FREUD IN CAMBRIDGE

Freud may never have set foot in Cambridge – that hub for the twentieth century's most influential thinkers and scientists – but his intellectual impact there in the years between the two World Wars was immense. This is a story that has long languished untold, buried under different accounts of the dissemination of psychoanalysis. John Forrester and Laura Cameron present a fascinating and deeply textured history of the ways in which a set of Freudian ideas about the workings of the human mind, sexuality and the unconscious affected Cambridge men and women – from A.G. Tansley and W.H.R. Rivers to Bertrand Russell, Bernal, Strachey, and Wittgenstein – shaping their thinking across a range of disciplines from biology to anthropology; from philosophy to psychology, education and literature. *Freud in Cambridge* will be welcomed as a major intervention by literary scholars, historians, psychoanalysts and all readers interested in twentieth-century intellectual and scientific life.

JOHN FORRESTER (25 August 1949 – 24 November 2015) was Professor of History and Philosophy of the Sciences in the University of Cambridge and head of the HPS department for seven years. He was Editor of the journal *Psychoanalysis and History* from 2005 to 2014 and authored *Freud's Women* (1992) with Lisa Appignanesi, *Dispatches from the Freud Wars* (1997) and *Truth Games* (1997), amongst others. He published over fifty papers in scholarly journals, principally concerned with the history and philosophy of psychoanalysis. His work on cases as a genre and as a style of reasoning was posthumously published as *Thinking in Cases* (2016).

LAURA CAMERON is an Associate Professor of historical geography at Queen's University in Kingston, Canada. She is the author of *Openings: A Meditation on History, Method and Sumas Lake* (1997), and co-editor of *Emotion, Place and Culture* (2009) and *Rethinking the Great White North: Race, Nature and the Historical Geographies of Whiteness* (2011), and has published numerous papers on the history of fieldwork, psychoanalysis, ecology and sound.

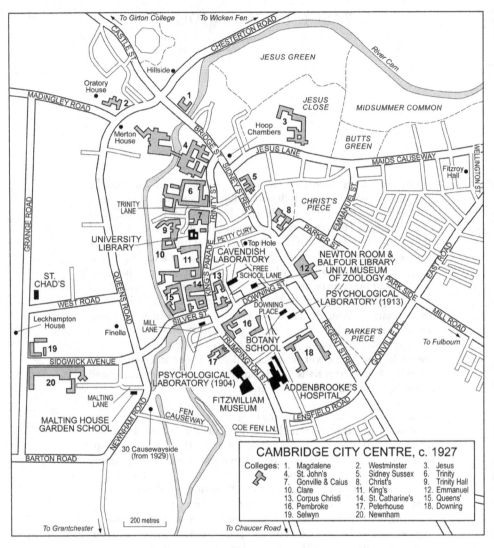

Cambridge city centre, *c.* 1927.

Other books by John Forrester
Thinking in Cases
Truth Games: Lies, Money, and Psychoanalysis
Dispatches from the Freud Wars: Psychoanalysis and Its Passions
Freud's Women, with Lisa Appignanesi
The Seductions of Psychoanalysis: Freud, Lacan and Derrida
Language and the Origins of Psychoanalysis

Other books by Laura Cameron
Rethinking the Great White North: Race, Nature and the Historical Geographies of Whiteness (co-editor) with Andrew Baldwin and Audrey Kobayashi
Emotion, Place and Culture (co-editor) with Mick Smith, Joyce Davidson and Liz Bondi
Openings: A Meditation on History, Method and Sumas Lake

FREUD IN CAMBRIDGE

JOHN FORRESTER AND LAURA CAMERON

CAMBRIDGE
UNIVERSITY PRESS

CAMBRIDGE
UNIVERSITY PRESS

University Printing House, Cambridge CB2 8BS, United Kingdom

One Liberty Plaza, 20th Floor, New York, NY 10006, USA

477 Williamstown Road, Port Melbourne, VIC 3207, Australia

314-321, 3rd Floor, Plot 3, Splendor Forum, Jasola District Centre, New Delhi - 110025, India

79 Anson Road, #06-04/06, Singapore 079906

Cambridge University Press is part of the University of Cambridge.

It furthers the University's mission by disseminating knowledge in the pursuit of education, learning and research at the highest international levels of excellence.

www.cambridge.org
Information on this title: www.cambridge.org/9780521861908

© John Forrester and Laura Cameron 2017

First published 2017

A catalogue record for this publication is available from the British Library

ISBN 978-0-521-86190-8 Hardback
ISBN 978-0-521-67995-4 Paperback

Cambridge University Press has no responsibility for the persistence or accuracy of URLs for external or third-party internet websites referred to in this publication, and does not guarantee that any content on such websites is, or will remain, accurate or appropriate.

Contents

Illustrations

Preface

This book is the product of research jointly sustained, mostly over long distances, for over eighteen years. John Forrester, whose sweeping knowledge of the human and physical sciences, as well as the history of psychoanalysis, was essential to the envisioning and final completion of this book, died six weeks after our manuscript was submitted to Cambridge University Press.

In our last conversation, John said we still needed to say something about how this enterprise began. We did not consider that I would be telling the story without him, and although I do so now with deep sadness, the beginning also underscores John's astonishing character. In a project whose hallmarks throughout were surprise, tenacity and inexhaustible excitement, it was Arthur Tansley's dream, but above all John's intellectual generosity, that set things in motion back in the spring of 1997.

As a postgraduate in historical geography, my studies focused in part on Sir Arthur George Tansley, the British ecologist who introduced the term 'ecosystem' and whose papers were then housed in some drawers at the Department of Plant Sciences, University of Cambridge. Although Tansley had written a book about psychoanalysis in 1920, this aspect of his life was largely unexplored. On a lead from Michael Molnar of the Freud Museum, I had corresponded with Kurt Eissler for permission to view the Tansley files held in the notorious ZR Section of the Sigmund Freud Archives at the Library of Congress. Eissler's eight word reply 'I do not recall an interview with Tansley' was as close as I got to the Freud Archive until I contacted John Forrester, then Reader in the History and Philosophy of Science, about the psychoanalytic papers that I had been examining over in Plant Sciences. One document appeared to be a letter from Freud to Tansley concerning the first patient of psychoanalysis, Anna O. With this 'find' which we published in the *International Journal of Psychoanalysis*, John helped me engage Eissler in further negotiations. Within a few months, I received notice that the material was no longer restricted: thanks to John, I was in.

xi

Tansley's contribution to the Freud Archive, as it turned out, was his own dream and self-analysis. His dream story made a fascinating case study in the significance of dreams in history, enabling an examination of the part they might play in an individual's life. It was also a jolting reminder of a time and a place when psychoanalysis was recognized as a science, when it was a marker of scientific modernity to be psychoanalysed, and when a dream, as a matter of course, had the radical potential to change an academic's life. We published joint papers on Tansley and his psychoanalytic networks: harbinger of things to come, John's inaugural lecture in May 2002 was entitled 'Freud in Cambridge'. Our massive research and email files continued to expand as I returned to Canada with my family to take up a position at Queen's University. We then planned a book, based on our earlier papers as well as research we each had been pursuing on related topics, including John's study of the life of W.H.R. Rivers (see Chapter 3) and my work on the Malting House School (see Chapter 7). Administrative loads, competing projects and health challenges slowed us down but the research continued and the book nonetheless grew, fleshed out over the ensuing years, and enlarged, most substantively so by John (Chapters 5 and 9) once he had leave from being Head of the Cambridge Department of History and Philosophy of Science in 2013. Although we worked collaboratively, his brilliance and iron will pulled it all together.

In helping us realize – and sustain – this work, we have been assisted by numerous people. As the research took place over two continents and nearly two decades, I am certain to forget someone and I apologize now for any omission. Sincere thanks, first of all, to John's wife, Lisa Appignanesi, who provided excellent insights and editorial suggestions all along the way and who has come to know this book and its making so well from beginning to end. Daniel Pick has been a close and careful first external reader and his enthusiasm buoyed John at the last. John would also have liked to thank the many colleagues who sustained him in his work, including Simon Schaffer, Liba Taub and Jim Secord, as well as the excellent Tamara Hug. He was particularly grateful to the Department of History and Philosophy of Science for supporting the research, long gestation, and publication of this book. The King's College Research Centre, then run by Simon Goldhill, now Director of CRASSH, hosted three work in progress seminars in May 2011 on different sections of the book. These were enormously helpful. We are both also grateful to Sarah Caro, our commissioning editor at Cambridge University Press, and to her successors, Richard Fisher, and most recently, Lucy Rhymer, for seeing the work through to publication. We are thankful, as well, for the coordinating efforts of Cassi Roberts and the very fine and thoughtful copyediting skills of Frances Brown.

We have drawn upon many sources for the book, chiefly from archives. For help with materials from the A.G. Tansley Collection when it was still based in Plant Sciences, we thank David Briggs, and the late poet-librarian, Richard Savage. For assistance in navigating other Cambridge and UK collections, we are grateful to: Jonathan Harrison, The Library, St John's College; Jacqueline Cox and Rosalind Grooms, King's College and the University Archives; Joanna Ball, John Marais and Jonathan Smith, Wren Library and Archives, Trinity College; Anne Thomson, Newnham College; the Archive Centre, King's College; the Department of Psychology Archives; the staff of the Cambridgeshire Record Office; and Mike Petty and Chris Jakes, the Cambridgeshire Collection, Cambridge Central Library. We thank also Michael Molnar and the Freud Museum, London; Ken Robinson, Allie Dillon, Gina Douglas, Jill Duncan and Polly Rossdale, Archives of the British Psycho-Analytical Society; Archives and Manuscripts of the British Library; the British Psychological Society; J.W. Belsham and Norman Leverets, Spalding Gentlemen's Society; the staff of the Wellcome Library; Sarah Aitchison, Institute of Education, University of London; Steve Roud, Croydon Local Studies Library and Archives Service, Central Library; Paul Rowan, Croydon Natural History Society.

Farther afield, we are grateful to: the late Lydia Marinelli, Daniella Seebacher and the Sigmund Freud Museum in Vienna; the Department of Library Services, University of South Africa; Lesley Hart, University of Cape Town Libraries; Harold Blum and the late Kurt Eissler, The Sigmund Freud Archives, Library of Congress, Washington, DC; Leslie Shores, American Heritage Centre, University of Wyoming; Jennifer Morrow, Hiram College Archives, Ohio; Stephen Yearl, Yale University Library; and, in Canada, Kathy Gray and Ken Blackwell, who assisted with The Bertrand Russell Archives and the CK Ogden Fonds at McMaster University.

Many people opened their homes and/or private archives to us. We appreciate the hospitality of Margaret Lythgoe-Goldstein, David Owers, Adrian Pyke, Janet Pyke and David Wills. Oral interviews also inform this book and we are grateful to the late Frances Barnes, the late Anna Dickens, Richard Grove, the late Lady Bertha Jeffreys, the Hon. Anne Keynes, Dan McKenzie, the late Frances Partridge, Janet Pyke, the late Martin Tomlinson, the late Helen Thompson and the late Richard West.

Research for this book was funded in part by the Social Sciences and Humanities Research Council of Canada, the Junior Research Fellowship at Churchill College, Queen's University and the Canada Research Chairs Program. We experienced enormous generosity from the research community and our writing and thinking have benefited from discussion with colleagues over the years in numerous conferences and seminars. In addition to those recognized in 'Acknowledgements', we would like to thank the Master of Sidney Sussex College, the late Sir Gabriel Horn, and the Librarian of the College for providing information on C.R.A. Thacker. Many thanks to Roy Foster, Colm Toibin, the late June Levine and in particular Mitch Elliott for information concerning Jonty Hanaghan. Thank you to Geoffrey Batten for information on Lella and Philip Sargant Florence. For help in the research process we also thank Bill Adams, Lady Lucy Adrian, Peder Anker, Peter Ayres, Alan Baker, Tim Bayliss-Smith, Drew Bednasek, German Berrios, Liz Bondi, Pete de Bolla, Mike Brearley, Andrew Brown, John Burnham, Gabriel Citron, Peter Cunningham, Mary Daniels, Joyce Davidson, Elizabeth Dougherty, Felix Driver, Mary Jane Drummond, Willem van der Eyken, Elizabeth Gagen, Peter Goheen, Paul Harris, Mike Heffernan, David Howie, Sir Michael Holroyd, Sarah Igo, Mary Jacobus, Edgar Jones, Heike Jöns, Gerry Kearns, Martin Kusch, Paul Kingsbury, Denis Linehan, Marin Levy, Roger Lohmann, Katrina Lythgoe, Elizabeth Lunbeck, David Matless, Hugh Mellor, Andreas Mayer, John Mollon, David Palfrey, Bronwyn Parry, Ian Patterson, Steve Pile, Jane Reid, the late Paul Roazen, William Rowley, Janet Sayers, John Sheail, Karl Snyder, Philip Stickler, Deborah Thom, Caroline Thomas, Edward Timms, Steve Trudgill, Andrew Webber, Paul Whittle and Jack Whitehead. *Bloomsbury/Freud*, edited by Perry Meisel and the late Walter Kendrick, was a continual inspiration.

John would have wanted me to express his abiding thanks not only to Lisa, but to his children and their partners, Katrina Forrester and Jamie Martin, Josh Appignanesi and Devorah Baum, as well as the grandson who filled his last years with joy, little Manny. On my side, thanks to Matthew Rogalsky, my partner in life for over thirty years, and our son Arden: their love and music infused the project throughout. Lastly, my eternal gratitude to John, himself, for sharing in the finest of archival adventures.

LAURA CAMERON

Acknowledgements

An earlier version of Chapter 2 appeared as Laura Cameron and John Forrester, '"A nice type of the English scientist": Tansley and Freud', in *History Workshop Journal* 48 (Autumn 1999), 64–100 and D. Pick and L. Roper (eds.), *Dreams and History*, London: Routledge, 2004.

Parts of Chapter 3 have appeared in Laura Cameron and John Forrester, 'Freud in the field: psychoanalysis, fieldwork and geographical imaginations in interwar Cambridge', in P. Kingsbury and S. Pile (eds.), *Psychoanalytic Geographies*, Farnham: Ashgate, 2014, and in John Forrester, 'The English Freud: W.H.R. Rivers, dreaming and the making of the early twentieth century human sciences', in Sally Alexander and Barbara Taylor (eds.), *History and Psyche: Culture, Psychoanalysis, and the Past*, London: Palgrave, 2012, pp. 71–104.

Parts of Chapters 4 and 6 have appeared in Laura Cameron and John Forrester, 'Tansley's psychoanalytic network: an episode out of the early history of psychoanalysis in England', *Psychoanalysis and History* 2(2) (2000), 189–256. Parts of Chapter 4 also appeared in John Forrester, 'The psychoanalytic passion of J.D. Bernal in 1920s Cambridge', *British Journal of Psychotherapy* 26 (2010), 397–404. Earlier versions of parts of Chapter 6 were also published in John Forrester, 'Freud in Cambridge', *Critical Quarterly* 46(2) (2004), 1–26.

An earlier version of Chapter 7 appeared as Laura Cameron, 'Science, nature and hatred: "finding out" at the Malting House Garden School, 1924–29', *Environment and Planning D: Society and Space* 24 (2006), 851–72.

An earlier version of Chapter 8 appeared in John Forrester, '"A sort of devil" (Keynes on Freud, 1925): reflections on a century of Freud-criticism', *Österreichische Zeitschrift für Geschichtswissenschaften* 14(2) (2003), 70–85.

We thank the late Anna Dickens and the late Martin Tomlinson for permission to quote from the materials which their grandfather, Sir Arthur Tansley, submitted to the Sigmund Freud archives; Joan Godwin for permission to quote from Tansley's published work; Sigmund Freud Copyrights for permission to quote from unpublished letters of Sigmund Freud; Tom Roberts of Sigmund Freud Copyrights for making Ernst Falzeder's transcriptions of Freud's letters to Rickman available to us; Ernst Falzeder for making available a series of transcriptions of unpublished letters, including Freud's letters to Abraham, the Rundbriefe, and other letters amongst the Committee's membership; Michael Young for making available unpublished materials relating to Malinowski; Adrian Cunningham and Lord Layard for sharing unpublished materials by John Layard; Janet Pyke who gave access to the Pyke Archive and who has given permission to cite, quote and reproduce several documents; Karina Williamson for permission to cite and quote from manuscripts and documents by Susan and Nathan Isaacs relating to the Malting House School.

Abbreviations

AA	Lord Adrian Archives, Trinity College Cambridge Archives
ABB	Andrew Brown, *J.D. Bernal: The Sage of Science*, Oxford: Oxford University Press, 2005
AF	See FA
AFSF	See SFAF
AGTSFA	*Three Contributions by Sir Arthur Tansley, F.R.S.*, The Sigmund Freud Archives, Manuscript Division, Library of Congress, Washington, DC
BAAS	British Association for the Advancement of Science
BF	Perry Meisel and Walter Kendrick, *Bloomsbury/Freud: The Letters of James and Alix Strachey 1924–1925*, London: Chatto & Windus, 1986
BIPA	*Bulletin of the International Psycho-Analytical Association*
BJMP	*British Journal of Medical Psychology*
BLSP	British Library Strachey Papers
BMFRS	*Biographical Memoirs of Fellows of the Royal Society*
BMJ	*British Medical Journal*
BPaS	Archives of the British Psycho-Analytical Society
BPCUL	Bernal Papers, Cambridge University Library
BPS	Archives of the British Psychological Society
BR	The Bertrand Russell Archives, McMaster University
BRA	Bertrand Russell, *Autobiography*. Followed by volume number: *The Autobiography of Bertrand Russell*, Vol. I, *1872–1914*, London: George Allen & Unwin, 1967; *The Autobiography of Bertrand Russell*, Vol. II, *1914–1944*, London: George Allen & Unwin, 1968; *The Autobiography of Bertrand Russell*, Vol. III, *1944–1967*, London: George Allen & Unwin, 1969

BRCP	Bertrand Russell, *The Collected Papers of Bertrand Russell*, London: Routledge, 1983– followed by volume number and page
CD	W.H.R. Rivers, *Conflict and Dream*, London: Kegan Paul, Trench, Trubner & Co., 1923
CM	*The Cambridge Magazine*
CR	*The Cambridge Review*
Crampton	Colin Crampton, 'The Cambridge School. The life, work and influence of James Ward, W.H.R. Rivers, C.S. Myers and Sir Frederic Bartlett' (PhD, University of Edinburgh, 1978)
CRO	Cambridgeshire Record Office
CU	Cambridge University
CUEP	Archives, Department of Experimental Psychology, University of Cambridge
CUL	Cambridge University Archives, University Library
CUR	*Cambridge University Reporter*
FA	E. Falzeder (ed.), *The Complete Correspondence of Sigmund Freud and Karl Abraham 1907–1925*, London and New York: Karnac, 2002. FA = Freud to Abraham; AF = Abraham to Freud. Followed by letter date and page number
FEC	Frederic E. Clements Papers, American Heritage Center, University of Wyoming
FJ	R. Andrew Paskauskas (ed.), *The Complete Correspondence of Sigmund Freud and Ernest Jones 1908–1939*, introduction by Riccardo Steiner, Cambridge, MA: The Belknap Press of Harvard University Press, 1993. FJ = Freud to Jones; JF = Jones to Freud. Followed by letter date and page number
FK	Pearl King and Riccardo Steiner (eds.), *The Freud–Klein Controversies 1941–45*, London and New York: Tavistock/Routledge, 1991
FPD	Frances Partridge, *Diaries, 1939–1972*, London: Phoenix, 2001
FPM	Frances Partridge, *Memories* (1981), London: Phoenix, 1996
FRP	Frank Ramsey Papers, King's College, Cambridge, Modern Archive
Holroyd	Michael Holroyd, *Lytton Strachey: The New Biography*, New York and London: Norton, 1995

HJPC	Papers and Correspondence of Sir Harold Jeffreys, St John's College, Cambridge Archives
HR 1910	J.R. Tanner (ed.), *Historical Register of the University of Cambridge, being a Supplement to the Calendar with a Record of University Offices, Honours and Distinctions to the Year 1910*, Cambridge: Cambridge University Press, 1917
HR 1911–20	G.V.C. (ed.), *Historical Register of the University of Cambridge, Supplement, 1911–1920*, Cambridge: Cambridge University Press, 1922
HR 1921–30	University Registry, *Historical Register of the University of Cambridge, Supplement 1921–1930*, Cambridge: Cambridge University Press, 1932
IJP	*International Journal of Psycho-Analysis*
IPA	International Psycho-Analytical Association
IPE	International Phytogeographical Excursion
IRP	*International Review of Psycho-Analysis*
IU	Rivers, *Instinct and the Unconscious*
JF	See FJ
Jones	Ernest Jones, *Sigmund Freud: Life and Work*, 3 vols., London: Hogarth Press, 1953–57, followed by volume and page number
JPBPaS	Ernest Jones Papers, Archives of the British Psycho-Analytical Society
KMFF	Kingsley Martin, *Father Figures: A First Volume of Autobiography, 1897–1931*, London: Hutchinson, 1966
Lighthouse	Jean MacGibbon, *There's the Lighthouse*, London: James & James, 1997
LLS	Paul Levy (ed.), *The Letters of Lytton Strachey*, London: Viking, 2005
LoCAF	Anna Freud Archives, Manuscript Division, Library of Congress, Washington, DC
LoCSF	The Sigmund Freud Archives, Manuscript Division, Library of Congress, Washington, DC
MBR I	Ray Monk, *Bertrand Russell: The Spirit of Solitude*, London: Jonathan Cape, 1996
MBR II	Ray Monk, *Bertrand Russell: The Ghost of Madness, 1921–1970*, London: Jonathan Cape, 2000
MLW	Ray Monk, *Ludwig Wittgenstein: The Duty of Genius*, London: Jonathan Cape, 1990

MST	Moral Sciences Tripos
NST	Natural Sciences Tripos
ODNB	*Oxford Dictionary of National Biography*, Oxford: Oxford University Press, 2004
OP	C.K. Ogden Papers, McMaster University Libraries
PC	I.A. Richards, *Practical Criticism: A Study of Literary Judgment* (1929), London: Kegan Paul, Trench, Trubner & Co., 1930
PP	Lionel Penrose Papers, University College London Manuscript and Rare Books Room. Numbers immediately following 'PP' refer to the Box number. Now digitised at the Wellcome Library: http://wellcomelibrary.org/collec tions/digital-collections/makers-of-modern-genetics/digi tised-archives/lionel-penrose/
Pyke Papers	Pyke Papers (held in the private residence of Janet Pyke, London)
RPBPaS	John Rickman Papers, Archives of the British Psycho-Analytical Society
SE	*The Standard Edition of the Complete Psychological Works of Sigmund Freud* (24 volumes), edited by James Strachey in collaboration with Anna Freud, assisted by Alix Strachey and Alan Tyson, London: The Hogarth Press and the Institute of Psycho-Analysis, 1953–74, followed by volume number and page number
SFAF	Ingeborg Meyer-Palmedo (ed.), *Sigmund Freud–Anna Freud: Correspondence 1904–1938*, Cambridge: Polity, 2013. SFAF = Sigmund to Anna; AFSF = Anna to Sigmund
Skid	Robert Skidelsky, *John Maynard Keynes*, Vol. I, *Hopes Betrayed 1883–1920*, London: Macmillan, 1983; *John Maynard Keynes*, Vol. II, *The Economist as Saviour 1920–1937*, London: Macmillan, 1992, followed by volume and page number
SPR	Society for Psychical Research
SSP	Sebastian Sprott Papers, Modern Archive, King's College, Cambridge
TA	Arthur Tansley Archives, Cambridge University Library
VWD	Virginia Woolf, *Diaries*, followed by volume number and page. *The Diary of Virginia Woolf*, Vol. I, *1915–1919*, ed. A.O. Bell, New York: Mariner Press, 1979; *The Diary of*

	Virginia Woolf, Vol. II, *1920–1924*, ed. A.O. Bell, New York: Mariner Press, 1980; *The Diary of Virginia Woolf*, Vol. III, *1925–1930*, ed. A.O. Bell, New York: Mariner Press, 1981; *The Diary of Virginia Woolf*, Vol. IV, *1931–1935*, ed. A.O. Bell, New York: Mariner Press, 1983
VWE	Virginia Woolf, *Collected Essays*, Vol. III, New York: Harcourt, Brace & World, 1967
VWL	Virginia Woolf, *Letters*, followed by volume number and page. Woolf, *The Letters of Virginia Woolf*, Vol. II, *1912–1922*, ed. N. Nicolson and J. Trautmann. New York: Harcourt Brace Jovanovich, 1976; *The Letters of Virginia Woolf*, Vol. III, *1923–1928*, ed. N. Nicolson and J. Trautmann. New York: Harcourt Brace, 1977

CHAPTER I

Introduction

That Sigmund Freud became a major intellectual presence in twentieth-century culture is not in doubt. Nor is there any doubt that at all times there was both fervent enthusiasm over and bitter hostility to his ideas and influence. But the exact means by which Freud became, despite this hostility, a master of intellectual life, on a par, already in the 1920s, with Karl Marx, Albert Einstein, Marie Curie and Bertrand Russell, has not been sufficiently explored. Strikingly, Freud emerged as a twentieth-century icon without the endorsement and support of an institution or a profession (in contrast to Einstein, Curie and Russell). Where are we to look for the details of this story of an emergent – and new – figure of immense cultural authority? One of the principal aims of this book is to show how this happened in one local, parochial yet privileged, site – Cambridge, then as now a university town stranded in the English Fens with a relatively small fluctuating population (59,212 in the 1921 Census, a 48 per cent increase since 1911).[1]

So this book contributes to the history and geography of psychoanalysis, but in an unusual fashion. Most histories of psychoanalysis start either in Paris, glittering metropolis of the nineteenth century, or in Vienna, capital of a doomed polyglot empire; this one starts in Grantchester, a picturesque village two miles outside Cambridge, the traditional destination of after-noon strolls across the Meadows for dons and students. Most histories of psychoanalysis assume a diffusionist model, with Freud's principal disci-ples functioning as essential relays for the transmission of doctrine and practical techniques, with the founding of local psychoanalytic societies

---

[1] The population of Cambridge was estimated at 6,490 in 1587 (which includes 1,500 members of the University), 7,778 in 1728 (including 100 college servants and 1,499 members of the University) and 10,087 in 1801. ('The city of Cambridge: economic history', in *A History of the County of Cambridge and the Isle of Ely*. Vol. III, *The City and University of Cambridge*, ed. J.P.C. Roach, Oxford: Oxford University Press, 1959, pp. 86–101, www.british-history.ac.uk/vch/cambs/vol3/pp86-101, accessed 6 May 2015.

and eventually orthodox training programmes as the key stages in the establishing of authorized psychoanalysis in different countries and regions; this one is inescapably full of surprising figures, loose connections between events and institutions, informal encounters. Most histories of psychoanalysis have been overly influenced by two crude models: the 'Great Man' model, in which specific individuals have decisive influence in turning history their way; and the bureaucratic transplant model, in which the oversight of the International Psycho-Analytic Association (IPA) and its sub-committee the International Training Committee (ITC) determined the forms and procedures for establishing psychoanalysis throughout the world. The two accounts come together for the British instance in locating Ernest Jones as the individual who, through his campaigning, through his writings and through his incessant organizing, created the British Psycho-Analytical Society (BPaS) in 1919 and founded the Institute of Psycho-Analysis in London in 1926.

In contrast, this account of the early history of psychoanalysis in England will be relatively Jones-free and will not be centred on the British Society. Instead, it purposely takes an Absent Great Man – Sigmund Freud – and looks not to a specific set of psychoanalytic institutions but to a flurry of activity in loose networks, some attached to the University, others not, yet associated with Cambridge. Freud the physical individual never came to Cambridge. This book is the story of his non-arrival. What Freud stood for – that is a different matter. 'Freud' did stand for a set of therapeutic practices that were deployed increasingly as the Great War dragged on and on. After the War, 'Freud' also stood for a revolution in psychology – the 'New Psychology'. For some Cambridge scientists, as we will see, 'Freud' stood for a revolution in thought quite the equal of those associated with Newton and Darwin. And he also stood for outrageous and immoral fabricated views on children, and on the importance of sexual life in general; his name was often shorthand for the pollution of the mind and society created by modernity.

1922 was the year of Cambridge in Freud's consulting room. James Strachey, Trinity graduate and Apostle, Bloomsberry, literary dilettante, had started analysis with Freud in October 1920 and finished at the end of June 1922; Alix Strachey, graduate of Newnham in modern languages, wife of James, had started at the same time and left in 1921. John Rickman, Quaker graduate of King's, doctor and enthusiast, fresh from a stint as a psychiatrist at Fulbourn Hospital just outside Cambridge, had begun analysis in April 1920 and completed at the end of June 1922. Joan Riviere, grande dame and intellectual, niece of Arthur Verrall, Apostle and first

Edward VII Professor of English Literature in the University, had moved from Jones's couch to Freud's in early 1922, returning to London in December. Arthur Tansley, University Lecturer in Botany, author of a psychoanalytic bestseller of 1920 entitled *The New Psychology and its Relation to Life*, began analysis with Freud on 31 March 1922, completing his first stint in June, and resigned his Cambridge lectureship in 1923 to come back for a more seriously sustained second analytic stint in late 1923 up to the summer of 1924. A Cambridge undergraduate, Roger Money-Kyrle, started analysis with Freud in the autumn of 1922, remaining in Vienna till 1926. In 1979 he described the milieu in which he moved:

> In Vienna, we met several people from Oxford and Cambridge, nearly all subsequently famous, who were more or less secretly in analysis. And I did not know till many years after that a half-uncle of my wife, a Fellow and Lecturer of Trinity, Cambridge, had spent one long summer vacation travelling Europe in analysis with James Glover, who was himself simultaneously in analysis with Abraham. Shades of the Peripatetic School of Athens in the third century B.C.! Incidentally, of course, I never mentioned psychoanalysis to [my doctoral supervisor Moritz] Schlick till I left, and then discovered that he himself was extremely interested in, but never spoke of it.[2]

So from March to June 1922, Riviere, Strachey, Rickman and Tansley were all in analysis with Freud, thus making up 40 per cent of his patient load.[3] What were they all doing in Vienna? Each had their own symptoms, their malaise in life, of course, but they were not ordinary patients. They and others like them were the means by which psychoanalysis became disseminated as a theory, as a vision of the world, as cocktail party chat, as a practice – and perhaps even as a form of knowledge suitable for inclusion in the teaching and research of an ancient university like Cambridge. By the summer of 1922, after listening for four hours a day, six days a week, for several months to a gaggle of elite Cambridge graduates, Freud must have known a lot about Tripos nerves, High Table backstabbing, the intricate family dynamics of large and eminent Victorian families and the sex lives of the English. He clearly knew what it meant to

---

[2] R. Money-Kyrle, 'Looking backwards – and forwards', *International Review of Psycho-Analysis* 6 (1979), 265–72 at 266.

[3] From the beginning of the decade, both Americans and English were making the pilgrimage to Vienna to be analysed by Freud. In the American cohort of – roughly – 1920–22 were Albert Polan, Clarence Oberndorf, Leonard Blumgart, Monroe Meyer and Abram Kardiner. See Kardiner Oral History Interviews, interviews conducted by Bluma Swerdloff, 1963, Columbia University Oral History Project, New York, p. 102.

be the favourite of Lytton Strachey and an intimate of Maynard Keynes – it meant being part of refined homosexual Cambridge culture. Freud was certainly an expert on Cambridge. But what of Cambridge – what did it make of Freud?

This study is also, inevitably, a contribution to the history of Cambridge – principally the University, but also the city in which the University is located – at a key period in its history, 1910–30. The nineteenth-century reforms, which included the introduction of specialist honours degrees, the removal of religious tests, the expansion of the sciences and the broadening of the social intake of undergraduates, including crucially women, were followed in the 1920s with the putting in place of a new, thoroughly modern and still existing structure of faculties and departments, of career paths for lecturers and researchers, of scholarships for poorer students and essential interlocking with state educational policy. So the period of the reception of psychoanalysis was also the moment in the history of the University when it fully recognized that, in the words of the Asquith Commission of 1922, 'the growth of science at Cambridge since the era of the Royal Commissions [the 1850s] has been perhaps the greatest fact in the history of the University since its foundation'. This is also the period of Cambridge 'High Science', a term by which Gary Werskey meant, amongst other things, first, the period of the supremacy of 'pure science', uncontaminated by applications or by necessary alliances with industry or government; second, the period when this corner of science was still dominated by the traditional British elites and classes; and third, the period of Cambridge's first fully self-conscious scientific glory. In the judgement of Eric Hobsbawm, Cambridge 'virtually monopolized top-level British scientific achievement in the first half of the twentieth century'.[4] This study of the reception of psychoanalysis in the foremost science-oriented university in Britain and its surrounding elite culture in the early twentieth century therefore gives insight into the development of science-based knowledge institutions in Britain and the place of psychoanalysis within them. At a time of transformation in British universities, when state funding is being withdrawn from both the universities and the poorer students attending them, while at the same time the prestige of the sciences, technology and medicine has never been

---

[4] Gary Werskey, *The Visible College: A Collective Biography of British Scientists and Socialists of the 1930s* (1979), London: Free Association Books, 1988, esp. pp. 19–42; Eric Hobsbawm, 'Preface', in Brenda Swann and Francis Aprahamian (eds.), *J.D. Bernal: A Life in Science and Politics*, London: Verso, 1999, p. xi.

higher, a study of the creation of new disciplines within the newly state-funded older universities is salutary.

Yet 'Cambridge' is not only a university peculiarly well configured for the development of the sciences, but also a traditional key node in the system of elite institutions sustaining British life, through the education of the next generation's elite. And 'Cambridge' is also the town of Cambridge, located in what was in this period an economically backward, non-industrial relatively poor part of England – transformed subsequently, from the 1960s on, by the development of science-based satellite industries closely associated with the University.

While this is a study in the dissemination of psychoanalysis, it does not directly concern its popularization, since the groups and cohorts examined are undoubtedly part of the educated and cultural British elite of the period. Not without a series of extensive and extended struggles, science became an integral part of elite culture – perhaps now at its very centre – and much of the account of psychoanalysis given in this study is of its interaction with, and its interrogation, absorption and repudiation by this elite culture. But it is also, almost by accident, a study of the reception of Freud's ideas by some of the key British intellectuals of the twentieth century (who all happened to be associated with Cambridge): Bertrand Russell, J.M. Keynes, Virginia Woolf, J.D. Bernal, Joseph Needham (from the point of view of the Chinese, Needham is the most important Briton of the twentieth century).

A question, however, certainly does hang over this study: why Cambridge? The first and obvious answer to that question is a straightforwardly empirical one: the remarkable range and number of Cambridge men and women engaging with psychoanalysis from roughly 1910 on was unmatched by any comparable cohort in Oxford, Manchester, London, Edinburgh or any other of the cultural and university centres of Britain. The question as to *why* this was the case is best addressed after taking the full measure of this varied and surprising engagement.

Much of the material that makes up this book is not well known to historians, or else – and this is a crucial point – is known under a different description. Placing the history of disciplines within the local Cambridge context brings out the fluidity of interchange and surprising cross-influences in their development – the advantages that 'local history', history in its place, has brought elsewhere. The study of the dissemination of ideas within Cambridge encouraged us to develop a prosopographical method: a collective study of the lives of a group, a population, a cohort. The links between these multiple 'life-lines' has proved not only fertile but surprising.

There is, however, one over-riding reason why the story this book tells has been overlooked till now: the episode recounted here *came to an abrupt end*. Historians committed to a continuist methodology (as most historians often inadvertently are) will naturally therefore be inclined to overlook it or at least downgrade its importance, if judgements of importance are based on relevance (to today, looking backwards) or 'influence' (on today, looking forwards). It would be too brutal to call this episode a blind alley of history with no progeny or issue of any kind, but it has certainly been overlooked by those seeking to find the sources of the present in the past, to tailor history to 'presentist' concerns and strictures.[5] Since disciplinary histories – whether of physics, literary criticism or psychoanalysis – are by definition committed to presentism, since they take the existence of the discipline as a given (though usually with very great concern about its place and date of birth), they will often find themselves at a loss with episodes, ideas, figures who do not immediately conform to the boundaries established later by those disciplines. To take one example from the stories told in this book: why did James Strachey, in drafting during the Second World War one of the most fateful documents in the history of British psychoanalysis, his Memorandum on Training, suddenly evoke the teaching of geophysics to psychoanalytical candidates? It is only by tracing the whole of the history of Freud in Cambridge that we find the answer.

There may also be another reason for the forgetting of the enthusiasm for Freud in Cambridge in the 1920s. First loves and youthful enthusiasms, particularly those that are tied to strong emotions and sexuality, are often later re-described by historical actors themselves in reproving and jocular terms. A youthful enthusiasm for psychoanalysis may be described in the same sort of terms, and with the same sort of attention to historical accuracy, as the stories many happily married middle-aged parents tell their children of their first loves. This analogy will remind us that in writing the history of psychoanalysis, passions, secret loves and deep inner troubles will play, even in Cambridge, as important a part as the architecture of scientific theory, the foundations of a scientific discipline or the proper way to educate the next generation. Psychoanalysis began with the emergence of the dream. A common thread weaving through our stories of Cambridge lives, the dream is, appropriately, where we too must start.

---

[5] George W. Stocking, Jr., 'On the limits of "presentism" and "historicism" in the historiography of the behavioral sciences', *Journal of the History of the Behavioral Sciences* 1 (1965), 211–17. Stocking's article is not the first work to use 'presentist' in roughly the same sense as Butterfield's *Whig Interpretation of History* (1931), but it appears to be the start of its more recent use.

CHAPTER 2

Tansley's Dream

'A nice type of the English scientist': Tansley and Freud

I dreamed that I was in a sub-tropical country, separated from my friends, standing alone in a small shack or shed which was open on one side so that I looked out on a wide open space surrounded by bush or scrub. In the edge of the bush I could see a number of savages armed with spears and the long pointed shields used by some South African native tribes. They occupied the whole extent of the bush-edge abutting on the open space, but they showed no sign of active hostility. I myself had a loaded rifle, but realized that I was quite unable to escape in face of the number of armed savages who blocked the way.

Then my wife appeared in the open space, dressed entirely in white, and advanced towards me quite unhindered by the savages, of whom she seemed unaware. Before she reached me the dream, which up to then had been singularly clear and vivid, became confused, and though there was some suggestion that I fired the rifle, but with no knowledge of who or what I fired at, I awoke.

Sir Arthur Tansley, FRS, 'The Dream'

The Cambridge scientist Arthur Tansley had this dream some time during the First World War, when he was working at the Ministry of Munitions in London.[1] It was, he later made very clear, one of the major turning points in his life. From this dream came his interest in psychoanalysis.

On 6 April 1922, Sigmund Freud wrote to Ernest Jones in London: 'Tansley has started analysis last Saturday. I find a charming man in him, a nice type of the English scientist. It might be a gain to win him over to our science at the loss of botany.'[2] Such information was the staple of the

---

[1] Sir Harry Godwin, 'Sir Arthur Tansley: the man and the subject, The Tansley Lecture, 1976', *Journal of Ecology* 65 (1977), 13: 'Tansley undertook a more or less routine clerking post in one of the Ministries, where his powers were barely called upon.'

[2] FJ, 6 April 1922, p. 468.

Fig. 2.1 Arthur George Tansley, International Phytogeographical Excursion (IPE) 1949, Newbridge Fen, Co. Wicklow, Ireland, by Eric Fulton.

correspondence between Jones and Freud that comprised some 671 letters over a thirty-year period to Freud's death. Implicit in such exchanges was the sustaining of the joint project that kept these two men, never soul mates, bound together – the fate and future of psychoanalysis – as a theory, a therapy and an institutional movement.

By following the trail revealed by this little snippet about an analysis begun in Vienna in the spring of 1922, we will discover that the early history of psychoanalysis in England was by no means confined to the professional and institutional lines that Jones, and even Freud, had in mind. By focusing on Tansley, we gain a more balanced and intriguing sense of the intellectual vitality and novelty of the set of ideas and practices spawned by Freud. Tansley also draws us into speculating about the historical significance of dreams and their interpretation, which, following Freud, many in the twentieth century have come to regard as 'the royal road to the unconscious'.[3]

[3] Sigmund Freud, *Five Lectures on Psycho-Analysis, SE* XI, 33.

It is the very implausibility of Tansley's involvement in psychoanalysis that, oddly enough, makes him so representative. He was, as Freud endearingly described him in his eccentric but precise English, 'a nice type of the English scientist' – and a distinguished one at that. Born in central London on 15 August 1871, Arthur George Tansley was the second child and only son of Amelia Lawrence and George Tansley – the 'exceptional people' to whom, at the end of his life, he would attribute the fact that his own Oedipus complex was 'almost negligible'.[4] George had a good business organizing society functions, and he also taught at the Working Men's College, where his real heart and enthusiasm lay. Arthur was educated at Highgate School; he went on to University College London to study the sciences, and in 1890 entered Trinity College, Cambridge, where he would obtain a double first in the Natural Sciences Tripos in 1893–94.

Tansley recalled that at Cambridge, besides advancing his knowledge of botany, zoology, geology and physiology, he took part in the 'usual interminable discussions on the universe – on philosophy, psychology, religion, politics, and sex'.[5] Writing to his mother in his second year, he described long talks on Shakespeare's tragedies with fellow undergraduate Bertrand Russell. In a postscript he added, 'Went to a meeting of the Psychical Society last night . . . and heard Mr. Myers[6] discourse on "subliminal self".'[7]

Russell, who was (as Tansley put it, writing in the third person) 'the most penetrating mind with which he came into contact, and who was his favourite companion in midnight talks',[8] became a close friend, working with Tansley on a student journal, *The Cambridge Observer*, forming a new society called the 'Mathetics' and travelling with him in Europe. Russell described Tansley in a letter to his wife-to-be Alys Pearsall Smith in January 1894:

> He is a man I have always made more confidences to than to any one else up here: the consequence of wh[ich] is that (being a quite a good judge of character and very sympathetic) he knows me better than any other man does. I once travelled in Italy with him but although I liked & still like all the main elements of his character I got so much annoyed with his ugliness

[4] AGTSFA, Interview with Kurt Eissler, Summer 1953, p. 9.
[5] Laura Cameron, 'Histories of disturbance', *Radical History Review* 74 (1999), 2–24 at 6.
[6] F[rederic].W.H. Myers, classicist, poet, philologist and co-founder of the Society for Psychical Research, discussed more extensively in Chapter 5.
[7] Peter Ayres, *Shaping Ecology: The Life of Arthur Tansley*, Chichester, UK and Hoboken, NJ: Wiley-Blackwell, 2011, p. 45. Note it is more likely that the date of the letter is 6 November 1891, not 1890.
[8] H. Godwin, 'Arthur George Tansley 1871–1955', *BMFRS* 3 (November 1957), 227–46 at 229.

Fig. 2.2 Arthur Tansley as a Cambridge undergraduate, *c.* 1893.

(wh[ich] is very great) & with his manners & laugh (wh[ich] are very bad) that I have not seen so much of him since, tho' whenever I have I have spoken very confidentially to him, as he is more sympathetic than any other man I know. He is not sentimental, but has been interested in watching the development of my character, as much I think from a biological as from any other motive (being a student of biology), & appears by what he said to have thought love the only thing that w[ould] make me calm down from my former foolish fancies.[9]

As Russell's letter indicates, the intimate friendship between Russell and Tansley seems to have ended before Russell's marriage to Alys. But two days prior to the wedding, Tansley felt compelled to write to Russell: 'You have been more to me as a friend than any other single individual, during

[9] BR 300038A, Russell to Alys Pearsall Smith, 24 January 1894.

the two short years I knew you intimately. And although it is the emotional side of our friendship that rises first into one's mind, there is a great deal besides. I suppose <u>because</u> of the attraction there was between us, I have learnt more from you than from anyone. It was by your help that I got first to some sort of intellectual standpoint, and it was from your talks (and life) that I first realised what morality was.'[10]

Although Tansley would continue to focus his academic energies on botanical subjects, he never lost sight of the wider vision, opened up to him by Russell and undergraduate experience, that encompassed – and made him vulnerable to feelings of being torn between – natural and moral sciences. During his final year at Trinity College, Tansley assisted his first teacher, the botanist F.W. Oliver, in teaching and research at UCL. Oliver aroused Tansley's interest in fern-like plants and shared Tansley's interests in the new subject of ecology. Tansley taught and researched at UCL for the next twelve years with Oliver and other colleagues, such as F.F. Blackman and Marie Stopes, with whom he would form long-lived associations. While at UCL, Tansley taught himself German and thus could read the 1896 German translation of Eugenius Warming's *Plantesamfund* and A.F.W. Schimper's 1898 *Pflanzen-Geographie auf physiologischer Grundlage*. Tansley felt these books laid the foundations for plant ecology as they developed concepts of plant communities and described the relations between plants, soils and climates. Tansley also was able to read Wilhelm Max Wundt's 1874 *Grundzüge der physiologischen Psychologie*. In 1896, Tansley submitted an essay entitled 'Natural selection considered as a special example of the general principle of evolution', in his bid to win the first Arnold Gerstenberg Studentship at Cambridge, an award 'with the object of promoting the study of Moral Philosophy and Metaphysics among students of Natural Science, both men and women'.[11] In his seventy-three page essay, redrafted after receiving suggestions from Russell, Tansley turned a critical eye to Wundt and regretfully noted his own inability to 'do more than suggest the possibility of the application of the idea of natural selection to mental processes, a possibility which seems to present a most fascinating field of research'.[12]

---

[10] BR 710.056798, Tansley to Russell, 11 December 1894.   [11] HR 1910, pp. 280–1.

[12] Arthur Tansley, 'Natural selection considered as a special example of the general principle of evolution', unpublished essay, TA. Russell's assistance is noted in a letter from Arthur to his sister Maud, dated 18 September 1896 and held in Branscoll (a collection of family letters found at Branscombe, south Devon); Peter Ayres kindly shared some of these letters with us.

Tansley held the Studentship jointly with C.S. Myers[13] but retained, for the time being, his lectureship at UCL. In 1900–1 he travelled to Ceylon, the Malaya Peninsula and North Africa. His diary during this time describes human, animal and plant activity with insight and humour.[14] In 1903, he married one of his correspondents, his former student at UCL and co-author, Edith Chick, who was the daughter of a lace merchant, Samuel Chick. The honeymoon trip to the south of France and Italy loosely traced a familiar route: Tansley wrote to his mother that in Pisa 'we stayed in the same hotel that Bertie and I stopped at 12 years ago'.[15] Arthur and Edith were to have three daughters, who were to become a physiologist, an architect and an economist. In 1906, he returned to Cambridge on his appointment to a University Lectureship in Botany.

Tansley had by this time already demonstrated two of his most salient intellectual characteristics: his willingness to assist an admired intellectual figure in a seemingly subordinate position without loss of dignity or standing and his gift for organizing and leading scientific projects as one of a group of like-minded enthusiasts. An admirer of Herbert Spencer's scientific philosophy, he oversaw the sections on plant morphology and physiology in the 1899 revised edition of his *The Principles of Biology*.

A Fellow of the Linnean Society, Tansley was pivotal in yoking the concerns of professional botanists to the activities of naturalist societies in the national survey projects of the British Vegetation Committee which he co-founded in 1904. As the scope of these necessarily collaborative survey activities was broadened to include botanists from outside Britain, Tansley founded the International Phytogeographical Excursion (IPE), hosted first by the British botanists[16] and subsequently by the Americans in 1913. The IPE, an organization perhaps rather similar to the International Psycho-Analytic Association in the latter's early years, became a thriving institution (the last excursion was held in Poland in 1991), meeting every two or three years in a different country, with its headquarters at the Geobotanical Institute of the Swiss Federal Institute of Technology in

---

[13] This Studentship and others were one of the ways in which Cambridge turned itself into a (Germanic) research university, by providing fundings and thus potential careers to scientific researchers on a broad basis. The fact that Tansley and Myers were the first recipients indicates that they were amongst the early cohort of such researchers.

[14] Laura Cameron, 'Sir Arthur George Tansley', in N. Koertge (ed.), *New Dictionary of Scientific Biography*, Vol. VII, London: Thomson Gale, 2008, pp. 3–10 at 3.

[15] Ayres, *Shaping Ecology*, p. 57.

[16] Laura Cameron and David Matless, 'Translocal ecologies: the Norfolk Broads, the "Natural", and the International Phytogeographical Excursion, 1911', *Journal of the History of Biology* 44(1) (2011), 15–41.

Zurich. To acquaint the non-British scientists with local vegetation, of which they knew virtually nothing, Tansley edited and wrote *Types of British Vegetation* (1911) for the IPE. This was the first systematic account of British vegetation, and immediately found a large home market.

In 1913, the British Vegetation Committee became the British Ecological Society, the world's first ecological organization. Tansley was its first president. Already editor of a botanical journal, *The New Phytologist*, begun in 1902 and funded by his private income and (with shades of things to come) entirely independent of universities and the scholarly presses,[17] Tansley also acted as editor of the new Society's *Journal of Ecology* from 1917 to 1938. In 1915, he was elected Fellow of the Royal Society; in later years, affirming that this was the recognition that counted, he would always add the letters 'F.R.S.' to his signature.

Fig. 2.3 Arthur and Edith Tansley, IPE 1913.

17 Godwin, 'Arthur George Tansley 1871–1955', p. 232.

Concerned with effective teaching of the new ecology, Tansley used his editorial authority to advocate a new curriculum. The key term for early proponents of self-conscious ecology like Tansley and the American plant ecologist, F.E. Clements was 'dynamic'.[18] This was a departure from static morphology and biogeography, the prevailing focus on structure over function, and what ecologists derided as mere 'descriptive' botany with its emphasis on species lists. The 1917 so-called 'encyclical' in *New Phytologist* (signed by Tansley, Oliver, Blackman and two others) pleaded for a vitalized and practical curriculum, to be based on plant physiology and ecology alongside, rather than subordinate to, the currently dominant (and, in their opinion, static and dull) morphology.[19] Tansley's ideas for reform were denounced as 'Botanical Bolshevism' by Frederick Bower, the Regius Professor of Botany at Glasgow, and received a similarly chilly response in the Cambridge Botany School. They may have been a significant factor in his not being elected to the Sherardian Chair of Botany at Oxford, for which he was a candidate in the autumn of 1919[20] – a professional setback that may have had profound inner consequences, with reverberations to which later his conversation with Freud in 1928 may have been alluding. An additional source of frustration may have been Tansley's ultimately unsuccessful attempt, beginning in late 1917, to create a Scientific Research Association for the promotion of pure research (an initiative supported by Ernest Jones, Bernard Hart, C.S. Myers, T.H. Pear, William McDougall and W.H.R. Rivers amongst others): the SRA was dissolved in December 1919. Tansley had complained to Frederic Clements in 1918, 'I've been getting some experience in the "Gentle art of making enemies" lately ... Reactionary forces are pretty strong here, and it will be a hard struggle to get anything progressive done. But I am going to have a good try.'[21]

By this time, Tansley was looking elsewhere than the Cambridge Botany School or even the international ecology movement for his intellectual direction forward. A key influence since the early years of the century was his own former student Bernard Hart. Hart, working as a doctor in

[18] Robert McIntosh, *The Background of Ecology: Concept and Theory*, Cambridge: Cambridge University Press, 1985, p. 69.

[19] 'The reconstruction of elementary botanical teaching', *New Phytologist* 16(10) (1917), 241–52; A.D. Boney, 'The "Tansley Manifesto" affair', *New Phytologist* 118(1) (1991), 3–21. Psychological language infuses the piece and some of the ensuing correspondence.

[20] F.E. Bower, 'Botanical Bolshevism', *New Phytologist* 17(5/6) (1918), 105–7. Boney, 'The "Tansley Manifesto" affair', pp. 17–18.

[21] Letter dated 18 December 1918, FEC, cited in Frank Golley, *A History of the Ecosytem Concept in Ecology*, New Haven: Yale University Press, 1993, p. 208.

asylums near London, would often entertain Tansley with asylum lore. Through him, Tansley came to have an unbookish and hands-on experience of mental disturbances. Hart's interests were in the psychology of insanity – the title of his very influential short book published in 1912. Hart was eclectic – absorbed first by Janet's ideas, then by Freud's, and in turn by Jung's – and he impressed Freud as well as Tansley. Writing to Jones in 1910, Freud called Hart's essay on the subconscious 'the first clever word upon the matter'.[22] Jones, always alert to any danger to his position as first among English Freudians, replied: 'He was one of my best pupils in England, although I had at first some difficulty in getting him to take up your work. Ultimately he said "Freudism is strictly speaking a religion; you can't *prove* it, but you have to accept it because 'it works'", which was quite clever.'[23] Hart played a considerable role in the integration of Freudian and non-Freudian psychotherapeutic practitioners during the Second World War, when he was in charge of psychotherapy coordination for the Emergency Medical Services, first in London, and then throughout Britain.

Tansley was clearly quite intrigued by the new theories in psychopathology before the war, but, by his own account, his knowledge owed more to conversation than study or research. What then happened to him was curious and was, according to him, the reason why the second half of his professional life became intertwined with the early history of psychoanalysis in England. In 1916 or thereabouts, aged forty-five, married with three daughters, secure (though restless) in his profession and having recently attained the pinnacle of a scientist in early twentieth-century Britain and with further successes and achievements in his chosen field undoubtedly ahead of him, he had his decisive dream.

> [The dream and my analysis of it] impressed me very deeply and led to a resolve to read Freud's work. This I did in the months that followed, beginning with the Traumdeutung, and following with the Drei Abhand[lungen] zur Sexual theorie, and some others. The latter – the Sexual theorie – interested and excited me immensely. I felt that it was an extraordinarily able and illuminating work, and, after having read far more widely in Freud since then, I still think that in some respects it is his most outstanding contribution – a daring and successful synthesis clearly and admirably expounded. My interest in the whole subject was now thoroughly aroused, and after a good deal of thought I determined to write my own picture of it as it shaped itself in my mind.

---

[22] FJ, 10 March 1910, p. 48.   [23] JF, 30 March 1910, p. 49.

The above was what Tansley wrote in 1953, when setting down for Kurt Eissler of the Sigmund Freud Archives, later sited at the Library of Congress, his memories of his involvement with Freud and psychoanalysis.[24] The 'picture' Tansley refers to is his book – *The New Psychology and its Relation to Life*, completed in January 1920 and published in June. It was reprinted twice within eight months, ten times in four years, selling over 10,000 copies in the UK and over 4,000 in the USA in the same period,[25] and eventually its publishers printed 19,000 copies (compared with 4,084 of Barbara Low's *Psycho-Analysis* of the same year). The book was translated into Swedish and German, and reviewed more often than any other psychoanalytic book of its time.[26] Tansley had caught the post-war wave of enthusiasm and fascination with Freudianism. His book was an attempt, he said, to capture for the general reader the 'biological' view of the mind with the concepts taken from the work of 'the great modern psychopathologists, Professor Freud and Dr. Jung'.[27] Modestly, Tansley assured his reader that it was neither a treatment of 'psychopathology proper' nor a comprehensive review of the literature – the book is simply 'an outline picture of the subject as it shapes itself in the mind of the author'. (This, we might note, is exactly how Tansley had allowed his own dream to 'interpret itself' – almost 'automatically', as he put it to Eissler.)

By his own admission, Tansley was disconcerted by the response to his book. Not only did he have a best-seller on his hands, but his book instigated critical correspondence with eminent figures, including two old colleagues from his Cambridge days, psychologist William McDougall and physician Walter Langdon-Brown.[28] He also received 'a good many letters from strangers asking all sorts of questions, many of which I did not feel I could answer adequately without a much more

---

[24] The attempt to gain access to the Tansley file is detailed in Laura Cameron, 'Oral history in the Freud Archives: incidents, ethics and relations', *Historical Geography* 29 (2001), 38–44.

[25] William Cooper, 'Sir Arthur Tansley and the science of ecology', *Ecology* 38(4) (1957), 658–9, quoting a letter from Tansley dated November 1923.

[26] D. Rapp, 'The reception of Freud by the British press: general interest and literary magazines, 1920–1925', *Journal of the History of the Behavioral Sciences* 24 (1988), 191–201 at 201n80. The book was, for instance, reviewed favourably by Havelock Ellis in the *Daily Herald*, 22 December 1920 and G. Stanley Hall in the *American Journal of Psychology* 33(2) (1922), 286–7.

[27] A.G. Tansley, *The New Psychology in its Relation to Life*, London: George Allen & Unwin, 1920, p. 6.

[28] 'I think there is peculiar advantage in having such a book written from a wide biological standpoint and not merely from the medical point of view. I can think of no one more capable of doing this than yourself because you have always been interested in these problems, you have the necessary biological equipment and have been constantly in touch with the medical point of view through Bernard Hart and others. I sometimes feel that the only permanent advantage that will be derived from the war is that we know far more how man's mind really works' (Walter Langdon-Brown to A.G. Tansley, 9 September 1920, TA 'Press cuttings, letters, etc.').

extensive knowledge of psycho-analysis'.[29] Like his old friend and colleague Marie Stopes, whose *Married Love* (1918) was an even bigger seller than
The New Psychology, and like both Krafft-Ebing and Havelock Ellis before
them, whose writings on sexual idiosyncrasy expanded enormously in their
later editions from the weight of unsolicited private confession and testimony, Tansley found himself addressed as an expert by numerous individuals in need.[30]

> Accordingly in 1921 I asked Dr Ernest Jones to give me an introduction to
> Freud, to whom I then wrote asking if he could receive me for an analysis.
> On his consenting to do so I arranged to spend three months in Vienna,
> from March to June, 1922.[31]

This cool account of Tansley's journey from book to couch is hardly the
whole story: by the spring of 1923, after all, Tansley had resigned his
position at Cambridge. Even before he had taken this decisive step,
Freud, Jones and some others had begun to follow Tansley's psychoanalytic progress with some interest.[32]

Freud found a place for Tansley on the last day of March 1922. Four
weeks before, on 4 March 1922, Tansley had given a talk to the Cambridge
University Natural Science Club, of which he had been a member since
1893: this time his topic was 'Dream and day dreams'.[33] Once in Vienna, he
began his analysis in German, but was soon obliged by the difficulties of
communicating his innermost thoughts to switch to English.
The 'indefatigable' Rickman, as Freud called him,[34] at the Professor's
behest, found suitable lodgings – extremely suitable, given the symbolism –
in Vienna for Tansley in the house of the famous, recently deceased
botanist, Wiesner (whose lectures on Plant Physiology Freud himself had
attended as a student in 1876):[35] Freud was pleased Tansley would be able

[29] Tansley, 'The impact of Freud's work and personality on a non-medical biologist', AGTSFA, p. 2.
Tansley's scrapbook of reviews and correspondence regarding *The New Psychology* is held in the TA.
[30] See Ruth Hall, *Marie Stopes: A Biography*, London: Deutsch, 1977, and Ruth Hall (ed.), *Dear
Dr Stopes: Sex in the 1920s*, London: Deutsch, 1978.
[31] Tansley, 'The impact of Freud's work', p. 2. Jones mentioned to Freud that Tansley had approached
him in a letter dated 6 May 1921 and Freud responded on the 19 May informing Jones that he had
received Tansley's letter (see FJ, p. 424).
[32] See JF, 6 May 1921, p. 421, and Rundbrief by Jones, 19 October 1920, Otto Rank Archives, Columbia
University, New York.
[33] *Cambridge University Natural Science Club. Founded March 10th, 1872*. Cambridge: Cambridge
University Press, 1982. Tansley was an honorary member and gave eight talks across fifty-eight years:
his first talk to the Club was in 1893 on 'The development of the theory of organic evolution' and his
last in 1951 was on 'The Nature Conservancy'.
[34] SFAF, 19 March 1922, p. 256.
[35] Siegfried Bernfeld, 'Sigmund Freud, M.D., 1882–1885', *IJP* 32 (1951), 216.

to make use of Wiesner's library. Obviously well informed on Tansley's journey to Vienna, Jones enquired almost too eagerly the day after Tansley's analysis began: 'Has Tansley started yet? I think he is a very able and careful thinker, and shall be glad to hear your impressions of him.'[36] Freud's opinion of Tansley chimed with Jones's, and their joint effort to catch this big fish is palpable.

Once Tansley's resistances were mobilized, both Freud and Tansley agreed that the three-months analysis that ended in June 1922 was woefully incomplete. Tansley was obviously intent on returning to Freud, but his duties in Cambridge kept him from Vienna in the academic year 1922–23. In England, Tansley played a major role in a symposium on the relations of complex and sentiment for the July 1922 meeting of the British Psychological Society. In contrast to the positions taken by W.H.R. Rivers and Alexander Shand (whose language of 'sentiments' was the homegrown English competitor with the vocabulary of 'complexes'), Tansley argued that 'complex' was a key connecting term for normal and abnormal psychology and should not be limited to the latter field. As he stressed here and later in a 1923 letter to Clements outlining his view of the central issues in the field of psychology: 'The question of the applicability of Freudian method to the "normal" mind is doubtless the crucial question.'[37] Along with the notion that all energy, both physical and psychical, tended towards a state of fragile equilibrium, this focus on 'normal' and 'abnormal' provided conceptual links for his thinking in both psychology and ecology. As he explained to Clements, 'The limiting conception of "normality" is an abstraction – never seen in concrete form. We use the word "normal" in practice to cover quite large deviations from a theoretic balanced mean, just as we do with species.'[38]

In March 1923, Tansley was voicing his intention to resign to colleagues:

> Probably I shall cease to be a professional botanist after the [University] term, though for the present, at least, I shall continue to edit the two journals ... Adamson is going to the Cape and will be a terrible loss to me – I need a good 'florist' at my elbow. Together with the 'conservatives in authority' his departure will help make me spend more time at psychology and less at ecology. The last year or two I have been pursuing both, and

[36] JF, 1 April 1922, p. 467.
[37] Letter from Tansley to Frederic Clements, 12 July 1923, FEC, cited in Cameron, 'Sir Arthur George Tansley', p. 6.
[38] *Ibid.* See Cameron, 'Histories of disturbance', and Ayres, *Shaping Ecology*, pp. 114–16, for explorations of how Tansley may have applied this abnormal/normal linkage to the vegetational dynamics of deflected succession.

though my power of work is much better than it was, largely I think to the release of powers through emotional clarification – the double pull is a considerable strain.[39]

His increasing involvement with psychology did not stop him from publishing substantial works in botany;[40] in addition, he was President of the Botanical Section, British Association for the Advancement of Science in 1923, and spent part of the summer months doing research at Wicken Fen near Cambridge, a site of special scientific interest for him and his botanical co-workers.[41]

By the late spring, his decision was made. His colleagues, particularly his American rival and friend, the plant ecologist F.E. Clements, expressed consternation, though mixed with some understanding, about Tansley's resignation and the career and intellectual cross-roads it represented. They were obviously aware of the profound shift in Tansley's vision of his future. Clements voiced an ambiguous fear: 'I am not at all sure that your new field may not have greater opportunities for distinct and distinguished services.'[42] In May 1923, having resigned from Cambridge, Tansley told Clements of his plan to have more months of analysis with Freud in October – 'but for the present, at least, I shall continue to edit the two journals. It is likely that I shall take my whole family with me to Vienna.'[43] In July he informed Clements, 'if, as is quite possible, I become more and more absorbed in psychological research I may gradually drop plant ecology from sheer lack of time'.[44]

Move the family did. But Freud was not able to restart the analysis; he was undergoing the first of his many major operations for cancer in the autumn of 1923. Tansley waited in Vienna, and Freud recovered sufficiently to start work again at the end of December. This second slice of analysis lasted a further six months.

[39] Letter from Tansley to Frederic Clements, 8 March 1923, FEC, cited in Golley, *A History of the Ecosystem*, p. 209.

[40] Including *Elements of Plant Biology* in 1922 (based on the lecture course he gave to first-year medical students), *Practical Plant Ecology* in 1923, and a co-edited volume *Aims and Methods in the Study of Vegetation* in 1926. See Godwin, 'Arthur George Tansley, 1871–1955', p. 232.

[41] Wicken Fen, purchased in 1899 by the National Trust for Places of Historic Interest and Natural Beauty, is the oldest of Britain's nature reserves. Cameron's 1999 article, 'Histories of disturbance', examines Tansley's dual interests in psychoanalysis and ecology, particularly with regard to fieldwork carried out in this place/period.

[42] Letter from Clements to Tansley, 12 January 1923, TA, cited in Golley, *A History of the Ecosystem*, p. 209.

[43] Letter from Tansley to Clements, 30 May 1923, FEC, cited in Golley, *A History of the Ecosystem*, p. 209.

[44] Letter from Tansley to Clements, 12 July 1923, FEC.

Freud galvanized other analysts into welcoming Tansley to his new profession. On 14 March 1924, he wrote to Karl Abraham, then Secretary of the International Psycho-Analytic Association and convenor of the Congress to be held at Salzburg in April 1924, describing Tansley's book as having

> done a great deal for the spread of psychoanalysis, although it shows him still in a phase of development before being completely an adherent. He is now in analysis with me for the second time, and I hope to make considerable progress with his convictions. He is a distinguished, correct person, a clear, critical mind, well-meaning and highly educated.[45]

Tansley was welcomed with open arms by the analysts, just as Ernest Jones's somewhat diffident review of *The New Psychology* in late 1920 did not miss the opportunity to emphasize how distinguished Tansley was.[46] From Jones's point of view, Tansley appeared to be a godsend, who would help secure the biological flank of psychoanalysis and, if need be, stem the tide of biological speculation to which so many analysts, including Freud, were prone. Thus, on 19 October 1920, Jones announced in the Rundbrief to the Committee that:

> A. G. Tansley, Professor [*sic*] of Botany at Cambridge University, who has just written a good book called *The New Psychology*, read a paper on Oct. 13th on Freud's Theory of Sex from a Biological Point of View, before the British Soc. for the Study of Sex Psychology. He regretted the Ablehnung of biology in the preface to the Drei Abhandlungen, and asked me the meaning of the passage, which I should be glad to hear from Professor himself. T. was enthusiastically in favour of the theory, which he declared to be throughout essentially sound from a biological point of view and supported by much evidence from that science.[47]

Tansley and Jones were referring to the following passage from the Preface to the 1915 edition of Freud's *Three Essays*:

> [These essays are] deliberately independent of the findings of biology . . . my aim has rather been to discover how far psychological investigation can throw light upon the biology of the sexual life of man . . . there was no need for me to be diverted from my course if the psycho-analytic method led in a number of important respects to opinions and findings which differed largely from those based on biological considerations.[48]

---

[45] FA, 14 March 1924, p. 489.   [46] Ernest Jones, 'Review, *The New Psychology*', *IJP* 1 (1920), 480.
[47] Rundbrief written by Jones, 19 October 1920, Otto Rank Archives.
[48] Freud, *Three Essays on the Theory of Sexuality*, SE VII, 131.

To the 'modern, deterministic, empiric and dynamic' minds of Jones and Tansley, such a disdain for or even distancing from biology might not be the way in which to develop a truly scientific psychoanalysis. Two years later, there are hints in Jones's worrying to Freud about biology that he longs for Tansley to be able to take over the biological side completely, and to correct some of the errors to which Freud was inclined:

> I am not happy about our recapitulation theory and wish we could enlist the services of a good modern biologist. If Tansley were more advanced or experienced I would discuss it with him, but he is not yet sure of the ontogenetic side of the Oedipus complex, let alone the phylogenetic or prehistoric.[49]

Jones's perception was correct: Tansley was to remain resolutely agnostic on the question of the universality of the Oedipus complex and to employ lofty irony for attempts to employ a neo-Lamarckian theory of use-inheritance to underpin psychoanalytic findings.[50] But Jones continued to use Tansley as a secret weapon against the more speculative biological theorists, most prominently Ferenczi. Writing to Freud in September 1924, he adopted an almost threatening tone:

> I do not trace any suspicion of anti-analytic tendency in [Ferenczi's] work, but cannot refrain from the diagnosis of narcissism combined with poor judgement. No doubt you saw Tansley's review of his work in the *British Journal of Medical Psychology*.[51]

Tansley's review had criticized Ferenczi's neo-Lamarckian argument that evolution was the result of impulses and strivings on the part of organisms (an argument we now know was worked out in tandem with Freud in the period of the Great War and after) and had caustically undermined Ferenczi's view that psychology should be used to illuminate biology by asserting that science can only progress through the understanding of the complex on the basis of the simpler.[52] In his campaign against the 'romantic biological' inclinations of both Ferenczi and Freud, Jones was eager to make use of Tansley as ally and authority.

Coinciding roughly with the end of Tansley's analysis, back in London on 22 May 1924 John Rickman nominated him as an Associate Member of the BPaS, which approved the motion.[53] Freud had recommended that

[49] JF, [?] October 1922, p. 501.
[50] Tansley, 'Review, C.J. Patten, *The Memory Factor in Biology*', *IJP* 8 (1927), 292.
[51] JF, 29 September 1924, p. 555.
[52] A.G. Tansley, 'Critical notice of *Versuch einer Genitaltheorie* by Dr. S. Ferenczi', *BJMP* 4 (1924), 156–61.
[53] BPaS, FAA/134 Committee Minutes.

Tansley take on a psychoanalytic case, to acquaint himself fully with the technique and the findings of the discipline. At some point, probably starting in late 1924 or 1925, Tansley did so – 'an experimental analysis, lasting nearly two years, on an obsessional neurotic'.[54] His reappearance in Cambridge having been 'psychoanalytically trained' by Freud attracted at least another potential patient: James Strachey reported to Alix in late 1924 that 'Tansley was there last night and said that Sebastian [Sprott] had been making tentative approaches on the subject of being analysed by him. He didn't seem to think it'ld come to anything.'[55] And Tansley's daughter did remember young men coming as patients to the house in Grantchester.[56]

Even before he became a member of the British Society, Tansley was sought out as a powerful ally for psychoanalysis. At the Salzburg IPA Congress in April 1924, which he attended, it was agreed to hold the next Congress in Cambridge, a decision possibly connected with the curious fact that, by 1924, the BPaS had become the largest in the world, with 49 members (the Viennese Society had 42, the Swiss 40, the American 31, the Berlin 27 and the New York 26). It would have been very much in character for Jones to wish to carry on building his empire by holding the International Congress on British soil, in a city as welcoming of intellectual endeavour as was Cambridge, not to speak of the respectability that might as a result be conferred on psychoanalysis in English eyes. Once Tansley had become an Associate Member of the British Society in May 1924, a month after the Salzburg Congress, and completed his nine months of analysis with Freud, he was the obvious person to turn to as organizer of some kind of the Cambridge Congress. But in November Jones found himself obliged to write a letter to Abraham, the newly elected President of the IPA:

> Although Tansley promised me verbally to investigate the situation in Cambridge and I have written to him since reminding him, there is as yet no answer. I think that the delay is more likely to be due to personal inhibitions than to outward circumstances, but I will of course write to you the moment I hear anything.[57]

Newly freed from his teaching and academic responsibilities, Tansley may have felt somewhat uneasy about taking on similar responsibilities on behalf of his new psychoanalytic colleagues and institutions. Not being

[54] Untitled notes, probably delivered as an autobiographical introduction to a paper for Oxford's Magdalen Philosophy Club, 5 May 1932, TA.
[55] BF, 6 November 1924, p. 109.
[56] Interview with Helen Thompson, 21 April 2001, conducted by Laura Cameron.
[57] LoCSF, Jones to Abraham, 12 November 1924.

a Fellow of a Cambridge College – an increasingly common fate for practitioners of the burgeoning new disciplines, particularly in the sciences – he may not have had the base for organizing the beds, dinners and rooms that conferences require.[58] Or there may have been other, more personal inhibitions – his father, after all, had, although he despised them, spent his life organizing social functions for high society. Certainly Tansley seems to have been the only plausible Cambridge-based person for such organization – the two other British Society members resident in Cambridge were Susan Isaacs, newly arrived as head of the Malting House nursery school (scurrilously known as the 'pre-genital brothel'[59]), and Dr C.R.A. Thacker, physiologist and specialist on nervous diseases and shell shock, Fellow of Sidney Sussex College, by this time already suffering from the illness that was to kill him in 1929.

Whatever Tansley's ambivalences about specific involvement with the psychoanalytic movement, he was an intensely social being, a born scientific networker. On 7 October 1925, Tansley was elected to full membership of the Society, and for a year, until 17 November 1926, attended meetings frequently[60] and contributed at least one paper.[61] Free of those commitments at Cambridge which had become a burden and an incessant source of unprofitable struggle, he made new psychoanalytic links in at least two different directions: both informal, one within Cambridge (with the '1925 Group', see Chapter 6) and one within the field sciences milieu. An important member of the latter group was the botanist Ernest Pickworth Farrow, whose interest in psychology had been sparked by Tansley in 1912 when he brought the proofs of Hart's *The Psychology of Insanity* into his Cambridge botany class (for an account of Farrow's connection with psychoanalysis, see Chapter 4).[62] Another field scientist from Tansley's circle to acquire a life-long absorption with psychoanalysis was C.C. Fagg.[63] Born in 1883, Fagg was forced, because of family poverty,

---

[58] Tansley was eventually elected an Honorary Fellow of Trinity College, but only in 1944, well after his retirement from the Chair at Oxford.

[59] J.R. [John Rickman], 'Susan Sutherland Isaacs, C.B.E., M.A., D.Sc. (Vict.), Hon. D.Sc. (Adelaide)', *IJP* 31 (1950), 279–85. Susan Isaacs would 'read some notes upon Child Life' at a meeting hosted by Tansley at his Granchester home, 13 or 14 June 1925; see BF, p. 281. See Chapter 7.

[60] BPaS, FAA/161 Committee Minutes.

[61] 'A note on a masturbation phantasy'; see D. Bryan, 'Reports. British Psycho-Analytical Society', *BIPA* 7 (1926), 533.

[62] E. Pickworth Farrow, *A Practical Method of Self-Analysis*, with Foreword by the late Professor Sigmund Freud, London: George Allen & Unwin, 1942, p. 1.

[63] For more detailed analysis of Fagg's relations with Tansley, Freud and psychoanalysis, see Laura Cameron and John Forrester, 'Tansley's psychoanalytic network: an episode out of the

to work in a shop from the age of fourteen on. Seized with the desire for self-improvement, he took evening classes and was eventually admitted to the Civil Service in his early twenties, working as a Customs and Excise Officer until his retirement in 1941 at the age of fifty-eight. From 1906 to his death in 1965, he was a vigorous member, in fact the most stalwart member, of the Croydon Natural History and Scientific Society. He founded its Sociology Section in 1912, alongside a new Regional Survey Committee to help execute a survey of the Croydon area along the lines of other ongoing surveys inspired by the geographer-sociologist Professor Patrick Geddes. Tansley, one of Fagg's many University Extension Course lecturers, was extremely supportive of the regional survey move-ment in its formative years and came to speak to the Society in 1912 regarding aspects of vegetational survey. Given what Fagg says in his recollections,[64] it is quite possible that he was one of the correspondents who responded with questions to Tansley's *The New Psychology* of 1920. Soon after, *The New Psychology* became the textbook of the Sociological Section as the Section added 'Psychological' to its title and devoted itself wholly to the study of psychoanalysis. When Ernest Jones gave a lecture to the General Meeting of the Society on 16 May 1922, entitled 'Psycho-Analysis', over 400 people attended.[65]

Buoyed by such enthusiasm, Fagg would publish on psychoanalysis,[66] take patients and, between 1923 and 1933, intermittently correspond with Freud on issues of treatment and theory. What is most striking about Freud's interchanges with Fagg is how supportive he was of Fagg's work with patients. In 1928, Fagg wrote to Freud, prompted by the publication in the April number of the *International Journal of Psycho-Analysis* of the English translation of Freud's paper 'Fetishism', with its interpretation of the fetish object as representing the maternal phallus. Fagg clearly agreed with this, but wished to add a component concerning anal gratification,

early history of psychoanalysis in England', *Psychoanalysis and History* 2(2) (2000) 189–256, esp. at 235–43.

[64] Ms. Minute of meeting held 14 December 1959, Croydon Natural History and Scientific Society: Psychology and Social Science Section, Minutes of Meetings 22 January 1951–16 April 1962. Our transcription, in places uncertain.

[65] Minutes of meeting on 17 February 1958, signed 17 March 1958, Psychology and Social Science Section, Archives of Croydon Natural History and Scientific Society.

[66] C.C. Fagg, *Psycho-Analysis: A Select Reading List*, Croydon: Central Library, Town Hall, 1926; C.C. Fagg, 'Psychosynthesis, or evolution in the light of Freudian psychology' (read before the Medical Section of the British Psychological Society on 22 February 1933), *British Journal of Medical Psychology* 13 (1933), 119–42. His life's work, a book entitled *Sublimation or Perversion*, remained in draft form.

which Freud did not support. Nonetheless, in a Postscript, Freud reassured Fagg about the correctness of his unconventional analytic technique:

> Your following the obsessional patient into his privacy was by no means a sin against the orthodox technique. If we do not always the same lack of time is our only motive. In a similar case I had the patient watched by a person of my confidence with the same good results.[67]

It is probable that the patient referred to in Freud's next letter, written four weeks later, is this same obsessional:

> If your patient has succeeded in finding new outlets of social use for his pathological trends I think this effect should be considered as a recovery from his neurotic condition. We have no right to claim for more.[68]

And Fagg added a handwritten note at the bottom of Freud's letter clarifying what they had been discussing:

> A symptom of the obsessional patient referred to in the first paragraph was skoptophilia directed towards full-breasted women. Towards the end of my treatment of the patient this was displaced to watching birds in my garden.

Fagg's involvement with the psychoanalytic movement is intrinsically interesting for the light it throws on both Tansley's and Freud's broad-minded approaches to unorthodox practitioners and, with regard to the seriousness with which Freud took Fagg, Freud's high esteem for Tansley's judgement. The example of Fagg shows how an enthusiastic 'amateur' could develop a life-long commitment to Freud's ideas, publish papers developing them, and, without having any training of which we are aware, establish a psychoanalytic practice of some kind without encountering hostility from powers that be (such as Ernest Jones and the BPaS) and with continued encouragement from Freud himself – a practice which Fagg continued well into the 1950s. Fagg's relationship with Tansley not only gave him an entrée to Freud, but gave him at least one patient, when Tansley's own switching of professional direction required him to move on from his 'experimental' psychoanalytic practice. Some of the other links in Tansley's psychoanalytic network acquired a similar characteristic of being in large part hidden from history that necessarily accompanies the duty of clinical confidentiality. But, interestingly enough, Tansley's work to gal-vanize support for psychoanalysis was not confined to informal and private contacts. Throughout the summer of 1925, Tansley was engaged in IPE

---

[67] Freud to Fagg, 28 June 1928, Freud Museum, London.
[68] Freud to Fagg, 25 July 1928, Freud Museum, London.

ecological activities in Europe and at the same time in a public polemic
defending psychoanalysis in the correspondence column of *The Nation and
Athenæum*, as Chapter 8 will recount. Perhaps the quirky manner in which
he rounded off his robust defence pointed to the next step in his career:
'may I beg your correspondents' attention to the fact that I am not, and
never have been, a professor? Nor do I hold a doctor's degree.'[69] Plain Mr
Tansley he certainly was at that time, neither academic nor doctor, whether
of philosophy or medicine. But not for much longer.

At some point in 1926–27, Tansley's younger colleague Harry Godwin[70]
did some behind-the-scenes work in the botany world. In Godwin's later
judgement, the years 1923 to 1927 had been for Tansley years 'in the
wilderness so far at least as his relations to botanical science were concerned
and especially those with British botanists'.[71] Through Godwin's encouragement, Tansley accepted an invitation to apply for the Sherardian Chair
of Botany at Oxford. In an authoritative tone that betrays familiarity with
Tansley's relations with Freud and psychoanalysis, Godwin wrote: 'Not
until the end of 1926 did he complete what Freud had forecast for him, "the
return to the mother subject," ... He was elected in January 1927.
Indecision was abandoned.'[72] He took up the post in October 1927,
together with a Fellowship at Magdalen College. His lectures from then
on were on more conventional botanical subjects; he obviously felt the
need to devote himself to reforming teaching and research in botany at
Oxford. Whether he included discussion of psychoanalysis in those lectures is not known. He had done so in Cambridge, where, as Joseph
Needham, the Cambridge biochemist and sinologist, later recalled, it
had been Tansley who helped to generate an interest in Freud among
students during the early 1920s by mentioning him in his lectures.[73] While
at Oxford, he certainly did not leave psychoanalysis entirely behind him.

---

[69] A.G. Tansley, 'Freudian psycho-analysis', *The Nation and Athenæum*, 12 September 1925, p. 700.
[70] In addition to pioneering Quaternary research in Britain with his wife Margaret (who took up
pollen analysis at Tansley's suggestion), Godwin (1901–85, FRS 1945) was, like Tansley, a leader in
the Nature Conservancy. See R.G. West, 'Harry Godwin', *BMFRS* 34 (1988), 261–92.
[71] Godwin, 'Arthur George Tansley, 1871–1955', p. 236.   [72] *Ibid.*
[73] The source for Needham's reminiscence is a personal communication from Edward Timms (to
Laura Cameron, 17 March 1997), who asked Needham about interest in Freud in Cambridge in the
1920s. Charles Raven, botanist, historian of natural history and biographer of John Ray, probably
had the same teacher in mind when he noted that in the early years of the century ecology and
psychology were being drawn together – a comment which, without the particular research interests
of Tansley, would be somewhat mysterious. In a speech to the Cambridge Natural History Society
in 1957, Charles Raven said: 'at the beginning of the century, the tide had begun slowly but surely to
turn and though the gulf betwen the men of museums and the men of the open-air was still wide, the
liveliest minds in biology were already moving towards a denial of the antithesis, towards ecological
and psychological problems and to that sense of wholeness which is now influencing every dept.

In 1928, he initiated a further polemic in *The Nation and Athenæum*, seemingly stung by Vera Brittain's charge – which he must have read as a covert attack on psychoanalysis – that 'certain men of science have bestowed upon sexual gluttony a blessing which they would withhold in horror from any other form of immoderation'. Having pointed out to Brittain that the views of such 'men of science' on sexuality involved recommending moderation instead of abstinence – quite the opposite of advocating licensed gluttony, as in her image of little boys let loose on an unlimited quantity of jam – he moved on to his psychoanalytic point when he declared that a statement such as Brittain's 'that there is a danger of "over-estimating the importance of the part which sex plays in life" is a contention only ever, in my experience, made by those who seriously under-estimate that importance'.[74]

Although he contemplated writing a history of the early development of Freudian psychology[75] and drafted chapters for it, Tansley's main work of the 1930s was in his 'mother subject' of botany. This produced a concept of great significance for the future development of the discipline: the 'ecosystem'.[76] In this and other projects, there is substantial evidence of his continued interest and commitment to 'psychology', in particular manuscript materials relating to the project Godwin mentioned, a history of the development of Freudian psychology. Towards the end of 1932, Tansley wrote at least two papers on the early development of Freud's theories, which he intended to submit to the *British Journal of Medical Psychology*, focusing on the relation between psychoanalysis and biology, and he left an incomplete manuscript entitled 'The historical foundations of psychoanalysis', which may have incorporated those papers. Tansley was sufficiently immersed in this work to write asking Freud what became of the first patient of psychoanalysis,

---

from medicine to nuclear physics.' Cited in F.W. Dillistone, *Charles Raven: Naturalist, Historian, Theologian*, London: Hodder and Stoughton, 1975, pp. 174–5.

[74] *The Nation and Athenæum*: Brittain's letter is in the number dated 28 July 1928, p. 552; Tansley's first response, from Kielce in Poland, is in that for 11 August 1928, p. 618. Brittain's response is in the edition dated 18 August 1928, pp. 644–5 and Tansley's long and final rejoinder, from Saas Fee in Switzerland, is 15 September 1928, p. 757.

[75] Godwin, 'Arthur George Tansley, 1871–1955', p. 238.

[76] The influences on Tansley's thinking concerning the 'ecosystem' have been detailed in terms of physics, psychoanalysis, politics and philosophy. See, for instance: Peder Anker, 'The context of ecosystem theory', *Ecosystems* 5 (2002), 611–13; A.G. Tansley (prepared by Peder Anker), 'The temporal genetic series as a means of approach to philosophy', *Ecosystems* 5 (2002), 614–24; Laura Cameron, 'Ecosystems', in Stephan Harrison, Steve Pile and Nigel Thrift (eds.), *Patterned Ground: Entanglements of Nature and Culture*, London: Reaktion Press, 2004, pp. 55–7; Laura Cameron and Sinead Earley, 'The ecosystem: movements, connections, tensions and translations', *Geoforum* 65 (2015), 473–81.

Anna O.[77] In addition, there is a manuscript preface to 'a series of essays on various topics that have interested me specially in what I call the New Psychology'; the book, obviously a follow-up to his 1920 best-seller, was never completed.

Thus, even though Tansley published little work in psychology after 1927 until his final book, *Mind and Life*, an overarching synthesis of the twin preoccupations of his professional career, his interest in psychoanalysis did not diminish, nor did he lose his contacts with the BPaS.[78] *Mind and Life* (1952) was a direct continuation of his larger vision, first recognized by the Gerstenberg Studentship in 1896, well expressed in *The New Psychology and its Relation to Life*, for an overall vision of the integration of biology and psychology, psychology always conceived of as psychoanalytically based, together with characteristic English inflections (e.g. the instinct theory of McDougall; a properly Darwinian framework). In 1941, he provided the Royal Society with its obituary for Sigmund Freud.[79] Botany did, quite clearly, however, dominate the rest of his life. During the 1930s, he worked on his revision of his 1911 *Types*. Eventually completed after his retirement from Oxford in 1937, *The British Islands and their Vegetation* (1939), his magnum opus, was a vast survey of over 900 pages, the culmination of the phase of ecology which he had initiated. It was the first major book to employ the ecosystem concept systematically: vegetational communities are shown to be the result of the interacting processes of plants, climates and soils in a dynamic landscape lively with human and animal activities. In 1931, he handed over ownership and editorship of *The New Phytologist*. In 1938, he finally gave up editing the *Journal of Ecology*. The fifteen years following his retirement were very productive of publications.[80] In 1941 he took a guiding role in the planning of government postwar nature conservation which led to the foundation of the Nature Conservancy in 1949, of which he was the first Chairman, retiring in 1953.

He was also heavily involved (as President 1947–53) in the Council for the Promotion of Field Studies (later the Field Studies Council),

---

[77] See John Forrester and Laura Cameron, '"Cure with a defect": a previously unpublished letter by Freud concerning "Anna O."', *IJP* 80(5) (1999), 929–42.

[78] For instance, he gave a confidential opinion on a Prize Essay submission on 'The sense of injustice and its relation to oral sadism' in a letter to Sylvia Payne, 18 June 1941, BPaS, G06/BA/F04/03.

[79] A.G. Tansley, 'Sigmund Freud (1856–1939)', *Obituary Notices of Fellows of The Royal Society* 3 (1939–41), 247–75.

[80] *The Values of Science to Humanity*, 1942; *Our Heritage of Wild Nature*, 1945; *Plant Ecology and the School* (with E. Price Evans), 1946; *Introduction to Plant Ecology*, 1946; *Britain's Green Mantle*, 1949; *Oaks and Oakwoods*, 1952.

a voluntary organization which created and maintained resident field centres in various locations of ecological and geological interest (such as Flatford Mill in Suffolk), where students could explore natural history interests and painting. Such an interest in decentralized education and the nurturing of 'scientific curiosity'[81] resonated with his active joint leadership (with John Baker and Michael Polanyi) of the Society for Freedom in Science, an organization which, from 1940, fought strongly against the central planning of scientific research, the rising tide of 'Bernalism', being introduced as orthodoxy with the new bureaucratic forms and quasi-socialist ideals of the postwar settlement. Perhaps in part due to Tansley's recruiting activities, ecologists made up more than a quarter of the SFS membership.[82] In this, yet another of the new and extra-academic institutions which he had helped to found over some fifty years, Tansley felt free to express his views on psychology in a pamphlet, 'The psychological connexion of two basic principles of the SFS'. And, being now a distinguished knight and longstanding member of the BPaS, in 1951 Tansley was one of the signatories for an Appeal for £100,000 for an Institute and Clinic of Psycho-Analysis, along with Dr J.C. Flügel, Dr William Gillespie (its Chairman), Dr Ernest Jones, Professor L.S. Penrose and Professor F.R. Winton.[83] Tansley died on the 25 November 1955, aged eighty-four, in Grove Cottage, Grantchester just outside Cambridge where he had lived since 1906.

Tansley's Dream

> But the dream itself, the patent, the obvious content of it, is entirely harmless. Nobody outside the psychoanalytically trained could possibly tell what it meant, that it meant hell, deep . . .
>
> Professor Sir A.G. Tansley, FRS, 1953

Seen from the outside, Tansley's life places him as an intriguing and symptomatic character in the development of psychoanalysis in England. Coming from the non-medical sciences, inspired by his friends and

---

[81] A.G. Tansley, 'The psychological connexion of two basic principles of the SFS', *Society for Freedom in Science*, Occasional Pamphlet 12, March 1952.

[82] The figure, seventeen ecologists out of sixty-one members, comes from Stephen Bocking, *Ecologists and Environmental Politics: A History of Contemporary Ecology*, New Haven: Yale University Press, 1997, p. 27.

[83] On Winton, see Chapter 6, p. 391.

colleagues in psychiatry to immerse himself in Freud's work, he unexpectedly wrote a book – as much for his own satisfaction, it appears, as for any
other reason – that caught the spirit of the times and led him, dissatisfied
with his academic position and future, to engage seriously in psychoanalysis with Freud and other like-minded colleagues in Cambridge and
within his informal scientific community. As he himself put it, writing in
1932, in 1926 'it was touch and go whether I became a professional psychoanalyst' or took the Chair in Botany in Oxford. Oxford won out and
psychoanalysis suffered the loss of the 'nice type of an English scientist'.

But there are also a number of pieces of evidence that make it possible to
undertake a speculative reconstruction of Tansley's life from within. First
and foremost, he himself transcribed a dream which he regarded as a crucial
turning-point in his life. In addition, he left a number of autobiographical
works in which he tried to explain how and why his life took the shape it
did. With these materials, and with the benefit of hindsight, we can
venture a psychoanalytic reading in order to make clearer how his influential and, it would seem, fulfilling life came to have one shape rather than
another.

In the summer of 1953, when interviewed by Dr Kurt Eissler,
founder of the Sigmund Freud Archives, Tansley's account of why
and how he became involved with psychoanalysis centred entirely
around the dream. He found the interview procedure uncongenial for
recalling the events which Eissler was interested in, and instead promised to send Eissler a written account, which he did. Eissler, for his
part, sent a transcript of the interview to Tansley, who returned it
together with two other documents he had prepared. Thus the complete holding of the Tansley Section of the Freud Archives in the
Library of Congress consists of three documents: the first, a seven-
page account prepared by Tansley and entitled by him: '(1) The impact
of Freud's work and personality on a non-medical biologist'.
The second is a two-page type-written document he entitled
'THE DREAM'. The third is the Eissler transcript. Before returning
the documents to Eissler, Tansley added in pen a new title to the
package: 'Three Contributions by Sir Arthur Tansley, F.R.S.'

All three of these 'contributions' centre on Tansley's dream. In the
first, his written account of Freud's impact, he notes:

> At that time [the first decade of the twentieth century] I had read none of
> Freud's publications, and although I was intrigued by what I heard from
> Hart and his colleagues my interest was only vividly aroused as the result of

a dream which I had some years later, after I had moved to Cambridge. This dream and my 'automatic' analysis of its content are described in another contribution to the Archives (Ref. 2). I was so deeply impressed by this experience that I began to read Freud's works, notably the *Drei Abhandlungen zur Sexualtheorie* and Brill's translation of *Die Traumdeutung*, as well as Jung's study of Dementia praecox and his *Wandlungen und Symbole der Libido*. The *Sexualtheorie* I found particularly impressive and illuminating.

The third document, the transcribed interview, gives a similar, though less focused account of his initial serious interest in psychoanalysis as a result of the dream, which, as 'Ref. 2', thus forms the centre-piece of this triptych.

It is Tansley's whimsical afterthought for his submission that makes one pause, in Freudian style, for thought. To call these three documents, each centred primarily on Tansley's own dream, his 'Three contributions' is to allow them to mirror the work of Freud's that Tansley most admired, the *Drei Abhandlungen* – 'one of his most penetrating fundamental works, those *Three contributions*, as they have been called in translation'.[84] Yet the overall message of these three documents is: he learned little from Freud that he had not already discovered in this dream. To put it crudely: when Tansley was asked about Freud's influence upon him, he replied by saying that influence was minimal, and offered as proof the dream he had had long before meeting or even reading Freud.

The dream text Tansley submitted to the Sigmund Freud Archives opens this chapter. Tansley also included the following comments, associations and a page he entitled 'Interpretation':

> The dream was so vivid and dramatic and had made such a strong impression on me that I recounted it at breakfast with no notion that it had a hidden meaning. I was aware at the time of Freud's work, which had been described to me by a friend who was a psychiatrist, so that I knew roughly the nature of the technique of free association in the interpretation of dreams. At that time, however, I had read none of Freud's writings.
>
> Very shortly afterwards (I think the day following the night of the dream) I began to analyze this dream by seeking associations to the general picture and to the various dream images and sensations. This I did without conscious intention – my mind wandered, as it were, without conscious

[84] The *Drei Abhandlungen* were known in English translation either as *Three Contributions to the Sexual Theory* in Brill's translation of 1910 or from later editions on as *Three Contributions to the Theory of Sex* (1916, 1918, 1930, 1938). The first English edition to use the title *Three Essays on the Theory of Sexuality* was that of James Strachey in 1949. The volume of the *Standard Edition* in which the *Three Essays* appeared, Vol. VII, was published in 1953.

volition, while I was riding a bicycle, around the dream images and sensations. Gradually, but surprisingly quickly, with no notable resistance, the interpretation took shape and gained my complete conviction of its correctness.

These occurrences impressed me very deeply and led to a resolve to read Freud's work. This I did in the months that followed, beginning with the Traumdeutung, and following with the Drei Abhandlungen zur Sexual Theorie, and some others. The latter – the Sexualtheorie – interested and excited me immensely. I felt that it was an extraordinarily able and illuminating work, and, after having read far more widely in Freud since then, I still think that in some respects it is his most outstanding contribution – a daring and successful synthesis clearly and admirably expounded. My interest in the whole subject was now thoroughly aroused, and after a good deal of thought I determined to write my own picture of it as it shaped itself in my mind.

The New Psychology and its Relation to Life was published in 1920, and in 1922 I spent 3 months with Freud in Vienna, and in 1923–4 another six months.

Free associations to the setting, images and sensations of the dream

The sub-tropical scene – South Africa. Several of my old pupils had gone to that country, including a girl with whom I had fallen in love.

'Separated from my friends' – My old pupils, who were dispersed owing to the war of 1914–18, and especially the girl.

The savages. Pictures of Zulus on the warpath.

Their numbers and spears – my rifle – Overwhelming strength against me in spite of my superior weapon.

My wife's white clothing – 'Purity' in the sexual sense.

Interpretation

I was separated from my beloved and unable to take any active steps towards union with her because I was married and public opinion (of the 'herd'[85] in Wilfred Trotter's sense) symbolized by the savages would be unanimously against me. (Note that the 'herd' in this case was not of my own race and was regarded by me as intrinsically inferior.) Since I had a good reputation in 'my' world, this opposition was quiescent, only potential – there was no active hostility. I was in no danger where I was provided I remained there. But the barrier was impregnable, in spite of the fact that my mental

---

[85] The typescript has 'hero' here and in the next parenthetic phrase: apparently a typing error, either Tansley's or introduced by Tansley's assistant (if he had one) or by the Sigmund Freud Archives. The third possibility seems more probable as the typescript is in Pica 10 pt, unlike Tansley's first document, which is in Courier 12 pt. On the cover sheet for THE DREAM, a stamp specifies: 'THE SIGMUND FREUD ARCHIVES/Recollections/Sir Arthur Tansley's/self-interpretation of a dream./#832 (Copy).' Given the date, the word 'copy' probably refers to a re-typing, rather than to a photostat or other mechanical copying process.

equipment, symbolized by the rifle, was much superior to that of herd in quality, symbolized by the spears. The numbers and unanimity of the potential opposition made a successful escape impossible and my superior weapon useless. But the savages made no attempt to attack my wife or stop her coming to me – that she could legitimately be with me was a matter of course.

The end of the dream in confusion meant that my problem was insoluble. The suggestion of firing the rifle – quite vague and uncertain – I could not interpret. It occurred to me that the rifle might be thought to symbolize the male genital organ and the firing, orgasm. But this would not fit in with the rest, and I concluded that the only reason I thought of it was because any offensive weapon, especially an elongated one, is well known as a symbol of the penis, not because it had any such significance in the dream. A possible alternative is that I shot at my wife, but I cannot confirm this, perhaps because it was as impossible in the dream as it would be in real life.

What are the principal elements of Tansley's interpretation of his dream? He interpreted his dream as being a representation of an insoluble conflict between his desire to be with his 'beloved' and the overwhelming barrier of public opinion. It appears to be a dream in which the familiar themes of purity and pollution, the wife and the beloved other woman, are placed in the context of a struggle of conscience between 'civilized' sexual morality, as Freud called it, and the desires of the individual.

Yet, in a sense, the dream and Tansley's comments (written some forty years afterward) carry no sense of this struggle. Everything in the dream appears already decided: it is inconceivable for the dreamer to shoot his wife, in the dream as in real life; it is inconceivable for a single individual, no matter how gifted, to stand against public opinion, no matter how inferior. The game is lost before it has begun. There is an atmosphere of resignation to the inevitable in the dream. It is possible that this atmosphere stems in part from the tragedy unfolding in France, the victory of the 'herd' of public opinion over the intelligence of superior men. Certainly at the time of his dream Tansley had a striking example of that defeat close to hand, in the fate of his once close Cambridge friend Bertrand Russell, deprived of his Trinity Fellowship for his conviction under the Defence of the Realm Act in 1916. But we have no evidence that there is such a network of associations to the dream. Indeed, it is perhaps more important to note that Tansley did not share Russell's pacifism; excluded from active service due to the fused and twisted fingers of his left hand (perhaps the 'ugliness' that Russell had earlier seen in him) and his age, Tansley's work at the Ministry of Munitions entitled him to wear a uniform. Peter Ayres recounts the family story of Tansley in uniform walking into Trinity College and bumping into Russell who 'ostentatiously turned on his heel and without speaking left the room'.[86] One doubt does remain: the question of the firing of the rifle. Tansley could not interpret this element, but asserted quite definitively that it had neither a sexual, phallic meaning nor a murderously aggressive one. This curious gesture, of leaving the rifle neither sexual nor aggressive, makes the dream more mysterious than it appears at first sight.[87] It is this mystery, which underscores the sense of the dream's importance. After all, Tansley's life, and his account of the dream, lead us to believe that his life was transformed by the experience of dreaming it. We need to look in the dream for a mystery, a significance, which is worthy of such a thing as the transformation of a man's life. It's best to go slow in broaching such matters.

The first thing to note is that Tansley's discussion of his own dream is divided into four parts: the exposition of the dream; followed by a context for the dream (his prior knowledge of psychoanalysis, the circumstances surrounding the dream, including his initial complete lack of awareness of its meaning, the almost involuntary process of interpretation and its effect on him – his reading of Freud, his writing of *The New Psychology*, followed by his analysis with Freud). The third part is the set of associations, followed immediately by the fourth, the interpretation.[88] One would expect the content of the dream, and its interpretation, to throw some light on the connection between the preamble and the effect of the dream on his life. Yet we are immediately confronted with a mismatch. The centre of the dream appears to be about a moral conflict whose resolution is never in doubt, yet its effect on Tansley is in an entirely different sphere: in his relation to Freud and to the development of psychoanalytic ideas.

To put it crudely: where the dream appears to be about whether he should remain faithful to his sexually pure wife or disappear into the subtropical bush with his 'beloved', his account of its effect on him has him

---

[86] Ayres, *Shaping Ecology*, p. 107. The story of their difference of opinion over the war was significant enough for it to be a topic of conversation when Solly Zuckerman, whom Tansley got to know well in Oxford, came to stay at Grantchester in the early 1930s (Solly Zuckerman, *From Apes to Warlords: The Autobiography (1904–1946) of Solly Zuckerman*, London: Hamilton, 1978, p. 27).

[87] Tansley included an anonymized and considerably pared down version of his own important dream in *The New Psychology*, p. 130, including it as an example of extreme symbolism: 'The more extreme forms of symbolism are met with in the dreams of adults when the affects are very deep and very strongly repressed as the result of conflict. Very many such dreams are concerned with sexual relations.'

[88] In choosing this expository structure, Tansley used a method similar to Freud's own, as first set out in the analysis of his 'specimen dream', the 'dream of Irma's injection' – Preamble, Dream, Analysis. Whereas Freud fuses the associations and their interpretation, Tansley separates them, and inverts the order of Preamble and Dream.

disappearing into the bush of the new psychology with another new beloved, Freud. The mapping of the interpretation of his dream concerning his 'beloved' on to his new absorption in and by psychoanalysis is very close. Yet this is clearly a retrospective analysis of the dream, because on his bike-ride when he interpreted the dream, Tansley could not know that Freud and psychoanalysis would become his new beloved, a new affront to public opinion. In other words, his interpretation is not an interpretation in a Freudian sense, which recognizes dream-wishes as moulded into a 'perfect likeness [*Ebenbild*] of the past',[89] but rather more akin to a perfect likeness of the future. In interpretative terms, we smell a rat.

But let us go over this ground again, more slowly. Tansley's narrative of the dream and its immediate after-effect tells us that he recounted the dream over breakfast 'with no notion that it had a hidden meaning'. That is, even though he knew through Bernard Hart of some of Freud's views on dreams, he obviously did not take them to heart. He includes this detail in order to demonstrate that he believed thoroughly, at that time, in the innocence of dreams. But in the course of the next day, he underwent a process of automatic interpretation – a vision on the road to Damascus, except in all probability his was a vision on a bike-ride to Grantchester. Another epiphanic bike-ride to Grantchester was made by Bertrand Russell in the spring of 1902, when 'suddenly, as I was riding along a country road, I realised that I no longer loved Alys'.[90] Did Russell share this story with Tansley? Beyond Tansley's intimation that he and Russell argued about Russell's opposition to the war, there is little evidence that Russell – in Cambridge for most of the time till his dismissal by Trinity during the First World War in 1916 – and Tansley talked intimately after 1896.[91] In Russell's diary entry for 21 November 1902, he records seeing Tansley at a distance at a dinner meeting: 'gave me a bad conscience'.[92] Russell's autobiography was published only after Tansley's death. There remains the faint possibility that either Tansley or Russell used the other's recounted story of an epiphanic bicycle ride for his own purposes. Along with this possibility, whether or not they argued about war, or even Freud, there remain the strong emotions that Tansley had for Russell: their attraction,

---

[89] Freud, *The Interpretation of Dreams, SE* V, 621.

[90] BRA I, pp. 147–8; also commentary in MBR I, p. 145. It should be noted that Russell's *Autobiography* up to his marriage to Dora in 1921 was drafted in 1931 (MBR II, p. 125).

[91] Ayres in *Shaping Ecology* (p. 50) records a letter from Tansley to his sister Maud, 18 September 1896: 'I have been staying at Haslemere for a few days . . . in a cottage belonging to Russell's father-in-law, Mr. Pearsall-Smith. Russell was at Friday's Hill, Mr. Smith's house, part of the time, and several Cambridge men of our time . . . so you can imagine I had a splendid time.'

[92] BRCP 12, p. 11.

or what Tansley does not discuss, the pain of rejection and loss of intimate friendship with 'the most penetrating mind'.[93] The result of the epiphany was not a moral decision, but an intellectual certainty: 'my complete conviction' of the 'correctness' of this interpretation. This intellectual conviction then led to his absorption in Freud's work, principally the book on dreams and the essays on sexuality.

On the face of it, this is an odd response to a new intellectual enthusiasm. Tansley had few qualifications for this task – as he was to find out, after the event, when his readers wrote to him. Yet it repeats his own experience of his own dream: instead of resolving a moral dilemma, he emerged with an intellectual conviction of his own correct interpretation. In response to reading Freud, he did not engage in a moral or personal debate with Freud or any other worker in the field of psychoanalysis, but developed and then displayed his own intellectual convictions. In short, he responded pre-emptively with intellectual mastery, just as he had done with his dream. Faced by the confusion in his dream, which he knew signified an insoluble problem, he quickly arrived at 'complete conviction' of his own interpretation.

Tansley displayed his considerable intellectual virtues in this process: encountering an interesting problem, which led him to a profound conviction, he mastered the literature and provided a general, judicious and

---

[93] Russell was a recurring motif in Tansley's own story of his life. Harry Godwin relates Tansley's tale of meeting up with Russell in 1944 when Tansley was elected an Honorary Fellow of Trinity and Russell had been elected a teaching Fellow: 'Russell then made some amends by saying about that episode [of sharply differing on the war], "you were right then, Tansley, and I was wrong".' See Harry Godwin, 'Sir Arthur Tansley: the man and the subject'. *Journal of Ecology* 65 (1977), 19. A few months before he died, Tansley wrote Russell a letter, the entire text of which is: 'I trust your recent experience has not caused any of the leaves of the green bay tree to wither or wilt' (3 March 1954). The immediate occasion for the letter was Russell leaving hospital after a prostate operation (*The Times*, 24 February 1954, p. 8). But Tansley wrote the letter in a code only Russell would understand: the reference to the 'green bay tree' is to Petrarch's Sonnet 269: 'Rotta è l'alta colonna, e 'l verde lauro,/Che facean ombra al mio stanco pensero [Broken the column and the green bay tree/ That lent a shade to my exhausted thought].' On 11 December 1894, two days before Russell's wedding to Alys, Tansley had given Russell 'a copy of Fraticelli's *Dante*, and also a little pocket "Commedia", and Petrarch', recalling the holiday in Italy Tansley and Russell took together. Tansley's letter was effectively an envoi to their friendship: 'You have been more to me as a friend than any other single individual, during the two short years I knew you intimately. And although it is the emotional side of our friendship that rises first into one's mind, there is a great deal besides . . . And the past nothing can remove – you have made me permanently happier, and, I think, better.' After Tansley's death in 1955, Lord Adrian, then Master of Trinity College, specifically invited Russell to attend a celebration on site of the completion of the memorial to Tansley constructed at Chanctonbury Ring on the South Downs, which Russell declined. The Trinity College Report recalled Tansley's oft-quoted memory of Russell's comment to him on one of their vacation walks: 'It can be regarded as an axiom that any view which includes Chanctonbury Ring is a good view' (BR 0000526, 10 November 1957).

unbiased overview of a large and unstable field of antagonistic views. The very literary success of Tansley's book revealed something incongruous about his achievement: it was the first general account of the new findings in dreams, sexuality, psychopathology and the theory of the unconscious in English, when it came out in June 1920. Tansley had published the right book at the right time. Yet the response from readers and, we may speculate, residual doubts of his own, led him to view with some scepticism his right to stand before the English public as an authority in the field of psychology. Perhaps the problem that he regarded as insoluble in his dream still remained insoluble, despite his having successfully displaced it into the intellectual terms of the relations between biology and psychology.

One of the strangest effects, then, of his proclivity for intellectual mastery was him appearing as an authority on psychoanalysis. A self-taught authority, and therefore in danger of occupying Freud's position as authority. It is this independence that so clearly marks Tansley's relations with psychoanalysis from the start. He is astonished to make the independent discovery of the meaning of his dream; he acquires his authority on psychoanalysis entirely independently. He is, it would seem, rather like Freud.

The similarity between the two men should not be discounted. Tansley was older than most of those interested in psychoanalysis at this period. In 1920, he was forty-nine. Born in 1871, he was four years older than Jung, eight years older than Jones, and only fifteen years younger than Freud. More immediately relevant, perhaps, his scientific trajectory had two important similarities to that of Freud: he was identified by others as the founder of a new scientific discipline – ecology for Tansley, psychoanalysis for Freud; and his appetite for the organization of colleagues into newly minted institutions was, if anything, even greater than Freud's. Tansley's journals and his societies were commercial in the same way as Freud's were, not primarily affiliated with university departments, again like Freud's, and were astutely sensitive to a new wave of internationalism in the early years of the twentieth century. There is, then, a strange similarity between these two men, on the face of it so different. Tansley's tone in discussing Freud was predominantly that of an equal discerning the grounds for his undoubted admiration of the founder of psychoanalysis. It was Tansley who wrote Freud's obituary for the Royal Society, and, as Godwin shrewdly noted, nearly all of the gifts that Tansley described in Freud were ones that he 'unconsciously acknowledged' as being his own – or were at least those they had in common.[94] The key difference between them, though, and it was one Tansley would have readily admitted, was Freud's striking originality. Nonetheless, in talking to Eissler, Tansley remembered that Rickman's 'impression was that Freud and I discussed analysis rather like two sovereigns when we conducted Analysis'.[95] Tansley had a right to be regarded as a sovereign in his own discipline, and may himself have viewed his analysis like that (from where else could Rickman have received his impression?). Nonetheless, it is a telling observation about how Tansley approached analysis with Freud.

What happened in Tansley's analysis? When recollecting it in 1953, he was evidently disappointed that he had made no great discoveries of forgotten scenes from his childhood, and he was disappointed that Freud spent more time discussing theoretical questions than Tansley's own unconscious. Yet who exactly was responsible for this disappointment is not clear. Tansley made it plain to Eissler that he was not neurotic and did not need analysis:

> We never seemed to penetrate at all deeply into my 'Unconscious', and I think the main cause of this failure was probably that I had no marked neurosis, but a fairly stable mental and emotional equilibrium which was difficult to upset or penetrate, so that there was little unconscious material which could be brought to the surface. The analysis was thus of the nature of a 'Lehranalyse', and could not closely resemble the analysis of a neurotic patient.[96]

Thus Tansley gives the impression that nothing much happened – and that it was a mixture of Freud's fault for being too interested in theory, and his own fault for not being neurotic enough:

> from a personal point of view, as I say, I don't think it was a really good analysis. I think he departed /laughs/ from his own technical procedure.

---

[94] Godwin, 'Arthur George Tansley, 1871–1955', p. 243; see Tansley, 'Sigmund Freud', p. 272: 'Joined with his questing intelligence and his dominant passion for knowledge and truth was profound natural psychological insight. His intuitive judgments of men and their motives were exceedingly shrewd and very seldom at fault, and his remarkable self-knowledge enabled him to recognize his own motives with the same sureness as those of others. These gifts were the foundations of his success in penetrating to and analysing the deeper causes of human behaviour. Allied with them were his inveterate scepticism of current "explanations" which did not explain and his unvarying insistence on independent judgment of all the relevant facts, including psychological facts such as the images and fantasies produced by the neurotic and the normal mind, previously ignored or dismissed as unworthy of the psychologist's attention. Lastly there was his power of hard and continuous application to daily clinical work and his simultaneous unfailing interest in and constant consideration of every human activity that could throw light on the workings of the human mind.'
[95] Eissler interview, AGTSFA, p. 4.   [96] Tansley, 'The impact of Freud's work', pp. 4–5.

Because of course he recognized that I was not an ordinary patient. /laughs/
I wasn't, I wanted information to get to know more about the subject and so
on, rather than concentrating on my unconscious, as I say.[97]

Tansley gives two different sorts of reason for this unsatisfactory state of
affairs. Firstly, Freud didn't focus on Tansley's own personal life, being too
ready to discuss general questions. Secondly, Tansley implies that he was
a special patient, who was in analysis to learn about the theory, not cure
himself of a neurosis. The image of the two sovereigns discussing analysis
obviously appealed to Tansley; he liked to think of himself as Freud's equal
and hence independent of him. As he said to Eissler: 'The best dream that
I ever had I analyzed myself. And I told Freud what my interpretation
would be. And he said I was perfectly right.' The same held true for
Tansley's overall view of Freudian theory. Enormous respect and agreement – but somehow he was a special case: 'I think in my own case, the
Oedipus complex was almost negligible. Maybe, because my parents were
exceptional people. But I, I could not trace by any means the sort of effect
that the complex is supposed to have on one's life and emotions in my own
case.'[98] Equally telling was Tansley's aside to Eissler about his dream:
'An excellent example, surely, of a frustrated, not a fulfilled, wish! But
I did not say so to Freud.' Clearly, Tansley held himself back from the
analysis, not willing to hurt Freud's feelings or provoke a conflict; he
thereby preserved his independence from Freud by failing to engage with
him. Despite his sovereign distance, though, he became and remained
a great admirer of Freud's, perhaps even besotted with him. The most
striking piece of evidence for Tansley's high regard for Freud comes from
a story recounted by his close friend and colleague Harry Godwin: when
Tansley, while at an Oxford social function in the 1930s, was asked to name
the most influential man since Christ, he answered, without hesitation,
'Freud'.[99]

Yet these accounts still leave a mystery around the question: why was
Tansley in analysis with Freud in Vienna in the first place? Accepting for
the moment that the reason he gave – that he needed to acquire more
expertise about psychoanalysis in order to answer those of his readers who
had approached him – is insufficient as a motive for starting analysis with
Freud, the account he gave Eissler in 1953 made it transparently clear that
he was there because of something in his dream, or as a result of his dream,
which he analysed before he had read Freud and before his analysis: he was

[97] Eissler interview, AGTSFA, p. 3.   [98] *Ibid.*, p. 7.   [99] Godwin, 'Sir Arthur Tansley', p. 25.

still seeking with Freud something that was 'left over' from his dream. Although his account of his analysis indicates his disappointment – he appears not to have found what he was in search of – the consequences of the dream constituted such an upheaval in his life that we should follow them all out before returning to the question of what, exactly, the dream changed in his life and what the dream signified that he changed his life in order to find.

The obvious question to ask is: what did Freud think of Tansley's dream? Tansley mentions this in his account of the dream to Eissler and in the written version of 1953: 'I recounted the dream and my interpretation to Freud in 1924, and he said the interpretation was undoubtedly correct.' Maybe this was what Tansley was in analysis for: Freud's approval of the interpretation of his dream? Certainly something to do with dreams took place early on in his meetings with Freud.

On Maundy Thursday, 1922 – the day when a sovereign reverses social roles, washes the feet of his or her subjects and gives out special gifts – Tansley presented Freud with a copy of *The New Psychology*, in which he inscribed the words: 'Prof. S. Freud, from the author, 13 April 1922.'[100] The next day, Good Friday, Freud returned the compliment – he gave Tansley a copy of the sixth edition of *Die Traumdeutung*, and signed it: '14.4.22 Herrn Prof. Tansley zur freundlichen Erinnerung an den Verf' – 'To Prof. Tansley with friendly memories from the author.'[101] Two days later, on Easter Sunday, Freud reported to Jones on Tansley's analysis: 'Tansley is bringing up enormous resistance.'[102] To the commentator, it is no surprise that the giving of the gifts provoked something substantial in the way of analytic material. And it was, it appears, Tansley who began the cycle on the Thursday, only to be outmanoeuvred by Freud.

This was to be Freud's last communication to Jones concerning the analysis; as with other patients, once something 'analytic' started happening, his communications to those outside dried up.[103] We have no record of the next three months of Tansley's analysis and the few surviving letters to Clements during this period reveal little as to what happened in the second

---

[100] Copy in Freud's Library, Freud Museum, London.
[101] Copy in Library, Sigmund Freud Museum, Vienna. Special thanks to the late Lydia Marinelli, Curator, for her generous assistance concerning books held by the Museum that were once owned by Tansley.
[102] FJ, 16 April 1922, p. 474.
[103] For examples, see his analysis of Loe Kann and of Joan Riviere, discussed in Lisa Appignanesi and John Forrester, *Freud's Women*, London: Weidenfeld & Nicolson, 1992.

period of analysis in 1924.[104] Letters to his sister Maud at this time reveal that he was interested in learning details about his childhood, including the age at which the parents removed his cot from their bedroom.[105] What we do have, again, are a record of the gifts Tansley presented to Freud and Freud's responses to them. In the first week of 1924 Tansley restarted analysis with Freud, which came to an end with the summer break, when Tansley returned to England.[106] Tansley later admitted he did not have an extensive correspondence with Freud, but he did continue to send him gifts. The first was at the end of the year: a copy of A.A. Milne's *When We Were Very Young*, with the dedication: 'To Prof. Freud, A.G.T., Christmas, 1924.'[107] With this gift (the knowing reader in the early twenty-first century imagines with a smile Sigmund Freud's reactions to James James Morrison Morrison Weatherby George Dupree taking great care of his mother, though he was only three), Tansley revealed yet another surprisingly astute intuition, this time in a very different field. This book was the first of Milne's (1882–1956) publications for children – the first addressed to his son Christopher Robin Milne. It was published on 6 November 1924, six weeks before it arrived on Freud's desk.[108] The even better known *Winnie-the-Pooh* was first published in 1926, *Now We Are Six* in 1927 and *The House at Pooh Corner* in 1928. Milne's book would have been very little known in Vienna when Tansley chose it as an 'end-of-analysis' gift.[109] It is, of course, possible that Tansley was

---

[104] There is correspondence between Tansley and Clements during this period, including discussion of botanical topics, the international situation and the contribution of Freud. Tansley sent to Clements five letters from 11 November 1923 to 10 April 1924 from three different addresses in Vienna. The first flat, remembered fondly by Helen Thompson, Tansley's youngest daughter, though she was homesick, was Alserstrasse 21, the home of Clemens, Baron von Pirquet, the originator of a tuberculin skin test and the coiner of the term 'allergy', and a five-minute walk from Freud's flat. Pirquet was to have spent the winter in Minneapolis but found the situation so 'impractical' he returned home early. Tansley's family then moved to the 'more expensive and less comfortable' Pension Pohl at Rathausstrasse 20. (Letters from A.G. Tansley to F.E. Clements, 11 November 1923, 20 November 1923, 28 December 1923, 19 February 1924, 10 April 1924, FEC. Interview with Helen Thompson, 21 April 2001, conducted by Laura Cameron.) Helen recalled her father being 'very depressed' when he was undergoing psychoanalysis, as you do because 'all sorts of horrors from your childhood come up'.

[105] Ayres, *Shaping Ecology*, p. 113.

[106] Tansley's family returned to Cambridge at the beginning of April and Tansley stayed on at the home of Frau von Selig, Freiheitsplatz 4–5 until around 24 June. Letters from Tansley to Clements, 19 February 1924, 10 April 1924, FEC.

[107] Inscription in Freud's copy in the Freud Museum, London.

[108] Individual poems were published in *Punch* in the preceding years, with considerable immediate success, even provoking parodies, before the book was published. Personal communication, Ian Patterson, 26 February 1999.

[109] Freud did not make professional use of Milne's verses, but his daughter Anna did, quoting and interpreting a passage from *When We Were Very Young* in her *The Ego and the Mechanisms of Defence*

aware of Milne's work, and may even have been a friend of Milne, who studied mathematics at Trinity College, Cambridge, Tansley's own College, in the years 1900–3.[110]

Tansley's next contact with Freud came in 1928. Probably while attending the IPE meeting in July and August 1928 in Czechoslovakia and Poland, he paid Freud a visit in Vienna. Tansley remembered this as their last meeting.

> [S]eeing on the card I had sent in to him that I was now a professor at Oxford he immediately enquired: 'Ordentlich?' 'Jawohl,' I replied. 'Das ist gut,' he said. He had had enough experience of being 'ausserordentlich' himself to be acutely conscious of the difference, and was unfeignedly glad of my new academic status, which he was sure would be good for me psychologically. At the previous Christmas I had sent him a reproduction of Leonardo's cartoon of the Virgin, St. Anne and the infant Christ, and this he now showed me hung on the door of his study where he could always see it as he sat at his study table. I had known it was a gift he would appreciate because he was of course a great admirer of Leonardo, and it was unlikely that he had seen the original cartoon which hung in the Diploma Gallery at Burlington House in London. I was proud that my present was so much honoured, and flattered when he added, 'You do know how to give presents!' . . . I never saw him again and we exchanged very few letters.[111]

(1936), London: Hogarth Press and Institute of Psycho-Analysis, 1968 (revised edition), p. 83, concerning 'Denial in word and act'.

[110] In *The New Psychology*, p. 197, Tansley referred to another children's author to illustrate his point that 'the moral law holds for animal communities just as it does for human societies' – 'Mr. Kipling, one of our greatest modern exponents of the glories of herd life and the herd instinct, has brought this out in his delightful stories of the wolf pack and the bee-hive.' He specifically recommended 'Mowgli's brothers' in *The Jungle Book* (1894), 'How fear came' in *The Second Jungle Book* (1895), and 'The mother hive' in *Actions and Reactions* (1909).

[111] Tansley, 'The impact of Freud's work', pp. 3–4. In the interview, Eissler asked Tansley if Freud ever gave him a gift and he replied: 'No, he . . . Yes, he gave me a portrait, a photograph, and a copy of the *Traumdeutung*.' In Engelmann's photographs of Freud's study, taken in 1938, the Leonardo is visible from his desk, as Tansley noted, with evident pride, when he saw it on his 1928 visit to Freud. There is no record of the Pepys *Diaries* ever having been part of Freud's library, neither in the collection of books he took from Vienna to London, nor in the portion sold when Freud was forced to leave Vienna. See Nolan D.C. Lewis and Carney Landis, 'Freud's Library', *Psychoanalytic Review* 44 (1957), 327–561 and David Bakan's corrective 'The authenticity of the Freud Memorial Collection', *Journal of the History of the Behavioral Sciences* 11 (1975), 365–7. Tansley's recollection that this was their last meeting may not have been accurate. Freud entered Tansley's name in his Diary for the date Sunday, 12 April 1931 (see Michael Molnar (trans., annot. and ed.), *The Diary of Sigmund Freud, 1929–1939: A Record of the Final Decade*, London: Hogarth Press, p. 96); such an entry nearly always meant a personal visit. The date of the hypothetical visit was in the middle of Oxford's Easter Vacation; but there is no extant evidence of Tansley's movements or attendance at conferences for this period. It is not so much the fact of their meeting that is of significance; perhaps equally so is the fact of their not meeting in 1938 or 1939, when Freud was living in north-west London, where Tansley had himself grown up and gone to school. Tansley certainly did not visit Freud then.

What do we learn of Tansley's relationship to Freud from the four gifts he made him?[112] All four of them are very 'English' – the Leonardo because of its physical location, the others, being books, in a cultural rather than geographic sense. Tansley may have been careful only to send Freud things 'he would like and would be interested in', but he was also careful only to send Freud things that he didn't 'possess' already – because they were English. The Pepys *Diaries* and the A.A. Milne verses are telling: the first appears to contain a message that it was an Englishman, over two hundred years before Freud, who first conceived of the unravelling of the daily and inner life in a discipline of writing.

The Milne verses demonstrate an approach very different from Freud's but equally honouring to the inner life of the child – as if Tansley's message is that Freud may well be the great discoverer of the inner world of the child but his English contemporaries are pursuing similarly searching, if infinitely more light-hearted and less scientifically pretentious, projects of understanding the mind of the child. With hindsight, we can acknowledge how culturally alert Tansley was with this gift – it is not clear which of Christopher Robin or Little Hans should serve as the exemplary child of the early twentieth century. With these truly excellently chosen gifts, Tansley thus conveyed a curious message: a repeated attempt to reveal to Freud the autonomy of English culture with respect to psychoanalysis – its autonomy in the quest for self-revelation, its autonomy in the quest for knowledge of the inner world of the child. Yet again, it is the 'sovereignty' question, with Tansley, this 'nice type of the English scientist', implicitly cast as not required to submit to Freud's sovereignty.

Interwoven with the account of Tansley's gifts is his account of Freud's response to the news of his Professorship at Oxford. 'Ordentlich?' is Freud's response, and Tansley then explains how Freud knows very well – too well, Tansley implies – the difference between 'Ordentlich' (or 'Ordinarius') and 'Extraordinarius'. Tansley cannot resist letting Eissler – his reader – know what Eissler already knew very well: that Freud was never an 'Ordinarius' Professor. But Tansley here distorts rather severely the more usual view of Freud's academic status: that he struggled mightily, as an outsider, and against considerable prejudice, to be appointed Professor Extraordinarius; then having achieved it (in 1902) appeared perfectly content with his position, satisfied that it guaranteed him a place from

---

[112] The Leonardo used to hang in the Burlington, and thus off the beaten track – as Tansley recognized. Its acquisition by the National Gallery, London, was the occasion for a very public and impassioned campaign in the 1960s.

which to disseminate his teaching and the social recognition that went with the title of Professor.[113] In Tansley's implied version, Freud's position of Extraordinarius was the source of an acute sense of an inferior status, rather than the achievement of a long-wished-for goal. Here the question of sovereignty is being harped on with a vengeance – and strictly to Tansley's advantage: he implies that he had now achieved something that Freud himself had long wished for and had never succeeded in (at the date of the conversation, Freud was seventy-two, and thus beyond an age at which one could hope for such promotion). But it is clear that the person who cares about being Professor Ordinarius is not Freud but Tansley. Freud's question to Tansley – 'Ordentlich?' – may well have been one based on Freud's acute awareness of the relevant differences between ranks, but the question was not addressed because of Freud's *own* preoccupation with this question but most likely because Freud sensed Tansley's pride at having fulfilled what must have emerged in his analysis with Freud as a heartfelt ambition of his own.

Tansley's appointment to the Chair at Oxford does not only represent his assertion of independent, even higher, sovereignty in relation to Freud; he 'was unfeignedly glad of my new academic status, which he was sure would be good for me psychologically'. It is here that there is a sliding from the issue of sovereignty to an issue more closely tied to Tansley's analysis with Freud. It is not the fact that Tansley has become a Professor that is important to Freud, one might say; it is the fact that he has made what Godwin acutely called the 'return to the mother subject' of botany. What Tansley had done, in Freud's eyes, we speculate, was finally resolve the crisis in his life which had been initiated by his dream and the events it referred to. The intriguing – but not, finally, atypical – form of that life-crisis was Tansley's involvement with psychoanalysis and Freud. In this sense, we can stand back and see this psychoanalytic episode in Tansley's life as a protracted transference neurosis. To remind readers: the transference neurosis is that structure created during psychoanalytic treatment in its central phase, after the initial phase when the pre-existing symptoms have been interpreted and disappear – the transference neurosis is the

---

[113] Freud's desire to be a Professor is analysed in both Jones's and Gay's biographies of Freud and in the detailed studies of Joseph and Renée Gicklhorn and of K.R. Eissler. In 1924, Freud pedantically corrected Pfister on this biographical fact: 'Actually I was never a full professor of neurology and was never anything but a lecturer. I became a titular professor in 1902 and a titular full professor in 1920, have never given up my academic post, but have continued with it for thirty-two years, and finally gave up my voluntary lectures in 1918.' (Freud and Oskar Pfister, *Psycho-Analysis and Faith*, London: Hogarth Press, 1963, 9 June 1924, p. 95.) 'Titular' here means something between 'Associate' and 'Honorary'.

expression of the subject's neurosis entirely in terms of his or her relation to the analyst, to psychoanalysis, and to 'Freud'. Tansley's dream undoubtedly precipitated him towards an intense relationship with Freud. Despite the fact that his account of the dream in his deposition to the Library of Congress affirms the *achievement* represented by the dream – an achievement of interpretative mastery – it is more than likely, given what happened in the next ten years of Tansley's life, that it initiated a period of great confusion rather than clarification. 'Nobody outside the psychoanalytically trained could possibly tell what it meant, that it meant hell, deep', Tansley told Eissler. His moment of psychoanalytic understanding had revealed to him how his life had become hell. Thus precipitated into a period of emotional upheaval by the dream, we can regard his writing of the book, his resignation from Cambridge, his analysis with the man himself as the unravelling of this transference neurosis. And, Tansley plausibly intimates, Freud regarded the appointment to the Chair in Botany at Oxford as its final resolution.

One piece of evidence indicates that, no matter how Tansley had resolved his vacillation in the 'displaced' professional domain, his inner erotic preoccupations had not been resolved. We saw how the normally efficient Tansley failed to organize the 1925 International Psychoanalytic Congress in Cambridge; when the 1929 IPA Congress took place from 27–31 July at Queens' College in Oxford, Tansley, despite his being the sole member of the British Society who had a formal connection with the University, was at no point involved – the organizers were Joan Riviere and Sylvia Payne, and guided tours of the Colleges were led by Ernest and Mrs Jones, Edward Glover, J.C. Flügel and the psychiatrist W.H.B. Stoddart. Letters to Clements in this period indicate that Tansley was busy with writing and fieldwork and had already started plans for the following year's International Botanical Congress in Cambridge for which he was taking a major organizing role.[114] In a notebook from this period, there is the following undated paragraph:

> There is no 'armour' to protect one against such elemental hurts, I find. You must know this because the knowledge that we are sharing the pain may help. I am numb toward everything but these two days. But I cannot, cannot regret them, nor can I face absolute finality . . . It eases the pain to write but it is an indulgence, and I have hurt you too much already, my very dear.[115]

[114] Letters from Tansley to Clements, 11 June 1929, 8 July 1929, 21 August 1929, FEC.
[115] Undated entry in notebook entitled 'Psychology' which contains notes for 1929–36, TA.

These lines follow random notes on psychoanalysis, and vitamins, and quotes from Claude McKay's 1929 novel *Banjo: A Story without a Plot* (p. 252) under the header 'Psychology of sex. Negroes & whites', and an address delivered to the Chemistry section of the British Association for the Advancement of Science in Cape Town and Johannesburg, a meeting overlapping in time with the Oxford Congress (22 July to 3 August). While it is difficult to be certain to what this refers, this personal note set amongst references to South Africa stands as testimony to the continued strength of his feelings for the 'beloved'. The conflict of his dream from more than ten years was still as alive as ever. What is more, this theme of 'South Africa' may have infiltrated his professional work: one of Tansley's major ecological arguments in the 1930s was with John Phillips, a botanist based in South Africa, whom Tansley charged with drinking 'the pure milk of the Clementsian word':[116] the 1935 paper 'The use and abuse of vegetational concepts and terms' in which he introduced the concept of the ecosystem was largely an attack on Phillips who had been inspired by the holism of General Jan Smuts.[117] From the early 1910s on, South Africa was the scene of many important events and themes in Tansley's botanical work, and represented too the loss of close colleagues such as Charles E. Moss who left Cambridge for South Africa in 1917 and Robert S. Adamson, who in 1923 accepted the Harry Bolus Chair of Botany at Cape Town – and it probably continued to possess the secret emotional resonance stemming from his 'beloved's' past presence there.

We are thus reconstructing Tansley's 'neurosis' as having as its principal content the dual vacillation or splitting: splitting of his interest or commitment between botany and psychoanalysis, a vacillation or splitting that duplicated the stark choice in his dream between his love for a woman who was not his wife and submitting to the inferior but stronger forces of 'public opinion'. He is married to botany, but his beloved is psychoanalysis. Tansley himself used the terms splitting to describe his situation

---

[116] A.G. Tansley, 'The use and abuse of vegetational concepts and terms', *Ecology* 16(3) (July 1935), 285.

[117] Tansley argued that Phillips simply trotted out F.E. Clements's theory that vegetation constituted a 'complex organism' which could be studied in the manner that physiologists approached the individual organism. Although Tansley had initially adhered to this idea, he increasingly expressed dissatisfaction with Clements's organismal analogy as well as his 'monoclimax' – the theory that there was a single end point to vegetational succession in a given climatic area. In his 1935 paper, Tansley wrote that Phillips's three linked articles in Tansley's *Journal of Ecology* – John Phillips, 'Succession, development, the climax and the complex organism: an analysis of concepts I', *Journal of Ecology* 22 (1934), 554–71 and Phillips, '[Parts II, III]', *Journal of Ecology* 23 (1935), 210–46, 488–508 – invited 'attack at almost every point'. See Peder Anker, *Imperial Ecology: Environmental Order in the British Empire, 1895–1945*, New Haven: Yale University Press, 2001, pp. 118–56. No general debate appears to have followed the paper's publication and neither Phillips nor Smuts published a reply.

in 1923: 'I am doing psychological work here now [in Vienna], having resigned my lectureship at Cambridge ... I shall continue to edit the journals and also to work at ecology, but I do not know to what this splitting of interest will eventually lead.'[118] His resignation from Cambridge and his moves in the mid-1920s towards becoming a full-time psychoanalyst represent perhaps a more courageous defiance of 'public opinion'; yet it is recognizably a displacement of this defiance from one domain to another. And it is an interesting question whether the defiance of leaving botany for psychoanalysis was more or less stark than the alternative action of leaving his marriage for his 'beloved'.

It is natural, given this displacement, to seek understanding of his vacillation between botany and psychoanalysis in the more private vacillation of his familial and erotic life. But of this we know very little. His wife Edith figures as hardly a shadow in any account he or any others gave of his life. She was his student at UCL, but, apart from acknowledgements of indebtedness to her in some early index work and two early collaborative articles they wrote together, she is remarkably absent from his professional life.[119] James Strachey's letters to Alix of 1925 include a portrait of Tansley family life, but it says as much about Strachey and his tastes as it does about his hosts:

> Give me a well-off middle class household. Blazing fire in the bedroom, perfect bed, five-course dinner, excellent cooking, claret and port at dinner, hock at lunch, good coffee – what more can one desire? But besides these essentials Tansley himself is very nice & quite intelligent, Mrs. T is not too tiresome, and the girls most inoffensive though unluckily far from beauties. –They were all most affable; and last night Tansley went rather further, I think, than he'd intended, and poured out a good deal of his troubles: his life's interest hopelessly divided between his old love, botany and his new one, psychoanalysis.[120]

Prior to his resignation from the Botany School, Tansley was sparking interest in botany as well as psychology amongst young students, both male and female, and we might consider the various candidates for Tansley's 'beloved' amongst them. Appreciative students included Harry Godwin

---

[118] Cooper, 'Sir Arthur Tansley and the science of ecology', citing a letter dated November 1923.
[119] 'Notes on the conducting tissue-system in Bryophyta', *Annals of Botany* 15 (1901), 1 and 'On the structure of *Schizaea malaccana*', *Annals of Botany* 17 (1903), 493. Edith also created the index for *Types of British Vegetation*, Cambridge: Cambridge University Press, 1911, for which Arthur thanked her in the Preface.
[120] BF, p. 216, 23 February 1925. According to his granddaughter, Dr Anna Dickens, Tansley's daughters 'were well known for being attractive both in appearance and intellect'.

who would become, like Tansley, a key figure in British ecology, and
Ernest Pickworth Farrow, a young man who would later himself publish
work in both ecology and psychoanalysis and correspond with Freud (see
Chapter 4). Another was Eleanor Muriel Margaret 'Margot' Hume
(1887–1968), a pioneer in the field of nutrition research, and, perhaps, the
loved one associated with his dream.

After attending the Eastbourne Ladies College, Margot Hume entered
Newnham College and excelled in the Natural Sciences Tripos, completing Part II in 1910 with a Class I in botany.[121] Reminiscing of 'reading-
parties in the country in the Easter Vac, ostensibly to swot for the Tripos',
fellow Newnhamite Francesca Wilson recalled 'the beautiful Margot
Hume with her voice full of tears and her gay laugh. We didn't know
then what a distinguished scientist she was to become.'[122] She was

Fig. 2.4 E. Margaret 'Margot' Hume at Newnham College, 1909.

[121] *Newnham College Register 1871–1971*, Vol. I, *1871–1923*, Cambridge: [The College], 1979, p. 193.
The youngest of the Chick women, Dorothy Chick (1887–1919) was a classmate of Hume in NST
Part II. She became a surgeon, worked at the Royal Free Hospital in Gray's Inn Road and in 1915
was part of the first medical unit organized in Britain for the relief of the Serbian Front.
[122] *A Newnham Anthology*, ed. Ann Phillips, Cambridge: Cambridge University Press, 1979, p. 68.
Francesca Wilson later worked with displaced children and refugees in southern Spain during the
Spanish Civil War.

awarded the Bathurst Studentship and continued at Newnham as a senior research student. Hume was already familiar with Tansley through his course on Pteridophyta (ferns) which she had attended with her Newnham friend, Lucy Wills.[123] In 1911 they were both enrolled in Tansley's advanced plant ecology class. Hume was twenty-four, Tansley was forty, and it was likely during this period, while Hume held the research studentship, that she and Tansley began discussing Freud's work.[124] Over the next two years, Hume started to publish her research, one of her papers appearing in Tansley's *New Phytologist*.[125] Under Tansley's supervision, and at his suggestion,[126] she commenced new ecological experiments at the University Botanic Garden beginning in the autumn of 1911.[127] This research on the competition between *Galium sylvestre* and *Galium saxatile* in different soils was taken over by another of Tansley's students A.S. Marsh in autumn of 1913 when Hume went to South Africa. Marsh later enlisted and would be killed by a sniper in France in 1916, his loss sadly commemorated by Tansley in *The New Phytologist*. Tansley himself would take over and publish this work, the only time in his career that he was involved with small-scale experimentation.[128]

In November 1913 Hume accepted a Lectureship in Botany at the South African College in Cape Town.[129] Lucy Wills came to visit her there and when war was declared in August, volunteered for a few weeks as a nurse. Through discussion with Hume, Wills became deeply interested in 'Freudian teaching' to which Hume 'had been introduced by A.G. Tansley'[130] and, upon her return to England in December, Wills

[123] *The Newnham College Register*, Vol I, p. 203 records that Wills was at Newnham from 1907 to 1911 and received a Class II in Part II (Botany) in 1911. This is contrary to obituary information in *The Lancet* 283(7344) (30 May 1964), p. 1225, and commentary in Barry G. Firkin, 'Historical review. Some women pioneers in haematology', *British Journal of Haematology* 108 (2000), 6–12, both of which state she took the second part of her tripos in geology.

[124] Newnham University and College Lecture Lists, Newnham College Archives, Cambridge; 'Obituary: Lucy Wills', p. 1225.

[125] E. Margaret Hume, 'On the presence of connecting threads in graft hybrids', *New Phytologist* 12(6) (1913), 216–21. Other published work on botanical subjects includes: E.M. Margaret Hume, 'The history of the sieve tubes of *Pteridium aquilinum*, with some notes on *Marsilia quadrifolia* and *Lygodium dichotomum*', *Annals of Botany* 26 (1912), 573–87, pls. 54, 55.

[126] A.G. Tansley, 'On competition between *Galium saxatile* L. (*G. hercynicum* Weig.) and *Galium sylvestre* Poll (*G. asperum* Schreb.) on different types of soil', *Journal of Ecology* 5 (1917), 173–9 at 174.

[127] *Annual Report of the Botanic Garden Syndicate*, 3 June 1912, 4 June 1914, Cambridge University Botanical Garden Archives.

[128] Ayres, *Shaping Ecology*, p. 106.

[129] Professor W. Ritchie, *The History of the South African College 1829–1918*, Vol. II, Capetown: T. Maskew Miller, 1918, p. 612.

[130] 'Obituary: Lucy Wills', p. 1225.

enrolled in the London School of Medicine for Women where she initially planned to focus on psychiatry; however, by 1920 she was refocused on the field of nutrition research. She was later the first to realize that yeast extracts such as Marmite were effective in treating anaemia in pregnant women, and she discovered folic acid for which the term 'Wills' factor' remains a synonym.[131] Hume had accompanied Wills on her trip back to England, but would return to work at the South African College until 1916 when she went 'on leave to England and then had taken up munition work'.[132] At this time Tansley was engaged in war work at the Ministry of Munitions, and in 1916 Hume, perhaps through Tansley's recommendation, also became a temporary assistant at the Lister Institute of Preventive Medicine, working with Harriette Chick, Tansley's sister-in-law, on the problem of scurvy in troops in the Middle East and India.[133] In a lecture series delivered in the early months of 1918, both Hume and Tansley would speak at University College London on 'Biological problems of the day'.[134]

Hume, Wills, Harriette Chick and others formed a network of highly educated women and men with 'modern' patterns of relationship: long-held friendships, with affairs and marriages kept within the family of friends. The group had strong, but not exclusive, scientific interests and their university connections, like those of Tansley, wove through University College London and the University of Cambridge. The 'Chicks' proper were the daughters (seven in total) of London lace merchant Samuel Chick and his wife Emma.[135] Physically vigorous and

[131] Daphne Roe, 'Lucy Wills (1888–1964): a biographical sketch', *Journal of Nutrition* 108(9) (1978), 1379–83.

[132] Ritchie, *The History of the South African College*, p. 650.

[133] Harriette Chick, Margaret Hume and Marjorie MacFarlane, *War on Disease: A History of the Lister Institute*, London: A. Deutsch, 1971, p. 126.

[134] Hume's topic was 'Accessory food-factors (vitamines) in war-time diets', delivered in her capacity as Temporary Assistant in the Director's Department, Lister Institute of Preventive Medicine. The public series was published as a book, W.M. Bayliss, *Life and its Maintenance: A Symposium on Biological Problems of the Day*, Glasgow: Blackie & Son, 1919. Tansley's chapter is entitled 'Co-operation in food supply'. The Preface written by F.W. Oliver refers to Hume's research as 'the beautiful work on vitamines'.

[135] The Chick children had attended the Gower Street School which attracted students from nearby Bloomsbury: the school was run by two women keen to give their young charges a good grounding in literature and the arts. Rather than join her father in the lace business, the oldest Chick daughter, Edith, desired to continue her education to become a teacher. The intervention of Lady Stanley of Alderley, an advocate for women's higher education, led to her enrolment in the Notting Hill High School, where she was followed by her sisters. Here much of the teaching was by women graduates from Cambridge, Oxford and London. Five of the Chick daughters went on to become university graduates in botany, physics and chemistry, English and medicine. See Margaret Tomlinson, *Three*

intrepid, the Chick girls were known for their walking parties and for swimming even in rough seas: on one of these dramatic bathing occasions, a family friend described watching a huge wave crash on shore: 'it seemed to be full of Miss Chicks'.[136] In 1903, the Botany Department of University College London would initiate fieldwork expeditions quite in line with the experience of the Chick women but what others would regard 'as a bold and risky innovation'.[137] Tansley led the first excursion to the Norfolk Broads,[138] and perhaps not least of the boldnesses required for this venture was the fact that unchaperoned women were as welcome on these expeditions as the men.[139] The 'New Ecology' had its attractions for the 'New Woman', a term based in part on reality as well as the literary constructions of the term's inventor, Sarah Grand, the feminist novelist.[140] According to Tansley/Chick family lore, Arthur Tansley first fell in love with the vivacious dark-haired Harriette 'Harty' Chick,

Generations in the Honiton Lace Trade: A Family History, Exeter: Devon Print Group, 1983, pp. 64–79.

[136] *Ibid.*, p. 77.

[137] F.W. Oliver and T.G. Hill, *An Outline of the History of the Botanical Department of University College London*, Issued by the Department on the Occasion of the Centenary of the College, 18 June 1927.

[138] A.G. Tansley, 'The physical features of the Norfolk Broads: lecture introductory to Summer Ecological Excursion July 1903', unpublished lecture notes, Tansley Archive, Plant Sciences Library, University of Cambridge.

[139] By his own account, Oliver's department was supportive and inclusive of women academics. The prestigious Quain studentship (once held by Tansley), was, from 1890 to 1927, held by eight people, four of whom were women, including: Edith Chick (later Mrs Tansley), Agnes Robertson (who become Mrs Agnes Arber), Sarah Baker (who died in 1917 and in whose memory the Sarah Baker Memorial Prize was created) and Violet Anderson. At some point in the period 1888–1927, all of these women held positions on staff at UCL and were joined by a number of other women, including Dr Ethel N. Thomas, Dr Marie Stopes, Miss Winifred Smith, Miss B. Russell-Wells and Miss G. L. Naylor. Information regarding the 'unchaperoned nature' of these expeditions is from John Sheail (personal communication to Laura Cameron) who gathered this knowledge through his own interviews with early ecologists.

[140] Lucy Bland, *Banishing the Beast: English Feminism and Sexual Morality 1885–1914*, London: Penguin Books, 1995, p. 144. As Lucy Bland describes her, the hallmark of the 'New Woman' was personal freedom: intent to work with men on equal terms, she was 'given to reading "advanced" literature, smoking cigarettes, and travelling unchaperoned, often on a bicycle'. Charlotte Mew the poet, a 'New Woman' by such measures, and the same age as Edith, lived with the Chick family for a period when she was in her early twenties and travelled with Harriette to Brittany, an experience documented in Mew's poetry. Mew, who committed suicide in 1928 by drinking lysol, was romantically linked with the Chicks' oldest brother Samuel as well as the novelist May Sinclair, one of the first English writers to use the theories of psychoanalysis in her writing. See Suzanne Raitt, *May Sinclair: A Modern Victorian*, Oxford: Oxford University Press, 2000; Theophilus E.M. Boll, 'May Sinclair and the Medico-Psychological Clinic of London', *Proceedings of the American Philosophical Society* 106 (August 1962), 310–26; Leigh Wilson, 'May Sinclair 1863–1946', *The Literary Encyclopedia*, www.litencyc.com/index.php.

a student at University College London where he was a new Lecturer in 1894, but Tansley was rebuffed (the story has it that Harriette was disturbed by his depressive tendencies) and he subsequently married her older sister, Edith, also a talented botanist, in 1903.[141]

Harriette earned her D.Sc. from London University in 1904 and with the encouragement of Charles Sherrington became in 1905 the first woman to be appointed to the staff of the Lister Institute of Preventive Medicine where she would begin pioneering work on vitamins and nutrition.[142] In 1919 the Accessory Food Factors Committee of the Medical Research Council was formed with Harriette Chick as Secretary, and she along with Elsie Dalyell from the Lister Institute were sent to investigate nutrition deficiencies in Vienna. Margot Hume would join the team a few months later and it is clear from Harriette's diaries that both she and Margot met up with Arthur Tansley during his first period of analysis with Freud in the spring of 1922.[143] In a two-year study based at the Universitäts Kinderklinik, the team confirmed the important discovery that cod liver oil protected against the development of rickets as well as the discovery of K. Huldschinsky of Berlin that exposure to ultraviolet light had healing properties.[144] Their findings, published as *Studies of Rickets in Vienna, 1919–22* (1924) had wide impact in Austria and Britain, shifting the view that rickets was caused by microbes and poor hygiene to one that promoted awareness of diet and vitamins in preventing the condition. Chick was appointed CBE in 1932 and DBE in 1949. During the Second World War, the Lister Institute would relocate to Roebuck House in Cambridge (something of a laboratory animal equivalent for the children's Evacuation Programme), where Charlotte and J.B.S. Haldane had been living in the inter-war years.

[141] Katrina Lythgoe, 'Sir Arthur Tansley – the man', *Bulletin of the British Ecological Society* 36(1) (2005), 26. Harriette Chick received the Advanced Class Prize for Botany in 1894 and the Gold Medal for Botany the following year, but then transferred her interests to bacteriology.

[142] H.M. Sinclair, rev. David F. Smith, 'Chick, Dame Harriette (1875–1977), nutritionist', in *New Dictionary of National Biography*, ed. Andrew Roberts and Betty Falkenberg, 2005; Charlotte Mew chronology with mental, historical and geographical connections linking with her own words, Middlesex University resource available at www.mdx.ac.uk/www/study/ymew.htm.

[143] Harriette recorded for 17 April 1922: 'day walking, lovely day: met Arthur T and then later Sister [likely Margaret Chick] & Margot and AGT stayed the night'. Harriette Chick Papers, Wellcome Library.

[144] Sinclair, rev. Smith, 'Chick, Dame Harriette (1875–1977), nutritionist'. Another member of the Vienna team was Helen Mackay who produced key and enduring findings on iron deficiency anaemia in infancy. See David Stephens, 'Helen Marion McPherson Mackay (1891–1965), paediatrician', *New Dictionary of National Biography*.

Although a number of interlinked trajectories and professional and familial ties point to Margot Hume as the beloved of his dream, until a cache of papers confirms the identity of the 'girl with whom I had fallen in love' we are left solely with speculation. In detailing the 'Chick Group', Tansley emerges, in some sense, as a man who wedded himself, not just to Edith Chick, the one wife, but to a tight network of women scientists. According to his eldest granddaughter, 'there was a frankness and openness with the family, quite unusual for the times'.[145]

Thus, in attempting to understand the significance of Tansley's dream, we are left with fragments, rather than solutions. We have insufficient evidence to articulate Tansley's erotic and professional crises. We have been struck by the parallels between the figures of his wife and 'beloved', and his relationship to botany and psychoanalysis. Could the core wish of the dream – the truly 'impossible' thing to contemplate, 'as impossible in the dream as it would be in real life' – be the desire to murder his wife (botany) in order to be with his new beloved? If so, his ten-year dalliance with psychoanalysis was an 'acting out' of this core wish. But when we look to Tansley's work in botany, one of its most striking and original features is the *lack* of 'splitting' – between Man and Nature, between Nature and Culture. Tansley's principal contributions were, in contradistinction to American ecology, to emphasize the systemic interrelations of human activity and botanical phenomena – he sees no real difference between those ecosystems which are natural and those which are 'anthropogenic' (nature 'produced by man', as he glossed it in 1923).[146] The American and 'preservationist' theme of the 'wilderness', prior to and independent of human intervention, with its image of 'virginal nature' and its ethos of non-interference, was never Tansley's, in whose work there is very little talk of 'mother Nature'. What is ever-present there is the possibility of the human control or 'regulation' (Tansley's preferred term) of natural processes. Reading this posture alongside the dream, we note how the rifle is what symbolizes Tansley's knowledge, his 'superior equipment'. And, in terms of his life-choices, Tansley's quest for psychoanalytic knowledge continues his over-estimation of the importance of mastery embodied in knowledge, at the expense of the resolution of his conflict by an erotic choice. Tansley's principal responses to Freud are, after all, to emphasize

[145] Personal communication to Laura Cameron, Dr Anna Dickens, Cambridge, 23 June 1999.
[146] A.G. Tansley, *Practical Plant Ecology*, London: George Allen & Unwin, 1923, p. 48. For more detailed analysis, see Cameron, 'Histories of disturbance', pp. 4–24 and Cameron, 'Anthropogenic natures: Wicken Fen and histories of disturbance, 1923–1943', PhD dissertation, University of Cambridge, 2001.

his own mastery, his epistemic independence from his analyst. And it was this attitude that he reiterated in 1953 when interviewed by Eissler. More than that: this attitude helps us explain how Tansley felt it appropriate to submit his dream and its interpretation to the Freud Archives as the principal testimony concerning his place in the history of psychoanalysis. A strange decision: a dream which Freud never analysed, dreamt prior to any real knowledge of Freud or his writings. What Tansley's dream bears witness to is the superiority of psychoanalytic knowledge, and thus implicitly disdain for Freud himself. And, implicitly, also disdain for the 'hell' of the erotic conflict in which he had found himself at this time. It is the mastery which always seems to win out over the recognition of the conflict; again and again, we are tempted to see Tansley's involvement with psychoanalysis as an attempt to arm himself with a more powerful rifle so as to win out over 'the herd'. As he said to Eissler:

T   It was really this remarkable dream of mine that impressed me most deeply. I [was] sure that psychoanalysis is going to be a very important fundamental contribution to the general theory of psychology. And that was why I went to Freud, you see. But I have also been an amateur, I have never got a professional, I always have been an amateur if you like /laughs/

E   But you always were engaged in /unclear/ /laughs/ You wrote a book on the topic!

T   Oh yes, I wrote a book that's true! That book I wrote before I went to Freud. As a result of this stirring up of my interest and emotion about the thing . . .[147]

In this version, Tansley makes it clear that the book he wrote was a result of emotion stirred up by his dream and its interpretation.[148] Quite clearly, the book was an attempt at mastering these emotions. It may be fair to say that this was a desire to win the battle, as if that was the only way to resolve the conflict, and in that quest he may have later used psychoanalysis against 'the herd', such as Vera Brittain, who equated sexual freedom with sexual gluttony; but a truer resolution would be couched less in terms of the superiority of psychoanalytic knowledge than in his eventual return to his first love, botany.

When Tansley died in 1955, his papers passed into the care of Harry Godwin, and were eventually deposited in his old Department in

---

[147] Eissler interview, AGTSFA, p. 12.

[148] A number of passages in *The New Psychology* single out a married man who conducts 'a serious illicit love-affair' (p. 103) as peculiarly subject to intense conflict; see also p. 111.

Cambridge, now known as the Department of Plant Sciences. His wife Edith lived on to 1970, dying at the age of 101. As is well known, virtually nothing of Freud's presence in Vienna survived the Nazis. For years, the city of Vienna behaved as if it had successfully forgotten Freud. Early in the 1970s, Anna Freud was asked by the newly formed Sigmund Freud Gesellschaft in Vienna if she could help locate psychoanalytic books to furnish the beginnings of a Museum which was intended to occupy Freud's old apartment there. Some time after 1972, she made an appeal to members of the International Psycho-Analytical Association for contributions to the library.[149] Somehow, she located the psychoanalytic books of Professor Sir Arthur Tansley, FRS, which he had bequeathed to Harry Godwin, along with his other books, papers, pamphlets and copyrights.[150] As a result of Godwin's gift, the Freud Museum in Vienna acquired forty-seven books from his library, including fourteen books by Freud in German, some of them first editions. It is the books of this 'nice type of the English scientist' that make up a significant portion of the oldest of Freud's books to be found in Freud's old apartment that now houses the Freud Museum in Vienna, making reparations for the destruction of the Nazis.

One of the consequences of this scouring of the psychoanalytic community for early editions of Freud's work is that Tansley's own library contains no psychoanalytic books. The books on botany from his library were donated to the Department in Cambridge, where, for a time, they were housed separately from the main collection; more recently, they have been integrated with that collection, and have thus lost their unity as elements in a distinguished botanist's life-work.[151]

Even Tansley's edition of *Die Traumdeutung*, with Freud's own dedication to him, has completed the circuit it started out on, a few days after his

[149] Ms. letter, undated but clearly after 1972, addressed to Members of the International Psycho-Analytical Association, LoCAF.

[150] Information received from Dr Richard West, Godwin's colleague and executor, who donated much of Godwin's estate, including Tansley's papers, to the Department of Plant Sciences, Cambridge, and informed us (14 May 1999) that it was Godwin who donated the books to the Freud Museum in Vienna. Tansley's will was probated on 14 April 1956.

[151] Professor Peter Grub of the Department of Plant Sciences in Cambridge has a small private collection of Tansley's botanical works, received from Sir Harry Godwin. Tansley's own edition of Freud's works in English, prepared by his long-standing friend James Strachey, found its way into a second-hand bookshop in Cambridge in the 1970s, where it was eagerly acquired by Edward Timms, Emeritus Professor of German at the University of Sussex and a distinguished commentator on psychoanalysis. He retained the original invoices that came with the set, from which the following information is derived: Tansley subscribed to the *Standard Edition* on 2 April 1953, ordering it through the London bookseller's I.R. Maxwell of 4–5 Fitzroy Square. Before his death in 1955, Tansley certainly received two other volumes (Vol. XIII, 1953, and Vol. XVIII, 1955, dispatched to him in May 1955). The final volume must have been delivered to Tansley's estate in 1974.

analysis began on 31 March 1922, from Freud's hands to Grantchester, and now back on the bookshelves of Berggasse 19. This journey of Freud's dream book is a fitting allegory of Tansley's own journey, driven by his own dream which he interpreted himself, to Freud's books, then on to involvement with the psychoanalytic movement and the nine months he spent on Freud's couch, and back to his own field of ecology. Nonetheless, this circular journey of Freud's dream book, from Vienna to Cambridge and then back to its point of departure was in danger of erasing its own history. Tansley's dream, his professional crisis, his analysis with Freud, his bringing of psychoanalysis to Cambridge – all were in danger of being obscured by the completion of his analysis and his return to botany, symbolised by the vanishing of his psychoanalytic books back to the shelves of Berggasse 19.[152]

[152] While chatting in the Library of Berggasse 19 with the late Lydia Marinelli, Archivist of the Freud Museum in Vienna, John Forrester discovered Tansley's books on the bookshelves beside which he was sitting. As professional librarians and experts on the history of psychoanalysis, the Freud Museum staff had excellent records of the source of the books, but no knowledge of the identity of Sir Arthur Tansley, FRS.

CHAPTER 3

W.H.R. Rivers, the English Freud

In the 'General Preface' to *The Standard Edition of the Complete Psychological Works of Sigmund Freud*, James Strachey wrote: 'The imaginary model which I have always kept before me is of the writings of some English man of science of wide education born in the middle of the nineteenth century. And I should like, in an explanatory and no patriotic spirit, to emphasize the word "English".'[1] It is an amusing, seemingly pointless, question to ask: whom did Strachey have in mind as his model? And it leads to another parallel question: since Strachey actually did create, through his translation, an *imagined* 'English Freud', a man of science of wide education born around 1850, can we pinpoint an *actual* English man of science who corresponds closely to this figure? This 'English man of science' would have to have an inclination for bold speculation and adventure. He would have to be courageous, imaginative, and empirically immersed through firsthand experience in the construction of a new human science or sciences. Once one specifies these characteristics, the most plausible candidate to come into focus is W.H.R. Rivers.

Only eight years younger than Freud, Rivers qualified medically in 1886, the same year Freud set up his medical practice; he became the most eminent English psychologist and anthropologist of the first two decades of the twentieth century. William Halse Rivers Rivers (1864–1922) came from a solidly middle-class Kent family with naval, Church and Cambridge connections; his father entered the Church and married Elizabeth Hunt, whose brother James was an energetic churchman. While still a youth, James had developed an expertise in the treatment of stammering, authored an influential series of books, and developed an interest in ethnography, partly in connection with his study of speech. He joined the Ethnological Society in 1856, became its Secretary and in 1863 led

[1] Strachey, 'General Preface', *SE* I, xiii–xxvi at xix.

a breakaway group into founding a rival society, the polygenist Anthropological Society. When Uncle James died in 1869, young Will's father took over the work of producing a later edition of his brother-in-law's work on stammering.

Will Rivers went to school at Tonbridge and was preparing to enter Cambridge, as his father and uncle had done, when he was struck down with typhoid fever. His slow recovery required him to change his plans: he studied medicine at Bart's in London, graduating in 1886 at the very early age of twenty-two. The next years were spent acquiring an unusually varied education. He travelled the world as a ship's surgeon, then developed an interest in neurology and the mental diseases, and worked in London under or alongside the most interesting medical men of the era: Michael Foster, Charles Sherrington, Henry Head and John Hughlings Jackson. In 1892, Rivers spent four months in his generation's Mecca for scientists, Germany, studying at Jena with the psychiatrist-neurologists Binswanger and Ziehen. He concluded that 'I should go in for insanity when I return to England and work as much as possible at psychology.'[2] Prior to his German months, Rivers's approach to neurology in his papers on delirium, hysteria and neurasthenia had emphasized the organic basis of these conditions.[3]

His interest in insanity led him to take a post at Bethlem and to assist G.H. Savage, who was to be Virginia Stephen's (later Woolf's) principal physician, with his lectures on mental diseases. Before taking up his duties at Cambridge, Rivers spent the summer of 1893 working with Emil Kraepelin in his psychological laboratory at Heidelberg, where the two men co-authored papers on fatigue. Rivers would continue to make regular visits to German laboratories over the next few years, working on a number of occasions during the 1890s with leading scientists in Germany – Ewald Hering, Emil Kraepelin and Wilhelm Kühne. His two German visits had instilled an enthusiasm for modern psychology in Rivers, a psychology that stood in vivid contrast to the traditional materialism of English alienists, as represented by the foremost Victorian alienist, Henry Maudsley:

> Dr Maudsley endeavours at every opportunity to persuade his readers to disregard the psychological aspect of mental disease and to look solely at its material aspect as brain disorganization ... Dr Maudsley does not seem to recognise the possibility of a scientific psychology to be developed side by

---

[2] Richard Slobodin, *W.H.R. Rivers*, New York: Columbia University Press, 1978, pbk version, p. 13.
[3] Allan Young, 'W.H.R. Rivers and the war neuroses', *Journal of the History of the Behavioral Sciences* 35(4) (1999), 359–78 at 360.

side with our knowledge of brain structure, which may be mutually helpful, each advance on one side throwing light upon the other.[4]

Particularly in the area of the measurement of reaction times in response to a great variety of stimuli and circumstances, being pioneered by Kraepelin as part of his turn away from brain anatomy as the basis of scientific psychiatry towards an autonomous clinical and classificatory psychiatry closely allied to 'auxiliary sciences', in particular experimental psychology, there was the hope for real advances in the understanding of insanity. Rivers was adopting a position in favour of psychology not unlike that of Freud at the same time, and he would certainly have recognized the significance of the application of reaction-time experiments by which Jung would in 1907 conquer the psychology of dementia praecox for Freudian theory.

Rivers's position in Cambridge was initially as a lowly Fellow-Commoner; he had living and dining rights in St John's College and lectured to students in both Natural and Moral Sciences Triposes. His position did not change radically when he was appointed in late 1897 to the University Lectureship in the Physiology of the Senses and Experimental Psychology. Giving lectures on mental diseases at Guy's on his return from Germany, he was also invited in 1893 to teach experimental psychology and physiology at both University College London and Cambridge; in 1897 he established 'the first' laboratory in Britain in experimental psychology in both universities. From 1897 on, his principal residence was St John's College, Cambridge, where he became a fellow in 1902. He also participated in the Cambridge Torres Straits Expedition of 1898, which led to the addition of anthropology to his already substantial range of scientific interests. In the period 1899–1914, Rivers worked both at physiological psychology and at anthropology.[5] The latter field preoccupied him increasingly from 1906 onward, with expeditions to India and to Melanesia. Arriving back in England in 1915, he took on war work: first at the Great Eastern Hospital in Cambridge, then at Maghull Hospital near Liverpool, which was becoming the national centre for the development of psychotherapeutic treatment of the war neuroses. In late 1916, now commissioned in the RAMC, he moved to Craiglockhart Hospital for Officers, where he remained till late 1917. Then, and for the rest of the war, he was

---

[4] W.H.R. Rivers, 'Review. Maudsley, *The Pathology of Mind*; Kraepelin, *Psychologische Arbeiten*', *Mind* N.S. 4(15) (1895), 400–3 at 401.

[5] For further details of Rivers's contribution to the development of psychology in Cambridge see Chapter 5, 'Discipline formation'.

appointed as aviation psychologist at the Royal Flying Corps Central Hospital at Hampstead in London, once again requiring prolonged self-experimentation by extensive flying and acrobatic manoeuvres in the dangerous aircraft of the time.[6] He had resigned his university position at Cambridge in 1916[7], possibly because he was envisaging a move to Manchester as professor of comparative religion[8]; however, after the war he did return to Cambridge without a university position and, in the course of an attempted reform by St John's College of its system of teaching natural sciences, Rivers, conforming to the College's requirement that the new position be for a man 'whose position and influence are primarily due to his interest in research work',[9] was appointed Praelector in Natural Sciences, acting as a general tutor for all students taking the natural sciences. (St John's was perhaps following the precedent of Trinity, which, since its reforms in the late 1860s, had used the position of Praelectorship to aid exceptional leading scientists stranded without a university position.[10]) After the war, Rivers became a leading figure, both within the academic disciplines – psychology, anthropology, folklore – and as a public intellectual, agreeing to stand as the Labour candidate for the University of London in the 1922 general election – jokingly, he said, so as to psychoanalyse Lloyd George and because the House of Commons was in urgent need of a resident psychiatrist.[11] His sudden death of a strangulated hernia on 4 June 1922, at the age of fifty-eight, cut short an expansive new phase of his life and was a blow to the public life of England, a profound one for the academic communities of which he was the acknowledged leader.[12]

[6] Slobodin, *Rivers*, p. 66.   [7] HR 1911–20, p. 15.

[8] L.E. Shore, 'W.H.R. Rivers', *The Eagle* (1923), 2–14 at 9; Alan Costall, 'Dire Straits: the divisive legacy of the 1898 Cambridge Anthropological Expedition', *Journal of the History of the Behavioral Sciences* 35(4) (1999), 345–58; W.H.R. Rivers, *Conflict and Dream*, London: Kegan Paul, Trench, Trubner & Co., 1923, p. 86, abbreviated as CD.

[9] Peter Linehan (ed.), *St John's College, Cambridge: A History*, Woodbridge: Boydell Press, 2011, pp. 453–4.

[10] In Cambridge, a praelector is now usually a college fellow who formally presents students during their matriculation and the graduation ceremony (a strictly ceremonial role with no other duties). However, the post of Praelector at Trinity College was the subject of an important reform in the late 1860s, involving the issue of the possibility of a man retaining his Fellowship and his living even though married; see D.A. Winstanley, *Later Victorian Cambridge*, Cambridge: Cambridge University Press, 1947, pp. 244–5. See comments on Foster's and Hopkins's earlier praelectorships in Chapter 5. St John's College introduced a similar system on a more programmatic but equally episodic basis in 1919; the brilliant Paul Dirac, a notoriously silent and remote character, was appointed Praelector in Mathematical Physics in 1929 precisely to protect him from teaching duties (Linehan, *St John's College*, p. 510).

[11] Slobodin, *Rivers*, p. 80.

[12] For biographical information on Rivers, the best source is still Slobodin, *Rivers*.

Fig. 3.1 William Halse Rivers Rivers, *c.* 1917, by Douglas Gordon Shields.

Rivers's response to Freud has rightly been viewed as something of a litmus test for the receptivity of English science to psychoanalysis. Many historians have distinguished Rivers's views and methods from those of the psychoanalysts, principally Freud.[13] In this they have taken Rivers at his word, noting how he criticized the doctrines of the unconscious, of repression, of the importance of infantile sexuality – all shibboleths of psychoanalysis, according to Freud. Yet what is most striking in Rivers's work is, again taking him at his word, how impressed he is by the 'genius' of Freud, by the revolution in medical psychology wrought by Freud, and how under the spell of Freud he is – not only at the level of theoretical concepts, where he went out of his way both to acclaim and to criticize and disagree, but at the level of method. Indeed, the book *Conflict and Dream* would be best titled *A Dialogue with Freud in and on Dreams.* It is a book that is 'normal science' in the Kuhnian sense at its clearest: it

[13] Allan Young, *The Harmony of Illusion: Inventing Post-Traumatic Stress Disorder*, Princeton: Princeton University Press, 1995; and 'W.H.R. Rivers and the war neuroses', *Journal of the History of the Behavioral Sciences* 35(4) (1999), 359–78.

takes the exemplars of a great scientific achievement as its model and worries away at the puzzles the achievement of that model opens up for subsequent researchers.[14]

Siegfried Sassoon's semi-autobiographical accounts of his life-transforming encounter with Rivers in 1917 at Craiglockhart have given Rivers wider cultural resonance, while Pat Barker's celebrated *The Regeneration Trilogy* has elevated him to near-mythical status. Barker uses Rivers's own writings and historical materials relating to him to construct an imagined portrait of a humane but internally troubled physician, trying out the new methods of treatment made necessary by war, tracking his own way between the moral pressures of treating soldiers only so as to send them back to the hell of the front and his sense of patriotic duty.[15] In Barker's account Rivers is a good doctor because he uses an exploratory psychotherapy with his shell-shocked patients. He refuses the more direct method of suggestion and symptom-removal using physical, particularly electrical shock, techniques, cast by many historians as disciplinary intimidation. Barker explores fully the irony that Rivers's method, of civilized conversation seamlessly interwoven with cathartic psychotherapy, aided Sassoon to choose to give up his protest against the conduct of the war and return to the front line in France, where he was wounded once again in a 'friendly fire' incident.

In parallel with his work with the war neuroses of ordinary soldiers at Maghull in 1916 and officers at Craiglockhart in 1917, Rivers was seized with scientific enthusiasm for Freud's theory of dreams. Starting in 1917, he wrote an influential and self-revelatory series of papers on Freud's theories, Freud's method of treatment, and the concept of the unconscious, summed up in his *Instinct and the Unconscious* (1920). Three themes were woven together in this rejuvenating work: his real interest in psychotherapeutic work; his growing interest in Freud's psychological views in general, but particularly in dreams; and the 'personality change', perhaps a result of the first two, alongside the overall impact of the war. As Rivers stated:

> Though I had taken much interest in the general views of Freud before the war, I had not attempted to master his theory of dreams. I was more interested in the applications of his scheme to the explanation of psycho-neurosis and the anomalous behaviour of everyday life. When the war

[14] Thomas S. Kuhn, *The Structure of Scientific Revolutions*, Chicago: University of Chicago Press, 1962.
[15] Pat Barker, *Regeneration*, London: Viking, 1991; *The Eye in the Door*, London: Viking, 1993; *The Ghost Road*, London: Viking, 1995.

brought me into touch with dreams as prominent symptoms of nervous disorder and as the means of learning the real nature of the mental states underlying the psychoneuroses of war, it became necessary to study Freud's scheme of dream-interpretation more closely, and I read his book carefully.[16]

Before the war, Rivers had been interested in Freud's theories of the neuroses and this, along with the work of many others, including Hippolite Bernheim, Pierre Janet and Jules Déjerine, fed into the lectures he gave at Cambridge on, for instance, 'Physiological and pathological psychology' from 1909 on, giving rise to Tripos examination questions such as: 'Discuss the psycho-pathology of hysteria' (1911). With the establishment by Rivers and his student and then colleague C.S. Myers[17] of a Diploma in Psychological Medicine in Cambridge in 1912, the new Psychological Laboratory became for the first time a national centre for medical psychology; Rivers, despite his principal commitment at this time to ethnography, was fully part of that work. So Rivers was entirely aware of the growing interest in Freud's work from 1908 on.[18]

Myers's student T.H. Pear, psychologist at Manchester and early enthusiast for Freud's theory of dreams, gave a colourful version of the origin of Rivers's new interest in Freud's dream theory on his arrival at Maghull:

> Almost before he had unpacked, Rivers paid me an honour which I shall never forget. He said he would like to be regarded as a student who had been away from books for a long time (the outbreak of war had found him for the second time visiting Melanesia) and wanted to catch up. Would I direct his reading for the next few weeks, and on afternoon walks – Cambridge fashion – discuss it? His first desire was to grasp what Freud meant by the Unconscious, which Rivers thought the most important contribution to psychology for a long time.[19]

The record provided by Rivers's own publications indicates a less dramatic moment of revelation, but there is no doubting the marked intensity of Rivers's interest in Freud's work, in particular on dreams – but only once he began work on his own dreams.[20] The period of most intense dream

---

[16] CD, 5.

[17] Rivers's student Myers was co-founder of the Cambridge Laboratory; see John Forrester, '1919: psychology and psychoanalysis, Cambridge and London: Myers, Jones and MacCurdy', *Psychoanalysis and History* 10(1) (2008), 38–43, 45–7 and Chapter 5 of this book.

[18] L.S. Hearnshaw, *A Short History of British Psychology, 1840–1940*, London: Methuen, 1964, p. 180; *Cyril Burt, Psychologist*, Ithaca, NY: Cornell University Press, 1979, p. 26; Crampton, pp. 37–94.

[19] T.H. Pear, 'Some early relations between English ethnologists and psychologists', *Journal of the Royal Anthropological Institute of Great Britain and Ireland* 90 (July–December 1960), 227–37 at 232.

[20] John Forrester, 'Remembering and forgetting Freud in early twentieth century dreams', *Science in Context* 19(1) (2006), 65–85.

analysis began in early December 1916: 'I made no great progress in dream-analysis or in the clinical utilisation of dreams until I had a dream myself which went far to convince me of the truth of the main lines of the Freudian position.'[21]

Rivers became an authoritative spokesman for the importance of Freud's work, in particular with his paper 'Freud's psychology of the unconscious', given to the Edinburgh Pathological Club on 7 March 1917, and published in *The Lancet* on 16 June 1917. Among his many strong claims for the significance of Freud's work, Rivers emphasized that it ranged far beyond its uses in medical therapeutics:

> Freud's theory of the unconscious is of far wider application than the perusal of recent medical literature would suggest. It is true that Freud is a physician and that he was led to his theory of the unconscious by the study of disease, but his theory is one which [p. 160] concerns a universal problem of psychology. If it is true, it must be taken into account, not only by the physician, but by the teacher, the politician, the moralist, the sociologist, and every other worker who is concerned with the study of human conduct . . . It is possible, even probable, that the practical application of Freud's theory of the unconscious in the domain of medicine may come to be held as one of its least important aspects, and that it is in other branches of human activity that its importance will in future be greatest. I may perhaps mention here that my own belief in the value of Freud's theory of the unconscious as a guide to the better understanding of human conduct is not so much based on my clinical experience as on general observation of human behaviour, on evidence provided by the experience of my friends, and most of all on the observation of my own mental activity, waking and sleeping.[22]

Rivers here places himself on the front line: his conviction of the significance of Freud's theory stems not only from his authority as psychologist, anthropologist or practising military physician, but from his *personal* life and the private experience of his friends.

This celebrated paper was avowedly an attempt to tread a fine line between the warring Freud enthusiasts and detractors;[23] yet in brokering agreement in a disinterested manner, Rivers combined the tone of the

[21] CD, 7. Dated by one of the associations for the initial 'Presidency' dream to a copy of the *British Medical Journal* for 2 December 1916 (CD, 14n). Rivers discussed Freud's theory of dreams with patients, as Max Plowman's April 1917 letter to his wife attests – he lent him Hart's *The Psychology of Insanity* as a good guide to them (see p. 248 note 122).

[22] IU, 159–60.

[23] The particular battles Rivers may have had in mind in early 1917 were the seventeenth International Medical Congress held in London in August 1913, when Janet had attacked Freud, who was defended by Ernest Jones; and a tempestuous series of papers and rejoinders in January–May 1916 in *The Lancet* on the topic of psychoanalysis when senior alienists (Mercier, Armstrong-Jones and

impartial observer and judge with the enthusiasm and conviction of a patient who has succumbed to a series of personal epiphanies. Without entering too far into the vexed and still controversial question of the overall place of psychological theories and psychological therapies in the treatment of the tens of thousands of cases of 'war neuroses',[24] there is clear evidence that a large number of physicians became convinced of the efficacy and necessity of psychological therapies, ranging from suggestion and hypnosis via the cathartic cure derived from Breuer and Freud to a more 'psycho-analytic' therapy of free association, analysis of resistances, and dream analysis. Among these doctors, Rivers was regarded as a leader whose outstanding scientific reputation provided a useful cover of serious respectability for the scandalous sexual theories that emanated from a capital of the *Central Powers*. Rivers closed his *Lancet* paper by emphasizing: 'Are we to reject a helping hand with contumely because it sometimes leads us to discover unpleasant aspects of human nature and because it comes from Vienna?'

The principle of conflict, the theory of forgetting, the aetiological importance of infantile experience, the distinction between the manifest and latent content of dreams, the mechanisms of displacement, condensation and symbolization both in dreams and in the neuroses, the psycho-analytic theory of the unconscious – many of the key principles and claims of Freudian psychoanalysis feature in Rivers's papers up to 1920. The fact that he endorsed them made the discussion of Freud's controversial and scandalous views, so mercilessly attacked by some senior figures, amongst them asylum alienists, incomparably easier. As Myers puts it in his obituary of Rivers: 'He had the courage to defend much of Freud's new teaching at a time when it was carelessly condemned *in toto* by those in authority, who were too ignorant or too incompetent to form any just opinion of its undoubted merits and undoubted defects.'[25]

Rivers the Dreamer

Following the armistice in November 1918, the academic year in Cambridge was extended so that the many students starting

Donkin) had clashed with William Brown, Ernest Jones and G. Elliot Smith, a clash that produced the first usage in print we have been able to establish of 'psycho-anal-ists'.

[24] Tracey Loughran, 'Shell-shock and psychological medicine in First World War Britain', *Social History of Medicine* 22(1) (2009), 88.

[25] C.S. Myers, 'The influence of the late W.H.R. Rivers', in W.H.R. Rivers, *Psychology and Politics and Other Essays*, London: Kegan Paul, Trench, Trubner & Co., 1923, p. 169.

in January 1919 fresh from the forces could complete a full year by the autumn. From early August, Rivers gave nineteen lectures on 'Instinct and the unconscious'.[26] In 1920–21 and 1921–22 he lectured on dreams.[27] The first series of lectures was published as *Instinct and the Unconscious* (1920); after his sudden death in June 1922, the second series was published by his literary executor Elliot Smith as *Conflict and Dream* (1923). In both books Rivers revealed himself in new and unprecedented ways. In the first, returning to his original enthusiasm for a scientific psychology of the insane, he blended accounts of his treatment of the war neuroses and his championing of Freudian ideas with a neurophysiological theory of the biological basis of a fully Freudian unconscious, not that of the introspectionists, 'clinging to a *simulacrum* of the conscious'.[28]

> One who wishes to satisfy himself whether or no unconscious experience exists should subject his own life-history to the severest scrutiny, either aided by another in a course of psycho-analysis or, though less satisfactory and less likely to convince, by a process of self-analysis. It will perhaps be instructive if I give a result of my own self-analysis, which though at present incomplete, has done much to convince me of the reality of the unconscious.[29]

The curious example he gives, which he himself finds so convincing of the existence of the dynamic unconscious, is 'the completeness of the blank in my mind in connection with that upper storey'[30] of the house in which he grew up to the age of five, thereby emphasizing that a self-analysis must confront infantile experience and infantile amnesia, just as Freud asserted. What the publicly announced inconclusive retrieval of his childhood memories clearly demonstrates is that Rivers's own self-analysis of 1917, if we can call it that, did attempt to gain access to infantile material – and that he was convinced that it was essential to do so. *Conflict and Dream* was even more revealing of Rivers's internal world, since he followed Freud's example in both spirit and letter in using principally his own dreams to examine critically Freud's theory of dreams. It was, as Rivers fully recognized, inevitable that, in doing so, one expose one's inner life to the scientific public.

As had Tansley, Rivers's commitment to dream analysis was initiated by a nocturnal epiphany:

[26] The start date is mentioned by Karin Stephen in a letter to her mother, 9 August 1919, BPaS Archives: KS/40.

[27] However, these lectures were listed by the Board of Psychological Studies, not by the Moral Sciences Tripos, i.e. these lectures were offered more generally, including for any student not taking a Tripos but intending to take an Ordinary Degree.

[28] IU, 8.   [29] IU, 11.   [30] IU, 13.

In October 1916 I was transferred to a hospital for officers, where I soon began to obtain from my patients dreams of a less simple kind [than manifest wish-fulfilments], but I made no great progress in dream-analysis or in the clinical utilisation of dreams until I had a dream myself which went far to convince me of the truth of the main lines of the Freudian position.[31]

Rivers's 'Presidency' dream opened his book, just as Freud's dream of 'Irma's injection' had opened the substantive part of Freud's dream book. The dream of 'Irma's injection' established two principles: first that dreams could be interpreted if one followed the method of associating to each discrete element of the dream as remembered on waking; second that the overall dream, once it had been interpreted, was in the nature of a wish-fulfilment. Rivers confirmed both these principles in his 'Presidency' dream.

Set in a Cambridge College garden, the dream moved into a vague evocation of a meeting of the Council of the Royal Anthropological Institute, in which the only person the dreamer recognized was reading a list of names.

> When the reader was finished, he put the paper from which he was reading on the table, and I leaned over to look at it, in order to ascertain who had been nominated as President, for I knew that his name would appear at the head of the list of new members of Council. There I read
> (*S. Poole.*)

In Rivers's dream analysis, he was quickly led to a conflict between his desire to spend more time in Cambridge doing his own work and his desire to be elected President of the Institute. Via a series of names – Stanley Pool; Professor Lane-Poole; a young doctor named 'Temp. Lieut. Samuel Pool, M.B., R.A.M.C.' whose name Rivers had glanced at in *The Scotsman*; a bookseller's catalogue in which the form 'S. Lane-Poole' had occurred – Rivers inferred that the name was a disguised version of his own – the 'S' from 'RiverS', the 'Pool' from schoolboy pranks with 'river' – 'streams, waters', and so on: 'A wish that I should be chosen to be president of a society was disguised by the appearance of my name in a distorted form.'[32]

[31] CD, 7.

[32] CD, 16. Some of Rivers's psychoanalytic contemporaries were alert to the personal resonances of the theme of water for him, as can be seen in a letter from John Rickman to Géza Róheim, dated 2 April 1922: 'When you reply to Rivers On Baptism (note: <u>Rivers = Water-Birth</u>) please deal gently but firmly with him. I know from personal knowledge that <u>Rivers–Birth Phantasies</u> is a personal complex of his & one can be firm & kind without hammering on his raw nerves' (BPaS, CRR/FO8/30). Rickman was writing before the publication of *Conflict and Dream* (though he may well have heard much of its contents in lectures), in which two of the principal dreams analysed – one of Rivers

Rivers's analysis was visibly preoccupied with repeating the two founding principles of Freudian interpretation: first, the rigorous determinism of the elements that appeared in the dream, in particular the recent occasions on which he had glanced at newspapers or lists in catalogues and seen names in passing, which were then opportunistically utilized by the process of dream distortion to represent his own name in a disguised form. Implicit in this empirical demonstration of the exact determination of the name was the overall principle: dreams are capable of being interpreted – of being understood, of being shown to be the result of mental relations of cause and effect. This demonstration of mental determinism, of causal laws in the mental sphere, clearly appealed to the kind of scientist Rivers had always been: in search of causal laws through experiment. On 6 May 1917, a few months after Rivers, Bertrand Russell was undergoing the same revelatory experience:

> I am reading Freud on dreams, most exciting – I see in my mind's eye a great work on how people come to have the opinions they have – interesting scientifically, & undermining *ferocity* at the base (*unmasking*, I ought to have said) – because it is always hidden behind a veil of morality.[33]

Russell would express this appeal of Freudianism in his authoritative style in *The Analysis of Mind*: '[D]reams, as Freud has shown, are just as much subject to laws as are the motions of the planets.'[34] These two key figures in Cambridge science, Rivers and Russell, felt the same immediate enthusiasm for Freud's dream theory: for them, Freud offered an immense extension into the mind of the rule of scientific law.

Rivers also introduced the 'Presidency' dream as the first that convinced him of the 'main lines' of the Freudian position. Not only were undistorted dreams wish fulfilments, but so were more complex, more distorted dreams, once suitably interpreted. Having demonstrated the key principles of Freudian dream analysis and having shown how convinced a 'Freudian' he therefore was, Rivers immediately set out to develop criticisms of the Freudian theory of wish-fulfilment. Indeed, his book's 'plot' is a narrative unfolding, through a series of key dreams, most of them his own, of his critical modification of Freud's theory.

himself, one of his patient's – were focused on plays with Rivers's name. It should also be noted that Rivers's full name was William Halse Rivers Rivers – the repetition of 'Rivers' was said to be the result of a vicar's confusion at the baptismal font, so yet another 'watery myth'; the full context of the 'Presidency' dream includes the fact that his maternal uncle James Hunt was the founder and first president in 1863 of the Anthropological Society.

[33] MBR I, pp. 495–6.

[34] Bertrand Russell, *The Analysis of*, London: George Allen & Unwin, 1921, p. 138.

In the history of my attitude towards Freud's theory of the dream ...
a sceptical tendency was overcome by the experience of a dream arising
out of a latent desire to be President of a Society. One result of this dream
was to make me a temporary convert to the view that the dream expresses
the fulfilment of a wish.[35]

Rivers then recapitulated the development of his views beyond those of
Freud's wish theory, first in the hypothesis that the motive force of the
dream is *any* dominant affect, such as fear, anger, reproach or desire,
present in the dreamer prior to sleep, and second to his more stable final
position, that dreams are the expressions of recent conflicts in the dreamer's mental life.

While Rivers was explicitly and persistently critical of Freud's theories in
the lectures on dreams that became *Conflict and Dream*, the whole form of
his book replicated the model of *Die Traumdeutung*. With each chapter,
with each new concept, Freud had used one of his own dreams to illustrate
and demonstrate a new concept. Rivers did the same and added an
additional dimension: he made it quite clear that his views of Freud's
theories had changed significantly over the course of the months when he
was putting Freud to the test with his dream life, from December 1916 to
the autumn of 1917, so that each of his principal dreams was analysed to
show both the importance of Freud's specific theory and also the historical
unfolding, in the narrative of his dream life, of his differences from Freud
on each specific question. And not only did Rivers, following Freud,
intertwine the public analysis of his own dreams with the exposition of
his theory, he repeated Freud's warning to the reader to expect some, but
not too much, self-revelation:

One of the greatest hindrances to the psychological study of dreams, or
rather to the general discussion of its problems, is the fact that the dream is
continually revealing thoughts and sentiments of the dreamer which cannot
easily be made public ... One of the infantile characters of the dream-
consciousness is that it blurts out like a child just what it really thinks and
feels about persons and things. Thus, in the dream of my own already
recorded I was obliged to omit certain features for this reason ... Freud
himself has suffered greatly from this limitation.[36]

As with Freud, whose dreams and their analyses constitute the most
extensive evidence we have as to his inner life, so with Rivers: nowhere
else but in his dreams does he reveal his feelings about the family tradition
of anthropology; nowhere else does he reveal his sense that it may be

[35] CD, 117. [36] CD, 41.

immoral to receive his fellowship emoluments while being paid as a major in the RAMC during the war; nowhere else does he reveal the difficulties of treating soldiers who had been through extreme experiences in the trenches. Amid the swirl of themes that flood out of his dream life, there is one, the politics of the war, that clearly constitutes a central portion of Rivers's own self-analysis. And in Rivers's dream life, the politics of the war is summed up in one image: the *Cambridge Magazine*.

The *Cambridge Magazine*, started in 1912, grew out of the Heretics Society founded by undergraduate C.K. Ogden in 1909. The Society soon developed into the most adventurous forum for intellectual debate in Cambridge, lasting until 1930.[37] In addition to reviews of drama, sport and other topics of university interest, it printed talks given to the Heretics (which had included before the war addresses by Jane Harrison, Bertrand Russell, G.H. Hardy and G.B. Shaw, and Rivers's 'The primitive conception of death' in May 1911) and poetry. Once war was declared, the magazine gave conspicuous coverage to sceptical views about the value of the war, and to the formation of the Union of Democratic Control, which called for parliamentary control of the war and negotiations between the warring powers, and, in particular, printed Romain Rolland's famous denunciations of the war. From October 1915 on, it published a weekly section of 'Foreign opinion: a weekly survey of the foreign press', edited by Dorothy Buxton from newspapers across Europe, including the Central Powers, and outside Europe; it was the only source of such foreign press reports to be published in England throughout the war. As a result, the circulation of the *Magazine* during 1915–18 leapt to twenty thousand.

To understand the significance of the *Cambridge Magazine* for Rivers, we must recapitulate the principal dreams that Rivers analysed in the key period from the end of 1916 to his departure from Craiglockhart to Hampstead in autumn 1917 and the overall development of his views concerning psychoanalysis.

As we have seen, the dream that ignited Rivers's self-analysis and persuaded him of the cogency of Freud's theory he called the 'Presidency' dream. In March 1917 he was developing the view that 'dreams might be the expression of any affective state of which the dreamer had been the subject during the preceding day'[38] and not only wishes. During the night of 20–21 March 1917, Rivers dreamed that 'I was reading a letter

---

[37] Philip Sargant Florence and J.R.L. Anderson (eds.), *C.K. Ogden: A Collective Memoir*, London: Elek Pemberton, 1977; Damon Franke, *Modernist Heresies: British Literary History, 1883–1924*, Columbus: Ohio State University Press, 2008.

[38] CD, 117.

from a Cambridge friend' – 'the highly reproachful character of the communication was evident'.[39] The reproaches were connected with 'the general European situation at the time'. 'On the previous day', Rivers reported, 'I had also received the *Cambridge Magazine* of 17th March, containing an account of the attack which was being made on the *Magazine* at the time and of the measures by which the attack was being met.'[40] The first layer of background dream thoughts were preoccupied with the international politics and economics of continuing the war '*jusqu'au bout*'. But Rivers then turned his analysis away from politics to the immediate anxieties of his hospital work and arrived at an interpretation in terms of a self-reproach concerning a difficult disciplinary matter he had had to deal with, in which he had solved the problem by 'the application of a somewhat violent procedure' for which 'I had definitely reproached myself for what I counted as a failure'.[41]

Rivers was satisfied with this analysis of the dream, referring the reproach in the letter to the self-reproach concerning his violent disciplining of hospital patients, because it 'seemed to furnish striking confirmation of the view to which I was already being led, that dreams are attempts to express in sleep the affective state which is prominent in the dreamer's mind before going to sleep'.[42] But by 29 July 1921, when he was writing up this chapter, he had changed his theoretical position to the view that dreams are 'a solution or attempted solution of a conflict'.[43] Looking back, Rivers now revised his view:

> [T]he real factor determining the dream proper was a conflict arising in some way out of my attitude to the war ... At the time of the dream (1917) I was manifestly adopting the orthodox attitude, and any such pacifist tendency as might have been aroused by reading the *Cambridge Magazine* would have been repressed, thus providing exactly the conditions by which such a dream as that with which we are dealing would have been produced.[44]

That night, 29–30 July 1921, Rivers had a long and confused dream whose only clear content was 'that I was going to my bedroom to have a siesta after lunch, taking with me books to read'.

> On waking from this dream I found myself thinking about the problem of the day before [i.e., the political interpretation of the original dream], and

[39] CD, 118.
[40] CD, 120. The attack on the *Magazine*'s 'pacifism' began on 24 February and a vigorous defence was published on 17 and 24 March.
[41] CD, 121.   [42] CD, 122.   [43] CD, 122.   [44] CD, 122–3.

then remembered clearly what I had then completely forgotten, that I had
had a definite conflict in my mind at the time (*i.e.* March 1917) whether
I was right in subscribing to the *Cambridge Magazine*. The conflict was
between the view that it must be right to know the truth, to know what the
people of other nations, enemy or allied, were thinking, and the view that in
time of war nothing should be done to make people doubtful about the
absolute justice of the cause for which they were fighting. In such a conflict
there would be little question that the former attitude would appeal more to
my adult intelligence, while the second point of view would have appealed
to me in youth.[45]

So Rivers now revised his interpretation: instead of it being a continuation
into sleep of the self-reproaches over a medical decision, it became an
expression of a personal conflict, symbolized by the *Cambridge Magazine*,
between 'knowing the truth' at all costs, even that of failing in one's
patriotic duty, and 'winning the war' at all costs, even that of sacrificing
the truth.

The symbolism of the *Cambridge Magazine* went even further:

I had the impression that the books which I was taking to my bedroom were
connected in some way with the *Cambridge Magazine*, though this impression was vague. (Footnote: In connection with this, it may be mentioned
that the *Cambridge Magazine* undertook the sale of books.) . . .
In association with the act of taking these books to my bedroom I had the
idea that it might have been right to read the Magazine in private, but that it
was not suited for general circulation . . The bedroom of the dream thus
seems to have served as a symbol for privacy as opposed to publicity in
relation to this publication.[46]

What had prompted the lifting of repression concerning the conflict over
the *Cambridge Magazine*? This lifting took place in July 1921 as a result of
the second dream. But there was another dream analysed later in the
book, the 'Pacifist dream', which raised similar issues. Rivers gave no date
for this dream, but internal evidence places it at the end of July 1917.
The dream took Rivers back to the late 1890s when he was conducting
physiological-psychological research with Kraepelin and Kühne at
Heidelberg, a time when he had worked side by side in friendship with
German scientists. The dream's narrative 'not only implied peace, but
also the restoration of the friendly relations between the scientific men of
the two countries which existed before the war and was still more definite
in the student days twenty years ago'. In addition, the journal *Anthropos*,

[45] CD,123–4.   [46] CD,125–6.

whose yellow cover figured prominently in the dream, was edited and published in Austria in French, Italian and English, as well as German, and 'thus forms a fitting symbol of international peaceful relations'.[47]

However, the immediate instigator of the dream was 'Patient B.', with whom Rivers had lunched and spent the evening in conversation, and who 'was not suffering from any form of psycho-neurosis, but was in the hospital on account of his adoption of a pacifist attitude while on leave from active service'.[48] The patient was Siegfried Sassoon, who had arrived courtesy of Robert Graves at Craiglockhart by 24 July, when they together celebrated Graves's twenty-second birthday.[49] It is Sassoon's pacifism that gives the dream its title.

> Three evenings a week I went along to Rivers' room to give my anti-war complex an airing. We talked a lot about European politicians and what they were saying. Most of our information was derived from a weekly periodical which contained translations from the foreign Press ... All that matters is my remembrance of the great and good man who gave me his friendship and guidance. I can visualize him, sitting at his table in the late summer twilight, with his spectacles pushed up on his forehead and his hands clasped in front of one knee; always communicating his integrity of mind; never revealing that he was weary as he must often have been after long days of exceptionally tiring work on those war neuroses which demanded such an exercise of sympathy and detachment combined.[50]

Sassoon's recollection of the 'weekly periodical which contained translations from the foreign Press', the *Cambridge Magazine*, combined with Rivers's dreams of the *Magazine*, highlights its importance as the principal object of contention and reference point for the two men. The significance of the *Cambridge Magazine* would have been increased by a factor that neither Sassoon nor Rivers mentions: the first of Sassoon's war poems were being regularly published in the *Magazine*.[51] On 9 June, 'In an

---

[47] CD, 170.   [48] CD, 167.

[49] Richard Perceval Graves, *Robert Graves: The Assault Heroic 1895–1926*, London: Weidenfeld & Nicolson, 1986; Jean Moorcroft Wilson, *Siegfried Sassoon. The Making of a War Poet: A Biography (1886–1918)*, London: Duckworth, 1998.

[50] Siegfried Sassoon, *Sherston's Progress* (1936), London: Folio Society, 1974, p. 13.

[51] Sassoon's principal contacts with the *Magazine* were Edward Dent, the musicologist, who initially advised Sassoon in early 1916 to publish in the *Magazine*, and A.T. (Theo) Bartholomew, under-librarian at the University Library, both part of a homosexual circle in Cambridge. By June 1917, Sassoon was in regular contact with Ogden, who was fifth on the list of those to whom he sent his 'Statement' (numbering fourteen in all, only ten of which were ticked to indicate, presumably, that the statement had actually been sent to them), which was read out in the House of Commons, in early July 1917 ('sent off July 6th', Sassoon noted). Top of the list was Thomas Hardy; an afterthought was H.G. Wells; it included three MPs. (See Sassoon's Notebooks: http://cudl.lib.cam.ac

underground dressing-station' and 'Supreme sacrifice' were published in the *Magazine*; 'To any dead officer' was published by Ogden as a separate *Cambridge Magazine* pamphlet in August 1917, just when Rivers and Sassoon were engaged most concentratedly in their exchanges. These were very powerful anti-war poems, full of bitter satire for civilian non-combatants and the old men sending the young to their deaths, full of bleak compassion for the men who had died in the mud of Flanders, poetic grenades inducing 'the shock / Of ugly war brought home'.[52] In the analysis of the 'Pacifist dream', Rivers recalls how, as soon as he had arrived, Patient B. was directing his reading 'partly in order to help me to understand his position':[53] he finished reading Barbusse's *Under Fire* (the translation published in June 1917) and the evening before the dream had been reading an article in the *English Review*, which had persuaded Rivers more strongly than any previous piece of the virtues of peace through negotiation.

'Friendly relations between ... men':[54] this is the scene reproduced in the dream, most obviously in the 'student days twenty years ago' in Heidelberg, but also in the new and dramatically challenging relations with Sassoon. This immediate stimulus for the dream sees Sassoon and Rivers talking politics, reading through the *Cambridge Magazine* and other 'left-leaning' material that Sassoon thrust at Rivers, Rivers opening up to Sassoon on his life of fraternal collaboration with German scientists in the 1890s when he had lived with a professor and worked with him in the same building, a 'combination of dwelling-house and laboratory, which is unusual in England, [but] frequently occurs in Germany'.[55] What might result from such warm and unguarded conversations? '[W]hen I was reading the *Review* I had thought of the situation that would arise if my task of converting a patient from his "pacifist errors" to the conventional attitude should have as its result my own conversion to his point of view.'[56] '[O]ne of our jokes had been about the humorous situation which would

.uk/view/MS-ADD-09852-00001-00011/26.) Being a regular reader of the *Magazine*, Rivers would have seen the following poems first published in the *Magazine* (e.g., 'A subaltern', on 24 February 1917; 'The optimist', 21 April 1917; 'Base details', 28 April 1917; 'News from the Front', 2 June 1917; 'Supreme sacrifice' and 'In an underground dressing-station', 9 June 1917). Four poems published in the *Magazine* in 1917 were also published by Ogden in pamphlet form in January 1918 ('Dreamers', 'Does it matter?', 'Base details' and 'Glory of women'). Some of these poems were not published in the collections *Counter-Attack and Other Poems* (1918) and *War Poems* (1919) that made Sassoon immediately famous as a war poet. See Jean Moorcroft Wilson, *Siegfried Sassoon. The Journey from the Trenches: A Biography (1918–1967)*, London: Duckworth, 2003.
[52] Siegfried Sassoon, 'Supreme sacrifice', *Cambridge Magazine*, 9 June 1917, p. 691.   [53] CD, 168.
[54] CD, 170.   [55] CD, 167.   [56] CD, 168.

arise if I were to convert him to my point of view', Sassoon wrote.[57] Many later writers, the most prominent of them being Pat Barker in *Regeneration*, have been drawn to the inherent drama of this meeting between the humanely sympathetic, previously duty-bound Rivers and the hero and poet Sassoon struggling between his own sense of solidarity with the soldiers, both dead and alive, in France, and his revulsion at the dirty trick played on his generation by the generals, by the patriotic women, by the English.

Rivers initially portrayed his own pacifist leanings as 'egoistic' in character: peace would allow him to 'get back to my proper studies, which had been interrupted by the war'.[58] But as his analysis unfolded, he set up a stark contrast between his long-held 'fight to the finish' views and the revival, as represented in the dream, of the conviction he had held as a young man of 'the value of scientific co-operation as a step towards international friendship'.[59] Even though a crucial moment in the dream had seen him turning 'the wrong way', requiring his German host to call out to him that he must go to the left, not the right, the conflict remained unresolved. Him turning to the left he later defended as to be correctly understood as a political, not an ethical, symbolization:

> It seems far more likely that in this case 'right' and 'left' had reference to the customary means of denoting Conservative and Liberal tendencies, especially on the Continent. I was especially familiar at the time with the use of these expressions in the extracts from foreign journals published in the *Cambridge Magazine*, which I read regularly, and a movement to the left in such journals is a regular symbolic expression for Liberal tendencies.[60]

Rivers did indeed move to the Left as a result of his personal transformation during the war. He was sympathetic to Sassoon's immediate postwar wish to join the Labour Movement, advising him to study political economy at Oxford in order to do so effectively.[61] Postwar, the indefatigable Rivers increasingly spoke to meetings where non-elite audiences were to be found and, in his last months, took on as an assistant an uneducated Northern worker. In 1922 he himself agreed to stand as the Labour candidate for the University of London in the general election.

Rivers's final theoretical position on dreams was that they are 'attempts to solve in sleep conflicts of waking life'.[62] His account of the 'Pacifist dream' showed it to be a representation of a conflict (recently aroused by his conversations with Siegfried Sassoon) between newly awakened

---

[57] Sassoon, *Sherston's Progress*, p. 13.   [58] CD, 168.   [59] CD, 170.   [60] CD, 175.
[61] Wilson, *Siegfried Sassoon. The Journey from the Trenches*, p. 29.   [62] CD, 81.

pacifism, which could draw upon his egotistic motives, and his youthful internationalist ideals. Yet the dream did not solve the conflict, and nowhere in his discussion of the dream did he reveal his own feelings or views concerning Sassoon's pacifist position. Still frustrated, many years on, Sassoon himself felt this absence: 'Looking back from to-day [1936], however, I am interested, not in what my own feelings were, but in what Rivers had been thinking about the decision which he had left me so entirely free to make.' In his 'medical' relationship with Sassoon, and in his account of his own dreams, Rivers maintained a studious neutrality about his political position: he adopted the strictly neutral position of the psychoanalyst.

Yet the interpretations of the two dreams, the 'Reproachful letter' and the 'Pacifist dream', indicate very clearly that Rivers went through a process of political awakening in the course of the war, even if he held firm to the practical ideal of psychotherapeutic neutrality in his army medical work. What prompted this awakening? If Rivers had not met Sassoon, would he have remained imperturbable in his 'fight to the finish' attitude? This is quite possibly the case. Yet the re-analysis of the 'Reproachful letter' dream, centred on the revival of the symbol '*Cambridge Magazine*', indicates that, in July 1921, Rivers rediscovered this political conflict as active in his own inner life four months *before* Sassoon arrived at Craiglockhart. By rewriting his own autobiographical journey (in the revision of his analysis of the 'Reproachful letter' dream), he effectively took this pivotal role away from Sassoon. He would not allow that much power to one man.

Curiously enough, then, there is a parallel in Rivers's relationship with Freud and his relationship with Sassoon. Both relationships represented threats to the independence of his own judgement. The dream book is overtly an account of Rivers's shifting relations with Freud's dream theories. At first, he was enthusiastically under the spell of Freud's theories; then he drew back, critical scientist to the fore, proposing that any affective state could be the driving force behind the dream, then that dreams were attempts at resolution of the unresolved conflicts of waking life. He even fell into the trap into which many of Freud's readers are tempted: he thought he could revise Freud's theories by reinterpreting Freud's own dreams. The overall result was that Rivers's views on dreams were very close to Freud's, though this was often difficult to see because of his frequent minor – and confusing – changes of terminology and emphasis.[63] Rivers

---

[63] The most peculiar and confusing of these was Rivers's decision to reserve the term 'repression' for 'the process by which we wittingly endeavour to banish experience from consciousness' (IU, 17),

may have insisted there was considerable distance between his own views and those of Freud. But close readers of Rivers's work were more struck by their similarities than their differences and were irritated by his fussing, to no good purpose, over terminology.

> 'I propose,' [Rivers] says, 'to regard Freud's formula [of wish-fulfilment] as unduly simple, and suggest as an alternative the working hypothesis that the dream is the solution, or attempted solution, of a conflict which finds expression in ways characteristic of different levels of early experience.' This is not really so big a change as might appear, and the difficulty is in connexion with the term 'wish,' rather than with the underlying notion. A conflict must mean the presence of opposed tendencies having different ends. A solution or an attempted solution must mean that one of the ends, or a combination of the ends, is sought to the exclusion of the others. Any such seeking of an end to the exclusion of others is a 'wish' in the Freudian sense of the term.[64]

This acute observation, showing that Rivers had not fundamentally altered the Freudian theory but had only engaged in superficial terminological tinkering, did not come from an arch-Freudian. Far from it: the reviewer was Frederic Bartlett, Rivers's former student and, following his death, director of the Psychological Laboratory in Cambridge, writing in the journal *Mind*. Less surprisingly, perhaps, an even sharper observation came from Ernest Jones:

> [Rivers] has been able to confirm Freud's theory up to a certain point, though not beyond this where it concerns the deeper layers of the mind. Thus, he agrees with Freud that dreams are of great psychical significance; that the distinction between the manifest and latent content is of cardinal importance; that the latent content indicates a remarkable distortion (which he prefers to term transformation) before it is converted into the manifest content; that there are regular laws by which this transformation occurs; that the latent content is a repressed one – in short, all the more essential parts of Freud's theory. He dislikes Freud's term 'wish-fulfilment' and prefers to regard the dream as the expression of an attempt to solve some conflict; the difference here is in most cases verbal only.[65]

proposing the term 'suppression' for 'unwitting' elimination. In other words, Rivers interchanged the meaning of the (Freudian) terms 'suppression' and 'repression'. For '*unwissentlich*' – see W.H.R. Rivers, *Reports of the Cambridge Anthropological Expedition to Torres Straits*, Vol. II, *Physiology and Psychology*, Part I, *Introduction and Vision*, Cambridge: Cambridge University Press, 1901, pp. 6–7. For Jones's acute comments, see E.J. [Ernest Jones], '*Instinct and the Unconscious. A Contribution to a Biological Theory of the Psycho-Neuroses*: By W. H. R. Rivers, M.D., D.Sc., F.R.S., (Cambridge University Press, 1920. Pp. 252. Price 10s. 6d.)', *IJP* 1 (1920), 475.

[64] Frederic Bartlett, 'Review. W.H.R. Rivers, *Conflict and Dream*', *Mind* 33(129) (1924), 94.

[65] E.J. [Ernest Jones], '*Review, Rivers*, Conflict and Dream', IJP 4 (1923), 499.

These two well-informed reviewers agreed: Rivers's criticisms of Freud were essentially terminological; what he portrayed as major criticisms were in reality minor quibbles over nomenclature. In other words, he had persuaded himself that there was a greater distance between his views and Freud's than actually existed. He had, unwittingly, become more of a 'Freudian' than he believed he was. But, out of fear for his independence of mind, he magnified his own differences from Freud.

An Independent Mind

A similar dynamic is at work in Rivers's relations with the pacifists of the *Cambridge Magazine* and his medical charge, Siegfried Sassoon. Did Rivers become more of a 'Sassoonian' than he believed he was? The impact of Rivers on Sassoon is well documented. In his *Diaries*, Sassoon recorded on 26 July 1917: 'Rivers, the chap who looks after me . . . I am very glad to have the chance of talking to such a fine man.'[66] On 19 October, he reported: 'He says I've got a very strong "anti-war" complex, whatever that means',[67] and on 29 October: 'Rivers thinks my "Fight to a Finish" poem in the *Cambridge Magazine* very dangerous!'[68] ('And with my trusty bombers turned and went/To clear those Junkers out of Parliament.') Trustworthy Rivers protected Sassoon very effectively: 'The Board asked if I had changed my views on the war, and I said I hadn't, which seemed to cause surprise. However Rivers obtained, previously, an assurance from a high quarter that no obstacles would be put in the way of my going back to the sausage machine.'[69] Sassoon, strongly prone to idealization of authoritative figures in his life, recounted repeatedly in his autobiographical writings the crucial place that Rivers quickly came to have in his inner life. In his whimsical best-selling semi-autobiographical writings, classics of anti-war literature, he replaced all real names of people, places and houses with fictional names – with one exception: Dr Rivers always remained Dr Rivers.

While recuperating at his mother's house during May and June 1917, Sassoon decided to make his Public Declaration – the declaration that eventually led him into the care of Rivers at Craiglockhart and which was drafted in consultation with Bertrand Russell. Simultaneously he was

[66] Siegfried Sassoon, *War Diaries. 1915–1918*, ed. and intro. Rupert Hart-Davies, London: Faber & Faber, 1983, p. 183; Patrick Campbell, *Siegfried Sassoon: A Study of the War Poetry*, Jefferson, NC: McFarland, 1999, p. 152.

[67] Sassoon, *War Diaries*, p. 192. [68] *Ibid.*, pp. 193–4.

[69] Sassoon to Graves, 7 December 1917; Sassoon, *War Diaries*, p. 196.

already drafting one of his first powerful 'war poems', eventually published under the title 'Repression of war experience'. Six months later, on 4 December, having left Craiglockhart, Rivers gave an address to the Royal Society of Medicine in London entitled 'On the repression of war experience'.

Whose title was it? It seems unlikely that Sassoon would have chosen this title without Rivers's example. Most plausibly the poem's title is a tribute to Rivers. But it is a well-considered tribute, since the subject of Sassoon's poem is the precise subject of Rivers's paper. 'Many of the most trying and distressing symptoms from which the subjects of war neurosis suffer', Rivers wrote, 'are not the necessary result of the strains and shocks to which they have been exposed in warfare, but are due to the attempt to banish from the mind distressing memories of warfare or painful affective states which have come into being as the result of their war experience.'[70] Rivers's paper is a criticism of the policy of repression – the deliberate attempt to push horrifying memories and affects out of consciousness: '[W]e must not be content merely to advise our patients to give up repression, we must help them by every means in our power to put this advice into practice.'[71] Sassoon was the best example possible of someone who never let his 'war experience' out of his mind, and was undoubtedly 'healthy'. Walking in step with Rivers's argument, Sassoon's poem[72] enacted the effects of the doctrine of repression:

> – it's bad to think of war,
> When thoughts you've gagged all day come back to scare you;
> And it's been proved that soldiers don't go mad
> Unless they lose control of ugly thoughts
> That drive them out to jabber among the trees.

They were of one mind on this issue.

The preoccupation Rivers and Sassoon thus shared with the 'repression of war experience' helps us understand more clearly Rivers's idiosyncratic usage of the term 'repression', indicating a political dimension not immediately apparent in the Freudian usage. For Rivers and Sassoon, the injunction to 'stop thinking about the war', to 'turn one's mind to other, happier things', went hand in hand with the political position of refusing to recognize the reality of the war being fought in the mud of the trenches. Countering the policy of repression amounted to a politico-therapeutic

---

[70] IU, 187.   [71] IU, 204.
[72] Siegfried Sassoon, *War Poems*, London: Heinemann, 1919, p. 75; see Campbell, *Siegfried Sassoon*, pp. 151–61.

injunction to 'face the facts!' – the stark contrast to 'fight to the finish!' Dividing the individual mind (between soothing thoughts of pastoral peace and horrible recollections of war), dividing the geography of Europe (between the horrors of war in France and the continuity of ordinary life in England), dividing the mind of the nation (between those who knew what was going on in Europe through the *Cambridge Magazine* and those who were wilfully blind to everything but the military imperative, wishing to impose this blindness on others) – these divisions were all intertwined in Rivers's dream life. And the image that Rivers introduced to sum up his own version of these conflicts speaks of yet another division: his thought that it was improper to read the *Cambridge Magazine* in public since it sapped morale made the reading of the *Magazine* (and thus of Sassoon's poetry) a private act to be kept in the bedroom – like other private acts. The privacy of the bedroom for one's personal political life thus hovers as a metaphor for other private acts in the bedroom.

In discussions with Sassoon, Rivers ran the risk of being converted to his pacifist views. In discussions with Freud, Rivers ran the risk of being converted to his scandalous views about sex. 'Friendly relations between men' could turn into being overwhelmed by an alien mind. Any dissemination of views from the larger world (as in the Foreign Press section of the *Cambridge Magazine*) should take place only in private, as in a bedroom.

This distrust of being unwittingly affected was a deep theme in Rivers's identity, most obviously in his scientific identity. Trained in psychology at the historical moment when experimental introspection was the psychologist's principal and most vaunted special skill, Rivers's two most famous experimental studies were organized around the elimination of extraneous, disturbing influences on the introspected data. In his experiments on nerve regeneration with Henry Head, his role was as 'guide and counsellor'. His interest lay rather in the psycho-physical aspect of the work and he was impressed with the insecurity of this side of the investigation. Introspection could be made fruitful by the personal experiences of a trained observer only.[73] 'Insecurity' here meant the difficulty of securing reliable data – that is, reliable reports. The study of the influence of drugs employed the classic Wundtian and Kraepelinian apparatus; what was distinctive and entirely due to Rivers was the experimental design of introducing a blind control in order to eliminate the influence of the *experimenter's* interest upon the

---

[73] W.H.R. Rivers and Henry Head, 'A human experiment in nerve division', *Brain* 31(3) (1908), 324.

experimental data produced.[74] Once again, the objectivity and reliability of the *experimenter* (*Versuchsperson*), not the 'object' (*Reagent*), was at the forefront of Rivers's mind and led to his methodological innovations.[75] When Rivers became involved in Freudian dream analysis, he quickly became alert to the 'danger' that 'according to Freud the wish of a patient to prove or disprove the views of his physician can provide the leading motive of a dream, and this suggests the danger that the theories of the dreamer may influence his dreams'.[76] He even had a record of this thought, from 19 March 1917, stimulated by a rereading of the final chapter of *The Interpretation of Dreams*: '[T]he danger of having his dreams influenced by his theories must be especially great in the case of one who had formulated so definite a theoretical position as that of Freud. I had wondered whether it might be possible to find evidence for such influence in any dream of mine.'[77] The next night he dreamed the dream of 'The reproachful letter'. Here was a dream driven entirely, it seemed, by a reproach, not a wish. But, given his written note from the previous day, how could he be sure he was not dreaming a reproachful dream as a *refutation* of Freud's theory – just as Chapter 4 of Freud's dream book predicted he would?[78] If Freud's dreams were influenced by his theories, might not his, Rivers's, be? Notwithstanding this threat, he now took his re-analysis of the dream in terms of a personal conflict over politics rather than a self-reproach for medical failure to be evidence that the dream itself (of reading the letter) had *not* been influenced by his eagerness to confirm his own theory; only his first analysis was 'contaminated' by this desire: '[T] he theories of a dreamer may influence his self-analysis, but provides no evidence that they influence his dreams.' What his re-analysis reveals is 'one of the dangers of such self-analysis'[79] – the ever-present possibility of contamination of one's analysis by one's own pet theories. But the second dream could, of course, be a way of protecting himself from the fear that his very own dreams, not just his analyses, were contaminated. Rivers did not raise or answer this meta-objection.

---

[74] W.H.R. Rivers, *The Influence of Alcohol and Other Drugs on Fatigue: The Croonian Lectures Delivered at the Royal College of Physicians in 1906*, London: E. Arnold, 1908, p. 20.
[75] Ted J. Kaptchuk, 'Intentional ignorance: a history of blind assessment and placebo controls in medicine', *Bulletin of the History of Medicine* 72(3) (1998), 419.
[76] CD, 126.   [77] CD, 126.
[78] John Forrester, 'Dream readers', in Forrester (ed.), *Dispatches from the Freud Wars: Psychoanalysis and its Passions*, Cambridge, MA: Harvard University Press, 1997, pp. 138–83; John Forrester, 'Introduction', in Sigmund Freud, *Interpreting Dreams*, trans. J.A. Underwood, London: Penguin, 2006, pp. vii–liv.
[79] CD, 127.

The potential contamination inherent in psychoanalysis went even further. The second chapter of Rivers's book moved smoothly from his own 'Presidency' dream with its play with 'rivers' to two dreams of a suicidal patient, which also played with 'rivers'; the second dream dramatized a conflict between 'one desire to come to me for help and another desire to stand on his own feet and rely on his own strength'.[80] Rivers added immediately: 'I need hardly say that it was a regular part of my treatment to guard against the process known to the psycho-analysts as transference.'[81] The striking phrase in this sentence is the opening: 'I need hardly say'. It speaks volumes: it tells us quite unambiguously, because it is so unself-conscious, that Rivers assumed without a moment's second thought that transference – the development of a baseless, mistaken affective relationship between patient and doctor – was an evil that should at all costs be avoided. 'There had at first been rather violent resistance to [the] process [of gaining access to the patient's early experiences], followed later by a state in which I had recognised the danger of transference, and it had formed an essential part of my treatment to inculcate independence.'[82] No wonder, we might reflect, the patient felt this violent conflict: River's offer of help came in the form of exhortation to independence.

Yet it is entirely in character – the character of one of the most rigorously objective human scientists of his age – for Rivers to be wary of transference. With his patients – all male[83] – there was quite probably the additional complicating factor, seen most clearly in his relations with Sassoon. The positive transference would manifest itself in a form in which the 'paternal' transference and the 'homosexual' transference would be inextricably intertwined. For Rivers, a man whose homosexuality was covert and ill-defined, any response to this transference would have appeared potentially threatening and could have unpredictable effects on the patient. He had every reason to retreat from 'working with' the transference into the position he made explicit in his writings: avoidance at all costs. And, as a result of finding no way of working with the transference in a positive fashion, he would find no way of working with his own countertransference: the evidence we have takes the form of his wondering to what extent

---

[80] CD, 35.   [81] CD, 35.   [82] CD, 36.

[83] There is the striking and slightly mysterious brief appearance of a woman in Rivers's psychotherapeutic work: Mrs Edith Eder, David Eder's wife, is thanked warmly in CD, 15 for supplying corroboration of a fine point of dream analysis through one of her patient's dreams. Edith Eder was also a close friend of Ernest Jones (Brenda Maddox, Freud's Wizard, London: John Murray, 2006, p. 113) and is described with her husband by D.H. Lawrence in 1921 as 'the great Freudians in London'.

might Sassoon's pacifist views, whether in the form of public protest or poems, be undermining his formerly clearly held political views.

Placing these issues of transference and countertransference in the context of his understanding of science, we see Rivers, always aware of issues of objectivity and the interests of the observer, standing on the verge of a recognition that the observer's implication in the scene of inquiry was constitutive of the field of the human sciences. It was Rivers who mapped out the terrain in ethnology of 'intensive fieldwork', of the 'participant-observer', but never fully entered on it himself. In the field of psychotherapy, he hoped to escape from Freud's recognition that, both in dreams and in the neuroses, there was no avoiding the implication of the subject's own desires and the observer's relationship with his patients – indeed, this implication was being turned around by Freud into the foundation stone of the theory and therapy of the 'transference neuroses'. As the literary historian James Buzard astutely notes,

> [A]uthority derives from the demonstration not so much of some finally achieved 'insideness' in the alien state, but rather from the demonstration of an *outsider's insideness*. Anthropology's Participant Observer, whose aim was a 'simulated membership' or 'membership without commitment to membership' in the visited culture, went on to become perhaps the most recognizable (and institutionally embedded) avatar of this distinctively modern variety of heroism and prestige.[84]

There is the sense that Rivers's 'Faustian' character, restless, unrooted, playing all his parts perfectly – exacting experimentalist, roving explorer, sympathetic physician, punctilious and demanding methodologist – was made to stand on the edge, looking forward to the intermingling of these future sciences. In Rivers one senses all the possibilities of the relations of ethnology and psychoanalysis that provoked Michel Foucault to make these two 'counter-sciences' the master disciplines for modernity:

> Psychoanalysis and ethnology are not so much two human sciences among others; rather they span the entire domain of those sciences, they animate its whole surface, spread their concepts throughout it, and are able to propound their methods of decipherment and their interpretations everywhere. No human science can be sure that it is out of their debt, or entirely independent of what they have discovered, or certain of not being beholden to them in one way or another.[85]

[84] James Buzard, *Disorienting Fiction: The Autoethnographic Work of Nineteenth-Century British Novels*, Princeton: Princeton University Press, 2005, pp. 9–10.

[85] Michel Foucault, *The Order of Things* (1970 [1966]), London: Routledge, 2002, p. 413.

A fused project was very much alive in the early 1920s, what John Rickman, Quaker doctor, Rivers's student and Freud's analysand, in 1923 called 'the stormy seas of anthropo-ethno-analytico-sciences'.[86] As for sex and transference: these, in the end, were the topics where Rivers seemed most at odds with Freud. There are no records available that give explicit evidence concerning Rivers's sexual life. Many writers have assumed he was homosexual in orientation, and there is ample room for speculation that his relationship with Sassoon aroused homosexual currents, at the very least in the more general sense of relations of dependency and intimacy between men. It is sensible to assume that the repressive atmosphere of British public life when it came to homosexuality in the period from the Great War to the 1960s persuaded most of those who knew Rivers to remain silent about his sexuality. In his thorough study of Rivers as an anthropologist, Ian Langham wrote:

> As one might be inclined to predict for someone born in the middle of the Victorian era and having a clergyman for a father and two sisters who never married, Rivers seems to have been the victim of severe sexual repression, and this problem seems to have been compounded by what one of his close friends interpreted as Rivers's own tendencies towards homosexuality. Note: For reasons which, if they could be revealed, would be seen as perfectly understandable and proper, I am duty bound not to reveal the source of the suggestion, which was advanced as a clear and confident statement of fact, that Rivers was a closeted homosexual.[87]

It is even unclear if Rivers ever discovered the extent to which Sassoon's friendships were infused and traversed by homosexual currents and passions. There is the striking evidence – unreliable, because remembered sixty years on, and unreservedly and self-declaredly subjective – of Rivers's student John Layard, one of the earliest 'participant-observers' on Atchin, a small island off Malakula Island in Melanesia, in 1914–15,[88] suffering a nervous breakdown on his return, following Rivers to Craiglockhart where in late autumn 1916 he received daily psychotherapy in his nearby

---

[86] Rickman to Róheim, letter dated 22 January 1923, BPaS, Rickman Papers, CRR/FO7/12. On Rickman, see Chapter 5.

[87] Ian Langham, *The Building of British Social Anthropology: W.H.R. Rivers and his Cambridge Disciples in the Development of Kinship Studies, 1898–1931*, Dordrecht: D. Reidel, 1981, pp. 52 and 340n7.

[88] Jeremy MacClanchy, 'Unconventional character and disciplinary convention: John Layard, Jungian and anthropologist', in George W. Stocking, Jr. (ed.), *Malinowski, Rivers, Benedict and Others: Essays on Culture and Personality*, History of Anthropology 4, Madison: University of Wisconsin Press, 1986, pp. 50–71.

Fig. 3.2 Group in front of abandoned RC Mission house with Dr W.H.R. Rivers.
Photo by John Layard, Atchin, 1914.

hotel, which came to an abrupt end, according to Layard, when Layard
declared his (transferential) love to Rivers.

When it came to Rivers's understanding of the importance of sexuality
for understanding the war neuroses, we appear to be on firmer ground.
There is ample evidence that Rivers publicly disputed the importance given
to sexuality by Freud. Rivers's views were certainly trumpeted by many
who followed after as evidence of the sensible and necessary corrections
that the vast bulk of informed British medical writers made to the extreme
and unbalanced views of the Freudian psychoanalysts on the question of
sex. To give just one example, an often-quoted passage from Rivers's
Instinct and the Unconscious:

> It is a wonderful turn of late that just as Freud's theory of the unconscious and
> the method of psycho-analysis founded upon it should be so hotly discussed,
> there should have occurred events which have produced on an enormous scale
> just those conditions of paralysis and contracture, phobia and obsession,
> which the theory was especially designed to explain. Fate would seem to
> have presented us at the present time with an unexampled opportunity to test

the truth of Freud's theory of the unconscious, at any rate in so far as it is concerned with the part taken by sexual factors in the production of mental and functional nervous disorder. In my own experience, cases arising out of the war which illustrate the Freudian theory of sexuality directly and obviously have been few and far between. Since the army at the present time would seem to be fairly representative of the whole male population of the country, this failure to discover to any great extent the cases with which the literature of the Freudian school abounds might well be regarded as significant. If my experience is a trustworthy sample, it would seem as if the problem was already well on the way towards settlement.[89]

Rivers's strategy was a familiar and repeated one: he wished to bypass the issue of sexuality, regarding it as not essential, an 'unfortunate excrescence',[90] which could be excised from the more important real discoveries that Freud had made; he wished to rescue Freud certainly from the Freudians and, in this respect, from himself.

However, Rivers made other comments about sexuality, which reveal a more nuanced and less self-certain position. In his inaugural address to the newly formed Medical Section of the British Psychological Society in May 1919, an audience predominantly made up of the enormously expanded group of medical psychologists streaming back to civilian life from their wartime military duties,[91] he argued against underestimating the importance of sexuality now that peace was restored:

> You all know how the most prominent school of students of the psycho-neuroses believe that the instinctive tendencies which stand on one side of the battleground belong exclusively to the instinct of sex. However repellent this may be to the traditions of the medical profession, we must be prepared to face this problem honestly and without prejudice. In turning from the practice of war to that of peace we must expect to find a great increase in the part taken by the sexual instinct, for the simple and obvious reason that the conditions of our civilisation make this instinct the special object of its repressions and taboos. We have found reason to believe that sex plays but a little part in the causation of war neuroses, but it does not follow that this will hold good of the neuroses of civil life. On the other hand, we must be careful to hold the balance and not allow ourselves to give to sexual tendencies a prominence greater than they deserve. The sexual instinct is far from standing alone as the subject of the repressions and taboos of social tradition. It should be our working hypothesis that any instinct which needs repression in the interests of society may furnish that occasion for conflict which forms the essence of neurosis.[92]

---

[89] IU, 164.   [90] IU, 164.   [91] Forrester, '1919', pp. 52–6.
[92] W.H.R. Rivers, 'President's Inaugural address. Medical Section, British Psychological Society', *The Lancet* 193(4995) (24 May 1919), 889–92 at 892.

In preparing his book on dreams over the period 1920–21, Rivers gave a further, more personal twist to this counsel to be open-minded about the Freudian emphasis on sexuality. He admitted that the dreams he had analysed were unrepresentative because they came from only two sources: the dreams of traumatized soldiers and

> my own dreams, where sexual conflicts might perhaps hardly be expected to be as active as in the dreams of younger people. Moreover, most of my own dreams which have been analysed occurred at a time when, owing to the extreme interest of my work and my absorption in it, I was far more free than usual from the sexual conflicts which are generally believed to be active in dreams.[93]

And Rivers went on to make it quite clear that his own book was inevitably discreet about sexuality: 'If [my dreams] had dealt with sex-conflicts the analyses would probably have been full of passages which a natural reticence would have driven me to withhold or garble, thus interfering with the cogency of the demonstration.'[94] He made it pretty clear that the dreams he had used to demonstrate his theories were, 'wittingly' or 'unwittingly' (key terms for Rivers, as we have seen, a fusion of experimental psychological and psychoanalytical themes), the result of 'comparatively innocent conflicts'.[95] Make no mistake, he informed the reader: if they had not been 'innocent' he would not have analysed them in public.

So we see that Rivers's considered view was the war neuroses had demonstrated that the emphasis placed on sexuality in the aetiology of all neuroses by the Freudians was unwarranted. But Rivers did not turn his back on this problem: he warned that the peacetime neuroses might well manifest the force of the sexual instinct as described by Freud; he gave exceptional warning signals in his dream book that a number of factors – his age, his public diffidence, perhaps even his inner resistances – entailed that he was not well placed to draw any certain conclusions about the importance of sexual conflicts in dream life generally.

When it came to his own personal conflicts as revealed by his published writings, there was no question of anything sexual being explicitly – wittingly – mentioned. Professional ambition, personal reproach, Sassoon's protest and pacifism – these are some of the issues Rivers's published dream life is preoccupied with; but there is more. Two other sets of dreams (the 'London lectures' dreams and the dream of 'Hidden sources') indicate a conflict between Rivers's anthropological interests and his professional

[93] CD, 110. [94] CD, 111. [95] CD, 111.

psychotherapeutic duties, to which he had found an *intellectual* solution: what Rivers discovered was a way to integrate his anthropological and psychological interests. The programme he outlines had important consequences for the development of anthropology in the twentieth century. It represents the convergence of two major strands of Rivers's work, which he himself had thought at odds with each other. To appreciate fully the significance of this moment – for Rivers, for anthropology, for psychoanalysis – some background is needed concerning Rivers's part in the development of anthropology.

Psychoanalysis and Ethnography

There are two mythicized founding dates for social anthropology in Britain: the Torres Straits Expedition of 1898 and the publication in 1922 of both Malinowski's *Argonauts of the Western Pacific* and Radcliffe-Brown's *The Andaman Islanders*.[96] The second of these dates also marks the death of the leading anthropologist of the period 1898–1922: W.H.R. Rivers, whose ethnographic career began with the Torres Straits Expedition. Another way of putting this is to claim that what was started in 1898 – the survey fieldwork approach to understanding primitive cultures – came to fruition with the publication of two monographs by practitioners who had learned the lessons of the Torres Straits programme and put them into practice in the 1910s, and were as a consequence able to produce perfectly formed exemplars of intensive fieldwork and the understanding based upon it.

---

[96] Adam Kuper, *Anthropology and Anthropologists: The Modern British School*, 2nd revised edition, London: Routledge and Kegan Paul, 1983, p. ix; Michael North, *Reading 1922: A Return to the Scene of the Modern*, Oxford: Oxford University Press, 1999, p. 42; George W. Stocking, Jr., *After Tylor: British Social Anthropology, 1888–1951*, Madison: University of Wisconsin Press, 1995, p. 271. A.R. Radcliffe-Brown (1881–1955), who came up to Cambridge in 1901 and took the Moral Sciences Tripos at Cambridge in 1904 and 1905, was very much Rivers's student until their estrangement before the war: 'My own study of kinship began in 1904 under Rivers, when I was his first and at that time his only student in social anthropology, having for three years previously studied psychology under him' (A.R. Radcliffe-Brown, *Structure and Function in Primitive Societies*, Glencoe, IL: The Free Press, 1952, p. 50). Langham (*The Building of British Social Anthropology*, pp. 244–300) argues vigorously that Radcliffe-Brown's self-presentation as entirely indebted to Durkheim for his advocacy of the structural-functionalist method conceals the fact that, in the period 1912–14, he developed this method explicitly as an extension of Rivers's work. Rivers was then systematically excluded from the discipline-founding accounts by later students of Radcliffe-Brown's. Malinowski was formally Seligman's student, but looked to Rivers as the unquestioned leader of British anthropology (see Stocking, *After Tylor*, pp. 267–8; Michael W. Young, *Malinowski: Odyssey of an Anthropologist, 1884–1920*, New Haven: Yale University Press, 2004, pp. 88, 161–2, 165, 233–7, 245, 251, 265, 349–50, 369, 373, 408, 558–9.

Fig. 3.3 Rivers and the Cambridge University Torres Straits Expedition, 1898: Haddon (seated) with (l–r) Rivers, Seligman, Ray and Wilkin, Mabuiag.

This account places all the emphasis upon method: the ideal of the lone researcher going out, immersing himself in one very specific native culture, and reporting back – becoming a 'participant observer', as the neologism of the mid-1920s characterized it. Part of the eventual prestige of social anthropology would come to rest on the sheer arduousness of the labour involved and hardships undergone in this process and upon the fact that no other researchers were involved or had ever been there: the new fieldwork ideal was strangely solitary and naïvely empiricist in its conception of professional expertise. Yet there was an accompanying *conceptual* transformation in the period 1898–1922: where the centre of gravity of anthropology had been evolutionism and religion, the new social anthropology was non-evolutionist (because synchronic structural-functionalist) and preoccupied with social structure and order. The key area for understanding social structure became 'kinship'. For Olympian theoreticians such as Lévi-Strauss, it was *this* innovation that counted: 'Anthropology found its Galileo in Rivers, its Newton in Mauss ... Rivers, whose genius is largely ignored today, employed both types of interpretation [socio-psychological and historical] simultaneously; since his time, no one has said anything not already anticipated by that great theoretician.'[97] Haddon, Rivers and Myers together founded Cambridge ethnographic anthropology between 1898 and 1921: Haddon became lecturer (1900) and then reader (1909) in Ethnology; a Diploma in Anthropology was established in 1908 and the persuasiveness of the prestige of Rivers's achievements led the university to establish a new Tripos in Anthropology in 1913, taking its students for the first time in 1920–21.

Rivers's 'genealogical method' developed in his work with the Todas, the peoples of the Solomon Islands, Melanesia, and Polynesia during the first decade of the twentieth century. A new methodological ideal was promoted in the section Rivers wrote for the Royal Anthropological Institute's updated edition of *Notes and Queries*,[98] contrasting 'survey work' with what was now required – 'intensive work':

> A typical piece of intensive work is one in which the worker lives for a year or more among a community of perhaps four or five hundred people and studies every detail of their life and culture; in which he comes to know every member of the community personally; in which he is not content with generalized information, but studies every feature of life and custom in concrete detail and by means of the vernacular language.[99]

Originally introduced as a method of generating 'social statistics' in a Galtonian spirit, investigating the frequency of brother–sister exchange, polygamy and levirate marriage, Rivers's method required asking all members of a community what was their 'naming relationship' to all others; it was conceived of as a method for employing the supreme mastery of natives in the domain of 'concrete facts' to solve the 'abstract problems' that natives were not adept at addressing.[100] Gradually recognizing that the answers to the questions did not produce reliable data for drawing up patterns of 'biological' relationships, Rivers simultaneously realized that

---

[97] Claude Lévi-Strauss, 'Do dual-organizations exist?' (1956), in *Structural Anthropology* (1958), trans. Claire Jacobson and Brooke Grundfest Schoepf, London: Penguin, 1972, pp. 162–3.

[98] James Urry, '"Notes and Queries on Anthropology" and the development of field methods in British anthropology, 1870–1920', *Proceedings of the Royal Anthropological Institute of Great Britain and Ireland* 1972 (1972), 45–57.

[99] W.H.R. Rivers, 'Anthropological research outside America', in W.H.R. Rivers, A.E. Jenks and S.G. Morley (eds.), *Reports on the Present Condition and Future Needs of the Science of Anthropology*, Washington, DC: Carnegie Institute, 1913, pp. 5–28 at 7.

[100] W.H.R. Rivers, 'A genealogical method of collecting social and vital statistics', *Journal of the Anthropological Institute of Great Britain and Ireland* 30 (1900), 82; Langham, *The Building of British Social Anthropology*, pp. 64–93; Henrika Kuklick, *The Savage Within: The Social History of British Anthropology, 1885–1945*, Cambridge: Cambridge University Press, 1991, pp. 140–9.

the method was increasingly revelatory of the complex systems of social relationship between all individuals in a community. The method revealed the kinship structures that provided the social organization for 'primitive society'.[101] These kinships structures could very quickly become describable in formal terms, allowing basic templates and rules to be generated in a quasi-algebraic fashion. This was to be the principal method and contribution of Rivers's Cambridge school; this very Cambridge penchant for a baroque formalism built upon many years of painstaking data-gathering was why Lévi-Strauss called Rivers the Kepler of anthropology. (He of course was to be its Einstein.) It is this contrast between the 'survey work' of teams and the 'intensive work' or Haddon's term 'fieldwork'[102] of the single worker that prompts the pre-eminent historian of anthropology, George Stocking, Jr., to deem this enterprise and its associated 'genealogical method' the 'major methodological innovation associated with the Cambridge School'.[103]

At least for Rivers and Myers, the Torres Straits Expedition had initially been a project in comparative psychology. The findings, all hedged with qualifications, were most straightforwardly interpreted as demonstrating little if any difference between the perceptual capacities of white Europeans and Torres Straits islanders. These probably did form the starting-point for the anti-racism or non-racism, particularly of Rivers and Myers (in contrast to McDougall), and thus demonstrated a clear-cut break with and estrangement from the evolutionist assumptions of Victorian anthropology.[104] Rivers's surprising and infamous 'conversion' to diffusionism in 1911, developed at the same time as his call for 'intensive work' of ethnological researchers, certainly put explicit distance between his new views and the older traditions of evolutionism. Historical contingency replaced overarching universal laws of development. Diffusionism seemed in some sense to point away from the intensive work in one place by one

---

[101] 'in his field work Rivers had discovered and revealed to others the importance of the investigation of the behaviour of relatives to one another as a means of understanding a system of kinship' (Radcliffe-Brown, *Structure and Function in Primitive Societies*, p. 51).

[102] George W. Stocking, Jr., 'The ethnographer's magic: fieldwork in British anthropology from Tylor to Malinowski', in Stocking (ed.), *The Ethnographer's Magic and Other Essays in the History of Anthropology*, Madison: University of Wisconsin Press, 1992, p. 27; Anita Herle and Sandra Rouse, 'Introduction: Cambridge and the Torres Strait', in Anita Herle and Sandra Rouse (eds.), *Cambridge and the Torres Strait: Centenary Essays on the 1898 Anthropological Expedition*, Cambridge: Cambridge University Press, 1998, p. 17.

[103] Stocking, *After Tylor*, 112.

[104] Graham Richards, 'Getting a result: the expedition's psychological research 1898–1913', in Herle and Rouse (eds.), *Cambridge and the Torres Strait*, pp. 136–57; Graham Richards, *'Race', Racism and Psychology: Towards a Reflexive History*, London: Routledge, 1997.

worker that Rivers's methodological directives recommended: yet the task of tracking changes across geographical expanses, collating material techniques, hypothesizing intersections of peoples leading to hybrid institutions (in order to explain anomalous kinship organizations or amalgamated architectural styles, for example) required, in Rivers's eyes, yet more fine attention to detail, whether of vernacular nomenclature or the technological level of 'useful arts'. Rivers's magnum opus *The History of Melanesian Society* (1914) covered an enormous range of cultures, with complex histories and widely differing social structures, occupying an area of 30° of longitude and 25° of latitude – roughly 2.4 million square miles (almost the size of Australia), 'a vast Aegean under the lee of the continental mass of South-Eastern Asia', as Rivers's memorialist described it.[105] One could not be much further from the study of a circumscribed group of perhaps three hundred people who share one language and one way of life – the ideal Rivers offered his profession and which they roundly endorsed, particularly in word but also in deed.

Rivers's public support for Freudism in 1917 not only provided some sort of 'cover' for doctors engaging with Freud's theories for the purpose of psychological therapies with the war neuroses; it also promoted the very wide deployment of Freudian ideas for the use of 'the teacher, the politician, the moralist, the sociologist'.[106] So it followed that the next phase of Rivers's engagement with psychoanalysis should entail his extension of Freud's ideas into his other home terrain, that of ethnology. Was there a path from Freudian psychology to ethnology, or was there, as he felt, a conflict between his interests in each of these two areas ('my growing interest in the psychological problems suggested by war-neurosis began to compete and conflict with my interest in ethnology'[107])? This was the problem that, initially unbeknownst to the dreamer, was examined in his own dream life as he tested Freudian dream theory. In the dream of the night of 24–25 March 1917, 'hidden sources' referred both to Rivers's theory of the reconstruction of forgotten histories for Melanesian stone-carving practices and kinship terms and to the 'unconscious experience which bulks so largely in the Freudian psychology'.[108] The triumph of his dream analysis was that it

---

[105] J.L. Myres, 'W.H.R. Rivers', *Journal of the Royal Anthropological Institute of Great Britain and Ireland* 53 (January–June 1923), 16.

[106] IU, 160; see also W.H.R. Rivers, 'Sociology and psychology', *Sociological Review* 9 (1916), 1–13; and 'An address on education and mental hygiene' (1922), in *Psychology and Politics and Other Essays*, with a Prefatory Note by G. Elliot Smith and an Appreciation by C.S. Myers, London: Kegan Paul, Trench, Trubner & Co., 1923, pp. 95–106.

[107] CD, 132.   [108] CD, 131.

solved this problem, this personal problem. Rivers's revised dream theory would eventually reject the universal claim of Freud's wish-fulfilment theory and replace it with a view that dreams are 'the attempted solutions of conflicts'.[109] The particular conflict on which he based his revision of Freud's theory was his own conflict between the 'egoistic motive which urged me not only to go on with my ethnological work, but also to avail myself of an opportunity to demolish opponents' and the 'more altruistic motive that it was now my business [i.e., duty] to understand and apply the principles of psycho-therapy'.[110] In sleep Rivers paved the way for a resolution of the conflict that waking life could not arrive at:

> I came to see that there was no real conflict between ethnology and psychology, but that the two studies are mutually helpful, and that such knowledge of the two as had come to me formed an opportunity to be utilised, and later in the year I prepared the lecture *Dreams and Primitive Culture*, which forms the first of a series of papers in which I have dealt with the extensive border-region between psychology and ethnology.[111]

What diffusionism thus made important for Rivers was the subtle interpretation of minute details in order to facilitate the reconstruction of hypothetical historical movements. The ethnologist was to take the form of a weapon as it presented itself in a series, or the exact pattern of marriage rules, and reconstruct the interaction of peoples, the shift in social institutions as a result of recent warfare, of bride bartering, of commercial transactions of a precious material or a novel foodstuff. The grand scale of diffusionist speculation went hand in hand with an acute awareness of the subtle interpretative skills required of the ethnologist. This shift in the kaleidoscope of Rivers's scientific interests would provide the necessary background to create a new version of psychological anthropology, very different from the comparative experimentalism undertaken on the Torres Straits Expedition of 1898. In the lecture he mentions in his dream analysis (delivered in Manchester in April 1918), Rivers found the perfect confluence between psychoanalysis and ethnology: the dream mechanisms discovered by Freud – dramatization, symbolization, condensation and displacement – provide the tailor-made grid of interpretation for both psychoanalyst and ethnologist.

> Wholly independently of one another, two groups of students [students of early culture and Freud and his followers] concerned with widely different aspects of human behaviour have been led by the facts to adopt an almost

[109] CD, 136.   [110] CD, 134.   [111] CD, 134.

identical standpoint and closely similar methods of inquiry. Both agree in basing their studies upon a thorough-going determinism according to which it is held that every detail of the phenomena they study, whether it be the apparently phantastic and absurd incident of a dream, or to our eyes the equally phantastic and ridiculous rite or custom of the savage, has its definite historical antecedents and is only the final and highly-condensed product of a long and complex chain of events. In this matter of condensation we meet a fundamental problem of those sciences which deal with human behaviour, whether individual or collective.[112]

This confluence of the two disciplines is based on a thorough-going determinism coupled with the greatest respect for every detail of history, whether documented or reconstructed. Crucially Rivers had set out an architecture for the consilience of the two fields: from now on, empirical investigations in one could give indirect support to the other – and vice versa.

Rivers was not alone in this programme of rapprochement between psychoanalysis (or 'the new psychology') and anthropology, as George Stocking has magisterially demonstrated:

> As Rivers and others were so evidently aware, traditional British psychology had been called into question along a number of lines: the irrational, the instinctive, the abnormal, the pathological, the collective. Foreshadowed in the work of William James and in the psychic research movement, pre-figured in the social psychologies of Graham Wallas and William McDougall ... catalyzed by the experiences of World War I, which opened the way for a serious consideration of Freud, the reorientation of British psychology seemed to Rivers and others to provide the basis for a new relationship of anthropology and psychology.[113]

With Rivers's sudden death, this programme was unexpectedly transformed by his replacement as the intellectual leader of British anthropology, Bronislaw Malinowski who, despite his manifest and self-declared great debt to Freud, would later be recalled more for his *rejection* rather than incorporation of psychoanalysis on behalf of the discipline he did so much to professionalize. Malinowski was acutely aware of Rivers's legacy, not least because his former supervisor, Rivers's close colleague Seligman, fed him Rivers's work while he was in the field, with directives to start analysing the dreams of the natives.[114] Malinowski's forum for publication

---

[112] W.H.R. Rivers, 'Dreams and primitive culture. A lecture delivered in the John Rylands Library on the 10th April 1918', reprinted from the *Bulletin of the John Rylands Library* 4(3–4) (1918), 14.

[113] Stocking, *After Tylor*, pp. 243–4.

[114] George W. Stocking, Jr., 'Anthropology and the science of the irrational: Malinowski's encounter with Freudian psychoanalysis', in Stocking (ed.), *Malinowski, Rivers, Benedict and Others*, pp. 29–31.

was the Cambridge general psychological house-journal, *Psyche* (Ogden's journal once he had closed down the *Cambridge Magazine* in 1920), in the early to mid-1920s, where he was invited to replace Rivers as ethnographic expert as soon as Rivers had died. In bold contrast to Rivers's reflex diffidence towards sexuality, Malinowski's tribute to psychoanalysis was to place sexuality at the heart of the project of psychological anthropology. While recognizing, like Rivers, the genius of Freud and the fertility of psychoanalytic ideas, Malinowski was gleefully critical – again like Rivers but in his own triumphantly ill-mannered, decidedly swashbuckling style – of the Freudian project to found a universal anthropology on the Oedipus complex.[115] Malinowski repeated the ambivalence of Rivers: praising psychoanalysis for numerous correct fundamental innovations (its emphasis on sex, its account of infantile sexuality, its attempt to articulate a theory of the relation of biological instinct, psychology and social structure – the list was extensive and repeated) while ridiculing some of the fundamental psychoanalytic hypotheses – most famously, the centrality and cultural universality of the Oedipus complex. Both Rivers and Malinowski repeatedly acknowledged the 'genius' of Freud; both Rivers and Malinowski mounted fierce critiques, based on their ethnographic expertise, of key features of psychoanalysis. Their contemporaries, for their own reasons, heard only the criticisms, not the admiration. Once Malinowski had put the boot into psychoanalysis, the way was clear for Radcliffe-Brown to expunge all psychology from the nascent field of social anthropology, perfectly summed up in his suggestion to his publisher in 1931, when preparing the second edition of *Andaman Islanders*, that the word 'sociology' should be substituted everywhere he had previously used the word 'psychology'.[116]

The distinguished historian of medicine Erwin Ackerknecht observed in 1942:

> In 1911 as an established authority in anthropology, [Rivers] could only undermine his position when he professed diffusionism in his presidential address to the Anthropological Section of the British Association. After almost 30 years of successful work in the field of experimental psychology,

[115] Pear recalled that during the War Rivers preferred not to speculate about Freud's claims concerning the universality of the Oedipus complex until Malinowski had returned from the field (T.H. Pear, 'Reminiscences', typed with a few modifications and additions, from notes made for a recording kept in the archives of the British Psychological Society, 10 May 1957, p. 232).

[116] George W. Stocking, Jr., 'Books unwritten, turning points unmarked. Notes for an anti-history of anthropology' (1981), in Stocking, *Delimiting Anthropology: Occasional Essays and Reflections*, Madison: University of Wisconsin Press, 2001, p. 344.

he had a reputation to lose when he professed Freudian ideas in 1916. Nevertheless he did. The results of these conversions may be appreciated quite differently, but no one can resist the charm of the 'Faustian' attitude which is the very kernel and the goal of our own occidental culture.[117]

This arresting use of 'Faustian' points to the roving and unending quest for knowledge in so many fields which was Rivers's way. (Who else would give up experimental psychology the moment his close friend Myers had built him the best Psychological Laboratory in the country?)

Rivers's Faustian fearlessness has immediate echoes with Freud, who reflected in 1929 to Lou Andreas-Salomé: 'My worst qualities, including a certain indifference towards the world, have no doubt had the same share in the end result as my good ones, e.g., a defiant courage in the search for truth.' Rivers possessed this same courage in pursuit of the truth. But unlike Freud's, his 'search' required geographical wandering. His first medical job was as a ship's surgeon. In the 1890s he was a peripatetic university student and teacher, shuttling between London, Cambridge and the German universities. His almost accidental presence on the Torres Straits Expedition was the equivalent of Freud's self-analysis as a mind- and career-changing experience. Once introduced to ethnographic work in the Torres Straits, he shuttled between Cambridge and India (1901–2), the Solomon Islands and Melanesia (1907–8) and the New Hebrides (1914–15). Even during the war he was restless – from the South Seas to Cambridge then to Liverpool then to Hampstead via Edinburgh. The same restless adventurousness led him to roam the world, to open up new scientific lines of inquiry and alight on new problems, generating new techniques and conceptual schemata. Freud's declaration, in a moment of profound self-doubt in 1900, that he was 'nothing but a *conquistador* – an adventurer, if you want it translated – with all the curiosity, daring and tenacity char-acteristic of a man of this sort',[118] pointed to the fact that he would have wished to have had a life like Rivers's. In certain respects, he did. Freud's roving adventurousness manifested itself in the same encyclopaedic ran-ging over disciplines as Rivers's – expertise in neurology, psychiatry, neurophysiology, psychopathology, theoretical biology, folklore (*Volkskunde*) and ethnology were the shared ones. Most obviously and critically, both men were profoundly influenced by Hughlings Jackson's

---

[117] Erwin Ackerknecht, 'In memory of William H. R. Rivers, 1864–1922', *Bulletin of the History of Medicine* 11 (1942), 481.

[118] Sigmund Freud, 'Letter from Freud to Fliess, February 1, 1900', in J.M. Masson (ed.), *The Complete Letters of Sigmund Freud to Wilhelm Fliess, 1887–1904*, Cambridge, MA: Harvard University Press, 1984, p. 397.

programme in evolutionary neurophysiology, so that Rivers's distinction between the 'primitive' protopathic and the 'higher' epicritic nerve-sensibilities (to be discussed further in Chapter 5) is immediately recognizable as a close cousin of Freud's distinction between primary and secondary processes.[119] For Freud, there was also sexology and a deep acquaintance with world literature; for Rivers, there was the geographical specificity inherent in ethnographic fieldwork, as well as sensory physiology, especially vision. Yet there is one glaring difference between the two men: while Rivers roamed in the style of European adventurers, taking full advantage of the opportunities supplied by British imperial hegemony, truly putting a girdle round about the earth, Freud was comparatively less mobile, voluntarily chained to his medical practice, his family and his consulting room. The Freudian adventure is a '*voyage autour de ma chambre*'.

Despite his enthusiasm for and courage in championing the Freudian cause, after the war Rivers had distinctly mixed views of psychoanalysis. He insisted that some of his friends in trouble seek out a psychoanalyst (e.g., Mansfield Forbes, part of the same Cambridge circle as Sassoon, who complained 'I rarely go to bed without apprehensions of this tormenting vitality, the only release from which is liver-deadening enlivening drink? or sexual fantasies leading to orgasms'[120]). He set John Rickman on the path to being an analyst by insisting he seek analysis with Freud.[121] Rivers and Jones worked closely together in the Medical Section of the British Psychological Society.[122] In April 1919, Rivers was the first person to become an associate member of the BPaS, having invited himself, and he expected Freud to respond positively to his invitation to come to Cambridge in September 1920. Rickman mediated a fruitful relationship between Rivers and the psychoanalytic anthropologist Géza Róheim.[123] But Rivers was very disappointed by the 'mythical' instinct-theory of

[119] John Forrester, *Language and the Origins of Psychoanalysis*, London: Macmillan, 1980; R.G. Goldstein, 'The higher and lower in mental life: an essay on J. Hughlings Jackson and Freud', *Journal of the American Psychoanalytic Association* 43 (1995), 495–515; Allan Young, *The Harmony of Illusions: Inventing Post-Traumatic Stress Disorder*, Princeton: Princeton University Press, 1995.

[120] Hugh Carey, *Mansfield Forbes and his Cambridge*, Cambridge: Cambridge University Press, 1984, p. 42. Rivers's diagnostic view and recommendation were endorsed by Ffrangcon Roberts, appointed as a Junior Demonstrator in Physiology in 1919 at the same time as Adrian and Thacker, the only three University appointments in the field made at that time, and all to young medical scientists interested in psychoanalysis; Roberts, incidentally, had given a talk during the war to Clare College Society, 'The dilettanti', explaining psychoanalysis (see H. Godwin, *Cambridge and Clare*, Cambridge: Cambridge University Press, 1985, p. 30).

[121] Sylvia Payne, 'Dr. John Rickman', *IJP* 33 (1952), 54.   [122] Forrester, '1919', p. 52.

[123] See the Rickman–Róheim correspondence, BPaS, Rickman Papers, CRR/FO7 and CRR/FO8.

Beyond the Pleasure Principle[124] and, as he explained to Jones, his long-standing aversion to transference had by 1921 taken on increasing weight:

> I am becoming progressively more and more doubtful [what to make] of psychoanalysis, as at present [conducted], especially as to transference [?] and its production of undue dependence and loss of critical faculty, and consequently have taken a great dislike to undertaking any kind of responsibility in connection with it.

He had expressed similar reservations to Karin Stephen earlier that year: 'Tremendous talk with Rivers about Psychoanalysis. He thinks it is very dangerous though a few people might come out of it all right!'[125] Rivers was in flight again, but still committed to the psychoanalytic project: he was intending to attend the IPA Congress in Berlin in September 1922; Rickman hoped that meeting Freud there would entice Rivers into having an analysis with him.

When C.S. Myers, Rivers's longtime colleague and builder of Cambridge psychology, gave an overview of Rivers's career following his sudden death in 1922, he noted that Rivers had given up teaching experimental psychology in 1907, leaving Myers a free hand in that field, and had devoted himself to his ethnology (and teaching abnormal psychology) until he arrived at Maghull, where he became a transformed man:

> The period beginning in July 1915 is characterized by a distinct change in Rivers's personality. In investigating the psychoneuroses he was fulfilling the desires of his youth. Perhaps because of the gratified desire of an opportunity for more sympathetic insight into the mental life of his fellows, he became another and far happier man. Diffidence gave place to confidence, hesitation to certainty, reticence to outspokenness, a somewhat laboured literary style to one remarkable for its ease and charm. Not only did he win the gratitude and affection of numberless nerve-shattered soldier patients, but he attracted all kinds of people to this new aspect of psychology. Painters, poets, authors, artisans, all came to recognize the value of his work, to seek, to win, and to appreciate his sympathy and his friendship.[126]

Myers gives as broad a hint as he felt able that, in his Freudian moment, Rivers had found what he had been seeking all his life. The 'new psychology' – Freudian psychology – not only was the fulfilment of his youthful desires but was also transformative in the present. Rivers underwent something like a self-analysis – in his dreams and in his relations with his soldier-patients.

---

[124] Rivers to Jones, 21 November 1921, BPaS, Jones Papers, CRB/Fo1/o1.
[125] Karin Stephen to Jones, 14 June 1921, BPaS, KS/27.
[126] Myers, 'The influence of the late W.H.R. Rivers', pp. 167–8.

For two years in his maturity (1916–17), at Maghull and Craiglockhart, Rivers practised as a physician and allowed himself to adopt the scientific style of Freud – an adventurous researcher engaged in intense therapeutic relationships with patients. But Rivers did not stay long, seduced away at the end of 1917 by Henry Head – again – into another great adventure of the age, flying. For Rivers, in the end, the adventure would always be outer-directed. And in clear contrast to that archetypal hedgehog Freud, Rivers was always a fox.

Charles Myers may have been asking himself the question, 'Could Rivers have become Freud?', when in 1926 he gave a sweeping, insightful and waspish overview of Freud's scientific contributions. Full of admiration for Freud's genius, aghast at his incessant speculative forays, he measured Freud against his own – and Rivers's – experience of psychology:

> Freud himself is not a trained psychologist. Had he been, it is quite possible that his developing genius would have been so stunted by the (now superseded) psychological doctrines of his day, that his brilliant contributions to the subject – of such varying worth – would have been altogether denied to us. Contempt is often nature's cloak for ignorance, but it is not altogether unjustifiable, because of the errors of its youth, that Freud despises or ignores current psychology. Of him the psychologist might say what Dostoievsky, so I have read, once said of himself: 'I am called a psychologist. It is wrong. I am simply a realist in the higher sense of the word; that is, I depict all the dim recesses of the human soul.'[127]

Implicit in this meditation is the counterfactual: what if, like Freud, Rivers had avoided the embrace of 'the (now superseded) psychological doctrines', 'the older academic, "abstract", Teutonic, experimental psychology', whose doctrines concerning memory were totally vitiated by its disdain for emotion?[128] Freud escaped from academic psychiatry and overly mechanistic neurophysiology by a daily confrontation with suffering patients, while Rivers's escape route was to throw himself into the embrace of 'primitive' peoples. Rivers's eventual return and qualified espousal of Freudianism showed what might have been; it demonstrated what happened to the English Freud.

---

[127] C.S. Myers, 'A lecture on Freudian psychology', *The Lancet* 207(5364) (19 June 1926), 1183.
[128] C.S. Myers, *Present-Day Applications of Psychology, with Special Reference to Industry, Education and Nervous Breakdown*, London: Methuen, 1918, p. 28.

CHAPTER 4

Becoming Freudian in Cambridge: Undergraduates and Psychoanalysis

One day I was walking through Pump Court in the Temple in London, when suddenly I felt extraordinarily happy. Then I recognized that the feeling of happiness was associated with the noise of a can being filled at a stand pipe. Then I knew that it recalled the noise my college servant made filling the water-can for the bath that awakened me when an undergraduate at Cambridge, when to awaken was to anticipate another delightful day.[1]

Sir Walter Langdon-Brown, 1938

Walter Brown, as he was then called, was an undergraduate at St John's College in the early 1890s. By the time he wrote these words in 1938, he had a claim to be the founder of modern clinical endocrinology in Britain. He was also well known as a champion of the application of Freud's ideas to clinical medicine, had been elected President of the Medical Society of Individual Psychology, and had recently retired as Regius Professor of Physic at Cambridge. His Proustian evocation of an undergraduate awakening to a new morning is a very knowing one.

Looking back with nostalgia to an idyllic pre-First World War Cambridge when no hint of doom and darkness lay on the horizon was inevitably the fate of those who had been undergraduates in the early years of the century. For that golden generation, the angel of history certainly flew with his back to the future. Lord Adrian, undergraduate at Trinity from 1908 to 1911, physiologist and Nobel Prize-winner in 1932 and later Master of Trinity, was, not surprisingly, given his eminence, a frequent after-dinner speaker in the 1960s. He had perfected the art of evoking both the bliss that it was to be alive, young and at Cambridge in the dawn of the twentieth century, and the brusque irrevocability of its destruction. Invited to speak at a King's Feast in 1962, he evoked his awe for the College as an undergraduate and reminisced:

---

[1] Walter Langdon-Brown, "'Just nerves'", in Langdon-Brown, *Thus We Are Men*, London: Kegan Paul, Trench, Trubner & Co., 1938, pp. 77–92 at 81.

Fig. 4.1 Edgar Douglas Adrian, 1st Baron Adrian, by Elliott & Fry.
In the laboratory, *c.* 1935.

I want to tell you how much we all owed to King's in those days. I expect I
see them in a rosy light, for it was before the first war, when we could take
our standing in the world for granted and spend our time talking about the
Principles of Ethics,[2] or the Irish Theatre,[3] or the remarkable views about
our dreams attributed to a doctor in Vienna called Freud. I expect we were
much too care free and pleased with ourselves, but the Cambridge men of
my generation were genuinely proud of King's.[4]

---

[2] I.e. G.E. Moore's *Principia Ethica.* [3] I.e. Shaw and Yeats.

[4] AA, Box 10, 'Trinity College materials and speeches and other talks', King's College Founder's Day,
6 December 1962. Why did he hold King's in awe? King's College went through a period of dramatic
reform in the late nineteenth century, first changing its Statutes to allow non-Etonians to matriculate
and shortly after requiring all undergraduates to take the Tripos – the first Cambridge college to
require this. See L.P. Wilkinson, *A Century of King's, 1873–1972*, Cambridge: King's College, 1980.

Freud here is part of the carefree pre-War undergraduate's world; he occupies a somewhat different place when Adrian speaks in 1968 to his own Trinity College at a Feast in his honour celebrating his sixty years in the College:

> I had always intended to be a doctor & I wanted to learn more science before I went to hospital. So I came to Trinity. It was already the leading College for Physics & Physiology, it was starting Biochemistry with Hopkins; it had J.G. Frazer, the father of Social Anthropology (I went to some of the lectures he gave in College); Moore & Russell were making us read philosophy & we could still believe in the human virtues for by 1908 the shadow of Sigmund Freud had scarcely fallen on England.

For Adrian, the name 'Freud' marked a loss of innocence, both personal and for English culture as a whole. 'Freud' marked the end of the golden Edwardian years, the end of honest virtue, the end of illusions about humanity and about oneself. In 1953, the trope was already present: when celebrating the retirement of his close friend Sir William Bragg, he evoked the historical era they had shared in their youth:

> The change had begun of course: there were cars for the wealthy instead of carriages and motor bicycles for the enterprising ... we carried lumps of gold about in our pockets instead of paper and there was still an ether to conduct the light waves, at all events in Part I of the Tripos – I can recall J.J.[5] demonstrating smoke rings in his lectures on the properties of matter to illustrate one of the theories of the atom.
>
> We have gone a long way since then with two wars and Einstein and Freud to keep us in our places.[6]

Earlier, in 1946, when Adrian was asked by the BPaS to give the first Ernest Jones Lecture – an invitation that his former student at Cambridge, John Bowlby,[7] was almost certainly behind, knowing Adrian's long-standing interest in psychoanalysis – he had opened by placing the Great War both as watershed and as marking an irreversible change in his relation to Freud, to Jones and to psychoanalysis:

> Before the war of 1914 I was a medical student at Cambridge with two friends who were particularly interested in disorders of the mind. Nowadays these things have become the stock-in-trade of the entertainment industry

---

[5] J.J. Thompson, Head of the Cavendish Laboratory, discoverer of the electron.

[6] AA, Box 11, 'After-dinner speeches', Farewell dinner for (William) Lawrence Bragg, Willy and Alice, 18 December 1953, f. 2.

[7] In the wake of the Controversial Discussions, John Bowlby took on important organizational functions within the BPaS, replacing James Strachey as Training Secretary in 1944.

and Hollywood spends millions in depicting the more respectable aspects of the unconscious; the neuroses must have lost some of their glamour for the medical student, but then they had all the attraction of a new and mysterious field out of relation with anything which we were taught in our laboratories. The older generation showed little interest in the subject and they could scarcely be blamed, for medical science had advanced so positively in other fields, in surgery and bacteriology, for instance, whereas if the neuroses were better understood and treated it was by a few specialists each with his own method and outlook. We dabbled in hypnotism; at that time there was one of the recurrent waves of interest in hypnotism in Cambridge and medical students were constantly assuring one another that their eyes were growing heavier and heavier. But the theories led nowhere. We could agree that the hypnotized subject, like the hysteric, had a restricted range of consciousness and was unduly suggestible, but we were little the wiser.

Then one day one of my friends came to read Dr. Ernest Jones's papers on Psycho-analysis, published in 1913. He made us read it too, and we found it disturbing stuff. We were naturally repelled at the thought of the fantastic tricks our minds were supposed to play on us, but we were young and curious and it could not be denied that Freud's ideas were on quite a different plane from any of the others we had come across. Freud's seemed incredible, but they led to definite conclusions not only about neurotic symptoms, but about memories and dreams and normal behaviour. Unlike the others, this theory went far beyond a single range of facts; it showed or tried to show quite unexpected relations between different fields, and it made assertions which should have been open to direct confirmation or disproof. Although we did not know what to make of it we were sufficiently excited to decide that we would try to get into touch with Dr. Ernest Jones as soon as we went down from Cambridge to see if he could resolve our doubts. None of us did so, for the war intervened. My friends were killed and I returned, regressed perhaps, to physiology and have never been more than a spectator of the early struggles of psycho-analysis and of the gradual acceptance of most of its principles.[8]

Another way of putting Adrian's trope is the following: before the war, undergraduates – but not only undergraduates, the entire literary and scientific culture perhaps – could have an interest in psychoanalysis. Yet the Freud that pre-war England read was not necessarily the later Freud. This later Freud destroyed all one's illusions in humanity and simultaneously offered a way forward out of death, destruction and the mud: it was only after the war that Freud became 'Freud', a dark figure of yearning and disappointment who presided over the destruction of illusions about human virtue. This post-war 'Freud' somehow became identified with

[8] E.D. Adrian, 'The mental and the physical origins of behaviour', *IJP* 27 (1946), 1–6 at 1–2.

the bracing reality of the Great War, for good and for ill.[9] Reviewing the
first volume of Jones's life of Freud in 1953, now as Dr E.D. Adrian, OM,
Nobel Prize-winner, President of the Royal Society and Master of Trinity
College, Cambridge – in other words the most important and prestigious
scientist in Britain – Adrian wrote:

> It is difficult now to believe that we ever held such lofty views of human
> nature and that alone is a measure of the change of outlook which has taken
> place. How long the present elaborate doctrine of psycho-analysis will
> survive is another matter, but whatever happens to it we must accept
> Freud as one of the most important scientists of our time.[10]

Today it is generally acknowledged that psychoanalysis deals with issues that
are often unbearably alive for young people in the process of becoming adults.
As such, psychoanalysis is bound to be an object of fascination and often fear
for the young who are emotionally in flux and intellectually alive. The
historical question then becomes which generation was the first to experience
psychoanalysis in this way. In Britain, the evidence points to the fact that it
was the generation who arrived in Cambridge in the wake of the Great War.

Ogden's Cambridge

The Heretics

The 'revolution of the dons' of the late nineteenth century concerned,
amongst other things, the transformation of the universities consequent
upon three sources of change: science, women and religion.[11] The

---

[9] It is this 'morbid' Freud who looms so large over the entire inter-war period in Richard Overy's
important book *The Morbid Age: Britain and the Crisis of Civilization*, London: Allan Lane, 2009.
[10] E.D. Adrian, 'Discoveries of the mind', *The Observer*, 28 October 1953. Adrian had had such a
conviction of Freud's importance that in January 1936, exercising his newly fledged right as a Nobel
Prize-winner (awarded 1932), he had confidentially but unsuccessfully nominated Freud for the
Nobel Prize in Physiology or Medicine (Carl-Magnus Stolt, 'Why did Freud never receive the Nobel
Prize?', in Elisabeth Crawford (ed.), *Historical Studies in the Nobel Archives: The Prizes in Science and
Medicine*, Uppsala: Universal Academy Press, 2002, pp. 95–106 at 99). Adrian's position on psycho-
analysis was decidedly nuanced, at least in Ernest Jones's eyes writing to Heinz Hartmann in 1955
about the approaching celebrations for Freud's centenary: 'I am doubtful about approaching Adrian
as he has no knowledge of psychoanalysis, and he also refused to support my nomination for
Fellowship of the Royal Society' (Jones to Hartmann, 15 February 1955, CHB/FO2/06). There is, of
course, no change of view implied concerning psychoanalysis in judging Freud worthy of a Nobel
Prize in 1936 and Jones unworthy of an FRS in 1953 (when a campaign was in train), nor is it safe to
infer that Adrian knew nothing of psychoanalysis because he thought Jones unworthy of an FRS.
Adrian was President of the RS, 1950–55.
[11] Sheldon Rothblatt, *The Revolution of the Dons: Cambridge and Society in Victorian England*, London:
Faber & Faber, 1968, reprinted Cambridge University Press, 1981

Universities Tests Act of 1871 opened the universities to non-Anglicans. This was an essential element in the transformation of Cambridge from a community dominated by quasi-monastic, autonomous colleges of bachelor Anglican dons to a university in which women and 'others' might participate. Not least amongst the effects was the rise of the don's family. It brought in train a distinctive new set of social and economic relations amongst women, servants, families, the city and the university – not least as a result of dons constructing grand red-brick family houses in open land to the west of the colleges.

John Maynard Keynes was a product of this new configuration of personal life: his father was an undergraduate and Fellow of Pembroke, then lecturer in Moral Science and University Registrar; his mother, Florence, a Newnham graduate, was the first director of the Cambridge Labour Exchange, later Mayor of Cambridge. She was also sister to Walter Langdon-Brown, Regius Professor of Physic in the 1930s, who remembered the can being filled from the St John's College standpipe. Keynes pinpointed the link between Moral Sciences and Newnham as a key element in the development of the distinctively agnostic high culture of progressive Cambridge – a link which, as we will see, included the culture which supported the heroic age of the Society for Psychical Research, the society, based at Trinity College, which sponsored research into psychical phenomena:

> In the first age of married society in Cambridge, when the narrow circle of spouses-regnant of the Heads of Colleges and of a few wives of Professors was first extended, several of the most notable dons, particularly in the School of Moral Science, married students of Newnham. The double link between husbands and between wives bound together a small cultured society of great simplicity and distinction. This circle was at its full strength in my boyhood, and, when I was first old enough to be asked out to luncheon or dinner, it was to these houses that I went. I remember a homely, intellectual atmosphere which it is harder to find in the swollen, heterogeneous Cambridge of to-day.[12]

This newly married and suddenly domestic Cambridge, created out of the recent suppression of medieval celibacy, was, as Rivers observed in 1904, 'only thirty years from the Middle Ages'.[13] As depicted by Gwen Raverat in her memoir *Period Piece*, the archetypal Victorian don of the late

[12] John Maynard Keynes, 'Alfred Marshall' (1924), in *The Collected Writings of John Maynard Keynes*, Vol. X, *Essays in Biography*, London and Cambridge: Macmillan and Cambridge University Press for the Royal Economic Society, 1972, pp. 161–231 at 213; see also Skid I, 68–9.
[13] Peter Linehan (ed.), *St John's College, Cambridge: A History*, Woodbridge: Boydell Press, 2011, p. 479.

nineteenth century, scion of the Wedgwoods, Darwins, Sidgwicks, Balfours, Stephens, Maitlands, Keyneses and the other families who were part of the 'intellectual aristocracy',[14] was eccentric, high-minded, public-spirited, somewhat ascetic, supremely self-confident, sceptical of religion and deeply antagonistic to the hold that the Church and the College ecclesiasts had over the University[15] – 'he had clerical friends, but God was his personal enemy', C.P. Snow, chemist, novelist, politician, chronicler of Cambridge mores and inventor of the 'two cultures' thesis, reminisced of G.H. Hardy, outstanding Trinity mathematician.[16] The generation of undergraduates which they helped foster might not be quite as peremptory in their manners as James Strachey, who would leave the room if a Christian started to hold forth, but religion-baiting became a favourite undergraduate sport in the early years of the new century. Strident irreligiosity was to be a hallmark of many Freudians and many, though not necessarily all, of those attracted to psychoanalysis. So it is fitting that the first rumours of Freudian ideas would be heard in an undergraduate Cambridge culture distinguished by an innovation designed to put an end to the compulsory chapel endured for centuries by students.

Earlier in the Michaelmas Term of 1909, W. Chawner, recently Vice-Chancellor of the University and Master of Emmanuel College, a Cambridge college with a great tradition of evangelical Christianity, had printed a paper, *Proved All Things*, with commentaries, which recorded his recent sudden loss of faith and constituted a vitriolic attack on the claims of Anglican orthodoxy. His attempt to dissuade undergraduates from attending chapel precipitated a conflict between the Master and the Governing Body of the College only brought to an end by his death in early 1911. Twelve undergraduates requested copies of *Proved All Things*, and it was

---

[14] Noel Annan, 'The intellectual aristocracy', in J. Plumb (ed.), *Studies in Social History: A Tribute to G. M. Trevelyan*, London: Longman, 1955, reprinted with light revisions in Noel Annan, *The Dons*, London: HarperCollins, 1999, pp. 304–41. See Stefan Collini, *Absent Minds: Intellectuals in Britain*, Oxford: Oxford University Press, 2006, esp. p. 140.

[15] A later scion of this Cambridge culture, Gregory Bateson, could find no better way of recommending his new fiancée, Margaret Mead, to his mother in the late 1930s than by writing: 'Her family are "rationalistic, agnostic, Spencer-reading, New England Puritans", and when I boasted of my five generations of atheism, she capped the tale with a statement that her great-grandmother was "read out of" the Unitarium [*sic*] Church for heresy ... [she has] a good sound plain intelligent – almost female Darwin face' (D. Lipset, *Gregory Bateson*, Englewood Cliffs, NJ: Prentice-Hall, 1980, p. 150, quoted in Jane Howard, *Margaret Mead*, New York: Fawcett Columbine, Ballantine, 1984, pp. 184–5).

[16] C.N.L. Brooke, *A History of the University of Cambridge*, Vol. IV, *1870–1990*, Cambridge: Cambridge University Press, 1993, p. 117.

Fig. 4.2 Charles Kay Ogden, *c.* 1916.

these twelve who constituted the first members of the Heretics. The power behind this innovation is a figure who will recur often in these pages: Charles Kay Ogden (1889–1957), the foremost English intellectual entre- preneur of his generation and one of those figures who, in his early years at least, was always in the right place at the right time. A second-year under-graduate at another Cambridge college, Magdalene, reading classics, he decided to found a society to be called the Heretics which sponsored two kinds of meetings. The first, restricted, was an occasion for every member to have a chance of speaking – the meetings were held first above the Pepys Library in Magdalene College, later at Ogden's hideaway, 'Top Hole' above the fishmongers in Petty Cury.[17] The second were public meetings with invited speakers, scheduled to clash with Sunday evening chapel.

The society's principal aim was to discuss 'problems of religion, philo-sophy and art'. After heated discussion, Ogden persuaded the members to accept Rule 4 of the Society: 'membership of the society shall imply the rejection of all appeal to Authority in the discussion of religious ques-tions'.[18] The inaugural meeting took place in December 1909, addressed by the classicist Jane Harrison and the philosopher Dr J.E. McTaggart, for whom the occasion must have brought back memories of old battles, since he had been expelled from his prep school for declaring he did not believe in God.[19] Senior members – McTaggart, Keynes, Verrall, G.M. Trevelyan, G.H. Hardy and the obligatory Darwin (Francis) – were elected Honorary Members. In the next few years, when the Heretics came to discuss religion, the light satirical mode was well represented by Francis Cornford (author of an anonymous 1904 pamphlet about compulsory chapel and the famous satire *Microcosmographia Academica*), who spoke in 1911 on 'Religion and the university', recommending that each college should have a mosque and a temple in addition to a chapel – a proposal as anodyne and *bien-pensant* now as it was hilariously scandalous then. The reflex connection between frivolity and religion on another occasion brought out one of Bertrand Russell's best *bon mots*: 'The Ten Commandments were like the customary rubric for a ten-question exam-ination paper: "only six need be attempted".'[20] While the Heretics mocked, the issue of compulsory chapel was fought out in bitter conflicts characteristic of college politics: the party of Cornford, Russell and Hardy, all Fellows of Trinity, joined by the young Edgar Adrian elected in 1913,

---

[17] Later part of the Macfisheries chain started by Lord Leverhulme in 1918. Ogden's lair later became the clubroom of the Cambridge University Socialist Society in the 1930s and then the home for the Footlights (Raymond Williams, *What I Came to Say*, London: Hutchinson Radius, 1989, p. 6).

[18] P. Sargant Florence and J.R.L. Anderson (eds.), *C.K. Ogden: A Collective Memoir*, London: Elek Pemberton, 1977, p. 228.

[19] See C.D. Broad, 'McTaggart, John McTaggart Ellis (1866–1925)', rev. C.A. Creffield, *ODNB*, www.oxforddnb.com/view/article/34824, accessed 19 April 2006.

[20] Sargant Florence and Anderson (eds.), *C.K. Ogden*, p. 229.

forced through a change in college policy just before the war. And, as so often in Cambridge, where Trinity goes, the rest eventually follow.

The leading lights of Cambridge philosophy – McTaggart, Russell, Moore – all gave talks to the Heretics before the war; after the war Broad, Moore and Wittgenstein[21] all spoke to this community well socialized into the Mooreian style of philosophical address: 'What *precisely* do you mean by ...?' The Heretics was created in time to provide Wittgenstein with a forum he would regularly attend; Ogden's ability to be in the right place at the right time meant he was even present on the occasion when Wittgenstein and Russell first met – Wittgenstein announced himself on 18 October 1911, as Ogden and Russell were discussing Heretics business over tea.[22] For many students, it was the opening on to the arts that made the Heretics so exciting; talks on Wilde, Poe, Hardy ('the Poet of Heresy'), Rupert Brooke on 'Drama present and future'. Ogden, as befitted a Germanophile intellectual, was creating a Cambridge version – polite and self-consciously heterodox – of a European idea of intellectual debate, half political meeting, half research seminar. What was epoch-making about the Heretics was not only the format (lecture and bracing debate) and the consistent distinction of the invited speakers – from both within Cambridge and, increasingly after the war, from outside – but also the long-term stability of what was initially an undergraduate society. The muscle behind the conception, the style and the stamina of the Society was the commitment of its founder; but one should not overlook the long-serving group of committed helpers, chief amongst whom were the Sargant Florence family: Philip Sargant-Florence, later economic historian and statistician, his mother Mary, feminist, pacifist, organizer of chaperone-substitutes, co-author with Ogden of *Militarism and Feminism* (1915), Alix Sargant-Florence, Philip's sister, later Alix Strachey, a psychoanalyst, author and Freud's translator, and then, after the war, even Lella Secor Florence, Philip's wife, who was to

[21] According to the editors of its posthumous publication, this address to the Heretics was the only public address Wittgenstein ever gave (Ludwig Wittgenstein, 'A lecture on ethics', *Philosophical Review* 74(1) (1965), 3–12); the occasion was organized by one of the last Secretaries of the Heretics, William Empson (see John Haffenden (ed.), *Selected Letters of William Empson*, Oxford: Oxford University Press, 2006), though Empson was forced to leave Cambridge in the course of this organization; Wittgenstein in the lecture refers to 'the honour' of originally being invited by 'your former secretary' and Ray Monk is probably right to infer that he is referring to Ogden (MLW, p. 276), so that the unique acceptance of the invitation to give a public address is a measure of the debt and friendship he owed both to Ogden and to the Heretics, part of the pre-war past he returned to in early 1929 'as though time had gone backwards' (MLW, p. 255).
[22] Brian McGuinness, *Wittgenstein: A Life. Young Ludwig (1889–1921)*, London: Duckworth, 1988, pp. 74, 95; MLW, pp. 38–9.

become the first Director of and research scientist associated with the Cambridge Birth Control Clinic (1925–74).

The Heretics continued to meet throughout the war and flourished throughout the 1920s. William Empson, student of Richards and later an eminent literary critic, was its Secretary in 1929. He was forced to hand over to Julian Trevelyan, later surrealist painter and poet, having been expelled from Magdalene College in the summer 1929 when his bedmaker read his correspondence and discovered condoms in his rooms.[23] The Heretics was the most prominent forum for public intellectuals to speak at in Cambridge: in anthropology, Haddon (1910), Rivers (1911), Elliot Smith (1922), Malinowski (1926); in the sciences during the 1920s, R.A. Fisher, William Bateson, Julian Huxley, A.S. Eddington and J.B.S. Haldane. The psychoanalysts were regularly invited: Ernest Jones (November 1922 on 'Narcissism'), James Glover (May 1925 on 'Biological lying'), Edward Glover (March 1928, 'Forerunners of conscience'), Adrian Stephen (February 1930, 'A description of Freudian analysis;) and the rebel Alfred Adler, provoking a vigorous debate with resident Freudians. If Freud had accepted the invitations from Rivers or from Sprott on behalf of Keynes, the forum at which he would have spoken would most probably have been the Heretics, just as *Daedalus*, Haldane's famous utopian fantasy about artificial wombs, and Virginia Woolf's equally famous 'Mr Bennett and Mrs Brown' were first delivered to that audience in the early 1920s.

The Heretics Society was a distinctive and original academic and cultural initiative, but one that was entirely in keeping with the idiosyncratic habits of Cambridge. While the University had existed as an institution for centuries, it was dominated until the twentieth by the wealthy and entirely autonomous colleges; as a result, the academic life of Cambridge was always dispersed and incapable of central coordination. This had obvious disadvantages; the less obvious advantages were that independent initiatives, separate from both colleges and University, were not only entirely in keeping with tradition, but were one of the principal means by which new subjects and ideas could enter into the culture of the small university town. One such independent initiative can stand for many others, but it is a very important one: the Cambridge Philosophical Society, founded in 1819 by Henslow and Sedgwick, Darwin's teachers in botany and geology. This society had at least four important functions. It served as a general 'scientific society' for

[23] John Haffenden (ed.), *Selected Letters of William Empson*, Oxford: Oxford University Press, 2006, pp. 8–11. The very first letter in this collection is concerned with aligning Empson's own readings of Hopkins with 'Freud and ambivalence'. Empson (and Richards): between the Heretics and Freud – this is an accurate portrait of his Cambridge milieu.

members, akin in this respect to the Royal Society in London, but with a crucial difference: the membership of the Philosophical Society was not as restrictive as the Royal Society, indeed local members were welcome. Secondly, it published *Proceedings*, thus providing an outlet for its members' writing and research. Thirdly, it hosted regular lectures and meetings, providing its members with a meeting place (and a club, inevitably – this is England). Finally, it employed its membership dues, and most importantly the mutually agreed reciprocal exchange of its *Proceedings*, with other learned journals world-wide, to provide a library service for its members.

The Cambridge Philosophical Society became perhaps the principal vehicle for the local advancement of science in Cambridge and it was very broad-ranging in the sciences it welcomed – observational, field sciences, experimental sciences, mathematical contributions. Its Library and Reading Room became the principal resource for scientific literature in Cambridge, so much so that in 1865 the University became fearful of its dispersal and provided a room in the University for it. From then on, the Philosophical Society, although always independent and always owned and controlled by its members and serving their interests, had a permanent alliance with the University, cemented in 1881 when the University agreed to house and staff the Library, in return for which they requested that all members of the University be able to make use of it. So over the years, the libraries of the Society and the University came to fuse and become unified. Yet the Philosophical Society is still, in 2015, the most influential *local* forum for general scientific lectures and papers in Cambridge – and remains apart from the University.

Such an odd arrangement between independent institutions founded on the initiative of members of the University – and others – was entirely characteristic of the higgledy-piggledy manner in which Cambridge developed as an institution of learning and research. Colleges themselves as clubs; scientific, literary, antiquarian, archaeological, musical or dramatic societies as a nether layer of clubs (the Marlowe Society; Footlights; Tansley's Ecology Club; the Society for Psychical Research; the Natural Science Club; the Committee that guarded Wicken Fen). Somewhere in the middle of these Chinese boxes of groups, clubs and gatherings, there is a University, formally authorized, and always behind the times. The club function should never be underestimated: the annual dinner of the Philosophical Society in 1919 had 'nine courses and ten speeches'.[24]

---

[24] A.R. Hall, *The Cambridge Philosophical Society: A History 1819–1969*, Cambridge: Cambridge Philosophical Society, 1969, p. 32.

So the Heretics was a classically configured Cambridge institution, doing what neither the University nor the colleges would or could do: hosting streams of thought and debate, in the arts, literature, the social sciences and the natural sciences, that are recognizably part of modernist movements in those domains that were so influential in the period 1910–30. Ogden, its inspirer, was also a recognizable figure: an encyclopaedic scholar who networked in Cambridge and across Europe and America and built a publishing arm to accompany the club, talk and debating forum. And this publishing arm was to be centred on modernist psychology.

The Heretics gracefully died in 1931, not long after Prince Mirsky, a son of a Tsarist Minister and a critic active in the best circles in literary London in the twenties, spoke on 'Dialectical materialism' – presaging his brave and foolhardy return to Soviet Russia, where he penned his astute survey *The Intelligentsia of Great Britain* before losing his life in the Stalinist purges in 1939.[25] With the shift to the political thirties, well encapsulated in Mirsky's topic and his life-choices, the spirit of the Heretics was no more.

The Cambridge Magazine

Ogden was not content with having created an institution which, as things turned out, functioned for over twenty years as a forum for the most advanced ideas in academic, political and artistic life. In January 1912, the first issue of his next invention, a weekly newspaper called the *Cambridge Magazine*, was sold for one penny in the shops of Cambridge. At first the magazine had its fair share of prominent reports of rowing and rugby, amateur dramatics and exquisite in-jokes about local celebrities and friends of the editorial team, but its explicit aim of being an intellectual forum soon came increasingly to the fore. Within a month of its foundation it devoted articles to the cult of Bergson and the new vitalism; in March 1912, the magazine gave a blow by blow account of a vigorous debate started by the psychologist C.S. Myers's talk to the Heretics on 'The new realism', with Moore and Russell attacking his views. In May, G.H. Hardy's talk on 'The new, new realism' – the philosophy of Moore and Russell – was printed.

By June the centrepiece was J.S. Haldane on vitalism with a response from the German biologist Hans Driesch. In the autumn of 1912, one of

---

[25] There are references to Mirsky sprinkled throughout Virginia Woolf's diaries and an interesting analysis of his book on the intelligentsia in Collini, *Absent Minds*, pp. 126–30.

Ogden's preoccupations, industrial training and progressive education, provided the theme, with a report on 'the ideal school' from Wickersdorf in Germany – a school where 'children may feel themselves happy in a life which suits their natural instincts'.[26] In neighbouring issues Mr Ivor Tuckett launched a sceptical attack on the research programme of the Society for Psychical Research, in particular the telepathic claims of Mrs Verrall, who had spoken to a packed audience at the Heretics in February 1912. Late in 1913, Geoffrey Pyke reported on Georg Brandes's lectures in Cambridge, culminating in the claim that the 'history of literature . . . is the history of European psychology'.[27] In early 1914, Russell's Heretics talk on 'the external world' was reported side by side with a review of Durkheim on *Totemism*, followed four months later by a lengthy article on four 'unknown German philosophers: Mauthner, Meinong, Rickert, Husserl'.[28] For those interested in the arts, there was the home-grown first publication of a part of Rupert Brooke's new satirical poem written in Berlin, 'The Old Vicarage', and excited anticipation of the futurist Marinetti's talk to the Heretics in May Week, 1914.

The first mention of Freud in the *Cambridge Magazine* came in February 1912 in a typically erudite review by Ogden of a new study of Aristotle's poetics by the Professor of Arabic at Oxford, David Samuel Margoliouth. The undergraduate Ogden concluded a complimentary review with regret at the professor's failure to incorporate some of the recent advances in classical scholarship:

> Admirers of Butcher's great work on the *Poetics* will be surprised at the result of a search for evidence that Professor Margoliouth is acquainted with it, – still less indebted thereto; and one looks in vain as usual for a reference to Steinthal's monumental work on the early history of linguistic theory: perhaps too Naumann's *Geschlecht und Kunst* (page 142, etc.) has been rendered obsolete by the more recent studies of Freud and his school.[29]

Here Ogden writes as if every Cambridge reader, even an Oxford professor renowned for his polylingual scholarship of the Ancient Worlds, is of course informed about the importance of the Freud school!

As an undergraduate publication, the *Cambridge Magazine* appeared only in the brief eight-week terms, with a long gap over the summer. Some

---

[26] *CM* II No. 7, 30 November 1912, p. 160.   [27] *CM* III No. 8, 29 November 1913, p. 194.
[28] *CM* III, April–May 1914.
[29] *CM* I, 3 February 1912, pp. 80–1. This passage particularly appeals since the final chapter of Forrester's *Language and the Origins of Psychoanalysis* pointed out the importance, so often overlooked, for the development of psychoanalysis of Steinthal's linguistics (see John Forrester, *Language and the Origins of Psychoanalysis*, London: Macmillan, 1980, pp. 175–8).

attempt at retaining its pre-war character was made in October 1914, but the effects of the war were immediate: reports of the deaths of students and calls for support for the plight of the displaced Belgian students now resident in Cambridge sat alongside an article on 'What Nietzsche thought about Germany' (defending the philosopher 'against the charge that he is the root of all our troubles'[30]) and a wittily laconic report from 'Miss A. S. Florence, of Newnham', beached high and dry by the war in Petrograd, about a Russian professor being robbed of his notes on Parsifal while stranded in Leipzig at the declaration of war. In November 1914, there was an intimation of how the magazine would develop: a report from the only Cambridge man, a Scholar of King's who was Armenian (and therefore not subject to internment by the German authorities), who had spent August and September in Berlin; it was followed by a reprint of Romain Rolland's 'Above the battlefield. An appeal to the young men of all nations' translated from the *Journal de Genève*. In January a local branch of a new Society, 'Union of Democratic Control', an anti-government grouping formed by Liberal and Labour politicians critical of the war, was announced; two weeks later, a UDC statement printed in the *Magazine*, signed by prominent younger dons, amongst them Dickinson, Cunningham, Hardy, Russell and Sedgwick, called for the recognition of independent opinion on policy and the war. The UDC's programme was to work for parliamentary control of foreign policy and a moderate peace settlement, on the basis of the widely shared suspicion that Britain had been dragged into the war because of secret military agreements with France and Russia. A sign of the lack of enthusiasm for the war displayed by the small number of students still in residence in February 1915 was the outcome of a Union Society debate on the motion 'that Great Britain is seen at her best during war time' – lost by 29 to 43.[31]

Increasingly the *Magazine* published material concerning the war and, what is more, material from as far afield as they could find: an article by Pigou, the Professor of Economics, on the question of peace terms and the virulent storm of debate this elicited, was followed by a report from a Cambridge man in Serbia. In March 1915, a meeting of the Wounded Allies Relief Committee was addressed by Archdeacon Cunningham, who made an attack on pacifists and advocates of peace and neutrality, quoting – approvingly – an Austrian minister who said that liberals and pacifists were 'worse than Jews'.

---

[30] *CM* IV, 10 October 1914, p. 27.   [31] *CM* IV, 6 February 1915, p. 234.

Whereupon between 100 and 150 members of the audience rose and left the hall ... Outside the hall an indignant throng was gathered, and cries of 'Down with Bill Cunningham,' 'Disgrace to the University' were loudly raised when the meeting finally broke up.[32]

In the same number, Alix Sargant-Florence's piece 'Petrograd in war-time' recounted how news of the war had reached her during her six months in Petrograd. In May the shock of Rupert Brooke's death occupied many pages; in June Romain Rolland's article on 'War literature' and the response of writers across Europe to the war was translated. Alix's husband-to-be, James Strachey, also was connected with the *Magazine*. At some point as it expanded, he wrote to Ogden from his Hampstead home offering to write a weekly *London Letter*: 'I suppose it'd be partly political and partly art etc., etc., – all of it viewed from a rather Cambridge standpoint. But the question is whether it could be made respectable without being dull.'[33] In October 1917, he suggested to Ogden that he might write an article on a proposed investigation of the cinema by the National Council for Public Morals.[34] In January 1916, the ever enterprising Ogden moved into book publishing, when he translated and published a collection of Romain Rolland's appeals and addresses on the war, *Above the Battle*. What had been an undergraduate magazine had mutated, under the pressure of the war, into something much more serious, something affecting national life.

What transformed the *Magazine* was, starting on 28 October 1915, a separate section devoted to 'Foreign opinion: a weekly survey of the foreign press' edited by Dorothy Buxton (1881–1963). Sister of Richard Jebb (1841–1905), Regius Professor of Greek and Tory MP for Cambridge in the 1890s, she was thoroughly well connected with the Darwins and the Stephens. Together with her political husband and brother-in-law, she was active throughout her life in social work on behalf of the poor in London pre-war and starving children in Europe post-war: she became the co-founder of the future Save the Children Fund. From the start of the war she had been determined to publicize the internal opposition to the war that existed within Germany; the Board of Trade, perhaps not initially loath to have both its intelligence work and its useful propaganda work done for it, allowed her to import twenty-five enemy papers from Scandinavia. As a

---

[32] *CM* IV, 13 March 1915, p. 333.
[33] OP, Box III F.12, James Strachey to Ogden, no date, probably after Strachey had been fired from *The Spectator* for his pacifist views, so probably in 1916.
[34] OP, Box III F.12, James Strachey to Ogden, 12 October 1917.

result the *Cambridge Magazine* became the only printed source in Britain for material from the foreign press. By 1917 the national circulation of the *Magazine* had risen to over 20,000 and it had become an essential part of the intellectual and political life of the country. When Siegfried Sassoon, as was earlier noted, issued his statement against the war in July 1917 and was bundled off to Craiglockhart into the care of Rivers, his 'treatment' consisted of a discussion of the *Cambridge Magazine*. Rivers's accompanying self-analysis revolved around the emotions and attitudes aroused by the *Magazine*.

The success of the *Magazine* created tremendous new problems of management for Ogden, which he overcame in his inimitable style. He acquired Cambridge shops, empty because of the war, and converted them into *Cambridge Magazine* bookshops – one was officially called the Cambridge Magazine Shop, another was called the Other Shop.[35] At one point he had three open on King's Parade. Doing a deal with the well-known second-hand booksellers David's, he would buy up large caches of second-hand books, and house them temporarily in the shops while he read through them deciding which to pulp for paper for the *Magazine*. Ogden took on another assistant: Geoffrey Pyke, a close friend and walking-companion of Philip Sargant Florence before the war and an early contributor, who had been captured in Berlin while working as a correspondent for the Chronicle and placed in the Ruhleben internment camp near Charlottenburg.[36] After he made what he called a 'scientific escape' in daylight and wrote his tale *To Ruhleben – and Back; a Great Adventure in Three Phases*, Pyke became something of a national celebrity. Back in Cambridge, he was entrusted by Ogden with scouting out extremely scarce newsprint; he also became the London advertising manager for the *Cambridge Magazine*. It was Pyke who placed advertisements in *The Times* quoting a string of laudatory comments from MPs and writers about the *Magazine*: 'the extracts ... enable one to see England bare and unadorned – her chances in the struggle freed from distortion by the glamour of patriotism' (Thomas Hardy).

> Some day the Government will be forced by military necessity to propose some step for which the nation is unprepared and against which it is violently prejudiced. The Government will be overborne by public opinion; and the military authorities will be compelled, not for the first time, to

---

[35] Marjory Todd, in Sargant Florence and Anderson (eds.), *C.K. Ogden*, p. 113.
[36] Geoffrey Pyke later became the founder and driving force behind the Malting House School (1924–27), to which Chapter 7 is devoted.

ignore the Government and to coerce the nation. The only available pre-
caution against such a schism at present is a conscientious study of the
Cambridge Magazine. (G.B. Shaw)[37]

As a result of Pyke's great success at selling space, advertisements in the
Magazine now became the subject of parliamentary questions concerning
whether it was proper for the government to place advertisements, as they
had done, in a seditious publication. Later in the war, the senior member
supporters of the *Magazine* (Bury, Quiller-Crouch, Dent, Manny Forbes,
Johnson, J.M. Keynes, Tansley et al.) rallied to its defence.[38] Armistice Day
brought another kind of reckoning for Ogden: his shop on King's Parade
was smashed up and his pictures by Duncan Grant, Vanessa Bell and Roger
Fry, his *objets d'art* and his two grand pianos were thrown into the street.[39]
Visiting his tenant I.A. Richards at 1 Free School Lane that night to inquire
if he had identified any of the hooligans, the two men engaged on the stairs
in a lengthy conversation in which they set out the plot for their landmark
study of language, *The Meaning of Meaning*. These two simultaneous
events – the end of the war and the beginning of the collaborative *The
Meaning of Meaning* – spelt the beginning of the end for the *Cambridge
Magazine*. Peacetime meant the need for an undergraduate magazine to
carry the burden of unofficial national opposition to the war no longer
existed. In January 1919, another undergraduate magazine, *New
Cambridge*, was founded explicitly in repudiation of the *Cambridge
Magazine* and its stand: in its second issue it reprinted a statement of
protest signed by 120 ex-servicemen undergraduates against the *CM*. But
this patriotic mood was not to last, not with the likes of Maurice Dobb,
later a major Marxist economist, and Rosamund Lehmann, later a modernist novelist, taking over its editorial board in 1920. And by 1920 the
Great War was falling under erasure in undergraduate circles. Two years is
a geological era in undergraduate life.

Having intended to expand and convert the Foreign Press section into a
section on 'International Understanding', Ogden quickly found this unsatisfactory: the urgent demand for uncensored views from outside Britain was
now less urgent. So in February 1920 Ogden transferred the Foreign Press
section to the *Manchester Guardian* and converted the *Magazine* into a
quarterly. From the summer of 1920 on, he used it principally as a vehicle
for publishing his joint work with Richards on language and symbolization –
The Meaning of Meaning. But the magazine also contained correspondence

---

[37] *The Times*, 1 November 1916, p. 4.
[38] OP, Box 112, F.1, Tansley to Ogden, 17 November 1917.   [39] KMFF, p. 106.

MODERN PSYCHOLOGY

The following recommended books are always kept in stock.

BAUDOUIN. *Suggestion and Autosuggestion.* 10s. 6d.
BERGSON. *Matter and Memory.* 12s. 6d.
BLANCHARD. *The Adolescent Girl.* 7s. 6d.
BRETT. *History of Psychology.* Vol. I. Ancient. 16s.
          Vol. II. Mediæval. 16s. Vol. III. Modern. 16s.
BROWN. *Psychology and Psychotherapy.* 8s. 6d.
BRIFFAULT. *Psyche's Lamp.* 12s. 6d.
CABOT. *Seven Ages of Childhood.* 12s. 6d.
CARRITT. *The Theory of Beauty.* 7s. 6d.
DEWEY. *How We Think.* 4s. 6d.
DREVER. *Psychology of Everyday Life.* 6s.
FREUD. *Psychopathology of Everyday Life.* 12s. 6d. *Interpretation of Dreams.* 15s. *Totem and Taboo.* 10s. 6d.
GREEN. *Psycho-analysis in the Class-room.* 7s. 6d.
HART. *Psychology of Insanity.* 3s.
HOBHOUSE. *Mind in Evolution.* 12s. 6d.
HOLT. *The Freudian Wish.* 5s. 6d.
JAMES. *Principles of Psychology.* 2 vols. 31s. 6d. *Varieties of Religious Experience.* 12s. 6d.
JONES. *Papers in Psycho-analysis.* 25s.
JUNG. *The Psychology of the Unconscious.* 25s.
MORGAN. *Instinct and Experience.* 6s.
McDOUGALL. *Body and Mind.* 12s. 6d. *Introduction to Social Psychology.* 7s. 6d. *The Group Mind.* 21s.
MUSCIO. *Lessons in Industrial Psychology.* 6s. 6d.
MITCHELL. *The Structure and Growth of the Mind,* 12s. 6d.
O'SHEA. *Mental Development and Education.* 12s.
PRINCE. *The Dissociation of a Personality.* 16s.
RIBOT. *The Psychology of the Emotions.* 7s. 6d.
RIVERS. *Instinct and the Unconscious.* 16s.

SCHRENCK-NOTZING. *The Phenomena of Materialisation.* 35s.
SEMON. *The Mneme.* 18s.
SHAND. *The Foundations of Character.* 20s.
SMITH. *The Foundations of Spiritualism.* 3s. 6d.
TANSLEY. *The New Psychology.* 12s. 6d.
TROTTER. *The Instincts of the Herd in Peace or War.* 8s. 6d.
TERMAN. *The Measurement of Intelligence.* 7s. 6d.
URBAN. *Valuation.* 12s. 6d.
WATSON. *Psychology from the Standpoint of a Behaviourist.* 12s. 6d.

———

Fig. 4.3 The range of psychoanalytic literature available at the Cambridge Magazine bookshops. Advertisement, 1921, from the inside cover of *Cambridge Magazine,* decennial number 11(1) (1912–21).

from enthusiasts heading into the field: 'Psycho-analysis continues to attract attention in University circles, and not only amongst medicals proper as the recent essays of Dr. Rivers [*Instinct and the Unconscious*] shew. Two of our contributors, Dr. Rickman of King's and Dr. Baynes of Trinity, have just left for the continent; the former to work with Freud in Vienna, the latter with Jung in Zurich.'[40] In January 1922 it provided a space for publication of the philosopher prodigy Frank Ramsey's razor-sharp demolition of Keynes's *A Treatise on Probability.*[41] But later that year, the *Cambridge Magazine* ceased publication.

Ogden was now over thirty and, although he remained President of the Heretics until 1924, he wished to play on a larger stage than that provided by an undergraduate magazine. In 1921, he shifted his base of operations to Bloomsbury, sleeping in his bookshops when in Cambridge – as would his academic guests, such as Malinowski, who delighted in Ogden's collection of pornography.[42]

The Heretics were originally heretical in a self-consciously religious sense, opposing the expectations of religiosity, in conventional ritual if in nothing else, of Edwardian Cambridge. In the 1920s, the underlying assumption broadened: what was now heretical was the refusal of the assumption of rationality as underlying human life and historical progress. The hereticism of the Heretics and the subjectivism, aestheticism and commitment to truth above all other values of that other elite group, the Apostles, the secret conversation society with close connections to Bloomsbury from the early years of the century on,[43] all defining characteristics of pre-war Cambridge, were now transformed, across the abyss of the war, into a new preoccupation with the subjective, inflected by awareness of the primacy of emotion and instinct. Above all, these new student intellectuals were concerned with the fundamental irrationality of human beings.

Undergraduates and Psychoanalysis before the Great War

Was Ogden's throwaway line in an early number of the *Cambridge Magazine* about 'Freud and his school' comprehensible to the ordinary

[40] *CM* 10 No. 1 Double Number, Summer 1920, p. 5.
[41] *CM* 11 No. 1, January 1922, pp. 3–5, reprinted as F. Ramsey, 'Mr Keynes on probability', *British Journal for the Philosophy of Science* 40 (1989), 219–222. See Skid II, 70–3.
[42] Richards's recollections in Sargent Florence and Anderson (eds.), *C.K. Ogden*, pp. 103–4. See also I. A. Richards to Helena Wayne Malinowska, 25 January 1973.
[43] See R. Deacon, *The Cambridge Apostles: A History of Cambridge University's Elite Intellectual Secret Society*, London: Robert Royce, 1985, esp. p. 92 with regard to the 1920s: 'psychology was the magic word at this time, though sociology was regarded with some contempt'.

Cambridge undergraduate in 1912? Available documentation concerning the interests of undergraduates varies hugely. Later in this chapter we will examine in detail the materials available relating to five undergraduates from the immediate post-war period. But what of the pre-war period?

If Ogden could, in his all-knowing style, refer blithely to the Freud school, Lytton Strachey – true, no longer an undergraduate – could compose a spoof examination paper on Sex Education for those who still were:[44]

Elementary

Describe the reproductive system of the periwinkle.
Write a short biography of Oedipus.
'When the hens are away, the cocks will play.' Illustrate this.
What do you understand by 'the language of flowers'?

II. For Higher Forms

Distinguish clearly, with diagrams, between the clitoris and the vagina.
Write a short essay on

EITHER   Shakespeare's Sonnets, psycho-pathologically considered.
OR   The history of French Letters.

Advanced. (If the Essay is chosen, not more than 3 other questions should be attempted.)

Trace the connections between Sadism and Masochism with examples from history, or from cases privately known.

Elucidate, on the basis of Dr. Freud's teaching,
The Conversion of St. Paul.
The Channel Tunnel project.
The European War.
The growing popularity of tooth-picks in the United States.
Bestiality: should it be encouraged? And if not, why not?
What evidence of inversion can you point to in the works of
Either (a) Sophocles
Or (b) Rupert Brooke?

6. Subjects for an essay. (One only to be chosen.) –

---

[44] Lytton Strachey, 'According to Freud', in *The Really Interesting Question, and Other Papers*, ed. and with intro. and commentaries by Paul Levy, London: Weidenfeld & Nicolson, 1972.

Anus v. Vagina.
The Influence of the Stool upon Social Institutions.
The Pleasures of a Single Life.
Dr. Freud, analysed by himself.

The dating of this curio is disputed[45] – 1914–15 seems to be the best bet. Non-spoof examination questions on psychoanalysis were set in Cambridge at this time and considerably earlier, but Lytton's ready wit with Dr Freud is indicative of what a very well informed undergraduate just might have been expected knowingly to laugh at, if not actually to know about. Lytton's close friend Leonard Woolf read the English edition of *The Interpretation of Dreams* on 12 May 1914, in preparation for his respectful and laudatory review of *The Psychopathology of Everyday Life* for the *New Weekly*:

> Whether one believes in his theories or not, one is forced to admit that he writes with great subtlety of mind, a broad and sweeping imagination more characteristic of the poet than the scientist or medical practitioner . . . There can be no doubt that there is a substantial amount of truth in the main thesis of Freud's book, and that truth is of great value.[46]

Woolf's approval was complemented by the pseudo-omniscience and lofty tone of the reviewer (Ogden?) of *Psychopathology* in the *Cambridge Review* in late 1914:

> The translation of the fourth edition of Professor Freud's most popular work should do much to make it very much more widely known. The work

---

[45] The editor of the collection, Paul Levy, surmises 1902, which is nigh on impossible, since Rupert Brooke, who is referred to, would have been fifteen. There is also a reference to the Channel Tunnel, which was much discussed in England in 1906–7; but this date too seems much too early – Freud's theories referred to in it had been published only very recently in German at that time. The reference to the 'European War' would seem to indicate a date in 1914 or later; it is also unlikely that even Lytton Strachey would have been quite so flip about Rupert Brooke in the period after his death in April 1915, when he had become a national emblem of a generation's heroic sacrifice. In addition the sprinkle of Freudian references implies some knowledge across a reasonably broad range of Freud's works, but particularly the *Drei Abhandlungen zur Sexualtheorie* (1905). Not being a reader of German, Lytton's source would most probably have been the translation by A.A. Brill published in 1910 (*Three Contributions to the Sexual Theory*, New York: Journal of Nervous and Mental Disease Publishing Company (Monograph Series 7)). In addition, the question 'What do you understand by "the language of flowers"?' is a reference to a dream analysis in *The Interpretation of Dreams*, which was only published in English translation in 1913 (see *SE* V, 375), in which Freud discusses the sexual symbolism of 'the language of flowers'. The most likely date, therefore, for its composition is between 1913 and the early months of the war, in 1914–15.

[46] Leonard Woolf, *New Weekly* (13 July 1914); see Victoria Glendinning, *Leonard Woolf: A Life*, London: Secker & Warburg, 2006, pp. 184–5 and Dean Rapp, 'The early discovery of Freud by the British general educated public, 1912–1919', *Social History of Medicine* 3 (1990), 217–43 at 224.

itself does not call for criticism. The positions upheld in it are familiar to every pathologist, and it only remains to observe that the translation is rather intended for American than for English readers. It could also be improved. Moreover, the references in the index have not been corrected. But these are minor details. The translation will be extremely useful. The Introduction is negligible.[47]

This says a great deal: Freud's positions 'are familiar to every pathologist'. Not because they are run of the mill, but because, it is implied, Freud's views are so widely disseminated. The author also places himself in a position of knowing which of Freud's works are the most popular. Philip Sargant Florence remembered that his sister Alix first became acquainted with Freud's work while a student before the war.[48] As we have seen, Edgar Adrian's interest in psychoanalysis was roused well before the war; 1912 was the date that James Strachey remembered as his first reading of Freud: the 'Note on the unconscious in psycho-analysis' was published in the *Proceedings of the Society for Psychical Research* in November 1912. It very much looks as if the steady flow of Freudian commentary in the general press so well documented by Rapp had certainly reached Cambridge: by 1914, Freud was an author 'everybody knows'. Which of course means: 'everyone is *supposed* to know'.

Two Psychoanalytic Organizers

Of the pre-war cohorts at just one Cambridge College, King's, three undergraduates later had much to do with the development of psychoanalysis in Britain: John Rickman, J.R. Rees and John Layard. Rickman, together with Ernest Jones, was the key architect of the BPaS from the early 1920s to his death in 1951; his career will be discussed at length later.

J.R. Rees (1890–1969), born in Leicester from a Methodist background, took a Third (like Rickman in 1913) in the Natural Sciences Tripos in 1910.[49] Qualifying in medicine at the London Hospital in 1915, he served with the Friends' Ambulance Unit, then the RAMC on the Somme, in Mesopotamia and India, the experience persuading him to specialize in psychological

---

[47] [Anon.] Review, '*Psychopathology of Everyday Life*. By Prof. Dr. Sigmund Freud, LL.D. Authorized English Edition with Introduction by A.A. Brill, Ph.B., [*sic*] M.D., T. Fisher Unwin. 1914, 12s 6d. net', *CR*, 4 November 1914, p. 62.

[48] BF, 'Introduction', p. 11.

[49] On Rees, see Malcolm Pines, 'Rees, John Rawlings (1890–1969)', *ODNB*, www.oxforddnb.com/view/article/54198, accessed 12 October 2015; H.V. Dicks, *Fifty Years of the Tavistock Clinic*, London: Routledge and Kegan Paul, 1970; Nikolas Rose, *Governing the Soul: The Shaping of the Private Self*, London: Routledge, 1990.

medicine. Back in London after the war, he worked closely with Hugh Crichton-Miller in helping to set up in Bloomsbury the Tavistock Clinic, the first out-patient clinic in Britain principally devoted to medical psychological patients. Around this time he had analysis with Maurice Nicoll (1884–1953).

Nicoll's and Rees's backgrounds were very similar: both raised in the Manse (Nicoll in Kelso on the border of Scotland with England), Nicoll entered Gonville and Caius College, Cambridge and took a First in the NST in 1906,[50] before training in medicine at Bart's, following which he spent time in Vienna, Berlin and Zurich.[51] In 1913, he was one of the founder members of the London Psycho-Analytical Society, with the elite medical address of 114a Harley Street; however, Jones reported to Freud that in July 1914, Nicoll, 'a rather promising young fellow, but influenced by Dr. Constance Long',[52] was travelling to Zurich to work with Jung. At the outbreak of war he joined the Friends' Ambulance Brigade and then the RAMC, stationed in France (seeing action in the Somme in 1916), then in Mesopotamia and India. Assisting J.T. MacCurdy in the preparation of his influential book *War Neuroses* in 1918[53] and debating the same year with Rivers and Jones at the British Psychological Society on the 'Unconscious', he had already published *Dream Psychology* in 1917, 'the first didactic presentation of Jung's psychology'[54] – which was trenchantly critical of Freud's theory of dreams in favour of the future-oriented, 'prospective function' that Jung encouraged. He too was a member of the core staff that Crichton-Miller brought together to found the Tavistock Clinic, giving lectures on Jung's analytical psychology in 1921.[55] From the 1920s on, he became an enthusiast for the spiritual teaching of Gurdjieff and Ouspensky.

As Crichton-Miller's right-hand man at the Tavistock Clinic in the 1920s, the 'Parson's clinic' as it was sometimes called, Rees endorsed the development of its famously eclectic stance, recognizing the importance of the Freudian theories and practices, while reserving the right to scepticism and to broad-mindedness towards other approaches, whether Jungian, Adlerian, suggestive or 'English'. He was critical of the concern with

[50] HR 1910, p. 791.
[51] Beryl Chassereau Pogson, *Maurice Nicoll: A Portrait*, London: V. Stuart, 1961.
[52] JF, 1 July 1914, p. 288.
[53] J.T. MacCurdy, *War Neuroses, with a Preface by W.H.R. Rivers*, Cambridge: Cambridge University Press, 1918, p. 66.
[54] Sonu Shamdasani, *Jung and the Making of Modern Psychology: The Dream of a Science*, Cambridge: Cambridge University Press, 2003, p. 147.
[55] Dicks, *Tavistock Clinic*, p. 30.

doctrinal purity developing behind the high walls of the BPaS.[56] The Tavistock fell into rivalry with both the orthodox psychoanalysts and the orthodox psychiatrists based at the Maudsley Clinic, the other principal centre for out-patient psychiatry in London in the inter-war years, who were far more interested in the somatic therapies being developed than in the psychotherapeutic revolution which spread widely after the Great War. When Rees became the Tavistock's Director in 1932 following a 'Palace Revolution' demanding a more democratic structure for the institution, the eclectic ethos was reaffirmed and extended as his organizational energy led to a considerable expansion of its activities, particularly into social work and child guidance. Surprised at his appointment as Chief of Army Psychiatry at the outbreak of the Second World War, and meeting with an uncertain and less than helpful response to begin with, Rees nonetheless oversaw an enormous expansion of psychiatric expertise and its functions, famously in the introduction of aptitude testing and the War Office Selection Boards in 1942, developed by Jock Sutherland, W.R. Bion and Eric Trist. Rees transplanted much of the staff from the Tavistock to the Army and was one of the principal benign senior influences, with his colleague Ronald Hargreaves and their close alliance with General Sir Ronald Adam, to foster the experimentalism of these younger men. This included the famous series of experiments with groups of invalided soldiers and officers at Northfield Hospital, first by Bion and Rickman, later by S.H. Foulkes.[57]

The return of staff from war service to the Tavistock created a new dynamism, but also a new enthusiasm for psychoanalysis, amounting to a psychoanalytic over-riding of the eclectic ethos. Rees was out of sympathy with this new mood and the younger men promoting it – Sutherland, Bowlby, Bion and many others. So he resigned the Directorship in 1947, having successfully prepared for the Tavistock being absorbed into the new NHS on 5 July 1948. He then turned his galvanizing energies to international collaboration, being the driving force for many years behind the World Federation of Mental Health. In parallel, John Rickman stood down from his many years of editing the *BJMP*. Although hampered by poor health, he took on the Presidency of the BPaS from 1947 to 1950, in

---

[56] Rees maintained some connection with Cambridge through this time, giving lectures in the Chetwynd Lecture Rooms at King's College on 'Psychology and social work' in 1922. See *CR*, 13 October 1922, p. 5.

[57] Ben Shephard, *A War of Nerves: Soldiers and Psychiatrists 1914–1994*, London: Jonathan Cape, 2000, *passim*; Tom Harrison, *Bion, Rickman, Foulkes and the Northfield Experiments: Advancing on a Different Front*, London: Jessica Kingsley, 2000.

the course of which he galvanized the snail-like progress of the 'Memorial Edition' of Freud's Works (soon to be renamed the *Standard Edition*) and moved the Society into the stately residence in New Cavendish Street which became its home for the next fifty years. These two Cambridge men of the pre-war era, Rickman and Rees, thus stood in the centre of the development of psychoanalytic psychotherapy from the First World War to the post-World War II era. They were key organizers as well as animators of the English psychoanalytic and psychotherapeutic culture that developed from 1920 to 1950.

Two Wild Analysts: Layard and Farrow

A third undergraduate at King's pre-war was rather different. The grandson of Austen Henry Layard, who excavated Nineveh and Babylon in the 1840s, John Layard, Bedales-educated, took Medieval and Modern Languages in 1912 and 1913 in French, Spanish and German. Attending Heretics meetings,[58] he became smitten with W.H.R. Rivers, who persuaded him to stay at the University after his Tripos examinations to study anthropology. Layard did not take the only degree available in the subject at the time, the Diploma, but he was being nurtured by Haddon and Rivers, the two most prominent figures in the field, to become a researcher; they found him funds and took him along to the BAAS meeting in Sydney in 1914. As the war got under way in Europe, Rivers took Layard to Melanesia, soon leaving him alone on Atchin to do his research.[59] Layard, alongside Malinowski at exactly the same time, thus became one of the first intensive fieldworkers in British anthropology to proceed in the spirit of the recent methodological recommendations Rivers had penned for the Royal Anthropological Institute's *Notes and Queries*. In his autobiographical writings and interviews, Layard portrays himself as an ignorant fool in love with Rivers's charisma, but he inadvertently did the work he was destined for, throwing himself into native life, the happiest time of his life, thanking his lucky stars that he'd been circumcised as an infant so he would avoid that particular *rite de passage*.

On his return to wartime England in 1915, escaping from his 'mad' family, as he described them, he installed himself back in King's College, which was now a base for the Officer Training Corps. To aid his family, he had called in Rivers, who bundled Layard's father off to a private mental

---

[58] Ian Langham, *The Building of British Social Anthropology: W.H.R. Rivers and his Cambridge Disciples in the Development of Kinship Studies, 1898–1931*, Dordrecht: D. Reidel, 1981, pp. 202–4.
[59] See the photograph of Rivers by Layard taken before Rivers left Atchin, Fig. 3.2, p. 85.

nursing home. In his own deep distress, Layard again wrote to Rivers, who invited him to join him on a holiday in Devon, where Layard's condition only worsened. Rivers departed and Layard dreamt 'that I was asleep in this top storey room of mine in the house, that I heard a noise in the staircase coming up to it. I opened the door and there was Rivers carrying up the body of an insane young man, who when I looked at it, was me.'[60] Rivers extracted Layard from the Army, and spent more time on holiday in Salcombe with the troubled young man, who then moved to Edinburgh to be close to Rivers in his new job at Craiglockhart; there Rivers would come each day to his lodging house to conduct his 'analysis'. The still rather disturbed Layard wrote a letter to his one close friend from his undergraduate years, Robin Bevan-Brown, which he felt, with consternation, was a love letter.

> So next time Rivers came, I told him about this. I said, 'I love you, but you're sending me off my head. I can't see you any more.' I remember Rivers almost trembling and blanching when I said that. He went, and that was the last time I ever saw him.
> There's no doubt in me now at all that 'the most important thing,' which I had never told Rivers was the transference and the love which I must have had for him, and that he didn't appear to know consciously either or anyhow never spoke of.
> He did the very worst thing he could about it. He wrote to Robin and told Robin that it was better for him not to reply to me. This was Rivers' anti-sex in every direction . . . So my last and only male link with the world was gone, and that was obviously bound up with my dismissing Rivers. Because Rivers had obviously not recognized the whole homosexual content of our relationship, probably both sides.[61]

This passage presents a rare instance of what the messy business of doing proto-psychoanalysis was like in the autumn of 1916. Rivers was learning his own trade; Layard was a young man in trouble and would remain a man in trouble throughout his extraordinary life. After a year or more of a severe breakdown in which he was looked after by the Bagenals, friends he had made during his convalescence at King's College, he heard of

---

[60] John Layard, 'Autobiography: history of a failure', *c.* 1967– (Ms., John Willoughby Layard Papers, 1897–1974, University of California at San Diego, Mandeville Special Collections Library), pp. 115–16. There are substantial Layard archives at UCSD. The most informative published papers on Layard are by Jeremy MacClanchy: see his entry on Layard in the *ODNB* and 'Unconventional character and disciplinary convention: John Layard, Jungian and anthropologist', in George W. Stocking, Jr. (ed.), *Malinowski, Rivers, Benedict and Others: Essays on Culture and Personality*, History of Anthropology 4, Madison, WI: University of Wisconsin Press, pp. 50–71.
[61] Layard, 'Autobiography', pp. 115–16.

the eccentric and brilliant self-appointed revolutionary Freudian thera-pist Homer Lane, whose treatment extricated Layard from his quasi-psychotic regressed state. In the 1920s, Layard spent some time working in Cambridge at his ethnographic collections before moving to Weimar Berlin, where his flamboyant bisexuality found full expression. Lane having been brutally persecuted by the British authorities till he died a broken man, Layard moved on to a succession of other psychoanalysts: first Maurice Nicoll and then Paul Bousfield in London, briefly with Wilhelm Stekel and Fritz Wittels in Vienna in 1926, then Siegfried Bernfeld in Berlin. In 1929 Layard, partly out of love, partly out of rivalry for his young friend, sometime lover and acolyte, W.H. Auden, tried and failed to blow his own brains out.[62] Moving back to England, working desultorily in anthropology, he shuttled between London, Oxford and Cambridge, living in Finella on Queens' Road with its builder Manny Forbes, co-founder of the English Tripos. After a disastrous psychiatric encounter with Bernard Hart at UCL, he took up analysis with a con-temporary from Cambridge, H.G. Baynes, also a Heretics enthusiast, who had become one of the principal Jungians in Britain; after 1936, Layard eventually had analysis with Jung himself – an encounter that led to much friction and eventual mutual hostility. Finally settling in Oxford and practising as a therapist, during the Second World War he completed his anthropological study, *Stone Men of Malekula* and published an idiosyncratic case-history, *Lady of the Hare.*[63] Layard continued to prac-tise as a therapist, had episodic relations with the Oxford Department of Anthropology, engaged in lengthy affairs with two other women thera-pists (one Swiss in Zurich, when he returned to analysis with Jung; the other English in Oxford) and continued his analytic odyssey. By the 1960s, he was in therapy with R.D. Laing while living at Kingsley Hall.[64] In 1967, at a College feast for survivors of the undergraduate cohorts from the Great War and before, he proposed to Martin Bernal, son of his old Cambridge friend Margaret Gardiner, that he be appointed 'college

---

[62] There are two extensive accounts of this episode, one by Layard in his 'Autobiography', the second by Margaret Gardiner, Layard's close friend, who made an emergency visit to Berlin when told of his suicide attempt, in Margaret Gardiner, *A Scatter of Memories*, London: Free Association Books, 1988, p. 135 ff., part of the chapter 'Auden. A memoir', pp. 135–61. See also Humphrey Carpenter, *W.H. Auden: A Biography*, London: George Allen & Unwin, 1981.

[63] Reviewed rather sceptically by Wisdom in *Mind* 55 (1946); see John Wisdom, *Philosophy and Psycho-Analysis*, 'Critical notices of Howe, Glover, Layard and Sears', Oxford: Blackwell, 1953, pp. 182–94.

[64] Joseph H. Berke, 'Trick or treat: the Divided Self of R. D. Laing', www.janushead.org/4–1/berke.cfm, accessed 25 November 2005. See also Daniel Burston, *The Wing of Madness: The Life and Work of R.D. Laing*, Cambridge, MA: Harvard University Press, 1996, pp. 91–2.

psychologist' – not academic psychologist, but a therapist tutor, a special legacy of Freud in Cambridge.[65]

In the 1930s, while working in London at UCL's Human Anatomy Department, Layard had met Doris Dingwall, who somewhat stabilised his life and gave birth in 1934 to their son, Richard, who, following his father to King's College, Cambridge, became a well-known economist and New Labour's Happiness Tsar in the early twenty-first century. No doubt Richard Layard was acutely aware of his father's life-long analytic odyssey, perhaps the most star-studded such journey of the century, from Rivers via Stekel and Jung to Laing. It would not be surprising to learn that his family history informs the resolute support he gives to supposed evidence-based therapeutic interventions, such as cognitive behavioural therapy, in the National Health Service.

Another pre-First World War student had an equally singular, while completely contrasting, psychoanalytically infused life. E. Pickworth Farrow (1891–1956) gained a Diploma in Agriculture at Trinity College, Cambridge in 1912 and had his dissertation 'On the ecology of the vegetation of the Breckland Heath' approved in 1915. His interest in psychology was aroused in 1912 when Tansley brought the proofs of Hart's *The Psychology of Insanity* into his Cambridge botany class.[66] Hart's account of Jung's experimental investigations of the reaction times of associations, particularly when conducted on himself, persuaded him that there was something in psychoanalysis. He read *The Interpretation of Dreams* as soon as the translation was published in 1913.[67]

In the Great War, he volunteered for Army service, and in 1917 had a breakdown treated by a conventional doctor (possibly Sir Frederick Mott) who viewed the past solely as 'spilt milk', which left him with selective amnesia and slight depression – an 'army neurosis'.[68] Singled out by

---

[65] Martin Bernal, *Geography of a Life*, Xlibris, 2012, p. 279. The modern-day University Counselling Service, very much a product of 1960s attentiveness to the vulnerabilities of undergraduates, was founded around the time of Layard's conversation with Martin Bernal in the late 1960s. In the early 2000s, its website displayed the following: 'If you have any comments you'd like to make about the site please email us at: sigmund@counselling.cam.ac.uk.'

[66] E. Pickworth Farrow, *A Practical Method of Self-Analysis, enabling anyone to become deeply psycho-analyzed without a personal analyst, with some results obtained by the Author from early childhood, the earliest memories going back to the age of six months: also recounting the Author's personal experiences with two Psycho-Analysts*, with Foreword by the late Professor Sigmund Freud, London: George Allen & Unwin, 1942, p. 1.

[67] E. Pickworth Farrow, 'Experiences with two psycho-analysts', *Psyche* 5 (January 1925), 234–48 at 234.

[68] *Ibid.*, pp. 234 and 240n1; the identification of his war doctor as Mott is a surmise based on a mention of Mott's views in conversation concerning the effects of 'real experience'; see E. Pickworth Farrow, 'On the view that repressed fear of severance of the genitalia is solely caused by external reality and is not inherited', *IJP* 26 (1945), 161–8 at 163.

Fig. 4.4 Ernest Pickworth Farrow, photo by Harold Jeffreys, *c.* 1916.

Tansley[69] for being the first to call general attention to the influence of the biotic factor on British vegetation, Farrow kept in touch with his teacher, who later put him in contact with Freud.

In 1922, Farrow's interest in psychoanalysis was rekindled by a dream provoked by business worries – he was now working in the family engineering firm. Early in 1923, he 'decided with some reluctance (or resistance) to be analyzed'.[70] His first analysis lasted three months, when he moved on, at his analyst's suggestion, to another, with whom he had two-hour sessions five days a week for three and a half months. In commencing and then in persisting with analysis, he was very much indebted to 'friends' who had recounted the success of their own analyses and urged him to try it, in particular one, probably Tansley, 'who had been analysed and who argued that he had become much happier as a result'.[71] By the end of 1923, he had done with these two analysts and began his self-fashioned self-analysis. Clearly Tansley was his mentor in this field, since he prepared two papers with his encouragement and in the autumn of 1924 Farrow sent the first to Ogden for anonymous publication in *Psyche*, the journal Ogden was

---

[69] A.G. Tansley, *The British Isles and their Vegetation*, Cambridge: Cambridge University Press, 1939, pp. 129, 136–9, 500, drawing attention to the impact of rabbit attacks on heathland.
[70] Farrow, *A Practical Method of Self-Analysis*, p. 9.
[71] Farrow, 'Experiences with two psycho-analysts', p. 236.

now editing – adding in a postscript how well he remembered the old days of the Heretics of which he had been a member:

> My friend Mr AG Tansley is rather anxious that the enclosed paper 'Experiences with two Psychoanalysts' should be published and he considers that *Psyche* would be the best paper to publish it … Tansley was good enough to say that he thought it an interesting account of analysis from the point of view of the patient, & there have been practically no accounts like this yet published. Also Tansley thinks it is a useful criticism of some of the methods adopted by some analysts & that, on this account, its publication would be useful. He showed it to Freud who liked it a lot & who agreed with the criticisms & would like to attach his support to these but fears this might arouse hostility amongst some of his supporters in England.[72]

A few weeks later, Farrow, again with Tansley's guidance, sent his second paper to Ernest Jones, who wrote to Freud in late 1924 in one of his more inflexible and narrow-minded moods: 'A Mr. Farrow, with whom you are in correspondence, has sent me a rambling auto-biographical article. From the content of it I should suspect him of suffering from dementia praecox, though Tansley who knows him does not think so. I am trying to get the article re-written in a form possible of presentation.'[73] Freud replied defending the eccentric and maverick Farrow – and implicitly endorsing Tansley's judgement over Jones's:

> You must not take Mr. Farrow for a fool. I know him through Tansley and from a personal conversation. He is an odd man, but a very able, 'shrewd'[74] one, who had no luck with two analysts and has since then undertaken a self-analysis and is coming up with quite serious findings. To be sure, he is a bit of a grumbler, but both analysts (near you) really made technical mistakes with him.[75]

Clearly what had unsettled Jones was not simply the fact that Farrow had criticized the methods of his two analysts (both members of his recently formed BPaS) and was attempting to enlist Freud as a patron of self-analysis. What made Jones think of a diagnosis of psychosis was Farrow's insistence on the terrifying *reality* of the castration threats in childhood that his self-analysis had reconstructed: a scene from his early childhood in which an adult female cousin brought a pair of scissors closer and closer to his body while a collaborator held a cooking bowl to catch the penis as it

---

[72] OP, Box 19, F.2, Letter n.d.
[73] JF, 7 November 1924, p. 560; see E. Pickworth Farrow, 'A castration complex', *IJP* 6 (1925), 45–50.
[74] In English in the original.   [75] FJ, 16 November 1924, p. 562.

was lopped off.[76] The scene as described (in the version eventually published with extensive revisions by Tansley[77]) has a feeling of 'hyper-reality'. Farrow continued to find sympathy and support in Freud; Freud never did tell Jones that he had agreed to take Farrow into analysis in the near future – although in the event it came to nothing.[78] However, Freud did give his public imprimatur to a third paper of Farrow's – concerning a memory recovered from his seventh month – published in the *International Zeitschrift für Psychoanalyse* in 1926[79], in which, as if to challenge Jones's vision of what counts as authorized psychoanalysis, he commended Farrow's unorthodox efforts and assured the readers of the reliability of the agent and his results:

> The author of this paper is known to me as a man of strong and independent intelligence. Probably through being somewhat self-willed he failed to get on to good terms with two analysts with whom he made the attempt. He thereupon proceeded to make a systematic application of the procedure of self-analysis which I myself employed in the past for the analysis of my own dreams. His findings deserve attention precisely on account of the peculiar character of his personality and of his technique.[80]

Notwithstanding Freud's support, there was a flurry of nervousness about what might be seen as whistle-blowing. When the April 1925 number of *Psyche* came out, James Strachey immediately sent it off to Alix in Berlin, pointing out the highlights: 'another article by Malinowski and also an article "My experiences with 2 psychoanalysts" by a Patient. The man is known to Tansley & Rickman, & I gather the analysts were Eder & Co. He certainly gets in some hard blows at them.'[81] Farrow had also emphasized to Ogden the delicacy of the matter: 'as I mentioned to you before, please keep [Tansley's] name out of it as it may be rather controversial'.[82] However, now with both Tansley and Freud as patrons, Farrow was invited to speak to the BPaS in November 1927 on 'Conventional methods of scientific research in relation to psycho-analysis.'[83]

---

[76] Farrow, 'A castration complex', p. 45.

[77] JF, 24 November 1924, p. 563, and Farrow, 'A castration complex', p. 45.

[78] OP, Box 19, F.2, Farrow to Ogden, Letters 16 October 1924 and 8 January 1925.

[79] E. Pickworth Farrow, 'Eine Kindheitserinnerung aus dem 6. Lebensmonat', *International Zeitschrift für Psychoanalyse* 12 (1926), 79; see also Farrow, 'A castration complex'.

[80] Sigmund Freud (1926), 'Prefatory note to a paper by E. Pickworth Farrow', *SE* XX, 280.

[81] BLSP, Add. 60714, Vol. LX, October 1924–February 1925, Letter dated 27 February 1925. The phrase 'Eder & Co.' indicates that David Eder was one of Farrow's two analysts; the other is quite likely to have been Eder's sister-in-law Barbara Low or possibly Edith Eder, who was apparently practising as an analyst and had collaborated with Rivers at the end of the war.

[82] OP, Box 19, F.2, Farrow to Ogden, 7 April 1925.

[83] D. Bryan, 'Report. British Psycho-Analytical Society', *Bulletin of the International Psycho-Analytic Association* 9 (1928), 276–7.

Farrow's long-term study on the Breckland Heath, conducted first under Tansley's supervision at Cambridge, utilized pioneering fixed-point photography and experimental methods in the field. Farrow analysed rabbit dung in the laboratory and used rabbit-proof cages and fences in the field to test the hypothesis that the degeneration of *Calluna* (heather)-heath to grass-heath could be attributed chiefly to rabbit attack. In his *Plant Life on East Anglian Heaths*, a revised collection of several articles that had appeared from 1915 to 1924 in the *Journal of Ecology*, Farrow stressed the urgent need for such experiment so that the new science of ecology could provide something more than superficial and uncertain conclusions.[84] Similar concerns had led him into psychoanalysis: as he recalled, he was 'used to trying to solve certain kinds of problems in the external world, but here were some very interesting problems inside his own mind urgently calling for solution'.[85] In a pamphlet published on behalf of the National Trust's Blakeney Laboratory in 1926 to stimulate readers to start on their own ecological research work, Farrow related that the discovery of this 'Biotic Zonation of Breckland' due to rabbits came to him one night in a dream.[86]

Just how strongly resonant were the problems of outer and inner worlds (Farrow's 'real' world of heather and voracious rabbits and his equally 'real' world of small boys and scissors) is an interesting consideration. Does it take a man who has been 'really' threatened with castration in his childhood to discover the causal agency of rabbits' teeth in the ecology of grasses? Or vice versa? Does it take a man who has recognized the causal agency of rabbits' teeth to discover that castration threats might exist in reality? In later reflections, Farrow would note the successful planting of forests due to a management regime which heeded his discovery that areas bereft of vegetation 'will grow and develop adult trees quite satisfactorily, provided the young seedlings are protected from certain extremely detrimental influences to which they are particularly susceptible'.[87] He was struck by the 'curious coincidence that, in an apparently quite different form of biological research, he should have been caused to investigate in some detail detrimental influences upon young humans which similarly

---

[84] E. Pickworth Farrow, *Plant Life on East Anglian Heaths, Being Observational and Experimental Studies of the Vegetation of Brecklands*, Cambridge: Cambridge University Press, 1925.

[85] Farrow, *A Practical Method of Self-Analysis*, p. 5.

[86] E. Pickworth Farrow, *The Study of Vegetation*, London: Blackie and Son, 1926, p. 16. Sold for two shillings per copy, this pamphlet was reprinted, with additions, from *Discovery*, September 1925.

[87] E. Pickworth Farrow, *Psychoanalyze Yourself: A Practical Method of Self Treatment*, New York: International Universities Press, 1948, p. xx.

prevent them from growing into healthy adults'. For Farrow, the highly dynamic mind of the child was to be regarded as a 'rapidly growing tree'. An adult who hits or frightens a child 'risks pushing the young tree's roots unnaturally far down into the ground, or lopping off various branches as it were, and may easily do permanent damage which will prevent it from ever being able to grow into a satisfactory adult tree unless the damage is eventually removed and repaired by analysis'.[88]

After the publication in 1925 of his *Plant Life on East Anglian Heaths*, Farrow became less involved in ecology. Indeed, apart from a short comment for the *Journal of Ecology*, 'Notes on vegetation on Cavenham Heath, Breckland' in 1941, he seemingly dropped out of professional ecology altogether. In part he was devoting much of his energies to the family business in Spalding of E.W. Farrow and Sons Ltd., which sold surplus army supplies after World War I and later became an engineering firm dedicated to crop irrigation. He was a member of the Spalding Club and a very active member of the Spalding Gentlemen's Society (since 1913) which he addressed on 'Recent discoveries concerning the human mind' in 1927, the same year he spoke to the BPaS. For many years he was the secretary of the Spalding Chiming Clock and Carillon Committee, one of the prime movers in an attempt to restore the Corn Exchange's carillon. His gentleman farmer and industrialist's lifestyle did not prevent him from publishing another series of psychoanalytic papers, and, then in 1942, *A Practical Method of Self-Analysis*, which was translated into Spanish, then German and French,[89] and was republished twice in Great Britain with an American edition appearing in 1945 as *Psycho-Analyze Yourself*. This book, with a Foreword by Freud,[90] consisted mainly of his papers published in the 1920s stitched together with the help of 'his old friend, Professor Tansley',[91] who contributed an Introduction to the third edition of 1948. But the analytic experiences on which it was based were maintained by Farrow over two decades: 'This book includes the results of more than 2,800 hours' research work spread over a period of 18 years, and the production of more than 12,000,000 words of free-association.'[92] In his

---

[88] *Ibid.*, p. 145.   [89] Anon., 'Obituary [E.P. Farrow]', *Lincolnshire Free Press*, 5 June 1956.
[90] In the Preface to his 1942 book, Farrow translated the German original of Freud's Preface to his 1926 paper, published in translation in the *Internationale Zeitschrift*. By translating the simple 'Die Verfasser' as 'The author of this book' Farrow implied that Freud had seen the book. We cannot confirm that Freud did not approve its use – the book may have been completed and sent to Freud in, say, 1938/39. In addition, Strachey comments that in Farrow's book the Preface was printed 'with a statement that it was included by his [Freud's] permission. (This was, of course, some years after Freud's death.)', XX, 280n1. We have not been able to find such a statement in Farrow's book.
[91] Farrow, *A Practical Method of Self-Analysis*, p. xiii.   [92] *Ibid.*, p. xi.

revised and extended third edition, Farrow included excerpts from appreciative readers reporting unearthed repressions and new feelings of ease; he also quoted an undated letter sent by Tansley from Vienna (and almost certainly written in 1924) while the book was still in draft form:

> I have received Chapter 3 and have discussed it with Professor Freud. We were both greatly interested. I am particularly impressed by the vigour and success with which you bring out the correct technique of free association. I have never seen it so well and so vigorously done and when I drew Freud's special attention to this part of your work he replied, 'Ja, tadellos', meaning of course, 'Yes, beyond reproach' or 'Impeccable'; and this is very high praise from him.[93]

Farrow asserted that his *Practical Method* – noting down whatever comes into the conscious mind on fine writing paper (in five one-hour sessions per worrying matter) – had several advantages over work with an analyst: it was quick and comparatively inexpensive, and allowed shy people to keep thoughts to themselves, with the added benefit that the analyst could not be accused of reading the results into the mind of the patient. Above all, in proving the superfluity of the transference, the method checked the power of the (castrating) analysts who, in his experience, interfered too much and could not be induced to share in his spirit of experimental and democratic scientific inquiry. Farrow wrote: 'Being trained in what is probably a good school of scientific thought and method, he would have respected the analyst more if he had permitted him to experiment and for being an experimentalist like himself.'[94] Farrow also attempted to verify his findings concerning the reality of castration threats by interviewing the residents of Spalding: 'Careful observation and inquiries, extending over many years, in a typical English town have convinced the author that, on the average, not more than two, or at the most three, boys out of every ten escape some threat of this nature between the ages of 2 and 6 years.'[95]

Farrow's self-analysis brought up vivid memories of specific incidents of chastisement, allowing him to discover the lasting effects 'caused by physical blows and slaps which may be light in themselves but heavy to the child, in early infancy'.[96] After a session in 1924, Farrow traced the cause of his unconscious fear of the 'feminine body' to an incident that occurred sometime between the age of eleven and fourteen months. Hungry and having recently been weaned, he attempted to grasp the breast of a woman visitor to the house. She rebuffed him, he persisted and she

---

[93] Farrow, *Psychoanalyze Yourself*, p. xx.
[94] Farrow, *A Practical Method of Self-Analysis*, pp. 20–1.   [95] *Ibid.*, p. 120.   [96] *Ibid.*, p. 93.

eventually hit him: 'the feeling of injury, caused by a big strong woman to a defenceless infant, was enhanced no doubt because at the time she was regarded by the child as a substitute for his mother'.[97] Going back even further to the age of six months, Farrow remembered his father pulled him away from his mother's breast when he hadn't finished suckling and thinking something like: 'I'll teach this big object to move me away from my mother (or food). I'll kick him very hard and perhaps he will not do it again.'[98] He discussed the incident with Tansley, who suggested to Farrow the term 'instinctive foresight' to describe the kind of thought the six-month-old baby had had. Ten pages later, having just discussed how he had ventured too close to his puppy Sally while she was finishing off a nice bone, so that she expressed enormous 'anger and energy' in the 'most terrible and vicious doggy language', he comes back to Tansley's suggestion:

> While the author is extremely grateful to Professor Tansley's conception of 'instinctive foresight' as it seems to him to offer some reasonable explanation of his definite extremely early thoughts yet, nevertheless, on his thinking the matter over for several months, he rather doubts whether the foresight is really 'instinctive' in itself, although the power of the foresight is instinctive and inherited and foresight undoubtedly exists at a much earlier age in the human mind than is generally realized ... On this view the foresight would arise owing to a strong instinctive and inherited tendency towards inductive reasoning on the part of the minds of some of the higher vertebrates rather than to its being instinctive in itself.[99]

A loner, practising on himself, Farrow was not averse to biting the hand that fed him; not even the hand of Freud escaped Farrow's treatment.

In Farrow's last published article in the *International Journal of Psycho-Analysis*, then being edited by James Strachey, 'On the view that repressed fear of severance of the genitalia is solely caused by external reality and is not inherited' (1945), Farrow dissected a letter that Freud sent him (18 July 1939) just two months before Freud's death:

The manner I have come to look at castration-fear is the following:

(1) I am sure it is inherited as well as individually acquired.
(2) There is a well-known verse of Goethe giving expression to such a constellation: What thou hast inherited from thy father thou must acquire anew in order to possess.

---

[97] *Ibid.*, pp. 76–7.   [98] *Ibid.*, p. 81.   [99] *Ibid.*, pp. 91–2.

(3)   The intensity and pathogenic importance of the fear is based on the phylogenetic origin.

(4)   Few boys, if any, will escape the opportunities by which the inherited tendency is aroused, awakened.

(5)   In some cases the recent individual instigations may appear to have done the work themselves,

(6)   in other cases, where the threats have been too mild, insufficient,

(7)   the hereditary part will be called up to complete the effect.

(8)   The threat for itself could never be efficient if another condition be not fulfilled.

(9)   The boy must have had the visual experience of a female genital either before or after he is threatened.

(10) It is necessary that these two factors combine.[100]

Having numbered Freud's sentences, Farrow then considers – and, for the most part, attacks – each in turn. After commenting that 'It is important that the parents of a small boy should know that two women may, in certain circumstances, form a sort of "pack" in making extremely severe genital-severance threats against him – far more severe than either of them would make alone',[101] Farrow enlists the help of his old friend Tansley:

> With regard to point (1), Professor Tansley pointed out to the writer that it was difficult to see how a fear, as such, could possibly be inherited, and that the view that it could be seemed to him to be certainly founded upon confused thought. He said that what could be inherited was surely the tendency or disposition to feel a fear under the action of an appropriate stimulus. Perhaps what Professor Freud really meant when he said that the fear was inherited was that all males inherit such a mental disposition as to make the arrival of the fear inevitable under stimuli which they always, in practice, receive, or, which they can rarely or never escape. A strong tendency towards a primitive, or at least very early, feeling of omnipotence does, on the other hand, appear to be definitely inherited by the individual. This primitive feeling is, however, utterly opposed to any fear of loss of the genitalia.[102]

Exactly when Farrow and Freud met is unclear – it was most likely in 1924, shortly before Farrow informed Ogden he was beginning analysis with Freud; but in the third edition of Farrow's *Psychoanalyze Yourself*, there is the following passage:

---

[100] Farrow, 'On the view that repressed fear of severance of the genitalia is solely caused by external reality and is not inherited', p. 162.

[101] *Ibid.*, p. 163.   [102] *Ibid.*, p. 163.

The author well remembers Prof. Freud saying to him, when he had expressed doubt regarding the importance of the Oedipus complex; 'Ah! At one time you doubted the great importance of the castration complex. Eventually you will realize in the same way the great importance of the Oedipus complex.'[103]

Whether Farrow did or not is unknown. He died in Spalding in 1956, suffering from Parkinson's disease.

After the Great War

The four trajectories of Rees, Rickman, Layard and Farrow tell us very little about what they learned of psychoanalysis while students before the Great War. Layard's studied self-portrait as an ignorant fool includes his embarrassment when Rivers, in 1914, had to teach him what the word 'faeces' meant and, in 1916, introduced him to a wholly new word, 'neurosis'. With such ignorance, it is unlikely he had read Freud. Farrow's interest was roused by Tansley in much the same way Layard's was by Rivers, and their self-help Freudian education was sustained by the patronage and interest of these two important Cambridge figures. There is little evidence that Donald Winnicott, at Jesus College taking the Natural Sciences Tripos in preparation for medical training in 1914–17 and then going into the Royal Navy, came across Freud while in Cambridge, although by the end of 1919, studying at Bart's, he was fluently didactic concerning Freud, dreams and psychotherapy in a letter to his sister.[104]

Winnicott described himself at this time as 'an inhibited young Englishman, with few outlets in fantasy, except music'.[105] However, students taking moral sciences would undoubtedly have come across Freud, in particular his dream theory, in the years before the Great War. Lord Adrian's memory may have been quite accurate: in 1908, there was no more than a whisper of Freud to be heard. But by 1912, things had changed. And by 1914, many, but by no means all, knew what 'Freud' stood for.[106]

---

[103] Farrow, *Psychoanalyze Yourself*, p. 166.

[104] Donald Winnicott to Violet Winnicott, 15 November 1919, in F.R. Rodman (ed.), *The Spontaneous Gesture*, Cambridge, MA: Harvard University Press, 1987, pp. 1–4.

[105] Paul Roazen, unpublished manuscript 'Historiography', quoted in F.R. Rodman, *Winnicott: His Life and Work*, Boston, MA: Da Capo Press, 2004, p. 389.

[106] Another figure, of whom there is little known beyond the following, is Cecil Robert Allen Thacker, born 18 June 1889, who went up to Downing College, Cambridge in 1908, took Natural Sciences and specialized in physiology – a pupil of Sherrington's – in his final year (1912), completing his medical studies at St Bartholomew's Hospital; he received his MD in 1920. During the war he worked in military hospitals in Cambridge, and relinquished his commission in the RAMC to do work on the Special Medical Board for neurasthenia and shell shock, under the Ministry of Pensions. He was elected Fellow of Sidney Sussex College, Cambridge in June 1918, where he

Fig. 4.5 D.W. Winnicott, in uniform, Cambridge, c. 1917. J. Palmer Clarke.

The University of Cambridge did not close during the Great War, but numbers of College dons and University staff were greatly reduced and numbers of undergraduates even more so. It was the financial impact of the war years, with the heavy reduction in fee income, that contributed substantially to the emergency decision immediately after the war to request funds for the first time in the University's history from the government, that is, funds voted by Parliament. This triggered the creation of a Commission whose recommendations for reform, when implemented in 1926, would create the modern research University, with separate Faculties, salaries and pensions for University staff, and a commitment to a much broader intake of undergraduates. During those war years the tumult affected everyone – but in diverse and chaotic ways. Virtually every figure in this book, from Rivers and Russell to Lionel Penrose and Harold Jeffreys, from James Strachey and Virginia Woolf to Wittgenstein and Keynes, had the direction both of their lives and of their intellectual work dramatically changed by the war. But what of the effect of the war on the reception of Freud?

'There is no doubt that Freud's views have won a large measure of acceptance in England as a result of the war.'[107] Edgar Adrian again sets the tone, writing anonymously in October 1919. His war, spent first in a

taught till 1926, becoming University Demonstrator in Physiology for 1920–22. He was an associate member of the BPaS from 1920 to his death. Owing to ill health he resigned all his posts in December 1926, and went to the south of France, where he died in May 1929.

[107] Anon. [Edgar Adrian], Review, W.H.R. Rivers, 'Mind and Medicine', *Athenaeum*, 3 October 1919, p. 979.

lightning medical qualification at Bart's in 1914–15, then at the National Hospital for Diseases of the Nervous System in Queen Square and later at the Connaught Military Hospital in Aldershot, a military town in the English Home Counties, till his demobilisation in 1919, gave him considerable experience of the war neuroses, in the course of which he co-authored a striking paper with L.R. Yealland (used by Pat Barker in her trilogy as the contrasting sadistic medical figure to her sympathetic Rivers). Adrian and Yealland's method of treatment consisted in an amplification of all the doctor's suggestive resources, including the use of the magic and pain of electricity. At the same time, they recognized that this was, however successful, solely a symptomatic treatment and considered, with sympathetic scepticism, the promise of the more thorough-going treatment which was psychoanalysis. However, the urgent practicalities of treating soldiers in as short a time as possible contrasted severely with psychoanalysis, where 'the course of treatment may run into years if a serious attempt is made to cure the hysterical mind by purging it of all the accumulated filth of a lifetime'.[108] Less inhibited by his co-author and more at ease in familiar home territory once back in Cambridge, Adrian gave a talk to the Natural Science Club,[109] of which he had been a stalwart member since 1909, on 31 May 1919, entitled 'Freud without tears'. Fresh from military hospital, he reviewed the miracle cures promoted by some doctors, casting sober scepticism on the permanence of the removal of symptoms.[110] But his principal message was to communicate what his own experience had taught him about psychoanalytic claims.

> Adrian describes one case of a neurasthenic with a bad stammer whom he questioned in a 'hypnoidal state'. The man was a regular soldier who had been in France for two years, and remembered nothing of a series of traumatic experiences which took place between a shell burst and his 'coming round in hospital'. The case proceeded along classic lines. In a series of interviews Adrian uncovered experiences that even by the standards of the First World War seem particularly horrifying, as well as suppressed childhood memories of an incident that led to stammering. When all these memories were brought back to the patient's mind he had a considerable emotional discharge but in a day or two he began to improve and soon became an absolutely different man. Adrian concluded that those who have

---

[108] E.D. Adrian and L.R. Yealland, 'The treatment of some common war neuroses', *The Lancet* 189(4893) (9 June 1917), 867–72 at 868.

[109] Many of the Cambridge scientists in this study had been members of this long-lived club (1872–1982), including former members as honorary members. Amongst the members who figure in this book were Denis Carroll, F.H.A. Marshall, W. Langdon-Brown, Tansley, W. McDougall.

[110] Shephard, *A War of Nerves*, p. 80.

tried this method of treatment become convinced of two things: (a) that the
patient does seem quite honestly to forget a series of events of very great
emotional significance, which one would have supposed him to remember
to his dying day, and (b) that his neurasthenic symptoms often disappear
very rapidly after he has recalled these events.[111]

Adrian put forward direct personal evidence for the reality of repression in
the Freudian sense (i.e. amnesia which is maintained by constant inner
forces of which the patient is not aware). And he affirms what so many
others working with soldiers in the war affirmed: the striking efficacy of the
cathartic cure.

Rivers also gave lectures about his wartime experience in the summer of
1919, on 'Instinct and the unconscious'. Given that the more senior figures
arrived back in Cambridge after the war communicating their new-found
knowledge and experience in the war neuroses, perhaps it is not quite as
surprising as it seems on the face of it that Lionel Penrose arrived in
Cambridge in January 1919 intent on taking the Psychoanalysis Tripos.
Making do with second best, the Moral Sciences Tripos, his path for the
next few years was unusually single-mindedly intent on elaborating a
science of the mind through psychoanalysis, developing out from that
into the relations of psychology and logic, then of the brain and logic,
then into psychiatry and finally into his life's work of the genetics of mental
disease and defects.

Other new arrivals were probably less well informed about the revolutionary new science. But, just as the middle classes learnt about the facts of
life from jokes and games, not to forget gardening, so it was with the
dissemination of psychoanalysis. The 1919 *Cambridge Review* published the
following report from 'a correspondent':

> It may perhaps be worth the *Review*'s while to insert a brief protest against a
> dangerous 'game' which has recently come into fashion in certain quarters.
> The game goes by the high-sounding name of 'psycho-analysis' and is
> played as follows:– A word, *e.g.,* 'Scotch' is uttered by the analyser and the
> victim is expected to say at once the first word thereby suggested. Thus is his
> secret soul laid bare: if he says 'whisky,' he is a drunkard (possibly sup-
> pressed); if he says 'mist,' he is a nature-lover; if he says 'Liturgy,' he is a
> controversial theologian, etc., etc. All this may sound harmless enough, and
> excellent drawing- (or smoking-) room fun; and on the other hand no one
> will wish to minimise the successes scored by the use of 'psycho-analysis' for
> serious purposes by a responsible agent. But it is against the bye-products of
> this art that a protest should be made – against the decadent coteries that

[111] A. Hodgkin, 'Edgar Adrian', *BMFRS* 25 (1979), 1–74 at 19.

poke and pry for signs of 'tendencies' in others and delight in their own self-revelation. Such back-alleys exist in Cambridge and still lack the proper satirist.[112]

The author was not to know it, but that 'perfect satirist' lay very close by, living at No. 2, Trinity Lane from 1920 to 1922, an aristocratic Russian studying Modern Languages at Trinity: Vladimir Nabokov.[113] It has even been suggested that the anonymous author above was indeed the very one who later came to be the scourge of the 'Viennese quack'.[114] This seems unlikely since the little piece is not sufficiently well written.[115] But it is equally unlikely that Nabokov failed to notice the Freudian epidemic besetting the odd English town in which he happened to land after the Bolshevik Revolution. When in the mid-1930s he came – suitably planted astride a bidet in a one-room apartment in Paris – to write *The Real Life of Sebastian Knight*, his first novel in English, recounting the pursuit of the eponymous central figure by his narrator half-brother, Sebastian spent time in Cambridge in the early 1920s. In part as a result of this, he had already adopted a horror of Freud and other psychological modernisms, to the point of insulting and then changing his publisher, in 1925, when he was asked to tone down his exposure of these 'shoddy gods' in his first novel:

> the dark secret of ... success, which is to travel second-class with a third-class ticket – or if my simile is not sufficiently clear – to pamper the taste of the worst category of the reading public – not those who revel in detective yarns, bless their pure souls – but those who buy the worst banalities because they have been shaken up in a modern way with a dash of Freud or 'stream of consciousness' or whatever.[116]

Nabokov's Cambridge may not have been Freud's Cambridge: 'I was deeply in love with the country which was my home (as far as my nature could afford the notion of home); I had my Kipling moods, and my Rupert

---

[112] *CR*, 41 (1919–20), p. 111.
[113] He matriculated at Trinity College in Michaelmas Term 1919, as 'Nabokoff, Vladimir Vladimirovitch' (*University of Cambridge Calendar 1920*, p. 874) and was placed in Class II in June 1922, in Modern and Medieval Languages Tripos Part II, having taken the 'Examination in Literature and History'. Uncannily, alongside his name is a second Nabokoff: Nabokoff, S. of Christ's College, who also took the examination in Literature and History (*CUR*, 17 June 1922, p. 1141).
[114] Stephen H. Blackwell, 'Nabokov's wiener-schnitzel dreams: despair and anti-Freudian poetics', *Nabokov Studies* 7 (2002/3), 129–50.
[115] The best guide to Nabokov's Cambridge period 1919–22 is Brian Boyd, *Vladimir Nabokov: The Russian Years*, Princeton: Princeton University Press, 1990. Boyd (p. 260) places the first explicit reference in Nabokov's letters, diaries or writings to 'witchdoctor Freud' in 1926, when he had to listen to a talk about Freud in a Russian group in Berlin.
[116] Vladimir Nabokov, *The Real Life of Sebastian Knight* (1941), London: Penguin, 1964, p. 46.

Brooke moods, and my Housman moods', Sebastian Knight writes.[117] Yet throughout his career, Nabokov could not conceal the pleasure he got out of satirizing the Freudians. Two of his later masterpieces, *Lolita* and *Speak, Memory*, are strewn with gratuitous puzzles whose solution is both a laugh at the expense of and a serious undermining of Freudianism. Humbert Humbert loves to trifle with his psychiatrists, those 'dream extortionists',[118] while in *Speak, Memory* spoof Oedipal jokes abound and the narrator's mother has a lover who morphs into the ballooning Sigismund Lejoyeux, certainly a nod towards Sigismund der Freudige.[119] Nabokov may well turn out to have been, in the judgement of history, the most eminent of the Cambridge Freudians of the early 1920s.

How to Read Freud?

Apart from lectures, how might Cambridge undergraduates gain access to psychoanalytic ideas? The obvious answer is: through books, the very currency of a university. The story of Cambridge University Library (UL) book classification, although it may appear on the face of it an intricately tedious subject, is here both telling and fascinating. The inquiry is both historical – when were certain books restricted and for whom? – and geographical – where were they held, and by what rules? Undergraduates were not allowed to use the UL until the late nineteenth century: they had been kicked out for misbehaviour four centuries earlier. By 1925, an undergraduate's tutor was generally responsible for borrowing books for the students, and could take out up to five for each student of select classmarks. Undergraduates were only admitted during the last three hours that the UL was open, provided they wore academic dress.

The British state deployed devious means throughout the nineteenth and twentieth centuries to ensure active censorship of the press, book, magazines, the stage and, when it became an issue, which was very early, the cinema. The principle of the policy was to deny outright the policy of censorship by pointing to the fact that there was no government official (e.g. Home Secretary) who had a statutory function of censor and therefore censorship could not occur. The secondary tactics – the implementation of this denied policy – involved putting behind-the-scenes pressure on private (non-state) organizations (e.g. booksellers' organizations) as the means of

---

[117] *Ibid.*, p. 58.   [118] Vladimir Nabokov, *Lolita* (1955), London: Penguin, 1997, p. 34.
[119] Vladimir Nabokov, *Speak, Memory: An Autobiography Revisited* (1967), New York: Vintage, 1989, p. 156.

enforcement, followed by the use of red herrings to obfuscate the crude reality of behind-the-scenes government censorship, e.g. the question of artistic merit as a defence against a charge of obscenity (or pornography).[120] In the University of Cambridge, the internal disciplinary systems (Proctors, Courts of Discipline) performed this function.

The actuality of this disavowed but real and insidious censorship was made clear to F.R. Leavis. 'Having been awarded his doctorate in 1924, Leavis lectured for the English faculty as a freelance known for independent-mindedness: on one occasion he was in trouble with the police and university authorities because he wanted Joyce's banned novel *Ulysses* to be available for study.'[121] Following this incident, Manny Forbes, the founder of the Cambridge English Tripos and an influential don at Clare College who had hired both Richards and Leavis, 'entertained a friend for the week-end, who was apprehended by the Cambridge police for improper conduct. Perhaps fearing that he might suffer the indignity of being summoned before the Vice-Chancellor, as had happened to Leavis in 1925 ... he clandestinely dumped all the "unsuitable" books in his library in the Cam with the assistance of Sykes Davies, lest the police should find them on his shelves.'[122] So it is perhaps not surprising that tight control of books seeped into the policies of the University and its Library.

An undergraduate working with Wittgenstein from 1929 to 1931 recalled him recommending he read Weininger's *Sex and Character* and Wittgenstein's annoyance at finding that the book 'was in a section of the University Library which required a special procedure for borrowing: he thought that the implication was that it was in some way unfit for undergraduates and that that was nonsense'.[123] *Sex and Character* was in the notorious S3 category, 'S' meaning special. Over the years 'S' developed other associations for both library workers and users: 'S' for Sex and 'S' for Sin-bin. In the new University Library built in 1934, the S3 category books were stored up in the phallic tower, enhancing their reputation and mystique.[124] In terms of access to psychoanalytic literature, the general

---

[120] Lisa Z. Sigel, 'Censorship and magic tricks in inter-war Britain', *Revue LISA/LISA e-journal* [en ligne], Vol. XI – n° 1 | 2013, mis en ligne le 30 mai 2013, consulté le 29 mars 2015. http://lisa .revues.org/5211; DOI : 10.4000/lisa.5211.

[121] Ian MacKillop, 'Leavis, Frank Raymond (1895–1978)', *ODNB*, www.oxforddnb.com/view/article/ 31344, accessed 20 May 2015.

[122] T.E.B. Howarth, *Cambridge between Two Wars*, London: Collins, 1978, pp. 165–6.

[123] Desmond Lee, 'Wittgenstein 1929–1931', in F.A. Flowers III (ed.), *Portraits of Wittgenstein* (4 vols.), Bristol: Thoemmes Press, 1999, Vol. II, pp. 195–6.

[124] An S3 book remains unborrowable; a senior cataloguer noted to Laura Cameron, 'you don't want students just taking it out for a snigger'.

trend, in part mirroring the rise of interest in psychoanalysis and subsequent closing down, is one of increasing openness up to 1930 followed by increasing restrictions. The International Psycho-analytical Library series begun in 1920 was on open access until 1930. A student wanting to read *Future of an Illusion* (1928) in the year of its publication would seek it on the open shelves in the Goldsmith's Room. By 1930 the IPL books also went to closed access (closed to browsing and no longer borrowable), and in 1937 the IPL was shifted to a new 'S' classification (S180:01.c.1) specifically created for the IPL and all its books, including the Hogarth Press editions of Freud's work in translation. These remained long-term in the S category and had, and still have, to be consulted in one special room in the Library. In a world-class library famous for its liberal borrowing rules, they remain not borrowable to this day (2015).[125]

Five Undergraduates after the Great War

'An aristocrat of the highest family': Roger Money-Kyrle[126]

On first acquaintance, Money-Kyrle is a figure from an older era, an older England and an older Cambridge. Scion of a prominent English family, with a seat in Wiltshire since the thirteenth century, whose forebears included a Lord Chief Justice of the Court of Common Pleas in the early sixteenth century and one of the longest serving Chancellors of the Exchequer in the late seventeenth century, across the centuries the Ernles and Kyrles, later Money-Kyrles, bred high-ranking officers for the Royal Navy and the Army, the occasional barrister and parson, even servants of the East India Company. They were a classic family of well-established gentry. From one point of view, Roger Money-Kyrle's life trajectory honoured imperturbably the family traditions. Born in 1898, educated at home, then at Eton and Trinity College, Cambridge, a pilot in the Royal Flying Corps in World War I and a Squadron-Leader in the RAF in World War II, he resided at the family estate at Whetham throughout his life, starting in 1900 when it came into his father's possession. Like the gentleman he was, he never took up paid employment – with one brief notable

---

[125] Making a classification change was no small matter and a lot of work for the cataloguers. New restricted status for the IPL would reflect a decision not made lightly.

[126] A fine treatment of Money-Kyrle's pre-Kleinian career and work is Neil Vickers, 'Roger Money-Kyrle's *Aspasia: The Future of Amorality (1932)*', Interdisciplinary Science Reviews 34(1) (2009), 91–106.

exception – and was duly appointed Justice of the Peace for Wiltshire in 1935 and High Sheriff of the County in 1950.[127]

Then there was Roger Money-Kyrle's life without 'camouflage' – his own life, we might say, beyond the duties of a church-going grand family. Coming up to Trinity College in February 1919, he was shocked to discover how woeful was his preparation for studying his intended subject, physics, instead of taking a course in forestry, which would have been useful in running the family estates. Inner difficulties and guilt over war incidents also troubled him.

> It was then that I first heard of psycho-analysis (from E.F. Collingwood, a remarkable man who may have profited from analysis and who much later became a lecturer in mathematics at Trinity). I saw a psychotherapist recommended by him, and – although I went back later to take a 'Special in Economics' (which Collingwood also took, as neither of us dared face the Tripos) – I went down and started an analysis with Ernest Jones. This I camouflaged by going into a bank.

Having begun the analysis in the autumn of 1920, in May 1922 Jones proposed to Freud that he take Money-Kyrle for analysis, describing him as follows:

> Money-Kyrle, aged 23, an aristocrat of the highest family and, what is better a first class brain. Young as he is I regard him as a powerful thinker with a very clear grasp of essentials. His line is the modern application of mathematics to the principles of science and he may do much in defining the basis of our science in the future. He came to me because of Angst at using his voice, which proved to be based on an obstinate (manifest) masochistic perversion. He made good progress, but failed at the last when I gave him a terminus at Easter. After leaving he foolishly rushed off and married an older woman in a short time, against my advice.[128]

Money-Kyrle was not alone in establishing this project of 'defining the basis of our science in the future' through 'the modern application of mathematics to the principles of science': it is quite recognizably the psychoanalytic project that Ramsey and Penrose hoped to bring to fruition. (As James Strachey said of Ramsey in 1925: 'He is thinking of devoting himself to laying down the foundations of Psychology. All I can say is that

---

[127] As one of the many points of contrast with the Bloomsbury relation to the countryside, where they also resided, Bloomsbury (see Chapter 9) kept a distance from the social structure of country life, unlike Money-Kyrle, who was born to occupy its apex and did so.

[128] JF, 22 May 1922, pp. 479–80. Money-Kyrle married Helen Juliet Rachel Fox (1890–1980) on 8 May 1922.

if he does *we* shan't understand 'em. He seems quite to contemplate, in his curious naif way, playing the Newton to Freud's Copernicus.')

Money-Kyrle returned to Cambridge for Easter Term 1922 and took a Special Examination in Political Economy for the Ordinary Degree at the end of that term; he was placed in the Third Class.[129] This combination of psychology and political economy was the classic formula for Cambridge moral science of the old style. He then moved with his wife to Vienna and began his analysis with Freud; they were to live in Vienna for the next four years, with their first son born there in 1925. During his analysis, Jones, always curious as to the fate of his analysands in Freud's hands, inquired as to his progress; Freud was reluctant to say much. In November 1922, after two months of analysis, he wrote:

> As regards Money-Kyrle I can shape no judgment on his intellectual powers as long as he is exhibiting all his inhibitions and his resistance which he does like any 'greenhorn' in his first weeks. His history is clear but his extra-ordinary reactions to it are by far more conspicuous.[130]

Five months later, he said more:

> Mr. Money-Kyrle is progressing continuously. I think he has learned to listen and even to believe a little. The construction of his prehistoric case is complete and he is wont to lean on it for explanation of present situations. But the fact is, he has not yet penetrated to recollection, all his adherence is only [a] matter of reason and the practical effect of the treatment cannot be guessed. So I am ready to continue work with him, if he returns. He will leave on June 1st for an examination in Cambridge and much will depend on the result of it. I am sure he will call on you.[131]

He did leave Vienna for Cambridge, took his second 'Special', this time in psychology, in June 1923,[132] and then received his Cambridge degree. While in Vienna, he began work on a PhD in philosophy working with Moritz Schlick, the founder of logical positivism and the Vienna Circle. Money-Kyrle recalled:

---

[129] *CUR*, 17 June 1922, p. 1149. There were seventy-eight candidates, of whom seventy got Thirds. Regulations at the time required that students intending to take an Ordinary Degree were required to pass two 'Special Examinations'. University regulations excluded taking both a Special in Political Economy and a Special in Logic, whereas there was no such exclusion on taking both Political Economy and Psychology. (See *University of Cambridge Calendar 1920*, p. 175.) When Money-Kyrle took his second 'Special' (when there was only one other candidate), he again received a Third Class.

[130] FJ, 6 November 22, p. 502.   [131] FJ, 25 April 1923, p. 521.   [132] *CUR*, 12 June 1923, p. 1123.

Well I must confess unblushingly that, although I was genuinely interested in positivist philosophy, my Ph.D. was camouflage for psychoanalysis: and meeting Schlick was pure good luck. I had never heard of him before, and although I thought and still think he was in reality a brilliant and charming man, I no doubt transferred much of my positive transference, which could have no social outlet in analysis, from Freud to Schlick.

In other recollections, he wrote:

as my wife was sociable and popular (although she made no conscious effort to be either) we were soon on very friendly terms with Professor Moritz Schlick, a brilliant and delightful man, with whose help I eventually got my Ph.D. I remember I had suggested calling my dissertation '*Wirklichkeitslehre* (Theory of Reality)', but he most tactfully suggested '*Beiträge für Wirklichkeitslehre*' (Contribution to Theory of Reality). I kept my contact with Freud quite secret until the end when I discovered that Schlick, too, was most interested in Freud's work, but had also kept his interest secret! In Vienna we also met Frank Ramsay [*sic*], the mathematician, Lionel Penrose, later Professor of Genetics at London University, Adrian Bishop, a brilliant if somewhat erratic person who became headmaster of a Turkish school, and L.B., later Professor Sir Lewis, Namier – all also being secretly analysed.[133]

Money-Kyrle almost certainly finished his analysis with Freud in the summer of 1926. Frank Ramsey ran into him in Oxford at that time while delivering a paper on 'Mathematical logic' to the British Association for the Advancement of Science on 10 August: 'I spent yesterday evening with Money-Kyrle whom I met in the road. He is really very nice.'[134] This chance encounter was to have consequences. Money-Kyrle wrote a paper 'The psycho-physical apparatus: an introduction to a physical interpretation of psycho-analytic theory', published, no doubt courtesy of John Rickman, in the *British Journal of Medical Psychology* in 1928. In 1977 he added a note to its republication in his *Collected Papers*: 'This paper is almost purely behaviouristic but I leave it in because Frank Ramsay [*sic*] liked it.'

Money-Kyrle was not an especially original thinker. His first published paper was a variant on the principal themes of logical positivism as the Vienna Circle were developing it in the 1920s; his second paper, which Ramsey liked so much, was a hypothetical model of the 'neural apparatus' which could

[133] R. Money-Kyrle, *The Collected Papers of Roger Money-Kyrle*, ed. Donald Meltzer, Perthshire: Clunie Press, 1978, pp. xi–xvi at xiii. Money-Kyrle over-emphasized somewhat the 'secrecy' of his Oxbridge friends' analyses, perhaps to make them more like him, living in camouflage, than they in fact were. (One could also describe this, of course, as an instance of projective identification.)

[134] FRP, Letter to Lettice Baker, #26, undated but it can be dated by the dates of the meeting of the BAAS for that year (*Proceedings of the BAAS*, 1926, p. 342) as 11 August 1926.

Fig. 4.6 Roger Money-Kyrle, painting by Christian Schad, 1926.

explain the primary facts – the unconscious, repression, dreaming – of psychoanalytic psychology. His first book, *The Meaning of Sacrifice* (1930), based on his second doctoral dissertation produced under the supervision of J. C. Flügel at UCL, and published in the International Psycho-Analytical Library by Hogarth Press, received an excoriating review from Theodor Reik:

> Money-Kyrle gives an account of the analytical theory of sacrifice, together with some additions of his own, which are common knowledge to the analyst. He describes, but he originates nothing. What he describes is clear and correct, but we wait in vain for something fresh. The result is a very instructive, rather anæmic work, which affords much new knowledge to the anthropologist and the student of comparative religion, but to the psycho-analyst only a comprehensive account of the theory of sacrifice.
>
> The hypotheses which the author erects about the various kinds of sacrifice are largely, as has been said, unavoidable deductions from Freud's theories. Against the rest of the book there is nothing important to be said; but that is all that can be said for it.[135]

[135] T. Reik, 'The meaning of sacrifice', *IJP* 12 (1931), 370–1.

In his outer life, until he was nearly forty, Roger Money-Kyrle remained firmly on the path set by his family and social standing. But all was not entirely straightforward in the life of a gentleman of leisure and duty: in 1920 he had discovered, like so many others of his class after the Great War, 'that, instead of a comfortable income from two properties, I had a negative income of about £400 a year. I then sold Homme, for which I think the older tenants never forgave me.'[136] In the late 1930s he again had money troubles, and this may have been a significant factor in his eventual turn to psychoanalytic practice. But what is striking is that, even though he was a man with two PhDs, probably the first Englishman to come back from Vienna with a PhD in logical positivism (1927 – six years before Ayer's famous 1933 visit to Vienna), followed by a PhD in anthropology (1929), it never seems to have occurred to him to seek employment in a university.[137] Elected an Associate Member of the BPaS on 20 March 1929 and an Ordinary Fellow of the Royal Anthropological Institute on 19 November 1929, he was expressly asked by Ernest Jones not to practise as an analyst.[138] The new rules on training analyses were now in force. Until 1936 he remained a gentleman landowner with considerable intellectual interests in philosophy, psychoanalysis and anthropology.

In 1936, possibly when Money-Kyrle was again short of money and when he was involved in an intense and long-term affair, John Rickman persuaded him to undertake a training analysis with Melanie Klein[139] which lasted till the Second World War, then intermittently during it, while he worked in the Air Ministry, and for a brief period after. Working

---

[136] The online genealogy gives the date of sale of Homme House, Herefordshire as 1922. He sold the property to his uncle. Homme was the larger of his two properties.

[137] It is possible that he was deterred by the lack of an Honours degree from Cambridge; on the other hand, he did have two PhDs.

[138] Clarification of the status of 'associate member' is given in the discussion of Penrose's anomalous status as an associate member in Chapter 6, 'The 1925 Group' and also in the context of standard British scientific practices in Forrester, '1919', pp. 47–52.

[139] It is possible that Rickman's approach was not only on behalf of Melanie Klein, as it were, but also semi-officially on behalf of the BPaS. In 1935–36, two other individuals were approached with a view to conducting a research analysis as part of a British Society initiative: Dr Susan Isaacs and Dr William Stephenson. Isaacs had analysis with Joan Riviere, while Stephenson had analysis with Melanie Klein. As Stephenson later recalled: 'In 1935 I was chosen by Dr. E. Jones and a small committee to undertake psychoanalysis with Melanie Klein, not as a patient or trainee, but with the notion that I would undertake research of psychoanalytic doctrine in an academic (rather than clinical) framework. It was my understanding that Mrs. Susan Isaacs was given similar research significance. At the time few psychoanalysts were attached to universities in Britain – one could think only of Professor J. C. Flügel at University College, London' (William Stephenson, 'Tribute to Melanie Klein', *Psychoanalysis and History* 12(2) (2010), 245–71 at 245). Klein was by no means the most active training analyst; as she wrote to Sylvia Payne in 1942: 'I have actually had only four candidates since 1929' (FK, p. 671n1).

from then on as an analyst in private practice, he became one of the close circle of followers around Klein from the 1940s till his death in 1980. This included being a founding Trustee of the Melanie Klein Trust established in 1955,[140] which prepared the four-volume *The Writings of Melanie Klein* that he saw into print in 1975.

Kingsley Martin: '... mostly about God, Marx and Freud'

In the middle years of the twentieth century, Kingsley Martin was the most influential editor and journalist on the Left in Britain. Kingsley was born in Hereford in 1897, and raised from the age of sixteen in Finchley, London; his father was an outstanding Christian socialist Congregational minister with absolute pacifist convictions. Kingsley followed his father in many ways, including his opposition to the Great War, and registered as a conscientious objector while still at school, with the result that he spent the final eighteen months of the war, like Lionel Penrose, as an orderly with the Friends' Ambulance Unit. He came up to Magdalene College, Cambridge as soon as the war was over. On 25 January 1919 Patrick Blackett, future Nobel Prize-winner in Physics, creator of Operational Research during World War II and scientific advisor to Wilson's 1960s Labour government, arrived in the College in full naval uniform, let loose in Cambridge along with several hundred other officers by a special Naval scheme, and they immediately became firm and life-long friends.[141] From then on, it was the incessant talk that formed the backbone of undergraduate life:

> Conversation, argument, the search for the answers, were the stuff of Cambridge life. The claim of the old residential universities that the best part of their education lies in all-night discussion between men of all ages was in my day largely justified. We talked to one, two, or three in the morning, mostly about God, Freud, and Marx. I dismissed God in my first year ... In my second year I started to read and discuss sex, Freudian theories, and Socialism. The importance of Freud was that he engendered a new type of thinking. The discovery of the unconscious made Victorian thought seem childish. If people were driven by their unconscious it was foolish to blame them, and the world was much less easy to reform by reason than our fathers had imagined.[142]

---

[140] 'News, Notes and Comments', *IJP* 37 (1956), 514–17 at 515–16.

[141] Patrick Blackett, 'Boy Blackett', in Peter Hore (ed.), *Patrick Blackett: Sailor, Scientist and Socialist*, London: Frank Cass, 2003, p. 12. The *ODNB* is incorrect in stating he went up in October 1919; there is firm evidence that he went up, as did so many others returning from the war, in January 1919.

[142] KMFF, p. 110.

While 'the great event of our age was the Soviet Revolution', which led to a new form of optimism, the older form, the taken-for-granted possibility of an improvement in society, of inevitable progress, was dead: 'The war – and Freud – had killed any such optimism.'[143] As he wrote at the time:

> Many things have 'bust up' this [pre-war] picture. One: war, since all our progress was leading to the destruction of civilisation. Two: no one knew what end was in sight – perhaps society would always be like this? Three: a new psychology – perhaps we were not really 'directing ourselves' at all[144] . . . The end of the war meant that the theory of progress was dead.[145]

In 1925, he affirmed that 'the most devastating of all' fears was that promoted by Freud's psychology, which gives us solid grounds to believe that reason could no longer be relied upon to resolve the problems of the modern world.[146]

Besides the war and the Revolution, for Martin there were two more immediate great intellectual events in Cambridge: the publication of Lytton Strachey's *Eminent Victorians* in 1918 and the lectures Keynes gave on 'The economic aspects of the peace treaty' in Michaelmas Term 1919 that became his hugely successful and influential *The Economic Consequences of the Peace*:

> about Keynes's magnificent pamphlet everyone was agreed. I remember one Socialist complaining that the place should be called Keynesbridge. It was wonderful for us to have such a high authority saying with inside knowledge of the Treaty what we felt emotionally.[147]

Strachey's book set a somewhat different tone: 'Its satirical handling of the lives of people whom it was conventional to revere exactly satisfied the mood of revolt which was the common bond of Cambridge intellectuals. The style of my own first book was deeply influenced by it.'[148]

Martin's political commitments were evident from the start. He helped revive the Cambridge branch of the Union of Democratic Control and was vigorous in undergraduate socialist politics. He achieved a Double First in History in 1920–21 and in his next year, 1921–22, he grew ever closer to Rivers at St John's College. Enthused by the new psychology of his mentors, he gave a talk to the Heretics on 22 January 1922 on 'The

---

[143] KMFF, p. 112.

[144] Martin here referred to anti-representational art 'aiming only at emotional release'.

[145] KMFF, p. 112.

[146] Kingsley Martin, 'The war generation in England', published in *Revue des sciences politiques* 48 (1925), 211–30, Kingsley Martin Papers, University of Sussex Library, MS 11/24/1, draft, quoted in Overy, *The Morbid Age*, p. 145.

[147] KMFF, p. 103. [148] KMFF, p. 102.

psychology of the press'.[149] Rivers was not his sole well-informed source of Freudian ideas; he recorded in his diary further revelations about psycho-analysis from conversations with John Rickman, returned from analysis in Vienna and living at Grantchester:

> *August 17th 1922* I have got from him more than I ever had before, an idea of what Freudianism means. Most illuminating of all his observations is his analysis of factors no-one else would take into account concerning nations. For example – France's exaggerated nervous fear of Germany is partly due to the sexual institution in France of 'coitus interruptus', which produces excitement and nervous fear and therefore aggression for the individual. Explainable by the fact that connection of this sort satisfies only the physical craving and not the psychological. As for Russia, she has found a fixation in the oral stage, when the dominant characteristic seems to be the 'identifica-tion' seen in kissing, reaching agreement by the general will, i.e. complete identification of opinion in the total absence of any self . . . Rickman was interesting, too, on the psychology of the individual C.O.s and pacifists – the reasons are all sexual. I'm not sure whether 'love of humanity' is always homosexuality, but certainly usually. Another interesting example is the man who thinks the army is a bad thing because [he is] unconsciously afraid to live only with men . . . or again refuses to identify Germany with the Devil because to recognise the Devil would be to recognise God and he has a desire to suppress 'Father domination'. Astounding stuff! But I'm not inclined to scoff. [Yes, I am!]
>
> This is a curious entry into my diary! But I feel a definite need of a Diary now. It will help me in many ways, I believe – possibly sexually by allowing me to work things off.[150]

As is clear from this final phrase, discussions of psychoanalysis were always liable to intersect with a problem that confronted many of these young intellectuals: the problem of sex. Kingsley Martin would confront that problem with more pain a year later. But first he had to say goodbye to Cambridge and that meant remembering Rivers, who had died suddenly two months previously:

> My last day in Cambridge . . . I'm not at all sure just now whether I'm sorry to be leaving. I'm sorry to be leaving the Fellows' garden here, the Varsity library, this kind of life with its apparent leisure and good society. Also to be leaving certain individuals, though the number I mind about is rather small and Goldie [Lowes Dickinson] is far the most important. But I like seeing Richard [Braithwaite] often, and Frank [Ramsey] is better than anyone I

---

[149] Damon Franke, *Modernist Heresies: British Literary History, 1883–1924*, Columbus: Ohio State University Press, 2008, p. 228.
[150] The phrase, 'Yes, I am!' is written with a different pen, so probably at another time.

know; and yet their minds work so differently, really, that I doubt if I should find enough permanent satisfaction in them. I feel so very much older in many ways. Rivers, after all, was the person who mattered really most (except for Goldie). I think I remember our lunches together, our weekend at Dunsford, our casual good relationship, with as much pleasure as anything up here . . . I like to think of him last, lying out all the afternoon in the Fellows' Garden with me and RBB [Braithwaite], clearly rather tired and overdone, the feel of the grass on his back and the sun in his face.[151]

Martin left England to spend a year on a Procter Fellowship at Princeton while completing a dissertation for an Open Fellowship Competition at King's, but found he was beaten into second place by his close friend, later a Nobel Prize winner in physics, Patrick Blackett, who thus became the first non-Kingsman to be elected a Fellow; Frank Ramsey was to be the second, elected in 1924. Martin returned to Cambridge to a Bye-Fellowship at Magdalene and completed a book from his dissertation, on Palmerston and public opinion at the time of the Crimean War. The style of the book was much influenced by Lytton Strachey's approach to the Victorians. 'The book was not primarily about Palmerston', Martin later wrote,

> but a study of the part played by the press and by the personality of a minister in creating a public opinion in favour of war. It was in some ways a pioneer effort: it used the press not to establish facts about which other sources are more reliable, but to show how an illusory picture of the world scene can be painted on the public mind with the result that quite irrational actions seem morally imperative. It had a Freudian point though nothing of the sort was in my mind. I tried to analyse 'public opinion', explain what it meant and how it worked. I found that it reverted in 1854 to a former occasion and demanded a course of action which had been successful five years earlier though the situation was in fact quite dissimilar. It told a macabre story of how ministers of the Crown and influential public figures who had been bitterly opposed to the war gradually persuaded themselves that it was just and necessary.[152]

As Martin had noted in contemporary unpublished papers, 'one of its central assumptions is that there is no such thing as "public opinion" – there is only public emotion, which is manipulated by the newspapers so skilfully that in the excitement of a great crisis it can be mistaken for the *vox populi*'.[153] A striking, robust and cogent vision of the emotions of the

---

[151] C.H. Rolph, *Kingsley: The Life, Letters and Diaries of Kingsley Martin*, London: Victor Gollancz, 1973, pp. 77–8.
[152] KMFF, p. 132. [153] Rolph, *Kingsley*, p. 90.

masses for the future Editor of a national weekly, formed by the joint influence of Lytton Strachey and, unconsciously, of Freud.

After a year as Bye-Fellow at Magdalene, Martin gave up hope of a post in Cambridge. Moving more and more within the ambit of Maynard Keynes, who had taken over the *Nation* in the spring of 1923, installed Leonard Woolf as Literary Editor[154] and a Cambridge acolyte Dennis Robertson as nominal Editor, Martin got a chance to write for the new journal, now pumped full of Cambridge talent. Meanwhile his personal life grew complicated just as the question of his professional future became more urgent.

He had met Olga Walters in his family circle in 1921 before leaving for Princeton. She had come up to Cambridge to read anthropology at the same time as Martin's younger sister Peggy, and they became close. In March 1924, he attended a lecture in Cambridge given by Harold J. Laski, the great socialist intellectual of the era, where he met a young woman, a research student in the sciences, whom in his autobiography he calls 'Verity'. In the spring and summer, a love affair blossomed. But at the end of the summer, she decided to stay with the young man she was already engaged to. Throughout 1924 and 1925, Martin's passions vacillated between Olga and 'Verity'. At the same time, he moved to London, where once again Keynes engineered him an avenue to the future, this time an appointment at the LSE, where he grew close to Laski.[155] This was in many ways the most difficult and formative period of his life. Living in London, he struggled with the questions of love and sex. His urban and cultural location may not have helped: always describing himself as the outsider looking in, he lived 'on the edge of Bloomsbury'.[156] He sought out an analyst to help him make a decision about his love life. He felt bound to Olga but found himself seeking reasons for not marrying her; this was intertwined with a conviction that somehow his love for his father was also a cover for unconscious hatred and jealousy. His beloved mentor in Cambridge, Lowes Dickinson, a senior Apostle and generous friend, was suspicious of his desire for analysis: 'I wonder why you think you want psycho-analysis?' Lowes Dickinson wrote to him on 10 August 1925. 'I don't know whether it is aged stupidity which makes me so suspicious of its effects, but I am ...'[157] But Martin persisted and his diary follows his analysis and emotional tangles.

[154] See VWD II, 23 March 1923, p. 240.
[155] Martin would write a biographical memoir of Laski three years after his death in 1950.
[156] Rolph, *Kingsley*, p. 115.   [157] *Ibid.*, p. 106.

I think I am really better, but in spite of Olga, who has changed – very sure of herself, not as attractive as when she was full of shyness and admiration (here is the swank devil!). I think I shall do the analysis. I fear I can never get things right with women unless I do. I'm getting – like Ellis – furtive in meeting anyone's gaze. Oh God, I want full life and strength and work – and I *will* have them.[158]

In May 1925, he had written in his Diary:

Analysis proceeding all the time – I 'want to hurt Mother', apparently. It's my childish unconscious idea of love-making, and a jealousy of Father because I think he did this to Mother. I wonder why?[159]

Self-evidently, Martin was imbibing psychoanalytic theories as a direct script for his own personal history. His analysis lasted over a year; by the spring of 1926 he had fixed the end date as the summer of that year. But the urgency of his need to make a decision was at odds with the requirement of analysis to refrain from such action. On 1 June 1926 he wrote in his Diary:

Desire to 'down' analyst and get free.
  Desire to get 'something settled'.
  Desire not to run risk of making any kind of muddle with Olga liable to precipitate acts.
  Absurd, it seems, to go on with analysis, holding up complete relations with Olga, and yet absurd to break off before end of period arranged. Shall I tomorrow tell analyst I'm not coming any more? I have decided to ask Olga to marry and this involves my surrender – not a marriage such as I had imagined in the past. But a complete giving of myself.[160]

Further entries record his fury of indecision and agony over his feelings towards Olga and his mourning over the loss of his ideal love with Verity. But indecision was overcome and he married Olga on 7 July 1926. The marriage was not a success. Olga became more and more withdrawn as Martin became more and more a man of the journalistic and political worlds.

At the same time as he had been going through the agonies in his personal life, Martin was in permanent conflict with William Beveridge, the head of the LSE. To put an end to that conflict, he accepted the position of leader-writer from the renowned Editor of the *Manchester Guardian*, C.P. Scott. In Manchester, the marriage went from bad to worse. Martin's work was proving a disappointment too: his enthusiasm for democratic socialism was not welcomed by Scott's espousal of the cause of the Liberals led by Lloyd George. Let drop by Scott, in 1930 Keynes again came to the rescue.

---

[158] *Ibid.*, pp. 106–7.   [159] *Ibid.*, p. 110.   [160] *Ibid.*, p. 112.

Fig. 4.7 Kingsley Martin, far right, on the ENSA Brains Trust, 1943.

Hearing that Martin was on the job market, he successfully engineered a merger of *The Nation and Athenæum* with its great rival, Arnold Bennett's Fabian *New Statesman*, installing Martin as Editor of the *New Statesman and Nation* – the first issue appeared on 28 February 1931.[161] At last, things worked for Martin: he remained Editor of the *New Statesman* for the next thirty years. Remarkably, Keynes and Martin managed to get on tolerably over the next fifteen years till Keynes's death: Martin was allowed his editorial freedom. In the political thirties and forties, he made the magazine a principal voice of the Labour Left. His impracticality and his later notorious lack of political judgement – over Spain, over Stalin, over the Czech crisis in 1938 – did not seem to matter. Yet by 1934 he was considering a return to his other professional love, academic life. His marriage irretrievably broke down that year and his mother died; in these difficult times, he was offered the Professorship of International Relations at Aberystwyth; in March 1935 he once again recorded his agony of indecision:

[161] Skid II, 388–9.

I have often found decisions difficult, but only once – ten years ago – difficult in this absurd way ... There is no real difficulty except the horror that will come over me whatever job I take. Now if this horror were purely personal, no more than the common pathological symptom of a castration complex, it would not be worth while writing about. It might appear as a footnote in someone's textbook of psychoanalysis. The odd part to me is that I doubt if I am really a coward.[162]

Kingsley Martin somehow knew he was a classic psychoanalytic case, yet it seemed not to help him at all. In a way he had grown up with English psychoanalysis; it was clearly his way of representing himself to himself. In the event, he decided to stay with the *New Statesman*.[163] And about this time, he teamed up – in an open relationship – with the woman with whom he would share the rest of his life, Dorothy Woodman (1902–70), secretary of the Union of Democratic Control and later Asia Correspondent for the *New Statesman*.

In more ways than one, Kingsley Martin's life was lived in the long shadow of his Cambridge years. Dorothy Woodman recognized the connection when in her will she left money to create a trust fund in the University of Cambridge to support an annual Kingsley Martin Memorial Lecture 'on a South Asian topic, preference being given to a topic relating to Burma'[164] – a country she and Kingsley had visited often, leading to the publication of her anti-colonialist book in 1962, *The Making of Burma*.

J.D. Bernal: 'one of the greatest liberating discoveries of mankind'

J.D. Bernal arrived in Cambridge in 1919. The case of Bernal is instructive not only as an instance of an unprepared mind encountering Freudianism at exactly the historical moment when Freud gained such a grip on Western thought, but also because, in that encounter, we become aware why it was not for nothing that all those who met Bernal declared him a genius: for all the speculative grandiosity of his response, Bernal's intellectual excitement over psychoanalysis and the details of the Freudian revolution in philosophy and social thought he hoped to construct still strike one as authentic, impressive and seething with clever ideas – far more 'modern' than the encounters of Tansley and Rivers. What follows is almost

[162] Rolph, *Kingsley*, pp. 206–7.
[163] Skid II, 536 mentions Keynes having a part to play in persuading him to stay.
[164] University of Cambridge, *Statutes and Ordinances*, 2011, p. 854.

exclusively based on unpublished manuscripts from Bernal's archive; it is the hidden, youthful Bernal.

The shape of Bernal's highly successful professional life was not to be complicated. After studying science, he spent 1924 to 1927 working at the Davy-Faraday Laboratory at the Royal Institution, where he developed the techniques, both instrumental and mathematical, that made him one of the foremost X-ray crystallographers in the world. From 1927 to 1938 he returned to Cambridge, where his work and laboratory laid the foundations for the X-ray crystallographic study of larger organic molecules, including proteins, and viruses. After the revolution in molecular biology of the 1950s, many of the participants regarded Bernal as the principal inspirer of their discoveries, although he himself as well as many of his scientific contemporaries, realized he had somehow 'missed out' on making either of the major advances that were his due – notably the double helical solution for the structure of DNA and the elucidation of the structure of proteins. Taking the Chair in Physics at Birkbeck College, London vacated by Patrick Blackett in 1938, he remained there to the end of his professional life in 1968. From 1939 to 1945 he made many different scientific contributions to the war effort, in part as a favourite of the maverick Lord Mountbatten; after the war he was seen as a scientific ambassador on behalf of the Communist Party, spending much time in Eastern Europe and other 'socialist' countries. He built upon his already impressive reputation as one of the foremost British scientific intellectuals of his time, through his landmark books *The Social Function of Science* (1937) and the four volume *Science in History* (1954). Following a major stroke in 1963, his health declined considerably; he died, virtually incapacitated, in 1971.

Bernal was famous as a scientist, as a socialist intellectual and as a remarkable man: 'All who knew him were struck by the combination of intellectual brilliance, curiosity, bohemian culture and an erudition that leaped across the borders between science, history and the arts. Most of those who knew him thought he was a genius'[165] – such was his long-time colleague and fellow-Communist Eric Hobsbawm's summing up. Yet his unpublished papers reveal – and to an extent not documentable in any other case, not least because of the abundance of his archive – how caught up in the Freudianism of post-war Cambridge in the early 1920s this Communist scientific intellectual was. Half-jokingly planning his own biography in conversation with his long-term colleague and assistant Francis Aprahamian, Bernal pictured three

[165] Eric Hobsbawm, 'Preface', in Brenda Swann and Francis Aprahamian (eds.), *J.D. Bernal: A Life in Science and Politics*, London: Verso, 1999, pp. ix–xx at ix.

Fig. 4.8 John Desmond Bernal, by Ramsey & Muspratt, 1932.

strands: red for his politics, blue for his science and purple for his sexual life. Appropriately enough, his Freudianism, more hidden than his famously flamboyant sexual life, entwines all three.

Born 10 May 1901, John Desmond Bernal came up to Emmanuel College to study mathematics in October 1919 straight from Bedford School. From an Irish farming family on his father's side, he was descended from a Sephardic Jewish grandfather who had arrived in Limerick in the 1840s and converted to Catholicism. His American mother, with a cosmopolitan education in the United States and Europe, had met his father when he rescued her from what he took to be drowning, she swimming, on a beach in Belgium. From a Presbyterian background, she converted to Catholicism in order to marry. Her first-born and favourite, still resolutely Catholic when he arrived in Cambridge, Desmond, Des or, later, 'Sage', was already enthused by crystallography, the scientific field in which he would become the doyen. However, his academic career at Cambridge was not the success his later reputation as an undisputed genius – 'a man uniquely gifted ... in "shooting an arrow of original thought" into any target presented to him' – would lead one to expect.[166] He failed to get a

---

[166] *Ibid.*, p. ix.

First in the Mathematical Tripos of his first year, switched to the Natural Sciences Tripos, studying chemistry, geology and mineralogy, in his second year when he did achieve a First, but in his Part II, in Physics, taken in June 1922, again failed to get a First. This bumpy ride in the Cambridge system is not all that surprising given the way Bernal had thrown himself into social and intellectual undergraduate life as well as original and formidably exhausting work in the mathematic foundations of crystallography. He recalled: 'I never fitted in with my industrious co-scientists, but consorted with socialistically inclined economists or historians. Night after night, sitting over a fire and drinking more and more diluted coffee or strolling around moonlit courts, we talked politics, religion and sex.'[167] With his close friends Henry Dickinson and Allen Hutt, he was attending the Heretics and the Cambridge University Socialist Society as well as meetings of the many smaller societies with which the University and the Colleges teemed.

The general atmosphere of Freudianism seeped into Bernal from 1920 on; its influence, as with so many others, went hand in hand with his personal grappling with the problems of sex. In January 1920, he recorded a conversation with his friend Lucas, who 'told me the advice of a Harley St nerve specialist a freudian who puts everything down to sex in a rather material and cold blooded way. He also talked of personal affairs.'[168] The next day he read – or rather glanced at – McDougall's *Body and Mind: A History and a Defense of Animism* and W.J. Crawford's *The Reality of Psychic Phenomena, Raps, Levitations*, but discarded these to read with enjoyment that most Freudian of First World War novels, *The Return of the Soldier* by Rebecca West. Always a voracious and unsystematic reader, the rest of that year was taken up with Shaw, Wilde, Compton Mackenzie, H.G. Wells, Anatole France, alongside Keynes's famous book on the war ('moved') and Marie Stopes on *Married Love* ('mixed' was his verdict); the most psychological of these books read in 1920 was Havelock Ellis's *The World of Dreams*.

In February 1920, his friend Lucas 'psychoanalysed' him 'and suggested that "love was the cause of his despair"'.[169] A few weeks later he records: 'D stayed late in my bedroom discussing various aspects of personal ethics and egoism. We both feel better after dissecting our inner selves and

[167] Ann Synge, 'Early years and influences', in Swann and Aprahamian (eds.), *J.D. Bernal*, pp. 1–16 at 12.
[168] BPCUL, O.1.1 Diaries and Notebooks, Notebook 2 for 1920, entry Friday 16 January 1920.
[169] Fred Steward, 'Political formation', in Swann and Aprahamian (eds.), *J.D. Bernal*, pp. 37–77, quoting Diaries, 19 February 1920.

exposing the most diseased organs to each other's critical examination.'[170] At this point, Bernal was still a practising Catholic, so the rituals of confession certainly would have had a mixed pedigree for him. By the start of his second year, he was getting into his stride of making public declarative speeches and advancing extreme views in a convincing manner: 'I find myself more of a Freudian than any of the others, though I never read a word he wrote.'[171] A few days later, he wondered if he had an original mind, or if he were 'merely an undigested mass of Einstein and Freud with a top dressing of Wilde and Shaw'[172] – an appropriate mixture of Jewish scientist and Irish wordsmith. But he was not all Talmudic blarney: he recorded reading Bousfield's *The Elements of Practical Psycho-Analysis* on 29 September 1920.[173]

Back in Ireland for Christmas, psychoanalysis began to challenge politics as dinner-time conversation, although it was something like a more daring variety of parlour game at this stage: 'Freud had quite a vogue this evening. Psychoanalysis proves even more fascinating than Consequences but Mammy and Cuddie persist in calling it dangerous and childish in the same breath.'[174] In early 1921 began his first brief love affair, with his second cousin, Virginia Crawford, which came to an end in March. A few years later, he recorded – in a manuscript entitled 'Feminine influences' – the early stages of his sentimental education:

1. Adolescent. The romantic tradition. Clean fun.
2. Hopeless passion. Discussions on sex, the Modern attitude.

From then on, through the Socialist Society, he began to meet young women, serious about politics, free thinking in personal relations; three of them were to be constant companions in the next few months: 'Grey' (Dora Grey), Sylvia Barnes ('Barnes') and the politically active and experienced Eileen Sprague – the last of whom was not a student but worked at the secretarial agency Miss Pate's, where she had earlier typed the manuscript of Keynes's *Economic Consequences of the Peace* and may well have been typing up Wittgenstein's *Tractatus* at exactly this time – Richard Braithwaite remembered Frank Ramsey 'sailing into Miss Pate's typing office in Cambridge and dictating the translation with Ogden at his side'.[175] Another future life-long friend and colleague came into Bernal's life at this time: Maurice Dobb, undergraduate in economics at Pembroke,

[170] *Ibid.*, p. 46.
[171] BPCUL, O.1.1 Diaries and Notebooks, Notebook 2 for 1920, 27 October 1920.
[172] ABB, p. 35.   [173] Steward, 'Political formation', p. 46.   [174] *Ibid.*, citing 8 January 1921.
[175] McGuinness, *Wittgenstein: A Life*, p. 298.

later host and friend to Wittgenstein on his return in 1929, collaborator with Piero Sraffa, perpetual Communist and lecturer (later Reader) in Economics at Cambridge from 1924 to his retirement in 1967.

The talks Bernal gave to student societies indicate his growing preoccupation with combining the points of view of economics and psychology within a scientific theory of society; speaking to the Political Science Society he gives an account of humanity as 'a system of particles' with 'external forces (resources) acting on all the particles and internal forces (the new psychology) acting between every pair of particles'.[176] In May 1921, he committed himself to communism; a month later, with his exams over – always the most intense time of year for the undergraduate – on 3 June 1921, he read Freud for the first time: *The Interpretation of Dreams*,[177] which affected him profoundly. In his diary he recorded: 'I expound the new religion.'[178] And then came the following entry:

> Friday 10th June 1921
> I find a note inviting me to hear a psychoanalist [*sic*] on love in Dobb's rooms. Eat a hasty lunch abandon K[179] and go. Gr and B [Grey and Barnes] are there also Sprague, & religious people. We meet the stupendous Jonty Hannagan. He talks psychology rhapsodies & metaphysics and is immensely inspiring though he instils in me the spirit of contradiction. Hutt and I talk a lot. We go out inspired full of life and love. Walk to Barnes show and talk it over freely her & Grey.
> This at last is the new religion and I am in love with Grey.

In 1926, when writing 'Microcosm', his 'Autobiography' and intellectual testament of youth, Bernal gave the following summary of his personal development, in the form of a chronological list:

Father and Mother
The country
Ireland
Public School, Religion and The Phantasy, and The War
Cambridge
Jonty
Marriage
People

---

[176] Steward, 'Political formation', p. 47 but note gives this as 'The dynamics of human society', 1920, which can't be quite right for the date.

[177] BPCUL, O.1.1 Diaries and Notebooks, Notebook 1921, 'Books read', 3 June 1921.

[178] *Ibid.*, Diaries, 5 June 1921.

[179] Bernal's younger brother Kevin, to whom he had always been close, was staying with him for much of this period.

Clearly, then, this meeting with Jonty was to be of great importance for his development. Who was 'the stupendous Jonty Hanaghan'?

Jonathan Hanaghan (1887–1967), always known as Jonty, was born across the Mersey from Liverpool in Birkenhead, to an Irish father and a Scottish mother. Very little is known of his youth and early adulthood, but he became an enthusiastic Christian in his youth, a passion he never abandoned.[180] He came upon Freud by reading Wilfred Lay's *Man's Unconscious Conflict: A Popular Exposition of Psychoanalysis* (1917) sometime in 1918–19;[181] by 1920, when he wrote to Ernest Jones to enquire if he could become a non-medical member of the London Psycho-Analytical Society, he lists the books he has read (Freud, Jung, Pfister, Brill), gives a sketch of his interests and declares: 'I have tested the theory of Freud with some astonishing results in practise . . . Before coming across Freud I was groping after what he seemed to illumine with the light of genius.'[182] In 1922, he enquired again of Jones concerning the possibility of becoming a member of the International Psycho-Analytical Association.[183] In the summer of 1923, on the recommendation of Ernest Jones he began an analysis with Douglas Bryan (MD 1900, London) who had been honorary secretary of the Psycho-Medical Society, London, charter member and first vice president of the London Psycho-Analytic Society and a founding member and Honorary Secretary of the BPaS from its foundation in 1919 till 1930 and the Honorary Treasurer from 1930 to 1937; in 1922, Ernest Jones reminded Freud that Bryan and Flügel 'are certainly the two leading analysts in England'.[184] And the word-of-mouth history of Irish psychoanalysis attributes a meeting Jonty had with Ernest Jones when the Welshman issued an uncharacteristically hortatory instruction: 'It will take a Gael to start psychoanalysis in Ireland, so off you go!'[185] Hanaghan travelled to Ireland in 1917, but definitively moved to Dublin only in 1926, when he worked as a stage-hand and was married for the second time, to Mary Webb, whose

---

[180] Some sources indicate he moved to Dublin in 1917, but his heavily annotated copy of the 1921 reprint of *The Interpretation of Dreams* has an *ex libris* sticker: 'Jonty Hanaghan, "Kirkbean", Thingwall, Birkenhead'. The book is now in the possession of Tony Hughes, Dublin (email received from Hughes, 18 April 2006).

[181] JPBPaS, CHA/Fo3/02, Hanaghan to Jones, 2 May 1922.

[182] JPBPaS, CHA/Fo5/01, Hanaghan to Jones, 6 June 1920.

[183] JPBPaS, CHA/Fo3/02, Hanaghan to Jones, 2 May 1922.

[184] Jones to Hanaghan, 4 July 1923, text transmitted by David Hanaghan, 14 July 2006. On Bryan, see an interesting letter from Freud to Bryan, with brief editorial notes on Bryan in E. Zaretsky, 'An unpublished letter from Freud on female sexuality', *Gender and Psychoanalysis* 4 (1999), 99–104. JF, 10 April 1922, pp. 470–3 at 470.

[185] See Ross Skelton, 'Jonathan Hanaghan. The founder of psychoanalysis in Ireland', *The Crane Bag* 7(2) (1983), 183–90.

parents as a wedding present gave the couple a house at 2 Belgrave Terrace, which remained Hanaghan's home and later his consulting room until his death. The marriage produced two children.

Mary died in the early 1940s and very soon after Hanaghan, together with four colleagues, formed the Irish Psycho-Analytic Association (IPAA). He also founded the Runa Press which published his own writings, amongst them an edited volume, *Tidings* (1943), and *Society, Evolution and Revelation* (1957), *The Sayings of Jonathan Hanaghan* (1960) and *Freud and Jesus* (1966). His followers published a number of posthumous works, including *The Courage to Be Married* (1973) and *Forging Passion with Power, or The Beast Factor* (1979). Hanaghan remarried and this third marriage produced three children. In the 1940s, Hanaghan was in touch with Anna Freud, with whom he had a good understanding, but she was unable to persuade Ernest Jones to accept the IPAA's request to affiliate to the IPA.[186] At the Zurich Congress in 1949, Jones as President reported: 'An Irish Psycho-Analytical Society was founded in 1945. It appears to be doing some good work, but since none of its members have been analysed, let alone trained, we did not feel justified in recognizing them.'[187] In Jones's eyes, it seems, Hanaghan's analysis with Bryan did not count.

As leader of what became known as the 'Monkstown Group', Hanaghan developed an unorthodox, charismatic, radical Christian approach to psychoanalysis. The Society grew slowly; the heyday of Hanaghan's influence was the 1950s and early 1960s, and his charismatic imprint was left on every one of the frequent meetings. Freud was much admired and much read, but the overall tenor of the group was against the dry institutionalization characteristic, the group felt, of psychoanalytic societies. After Hanaghan's death, younger members felt the devotional adherence to his teachings prevalent in the Monkstown Group hampered their own development. The power and logic of the charismatic leader so endemic in psychoanalysis was obviously incarnate in the life and death of Jonty Hanaghan.

This was the man, then aged thirty-four, who came to spread the psychoanalytic word in Cambridge in 1921. Beginning with the meeting on 10 June 1921, his effect on Bernal – from the first day inducing intellectual intoxication, religious enthusiasm and love for all, most particularly for the young women in his circle – was overwhelming. At a meeting the next day to which he had taken Mrs Dobb, Jonty

[186] JPBPaS, CFF/F02/05, Anna Freud to Jones, 26 August 1945.
[187] Anna Freud, 'Report on the Sixteenth International Psycho-Analytical Congress', *BIPA* 30 (1949), 178–208 at 185.

deals more thoroughly with psychoanalisis [*sic*] and the great complexes. I argue more strongly. D makes admissions and his mother gets shocks. I become violently in love with Sprague and Grey. But I have to accompany Mrs D home. D loses an opportunity for a confession.

The next day, Sunday, is spent in conversation with 'Mrs D. mainly on manners and psychoanalysis finding out some interesting things'. On the Monday, 'Jonty gives D & I an interview and hears our confessions. He advised treatment for D very strongly, says so to his people but to me he has little to advise.'[188] By 17 June, Bernal has left Cambridge for London, and is still recording incessant talks about psychoanalysis as he prepares to depart with a friend for his first holiday abroad to Ostend, Brussels and then Paris. The only diary entry of intellectual substance, following a visit to the Salon de Printemps in Brussels, shows him assimilating his reading of Freud:

> Looking at a picture of a red staircase I began to realize the illuminating application of the *Traumdeutung* to pictures. A real picture (not made to order) must like a dream be made out of material taken from nature direct or through the memory but this material is chosen & formed unconsciously and in symbolic shape by the wish and the complete picture results. Now in looking at a picture we unconsciously analyse it in a similar manner again effecting a choice of material but not simple, as we get them from nature but in already formed bundles and if there is any aesthetic pleasure it is the sublimation or displacement of our pre-existing wishes as expressed on the canvas. A glorious theory, not original of course it is too simple a deduction from F for that. I would love to follow the threads.[189]

Back in London in July, between trips to the British Museum, he catches up with friends: 'I have tea miss my train and meet D. He tells me of psycho-analysis under Ernest Jones & other things.'[190] Returning to Cambridge for the Long Vacation term – lectures on colloid chemistry – he is given his friend Dickinson's diary to read, where the thread of Jonty and his effect on the intense group of friends is spelled out:

> Quite casually he gave me his to read I began with the Johnty [*sic*] affair . . . Gray and Hutt are modern disillusioned and thanks to Jonty they can be perfectly candid with each other, so that there are no promises and no expectations of the future. It will save them a lot of worry.[191]

A talk at the Science Club a few days later 'is very sound on psycho-analysis. but the most thrilly part was an exhibition of automatic writing of

---

[188] All quotes from Bernal's diary for 1921 in BPCUL, O.1.1 Diaries and Notebooks.
[189] BPCUL, O.1.1 Diaries and Notebooks, Notebook 2, 1921.
[190] *Ibid.*, 9 July 1921. D. is probably Maurice Dobb. [191] *Ibid.*, 15 July 1921.

hypnosis and autohypnosis. I left it in a very psychic state.'[192] Alongside the psychic enthusiasms, he has come to grips with Langmuir's theory of atoms, which he wishes to develop so as to determine the properties of simple inorganic compounds: 'It bears out my neo-Pythagorean philosophy that the universe is completely determined by the theory of numbers and the laws of chance. I won't rest till the whole is explained.'[193] At night he talks with Grey about her affair with Hutt, aligning it with Jonty's theories – and they kiss: 'A record day a new theory of the universe and a new love though really they are both second hand.'[194] But this new love affair is broken off: Grey senses she is in love with Bernal only out of pity. Serious talks with Sprague about religion prompted him to take a definite step: he told his Catholic mentor in College that he was leaving the Church. And he stuck by this decision for the rest of his life. His work on the theory of the atom develops apace. Tea next to 'Ogden and other intellectuals', walks with the other members of the complicated network of budding love affairs, all interlaced with serious psychoanalytic talk and analysing each others' dreams, 'singing while punting in the dark'. This was the formative summer for Bernal, culminating in the loss of one love affair ('Grey at Byrons Pool. Dream analyses. Farewell to love a last kiss bitterness.'[195]), the loss of his religion and a final walk through Cambridge:

> [Sprague has] really good old Christian morality broken into whenever it seemed impossible to conform. Most of this morality is of course unconscious. She is a normal person without scruples. I think I managed to persuade her of the truth of part of this ... Coming up Trinity St I explained the psychological standpoint 'I cannot tell you why you should do anything. I can observe you doing it and tell you why you do it ...' We shook hands warmly. To be continued next term.[196]

The summer ends with Grey and Sprague talking in Cambridge, while Bernal is in Ireland in love with both.

Jonty Hanaghan returned to Cambridge in the first term of Bernal's third year and conducted 'Seminars' for the group of friends, who were now calling themselves the Neurotics, no doubt in self-conscious parody of the Heretics.[197] These were 'people after my own heart, poets, theologians, mystics, fanatics and rationalists. I can talk (with them) to my heart's content, the most abstruse nonsense, and live up to my universally

---

[192] *Ibid.*, 20 July 1921.   [193] *Ibid.*, 21 July 1921.   [194] *Ibid.*, 21 July 1921.
[195] *Ibid.*, 15 August 1921.   [196] *Ibid.*, 18 August 1921.
[197] Unfortunately, the principal diary for late 1921 is missing (confirmed by Andrew Brown, email dated 28 January 2006, although it was read by Steward, 'Political formation'); the information gleaned here is from a remarkable diary of diaries which Bernal collated some years later.

acknowledged nickname, the sage.'[198] Meeting virtually every day, Jonty conducted seminars on fantasy, on Oedipus, on normal sexuality, while the liaisons and alliances in the group shifted from day to day: 'triangles have given place to the most complexly joined polygons and jealousy is by no means absent'.[199] Jonty's course came to an end on 1 November 1921: 'The "Neurotics" last lunch at the tea shop. Supper with J[onty] at Eileens.'[200] Many years later, Allen Hutt described Hanaghan, the instigator of these meetings, as:

> a sort of maverick missionary with a considerable load of what, in those early days of Freudian enthusiasm, passed for Freudianism . . . A whole lot of us, including Des, myself, Dick and Eileen gathered for what one could almost call, I suppose, a species of prayer meeting where this strange, rather magnetic man, Hanaghan, addressed us on the basic problems of what, as far as I can recall, were the relationships between the sexes, which we found very encouraging and mostly set us, as it seems to me now, on the right path at an early stage.[201]

The right path for Bernal lay with Eileen Sprague. At Christmas, he told his family he was intending to marry her. His mother gave him somewhat surprising advice: 'all these advanced girls believe in free love. Why can't you do what you talk about, live together without marrying and make a success of it.'[202] But Bernal and Eileen had made up their minds to get married. Strangely resigned to his fate, he was determined yet perplexed by his own determination:

> I act not from that conscious inward drive that I thought would carry me through life but for reasons I cannot glimpse at and am almost beginning to conceive them as reasons outside and beyond me. She who is woven into my thoughts will be woven into my life that is all. Why make a fuss about it.[203]

Returning to Cambridge, Eileen and Bernal met up again: '"You must perform the final rites" she had said.'[204] And married they were, before Bernal's results for his third-year exams – a First, as it turned out – were

[198] Synge, 'Early years and influences', p. 13.   [199] Bernal Diary, recorded in *ibid.*, p. 13.
[200] BPCUL, O.1.1 Diaries and Notebooks, Diary of diaries, 1 November 1921.
[201] Synge, 'Early years and influences', p. 13, quoting reminiscences of Allen Hutt (1901–73), who became a working journalist all his life and an active Communist, also interested in the history and design of newspapers, the history of journalism, trade unionism, and the French typographer Pierre Fournier.
[202] ABB, p. 43.
[203] BPCUL, O.1.1 Diaries and Notebooks, Diary 1922 (optics and geological sciences) then later 1924, 16 January 1922.
[204] *Ibid.*, 29 January 1922.

published, in June 1922. Eileen went flat-hunting, with an obvious port of call being C.K. Ogden, who not only ran Cambridge intellectual life, but rented out property to those in his circle.[205] In the next year, when Bernal was taking Part II of the Natural Sciences Tripos in Physics, their set naturally broadened; they became friends with the Sargant Florences[206], Bernal met Graham Wallas and Bertrand Russell and spent increasing amounts of time in London with the metropolitan Left – at the 1917 Club or lunching with his friends Henry Dickinson and Maurice Dobb.[207] The work he had first begun in the summer of 1921 on the complete account of the space lattice groups, drafts of which Eileen had typed up as long ago as November 1921, he submitted successfully for a College Prize in May 1923. But his results in his final Tripos exams in June 1923 were 'very gloomy'.[208] Bernal seemingly had no chance to continue work as a Cambridge scientist; but he did have support from his College mentor, Alexander Wood, who pleaded his case to William Bragg, just then leaving Cambridge for the Royal Institution in London. Bragg agreed to take him on. With a short-term grant from his College, help from his family and Eileen's wages, he started research in the autumn of 1923 in London.

Bernal's participation in the bohemian, socialist London world over the next four years established for good the three major threads of his life: science, politics and sex. The entries for Bernal's list of 'Feminine influences' relating to this period run as follows:

3. First kiss, engagement. Initial difficulties. Susceptibility.
4. learning. Problem renunciation.
5. Eileen. Intellectual attachment. The sexual act. M. Stopes.
6. Monogamy. Temptations.
7. First adultry [sic] – division of love.
8. The virgin, Sylvia. The occasion for jealousy.
9. Ivy, aupres de ma blonde. Lasting of passion. Married man and family.

The continual flow of women through Bernal's life would be unceasing from this point on.

Bernal had become a Marxist in the spring of 1921, a few weeks before he first read Freud. Amongst his unpublished papers from 1922–23 there is a

[205] OP, Box 3, Bernal to Ogden, 1922.
[206] Steward, 'Political formation', p. 52; see BPCUL, O.1.1 Diaries and Notebooks, Diaries, 29 March 1923.
[207] Ibid., Diaries, 7 April 1923.
[208] Ibid., Diary 1923, which E and Sage shared; most of it is in E's hand, including this entry.

document entitled 'Psychology and Communism',[209] based on his reading of Freud, Pfister, Jones, Flügel, Rivers (*Psychology and Politics*), Kollontai and Kolnai (*Psychoanalysis and Sociology*)[210]. The argument is put straight-forwardly at the outset:

> Ever since Marx, the chief scientific basis of communism has lain in economics. Economics has seemed the only basis on which communist theory could rest; but I think it is safe to predict that in future the place of economics will be taken by psychology . . . the whole theory of economics rests on a basis of psychological assumptions. The first economists were unfortunate in conceiving the psychological unity as the rational economic man, and except for the fact that the economics they produced appealed so strongly to bourgeois mentality, it would have been discarded years ago . . . To deal with economics without psychology is like using a motor car without any idea of mechanics.[211]

So what is the 'new psychology' on which Marxism will have to be based? The possibilities mapped out by Freud and his followers 'happen to be the only way of reaching an understanding that really goes beyond the appearances, and moreover the fact that they are almost universally condemned in respectable circles as subversive of morality is one that should make their study peculiarly attractive to communists'. While expressing the reservation that Freud's psychology is that of the bourgeoisie, he proposed that the best plan of attack was to 'divide humanity into types according to their dominant complexes'.

The main types of complex Bernal surveys are those associated with the anal-erotic and the Oedipus complex in two forms, one which generates abstract ideals and another which generates hatred of the father.

> In the large majority of cases, however, the hate aspect of the Oedipus complex is so completely repressed that it never appears in a direct form. Its place is taken by deference and submissiveness to the father and to all the father substitutes. For our purpose the two most important of these are the employer and the government.

---

[209] BPCUL, B.4.10 'Psychology and Communism'. Steward dates this to 1923. In the margin of the manuscript there are comments, sometimes signed 'HDD', i.e. H. Douglas Dickinson, along with Hutt, Bernal's closest friend at Cambridge. Some of the pages used have typed drafts of the space lattice paper on their reverse, so it probably dates from 1922 or early 1923. We regard it as safe to date the manuscript to his time in Cambridge before he left for London in late 1923.

[210] Aurel Kolnai, *Psychoanalysis and Sociology*, trans. Eden and Cedar Paul, London: George Allen & Unwin, 1921.

[211] Part of this passage is also quoted in ABB, pp. 76–7.

It is the capacity of the Oedipus complex to generate submissiveness that Bernal underlines; but 'on the other hand most of the psychological motive power of revolution is contained in the hate aspect of the Oedipus complex' – a thesis he backs up by evoking son myths and hero gods. If the source of revolution is to be found in the Oedipus complex, so also is the underlying dynamic of war, which promotes regression and the satisfaction of normally inhibited sadistic-cruel impulses.

> In an individual regression never takes place unless there is something that prevents his living a full life. Viewed in this way the last war is but another symptom of the insufficiency of the social system to satisfy the psychological needs of any of its classes. The bourgeois classes are prevented by their moral and economic codes from living a satisfactory sexual life; the proletarians on the other hand are more hampered in their ego development by their dependent condition. No attempts by reasonable methods, Peace Conferences, Leagues of Nations or the like, can possibly prevent war; they touch neither its economic nor its psychological sources. The end of war will come only with the solution of the social problem. In capitalist conscript wars there is yet another factor to consider. The great bulk of the combatants only enter the war in virtue of their dependent and submissive attitude towards their paternal governments. It is to the repressed hatred thus generated that Freud attributes the defeat of the German armies.[212]

To give more substance to this account, Bernal turns to a development of his theory of the anal-erotic, the 'root complex of capitalism', so dominant in the bourgeois classes. His focusing in on the Freudian theory of anality predated this essay; in 1922, in a paper entitled 'The Great Society', endeavouring to give an account of 'the hell of modern civilisation', he had written: 'It is money trite but true. Money is the magic power the mana of modern times ... Freud will give us enough to start on. Money = excrement.'[213] In 'Psychology and Communism' the anal-erotic complex's characteristic of retentiveness is manifest in collecting, then in the development of money, while its creative side is seen in the development of the plastic arts, with the unconscious identification of faeces and matter, of 'flatus with music and poetry and oratory'.

> There is little hope of eliminating the anal-erotic complex; but one of the aims of communist psychology must be to change the displacement on to something else, and this in itself involves the destruction of the capitalist system. Such a change is no new thing if we are to believe Herodotus. Cyrus,

---

[212] BPCUL, B.4.10 'Psychology and Communism', p. 9.
[213] BPCUL, O.1.1 Diaries and Notebooks, Diary 1922 (optics and geological sciences) then later 1924. Steward dates this passage to 1922.

he tells us, turned the gold-loving, warlike Lydians after conquering them into flute-players and artists, and they gave him no more trouble.

Bernal went on to identify a further type of complex, which manifests itself in three important varieties: the exhibitionist, with his desire of exposure mutating into the desire for admiration, producing actors, flashily dressed people 'and politicians of the L[loyd].G[eorge]. type', offering 'excellent subjects for identifications with ego-ideals by large numbers of proletarians'.

> Secondly we have the inspection aspect which manifests itself mainly in forms of curiosity and to which we owe a third enormously important type, the scientist. In the relation of these with communism, the same considerations apply as apply to the artist, and the same problems face the communists after the revolution; but as scientists are often retentive anal-erotics as well, their allegiance is even harder to gain.

Bernal then turned to more general considerations, heavily derived from Freud's *Group Psychology and the Analysis of the Ego*, which he calls 'probably the most important recent contribution to social theory'. While giving a faithful and detailed account of Freud's argument, he concludes with its adaptation to his own purpose: to find the appropriate psychology for Communism and in particular for the Communist Party, whose aim, as 'itself a free group, is to promote in the bulk of the proletariat this identification of themselves with each other and with a suppressed class struggling to be free. This is Class Consciousness.' However, Freudianism necessitates an inflection in the Hegelian-Marxist account of consciousness:

> The word class consciousness is however rather unfortunate; the binding strength of a group is emotional, not consciously rational. It is sufficient for the small nucleus of Communists to be class conscious in the full sense of the words, in order to guide the great proletarian group along the paths of constructive revolution; for without their guidance its emotions would lead it along regressive paths of anarchical or purely destructive nature.

Such a clear break with classical Marxism (in which identification of one's class interests would inevitably lead to identification with the class to which one belongs, to becoming 'conscious' – and vice versa) owes as much to the post-war conviction of the irrational forces that drive groups, nations, classes as it does to Bernal's growing acquaintance with psychoanalytic literature. But Bernal's focus on the importance of the unconscious – the most important concept, according to him, introduced by Freud to social theory – is then matched by his fidelity to the other great theme of psychoanalysis: sexuality. In considering the question of how to make work into a pleasurable activity, he detours in the style characteristic of thinkers of his time, including the psychoanalysts, via an imaginary ethnography: savages identify work and sexual activity, a residue of which one finds in games. Can such an identification be achieved with machines – or are they inherently antithetical to sexuality?

Bernal's conclusion underscores the primacy of the Freudian account of sociality, through a Freudian reading of the Leninist Party: 'The function of the Communist group in society is essentially the same as that of the psycho-analyst to the individual.' Modern civilization is in crisis because 'humanity had never learned to face itself'. The solutions to date are simply substitute symptoms, such as 'taboo, religion, nationality or the capitalist system', which are all breaking down: the repressed is returning. Bernal's climactic peroration evokes as its final terms of reference the struggle of therapy rendered global, not the struggle of classes:

> Is this then the fated end? Must man who has conquered nature be destroyed by that part of nature that lies in him? No! It remains for the communists to prevent this ... But just as in psychotherapy an analyst cannot cure a neurosis in an other, if it lies latent in himself, but must first himself be analysed, so the communist cannot help to cure society while he is ill himself. The communist must first face the facts of his inner life and it is here that the new psychology comes to his aid. He can learn the irrationality of his motives and the sources of his strength and can build upon the secure foundations of external reality.

The impression this youthful document gives – written when Bernal was twenty-two – is that, while he is a committed Marxist, his real intellectual enthusiasm lies with Freudianism and the object of criticism is Marxism. At every turn, the novel intellectual resources come from Freud, not from Marx. The initial intuition of a failure at the heart of Marxism, a failure shared by classical political economy, is of a psychological failure, and the development of a new account is drawn from the new psychology, that of Freud. Such an intuition of the psychological core of the present crisis – a crisis of civilization, of the economic arrangements of the imperialist and capitalist orders – was not Bernal's alone; the development of Keynes's diagnosis of its disorders also placed a psychological transformation at its centre. As an early reviewer, a close colleague of Keynes, described it, the ground of economic discussion shifted in his hands to 'issues of psychology, and not of figures', to 'what people feel and think'.[214] (Whether Bernal had much contact with Keynes during his undergraduate years in

---

[214] Dennis Robertson's 1920 review of *The Economic Consequences of the Peace*, quoted in Skid II, 29.

Cambridge is unclear: in his first term, he may have been amongst the hundreds who heard the famous lectures on 'The economic aspects of the peace treaty' in Michaelmas Term 1919, or in his second year, those on 'The present disorders of the world's monetary system' in Michaelmas 1920;[215] his close friend Maurice Dobb quickly became a member of Keynes's economics society (entry, Apostles-like, by invitation only) – the Political Economy Club, or Keynes Club[216] – and one imagines their conversations were full of Marxist economics and the latest in Cambridge economics, as well as of the explicit Freudianism of which we do have a record.) The young Bernal's political heart lay, clearly, with Freud.

During his time in London working at the Royal Institution, Bernal decided to write his autobiography. The resulting massive manuscript, entitled 'Microcosm', is a record of his personal development to date and of his vision of knowledge, the sciences and civilization in general. It opens with two chapters on psychology.

> In putting psychology first of the weapons of criticism I know that it is the most difficult and least known of the sciences, but in the absence of any psychology feeling would appear to turn directly into reason ... I must make clear at the beginning that I understand psychology neither in the sense of classical psychology (which I do not neglect but treat as part of the vocabulary of logic) nor in the sense of Experimental Psychology which lies at the juncture of Biology and Physiology, but of the autonomous clinical psychology of Freud. Whether it has the exclusive right to the name of Psychology does not concern me, but for the purpose of my method only that psychology is of any use which can be applied directly to processes of thought in myself and other people or to problems which implicitly contain such processes. To me only Freudian psychology can do this. Experimental Psychology and the Behaviouristic theory founded on it does not profess to attempt it while the Classical a priori introspective psychology seems a barren system of definitions and classifications ... The old psychology had the same relation to the infinitely various world of the mind as the Mediaeval bestiaries to natural history.

Here, again, Bernal asserts that Freudian psychology must underpin the great deterministic system of Marxism, 'much as the laws of atom mechanics supplement those of thermodynamics in physics ... Here again the Freudian approach towards such a group psychology seems the

---

[215] Skid II, 3, 44–5, where Skidelsky quotes Keynes on the attractiveness of the gold standard: 'the convention behind it is that it is ... disgraceful to tamper with gold'.
[216] Skid II, 5.

only possible one.' Bernal expresses his conception of the interrelation of the major disciplines of knowledge in a simple diagram:[217]

Psychology	→	Mathematics
↑		↓
Biology	←	Physics

Just as there is a surprising link at the heart of the radical economics of the 1920s between psychology and economics, so we see a comparable link between psychology and mathematics, also found in Ramsey's reflections and in Penrose's research projects. Bernal spells out why psychology – the new psychology, that of Freud – is the foundational discipline of the era:

> It might be safe to guess that the thought of the present age is more characterised by the importance of psychology than by any other movement. We have first the need for psychology in dealing with the secondary effects of the industrial revolution . . . the main advance [came] from clinical psychology. That this would have happened without the comprehensive genius of Freud and his unflinching courage is doubtful. For it must be remembered that Freudian psychology is an independent science arising directly from clinical observation, owing a little to medicine and biology by way of analogy and very little to philosophy or orthodox psychology. Very little of what Freud made use of was not already available to Aristotle. However it was Freud who took the essential step of demonstrating the mechanism of the unconscious mind through the technique of Psychoanalysis, of maintaining, in the face of universal execration, its essentially sexual and amoral nature and by pointing to the main directions in which these discoveries affected human conduct and knowledge.
>
> The laying bare of the unconscious basis for human desire and action will rank with that of the cosmologies of Copernicus and Newton and of the process of Evolution as one of the greatest liberating discoveries of mankind, successively taking from his illusory self importance and bringing him in return into closer contact with realities. It shows its greatness as a scientific theory by the number of separate fields it brings together – Philosophy, Psychology, Metaphysics, Ethics, Anthropology, Sociology, Politics, Economics, Medicine, Religion, Art, Education – all those in fact which have as their essence the operation of the human mind.
>
> The Freudian doctrine has not that simplicity which characterised the fundamental statement of Evolution or Newtonianism. To understand it

---

[217] Roger Penrose, the son of Bernal's contemporary at Cambridge, Lionel, mapped out a similar dynamic circuit with three rather than four terms: mind → mathematics → physics → mind. See Roger Penrose, *Shadows of the Mind*, Oxford: Oxford University Press, 1994.

requires as much personal observation as reading, so that it is not surprising that outside the small group of psychoanalysts its effect [*sic*] are only to be seen in a partial and distorted way. Among the most important are the revaluation of the emotional aspects of life ... This profoundly influences our whole attitude towards religious and ethical questions in a way very different from that of rationalism ... Religion remains not as an irrational and inexplicable human aberration but as an illusion embodying in itself the deepest of human desires.

Equally important, the hollowness of rationalism itself is exposed. Unsuspected motive becomes apparent beneath the most objective statements and a new mode of criticism appears, from which none of the orthodox philosophies escape unscathed. Unlike religion and philosophy, the processes of scientific thinking escape the most severe criticism and a mode of realistic thought is foreshadowed, based on observation and checked by a knowledge of the distorting motives to be expected.[218]

The unpublished *Microcosm* can be seen as Bernal's first stab at his global account of the sciences which would come to fruition in the monumental *Science in History* (1954). But here the foundations are laid by Freud's revolutionary science of psychology. Bernal asserts that all disciplines are transformed by its arrival, that only those who allow its effect to transform their inner lives can realize its full impact and that it is the new era's response to the twin failures of rationalism and religion. Yet even these effects on the sciences do not give an adequate measure of psychoanalysis's significance for modern civilization:

what is most characteristically Freudian, so much so that to nine people out of ten it is the only association in which his name is known, is the bearing of the doctrine of repression on morality, especially sexual morality. By showing that the universal amorality and sexuality of the human mind can not be extinguished by the most complete asceticism and rigorous discipline, but instead takes on forms more dangerous to the individual and to society, Freudian psychology has given a powerful impetus and a scientific sanction to the idea of a natural life, first romantically proposed by Rousseau. There appears for the first time the possibility of a natural morality, a guiding of instincts towards their greatest satisfaction rather than a set of arbitrary sanctions imposed by authority and tradition for their suppression. It is the revelation of the nature of the unconscious mind that brought about the profound hostility against which Freudian views have always had to fight. Having an ape for an ancestor was no more disturbing to a religious mind feeling itself akin to the angels than still possessing the ape's lusts and depravity would appear to those who considered we were evolving to higher

---

[218] BPCUL, B.4.1 *Microcosm*, 'Influence of psychology', pp. 2–4.

things. But to those who even here preferred to accept things demonstrated to things that should be, the realization of the nature of the unconscious made possible for the first time an objective unemotional facing of the evil nature in man.

More particularly we owe to Freud the removal of hitherto completely irrational fears and disgusts, such as those of incest or sexual perversions, by showing their universal nature and the mode of their development. At last nothing is *néfaste*[219] and by keeping our taboos we are only advertising our relation to savagery. At the same time our attitude to crime and punishment has been profoundly altered ... The vindictive aspect of punishment has now been exposed in its primitive sadism, in spite of its philanthropic disguises, which to the modern eye leaves very little to choose between justice and crime.

Equally if not more important is the emphasis now placed on the earliest years in the development of the individual. Under its stimulus there is a possibility of a real education ... without the realization of the sexual nature in the child, which we owe to Freud, all the good intentions in the world would not have prevented the same stunting of development that has hitherto marked the progress of education.[220]

This prodigious encomium pauses for a moment as Bernal catches his breath, confesses to only a layman's knowledge of the field – 'without practicing analysis or even being psychoanalsed [*sic*]'[221] – but asserts that he is supremely confident of its fundamental importance: 'it overshadows in importance every other branch of knowledge'. It is the application to one's own mental life that ensures this importance, in particular the never-ending permutations of the Oedipus complex: we 'see our lives developing under its ever present influence. We see the gradual growth and transference of our affections from object to object, with its corresponding negative reactions, and so grow to understand our loyalties and antipathies rather than to take them blindly for granted.'[222] The theory of the instincts allows us to 'see the duality of our desires themselves, drawing us on the one hand towards the satisfaction of violent action and on the other to the quest of death which is before birth'. And the anal-erotic complex, to which Bernal continues to attach great significance, 'gives us the basis of our creativeness or acquisitiveness, our joy in suffering or our cruelty'. Through the theory of groups, we understand the influence of the Church, the Army and the Party: 'The capitalist system appears as the expression of close anal-erotic individualism and Socialism as the heroic and guilty revolt of son against father.'

[219] 'nefas' in Ms.; Bernal almost certainly intended the French '*néfaste*' – 'wicked'.
[220] BPCUL, B.4.1 *Microcosm*, 'Influence of psychology', pp. 5–7.   [221] *Ibid.*, p. 7.
[222] *Ibid.*, pp. 7–8.

Displaying his orthodox attachment to Freudian theory, Bernal takes side-swipes at the heretics Jung and Adler, whose repudiation of sexuality 'naturally makes them important rather for well-intentioned than scientific people'. The encyclopaedic scope of the rest of Bernal's manuscript makes even clearer the fascinated importance he accorded to Freudianism: dealing with the new physics, he is less than impressed by Einstein's theories, which he regards as something of a blind alley, but is fully aware of the revolution in microphysics ongoing as he writes, coded with the name Schrödinger; physics was, after all, his scientific trade and one in which he was exceptionally gifted. But nothing in contemporary science matches the revolutionary impact of Freudianism. In his diaries of these London years, sparser than his student diaries, the topic is often on the agenda. A 1924 dinner party is recorded:

> The Russels [*sic*]. The Kapps arrive. Making room the extra glass of wine for our separation. Argument with Bertie on Psycho-Analysis. Intervals of talk with the lady. Max Lebens [?] and Naomi. A Cambridge group. Birth control at the other end of the table. The pronunciation of recondite 'We old stagers' the amende honorable. Ogden. Bertie on Ogden. Talking about life hard and loose living. The lady makes pessimistic contributions. Illness. I don't think anybody can help anyone. What a strange thing to say. Bertie on help and money. Kapp & assurance. The hour grows late.[223]

Other diary entries make it clear how the preoccupation with psycho-analysis is rubbing shoulders with the struggle to find a clear path in his sexual life:

> All the time I learn about what before I had theorised on. On promiscuity without secrecy.
> May destroy all happiness this way. But it is the only road he can go, because otherwise it would destroy his soul.
> I have strangely enough some faith in the road I follow. If it is a bad road for me others may see why where I fall how to avoid the snares. And maybe some day it will be a Royal road.[224]

And an undated page from *Microcosm*, probably from 1925, records his unsuccessful attempts to draw up a map of this road, 'each his own science',

---

[223] Note that these Kapps are not the electrical engineer Kapp mentioned in Chapter 6 below, but Edmond Xavier Kapp (1890–1978) and his wife Yvonne Hélène Kapp (née Mayer) (1903–99), the latter of whose DNB entry mentions Bernal as part of their Bohemian circle in the 1920s. Naomi is probably Naomi Bentwich, Eileen's fellow typist at Miss Pate's in Cambridge and one-time secretary of Keynes. See Skid II, 34–8, 51–5.

[224] ABB, p. 57.

by elaborating an algebra of triangular relationships based on the concepts
of 'connection', 'disconnection' and 'reconnection'.

> Peculiarities of sexual jealousy, sexual pride, desecration of intimacy. Sexual
> revulsion of B temporary or permanent. Timing of new sexual act in
> triangle ... Polygonal problems, systematic treatment impossible. Chains
> of triangles. The square, and the cross square problems. Homosexual
> problems.
>> Problems: money, alimony, concubinage. Law, blackmail and sexual life.
>> Difficulties.
>
> 1. Inner reactions, guilt. Transference and Don Juanism. Dissipation
>   leading to incompleteness of any relation.
> 2. Secondary difficulties. Jealousy and honesty.
> 3. Tertiary difficulties, attitude of society.
>
> Open or secret? 'One must be prepared to love everything without recrimi-
> nation. No property in love.'[225]

And a page entitled 'Present problems' sums up the tensions in Bernal's life
and thought at this time:

> Freudian vs. Communist communist vs. Freudian
> Scientific v. Communist communist v Scientific
> Scientific v Freudian Freudian v scientific[226]

Concealed in his 'Diary of diaries' is another factor at work in this moment
in the complex lives of Bernal – husband, Marxist, Socialist, promiscuous
radical, inventive scientist: in early 1925 he had 'meetings' with 'Bousfield',
almost certainly Paul Bousfield, a doctor practising psychoanalysis outside
the membership of the BPaS.

> 27 Jan 1925: A Bousfield meeting
>   17 Feb 1925: A walk on Putney Heath. Late for Maurice. The Bousfield
> meeting. Three whiskeys. An invitation to Brighton. Repentance & cure.[227]

Bernal made no other comments concerning these meetings. The
Bousfields were – and still are – a distinguished and numerous old
English medical family. Paul Bousfield, MRCS, LRCP (1916), was in
private practice in 1923 at 7, Harley Street and had published *The*

---

[225] BPCUL, B.4.1 *Microcosm*, Document: 'Problems. Limitations to personal problems, each his own
science'.
[226] *Ibid.*: 'Present problems'. Steward, 'Political formation', p. 60, dates this page to the end of the
1920s; its placing does not allow firm dating.
[227] BPCUL, O.1.1 Diaries and Notebooks, Diary of diaries.

Elements of Practical Psycho-Analysis (1920), which, in September 1920, was the first psychoanalytic book Bernal read.[228]

Bernal was also maintaining his connections with Cambridge circles; after dining with the Isaacs and the Russells in London, he spent a weekend in Cambridge, hearing a talk on sex differences and meeting J.B.S. Haldane at dinner at the Isaacs', where philosophy of science was the topic of argument.[229] A month later Bernal met Geoffrey Pyke for the first time, at a London party – a fateful meeting, since they were to work together very closely on some extraordinary projects during World War II. But early 1926 was a difficult time for the Bernals: Eileen was seven months pregnant and they were discussing separation; when Pyke came to them for dinner, Eileen and Pyke spent hours discussing Desmond's affair with Ivy.[230] But they resolved to stay together. For 28 April 1926, Bernal's diary entry reads: 'And now she lies in child bed fighting out life.' In the weeks that followed, however, Bernal was as preoccupied by the General Strike of May 1926, productive of so much bitterness and disenchantment on the British Left, as he was by his new paternity.

Bernal returned to Cambridge in early 1928 as Lecturer in Crystallography. The next year he published a small book, *The World, the Flesh and the Devil: An Enquiry into the Future of the Three Enemies of*

---

[228] For a note on Paul Bousfield's father, W.R. Bousfield, see Chapter 8; neither should be confused with Stanley Bousfield (MD Cantab. 1910), physician to the Legal & General Assurance Society in 1923, or E.G.P. Bousfield (physician to the London Neurological Clinic for the Ministry of Pensions in 1919), or Guy William John Bousfield (1893–1974), a general physician. (Thanks to Edgar Jones for help with the Bousfields: email 12 April 2006.) Paul Bousfield's book had been reviewed by Ernest Jones in his most severe and condemnatory style, in which errors of theory and technique and absence of the most basic knowledge of Freud's work had been exposed; the kindest thing he had to say was that 'the author displays a considerable talent for elementary presentation' – the rest of the sentence taking back anything that had been given – 'one which would be useful in dealing with a subject of which he had first made an adequate study' (J., E., 'The Elements of Practical Psycho-Analysis', *IJP* 1 (1920), 324–8 at 327–8). Jones also noted Bousfield's vigorously feminist (by which Jones meant 'denial of differences between the sexes') and heterodox views on female sexuality: 'The novel view is put forward that no shifting of excitability from the clitoris to the vagina takes place as a rule in normal women; when it occurs it is to be regarded as a regression to cloacal erotism. In 150 cases of apparently normal women, three were completely anaesthetic, fourteen felt pleasure chiefly referred to the vagina but without orgasm, and in the remainder, without exception, the glans clitoris was the essential seat of sensation (pp. 88, 89). This is, of course, the opposite of the psycho-analytical theory of sexual development.' It is quite likely that reading this passage in Bousfield's book was the instigator of Frank Ramsey's reaction, recorded in his diary on 17 January 1924: 'Curious reaction in morning on noticing by chance in U.L. a book on psychoan and opening it at a passage on sexual anaesthesia in women saying Freud wrong in supposing transference of sensibility from clitoris to vagina at puberty, statistics that majority remain sensitive mainly in clitoris: felt [inserted: rather] faint and dithering for quite some [mins.].' FRP, Diary 1924. (The final word of this passage is very difficult to decipher.)

[229] BPCUL, O.1.1 Diaries and Notebooks, Diary of diaries, entries for 13 and 16 January 1926.

[230] ABB, p. 59.

the Rational Soul in the highly successful 'Today and Tomorrow' series that C.K. Ogden span off from the Cambridge meetings of the Heretics. The first of the series to be riotously successful, the one that set its tone and established a new style, in tune with the outré 1920s, for writing about popular science – daring and outrageous, rather than worthy and superior – had been J.B.S. Haldane's *Daedalus or, Science and the Future* (1924), originally read to the Heretics in February 1923. Haldane's book has to this day retained its ability to tickle the imagination because of his insistence that the mastery yielded by physics would be followed by a comparable mastery yielded by a future biology, in which the process of reproduction would be taken out of the body of woman and conducted in the laboratory and clinic – a form of liberation.[231] When Haldane did talk of psychology, which was very little, he alighted on hypnotism and the possibility of systematic communication with spiritual beings in other worlds as promised by the spiritualists and psychical researchers: hypnotism gave a glimpse of the powers that might be liberated if such forces were controlled, as drugs have recently become, and the spiritualists, 'already Christianity's most formidable enemy', promise scientific verification for an indeterminable new world of psychic forces. His heart and his rhetoric were more exercised by the necessity of destroying religion, which inevitably is an obstacle to the advance of science: 'We must learn not to take traditional morals too seriously. And it is just because even the least dogmatic of religions tends to associate itself with some kind of unalterable moral tradition, that there can be no truce between science and religion.'[232] This was, after all, a talk to the Heretics.

A passage in Haldane's little book gave Bernal his cue and his blueprint: science 'is man's gradual conquest, first of space and time, then of matter as such, then of his own body and those of other living beings, and finally the subjugation of the dark and evil elements in his own soul'.[233] Bernal imagined the future of mankind resulting from the future developments in these arenas: the World (physics), the Flesh (biology) and the Devil (psychology). Bernal's vision for physics assumed that the destiny of mankind was to leave the encumbrance of gravity and the limited resources

[231] The most famous vehicle for the dissemination of Haldane's speculation was Aldous Huxley's *Brave New World* (1932) which pits Shakespeare and Freud against reproductive biotechnology and psychopharmacology. But the final pages of a feminist classic, Shulamith Firestone's *The Dialectic of Sex* (1970), are also infused with Haldane's imagined ectogenesis, the process by which pregnancy and childbirth are replaced with an artificial system.

[232] J.B.S. Haldane, *Daedelus; or: Science and the Future; a paper read to the Heretics, Cambridge, on February 4th, 1923*, New York: Dutton, 1924, p. 90.

[233] *Ibid.*, quote in ABB, p. 73.

of the earth behind to dwell in space; his vision of biology was equally de-centring: humans would free themselves of their bodies, becoming brains in vats, with the additional possibility that they would link together to form a network of brains that would be one Giant Brain, a Communal Brain. This was the most famous – or infamous – of his imaginings: 'the first properly immortal envatted human brain in literature', as Cathy Gere describes it – an 'emblem of our technocracy, a vision of scientists as immortality-bestowing gods and illusion-producing devils'.[234] True to his earlier claims in *Microcosm* about the foundational character of psychology in our era, Bernal executes an eminently Freudian opening gambit when he addresses the third area, that of the Devil – it is our inner resistance, our devil, that baulks at the future our science will bring us, refusing in the name of 'humanity' such horrors:

> Why do the first lines of attack against the inorganic forces of the world and the organic structure of our bodies seem so doubtful, fanciful and Utopian? Because we can abandon the world and subdue the flesh only if we first expel the devil, and the devil, for all that he has lost individuality, is still as powerful as ever. The devil is the most difficult of all to deal with: he is inside ourselves, we cannot see him. Our capacities, our desires, our inner confusions are almost impossible to understand or cope with in the present, still less can we predict what will be the future of them.

But try to understand our own unconscious motives we must. Taking his cue from Russell's pessimistic diagnosis of the impending end of science because of the fading of hope in the benefits it brings, he offers a psychological typology of humanity, consisting of 'a crop of perverted individuals capable of more than average performance, and a mass of people effective not so much by their number as by their secure hold on tradition'. The creative perverts are usually held in check by the mass; and it appears that 'we are approaching the close of the period of respectable comfort which puritanism demanded and mathematics and handicraft produced'. What may promote a new era of expansion is 'science, raised to power by industrialism', becoming the new dominant tradition. It is for this reason that the modern era is a critical moment in world history: we live in

> an age in which the nature of desire has been glimpsed at for the first time, and that glimpse enables us to see two very different possibilities. The intellectual life, both in its scientific and its æsthetic aspects, is seen no

---

[234] Cathy Gere and Charlie Gere, 'Introduction: the brain in a vat', and Bronwyn Parry, 'Technologies of immortality: the brain on ice', *Studies in History and Philosophy of Biological and Biomedical Sciences* 35C(2) (2004), 219–25 and 391–413 respectively.

longer as the vocation of the rational mind, but as a compensation, as a perversion of more primitive, unsatisfied desires. Now the question arises is this perversion in the line of evolution, or is it a merely temporary, patho-logical process? If by a sounder psychology, a way of living more in accordance with nature, it should be found that the satisfaction of purely human – or, as we might almost say, purely mammalian – desires is capable of absorbing all the energy that suppression now forces into scientific or æsthetic channels, then the human race may well find itself statically employed in leading an idyllic, Melanesian existence of eating, drinking, friendliness, love-making, dancing and singing, and the golden age may settle permanently on the world. On the other hand it may [be] that though the desire, the necessity to escape life on the paths of intellectual or æsthetic creation may be weakened by the application of an intelligent psychology, yet a corresponding freedom from the internal conflicts which now hinder both these forms of expression may more than compensate for what is lost, and we may find the capacity to live at the same time more fully human and fully intellectual lives. The latter alternative is more in line with the recent developments of Freudian psychology which divide the psyche into the primitive id, the ego which is its expression of contact with reality, and the super-ego which represents its aspirations and ideals. Rationalism strove to make the super-ego the dominant partner; it never succeeded, not only because its standard was too high to allow any outlet for the primitive forces, but because it was itself too arbitrary, too tainted with distorted primitive wishes ever to be brought into correspondence with reality. Naturalism, less definitely, aimed at giving the primitive wishes full play but equally failed because these wishes are too primitive, too infantile, too inconsistent with themselves to be satisfied even by the greatest license. The aim of applied psychology is now to bring, by analysis or education, the ideals of the super-ego in line with external reality, using and rendering innocuous the power of the id and leading to a life where a full adult sexuality would be balanced with objective activity. It is this alternative that makes the mechanical, biological progress that I have outlined not only possible but almost neces-sary, for a sound intellectual humanity will never be content with repeating itself in circles of metaphysical thinking like Shaw's Immortals, but will need a real externalization in transforming the universe and itself. Such a development could hardly leave unchanged the present types of human interests in art and science and religion.[235]

It is this passage, in Bernal's first published book, that places Freudianism on the cusp of the next fateful steps in human history. How to take the Next Step in the History of Humanity is uniquely envisaged by psycho-analysis; the alternative is a mere Golden Age, a finely worked State of Nature. It is clear that Bernal is fortified by his reading of psychoanalysis in

[235] J.D. Bernal, *The World, the Flesh and the Devil*, London: Cape, 1968, p. 57.

affirming the possibility, perhaps the necessity, of the march of Science taking humans away from Earth, away from their bodies, through the Ideal and the Real (super-ego and reality, sexuality and objective activity) merging and becoming One. Psychoanalysis, the only plausible account of the true nature of human desire, will show us how to make our Devil become our God.

How odd to find psychoanalysis as the principal enabler of an argument for detaching human beings from Mother Earth and from the Body to which Narcissus sacrificed his life! In terms of Bernal's publications for a general readership, this was a first triumphant sortie; but in terms of his relation to psychoanalysis, it was something of a swansong. The unpublished private, youthful passion he had for Freud was to wither away very soon; his scientism, his celebration of science as the great benefactor of humankind, especially when he became from the early 1930s on a committed exponent of the science of dialectical materialism, was to wax ever more strongly, and his attention to human desire, to the internal world from which the sources of human action flow, was to wane very markedly. With the Great Crash, the Depression, the rise of fascism, the politicization of intellectual culture of the 1930s, Bernal abandoned his Freudianism. When in 1931 he wrote an article 'What is an intellectual?', the field of psychoanalysis was swept aside: 'The whole complex of individualism, self-development and self-expression loses its overwhelming importance ... these are no longer advanced but reactionary aspirations.'[236] Even more peremptory was his call to order in a review in 1937, published in *Labour Monthly*, of Osborn's *Psychoanalysis and Marxism*. Freudianism, it was now clear, was just one more form of subjectivist philosophy and 'must be understood and rejected as such'. Psychoanalysis sought to 'set up the individual ... as the centre and measure of all things' and was thus 'a profoundly dangerous influence, paralysing action and tending to fascism'. Reality had made its demands and Bernal was now its spokesman:

> In recent years the Freudian wave has begun to recede. The effects of the world economic crisis of capitalism, and of the close menace of fascism and war, startled the intellectual strata into awareness of the objective world, and aroused a new wide interest in Marxism.[237]

This may well have been an accurate perception of how many people now saw things in 1937; it was certainly in keeping with the Stalinist line

---

[236] BPCUL, B.4.13; see Steward, 'Political formation', p. 62.
[237] Quoted in Steward, 'Political formation', p. 67.

emanating from Moscow which, from 1930 on, anathematized Freudianism absolutely.[238] By the late 1930s, with the publication of *The Social Function of Science*, Bernal was centrally involved in the radicalization and mobilization of British scientists: the earliest and most significant convert to Communism among noteworthy British scientists, he led the movement against war and for the organization of scientific workers and was the most influential prophet of the unlimited potential of science for progress.[239]

But what of the intellectual and emotional investment he had made in Freudianism? At one point, it looked as if Freudianism were the 'science' which underpinned Marxist politics and promised to generate formulae for understanding the sexual complexities of life, particularly the polygonal life that Bernal himself lived. Was his abandonment of Freud a sign that his own sexual radicalism was finally claimed by the moderation of age and wisdom?

Still living part of the time with Eileen and their two boys, Michael and Egan, whose base was more London than Cambridge, Bernal had become permanently involved with Margaret Gardiner, formerly of Bedales and Newnham College, Cambridge, close friend of John Layard, whom she rescued in Berlin when his attempt to blow his brains out failed. In 1937, Margaret gave birth to Martin, Bernal's son. When he moved to Birkbeck College, London as Professor of Physics in 1938, a difficult period ensued, with Bernal dividing his time between Eileen living on Clifton Hill in St John's Wood and Margaret, living on Downshire Hill in Hampstead. A close friend, Joan Malleson, sternly rebuked him, by reminding him that Eileen 'feels your silence to be flippant and your attitude casual . . . you are deserting her and slighting her'.[240] Joan suggested that he might try psychoanalysis. In his unpublished sketch for an autobiography, under the year 1939 he recorded: 'After inevitable difficulties between Downshire Hill and Clifton Hill, invited by Joan Malleson to try psychoanalysis, but failed to effect transfer.'[241] We do not know who his analyst was; it is more than implausible that she was Melanie Klein, Eileen's next-door neighbour. But the very fact that he tried is an indication not only that his personal life could cause him and those close to him much difficulty, but also that the ambitious attempt to find a theory of the personal and intimate life – the polygons, jealousies and striving for honesty – still had

[238] Alexander Etkind, *Eros of the Impossible: The History of Psychoanalysis in Russia* (1993), trans. Noah and Maria Rubins, Boulder, CO: Westview Press, 1997.
[239] Hobsbawm, 'Preface', p. x.   [240] ABB, p. 155.
[241] Andrew Brown, email, 28 January 2006, citing BPCUL, O.1.1.

a little life left in it. Freudianism had once seemed to fit hand in glove with his attempt to find an honest path for his personal life, free of the hypocrisies and conventions of the pre-Freudian era. But perhaps no longer.

The story of the rise and fall of Bernal's enthusiastic Freudianism is the clearest instance of the pattern we see emerging for the reception of Freud in Cambridge. In the immediate post-war period, the frenzied intellectual hunger of youth took up the new ideas at that age when young men and women are most receptive and are able to absorb such ideas with a speed that their elders envy – Keynes said of Ramsey's essays that they were 'a remarkable example of how the young can take up the story at the point to which the previous generation had brought it a little out of breath, and then proceed forward without taking more than a week thoroughly to digest everything which had been done up to date'.[242] A week is a long time in an undergraduate's intellectual and emotional life. By the mid-1920s Freudianism was ripe for a more serious and sober assessment, and Bernal, always attuned to the latest ideas, embedded them in a new vision of the world and of science. Bernal's *Microcosm* was the most ambitious – even, matching its author, the most grandiose – attempt yet to do this, parts of which were then distilled into his famously provocative futurological published essay. With the turning of the political, economical and cultural tide, so marked at the end of the 1920s, Bernal's version of Freudianism was left high and dry, drained of life by the wholesale repudiation of the political and intellectual search for the significance of the 'inner world', for the sources of cultural malaise in the inner nature of modern man. 'Science and society', 'Marxism and science', 'Fascism and war' were to be Bernal's great causes for the rest of his life, with hardly a trace of his early Freudianism. When in 1968 he had the pleasure of writing a Foreword to a second edition of *The World, the Flesh and the Devil*, there was only one thing he wished to change:

> The section in the book on the devil remains the most important but, looking at it now, I find it expressed too much in Freudian terms which are likely to be superseded. Many new ideas remain just under the horizon, notably those on memory which appears to have a chemical basis.[243]

---

[242] Skid II, 73. The point for Keynes was that he had worked for fifteen years intermittently on his *Treatise on Probability* and the nineteen-year-old Ramsey had read and demolished it in a week.

[243] Bernal, *The World, the Flesh and the Devil*; 'Foreword to the second edition', p. 10, dated November 1968; see Dorothy M.C. Hodgkin, 'John Desmond Bernal. 10 May 1901–15 September 1971', *BMFRS* 26 (November 1980), 16–84 at 64.

Late in life, Bernal may have wished to repudiate his earlier Freudian enthusiasm, but when his son Martin Bernal wrote his own autobiography in 2012, the first sentences of the description of his father read: 'My father was a bourgeois communist of the 1920s. He was a bohemian and, for a time, a Freudian.'[244]

The example of Bernal's Freudian passion of the 1920s helps clarify the combination of complicating factors conditioning the overall reception of Freud's ideas. For Bernal, the attractions of Freudianism lay in its unique ability to serve as a foundation for the transformation of many other disciplines – recall his prodigious list: 'Philosophy, Psychology, Metaphysics, Ethics, Anthropology, Sociology, Politics, Economics, Medicine, Religion'. Freudianism promised a revolutionary science of the human subject of knowledge, passion and action – whether the subject be a priest, a pilot or a physicist. Through this interrogation of the subject of knowledge and of action, a revolution in society would follow. Bernal's enthusiastic reading of Freud's work naturally leads to the development of a Psychoanalytic Movement, akin to that of Christian Science or of Socialism. Bernal's personal inclinations for such movements are clear. He was ready to join; he was ready to lead.

But Freudianism was not only a movement in Britain in the 1920s. The second aspect of Freud's work was being pushed to the fore: the practice of psychoanalysis as a therapy for the neuroses. Bernal had brushes with this aspect: through the informal analyses the Neurotics practised on each other in Cambridge in 1921, through his visits to Bousfield in 1925 and then through a late and failed encounter with a psychoanalyst in 1939. Bernal never engaged fully with the practical business of analysis, but others, particularly those setting up the BPaS, the Institute of Psycho-Analysis and the London Clinic in the 1920s, were principally concerned with that dimension of Freud's invention. To them, Bernal's revolutionary Freudianism, and other enthusiastic projects like it, were a danger and a distraction in the establishment of a Freudian professional practice.

Third and finally, there is an element of Freud's work with which the undergraduate enthusiasms of Kingsley Martin, historian, and J.D. Bernal, physicist, hardly made contact: the development of psychoanalysis as an academic discipline, a science, just like others, like physics or literary studies, like anthropology or philosophy – with its own object, its own methods, its own research and teaching programmes. What happened in Cambridge, in the first decades of the twentieth century, to this third

---

[244] Martin Bernal, *Geography of a Life*, p. 18.

dimension of Freud's new discipline? What did Cambridge make or not make of the establishment of psychoanalysis as one more new science amongst the several others? Some elements of an answer will emerge from examining the lives of two other undergraduates from the immediate post-war years: Joseph Needham and Sebastian Sprott.

Joseph Needham: 'Oedipus complexes, anxiety neuroses, penis envy, and Jungian archetypes'

Like those of his future collaborator J.D. Bernal, Joseph Needham's undergraduate years automatically included an education in Freud's psychoanalysis, which would lead to a life-long sympathy tempered but not altered by growing respect for the advances of psychopharmacology post-1945. Needham came up to Gonville and Caius College, Cambridge in 1918 to read Natural Sciences in order to study medicine; he was advised to take chemistry as well as biological subjects, through which he met and fell under the spell of Gowland Hopkins, the creator of Cambridge

Fig. 4.9 Joseph Needham, by Ramsey & Muspratt, 1937.

biochemistry. Already encouraged at school in his broad philosophical interests, these had developed into a wide-ranging acquaintance with the history of the sciences, which was to be made manifest in a variety of ways throughout his life. Needham had also developed a deeply religious conviction, such that, even while an undergraduate, he had taken up residence at the Oratory, founded in the early twentieth century by a group of young Cambridge Dons, trying to maintain a high standard of Catholic devotion and discipline amidst the luxury of University life: in 1920, they had created a house in which devotees could dwell. Needham's religious convictions and cloistered life did not prevent him being extremely active in his scientific research and in the social life of the University. With regard to Freud and psychoanalysis, he recalled:

> I always believe that I was fortunate in being a student when A.G. Tansley was expounding the ideas of Sigmund Freud to English readers . . . At the B.A.'s table at Caius[245] we talked exclusively of Oedipus complexes, anxiety neuroses, penis envy, and Jungian archetypes, and I believe that this familiarisation was profoundly beneficial to me; otherwise I could have been very alarmed by psychological phenomena that I experienced as I grew older. *The Psychopathology of Everyday Life* was assuredly a great help for many people. I believe that Freud, Adler, and Jung were men of the deepest insight, as revolutionary and liberating in their way as Darwin, Marx, and Huxley had been before them.[246]

The conviction that the new depth psychology, most prominently represented by Freud, was a profound and lasting contribution to the scientific understanding of human beings was of course one Needham shared with other Cambridge students of the post-war generation; again and again, Freud is compared with Darwin, with Marx, with Einstein as one of the key makers of the modern scientific world-view.

Needham's father died suddenly in 1920. The death ravaged the family's finances and emotions. In particular Needham's mother's mental instability became increasingly apparent; academic mentor, Gowland Hopkins, to a large extent had to take on the role of parent as well.[247] It is probable that

---

[245] The dining table set aside for those junior members of the College who had already been granted a BA – what would now be called 'postgraduates' or 'graduates' (research students).

[246] Therese Spitzer, *Psychobattery: A Chronicle of Psychotherapeutic Abuse*, with medical discussion by Ralph Spitzer and a Foreword by Joseph Needham, Clifton, NJ: Humana Press, 1980, p. ix. When asked in the 1970s by Edward Timms about interest in Freud in the 1920s, Needham again recalled Tansley lecturing on Freud in his botany classes of the early 1920s. (Personal communication, Edward Timms to Laura Cameron, 17 March 1997.)

[247] Gregory Blue, 'Needham, (Noël) Joseph Terence Montgomery (1900–1995)', *ODNB*, www.oxfo rddnb.com/view/article/58035, accessed 7 July 2011.

his mother's condition played a part in the direction his professional interests now took. Signs of considerable interest in medical psychology appeared; in 1922, he chaired a meeting at which the founder of the Tavistock Clinic, Hugh Crichton-Miller, spoke to undergraduates.[248] On graduating in June 1922, Needham spent some time in a research lab in Berlin but returned to Cambridge to take up, for 1922 and 1923, the Ben Levi studentship in biochemistry:

> when in the early twenties I was starting life as a research biochemist, I was greatly attracted to the biochemistry of mental disease. I followed the lectures for the Diploma in Psychological Medicine, and worked at Fulbourn Mental Hospital near Cambridge on the creatinine metabolism in catatonic patients suffering from what we used to call in those days dementia praecox. I published one paper (with T.J. McCarthy),[249] but my hopes soon faded, and when I read an excellent review on the subject which covered much literature, and ended by saying that biochemists had grown tired of 'fishing in distilled water for the causes of mental disease,' I realised that I had better find something more worthwhile. Eggs and embryos were the answer, and very worthwhile they were.

Needham's initial research project on the biochemistry of schizophrenia had followed the 'toxin' theories of mental disease which had originated around 1900. His informal taking of the courses provided for the Diploma (intended only for those with a medical degree) indicated the seriousness of his commitment to the field. The project may well also have been a combination of the fad for Freud so evident in the student world of the early 1920s, the burdensome intimate knowledge of his mother's mental condition and his growing commitment to his mentor's field of biochemistry. His good judgement in turning away from this project and turning instead towards eggs would determine the direction of his scientific research for the next twenty years. 'The question that primarily interested Needham was how changes in chemical composition take place in development, and how morphology can be interpreted in chemical terms.' Yet his wider concerns were already conspicuous, well before the unexpected emergence, from the 1940s on, of his world-changing interest in the history of Chinese science and technology. His first book, prepared while conducting his doctoral research in 1922–24, was an edited volume published in 1925, entitled *Science, Religion and Reality*, with contributions from

---

[248] Dicks, *Tavistock Clinic*, p. 30.
[249] Thomas J. McCarthy and Joseph Needham, 'The excretion of creatinine in nervous diseases', *Quarterly Journal of Experimental Physiology* 13 (1922–23), Suppl. Proceedings of XIth International Physiology Congress, Edinburgh, July 1923, pp. 175–6.

Arthur Eddington, Charles Singer, William Brown (on 'Religion and psychology') and Malinowski (on 'Magic, science and religion'), and a lengthy study by himself on 'Mechanistic biology and the religious consciousness'. His contribution to the book, while immensely learned, historically, philosophically, theologically and scientifically, does not venture into psychology or mental philosophy. Needham would never return to his interest in the science of mental disease.

Sebastian Sprott: 'the psycho-analytic giggle'

On 22 August 1922, Ernest Jones, at the time the most energetic advocate of psychoanalysis in England, wrote to Sigmund Freud from his country cottage on the borders of Sussex and Hampshire:

> I have just received today a letter from a Mr. Sprott telling me, to my surprise, that he arranged with you last July to lecture at Cambridge next autumn and asking me if I could arrange some public lectures for you to give in London. It would be wonderful to know that you were lecturing in England, but I must first inquire of you about the authenticity of the man, for perhaps he is nothing but a lecture agent. I know nothing about him.[250]

Jones always feared that the sixty-six-year-old Freud might make a fool of himself by acting on the trait of gullibility to which Jones also ascribed his scientific genius – Freud actually believed people![251] By 1922, Freud was famous throughout Europe and America, a scientific media star on a par with Einstein, Russell and Curie. Above all else, Jones was suspicious of other psychoanalytic enthusiasts taking control of the development in England of the young science and taking it out of his own hands. So he conjured up the scenario of his revered teacher in the hands of a lecture agent.

---

[250] JF, 22 August 1922, p. 498.

[251] The story Jones told to show Freud's gullibility was the following: 'Joan Riviere has related an extraordinary example of this combination of credulity and persistence. During her analysis Freud spoke very angrily one morning of an English patient he had just seen who complained bitterly of monstrous, and indeed fantastic, ill-treatment she had suffered at the hands of an English analyst in Ipswich – of all places. Mrs. Riviere's cool mind at once perceived that this was a cock-and-bull story, but she contented herself with remarking that there was no English analyst of the name mentioned, that there never had been an analyst in Ipswich nor indeed anywhere in England outside of London. That made no impression, and Freud continued his tirade against such scandalous behaviour. Shortly afterwards, however, he received a letter from Abraham saying he had recommended an English lady to consult him and that she was a wild paranoiac with a fondness for inventing incredible stories about doctors. So poor Abraham had been the wicked analyst in Ipswich!' (J II, 477–8).

He need not have feared, as Freud's rapid reply, written in his eccentrically interesting English from the Pension Moritz in Berchtesgaden, indicated:

> Mr. Sprott is a young man of excellent manners and good connections, a favourite of Lytton Strachey and friend of Maynard Keynes, a Cambridge student of psychology, who came to invite me for a course of lectures to be given at Eastertime (not autumn, as in your letter). I accepted for the case that I should feel so tired at Easter, that I had to give up work, and yet fresh enough for some other enterprise, which, as you see, is only a polite way of declining.[252]

One might say that Freud indicated his own excellent manners and good connections by bandying about the reassuring names of Lytton Strachey and Maynard Keynes – not 'John Maynard Keynes', nor 'Keynes', but 'Maynard Keynes' – Freud knew that in Cambridge there were two John Keynes, father and son, so one employed either Neville Keynes or Maynard Keynes to distinguish the two.

Freud never came to Cambridge. He had visited his relatives in Manchester in 1875 at the age of nineteen, and then, again with his half-brother, in 1908 he had spent a fortnight's holiday in Blackpool, Southport, Manchester and London.[253] The next time he arrived in England was in the spring of 1938, following the *Anschluss* – 'to die in freedom', as he put it. His daughter Anna did visit the University of Oxford on his behalf, when the IPA Congress was held there in 1929. By then Freud was too ill to travel and instead followed her visit with elderly eagerness, filling in the gaps out of his great love of England. As he wrote to his old friend Lou Andreas-Salomé:

> As to the accommodation, she telegraphed, typically enough: 'More tradition than comfort'. I expect you know that the English, having created the concept of comfort, then refused to have anything more to do with it.[254]

Freud's invitation from Sprott was not the first overture from Cambridge. Two years earlier Freud informed the Vienna Psychoanalytic Society that his close friend Professor Pribram had recently visited Cambridge and reported that 'Cambridge University is very keen on his work, especially *Totem and Taboo*.'[255] He also told Jones: 'My friend Prof. Pribram tells me I have been expected at Cambridge on Sept. 27th. But I never got a word of

---

[252] FJ, 3 September 1922, pp. 500–1. [253] J II, 58.

[254] Ernst Pfeiffer (ed.), *Sigmund Freud and Lou Andreas-Salomé: Letters*, trans. William and Elaine Robson-Scott, London: Hogarth Press and the Institute of Psycho-Analysis, 1972, p. 182, 28 July 1929.

[255] JPBPaS, CFC/FO5/04. Freud and Rank, Rundbrief, 14 October 1920.

invitation as you know. I once had a conversation with Rickman about my visit there but I dropped it hearing that my daughter could not be invited. That was all.'[256] Perhaps it was Jones who had let the invitation languish, as his immediate response gives some reason to think – one doesn't have to be a Freudian to think it significant that Jones forgot to transmit a message to Freud from Rivers: 'Perhaps I forgot to tell you that Rivers wrote to me in September expressing the hope that you would visit Cambridge, but it was of course no appointment.'[257] Knowing how important it was for Jones to retain control of access to Freud and of what counted as approved discussion of psychoanalysis, it is no surprise that he did not welcome this competing independent overture to Freud from Cambridge, to the point of 'forgetting' it entirely. In writing to his daughter Anna about Pribram's news that he had been 'expected in Cambridge on 27 September', Freud added, clearly irritated at having missed some kind of opportunity: 'Why didn't they write?'[258]

---

[256] FJ, 12 October 1920, p. 393.

[257] JF, 17 October 1920, p. 394. 'no appointment' is quite probably an English version of a German phrase Jones had in mind and is an indication of some kind of stress in Jones's prose not short of a cover-up for what was really going on. Perhaps he was thinking of the German phrase 'keine Bestimmung', which can sometimes be best translated as 'nothing definite'.

[258] SFAF, 126SF, 12 October 1920, p. 205. Part of the background to the invitation is to be found in the complicated travel plans the Freuds had had for the summer of 1920, whose highpoint was the first Congress of the IPA since the war, held in neutral The Hague, 6–11 September, attended by Sigmund and Anna (as a Guest). They planned to spend time in The Hague seeing Loe Kann and her husband Herbert Jones for the first time since the war – an important encounter, since Loe Kann had been supplying from The Hague the Freud family in Vienna with food and provisions during the war and in the post-war famine time. ('Loe is still supplying the greater part of our sustenance by means of her shipments from The Hague. We hope to see her there in the fall': Freud to Ferenczi, 15 March 1920, p. 13.) Although early in 1920 Ernest Jones was hoping Freud would come to England in September, it was only in late July that an invitation from Loe Kann and Herbert Jones fired up Anna Freud's imagination for the trip to England with her father; Freud, altogether more sceptical of its possibility, nonetheless alerted Ernest Jones and started to make plans for the September trip with Anna (possibly accompanying Loe and Herbert back to England after the Hague Congress), so they could meet up with their relatives (Sam Freud in Manchester). On 30 July, Freud and Minna Bernays went to Gastein; after a month there, Freud and Anna travelled to Hamburg (to visit the widowed Max Halberstadt and the two grandsons), where they were met by Eitingon and on 6/7 September the three of them went to the congress in The Hague. Freud was still in The Hague on 23 September when he wrote to Jones confirming that the trip to England had had to be cancelled because 'The Visum for Anna came too late for my dispositions' (FJ, 23 September 1920, p. 390). The likeliest reconstruction of the events is perhaps the following: John Rickman started analysis with Freud at Easter 1920 as one of the first English-speaking analysands upon whom Freud became financially dependent in the early 1920s. In October 1920, Freud recalled to Jones that 'I once had a conversation with Rickman about my visit there [Cambridge] but I dropped it hearing that my daughter could not be invited.' It is likely that Rickman, who was in England over the summer of 1920 and attended the Hague Congress as a 'Guest' (place of residence given as 'Cambridge'), had communicated with Rivers about the possibility of Freud visiting Cambridge, since he may well have known of this plan. Rivers in September communicated with Jones his interest in having Freud come to Cambridge (Jones:

The second invitation to Freud, in 1922, came again from well outside the Jones circle. Sprott was a quintessential young product of Cambridge: the Apostles, the Heretics, Bloomsbury and the Moral Sciences Tripos. Born in 1897 from a professional background, he came up to Clare College in 1919 to read moral sciences and achieved a double First alongside Lettice Baker and Frances Marshall (later Partridge) with their Upper Seconds.[259] He had a great gift for securing the love, affection and trust of eminent older men; he was also a great beauty. Elected an Apostle in October 1920,[260] from that same month on he was regarded by Bloomsbury as 'married' to Keynes.[261] During 1920 and 1921 he conducted an affair with Keynes, who introduced him to Lytton Strachey, with whom he quickly developed an intimate and permanent friendship.[262] He was also close friends with Frank Ramsey throughout the latter's short life. By early 1921, Sprott was Secretary of the Heretics with Lettice Baker as Treasurer and Frank Ramsey co-opted on to the Committee – their suggestions for speakers including Roger Fry, Walter de la Mare, Susan Stebbing, Keynes, Rutherford, Eddington and Whately Smith.[263] Sprott hosted Lytton Strachey when he spoke to the Heretics on 21 February 1921.[264] When Lytton and Sebastian holidayed together in Venice following Sprott's final examinations in June 1922, Sprott travelled on to Vienna to visit Wittgenstein and also, at Keynes's behest, to make contact with Freud and invite him to give a talk to the Heretics as well. Lytton got Freud's address for him from his brother James,[265] Ramsey got Wittgenstein's

'Perhaps I forgot to tell you that Rivers wrote to me in September expressing the hope that you would visit Cambridge, but it was of course no appointment' (17 October 1920, p. 394)). So Rivers, Rickman and others may have been waiting in Cambridge for Freud's arrival, not knowing that Freud had cancelled his trip to England from The Hague some time around the middle of September. (Why did Rickman not talk to Freud while at the Hague Congress, one might well ask.) It should be noted that 27 September 1920 was outside Cambridge term (which began on 1 October 1920 (*University of Cambridge Calendar 1920*, p. x)) – so a very unusual date for a University event, one which was clearly designed to fit in with Freud's (provisional) schedule. The evocativeness of 'Cambridge' persisted, at least in Anna Freud's mind: in July 1922, Loe Kann and Herbert Jones renewed their invitation to Anna, who was extremely keen to travel to England, and encouraged her father to respond positively to the invitation: 'I know of one pleasant travelling companion: you, if you accept the invitation to Cambridge' (185AF, 27 July 1922, p. 302). Whether or not Anna knew of Sprott's climb up to Bad Gastein (which Freud had mentioned to Ferenczi on 21 July 1922 but had not mentioned to Anna in preceding letters in July 1922 addressed to Göttingen, where Anna was visiting Lou Andreas-Salomé), and therefore of Keynes's/Sprott's invitation to Cambridge, she may well have thought the invitation of 1920 was still on Freud's table and knew that it was a temptation for Freud.

[259] *CUR*, 14 June 1921, p. 1138.   [260] LLS, p. 462.   [261] Skid II, 35.   [262] Holroyd, *passim*.
[263] FRP; Dora Black, '"My friend Ogden"', in W. Terrence Gordon (ed.), *C.K. Ogden: A Bio-bibliographic Study*, New Jersey and London: Scarecrow Press, 1990, p. 89.
[264] KMFF, p. 102; FRP *Diary*, 21 February 1921.
[265] LLS, Lytton to James Strachey, 22 July 1922, p. 518.

Fig. 4.10 From left to right: Walter John Herbert ('Sebastian') Sprott, Richard Bevan Braithwaite, and Mary Sprott, Sebastian's mother; early 1920s.

address for him from Ogden.[266] Sprott made his way from Venice to Berggasse 19, found Freud away on holiday and then hunted him down some 350 kilometers away in Bad Gastein. Freud mentioned his impending visit to Ferenczi: 'On Sunday an Englishman is supposed to come, who will negotiate with me in vain about a series of lectures in Cambridge, etc.'[267] Two weeks later, Sebastian himself was back in Norfolk and reported to Lytton:

> Freud sent his greeting to you – 'the greatest stylist of the age' – this is flattering because according to your brother he is rather proud of himself in that particular line – any way that is what he said – he said '*Eminent Vics*' was – 'fabelhaft', 'reizend' and 'wunderbar' and that your grasp of Psychology was most extraordinary. There! and I was very good about it too – I said that good though your looks were, charmingly though you wrote, acute though you appeared on paper – nevertheless he could form no sort of idea of what you were like in 'Realität' and I told him about Venice.
>
> He wasn't in Vienna – and after telegraphy and being answered I followed him to a waterfall in the Tyrol ... Freud was taking the baths and was with his sister. She is a gorgon if ever there were one. Cold, calculating, stubborn. She has a tremendous influence on Freud. She tells him all his plans, and he leans on her for all arrangements. He is [illegible: 'curing'? Not 'clearly', not 'cunning'] broken and depressed, I thought. Constantly throwing out bright ideas – rather like Maynard – but at the bottom bitter and disappointed. He was flattered by my request, and, under certain circumstances, he may come to Cambridge. His sister is against it, which is a bad sign. He is clearly far more sensible than most of them, and, though he hates disagreement – he does not brush it aside. To look at, he reminded me at once of the picture of Butler, in Festy Jones's life.
>
> Vienna was lovely – well, when I say 'lovely' – perhaps that's an exaggeration, but anyway it was amusing.[268]

Two weeks later, Sprott wrote to Ernest Jones, provoking the latter's anxious enquiry to Freud. Jones responded to Sprott the same day, agreeing without further ado to talk to the Heretics – which he did on 26

---

[266] OP, Box III, F.1, Ramsey to Ogden, ?? June 1922: 'I wonder if you could send me Wittgenstein's address, because I promised to send it to Sebastian who is going to Austria and wants to go and see him with an introduction from Keynes. Could you let me have it as soon as possible? Sebastian is also going from Keynes to try to persuade Freud to come and give lectures in Cambridge.'

[267] S. Freud, 'Letter from Sigmund Freud to Sándor Ferenczi, July 21, 1922', in *The Correspondence of Sigmund Freud and Sándor Ferenczi*, Vol. III, *1920–1933*, ed. Ernst Falzeder and Eva Brabant, Cambridge, MA: Belknap Press, 2000, pp. 83–5.

[268] BLSP 60699, Vol. XLV, Sprott to Lytton Strachey, 8 August 1922. The 'sister' to whom Sprott refers is undoubtedly Minna Bernays, Freud's sister-in-law.

November 1922, on the topic of 'Narcissism'. But he was worried about the invitation to Freud:

> As to Freud I should think it would be possible to arrange public lectures in London, though I know nothing about the technique of going about such things. What I am sure is that a large number of people would pay to hear him, but to organise the affair would be a serious undertaking. One practical difficulty is that he has a rooted objection to speaking before large or public audiences. He hates being lionised and only likes talking to people who understand and want to learn more. I doubt if he means to come to Cambridge unless the inviatation [sic] is a special honour; at all events it is strange that he hasn't mentioned it to me. I had better write to him on the subject.[269]

Jones was clearly still fretting over the prospect of Freud's arrival in Cambridge so he wrote a brief follow-up note to Sprott on 25 August:

> I quite inferred that Freud was coming this October. Was any particular month mentioned? Is his lecture to be before your Society or is it a University invitation?[270]

And then in another paragraph, concerning his own talk:

> Would you please let me know if I may expect any knowledge of psychoanalysis in the audience. Will it be confined to students of medicine and psychology?

Jones was asking pertinent questions. Who was inviting Freud to Cambridge and to whom would he be talking? Would an audience of 'heretics' – sceptical undergraduates from all disciplines – be an appropriate audience? Jones may well have been unaware that the most provocative and intellectually progressive lectures in Cambridge had been given for some ten years to Ogden's improvised Society, bypassing the formal University and College institutions.

In January 1923, Sprott contacted Freud about the impending visit and received the predictable reply:

> You remember what our stipulations were, that I should come to Cambridge if I felt too tired to continue my habitual work and yet able to undertake some easier job. Now this case is not likely to occur. I am still active and the times are such as to command our outmost efforts to keep up work. Besides the other difficulty my inability to give lectures in another

[269] SSP WJHS/52, Jones to Sprott, 22 August 1922.
[270] SSP WJHS/52, Jones to Sprott, 25 [August?] 1922.

language than German whereby I was sure to find little sympathy and less hearers is not overcome and not likely to be.

So let us give up the phantasy of my visit to Cambridge for another year.[271]

Sprott was not strongly attached to the project of bringing Freud to Cambridge. He was never a full-blown psychoanalytic enthusiast like his friends Penrose or Ramsey. It had, in any case, been Keynes's idea in the first place. By 1923, Sprott had been appointed Demonstrator in the Psychological Laboratory at Cambridge and had lectured on 'Theories of perception with special reference to experimental work'.[272] He had excellent manners and contacts; he had great facility with languages – Lytton Strachey admired the insouciant display of his 'mind of iron' in reading Proust throughout his Final Examinations,[273] and from the beginning of the 1920s his forays into the humming low-life of Berlin were well-known to the Stracheys, prompting James to ask him in August 1924 for a list of the 'amusing Sehenswürdigkeiten' and 'best cabarets', where 'Alix might easily pass as a Transvestite gentleman.'[274] His expertise in German commanded sufficient respect from James for him to send Sebastian a copy of the newly completed *Glossary for the Use of Translators of Psycho-Analytic Works*,[275] asking for 'any emendations or additions you can suggest to the *Glossary*. There are a certain number of misprints, etc. in it; and a revised version or appendix or something is under way. Can you suggest anything for *Das Es*?'[276]

In 1924, Ogden had signed up Sprott to translate Kretschmer's *Physique and Character* for the International Library, which he had delivered by the end of the year.[277] Sprott's fluent German made him the obvious chaperone for Alfred Adler when he arrived in Britain for the first major international scientific meeting since the war, the Seventh International

---

[271] SSP WJHS/30, Freud to Sprott, 30 January 1923.   [272] Crampton, p. 276.

[273] LLS, Lytton Strachey to Sprott, 6 June 1922, p. 514.

[274] BLSP 60699, Vol. XLV, James Strachey to Sprott, 31 August 1924.

[275] Ernest Jones and others (eds.), *Glossary for the Use of Translators of Psycho-Analytic Works*, IJP supplement No. 1, London: Published for the Institute of Psycho-Analysis by Baillière, Tindall & Cox, 1924.

[276] BLSP 60699, Vol. XLV, James Strachey to Sprott, n.d. but it is probable this letter was written in the summer of 1924, since Sprott had asked Strachey for Lytton's address, which JS gave as 'Ham Spray House', to which Lytton and his ménage had moved over the summer of 1924. Indirect corroboration for this date is a letter of 9 October 1924 (BF, p. 83) in which James informs Alix of the possibility that 'das Es' may be translated as 'the id', despite its unfortunate inevitable degradation into 'the Yidd'.

[277] OP, Box 69, F.2, Sprott to Ogden, letters dated 1924? and 1 January 1925. See E. Kretschmer, *Physique and Character: An Investigation of the Nature of Constitution and of the Theory of Temperament*, London: Kegan Paul, Trench, Trubner & Co., 1925.

Congress of Psychology, held at Oxford in late July and early August 1924.[278] Of equal interest to Sprott was gestalt psychology: he invited Karl Koffka to Cambridge in early 1924 to speak to the Heretics and then visited the Koffkas at Giessen that summer, later corresponding about experiments Koffka was suggesting he perform in the Cambridge Lab.[279] His intention of writing a book on perception was probably linked to this experimental work and to his friendship with the Gestalt School represented by Koffka and Köhler, then being promoted as the answer to American behaviourism.

Reviewing three books in philosophy and psychology for the *Cambridge Review* in November 1923, Sprott's blithe good nature and attractive super-ficiality are evident when he sums up Herbert's *The Unconscious Mind*:

> And now we have yet another book on Psycho-Analysis. The 'little red books' come tumbling in, and we have a sinking feeling as the pile rises. But Dr Herbert provides a relief. His book is admirable. He is not so dismally certain as most writers on this subject; he is eminently sensible, and a particularly clear writer. He is not infected with the psycho-analytic giggle. He speaks without that irritating nudge in the ribs about wedding rings and mispronunciations.[280]

James Strachey's familiar mode of character-assassination with rapier wit captures Sprott's attitude to psychoanalysis and much else in September 1924:

> Sebastian thinks he is practically converted to Prof. Freud's views – only one or two small difficulties still remain, such as Infantile Sexuality and the Unconscious and Repression. He is now most anxious to be analyzed, though of course it is unfortunately quite out of the question for him to manage it – for purely practical reasons. – It's really extra*ord*inary how little he knows on the subject. He's only just (last month) read the Drei Abhandlungen.[281]

The same urge to be analysed and its petering out in the difficulties of Sprott's life reappeared a few weeks later when, meeting up with Tansley at the BPaS, Strachey learned that 'Sebastian had been making tentative approaches on the subject of being analysed by him. He didn't seem to think it'ld come to anything.'[282] Nor did it.

[278] H.S. Langfeld, 'The Seventh International Congress of Psychology', *American Journal of Psychology* 35(1) (1924), 148–53; BLSP 60699, Vol. XLV, Sprott to Lytton Strachey, 13 July 1923.

[279] SSP GBR/0272/PP/WJHS, File 60, six letters from Koffka between 8 May 1924 and 22 December 1925.

[280] *CR*, 30 November 1923, p. 111.   [281] BF, James Strachey to Alix, 27 Sept 1924, p. 71.

[282] BF, James Strachey to Alix, 6 November 1924, p. 109.

In early 1925, Sprott learned that Bartlett would not keep him on in the Department of Experimental Psychology at Cambridge; instead, he was offered a post in the Department of Philosophy at Nottingham University – 'did you know there was such a place?', James Strachey asked Alix.[283] Dr H. Banister, who had completed a doctoral dissertation under Hartridge on binaural localization, a major area of research in the Lab under Myers and after, joined the Department as the new Demonstrator on 25 March 1925, replacing Sprott. When the new University Statutes were introduced in 1926, the list of staff positions in Psychology was agreed as '1 Reader, 3 Lecturers and Demonstrators of whom not less than 2 be Lecturers'.[284] Banister was appointed Lecturer.[285] Bartlett, MacCurdy and Banister were the backbone of the Cambridge Department for the next twenty years and more. No other positions were added in all that time.

Sprott accepted his banishment from Cambridge and the elite worlds he had grown accustomed to. As Virginia Woolf, uncannily astute as always, had noted in her diary in March 1923:

> Sprot [*sic*] & I lunched at Mary's; then, tipsy with echoing brains, went to tea at Hill's in [Kensington] High Street. Infinitely old I felt & rich; he is very poor . . . I dont know why his experience seemed to me so meagre . . . He is hungry as a wolf, & snapping up delicacies in an alarming way.[286]

Sprott remained at Nottingham the rest of his life, becoming Reader (1928) then Professor (1948–60) in Philosophy. His *General Psychology* appeared in 1937 and he then shifted his attention towards the even more unpopular field of sociology, with three books: *Sociology* (1949), *Social Psychology* (1952) and *Science and Social Action* (1954).[287] Together with his friend J. R. Ackerley, with the two of them tutoring E.M. Forster in the double life of homosexual London in the 1920s,[288] he developed a taste for strapping young working-class men and became a faithful patron and saviour of such

[283] BF, James Strachey to Alix, 17 February 1925, p. 206.
[284] CUL, Mins, Board of Biology B, 17 October 1927, p. 42.
[285] A position he held until his retirement in 1948. See Zangwill's letter to Crampton, 13 February 1979.
[286] VWD II, 17 March 1923, p. 239.
[287] He counts as one of the founder figures of British sociology and was one of the six holders of a post in sociology who, with others, called in 1950 for the foundation of a British Sociological Association (see M. Bulmer, 'Sociology in Britain in the twentieth century: differentiation and establishment', in A.H. Halsey and W.G. Runciman (eds.), *British Sociology seen from Within and Without*, Oxford: Oxford University Press for the British Academy, 2005, pp. 36–53 at 39). There is a nice portrait of him as a founder sociologist in A.H. Halsey, *A History of Sociology in Britain: Science, Literature and Society*, Oxford: Oxford University Press, 2004, pp. 26–8.
[288] Wendy Moffat, *E.M. Forster: A New Life*, London: Bloomsbury, 2010, p. 211.

men in trouble with the law. But he was academically sufficiently in the swim of things to be a member of Solly Zuckerman's scientific activists' Oxford dining society, 'Tots and Quots' (along with Hogben, Bernal, Haldane, Julian Huxley).[289] Fittingly, in Nottingham he dropped his Bloomsbury 'Sebastian' and became 'Jack';[290] equally fittingly, from 1952 to 1954 he headed up a Rockefeller Foundation funded project on 'The social background of delinquency'. On Sprott's death, the classicist E.A. Thompson recalled the small terraced house 'in a slum street of gruesome squalor' where he lived, entertaining a Master of the Rolls, a Chancellor of the Exchequer, E.M. Forster or his array of young men in difficulty. 'I have myself heard his quick footsteps hurry to the hall door in answer to a ring, and his warm greeting. "Come in, dear boy, I *knew* you were to be released this afternoon."'[291] When E.M. Forster died in 1970, he left the bulk of his estate on trust for life to Sprott, on whose death the next year it passed to King's College, Cambridge.

In 1941, writing in her diary, Frances Partridge, in awe of Sebastian when they were fellow-undergraduates taking moral sciences, remembered him as having become a dear friend from the early 1930s on: 'He likes telling stories, and listens to them with equal concentration. But general ideas? There are a few, like psychoanalysis, he will discuss.'[292] He had clearly kept up some interest in psychoanalysis, or at least some desire that was caught by the idea of 'psychoanalysis' – the Stracheys invited him to stay for the weekend to meet Melanie Klein at dinner in 1930;[293] he visited Freud in Hampstead in the last year of his life.[294] But it was only in his translations that this interest substantively expressed itself – his own books had little to say about it. In 1931, he was very keen on translating Aichhorn's *Verwahrloste Jugend* (1925), corresponding with Jones and Strachey, only to discover an American translation, under the sentimental title *Wayward Youth*, was already in the pipeline.[295] In the same year, Jones had vetted Sprott's capabilities as a translator by requiring a draft translation of a chapter and had been extremely impressed. So Jones and Strachey asked

---

[289] Zuckerman, *From Apes to Warlords: The Autobiography (1904–1946) of Solly Zuckerman*, London: Hamilton, 1978, p. 370. Penrose was a semi-detached member of the club.

[290] A.H. Halsey, 'Sprott, Walter John Herbert [Sebastian] (1897–1971)', *ODNB*, www.oxforddnb.com/view/article/62699, accessed 12 October 2015.

[291] E.A. Thompson, 'W.J.H. Sprott', *The Times*, 11 September 1971, p. 14.

[292] FPD, 20 December 1941, p. 69.

[293] BLSP 60699, Vol. XLV, Alix Strachey to Sprott, 12 June 1930.

[294] Freud Museum Archive, LDFRD 5238, List of Visitors, Dark blue notebook.

[295] The best translation would be *Neglected* (or *Unkempt*) *Youth*. See BLSP 60699, Vol. XLV, James Strachey to Sprott, 21 February 1931, and SSP WJHS/52, Jones to Sprott, 9 July 1931.

Sprott – 'a very good translator',[296] Jones told Freud – to take on Freud's *New Introductory Lectures* when it appeared in German in January 1933.[297] Strachey was due to visit Sprott in Nottingham in June, probably to help, but cancelled because he had to write a paper for the British Society – 'The nature of the therapeutic action of psycho-analysis', Strachey's most influential strictly psychoanalytic paper.[298] That summer, as agreed with Jones, Strachey produced revisions of Sprott's translation, apologising somewhat for the extent of these due to 'an irresistible itch for making changes which comes over me when once I get started'.[299] The amicable working relationship resulted in Sprott acknowledging Strachey's help in print when the book came out in late 1933; but even Strachey's considerable supervision of Sprott did not stop him from preparing what he called a 'new' translation for the *Standard Edition* in 1964. So even Sprott's significant contribution to the psychoanalytic literature was superseded by the time of his death.[300]

Sprott was perfectly placed to become the representative of psychoanalysis in Cambridge's Psychology Laboratory. He had all the contacts and friends, both senior and junior. He had all the languages and charm. But in the end, unlike his brother Apostle James Strachey, he passed through the psychoanalytic enthusiasm of the early 1920s in Cambridge with little of it actually sticking to him. More importantly, perhaps, he was deliberately rejected by the new Cambridge psychology of the 1920s, which excluded not only psychoanalysis and gestalt psychology, but also any fundamental engagement with educational and medical psychology – key elements of the 'psychological complex' that arose between the wars.[301] Sprott's eventual turn away from general psychology to a social psychology infused by sociology and cultural anthropology[302] was an indictment in its own way of the narrowness of mid-century British psychology; it is also an indication of all the psychology that embattled Cambridge now felt it

---

[296] JF, 10 January 1933, p. 711.

[297] BLSP 60699, Vol. XLV, James Strachey to Sprott, 12 February 1933.

[298] Anon., 'British Psycho-Analytical Society', *BIPA* 14 (1933), 517–18; BLSP 60699, Vol. XLV, James Strachey to Sprott, 7 May 1933. J. Strachey, 'The nature of the therapeutic action of psychoanalysis', *IJP* 15 (1934), 127–59.

[299] BLSP 60699, Vol. XLV, James Strachey to Sprott, n.d., internal evidence indicates latter half of August 1933.

[300] *SE* XXII, 3. Strachey distinguished clearly between preparing new translations and using previous translations as a starting point for the definitive version of the *Standard Edition*; he would use the inimitable phrases 'somewhat modified', 'a considerably modified version' or 'based on that published previously' to indicate lighter changes.

[301] Nikolas Rose, *The Psychological Complex: Psychology, Politics and Society in England, 1869–1939*, London: Routledge and Kegan Paul, 1985.

[302] W.J.H. Sprott, *Social Psychology*, London: Methuen, 1952, p. x.

should ignore. Psychoanalysis was only a part of that vast excluded field. Sprott's eventual special interest was delinquency, consonant with his own erotic dispositions. Unlike many others in the 1930s, in particular the initiatives from Cyril Burt and the Institute for the Scientific Treatment of Delinquency (ISTD), founded in the early 1930s, principally by psycho-analysts, this did not become a bridge for Sprott back to psychoanalysis.

The early enthusiasm for psychoanalysis amongst the pre- and post-war Cambridge students did not often sustain itself into their mature careers. But not all of them wished to expunge it, like Bernal, from their lives. Some, like Needham, retained or at least respected that youthful enthu-siasm, even though their main interests may have taken them far from psychoanalysis. Others, marked by their youthful views, kept faith with the psychoanalytic promise, even into the Cold War era. In his address in 1948 to the Annual Congress of the BMA when he was President of the Neurology and Psychiatry Section, Lord Adrian declared:

> medical science has progressed so rapidly in the past hundred years that in another hundred there should be very little bodily disease that we cannot prevent. That side of the picture is bright enough; but there is the other, that of the diseases of the mind not so much the gross disorders of thought but the conflicts which all of us suffer to some extent. We ought to aim at making men not only free from bodily pain but also as contented as the circumstances will allow them to be. In this field the advance is likely to be slower and the prospects are less certain. Disorders of the spirit seem to involve theories and ways of prevention and treatment which must take us into quite new fields, away from the laboratory and the hospital ward and into the psycho-analyst's parlour or even into the psychosurgeon's operat-ing-theatre.[303]

But none of this added up to the forming of a new discipline in Cambridge. Why was there no Department of Psychoanalysis at a university where so many influential figures had been marked by an experience of Freud? Was an interest in psychoanalysis for the young raised in Edwardian households merely akin to a student awakening, a rebellious, extra-curricular immer-sion in sex, dreams, and a revolutionary, post-religious social vision?

---

[303] E.D. Adrian, 'The aims of medicine', *The Lancet* 252(6539) (25 December 1948), 997–1001 at 997.

CHAPTER 5

Discipline Formation – Psychology, English, Philosophy

When Lionel Penrose, aged twenty-one, and fresh from wartime ambulance duties, arrived in Cambridge in January 1919, he was disappointed to discover that there was no Psychoanalysis Tripos for him to pursue. These may have been the years of the 'wild rise of psychoanalysis', as C.S. Myers, the founder of the Department of Psychology at Cambridge, called them. But neither then nor at any point in the twentieth century could psychoanalytic enthusiasts take a degree in Psychoanalysis. Two immediate questions arise: What led Penrose to expect such a course of study was on offer? And why, throughout the century, would Penrose and others like him always be disappointed?

To answer these questions the structure of subjects and disciplines in Cambridge needs to be examined. What is instantly clear is that they were undergoing rapid change in the immediate post-war period. Indeed, this is precisely the period in which the structures of teaching, research and administration in the University were transformed into the form that would endure throughout the twentieth century. The period also saw the rise of important new disciplines, unimaginable under the Victorian dispensation of knowledge. Psychoanalysis did not turn out to be one of them.

The Reform of an Ancient University

What is a Tripos? The word derives from the three-legged stool on which sat a bachelor of arts appointed to dispute, in a humorous or satirical style, with the candidates for degrees at 'Commencement' or graduation. In the eighteenth century, this person, the 'Universitie Buffoone' as John Evelyn called the equivalent figure at Oxford in 1669, was known as the Tripos and published his humorous verses; on the reverse side of the paper was the list of candidates qualified for the honours degree in mathematics. This developed into the Tripos List, applicable particularly to Mathematics, the principal Tripos field in the first half of the nineteenth century. From there, the Tripos grew into the course of study for an Honours, as opposed to an Ordinary, degree. In the late nineteenth century the Boards of Study established in mid-century as responsible for teaching in the University began to divide the Tripos into two, so that a candidate might take both Part I and Part II, thus in effect deepening the subject's specialization. By the late nineteenth century, the serious, Honours or 'advanced' students, still very much a minority, were taking Parts I and II of a Tripos, but this would not necessarily exclude taking another Tripos as well. And it was expected that candidates for Tripos would have already taken an ordinary BA degree or its equivalent. The Tripos thus became the mechanism by which the ideals of research were developed in Cambridge – a typical example of organic English reform within a traditional framework, covering itself by proclaiming that no changes were taking place. Combinations of the prestigious Mathematical Tripos and the Natural Sciences Tripos became a path followed by committed research scientists. It also became a standard pattern that students who had studied at other universities (predominantly the University of London, which granted degrees nationwide for several geographically scattered institutions for much of the nineteenth century) would then go to Cambridge to study and take a Tripos examination.

The reform that created the precedent for later invention of Triposes and was most durable was given form by William Whewell, neologizer extraordinary, Professor of Mineralogy (1828–32) then Knightbridge Professor of Philosophy (1838–55), the polymath or 'Omniscientist' Master of Trinity College from 1841 to his death in 1866. In an attempt to stave off more radical reforms of the existing 'liberal education' based on mathematics and classics which was under criticism from a diverse range of opponents led by the Prince Consort and including many of Whewell's own friends in the elite of natural philosophers, such as Charles Darwin, Charles Babbage and William Makepeace Thackeray, the Master of Trinity College found a compromise. In 1848 Whewell drafted a report recommending that two new Triposes be created, one in Natural Sciences, one in Moral Sciences. The recommendations were implemented immediately. The conservative impulse was translated into the requirement that *only* those with a BA could take these new Tripos subjects: the essential core disciplines of mathematics[1] and classics were thereby preserved. At the same time, advanced studies at the University[2] were now open to a wide range of new fields of knowledge, Natural and Moral Sciences.

These innovations were not sufficient to prevent the Lord John Russell, Prime Minister and Bertrand Russell's grandfather, from setting up a Royal Commission on the Universities of Oxford and Cambridge. The Graham Commission made recommendations – drafted after much resistance from the Universities, and in particular from the Colleges – that were intended to prevent the ancient Statutes of the Colleges and the archaic forms of University governance from stymieing the development of a modern University – a German research University – responsive to the reform spirit of the day. The elimination of religious tests; the establishing of Special Boards of Studies (the antecedents of the Faculties recommended by the 1922 Royal Commission and established in 1926); the addition of more new Triposes, in languages, law, engineering and theology, with ordinary examinations in each of these fields of study in addition to the traditional classics and mathematics; the expansion of University teaching staff (new Professors and University lecturers); the taxing of Colleges to provide the University with funds; and the creation of a general Council overseeing the governance of the University: all these recommendations were eventually implemented, although in classic Cambridge style, resistance and inevitable compromise resulted in a piecemeal rather than wholesale implementation.[3]

---

[1] On mathematics as the core of 'permanent studies', see Martha McMackin Garland, *Cambridge before Darwin: The Ideal of Liberal Education, 1800–1860*, Cambridge: Cambridge University Press, 1980, esp. pp. 28–51.

[2] This conservative impulse would endure, indicating the haphazard manner in which teaching and research were being reconfigured in nineteenth- and twentieth-century Cambridge. Well into the twentieth century, many creators of new triposes (e.g. Tillyard in English) were horrified at the thought that the new students they had worked so hard to attract might not have a prior education in a more established, 'foundational' subject, such as classics or mathematics – Tillyard's horror focused on the possibility that English students at Cambridge might already have concentrated on 'English' while at school, having had a 'premature aestheticism' imposed upon them (E.M.W. Tillyard, *The Muse Unchained: An Intimate Account of the Revolution in English Studies at Cambridge*, London: Bowes & Bowes, 1958, p. 105) or take the English Tripos direct from school. In the 1940s Leavis, perhaps more surprisingly, similarly thought it entirely appropriate that 'English' be the culmination of a prior university education, including the sciences. See his *Education and the University: A Sketch for an English School*, London: Chatto & Windus, 1943. Less surprisingly, perhaps, English dons Lucas and Rylands at King's College held similar views (see Noel Annan, *The Dons*, London: HarperCollins, 1999 p. 71).

[3] See Peter Searby, *A History of the University of Cambridge*, Vol. III, *1750–1870*, Cambridge: Cambridge University Press, 1997, esp. p. 507 ff. and Christopher Brooke, *A History of the University of Cambridge*, Vol. IV, *1870–1990*, esp. pp. 293–7 on the Tripos system.

The medieval university, once devoted principally to the training of priests, appeared to be in inevitable decline, perhaps even approaching disappearance, in the early nineteenth century, which was the fate of all but a hundred European universities as a result of the French Revolution and its aftermath.[4] While Cambridge eventually followed the example of the German states in becoming 'modern', it did so while preserving its autonomous, corporate medieval constitution, independent from the state. While universities would always remain the training-grounds for elites – whether those elites were conceived of as state servants (as in France and the German states) or as aristocrats, clergy and 'gentlemen' in England – a genuine and revolutionary innovation came in the ideal of combining 'Wissenschaft' (science) and 'Bildung' (formation or education). It was this combination that gave birth to the institutional model of a university as fostering both research and teaching in the sciences. For Friedrich Schleiermacher, influential in the founding of the University of Berlin, the prototype of the modern university, in 1808 the aim of the University was 'to stimulate the idea of science in the minds of the students, to encourage them to take account of the fundamental laws of science in all their thinking'.[5] 'Sciences', or *Wissenschaft*, here means the full range of areas of learning and inquiry to be found in the reformed 'Faculty of Philosophy' of the German universities: philosophy, philology, history, languages, mathematical and experimental sciences. This is the source of the 'liberal arts' model of a university. Though the myth of an eternal debate between 'two cultures' permeates the English-speaking world, we should recognize, as ethnographers, that the pertinent nineteenth-century contrast was certainly never between the sciences and the arts, but between three different ideals: the ideal of education and learning for its own sake; that of vocational and professional formation; and the one best exemplified in Oxford (and to a lesser extent Cambridge) of a moral education for gentlemen who would become the leaders of Church, State and Empire.[6]

---

[4] Walter Rüegg (ed.), *A History of the Universities in Europe*, Vol III, *Universities in the Nineteenth and Early Twentieth Century*, Cambridge, Cambridge University Press, 2004, p. 405.

[5] Quoted in *ibid.*, p. 5.

[6] This division between liberal arts and vocational/professional educations has always been the backbone structure of American universities; Harvard, a good example and a long-term leading institution, has several faculties, including the principal traditional vocational faculties of Divinity, Law and Medicine, together with more modern ones such as Education, Design, Engineering, Dental Medicine and Government. The liberal arts are housed together in the 'Faculty of Arts and Sciences', which is, of course, at least in terms of expenditure and capital accumulation, if not necessarily in terms of faculty numbers or student numbers, dominated by the 'sciences'.

German reforms had promoted the authority of 'Philosophy' at the expense both of the training of priests (Theology) and of the other ancient vocational Faculties, Law and Medicine. The University was to become secular and avocational in its orientation. In Germany however, the new University would very quickly and unashamedly become a branch of the state, with education and research as public civic duties, financed by the state, even when driven by the ideals of 'Wissenschaft' and 'Bildung'. In France, with the Revolution, the medieval university was displaced by professional and vocational institutions – state museums, hospitals, professional schools (the *Grandes Écoles*) and other specialized educational institutions; the university, when revived, also became a function of the secularized state.[7] Both these models of higher education as part of the growing state's functions, the German and the French, would remain anathema to key groups of Cambridge and Oxford dons. In Oxbridge, the emphasis on purity of knowledge, separate from its employment and sullying, would converge with the aristocratic and gentlemanly ethos of the two universities to produce an unlikely new hybrid. The pioneering scientific spirit infiltrated the traditional university, complete with its own corporate autonomy, and coalesced around a new ideal in which the ancient freedom of its members was now geared to teaching, study and research.[8] It was only because the two English universities had escaped the turbulence of the Napoleonic wars with their property and medieval freedoms intact that they could continue to defend fiercely their autonomy – and their financial and class interests. The contrast was stark: on the Continent, the Professor became a civil servant of the lay and bureaucratic state;[9] in Oxbridge, the University Professor was often entirely peripheral to the life of College Fellows and students in the first half of the nineteenth century. Until the 1870s, religious tests were still in full force: these excluded non-conformists from the student body and required the compulsorily unmarried dons to accept Anglican dogma.

Pressure for reform came from within Oxbridge and from without. The structurally inchoate organization, with a central University and a score or so of independent Colleges, sometimes facilitated reforms: if agreement for wholesale reform was all but impossible, partial innovation in one tolerated corner followed by another and then another eased piecemeal change. An initiative such as the Cambridge Philosophical Society, created by natural

---

[7] Peter J. Bowler and John V. Pickstone (eds.), *The Cambridge History of Science*, Vol. VI, *The Modern Biological and Earth Sciences*, Cambridge: Cambridge University Press, 2008, p. 6.
[8] Rüegg, *A History of the Universities in Europe*, p. 14. [9] *Ibid.*, p. 7.

philosophers (Henslow and Sedgwick, Darwin's teachers) outside the University, but eventually in coordination with it, created the conditions for the expansion of studies first in botany, zoology and geology. From such initiatives came the mid-Victorian era of the Museums, the first local spaces and buildings devoted to the sciences. These eventually led to the foundation of the Natural Sciences Tripos, when pressure for more general reform to allow the representation at Cambridge of new middle-class, manufacturing, and often Dissenting interests had built up sufficient momentum.[10]

The relatively easy bedding down of the natural sciences was facilitated by the fact that most of the professorships established in the eighteenth century were in the new sciences, despite the fact that Cambridge then was at its most corrupt and peripheral to English scientific and cultural life, which went on elsewhere – in the Dissenting academies, the Scottish universities, and the new private London medical schools established by Scottish graduates such as the Hunters. The Chairs established at Cambridge from Newton to Whewell were mostly in mathematics, natural philosophy or the Baconian sciences.[11] So the organization of the natural sciences into a large group had considerable university resources to draw upon. The Natural Sciences Tripos has retained an overarching structure to this day, giving Cambridge 'science' a distinctively catholic and broad-based character, since new disciplines were readily brought under its generous umbrella – first geology, mineralogy, chemistry, zoology, botany, physiology, then physics, pathology, biochemistry, pharmacology, genetics and others.[12] This is not to overlook the fact that some subjects chose or were forced to make their own independent way: in particular, at the end of the nineteenth century, the 'mechanical sciences'(1894) and Agriculture (1900) were organized separately from the 'natural sciences'.

[10] Roy MacLeod and Russell Moseley, 'The "Naturals" and Victorian Cambridge: reflections on the anatomy of an elite, 1851–1914', *Oxford Review of Education* 6(2) (1980), 177–95.

[11] The complete list for 1650–1851 is: Mathematics (1663), Moral Philosophy (1683), Music (1684), Chemistry (1702), Astronomy and Experimental Philosophy (1704), Anatomy (1707), History (1724), Botany (1724), Geology (1728), Astronomy and Geometry (1749), Divinity (1777), Natural Philosophy (1783), Laws of England (1800), Archaeology (1851).

[12] Nor by any means was the development of sciences without great unevenness: for example, important developments in genetics in Britain (in both Cambridge and London) took place in the first ten years of the twentieth century (including the creation of the term by William Bateson, one of the key promoters of Mendelism, in Cambridge). However, a Department of Genetics scarcely existed till the 1950s and a Part II within the NST was created only in 1951; it never had more than three students until the 1960s. This was the apparently parlous state of affairs alongside the discovery of the molecular structure of DNA at the Cavendish Laboratory for Physics, also in Cambridge.

With the introduction of new subjects to be examined, reform of the relations between the Colleges and the University became indispensable. The Colleges were required to pay what was effectively a tax to the University to sustain the growing expenditure called for by the new subjects on new lecture rooms, museums and eventually University laboratories alongside the Colleges' own laboratories. Given the parlous state of College finances in the agricultural depression of the late nineteenth century, the University for the first time began actively to seek external patronage and donations and thus to offset the power of the Colleges.[13] The most momentous change of all, perhaps, began in the 1860s with the gradual and halting appearance of women in Cambridge. Two women's Colleges, Girton and Newnham, were founded. Their students attended lectures (if permitted to do so by the lecturers) and had, to start with, to take unofficial examinations, courtesy again of the good will of examiners. In 1881, women were granted the official right to sit the Tripos exams. Newnham insisted from that year on that all its students do so. The first men's College to insist on the Tripos had been King's in 1873.[14] The women students were encouraged to take their studies more seriously than the average man; their degree results, consistently superior to those of men, were the proof that they did, and proof of much else that for much of the twentieth century remained politically charged.

The arrival of women was in part due to external agitation for change: the early feminist campaigner Emily Davies, author in 1866 – the year of John Stuart Mill's petition for women's suffrage to Parliament which she, of course, signed – of *The Higher Education of Women*, was one of the founders of Girton College and its first mistress. She insisted her women sit the Tripos even though it had to be an unofficial one. The foundation of Newnham College, on the other hand, was in large part due to the Cambridge insider Henry Sidgwick, one of the 'Trinity reformers' of the 1860s and 1870s, a leader of the 'revolution of the dons'.[15]

Sidgwick embodies in his life and career many of the forces transforming late Victorian Cambridge. Born in 1838, the son of a teacher-clergyman, educated at Rugby and Trinity College, Cambridge, Senior Classic (ranked

[13] J.P.D. Dunbabin, 'Oxford and Cambridge college finances, 1871–1913', *Economic History Review*, N. S. 28(4) (1975), 631–47.

[14] D.A. Winstanley, *Early Victorian Cambridge*, Cambridge: Cambridge University Press, 1940, pp. 220–33.

[15] S. Rothblatt, *The Revolution of the Dons: Cambridge and Society in Victorian England*, London: Faber & Faber, 1968.

top of the Tripos List) and Wrangler (a First in Mathematics), he was elected into a Fellowship at Trinity College in 1859, the year of his graduation, in classics. In turmoil over ethical and theological issues throughout the 1860s, he became, with his *Methods of Ethics* (1874), the pre-eminent moral philosopher of his time, still regarded as the first modern philosopher in the field. The leading theme of his life was, perhaps, a full recognition that the great advances of the age had been made in the natural sciences, so that, instead of compulsively pursuing edification, ethics would benefit greatly from adopting 'the same disinterested curiosity to which we chiefly owe the great discoveries of physics',[16] including, crucially, the same sceptical and individualist relation to authority. Wrestling with theological doubts – he had already adopted a Theist position – at the height of discussions in Cambridge about the removal of the Test Acts, he resigned his Fellowship at Trinity in 1869 since he could no longer subscribe to the Thirty-Nine Articles of the Church of England. Already indebted to Sidgwick as a principal driving-force behind reform and recognizing his philosophical excellence, Trinity immediately appointed him to a position which did not require 'subscription' (to the Thirty-Nine Articles) – a Lectureship in Moral Science. Alongside his theological struggles, he pushed, successfully, for College reforms, firstly by encouraging both the College and the natural sciences, through the provision of scholarships for students, to elect a Fellow every three years in the natural sciences, secondly by opening up the office of Praelector to a married man, and giving the post some security of tenure.[17] Trinity moved to appoint an important figure in the sciences, first offering a Praelectorship to William Thomson (later Lord Kelvin) who declined, and then, at the behest either of George Eliot and George Henry Lewes or of T.H. Huxley, decided upon physiology and Michael Foster as the person to invite – who duly accepted.[18] Foster's arrival in Cambridge led to the establishment of the 'Cambridge school of physiology', recognized nationally then internationally, and which became a regular supplier of Nobel Prizes in the twentieth century (A.V. Hill, C.S. Sherrington, E.D. Adrian, Henry Dale).

Trinity's connections with Cambridge physics was no less close than with physiology. The Cavendish school became the prerequisite for, and closely identified with, the 'long-standing project to support Maxwell's

[16] Henry Sidgwick, *The Methods of Ethics* (1874), London: Macmillan, 1907, p. vi.
[17] Winstanley, *Later Victorian Cambridge*, pp. 246–9.
[18] Gerald Geison, *Michael Foster and the Cambridge School of Physiology*, Princeton: Princeton University Press, 1978, pp. 75–6.

theory via electrical metrology';[19] its successive Cavendish Professors of Experimental Physics, Maxwell, Rayleigh, Thomson, Rutherford, Bragg, were all Fellows of Trinity (and all, save Maxwell, who died before Nobel Prizes were instituted, were awarded Nobel Prizes (1904, 1906, 1908, 1915 respectively)).[20] A similarly critical intervention in the development of Cambridge biochemistry also came from Trinity: Gowland Hopkins, hired on spec by Michael Foster as a 'chemical physiologist', was struggling to survive as an Emmanuel Fellow, then Reader, teaching medical students; indeed he succumbed to a breakdown in 1906. Trinity College rescued him to the post of Praelector in 1910[21] – the same research-privileged College post Foster occupied. Hopkins was elected into a Professorship of Biochemistry in 1914, but Trinity continued to support him since the University paid no stipend.[22] The success of Trinity's research ethos can be seen in the fact that the six Presidents of the Royal Society from 1915 to 1950 were all Cambridge-educated in this period, all Nobel Prize-winners and, with the exception of Sherrington (Caius College), all Trinity men. So a string of 'Cambridge schools' in the sciences were established, the most famous of which were physics, physiology and biochemistry. It was the wealth, authority and increasingly unambiguous support of Trinity for the sciences, from the era of the 1860s Trinity reformers on, that mattered a great deal in these developments, even, so historian of physiology Gerald Geison speculated, to the point of marking the principal difference between Oxford and Cambridge in the late Victorian era.[23]

Sidgwick's famous labours to create more educational possibilities for women in Cambridge were, again, by no means solitary – he was nothing if not a networker. But in the establishment of Newnham College he was the acknowledged leader: organizing lectures, inviting Anne Clough, early suffragist and promoter of higher education for women, to be responsible

[19] Andrew Warwick, *Masters of Theory: Cambridge and the Rise of Mathematical Physics*, Chicago: University of Chicago Press, 2003, p. 336; see also p. 288.
[20] Geison, *Michael Foster*, pp. 113–14.
[21] See Mark Weatherall, 'Bread and newspapers: the making of "A Revolution in the Science of Food"', in Harmke Kamminga and Andrew Cunningham (eds.), *The Science and Culture of Nutrition, 1840–1940*, Amsterdam and Atlanta, GA: Rodopi, 1995, pp. 179–212. See also Harmke Kamminga, 'Hopkins and biochemistry', in Peter Harman and Simon Mitton (eds.), *Cambridge Scientific Minds*, Cambridge: Cambridge University Press, 2002, pp. 172–86.
[22] 'The University of Cambridge: the modern university (1882–1939)', in J.P.C. Roach, *A History of the County of Cambridge and the Isle of Ely*, Vol. III, *The City and University of Cambridge*, Oxford: Oxford University Press, 1959, pp. 266–306, www.british-history.ac.uk/report.aspx?compid=66634, accessed 1 April 2008.
[23] Geison, *Michael Foster*, p. 357.

for the young women students, acquiring an initial house on Regent Street in 1871, then moving the young women to Merton House on Queens' Road while a purpose-built hall in Newnham village was under construction. It opened in 1875. Sidgwick had used his own money to accomplish much of this, and had organized the supporters in and outside Cambridge to lecture, subscribe and donate. The college was formally established in 1880, with Clough as Principal.

In this same period, his long-term interest since the 1860s in spiritualism led him to attend séances at the house of his former student Arthur Balfour, where he befriended and became engaged to Balfour's older sister, the remarkable Eleanor Balfour. Their marriage in 1876 saw Sidgwick marry not only into the aristocracy but also into one of the richest families in England (through the railways) and one central to Tory politics for the next forty years. Eleanor's uncle was Lord Salisbury, Prime Minister 1885–92, and again 1895–1902, when he was succeeded by her brother Arthur until 1905. Eleanor was a match in every respect for Henry. A fine mathematician, she later assisted J.W. Strutt (Lord Rayleigh, who married her sister) in the Cavendish Laboratory with electrical standards experiments, became Deputy Principal and then Principal of Newnham College and was in later years the scientific-methodological backbone of the Society for Psychical Research. Henry Sidgwick became the academic politician par excellence, finding the compromises which gradually allowed the University to share in the Colleges' financial resources.

Wealth, power and influence: all these were visible in Sidgwick's life, even before the marriage, but he was a master at deploying them increasingly from this point on. To give one small example: he again supported Michael Foster's School of Physiology by personally paying £1,000 for his laboratory costs in 1889, and a further £500 in 1891, when the University's finances were in crisis.[24] The Balfour generosity to Cambridge was also constant and long-term: Arthur gave (or organized an anonymous donation of) £10,000 to establish a Professorship in Genetics in 1910. The Sidgwicks, the Balfours and the Rayleighs epistomized the patrician reformist period of Victorian Cambridge with its unambiguous support for the rise of 'modern subjects' (in particular the sciences). In the late 1860s, though himself a classics don, Sidgwick had originally hoped that the women students coming to Cambridge to study at what became Newnham College would be able to avoid learning Latin and Greek, since in his over-optimistic vision these subjects would very soon cease to

[24] *Ibid.*, p. 309.

be compulsory for men as well. It turned out that that modernizing battle would only be won fifty years later, during the First World War.[25]

The Trinity Reformers' ideal of the modern University – without religious restrictions and no longer dedicated to the training of the clergy, inclusive of women, increasingly excluding those who did not have a vocation of teaching and research, gradually opening up to the manufacturing, business and professional classes – was realized best in the expanding Natural Sciences Tripos. Students from northern families, from the manufacturing and new middle classes, increasingly took the NST without fear of being without a career, since professional scientific posts opened up from the 1870s on. The NST thus became a key factor in making Cambridge an agent of social and occupational mobility, through the opening up of new scientific professions, as well as the transformation in medicine. At the same time, following Trinity's lead, more Colleges created Fellowships for those teaching and researching in the sciences – and in the process created a research-oriented ethos in the University.

The shift away from the Church as the chief destination for Cambridge graduates was in effect a major step in the progressive secularization taking place not only in Cambridge but across Europe. A steadily increasing number of graduates would become members of the other professions, always most frequently teaching, but with significant numbers going into the Home and Indian Civil Services (for which the Triposes in Semitic Languages (1878) and Indian Languages (1879) had been created). By the turn of the century, many would go into the previously shunned fields of business and manufacturing.[26] Most importantly, a steadily growing proportion of students began to take the Tripos – the advanced courses which, in Cambridge, doubled up as preparation for research and a possible career in an academic speciality, and as an Honours degree.

Nonetheless, the principal career for NST graduates became and remained medicine: over 50 per cent of all undergraduates who took the NST between 1851 and 1916 did so as a preparation for their eventual medical practice.[27] The peculiarity of Cambridge's relation to medicine was that its ancient special privileges (with Oxford) granted by the Royal College of Physicians had decayed virtually to zero, in terms of medical graduates, across the eighteenth century. In their place, the new London medical schools had, with Edinburgh, dominated medical training

---

[25] A. Sidgwick and E.M. Sidgwick, *Henry Sidgwick: A Memoir*, London: Macmillan, 1906, pp. 208, 212, 351, 510.

[26] L.P. Wilkinson, *A Century of King's, 1873–1972*, Cambridge: King's College, 1980, pp. 158–67.

[27] MacLeod and Moseley, 'The "Naturals" and Victorian Cambridge', pp. 186–7.

throughout much of the nineteenth century and would continue to do so in the twentieth century. Cambridge created a new position of privilege for itself by subordinating clinical medicine to the ideal of non-vocational science teaching and research that the NST stimulated. Despite numerous attempts at reform, the efforts to bring clinical training to Cambridge met with failure from the 1860s to the 1950s, even during the period of the ill-fated Clinical School from 1880 to 1920.[28] Medical students there increasingly were, but all agreed that they would learn medical science, above all from the physiologists, but clinical medicine would not figure in Cambridge medicine. The sciences from which medicine might and increasingly must learn became the pride of the University, with clinical medicine a low-status unspeakable quasi-artisanal skill to be acquired elsewhere. No Medical Science Tripos was created. When Archie Cochran entered King's College in 1927 to take NST Part I in physiology, biochemistry and zoology and told his tutor he planned to study medicine, the response was: 'Cochrane, if you wish to study a trade, you must do that in your vacations. You have come here to be educated.'[29] The long-term hostility of Cambridge science to clinical medicine of any sort would have a significant effect on the possibility of psychoanalysis gaining recognition at Cambridge.

In 1870, the numbers of students taking Tripos examinations were as follows: Mathematics (115), Classics (66), Moral Sciences (15, of whom 9 were from Trinity or St John's), Natural Sciences (17) and Law and History (23). In the years after 1870, the numbers taking Classics and Natural Sciences increased rapidly. In 1900, 133 students took Part I of the NST and 25 took Part II (of those 25, 14 gained Firsts – indicating that this was an examination geared not to normal distributions but to sifted expertise); 127 students took Part I of the Classics Tripos and 12 took Part II (of those 12, 4 gained Firsts).[30] Not the least important factor in sustaining the development of these advanced courses was the growing influence of women students; in certain subjects, particularly the newer ones, in Medieval and Modern Languages (MML) and in Economics, women formed a significant proportion of Tripos candidates, far higher than their very limited absolute numbers within the University would indicate: in MML in 1895, 11 men and 17 women took the Tripos; 6 of the women

---

[28] Mark W. Weatherall, *Gentlemen, Scientists and Doctors: Medicine at Cambridge, 1800–1940*, Cambridge: Cambridge University Library, The Boydell Press, 2000.

[29] Archibald L. Cochrane with Max Blythe, *One Man's Medicine: An Autobiography of Professor Archie Cochrane*, London: BMJ, The Memoir Club, 1989, p. 8.

[30] HR 1910, *passim*.

got Firsts; in 1910, there were 29 men and 33 women, of whom 10 got Firsts. In Economics Part I in 1910, there were 11 men candidates and 7 women. By 1912, over 70 per cent of all candidates (1,066) taking Tripos were taking four out of the eleven examinations: History (21.4%), Natural Sciences (20.1%), Mathematics (19.3%) and Classics (10.5%); 12.6% of the candidates taking these four Triposes were women.

However, of degrees awarded in 1912 (necessarily only to men), 45% were Pass degrees (i.e. Ordinary degrees or those who had attempted Honours and failed).[31] Until well into the twentieth century, most Cambridge men took the Ordinary degree or left without a degree – some of these would have been in the tradition of the early nineteenth-century 'bloods and wastrels',[32] scions of the aristocracy, some rowing rowdies, since the prestige of rowing rose at the same time as the prestige of the Tripos. Even those who took 'Tripos' (i.e. for honours) might take a number of different Triposes. In effect, the Tripos was an advanced course of study – equivalent in certain respects to a research degree and regarded as such by the University. Even when a category of 'Advanced Student' was introduced in 1897, precisely to cater for students who wished to come to Cambridge to pursue research, one of the options offered them was to proceed to the Degree of BA or LLB by means of Tripos examinations. The fact that the Tripos examination was already framed as a 'research degree' for advanced students, many from other institutions such as University of London or Owen's College, Manchester, in part explains the late arrival to Cambridge (1919) of a distinctive feature of the German research university model – the PhD.[33] It also explains one of the oddest features of the relatively rapid shift from the mid-nineteenth-century norm of a degree consisting of a bare smattering of Greek, Latin and mathematics to rigid expectations of advanced specialization within the overall British educational system, a specialization completely out of sync with other countries' degree structures, which retained the generality of German 'Philosophy' (i.e. science, languages, etc.) or the American liberal arts curriculum (including physics, rhetoric and everything in between). The excessive

[31] Rita McWilliams Tullberg, *Women at Cambridge*, Cambridge: Cambridge University Press, 1998, p. 194, based on the data submitted to the Royal Commission and published in 1922.
[32] Sheldon Rothblatt, 'State and market in British university history', in Stefan Collini, Richard Whatmore and Brian Young (eds.), *Economy, Polity, and Society: British Intellectual History 1750–1950*, Cambridge: Cambridge University Press, 2000, pp. 224–42 at 227.
[33] And it also explains the relative reluctance of many subjects to adopt the PhD as the currency of disciplinary mastery. The introduction of the PhD in Cambridge was expressly designed to attract principally American students, hoping in the post-war era to stem their decades-long assumption that advanced study meant study in a German university for a doctoral degree.

specialization of British schoolchildren throughout the twentieth century until today is in large part due to the fact that, by the middle of the twentieth century, clever boys and girls from less affluent backgrounds, hoping to win State or Oxbridge College Scholarships in order to pay for their expensive and privileged education, were obliged to take a course of study leading to Tripos examinations which had originally been established as courses of advanced study for researchers. The tail of advanced research had inadvertently ended up wagging the dog of everyone's education.

The Oxford and Cambridge Act of 1923 for the first time brought the two ancient Universities under a certain amount of direct state control. They were virtually the last two universities in Europe to undergo this change of governance. (In the German lands, by 1880 all university income came from the state.[34]) Government control was inevitably exerted through the leverage of funding: money would only be forthcoming if specified reforms were introduced. The expansion of Cambridge throughout the nineteenth century had placed its finances under severe pressure. Generous private donations had began to take up some of the strain, particularly in the sciences. But the agricultural depression of the last thirty years of the century reduced College incomes dramatically, thus stymieing the new system of generating University income by taxing the Colleges. Some indirect state funding arrived with new subjects and indeed patterns of education. There had been considerable opposition to the Mechanical Sciences because of their 'trade' and vocational orientation; eventually founded in 1894, and renamed Engineering in the 1960s, they became the largest institution in the late twentieth-century University. With the crucial carrot of state funding, Agriculture had met with less opposition, perhaps because of its extreme relevance to the Colleges (and as a result the University), all dependent on agricultural rent from land which for a long time by law and also by traditional prudence they had been prohibited, and then were reluctant, to sell.[35] But the more significant and therefore contested form of state funding was from the Medical Grants Committee funnelling funds from the Medical Research Committee (later Council). In accepting the principle of state funding from the MRC pre-1914, the precedent was set for what followed post-1918 and continued until the 1980s. Confronted by bankruptcy, the University accepted a block grant from the newly founded University Grants

---

[34] Rüegg, *A History of the Universities in Europe*, p. 111.

[35] On the colleges' dependence on agricultural rents and the Board of Agriculture's consequent oversight of college finances, see the illuminating Robert Neild, *Riches and Responsibility: The Financial History of Trinity College, Cambridge*, Cambridge: Granta, 2008.

Committee. By committing itself to broadening the social background of the student body, it also accepted other indirect forms of state funding – for instance, granting entrance to increasing numbers of students with scholarships from poorer backgrounds. Financial deliverance came with conditions attached. The recommendations of the Asquith Commission (1922), which became the Statutory Commission, had to be implemented. The modern system of faculties and departments with concomitant centralization of fees, payments and a system of salaries and a pension scheme (with a compulsory retirement age) came into being.

An equally fundamental transformation had already taken place and was one of the reasons for Parliamentary intervention in the University's affairs: the broadening of the University from the education of clergy and gentlemen to the education of women, non-conformists and less wealthy sections of English and overseas society; and the establishing of the University as a research-dominated institution, particularly in the Sciences. As Asquith wrote in the Commission's Report: 'the growth of science at Cambridge since the era of the Royal Commissions has been perhaps the greatest fact in the history of the University since its foundation'. This was a fact distinctive to Cambridge. In contrast to Oxford which, well into the latter half of the twentieth century and despite its own considerable expansion of scientific activity, maintained an ethos centred on the education of the country's elite in government and public affairs, the leading edge of reform in Cambridge was from the 1850s on closely associated with the advance of the sciences, both in terms of teaching and in the establishing of new laboratories and facilities.

Was it the massed presence of scientists which created so receptive an environment for Freud's new science? Certainly this was one of the distinctive features of Cambridge compared with Oxford, dominated by the prestige of 'Greats' in place of mathematics and the NST. And it is striking that most of the figures in this book came from mathematics and the natural sciences (many because they intended to continue on to a medical degree in London) together with the other distinctive course of study invented in Cambridge in 1848: the Moral Sciences Tripos.

The Moral Sciences and the Revolution in Philosophy

The other pillar of mid-Victorian Cambridge examination reform, the Moral Sciences Tripos (MST), was to be consistently dogged first by a lack of focus and later by fissiparous tendencies. It was never altogether sure of its intellectual distinctiveness so its fledgling sub-disciplines ended up by finding their own feet through unilateral declarations of independence.[36] The term 'Moral Science' can be understood as an updating or division of 'Moral Philosophy', in parallel with the updating of 'Natural Philosophy' to 'Natural Science'.[37] The final section of John Stuart Mill's influential *A System of Logic* of 1843 (which Whewell specified as one of the set texts for the Tripos) was entitled 'On the logic of the moral sciences'. This gives some idea of the scope of the designation: it included ethics and moral philosophy, but also extended to psychological, political, sociological and historical domains, in short the laws of Mind and of Society. Whewell, whose promotion of the inductive sciences did much to foster science at Cambridge, but who was actually personally opposed to the educational innovations of the natural and moral sciences, in 1862 recorded how the new Tripos was a satisfactory response to the problems that had befallen the teaching of moral philosophy at Cambridge in the 1840s:

> after various attempts and changes, there was established a *Tripos* or Examination List, which may be regarded as an important era in the history of Moral Philosophy in England. The moral studies thus encouraged were

[36] It was even dogged by a peculiar indefiniteness in names: the committee that regulated teaching and research in the subject was officially designated the 'Special Board for Moral Science' or later the 'Board for Moral Science', whereas the examinations students took were in the 'Moral Sciences Tripos'; what hung on the difference between 'Moral Science' and 'Moral Sciences' was never made clear. The vestige that remains in the twenty-first century of the 'Moral Sciences' has opted for the plural – the Moral Sciences Club.

[37] Garland, *Cambridge before Darwin*, p. 58, bravely glosses Adam Sedgwick's 1832 critique of the teaching of 'Moral Philosophy' as 'in modern terms, the social sciences', but this anachronistic stab at 'translation' at least reminds us that both 'moral philosophy' and 'moral science' have complex and changing meanings. It's worth recalling that Whewell's neologizing propensity was expressed predominantly in the sciences, including the introduction of the term 'scientist' in 1833. The term 'moral sciences' has a complex history, not least in its peregrinations through other languages. Julien Vincent's comparative overview leads him to conclude that the most appropriate gloss on 'moral science' or 'sciences morales' is 'sciences mentales' or 'mental science' (see Julien Vincent, 'Les "sciences morales et politiques": de la gloire à l'oubli? Savoirs et politique en Europe au XIXe siècle', *Revue pour l'histoire du CNRS* 18 (automne 2007), 38–43). Mill's 'moral sciences' was translated into German as 'Geisteswissenschaften' (see J.S. Mill, *System der deduktiven und induktiven Logik: eine Darstellung der Principien wissenschaftlicher Forschung, insbesondere der Naturforschung*, 2 vols., Braunschweig: Vieweg, 1862, 1863) – the famous final section was translated as 'Von der Logik der Geisteswissenschaften oder moralischen Wissenschaften'. Despite misgivings about the term 'Geisteswissenschaften', Dilthey decided to employ it. Following his endorsement, it became a standard term in German philosophy, which is oddly enough often left by English-language commentators in German (as if it were untranslatable). See Frederick C. Beiser, *The German Historicist Tradition*, Oxford: Oxford University Press, 2012, pp. 325–31. Georg Simmel's youthful critique of Dilthey was entitled *Einleitung in die Moralwissenschaften* (written in 1892, eventually published in 1989), which indicates that the term 'Moralwissenschaften' did have some traction in German as well as in English. For the immediate Cambridge context, see D. Palfrey, 'Moral science at Cambridge University, 1848–1860', PhD dissertation, University of Cambridge, 2001; J. Vincent, 'Disestablishing Moral Science: John Neville Keynes, religion, and the question of cultural authority in Britain 1860–1900', PhD dissertation, University of Cambridge, 2006.

not only Moral Philosophy, but Mental Philosophy and Logic; and further, as another group of studies, History and Political Philosophy, Political Economy, and Jurisprudence.[38]

Shortly after its foundation, a Law Tripos was introduced, in part to respond to competition from University College London, which was making English common law part of a university education for the first time. In 1870, this became a short-lived Law and History Tripos, when in 1875 a separate History Tripos was founded. Because of this idiosyncratic development, with History first being introduced as closely bound to Law, History, originally part of the MST, became an autonomous discipline, at least for the purposes of teaching and examination. In 1902, Alfred Marshall's plea for the independence of economics depicted the history of the Tripos as follows:

> the first Moral Sciences Examination (1851–60) included ethics, law, history, and economics; but not mental science or logic. In 1860, however, philosophy and logic were introduced and associated with ethics; while history and political philosophy, jurisprudence and political economy formed an alternative group. In 1867 provision was made elsewhere for law and history; and mental science and logic have since then struck the keynote of the Moral Sciences Tripos.[39]

In Marshall's tendentious (but not necessarily inaccurate) view, the MST had suffered an early take-over by a narrowly conceived philosophy; clearly he thought it had never fully recovered from this take-over.

The Moral Sciences had by this time become principally the home of philosophy, politics, psychology and political economy, but it alone of the new Triposes failed to increase numbers of students; 4 men and 3 women took Part I in 1898; 5 men (plus one advanced student) and 2 women took Part II. In 1911, 6 men and 1 woman took Part I; 1 man and 3 women took Part II (all the women achieved Firsts that year). In this period, Marshall, who nonetheless shared Sidgwick's vision of the primacy of ethics, viewing economics as 'applied ethics', worked energetically to secure the autonomy and independence of his subject, noting how potential students were put off by 'the metaphysical studies that lie at the threshold of [the Moral

---

[38] William Whewell, *Lectures on the History of Moral Philosophy*, Cambridge: Deighton, Bell, 1862, Appendix: 'On the recent arrangements respecting Moral Studies in the University of Cambridge', pp. 278–9.
[39] Alfred Marshall, 'Plea for the Creation of a Curriculum in Economics', 1902, quoted in Keynes, 'Alfred Marshall', in *The Collected Writings of John Maynard Keynes*, Vol. X, *Essays in Biography*, London and Cambridge: Macmillan and Cambridge University Press for the Royal Economic Society, 1972, pp. 161–231 at 213.

Sciences] Tripos'.[40] So in 1905 an Economics Tripos was founded. (It was in this fledgling discipline that John Maynard Keynes was to take up one of the first positions in 1908, after following the Mathematics Tripos, spending much of his undergraduate time discussing philosophy and political questions while avoiding the MST within which his father, John Neville Keynes, taught Logic.) Such fission and separation were undoubtedly a means for the production of specialist courses of study, thus catering for the 'serious' students who, although still in a distinct minority, were growing in numbers. But the example of the Natural Sciences Tripos indicates there were other means available for this. The NST accommodated specialization under its umbrella, while undergoing stresses and strains, not least of which was the tension of its relations with the higher-status Mathematics Tripos, where mathematical physics of the Maxwellian variety was still to be found.[41] Whewell's original distrust of the NST, grounded in its divergence from his ideal of the command of knowledge in general, was in the twentieth century reconfigured as the promotion of the 'liberal scientific education' – and thus fused with the shadow of the gentlemanly amateur, providing a firm bulwark against any hint of vocational study. The requirement to take three (sometimes four) different sciences at Part I of the NST became, it was now argued, the best means to develop creative research scientists, in contrast with the narrow vocational scientist who had only studied one subject, his specialist subject, in depth.[42]

The more common method of allowing for specialization was the splitting of Triposes into Parts I and II while the most usual earlier course of study for serious students was to take more than one Tripos subject. In order to retain students and in order to provide them with opportunities for advanced courses, by 1910 most Triposes offered two Parts. A student could acquire an Honours degree (for which Tripos was the essential prerequisite) by taking first a Part in a Tripos, then later another part. New Triposes – such as English and Anthropology – would of necessity be founded with only one Part to start with.

At the turn of the century, before the departure of political economy to the new Economics Tripos, lectures in the MST covered in roughly equal measure psychology, logic, ethics and political economy; the more

[40] 'The present position of economics' (1885), reprinted in *Memorials of Alfred Marshall*, ed. A.C., London: Macmillan, 1925, p. 171.
[41] Warwick, *Masters of Theory, passim.*
[42] Alistair Sponsel, 'The Cambridge Natural Sciences Tripos, 1915–1949', MSc dissertation, Imperial College, London, 2001.

advanced Part II course introduced Politics and the History of Modern Philosophy (i.e. seventeenth- and eighteenth-century philosophy).[43] The dominant philosophical force in late nineteenth-century Cambridge, Henry Sidgwick, had wished to promote an ethics founded on reason as a viable alternative to theology and to make the MST into a 'nursery for intellectual statesman'.[44] But for G.E. Moore and Bertrand Russell, their most important influence was the idealist philosopher McTaggart (also, with them, and with Sidgwick, an Apostle). Their generation had different philosophical ambitions and gave birth in the first decade of the twentieth century to analytic philosophy, thereby professionalizing philosophy both in Britain and in America and leaving 'continental' philosophy to one side.

The German revolution of the universities was always a response to the rise of the sciences. Cambridge inflected this revolution differently. It had its own traditions and innovations. If Kantianism, neo-Kantianism and *Naturphilosophie* were distinctively German responses, and were followed by varieties of materialism as a rejection of the initial idealist waves, by the end of the century three different scientific programmes, developing across Europe, were putting in question the organizational hegemony of German philosophy. These were the programmes in mathematical and experimental physics; the programme in experimental psychology; and the various social sciences – in Comtean mode, in Weberian historical-economics-inspired sociology or in the tradition, strongest in Cambridge, of political economy. What price metaphysics in the face of the advances of physics in thermodynamics, in electromagnetic theory and later of relativity theory? What price the theory of knowledge when the psychology of perception, even a psychological account of the Kantian categories, was being advanced by trained introspectionists in the new psychology laboratories? What price logic when a 'science of thought' (George Boole), in which psychology and logic were seamlessly intertwined (Franz Brentano), was being developed from the mid-nineteenth century on? And what price ethics when a reformed utilitarianism was founded upon a psychology and an inductive logic (Mill)?

Physics, psychology and political economy were set to usurp, if not replace, the initially central place of philosophy in the new research university: the very name, 'moral sciences', gave this game away. Yet in Cambridge, philosophy's place had never been privileged. Nonetheless, it

---

[43] Students around this time referred to it as 'moral stinks'. E.g. Tansley to Bertrand Russell: 'I was very glad to see you & Sawyer got your fellowships all right. Rather a score for Moral Stinks isn't it?' (BR 710.056798, 12 October 1995).

[44] Skid I, 33.

was in Cambridge that a revolution, perhaps *the* revolution, creating modern philosophy took place. The key protagonists were G.E. Moore in ethics and Bertrand Russell in logic. Moore's rigorous distancing of philosophy from the naturalistic or reductionistic encroachments of psychology, sociology or 'the laws of matter' was the model for this revolution. The tool he wielded in order to fend off the claims of the sciences was to accuse all these other disciplines of committing the 'naturalistic fallacy'.[45] Instead, the process of 'analysis' ('A thing becomes intelligible first when it is analysed into its constituent concepts'[46]) and its accompanying concept of the 'unanalysable' (Sidgwick) revealed that 'there is a simple, indefinable, unanalysable object of thought by reference to which [Ethics] must be defined'.[47] A path was now set: 'analysis' could be deployed not only to fend off the claims of other 'sciences', but also to prevent philosophy itself committing the fallacies to which it was naturally disposed.

Russell's innovation was to fuse the traditional supremacy of mathematics in the University of Cambridge with the new autonomy of 'Philosophy' characteristic of the nineteenth-century German reforms. Logic and mathematics could form the basis for a philosophy, by the 1950s to be called 'analytic', that was characteristic of Cambridge. The principal focus of Russell's work was the project to place the foundations of mathematics on a new logic he developed, alongside, and following Gottlob Frege. The 'analysis' involved here centred on the concept of 'logical form'. Russell's project would be clearly esoteric and autonomous of other disciplines, in the fashion of the post-Kantian philosophical innovations of Germany. It would oblige philosophy to become technical and professional (here in the sense of 'exclusive' and 'esoteric'). Moore's *Principia Ethica* gave a name to Russell and Whitehead's *Principia Mathematica* and both were self-consciously modelled on the greatest Cantabrigian of them all, Newton, and his *Philosophiæ Naturalis Principia Mathematica*, another magisterial and monumental three-volume work – one also unreadable save by a few brave, brilliant souls.[48] In this fashion, as Randall Collins argues, 'within each national culture, as soon as it underwent a German-style university revolution, the academic

---

[45] G.E. Moore, *Principia Ethica*, Cambridge: Cambridge University Press, 1903, p. 10 ff.
[46] G.E. Moore, *Selected Writings*, ed. T. Baldwin, London: Routledge, 1993, p. 8.
[47] Moore, *Principia Ethica*, p. 21; see also p. 72.
[48] When the eighteen-year-old Frank Ramsey sat Part I of the Mathematical Tripos in 1921, he employed the symbolism used in *Principia Mathematica*; as a result, the Examiners (Arthur Berry, Samuel Lees, Ralph Howard Fowler, George Paget Thomson), not being familiar with it, had difficulty examining his scripts. (See FRP, A.S. Ramsey Papers, Part II.)

philosophers took over the center of attention from philosophers outside the university . . . wherever the university revolution occurred there was an upsurge of technical, metaphysically oriented philosophy'.[49]

> At Cambridge after the university reform [of the 1870s and 1880s], there occurred a confluence of all the major trends of British intellectual life: the algebraist-logicians; the Utilitarians; but also the Idealists, under whose auspices the newly reformed universities passed from religious to secular control.[50]

This revolution crystallized out of the wholesale repudiation of Hegelian Idealism, which had come late to England and later still, in the 1890s, to Cambridge, in the form of Bradley's *Appearance and Reality*; in Cambridge, the most vigorous representative of this tradition was McTaggart, slightly older than Russell and Moore, whom he had taught. McTaggart was still teaching a Hegelian thread in the MST into the 1920s. Moore and Russell each broke decisively with Idealism, the first in asserting an Absolute Realism intertwined with an adapted naturalistic ethics in his *Principia Ethica* of 1904 – the Bible of the generation of Apostles who grew up under his spell and later formed the nucleus of the Bloomsbury Group, in particular Lytton Strachey and Maynard Keynes. Russell's trajectory moved from a Pythagorean conviction of the ultimate nature of reality being mathematical, originally conceived within a Hegelian shell, through his great honeymoon period in which all the problems concerning the relationship between logic, mathematics and ultimate reality seemed solved, to the monumental technical achievement, co-written with Alfred North Whitehead, of *Principia Mathematica* of 1911–13 – the embodiment of the project of 'logicism'. Crucial to Russell's project was the separation of logic from 'psychology' – the distinction between what Frank Ramsey, with his limpid clarity, called logic as 'a symbolic system' and 'logic in the sense of the analysis of thought'.[51] The final inflection in this journey, foundational for the analytic philosophy of the whole twentieth century, came with Wittgenstein's arrival in October 1911 and his continuing pressure on Russell's edifice around the dual questions, 'what is logic?' and 'what is a proposition?' Wittgenstein's pressure resulted in the problem of logic being reconceived as a problem of language – a dismaying end-point for Russell. Logic, mathematics, language

[49] Randall Collins, *The Sociology of Philosophies: A Global Theory of Intellectual Change*, Cambridge, MA: Belknap Press, 1998, pp. 644–5.
[50] *Ibid.*, p. 709.
[51] Ramsey, 'The foundations of mathematics' (1926), in Ramsey, *The Foundations of Mathematics and Other Logical Essays*, ed. Richard Braithwaite, London: Routledge, 1931, p. 21.

and truth – these four terms and their interrelations were from then on to define the core field of Anglo-American philosophy. Under Wittgenstein's pressure, the linguistic turning of the great tanker of *Principia Mathematica* had begun.

The Development of Psychology

While Russell and Moore were beginning the transformation of philosophy and Marshall was doggedly lobbying to detach political economy from the 'metaphysics' irrelevant to the science of economics, there were developments taking place in psychology that would eventually transform its position within the Moral Sciences as well. 'Philosophical psychology' was the staple of the teaching for psychology from near the Tripos's inception. The principal teacher was James Ward, early student of Wundtian experimental psychology at Leipzig, Fellow of Trinity from 1875, first Professor of Mental Philosophy and Logic (1897), but others, in particular G.E. Moore, contributed well into the twenties and thirties to the bread and butter business of lecturing students on 'Psychology' for both Tripos and Ordinary examinations. However, the further development of psychology was not driven by its connection with philosophy nor the repudiation of Idealism. There was a triple motor of transformation at work in Cambridge: the introduction of the new laboratory methods of investigation in close connection with physiology; an alliance with the equally new field methods of investigation which would also transform anthropology; and the development of a genuinely 'medical psychology'.

More generally, the construction of 'psychology' as a discipline in Europe and the USA came about through three separate but eventually intertwining programmes of research at the end of the nineteenth century: firstly, the laboratory-based experimental practices, closely linked to emerging experimental physiology and the development of sophisticated techniques of introspection (pre-eminently Wilhelm Wundt at Leipzig); secondly, the observations and experiments conducted in a clinical context on 'abnormal' 'subjects' (patients) (J.-M. Charcot, leading neurologist at the famous Salpêtrière Hospital in Paris and Freud's teacher); thirdly, the counting of variations in a 'population' and the statistical (graphical or tabular) presentation of results (Francis Galton in London).[52] On the face of it, the

---

[52] This overview is derived from Kurt Danziger's seminal *Constructing the Subject: Historical Origins of Psychological Research*, Cambridge: Cambridge University Press, 1990; the best comprehensive survey is Roger Smith, *The Fontana History of the Human Sciences*, London: Fontana, 1997.

beginnings of psychology in Cambridge derive most directly from the first of these programmes. However, histories of psychology are still dogged by two assumptions from which they struggle to break free: firstly, that psychology became scientific *and* an identifiably separate discipline only once it became experimental; secondly that the essential narrative of the history of psychology has as its hero the University. However, each of these major programmes of research (experimental, clinical, statistical) emerged in the late nineteenth century in a dialogue, which turned into vigorous exclusionary boundary work, with 'spiritualism' and 'psychical research'. If there is one University in the world which is closely associated with the late Victorian tradition of psychical research that developed across Europe and America, it is Cambridge. This too is a crucial part of the development of psychology in Cambridge and the reception of psychoanalysis there.

The Society for Psychical Research

The wave of interest in spiritualism began first in upstate New York in the late 1840s, spreading vigorously in America and then to Britain and other countries. It was a middle- and upper-class movement, based principally in private homes and then in public gatherings and meetings. The core beliefs or concerns were the existence of spirits and the possibility of communication with them, in séances, through writing on slate, table-rappings, automatic writing and other means. Scientists such as Michael Faraday, Alfred Russell Wallace and William Crookes were early involved, both as sceptics exposing the fraudulent means so often used to produce miraculous effects and as converts deploying their esoteric theories and sophisticated experimental techniques to validate the apparitions and communications received from spirits.

In Cambridge in the 1860s, still torpidly in the grip of Anglican orthodoxy, the young were questioning both the dogmas of Christianity and the plausibility of the new spiritualist phenomena. Leslie Stephen resigned his Fellowship at Trinity Hall in 1862 because he realized he could not accept, let alone preach, the Anglican creed, and built himself a London life as a writer; Henry Sidgwick in 1869 also resigned his Tutorship at Trinity, which was adept enough to find him a position in the College immediately as Lecturer in Moral Science in order to retain him in the College – and Sidgwick's resignation was one of a number of factors that precipitated the College into dropping all religious test qualifications for Fellows.[53] About

[53] Sidgwick and Sidgwick, *Henry Sidgwick*, p. 196.

the same time, Sidgwick entered into what would be a life-long collaboration and friendship which became more systematic in the mid-1870s with
F.W.H. Myers, a younger Fellow in College; a third Fellow of Trinity,
Edmund Gurney, joined them when they began a more demanding
programme of research.[54] In 1881, the researches on 'thought-transference'
of William Barrett in Dublin caught the interest of the public, including
the Cambridge men. In early 1882 Barrett convened a conference which
immediately led to the foundation of the national Society for Psychical
Research, with Sidgwick as its President. 'A group of Cambridge friends
became convinced that the questions at issue could only be decided
through experiment and observation of contemporary phenomena. On
this basis the S.P.R. was founded.'[55] This was Myers's somewhat tendentious version of this history in 1901. The Society was committed to
investigating the purported 'mesmeric, psychical and spiritualistic' phenomena 'without prejudice or prepossession of any kind, and in the same
spirit of exact and unimpassioned enquiry which has enabled Science to
solve so many problems'.[56]

The Cambridge group of Sidgwick, Gurney and Myers dominated the
Library Committee of the SPR, devoting themselves to detailed examination
of all the extant evidence for historic cases of spirit activity and judging
harshly and authoritatively on a string of bogus claims. Sidgwick, Myers and
Gurney had already had many experiences of expectation followed by disappointment, as their method of investigation had uncovered fraud after
fraud amongst mediums. But they were never to be resolutely sceptical.
There would always be a glimmer of hope, which mattered greatly to
them, mattered even more as they became more experienced and older.
'The first definite and important point towards which all the evidence
converged was the thesis of Telepathy, the evidence for which was set forth
in *Phantasms of the Living*', wrote Myers in the historical review that opened
his posthumous *Human Personality and its Survival of Bodily Death* (1903).[57]
The vast work produced in 1886, *Phantasms of the Living*, dealt with 'crisis
apparitions' (hallucinations of persons within twelve hours of their
death) and was stuffed full of hundreds of cases expounded, examined

[54] Roger Luckhurst, *The Invention of Telepathy, 1870–1901*, Oxford: Oxford University Press, 2002,
    p. 53.
[55] Frederic W.H. Myers, *Human Personality and its Survival of Bodily Death*, 2 vols., New York:
    Longmans, Green, and Co., 1903, Vol. I, p. xxiv.
[56] Quoted in Alan Gauld, *The Founders of Psychical Research*, London: Routledge and Kegan Paul,
    1968, p. 138.
[57] Myers, *Human Personality and its Survival of Bodily Death*, Vol. I, p. xxiv.

and meticulously criticized by the principal researcher and author, Gurney, who framed his account of these events as thought-transference or 'telepathy' (Myers' coinage at the time of the foundation of the SPR). This choice of topic and approach was effectively a disciplinary power play and had that precise effect. Within a few years a substantial number of the 'spiritualists' (or 'spiritists', as the philosopher of science Ian Hacking calls them[58] to avoid confusion with explicitly religious groups) had resigned from the SPR, leaving the 'scientifically' inclined Cambridge group firmly in charge of the direction and ethos of the Society. This period, from the foundation of the SPR in 1882 to the deaths of Sidgwick (1900) and Myers (1901), spanning the publication of Gurney's magisterial work not long before his sudden death in 1888, Gurney's studies of hypnotism, the 'Census of Hallucinations', the Society's investigations of key mediums (Eusapia Palladino, Mrs Piper) and Myers's *Human Personality and Its Survival of Bodily Death* (1903), was the 'Heroic Age'[59] of the Society.

At its inception, the Society had linked up with American colleagues, the most significant amongst whom was William James at Harvard, long the principal transatlantic conduit, President of the SPR in 1894–95 and intensively engaged in the examination in Cambridge, Massachussetts of the famous case of Mrs Piper.[60] The most active members of the Cambridge group, particularly after the loss of Gurney, were Mrs Sidgwick and her assistant Miss Alice Johnson, 'Research Officer' of the SPR who also doubled up as assistant at Newnham College, even when Eleanor Sidgwick was Principal of the College in the 1890s. They developed the data and the protocols, deploying Eleanor Sidgwick's expertise as a mathematician and experimentalist garnered while aiding her brother-in-law John William Strutt's work on electrical resistance in the late 1870s. Richard Hodgson, an Australian law student who had taken the MST in the late 1870s and, under Sidgwick's influence, joined the SPR research team, supplemented the team with his youthful energy. He became a roving experimental researcher, first in India investigating Madame Blavatsky in 1884 (funded by Sidgwick), then in Boston investigating Mrs Piper (1887), a project involving ingenious experiments and exhaustive detection work which lasted for many years – indeed it occupied and dominated the rest of Hodgson's increasingly impecunious life in Boston

---

[58] Ian Hacking, 'Telepathy: origins of randomization in experimental design', *Isis* 79(3) (1988), 427–51 at 435.
[59] John Beloff, *Parapsychology: A Concise History*, London: The Athlone Press, 1993.
[60] See William James, *Essays in Psychical Research* (The Works of William James), Cambridge, MA: Harvard University Press, 1986.

till his death in 1905.[61] Last but not least, the centre of the network was F.W.H. Myers, whose omniscient reading and network meant that, within two months of the publication of Breuer and Freud's 'Preliminary communication' in early 1893, introducing the cathartic cure and proclaiming that hysterics suffer mainly from reminiscences, Myers was disseminating these important new findings in Britain. They corroborated and extended the lines of research that others, particularly the French, in the influential work of Pierre Janet, had contributed to the field of research and knowledge that the SPR had set out. And it was Myers who would propound positive theories based on their extensive research: the theory of the 'subliminal self', a self infinitely larger than consciousness, reaching down into oblivion and outwards towards who knows what.

What is striking about the research programmes of the Cambridge group and their network was the extent to which they incorporated the methods which were coming into currency elsewhere and which would later give foundation to the main strands of psychology outlined above. Experimental and observational methods, often in the service of revealing fraudulent mediums, but also in the service of developing evidence concerning the unknown forces operating in the spirit realm (chemist and spiritist William Crookes; telepathy), were a crucial part of the programme. But so also were the clinical methods of Charcot, Janet and the school of the Salpêtrière, deploying hypnotism, which itself became a field of intense inquiry, and also the clinical studies of dream, mesmerism and trance. The magisterial neurologist's style of individual case presentation perfected by Charcot and Janet, later by Théodor Flournoy and Morton Prince, was outdone by the devotion by psychical researchers to their mediums of hundreds of hours of séances and experiments, producing exhaustive documentation; there was no other field in which the prominence of the individual case was so crucial to the advance of the science, no other field so haunted by a figure such as Mrs Piper, almost a figure of scientific and personal destiny for William James and Richard Hodgson (who 'devoted much of the rest of his life to studying her'[62]). There were also statistical inquiries, sometimes, as with Richet's randomized studies of telepathy, the first deliberately randomized experiments,[63] but sometimes in the early Galtonian fact-gathering mode, accumulating and sifting large bodies of data (e.g. the 2,272 affirmations of hallucinations which were

---

[61] Janet Oppenheim, *The Other World: Spiritualism and Psychical Research in England, 1850–1914*, Cambridge: Cambridge University Press, 1985, p. 175; Gauld, *The Founders of Psychical Research*, pp. 202–3.

[62] *Ibid.*, p. 258.   [63] Hacking, 'Telepathy', pp. 437–40.

then winnowed down to provide a core of eighty instances of hallucinations within twelve hours of the hallucinated person's death[64]).

All histories[65] agree that the Cambridge group quickly came to dominate the SPR, indeed that the SPR's achievements and reputation coincided almost entirely with the twenty years of their leadership and intense activity. This led to some ambiguity about where the SPR was located. Its formal meetings were held in London, usually at Westminster Town Hall (later known as Caxton Hall); experiments were hosted in drawing-rooms by active members. Intense investigation often involved extended 'field trips' or inviting mediums to reside for the duration in the houses of the researchers – hence the prolonged visits of Madame Blavatsky, Eusapia Palladino and Mrs Piper, for which they were paid, to the grand houses that the Sidgwicks and Myers had had built in Cambridge (opposite Magdalene College on Chesterton Road ('Hillside') and Grange Road ('Leckhampton House') respectively). But the ambiguity is more telling than that: while the leaders of the SPR were a close group of friends, former Fellows of Trinity College resident in Cambridge, and there were ten other Fellows of Trinity who were members of the SPR in 1894, the SPR never developed any formal connection whatsoever with the University.

Given the close identification of the SPR with the Cambridge group led by Sidgwick, it is mysterious that no local institutional development was to come of their achievements, despite the research activities associated with the Cambridge group becoming internationally renowned. After all, the group was closely tied to and identified with Trinity College and was an integral part of the steady but immensely influential shift of Trinity College towards a scientific orientation in the last forty years of the nineteenth century. Why did Sidgwick, a core Trinity reformer of the 1860s and 1870s, a founder of a new College in Cambridge for women, a key figure in the modernization of the relations between Colleges and University, a member of the key ruling University Committee for nearly twenty years in the 1880s and 1890s, one of the principals of the Special Board for Moral Science, last but not least superbly integrated into the close circles of political and scientific influence, with his early friendship with Prime

---

[64] Gauld, *The Founders of Psychical Research*, p. 183.

[65] E.g. Luckhurst, *The Invention of Telepathy*, p. 53. The main sources drawn upon for this depiction of psychical research are Gauld, *The Founders of Psychical Research*; Oppenheim, *The Other World*; Luckhurst, *The Invention of Telepathy*, and James P. Keeley, '"The coping stone on psychoanalysis": Freud, psychoanalysis and the Society for Psychical Research', PhD dissertation, Columbia University, 2002. See also Andreas Sommer, 'Crossing the boundaries of mind and body: psychical research and the origins of modern psychology', PhD dissertation, University of London, 2013.

Minister Gladstone now supplemented with family ties to 10 Downing Street, where his brother-in-law lived from 1895 to 1905[66] – why did he not develop a university stake in the field of psychical research, maybe something akin to the 'Sidgwick and Myers Professor of Comparative Telepathy in the University of Cambridge'[67] imagined by C.S. Schiller, a leading psychical research enthusiast and idealist philosopher at Oxford, in 1902, in the wake of the deaths of Sidgwick and Myers?

The answer may lie in Sidgwick's personal diffidence. Despite the fact that he was President of the second International Congress of Experimental Psychology in London in August 1892, his natural reticence, almost half-heartedness, turned this honour and potential occasion for disciplinary expansion into 'the delicate and difficult task of persuading the orthodox psychologist to regard "Psychical Research" as a legitimate branch of experimental Psychology.'[68] What is most striking about the proceedings of the Congress is less this delicate tight-rope-walking act than the easy acceptance of Victor Horsley's experimental demonstration of localization in monkeys' brains alongside Mrs Sidgwick's meticulous experiments regarding thought-transference of images in seven hypnotised subjects and Ambroise-Auguste Liébeault's cure by suggestion of a case of monomaniacal suicide; Janet's studies of dissociation and synthesis in hysterics and Charles Richet's project for the linking of physiology and 'transcendant psychology' alongside Alexander Bain's depiction of the entire field of psychology, from psychophysics to comparative psychology.[69] As we will see, the initiative to make psychology in the MST in Cambridge more ambitious than book-learning did not come from Sidgwick at all, even though he was President of the International Congress of Experimental Psychology – in effect recognized as the leading 'psychologist' in England.

The question of the legacy of the Sidgwick, Myers and Gurney group, both in Cambridge and more generally, needs addressing. After his death, Eleanor Sidgwick published her husband's essays and, together with

---

[66] Arthur Balfour was First Lord of the Treasury from 1895 to 1905, normally the job taken simultaneously by the Prime Minister and conferring right of residence at 10 Downing Street, even though he did not become Prime Minister till 1902; the reason for this oddity was that his uncle Lord Salisbury combined the position of Prime Minister with Foreign Secretary and delegated the post of First Lord to Arthur – and in any case, Salisbury preferred to live at his own London residence across Green Park, close by the modern Ritz Hotel.

[67] F.C.S. Schiller, 'Telepathic communication' (April 1902), *Journal of the Society for Psychical Research* 10 (1901–2), 224.

[68] Sidgwick and Sidgwick, *Henry Sidgwick*, p. 513, letter of 16 February 1892.

[69] Arthur Macdonald, 'Congress Report', *Science* 20(511) (1892), 288–90.

A. Sidgwick, prepared a memoir, based on his autiobiographical writings and his correspondence; she also served as President of the SPR in 1908–9 (and in 1932). After Myers's death, his magnum opus, *Human Personality and its Survival of Bodily Death*, was prepared for publication by his surviving colleagues, principally Alice Johnson. As one might have expected, Myers's death was not his last word, at least for those surviving him. A new research programme developed, the cross-correspondences: fragmentary messages, received by four different mediums scattered across the world, containing allusions which, when coordinated, the interpreters were persuaded unmistakeably pointed, not least through the mesh of classical allusions and Victorian poetry in which they were couched, to a guiding intelligence. But one of the voices from beyond was, it appeared, that of a woman Arthur Balfour had loved and intended to marry, who died on Palm Sunday 1875. Among the mediums engaged in this global intelligence-gathering was Mrs Verrall, wife of the classical scholar Arthur Verrall, first Professor of English Literature at Cambridge; her daughter Helen also began to receive communications (deploying the established method of automatic writing). (Eventually Helen Salter, as she became when she married William Henry Salter, whom she brought into the activities of the SPR, became the administrative core of the Society for forty years.) This research project continued to create a stir in Edwardian Cambridge. The Society never lacked for august names to adorn its Presidency: Arthur Balfour, William James and William Crookes all served in the 1890s; in the period from 1900 to 1920, the Presidents included Oliver Lodge, Charles Richet, Gerald Balfour, Andrew Lang, Henri Bergson, Gilbert Murray and Lord Rayleigh. But for many, increasingly many, the Society for Psychical Research was beginning to give respectability a bad name.

This was the Society as James Strachey first encountered it. When in 1945 Ernest Jones prepared a draft of an overview of the early history of psychoanalysis in England, he passed over the SPR completely. Having been galvanized by Jones's insouciance into digging all the books out of the British Museum to check and supply him with quotes and footnotes, as was his forte, Strachey took him to task:

> [Your historical sketch] seems to me to be slightly unjust to the shade of Fred Myers ... I have a personal feeling about this S.P.R. episode, as that was actually my own road of approach to Ψα. The S.P.R. was still very lively (and not at all exclusively spiritualistic) at Cambridge when I was an undergraduate (1905–9), and, though I was never spooky, I was very much interested. I read a lot of the current literature on abnormal psychology – Janet, Prince, Flournoy

etc. But I remember quite well the impression made on me by Freud's paper in 1912 [published by the SPR] – which was the first thing of his I ever read.[70]

This Edwardian Cambridge world of the SPR was also that of another psychoanalytic pioneer of the 1920s: Joan Riviere, née Verrall, Arthur's brother's daughter. 'I knew Verrall and his wife quite well when I was at Trinity', Strachey recalled of his years as an undergraduate in 1905–9, 'and used often to go to their Sunday afternoon "at homes". And it was on one of these occasions, I like to believe, that I first met Joan, who was a frequent visitor at her uncle's . . . I still have a vivid visual picture of her standing by the fireplace at an evening party, tall, strikingly handsome, distinguished-looking, and somehow impressive.'[71] Strachey's recollections continued:

> Cambridge, and in particular the Verralls, were at that time the centre of the activities of the Society for Psychical Research . . . I used to read their published proceedings, not because I was much interested in the question of survival, but because that was almost the only place (apart from Janet's works) where I could read anything about abnormal psychology. Soon I became a member, and it was not long after I came down from Cambridge, in 1912, that Freud himself contributed one of his most profound short papers to those very S.P.R. *Proceedings*. This was my first acquaintance with Freud, and I was immediately fascinated. It seems to me likely that Joan Riviere, who was also connected with the S.P.R. through the Verralls, may have arrived at Freud by the same path.[72]

Strachey and others like him, he intimates, were interested in the SPR principally because it had been for many years a principal conduit of accessible information about the work of Janet, Flournoy, Prince and others working on abnormal states of mind, on dissociation, on hypnotism and psychotherapy – the field of medical psychology. And as historians such as Sonu Shamdasani and Philip Kuhn have demonstrated, there were many alternatives then being developed, some of them passing under the name of 'psychotherapy', not only in France, Germany and the USA, but also in Edwardian England,[73] particularly after the great enthusiasm for

---

[70] JPBPaS, CSD/F03/08, James Strachey to Jones, 18 July 1945. Like others from his background, James would of course read French but German only with difficulty. For a careful examination of other eccentric claims by Jones concerning the early history of psychoanalysis in England, see Philip Kuhn, 'Subterranean histories. The dissemination of Freud's works into the British discourse on psychological medicine: 1904–1911', *Psychoanalysis and History* 16(2) (2014), 153–214. The symbol Ψα (alternatively expressed as ΨA or Ψa) refers to psychoanalysis.

[71] 'Joan Riviere (1883–1962)', *IJP* 44 (1963), 228–35 at 228.   [72] 'Joan Riviere (1883–1962)', p. 229.

[73] Sonu Shamdasani, '"Psychotherapy": the invention of a word', *History of the Human Sciences* 18 (2005), 1–22; Philip Kuhn, 'Footnotes in the history of psychoanalysis. Observing Ernest Jones discerning the works of Sigmund Freud, 1905–1908' *Psychoanalysis and History* 16(1) (2014), 5–54.

hypnosis and suggestion of the years 1885–1905 waned dramatically. How these other currents affected the reception of psychoanalysis in Cambridge we will eventually see. But we will also see that the Society for Psychical Research, a vehicle for the dissemination of so much international work in the field of abnormal psychology, psychotherapeutics, dream and other altered states of consciousness, played surprisingly little role in the development of Cambridge psychology.

Early Developments of Psychology and Anthropology

The key figures in the development of psychology in Cambridge were W.H.R. Rivers and C.S. Myers, with backing from the older James Ward until a moment late in this story, after the Great War, when, aged seventy-six, he refused to follow the newest developments in psychology. Rivers was brought to Cambridge by Michael Foster, who had himself introduced laboratory teaching and research into physiology from 1870 on.[74] In December 1888, the Board of Biology and Geology was urging the University to appoint a Lecturer in the Physiology of the Senses, reaching out to the Moral Sciences to consolidate its case: the Lectureship should be 'adopted for those who are studying psychology (e.g. for the Moral Sciences Tripos) as well as for those who are studying physiology',[75] a proposal to which the Board of Moral Science gave its backing. As long ago as the 1870s, Ward, fresh from his successful Trinity Fellowship dissertation on German psychophysics, *The Relation of Physiology and Psychology* (1875), had urged the introduction of experimental methods into the teaching of psychology. But it was not till 1893 that the creaking wheels of implementation began to move, and only then because an External Examiner in Physiology had pointed out the insufficiency of teaching in this area.[76] Foster moved quickly to remedy this lack, asking Rivers – then working at Bethlem Royal Hospital, lecturing on mental diseases at Guy's Hospital and beginning to teach experimental psychology at University College London – to come to Cambridge to lecture on the physiology of the senses, while maintaining his London commitments. Four years later, on 19 May 1897, the proposal to establish formally a Lectureship in Physiological and Experimental Psychology was finally accepted by the

[74] Geison, *Michael Foster*, pp. 162 ff.   [75] Crampton, p. 64.
[76] It would have been a standard technique even then for a seasoned academic politician such as Foster to have mildly suggested to the External Examiner that he was aware of the serious lack of teaching in precisely this area, thereby indicating that a remark to that effect in his report would be entirely appropriate and beneficial.

General Board.[77] In that year Foster assigned Rivers a room for teaching, i.e. for doing practical work, as did James Sully, the Professor at University College London. So it came about that Rivers was the simultaneous 'founder' of both of the 'first' psychology laboratories in England.

In Rivers's first year of teaching at Cambridge, 1893, a bright young student taking the Natural Sciences Tripos Part II in Human Anatomy and Physiology came into his course: C.S. Myers.[78] The timing was ideal: for the next thirty years Myers would prove to be the perfect foil to Rivers in their joint expansion of scientific disciplines, both in Cambridge and on a broader national scale. Their gifts were complementary: Rivers the ideas man, roving, courageous, never tied down and always surprising in the turns his work would take; Myers the social animal par excellence, galvanizing hotch-potch collections of teachers and researchers, liable to found institutions and groupings in an afternoon and be able to reform sedate institutions where others in despair expected only genteel decadence and the ill will of the traditionalist.

Charles Myers (1873–1946) came from a Jewish family of merchants long-established in London; from his mother, he acquired a love of and great virtuosity in music. From City of London School he applied to Gonville and Caius College, Cambridge intending to study medicine; he took Parts I and II of the Natural Sciences Tripos in 1893 and 1895 and achieved a double First. Rivers became his mentor in physiology and psychology; even more of an inspiration was A.C. Haddon's teaching in anatomy. Already as an undergraduate he was interested in physical and racial anthropology and anthropometry, studying a recent find in Suffolk of ancient skulls,[79] alongside his already active urge to renovate institutions, transforming the moribund Cambridge University Musical Society, in which he played in a chamber music quartet.[80] Alfred Haddon was Professor of Zoology at Dublin, spending part of his time each year in Cambridge; in 1888–9 he had visited the Torres Straits to study marine biology and been more captivated by the natives than by the molluscs. At the urging of J.G. Frazer, the anthropologist Fellow of Trinity, he was planning a return expedition, this time anthropological and psychological, to record the lives of the inhabitants before they disappeared in the march of civilization. His first choice of companion

[77] Crampton, p. 64.   [78] No relation of F.W.H. Myers.

[79] C.S. Myers, 'Autobiography', in C. Murchison (ed.), A History of Psychology in Autobiography, Vol. III, Worcester, MA: Clark University Press, 1936, pp. 215–30 at 216.

[80] Crampton, p. 120 ff., based on the 'Personal records' Myers lodged with the Royal Society in 1942. It was their shared appreciation of music that drew Wittgenstein to Myers in 1912.

was Rivers, who declined, suggesting Myers, who accepted with alacrity. Myers suggested inviting William McDougall (1871–1938), another graduate of St John's College in Natural Sciences, also medically qualified, who had recently taken up a Research Fellowship at St John's and was moving into experimental psychology under the impact of reading William James's *Principles of Psychology* (1890). Haddon also invited a linguist, Sidney Ray (1858–1939) and another of his own students, Anthony Wilkin (1878–1901), to join the party as photographer; hearing that his students Myers and McDougall were joining, Rivers changed his mind.[81] An old friend of Myers's, C.G. Seligman (1873–1940) of London University, also medically trained, was so persistent in his request to join that he was eventually invited. They set out in March 1898 and straggled back at intervals until May 1899: Rivers had to come back in October 1898 to give his lectures in Cambridge. Anthropology at that time would not have been possible without steamships and trains that kept tight schedules from the Pacific to Southampton and from London to Cambridge.

As Haddon observed, the Expedition definitively seduced both Rivers and Seligman into anthropology. For the next fifteen years of his life, Rivers devoted much of his time either to India or to the Pacific: in 1901–2 he spent time in India, producing *The Todas* in 1904; in 1907–8 he went to the Solomon Islands with two younger researchers, Gerald C. Wheeler and Arthur M. Hocart, extending his trip to other areas of Melanesia and Polynesia, with a month on Hawaii and on Fiji – where Hocart was to remain for three years – and the eventual result was Rivers's magnum opus in ethnographic ethnology, *The History of Melanesian Society* (1914). But when he was in Cambridge, from 1902 on as a Fellow of St John's and Director of Studies in Moral Science,[82] Rivers was equally busy with experimental

---

[81] Wilkin was to die young, not long after their return from the Torres Straits. His parents set up in his memory a Studentship devoted to funding fieldwork in anthropology. As Langham remarks: 'Despite a singular lack of financial aid from the university, social anthropologists at Cambridge developed, under the guidance of Haddon and Rivers, into a surprisingly coherent and productive group. One symbol and cause of this coherence and productivity was the Anthony Wilkin Studentship which, being awarded to Radcliffe-Brown (twice), Armstrong, Barnard, Deacon and Bateson, did much to further the pursuit of fieldwork in the Rivers style' (Ian Langham, *The Building of British Social Anthropology: W.H.R. Rivers and his Cambridge Disciples in the Development of Kinship Studies, 1898–1931*, Dordrecht: D. Reidel, 1981, p. 314). The present-day regulations of the Anthony Wilkin Fund specify that 'the principal object of the Fund, namely, the furtherance of ethnological and archaeological research, preferably by fieldwork among the more primitive peoples, and in other lands than Greece, Italy, or Egypt, shall be maintained'.

[82] Frederic Bartlett, 'Cambridge, England: 1887–1937', *American Journal of Psychology* 50 (1937), 97–110 at 104.

Fig. 5.1 Charles Myers recording the sacred songs of the Malu ceremonies during the Torres Straits Expedition, 1898.

studies, two of which became landmarks.[83] The first was the study of the regeneration of peripheral nerves, begun on 25 April 1903. A surgeon severed two cutaneous nerves in the left forearm of Rivers's close friend Henry Head, rendering a largish area of skin on the hand and lower arm anaesthetic. 'For five happy years', says Head, 'we worked together on weekends and holidays in the quiet atmosphere of his rooms in St John's College'.[84]

'In addition, they made some control tests on Head's *glans penis*.'[85] The result of this study was the distinction Head and Rivers drew between protopathic and epicritic, so as to characterize

[83] Simon Schaffer, *From Physics to Anthropology and Back Again*, Cambridge: Prickly Pear Pamphlet 3, 1994, is an unrivalled study of Rivers and the question of the intersection of laboratory and field work.

[84] Henry Head, 'Obituary notice of William Halse Rivers Rivers', *Proceedings of the Royal Society* B77 (1923), i–iv.

[85] Schaffer is one of the few secondary sources to draw attention to the fact that this singular element of the experimental series was also crucial to its argument. In fact, the whole argument of the paper, concerning the development of epicritic and protopathic sensitivities, could have been developed and experimentally grounded without the severing of Head's nerve; the detailed examination of

Fig. 5.2 Henry Head and Rivers during an experiment in nerve division, *c.* 1903.

the existence of two definite stages in the return of sensibility. In one of these, the protopathic stage, the sensations are vague and crude in character, with absence of any exactness in discrimination or localisation and with a pronounced feeling-tone, usually on the unpleasant side, tending to lead immediately, as if reflexly, to such movements as would withdraw the stimulated part from contact with any object to which the sensory changes are due. At this stage of the healing of the reunited nerve there are present none of those characters of sensation by which we recognise the nature of an object in contact with the body … The second stage of the process of

Head's penis hanging down, usually with the foreskin rolled back, with Rivers testing the sensitivity of the glans, corona, neck and foreskin, would have been entirely sufficient in generating the relevant experimental data. (Being the scrupulous experimental scientist he was, Rivers of course compared the sensitivity of Head's penis with that of a circumcized subject.) However, the eventual reputation of the paper would probably have been rather different if it had been published as a study of Head's penis, rather than Head's radial nerve (*ramus superficialis nervi radialis*) divided at the point where it arises from the musculo-spiral (*n. radialis*). It is striking that Edgar Adrian's early physiological researches were initially focused on the protopathic–epicritic distinction and picked up immediately on the suitability of studying the sensitivity of the *glans penis* (see Edgar Adrian, 'The response of human sensory nerves to currents of short duration', *Journal of Physiology, London* 53 (1919), 70–85 and A. Hodgkin, 'Edgar Adrian', *BMFRS* 25 (1979), 1–74).

regeneration [the return of epicritic sensibility] is characterised by the return
of those features of normal cutaneous sensibility, such as exact discrimination and localisation, by means of which it becomes possible to perceive the
nature of an object in contact with the skin and adjust behaviour according
to this perception ... the two kinds of sensibility represent two distinct
stages in the development of the afferent nervous system ... two modes of
sensibility represent[ing] two stages in phylogenetic development.[86]

This distinction was directly derived from the neurological theories of
Head's and Rivers's teacher, John Hughlings Jackson, whose parallel
influence on Freud in his *On Aphasia* (1891) and *The Interpretation of
Dreams* (1900) was equally clear and avowed.[87] It was this common debt
that would help make Rivers's later assimilation and development of
Freudian theory uncomplicated but also eccentric.

While studying the function of the peripheral nervous system, Rivers
ran another series of experiments that also required mortification of the
flesh. Asking his experimental subjects, which included himself, to forgo
any intake of alcohol, caffeine and tea for something like a year, he studied
the effects of these drugs upon pure muscular fatigue; the design of the
experiment meant that Rivers and the other subjects were not aware if they
were taking the substance under investigation or a control substance or
mixture – a 'placebo' as it is now called. Rivers gave an account of these
experiments as the Croonian Lectures in 1906.[88]

The methodology of Rivers's experiments with Head and with his
colleagues on the effect of drugs conformed closely to the model developed
by Wundt and extended by Kraepelin; as described vividly by Kurt
Danziger, the roles of experimental subject (or 'reagent') and researcher
were fluid and interchangeable:

> in the earliest years of experimental psychology there simultaneously
> emerged two very different models of the psychological experiment as a
> social situation. These can be called the Leipzig model and the Paris model.
> The Leipzig model involved a high degree of fluidity in the allocation of
> social functions in the experimental situation, reflected in the lack of a

[86] IU, 23.
[87] See Allan Young, 'W.H.R. Rivers and the war neuroses', *Journal of the History of the Behavioral
Sciences* 35(4) (1999), 359–78; Young, *The Harmony of Illusion: Inventing Post-Traumatic Stress
Disorder*, Princeton: Princeton University Press, 1995, chapter 2, and John Forrester, *Language
and the Origins of Psychoanalysis*, London: Macmillan, 1980, pp. 15–29.
[88] W.H.R. Rivers, *The Influence of Alcohol and Other Drugs on Fatigue: The Croonian Lectures Delivered
at the Royal College of Physicians in 1906*, London: Edward Arnold, 1908 and Ted J. Kaptchuk,
'Intentional ignorance: a history of blind assessment and placebo controls in medicine', *Bulletin of
the History of Medicine* 72(3) (1998), 389–433 at 419.

uniform terminology to refer to experimenter and subject roles as such. Persons playing these roles at any particular time were more likely to be defined in terms of their relationship to the physical apparatus than in terms of their relationship to each other. At the same time, the subject function had a higher status than the experimenter function, though these functions could be assumed by any participant in the experimental situation. In the Paris model, by contrast, experimenter and subject roles were rigidly segregated, with the experimenter clearly being in charge and the subject serving as an object of study who underwent certain manipulative interventions on the part of the experimenter. With this went a more uniform terminology that unambiguously identified the subject as such. The two models were typically employed to investigate different aspects of psychological functioning: the Leipzig model to study aspects of normal cognition and the Paris model to study abnormal functioning.[89]

Rivers was still Director of the haphazard arrangements that passed for a Psychological Laboratory in Cambridge – from a room in the Physiology Laboratory they moved in 1901 to some rooms in St Tibbs Row, then in 1903, with an annual grant from the University of £50 for apparatus and expenses, to a cottage, 16 Mill Lane, made available by the Cambridge University Press. As Charles Myers described it for the benefit of the University authorities, it was 'damp, dark and ill-ventilated' – that 'hovel in Mill Lane', a student recalled, infested with unpsychological rats.[90] This was not principally a 'research laboratory' but firstly a 'teaching laboratory': 'All students reading moral science then did – or were supposed to do – four hours of experimental work weekly in the psychological laboratory.'[91] Indeed the idea of 'research' was a peculiarly ill-defined one at this time for most disciplines in Cambridge; advanced students and those engaged in research would agglomerate in an unsystematic manner rather than enter professional trajectories, given that there was no structure of higher degrees to acquire and there were very few professional positions to fill. At this time, the distinction between teaching and research did not exist. Rivers and others were beginning to put some elements of an organized discipline into place: in late October 1901 the Psychological Society was founded at a meeting in London by ten assorted university men, gentlemen of science and educationalists including Sully and McDougall from UCL, Rivers from Cambridge, three physicians from the

---

[89] Kurt Danziger, 'The origins of the psychological experiment as a social institution', *American Psychologist* 40 (1985), 133–40.

[90] Bartlett, 'Cambridge, England, 1887–1937'; Godfrey Thomson, *Education of an Englishman: An Autobiography*, Edinburgh: Moray House College of Education, 1969, p. 82.

[91] F.C. Bartlett, 'Autobiography', in Murchison (ed.), *History of Psychology in Autobiography*, Vol. III, p. 39.

LCC Pathological Laboratory at Claybury, Essex (dedicated to employing laboratory investigation for the prevention of mental disease[92]), including its Director, F.W. Mott and Robert Armstrong-Jones, as well as Mrs Sophie Bryant, Headmistress of North London Collegiate School for Girls.[93] (The model for this Society was the Physiological Society of Great Britain, founded in 1876 in response to the anti-vivisectionist campaigns – campaigns which had accelerated the formation of a 'professional' identity for physiologists.[94]) The members of the Psychological Society were to be restricted to those teaching and engaged in psychology, each with a recognized achievement in the field – essentially restricting the Society to the London and Oxbridge elite.[95] The Society was clearly the brain-child of those enthusiastic about laboratory methods in psychology.[96] Soon to follow in 1904, at the initiative of the Cambridge trio of Ward, Rivers and Myers, was the establishment of the *British Journal of Psychology*; Ward ceased being an Editor in 1911 and from 1913 to 1924 Myers edited it by himself, then handed the job on to his Cambridge protégé Frederic Bartlett, who edited what was now the *British Journal of Psychology (General Section)* until 1948. Meanwhile that organizational dynamo C.S. Myers took control of the British Psychological Society with important consequences.

From his early studies as an undergraduate, Myers showed himself as diverse in his interests as Rivers. He was enthusiastic about music, archaeology, anthropology, psychology and even medicine, his initial choice of profession. If there was anything approaching one strong thread binding his interests together it was music, so it is fitting that his first lectures at Cambridge, in Michaelmas Term 1902, were on 'Psychology and physiology of hearing'. With family wealth cushioning him throughout his career, Myers had no need of a professional income. Throughout his life, he devoted himself to the creation of structures which enabled the development of several professional academic disciplines. Myers had squeezed the bulk of his medical training into the time between leaving Cambridge in 1895 and heading for the Torres Straits in March 1898. Four years before

---

[92] Tatjana Buklijas, 'The laboratory and the asylum: the L.C.C. Pathological Laboratory at Claybury, Essex, 1895–1916', MPhil dissertation, University of Cambridge, 1999.

[93] Myers's absence at the founding meeting is explained by the fact that he was living in Egypt.

[94] Richard D. French, *Antivivisection and Medical Science in Victorian Society*, Princeton: Princeton University Press, 1975 and Geison, *Michael Foster*, esp. pp. 20 and 330.

[95] Sandy Lovie, 'Three steps to heaven: how the British Psychological Society attained its place in the sun', in G.C. Bunn, A.D. Lovie and G.D. Richards (eds.), *Psychology in Britain: Historical Essays and Personal Reflections*, Leicester: BPS Books, 2001, pp. 95–114 at 96–7.

[96] Galton was invited but declined to join; whether this was because the Society was not sufficiently committed to his statistical style of psychology or for other reasons is unclear.

Rivers, in 1896 he managed to get himself elected a Fellow of the Royal Anthropological Institute.[97] He returned to England in 1899 to a position at Bart's Hospital. A severe illness led him to spend 1900–2 in Egypt, when he did much anthropological work on hieroglyphics, tattooing and extensive anthropometry, all the while preparing his MD thesis, on the rare neurological condition *myasthenia gravis*. Back in Cambridge in 1902, he worked on all the anthropological and psychological materials he had collected, and helped Rivers run the practical classes in psychology, his only payment being fees collected from the individual students attending. In 1904, he married Edith Babette Seligman, who eventually bore three daughters and two sons. In 1904, an unpaid Demonstratorship in the Cambridge Psychological Laboratory was created for him and he became Secretary of the Psychological Society (renamed in 1906 the British Psychological Society). Knowing how to spot a winner, someone offered him the post of Secretary of the British Association for the Advancement of Science, but Myers turned it down. He began teaching part-time at King's College London. In 1906 he became part-time Professor at King's, a position he retained until 1909; at King's, one of his first students and then life-long friend was T.H. Pear, appointed in 1910 to a lectureship at Manchester and author of an enthusiastic account of Freud's dream theory in 1912.

In 1907, clearly as part of a Myers scheme, Rivers wrote to the Special Board of Moral Sciences in Cambridge complaining of an excessive workload and resigned part of his brief in Physiology and Experimental Psychology. The University responded by dividing his Lectureship in two: from October 1907 Rivers received a £100 stipend as Lecturer in the Physiology of the Senses and Myers was appointed to the £50 per year position of Lecturer in Experimental Psychology.[98] In 1908 Myers took over from Rivers as (unpaid) Director of the Cambridge Psychological Laboratory. In December of that year, Myers started a campaign for funds for a purpose-built Psychological Laboratory, using the Cambridge University Association as his institutional base – he had no College affiliation at the time; Myers toured Europe looking for best practice in new laboratories, and, at his instigation, the Special Board for Moral Science urged the University to support the proposal and persuaded the outgoing Vice-Chancellor to declare the present accommodation 'an actual disgrace'.[99]

---

[97] Richard Slobodin, *W.H.R. Rivers*, New York: Columbia University Press, 1978, pbk version, p. 21.
[98] Crampton; HR 1910, p. 124.
[99] CUEP, documents entitled 'Statement of the case for the establishment of a Laboratory for Experimental Psychology', dated December 1908, and 'Report of the Laboratory of Experimental Psychology Syndicate, Draft, Feb 2, 1911'.

Myers had his eye on the imminent construction of a new Physiology Laboratory, a scheme many years in the making, finally in sight of realization courtesy of a £10,000 donation from the Drapers Company, but the campaign for psychology made little headway until Myers announced that £3,000 from an anonymous donor had been 'offered to the University in November 1909 on the express condition that the Laboratory be erected without delay'.[100] Immediately in January 1910 a Sub-Syndicate dedicated to the new laboratory was set up, consisting of Ward, Rivers and Myers. By February 1911, the Drapers Company had agreed to a Joint Building, with a connecting staircase; the plans envisaged teaching rooms, laboratories, and a dark room and a sound proof room. Costings showed that the anonymous donation and small additional other sums covered the architect's estimate. By March a Report on the envisaged expenses and salary costs for the new building was drafted. Building started and it was complete a year before the adjacent Physiology Laboratory was ready to open. There was to be no other purpose-built Psychological Laboratory in England until 1942, when the Laboratory at Reading was completed.[101] Along the way, Myers's initiative had successfully shouldered to one side the competing claims of Gowland Hopkins's attempts to build up biochemistry within the Department of Physiology.[102]

In May 1913, the building was officially opened by the Vice-Chancellor, whose celebratory speech was reprinted in the *Cambridge Review*. Myers opened his formal reply by definitively situating the place of psychology in Cambridge within the Moral Sciences: 'That Psychology stands in the closest relation to Philosophy needs no demonstration.' In the typed manuscript of his speech, he continued:

> In its applications Psychology enters into relation with Biology, in the study of animal behaviour; with Education, in the study of the individual and general characteristics of the developing human mind; with Economics, in the study of the best methods of securing mental and muscular efficiency within the community, and of the relation between mental endowment and fitting occupation for the individual; with Anthropology, in the study of racial and mental differences; with Medicine, in the study of the disturbance of the nervous system and sense organs, and in the use of suggestion and hypnotism [*changed with pen to*: suggestion, hypnotism and psycho-

---

[100] CUEP, document entitled 'Report of the Laboratory of Experimental Psychology Syndicate, Draft, Feb 2, 1911'.
[101] A. Wooldridge, *Measuring the Mind: Education and Psychology in England c.1860–c.1990*, Cambridge: Cambridge University Press, 1994, p. 60.
[102] Robert E. Kohler, 'Walter Fletcher, F. G. Hopkins, and the Dunn Institute of Biochemistry: a case study in the patronage of science', *Isis* 69 (1978), 330–55.

analysis] as therapeutic measures; with Theology, in the study of the intellectual and emotional factors in religion; and with Art as a foundation for experimental Aesthetics.[103]

Thus, on 22 May 1913, at the inception of the new Psychological Laboratory, psychoanalysis found a place – as a late addition to the final draft – in the discipline's mission. It was, for Myers, the most recent in a line of therapeutic measures newly deployed in psychological medicine. This document gives us a very firm and accurate dating for the emergence of psychoanalysis as an item of significance for the expanding discipline of psychology in Cambridge.[104] It also gives us a standard of comparison for the disciplinary ambitions that were to be reconfigured in the turmoil after the Great War.

The expansion of psychology into medicine – or the interconnections of psychology and medicine – had already been part of the case for support for the new Laboratory Myers had put to the University in December 1908, where he had listed nine 'typical lines of research' to be carried out in the Laboratory: 1. 'sensations, normal and abnormal', 2. memory and learning, 3. attention, 4. perceptual illusions, 5. fatigue and work, 6. effect of drugs on efficiency, 7. 'mental characters of normal and defective children, primitive peoples and animals', 8. 'conditions of the aesthetic appreciation of form, colour, harmony, and rhythm' and, finally, 9. 'hypnotism, multiple personality, insanity and other abnormal mental states'.[105] In a revised version of this case, the Special Board for Moral Science had highlighted this last area of work in its plea for the Laboratory:

> It is hoped before long to offer courses in Psychology adapted to the needs of those who intend to devote themselves to teaching or to the study of abnormal mental conditions. The demand for such courses is now supplied at several Universities in this country, but it is quite impossible to hold them at Cambridge until adequate laboratory accommodation is forthcoming.[106]

---

[103] CUEP, Ms., Myers's speech, p. 4. The addition of 'psycho-analysis' is in Myers's hand. There are other additions elsewhere in the document in Rivers's hand.

[104] Close attention to the reception in the medical press of Freud's work should make this date for psychoanalysis's acceptance within psychological medicine less surprising; in a brief review in 1913, published in *Brain*, of the English translation of Freud's *Selected Papers on Hysteria*, the reviewer referred in passing to 'the well-known case of hysterical pains in the legs, "Elizabeth v. R."', indicating what well informed neurologists and other physicians would be expected to be casually familiar with (*Brain* 35 (1913), 324).

[105] CUEP, document entitled 'Statement of the case for the establishment of a Laboratory for Experimental Psychology', dated December 1908.

[106] CUEP, document entitled 'Statement of the case for the establishment of a Laboratory for Experimental Psychology', dated December 1908, revised draft.

Educational psychology and 'psychological medicine' were here being highlighted as future areas of growth. The timing of this expansion was well coordinated. From Michaelmas Term 1909 until 1913, Rivers taught a course each year on 'Physiological and Pathological Psychology' and the examinations set for students reveal the widening range of topics; in Part II for 1911, students were asked to write on 'Agnosia, apraxia and aphasia' and 'Discuss the psycho-pathology of hysteria.' In 1911, and then again in 1913, the general 'Essays' paper included the topic of 'Dreams'. Quite what the Examiners expected from psychology students writing on dreams is not documented. In his posthumous *Conflict and Dream*, drafted in 1920 and 1921, Rivers recalled that 'though I had taken much interest in the general views of Freud before the war, I had not attempted to master his theory of dreams. I was more interested in the applications of his scheme to the explanation of psychoneurosis and the anomalous behaviour of everyday life'.[107] And he also recalled:

> When I suggested a question on dreams in a University examination not many years ago, it was objected that the students would know nothing about the subject, which meant, of course, that they had been taught nothing about it. The consideration of the psychology of dreams was not deemed worthy of inclusion in a course of academic psychology. The great revolution in the attitude of psychologists which has since occurred is due to Freud – I think one can say entirely due to him. Among the many aspects of the vast influence which Freud has exerted upon psychology, none is more prominent than that concerned with dreams and their interpretation.[108]

Rivers's account post-war does not chime perfectly with the actual historical record: students *were* indeed asked to write on dreams. Was Rivers remembering the sceptical response of the Examiners but forgetting that he had successfully defended the appropriateness of 'Dreams' as a subject for examination in the Moral Sciences?

Diploma in Psychological Medicine

Closely interlinked with the steady rise to prominence of abnormal psychology in the Tripos teaching at Cambridge from 1908 on, there was another development inevitably fostering interest in psychoanalysis. In 1910, the President of the Medico-Psychological Association of Great Britain and Ireland had written to all medical examining bodies in Britain urging establishment of a diploma in psychological medicine; five

[107] CD, 5.   [108] CD, 2.

universities had responded[109] and Manchester (where T.H. Pear was based) established a Diploma in Psychological Medicine in that year.[110] A report to the University of Cambridge on 21 February 1911 by the Special Board for Medicine proposed:

> the institution of a Diploma in Psychological Medicine. They believe that such a diploma would improve materially the efficiency of those whose business it is to deal with Mental Disease and the care of the Insane. There already exist in connection with the University Diplomas in Public Health and in Tropical Medicine and Hygiene, which have proved successful in the highest degree.[111]

The number of candidates per year envisaged for the course's costs to break even was six.

Asylum and society doctor Maurice Craig,[112] close plotter with Rivers and Myers for this promotion of psychological medicine at Cambridge, returned to Cambridge especially to speak in the Senate in favour of the proposal, placing it in the context of the growing recognition that 'psychological medicine embraced practically the study of the whole mental life of the individual'[113] and the necessity for prophylaxis in the domain of mental illness just as in all other parts of medicine. The whole discussion was peculiarly slanted on account of the distinctive character of English laws on lunacy: because of the existence of such laws, 'psychological medicine' was naturally conceived of as part of 'State Medicine', along with two other areas (Public Health and Tropical Medicine). The Diploma was aimed at doing for lunacy – psychiatry, as modern men like Craig and

---

[109] Rosemary Stevens, *Medical Practice in Modern England: The Impact of Specialization and State Medicine*, New Haven: Yale University Press, 1966, p. 43n9.

[110] Alan Costhall, 'Pear and his peers', in Bunn, Lovie and Richards (eds.), *Psychology in Britain*, p. 197; *CUR*, 28 February 1911, p. 671.

[111] 'Report of the Special Board for Medicine on the Establishment of a Diploma in Psychological Medicine', *CUR*, 28 February 1911, pp. 670–2, followed by notes of a Discussion the following month led by Sir Clifford Allbutt (*CUR*, 21 March 1911, pp. 773–8). Myers recalled: 'I was largely concerned in establishing in 1912 the University Diploma in Psychological Medicine which was unique in requiring written and practical examinations both in psychology and in neurology – two most important adjuncts to the equipment of the thoroughly trained psychiatrist' (Myers, 'Autobiography', in Murchison (ed.), *A History of Psychology in Autobiography*, Vol. III, p. 221).

[112] Craig (1866–1935) was a close friend of Rivers from his student days, a specialist in psychological medicine. In the 1910s and 1920s Craig probably had the largest Harley Street practice in nervous and mental diseases. He was by default the specialist of choice for many of the Bloomsbury Group and their circle (e.g. Virginia and Leonard Woolf; Daphne Olivier's manic episode on hearing of Rupert Brooke's death in 1915; Harry Norton's breakdown in 1921). In Craig's Maudsley Lecture of 1922, he welcomed Freud's emphasis on repression and conflict, but was critical of the prominence he gave to sexuality. See H.C. Cameron, 'Sir Maurice Craig', *Guy's Hospital Reports* 85 (1935), 251–7.

[113] 'Discussion of a Report', *CUR*, 21 March 1911, pp. 772–8 at 775.

Myers began now to call it – what had been achieved since the 1880s in these two other areas of medicine. And there was an additional, geographical urgency, as with Public Health (cities) and Tropical Medicine (Empire): the lunatic asylums were, as Craig pointed out, 'placed in the heart of the country, in rural districts, away from the Universities, generally away from any place where they could study'.[114] The Diploma would make good the damage done by this isolation. In addition, the Diploma would help rectify, by dissemination of psychological knowledge, the parlous state of affairs dominant in the treatment of mental disease due to the legal restriction on any state (including local authority) funding of treatment before a patient had been certified as of unsound mind and confined to an asylum.

Relatively speedily the University approved the Diploma. The first Management Committee of the DPM consisted of the usual suspects: Myers, Rivers, Head, Craig and Sir Clifford Allbutt, Regius Professor of Physic (plus some others).[115] On 10 December 1912, Myers, inevitably the Secretary of the Managing Committee, published the detailed rules for its two Parts, in preparation for the first examinations in 1913: the first part was to be taken in June in Cambridge, the second part in July in London:

> the first part of the examination will consist of (1) a paper and (2) a practical and oral examination in the Anatomy and Physiology of the Nervous System, (3) a paper and (4) a practical and oral examination in Psychology ... [The second part] will consist of (1) a paper and (2) a clinical and oral examination in Neurology, (3) a paper in Psychiatry, Lunacy Law, and Asylum Administration, (4) a paper containing a choice of subjects for an Essay in Psychiatry, (5) a clinical and oral examination in Psychiatry.[116]

Included in the topics for examination in Psychology were: 'Personality and its disorders. Suggestion. Hypnosis. Sleep. Dream.' That tell-tale word 'Dream' is a reliable marker that psychoanalysis was envisaged: in the summer of 1912, a few months before this syllabus was published, T.H. Pear had given, with his mentor Myers's encouragement, the first paper on

---

[114] *Ibid.*, p. 776. See Chris Philo, *A Geographical History of Institutional Provision for the Insane from Medieval Times to the 1860s in England and Wales: The Space Reserved for Insanity*, Lampeter and New York: The Edwin Mellen Press, 2003.

[115] In October 1912, the State Medicine Syndicate appointed Members of the Managing Committee for the Examination in Psychological Medicine as follows: 'Dr. Humphry, Dr. Myers, and Professor Woodhead to serve until 31 December 1912; Dr. Anderson, Master of Gonville and Caius College, Professor Sir Clifford Allbutt K.C.B., M.D. and Dr. Hill to serve until 31 December 1913; and H. Head, M.D., M. Craig M.D., and W.H.R. Rivers M.A. of St. Johns College to serve until 31 December 1914.' *CUR*, 8 October 1912, p. 50.

[116] *CUR*, 10 December 1912, pp. 353–5.

Freud's dream theory to the British Psychological Society.[117] The English translation of Freud's *Die Traumdeutung* was published in 1913. Probably equally influential for jobbing students was the publication of Hart's *The Psychology of Insanity* in 1912, as it was for those who were lucky enough to get hold of proof copies. As we noted in Chapter 2, Arthur Tansley brought the proofs into his botany class.[118] For T.H. Pear, Hart's work was instrumental in sparking his interest in Freud:

> One lazy afternoon in the clinic library [at Giessen] I was idly riffling through journals and saw an article by a friend, Bernard Hart ... 'The Psychology of Freud and his School'. My second milestone. Neither in Würzburg nor in Giessen had I heard Freud's name. I read on: magic casements opened. Back in England, I wrote to Hart, who with quixotic generosity posted me the page proofs of his book, just about to appear, *The Psychology of Insanity*. Have complexes, logic-tight compartments, rationalisation ever been explained since in such simple, elegant English? I began to see light in many directions, and with that habit, which I fear must have irritated many of my readers and listeners, began to ask 'How does all this apply to me, and to people who are willing to describe their own experiences?'[119]

Thus on 4 June 1913, the first Diploma candidates sat an exam whose questions included the following topics: the classification of modes of consciousness and their value for study of insanity; the effects of adaptation on sensation and their importance for psychology generally; whether agnosia is a disorder of perception; the process by which we assign an event in personal experience to a given date; the aims and methods of experimental work on memory; the definitions and pathology of mood and

---

[117] 'The first addresses on Freud's theory of dreams, both at the British Psychological Society and at the British Association for the Advancement of Science, were given as a result of his encouragement': T.H. Pear, 'Charles Samuel Myers: 1873–1946', *American Journal of Psychology* 60 (1947), 289–96 at 296. Note that Bernard Hart had already, on 7 May 1910, given a paper to the BPS entitled 'The psychology of Freud and his School' (published in *Journal of Mental Science* 56 (July 1910), 431–52). Other former students of Myers such as Maurice Nicoll, graduating in NST in 1906 before going on to Bart's, soon to become a founding member of the London Psycho-Analytical Society, may have also been showing the Cambridge psychologists the way forward, as their interests in Freud, Jung and other psychotherapists grew.

[118] E. Pickworth Farrow, *A Practical Method of Self-Analysis*, with Foreword by the late Professor Sigmund Freud, London: George Allen & Unwin, 1942, p. 1.

[119] T.H. Pear, 'Reminiscences', typed with a few modifications and additions, from notes made for a recording kept in the archives of the British Psychological Society, 10 May 1957, p. 9. The first 'milestone' was 'a notice at King's [when studying physics at London in 1906] said that lectures on Colour Vision would be given by Dr. C.S. Myers', which led to his all-consuming interest in psychology and thence to his appointment as Lecturer at Manchester. Pear's lecture on dreams to the troops in France was the immediate catalyst for Lionel Penrose's interest in psychoanalysis.

sentiment; the nature of double personality and its relation to other examples of the dissociation of consciousness.[120] And sure enough, in the first Essay examination of the new Diploma in July 1913, one of the three questions was on: 'The value of psycho-analysis in the treatment of mental disease'.[121]

The speedy implementation of the new Diploma in Psychological Medicine, with psychoanalysis figuring prominently in its subject-matter, together with the conspicuous incursion of psychoanalytic questions into the MST – both probably employing Hart's pithy book as basic student reading[122] – indicates that the argument in Stone's classic paper on 'Shellshock and the psychologists' cannot be sustained:

> before shellshock the influence of [Freud's writings] on the mainstream of British psychiatry was marginal. Freud's redefinition of the pathological remained for all intents and purposes a literary event and the psychoanalytic study of the neuroses took place on the fringes of the medical world where it involved a handful of private practitioners dealing with a small number of upper- and middle-class patients.[123]

---

[120] CUEP, Examination paper, Diploma in Psychological Medicine, 3 June 1913, 2 p.m.–5 p.m.

[121] The other two were: 'Heredity and insanity' and 'The alleged increase of insanity'. The Examiners were Drs Batten, Craig, Sherrington and Rivers (*University of Cambridge Calendar 1913–14*, p. xcviii). A parallel influx of psychoanalytic exam questions took place in London in the same year: 'I have been coaching a doctor for the M.D. London, Psychiatry Branch. In the psychology paper four questions only out of six may be answered, and McDougall, the examiner, set four questions on ΨA. My man swept the board and I hope may get the gold medal. It will be some time before Berlin, Paris, or Vienna emulate the example of London University in this!' (JF, 11 December 1913, p. 248). Note that the instigators of these innovations were the close Cambridge group, Myers and Rivers together with their former student and research collaborator William McDougall.

[122] Hart's book was to remain the standard introductory textbook for several decades, thus defining the newly professional British approach to medical psychology as psychodynamic in orientation; as Rapp notes, Hart's book sold for only 1/- in contrast to the higher costs of books by Freud, Jung and Adler (8/6d to 25/-) and this in part explains why it went through six printings by 1919 (D. Rapp, 'The early discovery of Freud by the British general educated public, 1912–1919', *Social History of Medicine* 3(2) (1990), 217–43 at 220n6). For Hart's influence on the stabilization of a psychodynamic tradition of 'psychopathology' in Britain, see German E. Berrios, 'British psychopathology since the early 20th century', in German E. Berrios and Hugh Freeman, *150 Years of British Psychiatry, 1841–1991*, London: Gaskell, 1991, pp. 232–44. In April 1917, Rivers, while working at Craiglockhart, gave an officer patient, Max Plowman, Hart's book following a discussion on Freud's theory of dreams. 'I met one rather interesting man up here', Plowman wrote, 'A Dr — who's a Professor of Psychology at Cambridge. I was talking to him about Freud's book on dreams and he lent me Hart's *Psychology of Insanity* as an introduction to it' (D.L. Plowman (ed.), *Bridge into the Future: Letters of Max Plowman*, London, 1944, p. 65, quoted in Ben Shephard, *Head Hunters: The Search for a Science of the Mind*, London: Bodley Head, 2014, p. 201).

[123] Martin Stone, 'Shellshock and the psychologists', in W.F. Bynum, Roy Porter and Michael Shepherd (eds.), *The Anatomy of Madness: Essays in the History of Psychiatry*, Vol. II, *Institutions and Society*, London: Tavistock, 1985, pp. 242–71 at 266.

The advent of Freud may have been a 'literary event' – as the stir amongst novelists, avant-garde thinkers associated with Orage's *New Age* and in the pages of the *Times Literary Supplement,* the *Spectator* and other weeklies certainly makes evident – but it was *also* already, by 1912, an event in the sober scientific circles of the most important group of psychologists in Britain, those associated with the new Laboratory at Cambridge who, with the ambitious creation of the Diploma in Psychological Medicine, were intending that Cambridge become the national centre for the education of psychiatrists, as it already was for Public Health.

The title of a Cambridge MD dissertation for 1912, submitted by Guy Foster-Barham, indicates as much – 'The influence of emotional conflicts and repressed emotion in the causation of abnormal states of mind' (it was probably supervised by Bernard Hart).[124] A paper Foster-Barham published, also in 1912, 'Insanity with Myxœdema',[125] gave an account of the psychotic symptoms underlying a case treated conventionally and successfully for hypothyroidism; he argued that, as a result of his detailed and intimate conversations with the patient, underlying her delusional jealousy was a repressed erotic conflict derived from her deep love for her cousin. Her attempts at excessive intimacy with her doctor gave the clue to this source; in public discussion of the case, the other alienists immediately recognized, approvingly but with some reservations, that he was engaged in the psychoanalysis of asylum inmates.[126] Foster-Barham was no fringe doctor: he graduated MB from Gonville & Caius College in 1903, then working at Claybury, at that time the only asylum in England with a research laboratory attached, before taking charge, from 1907 to 1917, of the Women's Division with a thousand beds at Long Grove, the newest link in the chain of massive asylums put in place by the LCC round London. In 1917 he became Superintendent of Claybury, a position he held for many years. His Superintendent earlier at Long Grove, Hubert Bond, recalled in his 1944 obituary of Foster-Barham that 'those were the days when the effect of Freud's work tended to be dominant ... Foster-Barham spared himself no pains, often with striking success, in using [psychotherapy] in accessible psychotic states.'[127]

[124] Robert S. Wallerstein, *Lay Analysis: Life in the Controversy,* Hillsdale, NJ: Analytic, 1998, p. 8n2, claims that 'Theodor Reik's doctoral dissertation in Vienna (1912) in psychology was the first *psychoanalytic* doctoral thesis ever. It was a study of artistic creativity in Flaubert's *Temptation of St. Anthony.*' It would appear that Foster-Barham's MD dissertation would also have a claim to this title, and it was a clinical rather than literary dissertation.

[125] 'Myxœdema' is an older term for 'hypothyroidism'.

[126] G.F. Barham, 'Insanity with myxœdema', *Journal of Mental Science* 58 (1912), 226–35.

[127] H.B., 'Dr G. Foster-Barham', *The Times,* 17 October 1944, p. 6.

It was to train more doctors like Foster-Barham that Cambridge's Diploma in Psychological Medicine was founded. Parallel confirmation of the importance of the reception of Freud in medico-psychological circles and the alliance of new psychological approaches to mental illness is also to be found in the career of a seemingly obscure Lancashire pathologist, Dr R.G. Rows, who, from 1911 to 1914, shifted from the most organicist of views of mental illness to an enthusiastic espousal of psychotherapy employing the latest views of Freud and Jung – this was the man who was to take over the Maghull Hospital in December 1914 and, responding to the call of war, created a centre for the dissemination of Freudian views throughout the swelling ranks of medical psychologists.[128]

Beginning in 1905, Myers had also been instrumental in lecturing on animal psychology and bringing in lectures on educational psychology; he had also extended the reach of psychology beyond those few students taking the Tripos, by introducing a course in Psychology into the Ordinary BA.[129] In his speech in May 1913, he could point to all these and more expansive developments in the work of the Psychological Laboratory:

> The University has recently founded a Diploma in Architecture, in which Aesthetics is a subject of examination. Still more recently it has instituted a Diploma in Psychological Medicine, which includes a paper and a practical examination in Psychology; and if the proposed Tripos in Anthropology is established, a close connexion will be made between Psychology and Anthropology.

Myers could say this of the new anthropological Tripos with some confidence since he was, once again, a moving force behind its inception, being one of the six members of the Board of Anthropological Studies who had on 31 January 1913 signed a report recommending the establishment of this Tripos.[130]

---

[128] Ben Shephard, '"The early treatment of mental disorders": R.G. Rows and Maghull 1914–18', in G. E. Berrios and H. Freeman (eds.), *150 Years of British Psychiatry*, Vol II, *The Aftermath*, London: Athlone, 1996, pp. 434–64. On Maghull, see more generally Peter Barham's magnificent *Forgotten Lunatics of the Great War*, New Haven: Yale University Press, 2004, esp. pp. 44–6, and Peter Leese, '"Why are they not cured?" British shellshock treatment during the Great War', in Mark S. Micale and Paul Lerner (eds.), *Traumatic Pasts: History, Psychiatry, and Trauma in the Modern Age, 1870–1930*, Cambridge: Cambridge University Press, 2001, pp. 205–21, citing J.K. Rowlands, the local Liverpool historian, on p. 216.

[129] This was the course that Roger Money-Kyrle would be examined for in June 1923; see Chapter 4.

[130] Report of 31 January 1913 in *CUR*, 11 February, 1913, p. 632; 'In 1914, a Tripos in Anthropology was established, which I likewise initiated' (Myers, 'Autobiography', p. 221).

By all rights, the new Laboratory should have been called the Wolf Myers Psychological Laboratory. The anonymous donor was Charles Myers himself, using the legacy he had received from his father: 'My father had died just before this time and I devoted anonymously part of the money which he had left me to defraying most of the cost of the building in his memory.'[131] But it was not only his father's legacy that contributed; of the £3,565 collected in response to his call for funds,[132] £3,500 came from Myers or his family.[133] And the post of Director of the Laboratory that Myers had occupied since 1908 remained unpaid. Cambridge now had a new Laboratory which would form the institutional core for its domination of British academic psychology for the next sixty years.[134]

Rivers and Myers in the First World War

While Myers was creating in the years before the First World War the firm base for experimental psychology and for its extension out towards psychological medicine, education and animal psychology, Rivers was preoccupied with anthropology – with preparing his *History of Melanesian Society*, based on his survey work in 1906–7 and with the preparation of the fourth edition of *Notes and Queries*, the methodological rule-book of the Anthropological Section of the BAAS. He gave a talk to the Heretics in May 1911 on the 'Primitive conception of death'.[135] He took the proofs of the *History* with him on the ship to the BAAS meeting in Australia in July 1914, as well as his student John Layard. When news of the outbreak of war reached the hundreds of scientists attending the Meetings, he proposed to Layard that they should take the opportunity provided by the disruption of war to do some field research. He telegraphed Cambridge to extract himself from his teaching commitments and he and Layard, finding their gunboat now committed elsewhere, jumped on a boat heading towards Malekula in the New Hebrides.[136]

[131] Myers, 'Autobiography', p. 221.   [132] Crampton, p. 130.
[133] See also F.C. Bartlett, 'Remembering Dr. Myers', *Bulletin of the British Psychological Society* 18 (1965), 1–10 at 4, quoted in Costall, 'Pear and his peers', p. 189: 'He planned it, to a very large extent in detail he designed it, he himself, his family and his friends, mostly paid for it. With some air of reluctance the University accepted it.'
[134] The establishment of a comparable laboratory at Oxford also depended upon a benefaction – in 1936, from a grateful patient of William Brown, the Wilde Reader there and a practising physician (L.S. Hearnshaw, *A Short History of British Psychology, 1840–1940*, London: Methuen, 1964).
[135] P. Sargant Florence and J.R.L. Anderson (eds.), *C.K. Ogden: A Collective Memoir*, London: Elek Pemberton, 1977, p. 235.
[136] Layard, 'Autobiography', p. 18: 'To my amazement and delight, Rivers telegraphed to Cambridge suggesting that because of the war and all the young men joining up, could he have another term off

Myers too found he could not sit still when war was declared, but he reacted in the opposite fashion to Rivers.

> When the Great War came in August, 1914, I tried vainly to continue the work in which I was then engaged, of studying the unique records of Australian music which the late Professor Baldwin Spencer had presented to the large collection of phonographic musical records which I had gathered together from all parts of the uncivilized world. I was then 41 years of age and, on applying to the British War Office for service in France, I was informed that no medically qualified volunteer who was over 40 years old could be accepted. A few weeks after the outbreak of war, I resolved to journey to Paris *in mufti* and succeeded in persuading the Commandant of the Duchess of Westminster's Hospital, which was about to open at Le Touquet, to appoint me as Hospital Registrar.[137]

By early 1915, Army policy was more flexible and Myers received a commission in the RAMC in France, treating functional disorders in the BEF; by 1916 Myers was where one would expect him – Consultant Psychologist to the British Armies in France at General Headquarters. By 1917, quartered at Abbeville, he had 'a large staff of psychotherapeutic specialists under my supervision'.[138]

From early on in the War, Myers was publishing on the complex psychological picture that the condition known as 'shell shock' presented. It was his publications in 1915 that turned the term 'shell shock' from an informal category used by soldiers and their doctors at the front to a respectable and universally accepted, if for a brief time, medical category. In February 1915[139] he published an account of three cases scrutinized using stalwart techniques from the psychological laboratory – perimeter of visual field, visual acuity, reactions of taste buds to salt, sugar, quinine, ether, peppermint, iodine – deducing the 'functional' character of the soldier's condition from the lack of connection between the battlefield events and the bodily disturbances. Exceptionally noisy shells had little effect on the sense of hearing, but taste in particular was severely disturbed.[140] Myers

---

and be allowed not to go straight back to Cambridge but to go out and do some more field work in Melanesia. And that was accepted, and Rivers and I settled down to discussing where we should go.'
[137] Myers, in Murchison, *History of Psychology in Autobiography*, Vol. III, p. 222.
[138] *Ibid.*, p. 223. As late as the early 1920s, Rivers was referring to 'psychotherapeutists'. See Shamdasani, '"Psychotherapy": the invention of a word'.
[139] The first documented occurrence of the First World War usage of 'shell shock' is to be found in the *BMJ* 1(2822) (1915), 192, two weeks before Myers's.
[140] Charles S. Myers, M.D. Sc.D. Camb., 'A contribution to the study of shell shock. Being an account of three cases of loss of memory, vision, smell, and taste, admitted into the Duchess of Westminster's War Hospital, Le Touquet', *The Lancet* 185(4772) (13 February 1915), 316–20.

deployed hypnosis to aid recovery of memory of the soldiers' experiences, but it was the psychophysical examination that was clinching for him: it revealed 'the essentially psychological nature of this condition'[141] and demonstrated the exact character of the 'dissociated complex'.[142] The therapeutic measures he eventually adopted were 'by obtaining persuasively the recall of repressed memories, with or without the aid of light hypnosis'.[143] He found the use of hypnosis – these were the first cases on which he had ever used it – extremely disconcerting, warning others that it requires more self-mastery than one's first sight of a surgical operation.[144]

Later on in the war, Myers laid out the two extreme views of shell shock: on the one hand, there were those who viewed the sufferers as no better than malingerers, so that the more attention is paid to them, the worse they will become; on the other, there were those who 'would subject them to a prolonged course of psycho-analysis, or would tend indiscriminately to pamper them'.[145] Quite evidently, having drawn up the contrast in these terms, the truth for Myers lay between the two, but there was no doubt in his mind that the psychoanalytic enthusiasts were nearer to the truth than the neglecters, whom he compared to surgeons who blithely wave away appendicitis. Key to the treatment of shell shock were promptness, a conducive environment and psychotherapeutic measures, whose guiding principles 'consist in the re-education of the patient so as to restore his self-knowledge, self-confidence and self-control'.[146] With some cases it is necessary to engage in the 'analysis and elucidation of previous conflicts or of the dreams or strange ideas which force themselves on his notice, and to the revival of forgotten memories, if necessary under slight hypnosis'.[147]

While deriving much of his technique and conception of the psychotherapeutic process from a watered-down psychoanalysis, Myers was alert to keep his distance, assuring his reader that 'hypnosis will succeed in such cases where many weeks of psycho-analytic "free association" and "conversation" in the waking state may fail'.[148] But Myers is insistent that it may be necessary to revive long-forgotten conflicts and memories, often against much resistance on the part of the patient in order to overcome 'the strength of repression'.[149] Myers's theoretical framework also derived much from the French school, emphasizing dissociation and its consequences: a

[141] Myers, in Murchison, *History of Psychology in Autobiography*, Vol. III, p. 223.
[142] Myers, 'A contribution to the study of shell shock', p. 320.
[143] Myers, in Murchison, *History of Psychology in Autobiography*, Vol. III, p. 223.
[144] Charles S. Myers, *Shell Shock in France, 1914–1918*, Cambridge: Cambridge University Press, 2012, p. 57.
[145] *Ibid.*, p. 50. [146] *Ibid.*, p. 55. [147] *Ibid.*, p. 56. [148] *Ibid.*, p. 57. [149] *Ibid.*

trauma gives rise to a loss of consciousness, from which often arises the supremacy of an 'emotional' personality over the 'normal' personality. The apparently normal personality returns but this is an illusion: the 'emotional' personality is still always ready to make its presence felt in the symptoms of the soldier, which are the 'outward expression of ... the highly emotional "complex", as it works "subconsciously" through and beneath the "apparently normal" personality'.[150] Treatment consists in 'restoring the "emotional" personality deprived of its pathological, distracted, uncontrolled character, and in effecting its union with the "apparently normal" personality hitherto ignorant of the emotional experiences in question'.[151] Once this reintegration has taken place, all symptoms vanish: headaches and dreams disappear, the circulatory and digestive symptoms become normal, 'even the reflexes may change'.[152]

Myers found himself called upon to appear as an expert witness in numerous courts martial concerning desertion and serious infractions of discipline. This undoubtedly put a strain on his relations with the Army, since he could never wholly succeed in persuading the RAMC and other professional army officers that shell-shocked soldiers were not cowards deserving only of being shot. In addition, he was manoeuvred out of his position as principal psychological consultant to the Armies in France by a harder-nosed neurologist, Gordon Holmes.[153] Following the London Conference in October 1917 on shell shock, he gave up the struggle with the Army and decided to return to England permanently, taking up a new set of duties inspecting military shell shock hospitals in Britain: this would help the Army 'regulate', as his Army senior put it, 'the wilder spirits who lack a due sense of proportion'.[154] His first visit was to the hospital near Liverpool where many of his close academic colleagues were spending the war – the 'Maghull Academy'.

There is no question that Maghull was one of the most important locations for the intense discussion, exploration and dissemination of Freudian ideas in Britain in the second decade of the twentieth century;[155] amongst the physicians who worked there at some point were Rivers,

---

[150] *Ibid.*, p. 68.   [151] *Ibid.*, p. 69.   [152] *Ibid.*

[153] See Ben Shephard, *A War of Nerves: Soldiers and Psychiatrists 1914–1994*, London: Jonathan Cape, 2000, pp. 49–51.

[154] Myers, *Shell Shock in France*, p. 23; the phrase is repeated often in the later sections of the book.

[155] Some of the most vivid accounts of the atmosphere of Maghull were given by the young Freudian enthusiast T.H. Pear late in his long life, but following Pear's career is a good indicator of the progress of Freudian ideas within psychology in the whole of this period. See Costhall, 'Pear and his peers' and John Forrester, 'Remembering and forgetting Freud in early twentieth century dreams', *Science in Context* 19(1) (2006), 65–85 on Pear's dream analyses.

McDougall, Seligman, William Brown, Hart, Pear, Elliot Smith,[156] MacCurdy and Myers. (Even the grief-stricken Ernest Jones visited, to see Hart and seek solace, in the weeks after the sudden death of his first wife in September 1918.[157]) Filtered in part through Maghull, the two dominant figures in Cambridge psychology, Myers and Rivers, had more influence on the development of psychoanalytically informed practices and ideas about the treatment of shell-shocked soldiers than any other two individuals. And yet again, in their opposite and complementary fashions: Rivers, the man of ideas, who commanded immediate moral respect, the embodiment of the ideal character type of the distinguished gentleman of science, lending his moral and intellectual authority to the serious consideration of Freudian ideas; Myers, the man of organizations and reforms, always in the right place at the right time – at least for the first three years of the war – throwing his weight behind the growing movement for psychological treatment of soldiers just when that weight had considerable effect.

Psychology after the Great War

By the end of the war, the tide of enthusiasm for psychoanalysis was considerable. With the cessation of hostilities, key players in the development of psychology and psychoanalysis in Britain and in Cambridge switched their attention to the business of post-war disciplinary reconstruction. Charles Myers was in characteristic reformist mood. Having spent the war working with shell-shocked soldiers and the doctors who were drafted in to treat them, his vision of the future of psychology had been fundamentally altered. He set out to transform the institutions he had helped to create to make them fit for the transformed post-war landscape.

Colonel Myers operated on four principal fronts:[158] firstly, the reform of the British Psychological Society. Before the war the Society had been dedicated to those actually practising experimental psychology. Myers now knew that experimental psychology was becoming a failed backwater of

[156] While not a physician, Pear co-authored with Rivers's close friend Grafton Elliot Smith one of the most influential and polemical pro-psychological and pro-Freudian books of the period, *Shell Shock and its Lessons*, Manchester: Manchester University Press, 1917.

[157] Brenda Maddox, *Freud's Wizard*, London: John Murray, 2006, pp. 141–2.

[158] For further details of Myers's views and the actions to which they impelled him, see John Forrester, '1919: psychology and psychoanalysis, Cambridge and London: Myers, Jones and MacCurdy', *Psychoanalysis and History* 10(1) (2008), 37–94. For want of space, we have omitted detailed discussion, which can be found in that paper, of Myers's trajectory and the parallel trajectory of Ernest Jones in this period, of the details of the reforms of the British Psychological Society initiated by Myers and Jones, and of the parallel founding of the BPaS in February 1919.

psychology, so he loosened up the criteria for membership and created three sub-sections in the areas – all non-academic – in which psychology was burgeoning: medical, educational and industrial. The reward was immediate: hundreds of ex-RAMC doctors, trained up in the new psychological methods during the war, became members. For the next twenty-five years, the British Psychological Society would be less dominated by experimentalism, more open to the currents to which Nikolas Rose has given the name 'the psychological complex': social work, child study and guidance, criminology, as well as psychodynamic theories and therapies, educational psychology and Myers's own new pet field, industrial psychology and personnel management.[159] Secondly, following the Armistice, Myers's reforming activities for the British Society and in Cambridge, week by week, month by month, marched in parallel. Aiming to open up the BPS to the psychological therapists who had been trained up during the war, he also decided to transport Maghull to Cambridge, initially hoping to secure an appointment for Rows, the Director of Maghull, and put in place a clinic that would make Cambridge the national peace-time centre for the revolutionary new developments in psychological medicine. Like those to be found in Germany and the USA, the Clinic would be dedicated to the out-patient treatment of the neuroses, and thus form the 'clinic' wing to his expanded psychology empire.[160] He secured the support of the liberal Great and Good – Lady Darwin, some Heads of Houses, the Vice-Chancellor, Mrs Keynes and, of course, Rivers – in part by mobilizing the already existing organizational resources of the Cambridgeshire Voluntary Association for the Care of the Mentally Defective. Rivers actively assisted him, including locking horns with the inventor of the clinical thermometer and former Inspector in Lunacy, vociferous antagonist of psychoanalysis and probable model for George Eliot's Lydgate in *Middlemarch*, Sir Clifford Allbutt:

> The proposed clinic would therefore have two aspects, (i) the treatment of functional nervous disorders and incipient insanity, and (ii) research upon these subjects and the education, especially post-graduate, of students of

---

[159] Nikolas Rose, *The Psychological Complex: Psychology, Politics and Society in England, 1869–1939*, London: Routledge and Kegan Paul, 1985. Hearnshaw's history, still the only comprehensive history of psychology in Britain, concludes that British psychology between the wars was 'saved by its applications, educational, industrial and medical' (Hearnshaw, *A Short History of British Psychology*, p. 211). Rose's account puts flesh on Hearnshaw's overview, demonstrating the displacement of psychology in the inter-war years away from academic, experimental university-based psychology.

[160] For a fuller treatment of the story of the Cambridge Clinic, see Forrester, '1919', pp. 57–63.

medicine. Cambridge now has the chance of being first in the field in this most important movement.[161]

The Clinic did open, in September 1919. Prideaux, the physician attached to the Clinic in 1921, portrayed its activities as those of a network. Firstly, through the Voluntary Association for Mental Welfare, it provided 'visitors' so as to develop social services in association with the clinic, investigating home conditions, developing family histories, and finding work for patients when cured. The Clinic was also proving of extreme value to the general practitioners in the area. Schools, too, were able to pass difficult cases straight from the School Clinic to the Mental Clinic. 'Magistrates have taken to referring cases to the clinic for an opinion as to the mental condition of alleged criminals';[162] and finally, the Clinic functioned as a 'Research Centre', whose material passed directly to the University's Psychological Laboratory.

But this map was grander than the reality on the ground. Even with the imposing support of forward-thinking Cambridge, Myers failed to secure it permanent funding. Within a few years, the Clinic was most aligned in its operation with Fulbourn Asylum, then very much a traditional County Asylum just outside Cambridge, whose Director took over the management of the Clinic. With Bartlett, Myers's successor in 1922, suspicious of any psychology that was not experimental; with Rivers's successor, MacCurdy, effectively medically neutered and disdaining to teach in the specialist area in which he was supremely well qualified and for which he had been appointed; with insufficient funding; the plan to make Cambridge the national centre of the treatment of the neuroses went precisely nowhere. It was left to the Tavistock and particularly the Maudsley, both in London, to vie for dominance of the new field of operations which Myers had correctly identified.

Naturally, Myers wished to repeat in Cambridge what he was doing on the national scale – introduce to the University these larger areas of medical, industrial and educational psychology. Rivers had resigned his University position in 1916. Myers had secured funding from the Medical Research Committee to help reconfigure the Psychological Laboratory so that it could stand entirely independent of the Special Board for Moral

[161] Speeches at Special Meeting, Conference held at County Hall, Cambridge, convened by the Cambridgeshire Voluntary Association for the Mentally Defective, 13 January 1919, together with other papers, Cambridgeshire Record Offices, Ref. R84/23, Rivers CRO/SM 7–8, January 1919.
[162] 'Annual Meeting, Royal Medico-Psychological Association, 13th July 1921. Discussion', *Journal of Mental Science* 67 (October 1921), 525–32 at 529.

Science, securing permanent funding for the Director of the Laboratory and surrounding the Director, in scientific Professorial style, with the appropriate array of Lecturers and Demonstrators – he was successful in all of these.[163] He also campaigned to have a Board of Psychological Studies, comparable to and separate from the Board of Moral Science, established; again he met with success on this in 1920. He himself had finally been elected to a Fellowship at Gonville & Caius College in 1919, ahead of his award of the CBE in 1920.

But instead of 'Reader in Psychology', the Board of Moral Science forced him to adopt the title 'Reader in Experimental Psychology' – which was the last thing on his mind: his reformed idea of psychology involved relinquishing most of the brass instruments psychology of the turn of the century in favour of educational and industrial (testing) psychology, and above all of 'medical psychology', in which experiment played far less of a role than 'clinical observation'. In December 1920 Rivers jumped to Myers's defence in the debate in the Senate House over the name of the Readership being created:

> The experimental method took a relatively unimportant place in the science of Psychology, and to those acquainted with modern opinion it seemed calculated to bring this University into ridicule that the scope of a new post should be limited to the experimental aspect of the subject. It was possible that the General Board had been influenced in their decision by the fact that the funds for the post were provided by the Medical Grant Committee. If that was so, they had been misinformed, for the most important branches of Psychology in relation to Medicine depended on observation rather than experiment ... The title of his post would certainly not deter the new Reader from dealing with all aspects of Psychology, but it was perhaps well that there should be a protest against the out-of-date nomenclature adopted by the General Board.[164]

Rivers clearly knew this field better than anyone: he had personally founded the first two psychological laboratories in Britain; he had formally promoted and worked with the experimental methods of his teacher and German colleague, Kraepelin; he had participated in the famous five-year experimental studies of nerve regeneration with Henry Head; and he had engaged in methodically ground-breaking experimental studies of the effects of drugs on fatigue – all prior to his vigorous and influential work fashioning an independent English style of psychotherapy incorporating key elements of Freudianism during the war. It is a measure of the

---

[163] HR 1921–30, p. 26.   [164] *CUR*, 10 December 1920, p. 396.

transformation that ethnography and the war between them had effected in the vision of the future of psychology proclaimed by the foremost English experimental psychologist that he could declare that experiment was 'relatively unimportant' and that restricting the scope of the post to experiment by adopting 'out-of-date nomenclature' would 'bring this University to ridicule'. That this was Rivers's considered view, and not simply a response to a local academic spat, is clear from his survey of the whole of psychology he had delivered in May 1919, when he declared:

> There is now a widespread, and in my belief well-founded, opinion that this [experimental] movement has failed to come up to the expectations of its founders and has proved unfruitful as a direct means of increasing our knowledge, at any rate in so far as it confines its attention to the experimental investigation of the normal human adult. It cannot be said to have done much more than provide suggestions and clues for investigation on other lines.[165]

The teaching in psychology at Cambridge in this period bears out this broadened vision: in 1920–21, in addition to Rivers's lectures on dreams, Prideaux, Director of the new Clinic, lectured on abnormal psychology, Bartlett lectured on psychology and industry and Miss Lucy Fildes lectured on 'The psychology of the backward child'. The 1921 Tripos questions reflected the broad approach:

How far is 'wish fulfilment' an adequate principle for explaining dreams?

> Consider the relation of impulse, emotion, and instinct. Is it psychologically accurate to speak of a 'religious instinct'?
> Is religious conversion rightly regarded as psychopathological?
> To what extent have the phenomena of 'multiple personality' thrown light on the nature of self?
> Consider the view that all our experience is organised on a basis of different chronological levels. Has this view any special bearing upon the psychology of dreams?

On the Essay paper, one of the questions was 'Affective-tone in dreams'. So a student in 1921 might not have been foolhardy by braving the examination equipped solely with a detailed knowledge of contemporary dream theory, in other words, Freud and Rivers.

This moment perhaps marks the disciplinary pinnacle in Cambridge of psychoanalysis as it was interwoven into a broad understanding of what the field of psychology could cover. Myers already knew that his ambitions for the field were failing within the University in Cambridge: the failure of the

---

[165] W.H.R. Rivers, 'Psychology and medicine', *The Lancet* 193(4995) (24 May 1919), 889–92 at 890.

initiative of the Clinic; failure to secure funding from the Dunn Bequest for the project on 'The Study of the Human Mind'[166] (stymied, he later implied he had suspected at the time, by W.B. Hardy, Professor of Physiology in Cambridge); the sudden opposition to his expanded vision of psychology as expressed by the Board of Moral Science – these fuelled his sense of disgust at the conservative opposition in Cambridge. At a meeting in 1918, Myers had met a businessman with whom he quickly concocted the project of an Institute to put psychology in the service of the economic needs of the country. By 1921, they had founded the National Institute of Industrial Psychology. Myers announced his departure to direct the newly founded National Institute in December 1921, and left formally on 11 June 1922.[167] The NIIP became his main professional preoccupation until his retirement in 1939.[168] Seven days before Myers's departure, Rivers died.

[166] Kohler, 'Walter Fletcher, F. G. Hopkins, and the Dunn Institute of Biochemistry', pp. 341–2 includes details of a memorandum drawn up by William Bate Hardy (Biological Secretary to the Royal Society, Professor of Physiology at Cambridge, chairman of the MRC Food Committee), probably in October 1919, outlining areas in which the Dunn Trustees might deploy their considerable funds. 'The first, "The Study of the Human Mind," included ethnology and applied psychology and physiology. Hardy emphasized the relevance of anthropology to colonial government, and the public need for study of industrial efficiency, fatigue, vocational testing, and so on.' This is transparently a package drawn up with Hardy's colleagues Rivers and Myers in mind and was most probably drafted after discussion with Myers. Psychology lost out (and its loss led eventually to Myers's departure from Cambridge); the eventual very big winner was Gowland Hopkins, with the bequest of over £200,000 to found and build the Dunn Institute of Biochemistry in Cambridge. When originally constituted in 1913, the Medical Research Committee included Allbutt as one of its eight members (*Science*, N.S. 38(968) (18 July 1913), 79); by December 1918, the membership of the Committee included both Henry Head and Gowland Hopkins, both expert voices, neither disinterested, ready to speak to the relative merits of the psychology/physiology and the biochemistry proposals (*Science*, N.S. 48(1252) (27 December 1918), 639–40 at 639).

[167] Cf. his recollections: 'On demobilization I returned to Cambridge, fired with the desire to apply psychology to medicine, industry, and education and becoming increasingly disgusted, after my very practical experience during the War, with the old academic atmosphere of conservatism and opposition to psychology. I found that the wild rise of psychoanalysis had estranged the Regius Professor of Physic; I received little encouragement from the Professor of Physiology; and the Professor of Mental Philosophy, to my surprise, publicly opposed the suggested exclusion of the word "experimental" in the title, now about to be conferred on me by the University, of Reader in Experimental Psychology. Thus medicine, physiology and philosophy had little use then at Cambridge for the experimental psychologist.' (Myers, 'Autobiography', pp. 215–30, also quoted in Paul Whittle, 'W.H.R. Rivers: a founding father worth remembering', unpublished talk, now available online, December 1997.) Note the blatant contradiction in this passage: Myers resented Ward's insistence on appending the term 'experimental' to his title, but then described himself as being excluded precisely because he was an 'experimental psychologist'; this contradiction probably stems from Myers's wishing both to attribute continuity of professional identity to himself, while explicitly wishing to distance himself from 'old-style' experimental psychology.

[168] See Hearnshaw, *A Short History of British Psychology* and Stansfield, F.R. 'The growth years of the National Institute of Industrial Psychology, 1921–1930' (2005), unpublished paper presented at the British Psychological Society History of Psychology Centre, London, 2 March.

The structure of Psychological Studies at Cambridge was now decided by a University review. Bartlett's post of Assistant would be abolished and Bartlett would replace Myers as Reader and Director of Laboratory at the same salary of £650; the £175 p.a. freed up would be used for a new 'University Lectureship in Psychopathology in connexion with the Special Board of Medicine',[169] 'to ensure the future of the Diploma in Psychological Medicine'.[170] Myers's successor, Frederic Bartlett, successfully closed down all the avenues that Myers had looked to open up. He stabilized the discipline, against the grain of his very own inclinations and scientific judgement,[171] around a narrowly conceived core of experimental programmes, and exorcised all the possible new spirits of the age. A historic opportunity was lost; though some might say that Bartlett chose the right academic course and turned Cambridge psychology into the powerhouse of an *academic* discipline still struggling to establish itself. In contrast to Germany and particularly the USA, British psychology was decades late in establishing itself with numerous, as opposed to isolated, academic centres. When eventually, in the 1950s, it did, Cambridge was imprinted on it. As Eric Trist, the only student to get a starred First in Psychology between the wars, recalled of the 1930s, 'I knew Bartlett had control of all the appointments in psychology in England.'[172] In 1937 Bartlett boasted as much in a Report to the Vice-Chancellor of Cambridge, so as to reassure the University of the wisdom of accepting money from the Rockefeller Foundation for postdoctoral posts: 'if any post of a psychological kind is

---

[169] *CUR*, 23 January 1923, p. 580. In HR 1921–30, p. 34, the Lectureship is described as having been originally 'in connexion with the Special Board of Medicine' but then the 'Lectureship is now in the Faculty of Biology "B"', where Bartlett engineered to locate the Department of Psychology in the new Faculty system inaugurated in 1926, which created an umbrella system of Faculties within which Departments were located – although some Faculties (e.g. History) had no Departments within them.

[170] Crampton, p. 174.

[171] The great irony of Bartlett's life is that as a thinker he is remembered for his classic study, published in 1932, entitled *Remembering*, which proposed that memories are in fact mental reconstructions, influenced by culture, habit and personal idiosyncrasy. Such findings are of course an integral part of psychoanalytic theory; however, even though Bartlett recalled that, during the war, when working with 'shellshock' patients in Cambridge, 'I had discussed current treatments of "shell-shock" with Myers and Rivers and, like all psychologists of the period, I had read everything I could get of the work of Sigmund Freud' (Crampton citing Bartlett's (1969) unpublished autobiography, p. 168), it is difficult to find the explicit traces or influence of this absorption in the experimental and case work he began during the Great War which became *Remembering*. But certainly this classic study was completely out of keeping with the dominant Wundtian or behaviourist ideals of psychological experimentation. Yet Bartlett restricted his students, his colleagues and his Department to exactly the restricted ideal that his most important work repudiated.

[172] Eric Trist, 'Guilty of enthusiasm', in Trist, *Management Laureates*, ed. Arthur G. Bedeian, Vol. III, London: JAI, 1993, pp. 191–221.

created or falls vacant, whether academic or applied, anywhere in Great Britain, the very strong tendency is to turn first to Cambridge for somebody to fill the post'.[173] By 1939, there were only six Professors and thirty Lecturers in psychology in British universities; by 1957, there were sixteen Professors, of whom ten had been trained by Bartlett and Myers.[174]

Back in 1921 Myers had created a Rivers-shaped post, the Lectureship in Psychopathology, if Rivers were interested in university teaching any longer. Holding no university position immediately after the war, he had been lecturing on 'dreams' to huge audiences (often in the summer or on Saturdays, so that many, like Adrian and Karin Stephen, could come from London to attend), and on 'social organization', to help nurture the new Anthropology Tripos. With his death, there was an urgent need for a replacement in psychopathology. Who to appoint? There are some clues that Myers invited Emanuel Miller to apply.[175] Miller (1892–1970) had taken the Natural Sciences Tripos Part I at St John's College in 1913 (with a passion for zoology) and the MST Part II in 1914 (when he fell under the spell of Spinoza) before training in medicine at the London Hospital, qualifying in 1918. In 1920–21 he returned to Cambridge and, along with four others, took the Diploma in Psychological Medicine. Miller, as his later career attests, was the classic 'Cambridge Freudian', a youthful candidate to take over from Rivers and Myers. In February 1922, Ernest Jones reported to his colleagues in Vienna on a talk he had given to Tooting Neurological Society that the doctors were 'all in various stages of assimilating Psa, the most advanced being Miller, of course a Jew'.[176] The procedure for finding the first occupant of the new Lectureship in Psychopathology in connection with the Special Board of Medicine took place in the first three months of 1923, which is when Miller was probably invited to apply. He declined. At the end of that year he delivered a paper to the BPaS on 'Mythology and dreams',[177] though he never became either Associate Member or Member of the Society.

---

[173] CUEP, Frederic Bartlett, Report to VC, University of Cambridge, re Rockefeller funding, 22 February 1937.

[174] Crampton, chapter 13.

[175] H.G., 'Dr. Emanuel Miller', *The Lancet* 296(7668) (15 August 1970), 375: 'He did not accept Myers' offer of an academic career in psychology.' We have not been able to identify H.G., but given the content of the obituary it is certain he was a life-long close friend of Miller's and thus a reasonably reliable source. The only other post, much more in Myers's gift before he left Cambridge, was a Demonstratorship, to which Sebastian Sprott was appointed in 1922; it is not likely that the medically qualified Miller would have been offered a mere Demonstratorship.

[176] Maddox, *Freud's Wizard*, p. 169.

[177] D. Bryan, 'British Psycho-Analytical Society', *BIPA* 5 (1924), 250–3 at 251.

For the academic year 1924–25, Miller himself taught the DPM in Cambridge as a temporary lecturer[178] – a sign that things were already not going well with the DPM in Cambridge, since the new permanent Lecturer in Psychopathology, appointed in 1923, would have been expected to perform this duty, indeed was contractually bound to do so. In the mid-1920s Miller was one of a group of fifty psychiatrists invited to tour[179] the United States to become acquainted with American work in the field of child guidance; on his return, at the end of 1926 he became Honorary Director of the East London Child Guidance Clinic, the first Child Guidance Clinic in Britain, sponsored by the Jewish Hospital, Whitechapel, working closely with Sybil Clement Brown, a major organizing force in child welfare from the 1920s to the 1950s and close friend of Susan Isaacs.[180] He also taught for many years on the Diploma in Psychological Medicine course that succeeded where Cambridge's had failed, at the Maudsley in London.

Miller's approach to psychiatry was of Freudian inspiration without Freudian dogmatism. In his Presidential Address to the British Psychological Society, Medical Section, in 1935, he charted the development of his own psychological convictions from the 'calm groves of academe' of the moral sciences at Cambridge ('Ward, McTaggart, Russell and Moore') after which he passed through a period of scepticism before arriving at an open-minded eclectic position: 'with self-knowledge and the growth of experience of a very wide variety of patients, one was driven to the conclusion that after all there was quite a lot to be said for the theories of the many warring schools, although for my own part I found that perhaps the greatest security was discovered in the fundamental doctrine enunciated by Freud'.[181] Miller had broad interests: he was a good friend of Malinowski's, who thanked him in the 'Preface' (1927) to

---

[178] Edward Glover, 'In piam memoriam: Emanuel Miller, M.A., F.R.C.P., D.P.M., 1893–1970', *British Journal of Criminology* 11(1) (1971), 4–13 at 8–9.

[179] Mathew Thomson, 'The psychological body', in Roger Cooter and John Pickstone (eds.), *Companion Encyclopedia to Medicine in the Twentieth Century*, London: Routledge, 2003, pp. 291–306.

[180] The South African Meyer Fortes, completing a doctorate in psychology at UCL on culture-independent intelligence tests, was later to join them, working on the effects of sibling order on adolescent behaviour. Miller was probably instrumental in kindling Fortes's interests in psycho-analysis which would remain lifelong even when he became a social anthropologist, eventually Professor at Cambridge. 'I came into social anthropology after a training in experimental psychology and some experience of psychology under the tutelage of Emanuel Miller' (M. Fortes, 'Custom and conscience in anthropological perspective', *IRP* 4 (1977), 127–54 at 137); Fortes paid tribute to Miller by taking up once again his theory of the generations in a psychoanalytic-anthropological paper in 1974 (M. Fortes, 'The first born', *Journal of Child Psychology and Psychiatry* 15 (1974), 81–104).

[181] E. Miller, 'The present discontents in psychopathology', *British Journal of Medical Psychology* 15 (1935), 2–17 at 3.

his famous book which was equal parts use and abuse of psychoanalysis, *Sex and Repression in Savage Society*;[182] he was not only one of the leading child psychiatrists of the period 1925–60 but a co-founder in July 1931 of the Association for the Scientific Treatment of Criminals, together with Edward Glover, David Eder, E.T. Jensen, J.A. Hadfield, Marjorie Franklin and the originator Grace Pailthorpe, its founders coming from 'psychiatry with a psycho-analytical bias',[183] as he characterized it. In the early months of the Second World War, he joined up with Winnicott and Bowlby to urge that government plans to evacuate children be mindful of the damage that may be done by separating children from their mothers.[184]

On his death on 29 July 1970, he left a bequest to St John's College, Cambridge, to fund an annual prize for the best essay in the Philosophy of Science with special reference to the behavioural sciences (psychology and social science); two years later, his son, Jonathan Miller, paid an indirect tribute to one of the principal threads of his father's life by producing *Freud: The Man, his World, his Influence*.

In 1923, Emanuel Miller, for whatever reason, declined to envisage becoming the vehicle for the promotion of Freudian psychopathology in Cambridge. Nonetheless, the ideal candidate seemed to be at hand: John MacCurdy, a Canadian student of Ernest Jones and one of the few North Americans Jones trusted with the leadership of psychoanalysis there. He was one of the eight founder members in May 1911 of the American Psychoanalytic Association, and author (with a Preface by Rivers) of an influential book *War Neuroses* (1918), balanced but firmly Freudian in orientation. MacCurdy's private life had also become entangled with Ernest Jones's in pre-war Toronto and Vienna, turning into something of an *opera buffa* quintet.[185] However, by 1923, MacCurdy had published a substantial critique of both Freud and Rivers. This was the man appointed in March 1923 to replace Rivers and cover the new and exciting area of psychopathology, dream theory and medical psychology in Cambridge.

MacCurdy essentially took his appointment as an invitation to retire to the cloistered life of the Cambridge don in Corpus Christi College. Most

[182] Bronislaw Malinowski, *Sex and Repression in Savage Society*, London: Kegan Paul, Trench, Trubner & Co., 1927, p. 12.
[183] E. Miller et al., '1950–1970. Retrospects and reflections', *British Journal of Criminology* 10 (1970), 313–23 at 320.
[184] BMJ 2(4119) (1939), 1202–3.
[185] For a much more complete account of MacCurdy's career, as far as sources have so far allowed, see Forrester, '1919', pp. 72–89.

tellingly, he never regularized his medical qualifications so that he could practise in England. The Diploma in Psychological Medicine lost students, from five in 1920–21 to one in 1923–24 when Miller taught the course, and none for two consecutive years until it was abolished in 1927. A 1925 review of the University's Diplomas tellingly observed that the DPM had little influence on teaching or research in the University, though it did conclude that it was nonetheless recognized as an 'important qualification' by the mental health authorities.[186] A key part of the successful case for its abolition – that teaching for this national degree was not being supplied by the University – one can attribute very directly to MacCurdy's reluctance to teach what he had been appointed to teach. So the Diploma was abolished immediately.[187] However, viewed from a broader perspective, its demise was part of the overall demise of clinical medicine in Cambridge – its clinical school ignominiously folded, to the satisfaction of the laboratory sciences of physiology, biochemistry and pathology.[188] The disappearance

Fig. 5.3 John Thompson MacCurdy in his Corpus Christi College rooms.

[186] Weatherall, *Gentlemen, Scientists and Doctors*, p. 206.
[187] It was principally the Maudsley Clinic (rather than the Tavistock Clinic) that took up the slack and established itself as the national centre for postgraduate training in psychological medicine (i.e. psychiatry); see Edgar Jones, Shahina Rahman and Robin Woolven, 'The Maudsley Hospital: design and strategic direction, 1923–1939', *Medical History* 51 (2007), 357–78.
[188] A Clinical School that has lasted and excelled was eventually established in 1976.

of the DPM was self-evidently to the satisfaction of Bartlett and to the relief of MacCurdy.

So when in the early 1930s the Faculty Board of Biology 'B' felt it necessary to teach psychoanalysis to medical students, they again had to turn outside the University to Karin Stephen, former Newnham Fellow in Moral Sciences, now medically qualified and a psychoanalyst, for a course initially called 'Clinical aspects of psychoanalysis',[189] later 'Psychoanalysis and medicine', which she taught throughout the 1930s. But how the Department of Experimental Psychology viewed this course can be judged by the fact that it declined to acquire for its library the book she published on the basis of these lectures in 1933, which clearly formed a textbook for the subsequent courses she delivered until 1940.[190]

In 1927, the only year in which MacCurdy offered a course on 'Theory and practice of psycho-analysis',[191] he did have two important, though non-medical, students who earned his respect: Gregory Bateson and Ralph Pickford. The following year, John Bowlby took the Moral Sciences Tripos Part II; MacCurdy left no trace whatsoever in his recollections. In the next

[189] Listed in *CUR*, 23 March 1932, p. 777 under 'Psychology' in NST Biology 'B' as 'Mrs. Adrian Stephen, W.5, 15s' (15 shillings). It is noteworthy that this course was charged in the traditional manner (in large part done away with following the Statute reforms of 1926 with the introduction of career paths, i.e. the dons became salaried professionals) and that the *Reporter* declined to recognize her medical degree in their insistence on giving her her married name. (What were they thinking? That they couldn't call her Dr Adrian Stephen, since that would be ambiguous whether it was the husband or the wife who was lecturing, so they gave her an unambiguous gendered designation with the convention that she adopt both her husband's family and first name, so that students could distinguish between husband and wife? What would have been objectionable about calling her 'Dr Karin Stephen'?)

[190] The Experimental Psychology Library did acquire a copy of H. Banister, *Psychology and Health* (1935), a book written by the non-medically qualified University Lecturer in Experimental Psychology who took over Sprott's Demonstratorship in 1925; Banister (1882–1962) lectured on 'Psychology for medical students' in 1930–31. (Ernest Jones's deprecating review of Banister's book noted: 'The only puzzling thing about this class of book is that the author does not think it worth while to consult, before publishing it, someone who has studied the subject' (J., E., '*Psychology and Health*: by H. Banister, *IJP* 17 (1936), 382–3). Another review, by Louis Minski, an eclectic psychiatrist, noted: 'The psychopathological theories of Janet, Freud, Jung and Adler are described with a definite anti-Freudian bias. Some of the arguments against psycho-analytical theories seem to play into the Freudians' hands. For instance, on p. 57 it is stated: "It is a well-known fact that patients can produce dreams to please their analyst, and that many of the experiences they relate with minute detail are *unconscious* fabrications." Surely these phantasies are just what the analyst wishes the patient to produce' (Louis Minski, '*Psychology and Health*. By H. Banister', *Journal of Mental Science* 81 (1935), 694). The fact that Karin Stephen was asked to give her first set of lectures expressly for medical students in the year subsequent to Banister's probably indicates that his lectures were not thought to be adequate for the medical students. Stephen's lectures were not listed within Psychological Studies but only within the lecture list of the Faculty Board of Biology 'B', i.e. they were expressly for medical students, and moral science students would not be expected to follow them, even though Karin Stephen was herself a product of the Moral Sciences Tripos.

[191] *CUR*, 8 October 1926, p. 160.

twenty years, with his medical duties necessarily in abeyance, MacCurdy contributed little to Cambridge psychopathology and medical psychology. On his death in 1948, he left a substantial sum to his College for the MacCurdy Feast and his library to the University, including funds to acquire further books in abnormal psychology: it now forms the MacCurdy Psychopathology Library, properly sequestered from the main library of the Department of Experimental Psychology, and is a remarkable collection of early twentieth-century psychoanalytic works, with a substantial proportion in German.[192] But Rivers, MacCurdy was not.

War, the departure of Myers and the death of Rivers seem to have been the contingent and individual causes of the upheaval in Cambridge psychology. 'When the War was over psychology had, in a way, to be begun afresh in Cambridge',[193] Bartlett recalled in 1937. Myers had started this process by organizing a vast opening-up of psychology towards the new currents; his initiatives failed ignominiously and he departed in disgust. Students also recognized this fact: there were no candidates in option (c), the psychology option, in the MST, Part II for 1923. But Myers had prepared his succession well: on 12 June 1922, Bartlett at the age of thirty-six became the new Director of the Psychological Laboratory and therefore also the Reader in Experimental Psychology. In his unpublished autobiography written in 1969, Bartlett recalled his position:

> I now had to make a definite decision about how I would try to develop the Laboratory. Two possibilities seemed to be wide open. I could set to work to build a large teaching department. There was, for example, a rapidly expanding Medical interest in psychology, due in part to the popularity of the work of Freud, Jung and some of their associates, but perhaps still more to what had been done by Myers, Rivers, Sir Henry Head and others. The Government established Medical Grants [sic] Committee was outstandingly friendly and willing to help. Or I could decide to keep the Department rather small, highly selected, and principally directed towards the promotion of original research.[194]

It is surely indicative of his political choices that he remembered it as, in effect, a choice between popular demand for *teaching* in the arena of medical psychology symbolized by Freud and others and an elite and unflashy research Department. He chose the latter. So Bartlett's second

[192] The collection was for many years not adequately housed and some of the volumes in German were sold (private communication to John Forrester, Paul Whittle, email, 27 January 2000).
[193] Bartlett, 'Cambridge, England, 1887–1937', p. 108.
[194] Crampton, p. 265 citing Bartlett's (1969) unpublished autobiography 'What's the use of psychology?'.

new beginning went in the opposite direction from Myers's: eliminate most of the new developments and insulate Cambridge from their influence. This policy entailed that the University Department of Experimental Psychology would have nothing to do with the major innovation of the Malting House School (see Chapter 7), ten minutes' stroll away, when it was founded in 1924. Every single major psychologist in Britain, including Myers and Pear, was on the School's Advisory Council – with the exception of any representation from Cambridge itself.[195] Jean Piaget, Percy Nunn, Melanie Klein and many others travelled many miles to visit; Bartlett did not. He was developing a fortress mentality.

However, one does not even need to subscribe to the psychoanalytic doctrine of the return of the repressed to sense that fortresses do not always succeed in their exclusions. A remarkable retrospect of fifty years of psychology by the recently retired Bartlett in 1955 puts in a quite different perspective his retrenchments in the post-World War I period:

> [Freud's] fame reached its greatest heights during the 1920's. The odd thing is that the scientific psychologists, not only in this country, borrowed the clothes (if that is the right word to use – I doubt if it is) which Freud had provided, in their ardent quest for respectability. And the odder thing still is that it was this borrowed raiment which very largely got them what they wanted ... Whatever may be the ultimate verdict on the sombre picture Freud painted of human life and thinking, there is no possible doubt that his impact in experimental psychology was terrific, or that its chief effect was to make the whole subject more humane. When you look at a lot of experiments of the early classical period of experimental psychology, and at many of their direct derivatives down to the present day, you cannot escape the feeling that the experiments mainly provided strings for pulling puppets.[196]

The image of experimental psychologists stealing the Emperor's clothes in order to acquire respectability is a very odd one indeed. There are no details – no hint at which experiments or theories owed their existence or their interpretation to Freud's influence; nor is there a clear sense of what it means for experimental psychology to become more 'humane'. (Might he have meant that it became more realistic in its expectations concerning 'the human'? But then: the experimental psychologists, led by Bartlett, continued to pull the strings on their puppets.) Did the process of stealing Freud's clothes in order to become 'respectable' take place in Cambridge as well? Or

---

[195] *CR*, 23 October 1925, p. iii: 'Advisory council of the Malting House Garden School; Dr. P. Ballard; Prof. Cyril Burt; Dr. James Glover; Dr. C.S. Myers; Prof. T.P. Nunn; Prof. T.H. Pear; Prof. Helen Wodehouse; Directress Mrs. Susan Isaacs MA'.

[196] F.C. Bartlett, 'Fifty years of psychology', *Occupational Psychology* 19 (1955), 203–16 at 205–6.

did he still believe that Cambridge, under his guardianship, remained the only non-respectable (poor but honest) centre for psychology in the country? Given Bartlett's resolute refusal of psychoanalytic influence in Cambridge throughout his thirty years as Director of its Laboratory, the whole passage is mysteriously contradictory.

Psychology outside the University: *Psyche* and the International Library

Myers had left Cambridge in disgust at its conservative refusal to develop psychology beyond traditional experimental psychology, having attempted and failed to build a national centre for psychological medicine there. Bartlett's decision was to ally himself as necessary with the conservative forces and build up Cambridge as a centre for experimental, academic, non-flashy psychology – to exclude those trends, so active and fashionable after the war, associated with 'Freud'. But 'Freud' and psychoanalysis were inherently a multi-faceted movement, and Cambridge was not a unified university with coherent policies directed from its centre or centres. If initiatives failed in one corner they might find a suitable base somewhere else. A more general, expansive and contemporary development did take place in psychology; but it was not in the form of a conventional 'discipline' with its associated institutions. Instead, it came from a fresh set of schemes from the ever widening networks of C.K. Ogden.

Lancelot Whyte, post-war undergraduate beneficiary of Ogden's endeavours, summed up his achievement in his autobiography of 1963:

> In 1912 he had created *The Cambridge Magazine* which included, from 1915 onwards, what I believe was the first systematic attempt ever made to survey enemy opinion in war-time. In the history of war, Ogden's role is thus outstanding. But in peacetime he did almost the same thing: he brought European Continental thinking to insular Britain in his famous series: *The International Library of Psychology, Philosophy and Scientific Method* of some 160 volumes, of which about half came from outside Britain. This was surely the greatest single step in the education of British intellectuals since 1900.[197]

Ogden's friends – I.A. Richards, Philip Sargant Florence and others – speculated that from 1920 on he only kept the *Cambridge Magazine* going so that he could fill it with the essays he and Richards were producing, which appeared under a variety of pseudonyms. These friends, including Whyte, also transmitted Ogden's claim that he only began the

---

[197] Lancelot Law Whyte, *Focus and Diversions*, London: The Cresset Press, 1963, p. 49.

'International Library of Psychology, Philosophy and Scientific Method' at Routledge and Kegan Paul in order to find a place to publish their book, *The Meaning of Meaning*.[198] This is a typically witty half-truth – Ogden, Richards and their friends were extremely serious, without ever losing their whimsicality and penchant for ludicrosity and playfulness.[199] Ogden's Heretics Society was, after all, a Church for Atheists. After the Great War, the *Cambridge Magazine* was indeed destined to die, but Ogden transformed his interests into two immediate larger projects: the International Library and the journal *Psyche*. Both were direct products of the vision of the moral sciences Ogden, Richards and others shared and promoted – a direct descendant of the mapping found in J.S. Mill's *Logic*.

Like many others, Ogden fervently believed the new age was the age of psychology and he set out to establish the means for disseminating and promoting the new wave. In 1923, he took over editorship of the new Cambridge-based journal *Psyche*, which published on a quarterly basis; somewhat earlier, he became a roving commissioning editor for Routledge and Kegan Paul, eventually managing five series for them. They included the 'International Library of Psychology, Philosophy and Scientific Method' and 'History of Civilization'.

In 1920, Ogden's friend W. Whately Smith, graduate of King's in Moral Sciences, founded the *Psychic Research Quarterly*.[200] The aim was to provide a forum for the respectable scientific discussion of the lines of inquiry into the 'psychic' – hypnosis, unusual states of mind, telepathy, thought-transference, visions, sleep, dreams, etc. The upsurge of interest in shell shock during the war, with the new orthodoxy of psychogenesis and

---

[198] I.A. Richards, 'Co-author of the *Meaning of Meaning*. Some recollections of C.K. Ogden', in Sargant Florence and Anderson (eds.), *C.K. Ogden*, pp. 99–100. Also corroborated by Whyte's conversation with Ogden, recounted in Whyte, *Focus and Diversions*, pp. 48–9.

[199] Richards, 'Co-author of *The Meaning of Meaning*', p. 107. One might have expected that, thirty years later, the fifty-something Ogden would have grown more sedate, but 'when, after Winston Churchill's world-resounding Basic English speech at Harvard [in 1943], the journalists came to him for an interview, he gave it in a selection [from his collection] of masks, reappearing in different masks through unexpected doors'.

[200] Whately Smith (1892–1947) changed his name to Walter Whately Carington in 1933, and dedicated himself to research in psychical research, working with E.J. Dingwall (whose wife Doris later married John Layard). He was principally interested in psychical research; from the early 1920s on, this was his exclusive interest. He is known for attempts to introduce quantitative and more rigorous statistical methods into experiments testing the validity of claims of trance mediumship; the experimental methods he used were those introduced into the study of abnormal psychology by Jung (psychogalvanic reflex and word association tests). His methods and analysis were severely criticised by C.D. Broad, R.H. Thouless and Donald West, all significant Cambridge figures in the 1930s–1950s and all Presidents of the Society for Psychical Research. (Broad was Professor of Philosophy, Thouless Lecturer in Education and Donald West Professor of Clinical Criminology.)

the imperative towards psychological analysis, combined with the wave of preoccupation with communication with those beyond the grave that the death of so many beloved sons, brothers and husbands in the war had stirred. But by April 1921, with its fourth Editorial, as Whately Smith and Ogden began their collaboration, the journal was already changing tack, phasing out Whateley Smith's research agenda, an agenda close to the aims of the SPR, and adopting a new name: *Psyche*. Their sense of the new vistas open to psychology after the war ran very much in parallel with that of Myers and the other moral scientists with whom Ogden was in close contact: from now on, the journal would be principally concerned with the 'general applications of psychology' – medicine, education, law, industry and aesthetics. Its sub-title patently bore Ogden's encyclopaedic vision: 'A Quarterly Review of Psychology in Relation to Education, Psycho-Analysis, Industry, Religion, Social and Personal Relationships, Aesthetics, Psychical Research, etc.'[201] First and foremost amongst its aims would be finding an appropriate forum for the discussion of psychoanalysis:

> [Those interested in psychology] have seen and heard a good deal of war neuroses and ask whether analogous psychic mechanisms may not be responsible for all kinds of obstinate maladies in civil life. They especially desire a sane exposition of psycho-analytic doctrines in terms which bear some relation to their previous knowledge. The laity, also, are much intrigued by the smatterings of psycho-analysis which they have picked up from the mass of assorted verbiage which has recently been decanted upon them. Some of them have realised that although there is obviously 'something in it', the great proportion of accessibly published matter is unreliable, if not actively pernicious: but for the most part the public is either repelled, puzzled, or definitely misled, and there is an urgent need for the dissemination of sound views as an antidote to the sensationalism of the popular press.[202]

Certainly there was a need for an English forum for the *discussion* of psychoanalysis; by April 1921, the *International Journal of Psycho-Analysis* was starting its second volume, but was still very much a translation journal, with its centre of gravity in Central Europe. Ogden gradually took over *Psyche* and initially made it a forum for interesting, sometimes eccentric, sometimes critical, sometimes adulatory discussions of psychoanalysis – and a forum very much based in Cambridge. Starting with Volume 2, the journal devoted much space to psychoanalysis by members

---

[201] Later it became 'An Annual of General and Linguistic Psychology'.
[202] 'Editorial', *Psychic Research Quarterly* 1 (1920–21), 287–8.

of the University of Cambridge – in Psychology, Moral Sciences and elsewhere. J.P. Lowson (appointed as Demonstrator in the Laboratory of Experimental Psychology in 1920[203]) wrote an entirely orthodox Freudian account of dream-interpretation, with a lengthy discussion of a dream of an army doctor – 'a kind of story told in a sort of picture-writing'[204] – requiring free association to uncover the most intimate personal details of the dreamer. Lowson defended the Freudian emphasis on the seamy side of human life – 'Freud is not usually assailable on the side of facts observed'[205] – and only demurred somewhat at claims of universal dream symbolism.[206] The rest of the volume was filled with Ogden's other Cambridge contributors: Lowson's predecessor as the Physician in charge of the Addenbrooke's Psychological Clinic, E. Prideaux, discussing 'Criminal responsibility and insanity'; Hamilton Hartridge and Whately Smith on 'Sleep'; W.H.R. Rivers displaying his new political preoccupations with an essay on 'The instinct of acquisition' responding to the socialist-minded *The Acquisitive Society* of R.H. Tawney, the President of the London University Labour Party which had adopted Rivers as its parliamentary candidate.[207] Volume 3 (1922–23), understandably preoccupied with the death of the founder of Cambridge psychology, W.H.R. Rivers, included Myers's *éloge* to Rivers and a spiky anti-Freudian introduction by Grafton Elliot Smith to an unpublished paper by Rivers on ethnology and psychology; other papers included the prolific Swiss writer and psychoanalyst Charles Baudoin on the evolution of instinct from the standpoint of psycho-analysis, Cyril Burt on 'Causes and treatment of juvenile delinquency', Theodore Schroeder (the radical American promoter of obscenity and 'the erotogenetic theory of religion') on 'Behaviorism and psychoanalysis' and an essay on 'The psychology of faith' by a Cambridge doctoral candidate, R.H. Thouless (who would in the 1970s describe how his doctoral supervisor Rivers had encouraged him to employ a psychoanalytic approach to religion).[208]

---

[203] *University of Cambridge Calendar 1920*, p. 120.
[204] J.P. Lowson, 'The interpretation of dreams', *Psyche* 2 (1921–22), 4–12 at 7.   [205] *Ibid.*, p. 12.
[206] *Ibid.*   [207] Slobodin, *Rivers*, p. 185.
[208] On Thouless's memories of Rivers, see Crampton, p. 112. There is some indication of the lines of contestation within the differing approaches to 'religion' in the opposition that Thouless's dissertation, on 'The psychology of religion', faced in being approved for the doctorate: on 30 May 1923, for the second time of asking, the Moral Science Degree Committee voted in favour of granting the degree by 4 against 3. Those 'for' were Sorley, Keynes, Dawes Hicks and Bartlett, those against Ward, McTaggart and Moore (Minutes of Special Board for Moral Science, University Library Archives).

Ogden did not formally take over the editorship of *Psyche* until the July 1923 issue, Volume 4, No. 1,[209] exactly when the *Cambridge Magazine* closed definitively. With a seamless transfer of personnel between the two enterprises, Richards opened Volume 4 of *Psyche* with an essay on 'Psychology and the reading of poetry', which would become Chapter 11 of *Principles of Literary Criticism*, published in Ogden's International Library in 1925.[210] And this was the way the journal would develop, publishing the writings of his Cambridge friends and his ever widening circle of contacts in London and internationally. *Psyche* became the house-journal for Cambridge psychology. In doing so it competed directly with Ernest Jones's *International Journal of Psycho-Analysis*: Farrow published his 'Experiences with two psycho-analysts' in Ogden's *Psyche* in January 1925. And *Psyche* also competed directly with the *British Journal of Psychology*, edited out of the Cambridge Psychological Laboratory. The market and the interest in general psychology, of which psychoanalysis was the fashionable and controversial vanguard, would also be opened up by new initiatives that Ogden, with his impeccable taste in contacts and his disdain for intellectual focus, was now plotting.

Ogden's teaming-up with the London publishers Kegan Paul, Trench, Trubner allowed him to exercise his gifts on a larger scale. In early 1921 a proposal for offshoot book publications from *Psyche* was put forward; quite how and why the deal was done is not clear. The proposal must have been definitively accepted by the end of 1921,[211] Ogden having lined up the first four books in the course of the year. Dorothy Wrinch approached him about Wittgenstein's *Tractatus* soon after she received the rejection letter from Cambridge University Press dated 17 January 1921.[212] On 23 January 1921, the seventeen-year-old Frank Ramsey recorded in his diary:

> Moore talked to Heretics on Ethics; good paper; discussion not very good at first but then I asked some questions and got illuminating answers ... Moore, Ogden, Richards, Sprott, Martin and I adjourned upstairs where there was a most interesting discussion, in which Moore showed to great advantage. Sprott went soon; I talked a little. Moore seemed happy and wasn't overexcited. (It sounds as if he was a child I had care of). I think he's great.

[209] W. Terrence Gordon, 'Introduction', *Psyche* 1 (1920–21), reprint, London: Routledge/Thoemmes Press, 1995, p. xi.

[210] See John Paul Russo, *I.A. Richards: His Life and Work*, London: Routledge, 1989, esp. chapters 8 and 9.

[211] Moore's Preface for his *Philosophical Papers* is dated January 1922.

[212] Brian McGuinness, *Wittgenstein: A Life. Young Ludwig (1889–1921)*, London: Duckworth, 1988, p. 298n7.

Moore's talk was almost certainly 'The nature of moral philosophy', which is the final paper in the first book in Ogden's series. Ogden probably popped Moore the question at the time of this discussion – would he consider gathering his previously published papers together making a volume of *Philosophical Studies*? So, within the space of a week, Ogden had both Moore's and Wittgenstein's work on offer. In May 1921, Karin Stephen had submitted to Cambridge University Press her book manuscript, *The Misuse of Mind*, a study of Bergson, only for it to be rejected.[213] Probably over the summer, Ogden took on her book too. Also over the summer of 1921, Ogden was coming to an understanding with W.H.R. Rivers about publishing his next book,[214] which eventually appeared posthumously as *Conflict and Dream* in 1923. Ogden organized the translation of Wittgenstein, adeptly making use of Frank Ramsey, whose philosophical reading he had been guiding even while Ramsey was still at school. Ramsey had time on his hands. Ogden also accepted into the series his friend Whately Smith's book, *The Measurement of Emotion*. So by October 1922, the first four books in the series, now named the International Library of Psychology, Philosophy, and Scientific Method, all by Cambridge moral scientists of one complexion or another, Moore, Stephen, Whately Smith and Wittgenstein, were advertised in *The Times* alongside Malinowski's *Argonauts of the Western Pacific* as 'ready'.

Ogden had a problem of vocabulary. His authors – both for *Psyche* and the International Library – were principally Cambridge moral scientists. But outside Cambridge, the term 'moral science' would lead to confusion. So the explicit programme of the Library did not mention 'moral science'. Nor did it mention philosophy directly; instead its focus was on psychology bundled up with philosophy:

> The purpose of the International Library is to give expression, in a convenient form and at a moderate price, to the remarkable developments which have recently occurred in Psychology and its allied sciences. The older philosophers were preoccupied by metaphysical interests which for the most part have ceased to attract the younger investigators, and their forbidding terminology too often acted as a deterrent for the general reader. The attempt to deal in clear language with current tendencies whether in England and America or on the Continent has met with a very encouraging reception, and not only have accepted authorities been invited to explain the

---

[213] *Lighthouse*, p. 116. [214] OP, Box 60, F.4, Rivers to Ogden, 30 August 1921.

newer theories, but it has been found possible to include a number of original contributions of high merit.[215]

Ogden was not the only publishing entrepreneur responding to the 'remarkable developments which have recently occurred in Psychology'. The publication in 1921 of the first two volumes in Ernest Jones's International Psycho-Analytic Library, an obvious potential competitor, may well have galvanized Ogden into action.

When Rivers died suddenly in June 1922, Ogden organized the publication of his literary estate in his International Library, in three separate volumes, starting with the book Rivers had virtually completed at his death, *Conflict and Dream.* Over the course of 1922 and 1923, he cultivated a jovial friendship with Malinowski and it was their collaboration that may have been Ogden's single most consequential influence on the development of psychoanalysis in this period. Ogden and Malinowski plotted the bringing to fruition of the sustained dialogue between psychoanalysis and anthropology which had been embodied in Rivers.[216]

The first fruit was Malinowski's anthropological supplement to *The Meaning of Meaning,* followed by a series of papers in *Psyche* on the relations between psychoanalysis and anthropology. Ogden was more than willing to publish articles that viewed psychoanalysis from a critical distance; Malinowski's articles, appearing from late 1923 on, were products of his great admiration for, or perhaps more accurately his argumentative fascination with, Freud's work. With his trademark, swash-buckling verve Malinowski was more than prepared to take psychoanalysis off in directions distasteful to the faithful Freudians. Proposing that the Trobriand Islanders displayed a matrilineal variant of the Oedipus complex quite different from the Aryan patriarchal family type discussed by Freud[217] was

[215] From publicity material on dust-jacket of C. MacFie Campbell, *Problems of Personality: Studies Presented to Dr. Morton Prince, Pioneer in American Psychopathology,* London: Kegan Paul, Trench, Trubner & Co., 1925.

[216] See Chapter 3

[217] Bronislaw Malinowski, 'Psychoanalysis and anthropology', *Psyche* 4 (1924), 293–332. Most historians of anthropology ignore completely Malinowski's close working relationship with Ogden. The Malinowski Bibliography available on Yale Archive website is incomplete, omitting all of Malinowski's papers published in *Psyche.* Even the leading historian of anthropologists of our time, George W. Stocking, Jr., in the important paper in which he highlights the importance of *Psyche* as a forum for Malinowski's views ('Anthropology and the science of the irrational'), occludes the Ogden connection by not indicating their original provenance, instead citing some of Malinowski's *Psyche* essays in a rather scrappy 1962 volume, deficient from a scholarly point of view, put together by Malinowski's executors. So, for the sake of completeness of the record, the five papers by Malinowski that Ogden published in *Psyche* were: 'The psychology of sex and the foundations of kinship in primitive society', *Psyche* 4 (1924), 98–128; 'Psychoanalysis and

of a different order than jousting over the nature of symbolism in psycho-analysis and anthropology, which had up to then dominated discussion of the relations of the two fields. Most importantly, Malinowski was the first to implement Rivers's methodological innovation: intensive fieldwork. With his intention of performing self-analysis in the field, by which he recognized the 'undercurrent of desire' in himself, diaries and dreams became field-tools. This was a methodological break-through for professional anthropology. Fieldwork existed before Malinowski, but participant observation – participation with self-analysis – was a break with all earlier ethnographic research.

Malinowski also promised to work his material and ideas on sexuality up into a book for the International Library. Characteristically, he anticipated this being censored by the Lord Chamberlain ('Lord Chamberpot'), so taunted Ogden with potential scurrilous titles – *Melancholy Reflections on the Moral Naughtiness of our Degraded Brown Brethren* – *A Book for Methodist Virgins of Magdalene Institutions* – a dig at Ogden's Cambridge base of operations, Magdalene College.[218] Eventually the book was published in the International Library, using some of the material already published in *Psyche*, as *Sex and Repression in Savage Society* (1927). It was Malinowski's onslaught on psychoanalysis;[219] but, as many subsequent

---

anthropology', *Psyche* 4 (1924), 293–332; 'Complex and myth in mother-right', *Psyche* 5 (1925), 194–216; 'The role of myth in life', *Psyche* 6 (1926), 29–39; 'The life of culture', *Psyche* 7 (1927), 37–44. Graham Richards has also pointed out how Ogden has been omitted from the major histories of British psychology in 'C.K. Ogden's basic role in inter-war British psychology', *History and Philosophy of Psychology* 9(1) (2007), 56–65. See also Max Saunders, 'Science and futurology in the To-Day and To-Morrow series: matter, consciousness, time and language', *Interdisciplinary Science Reviews* 34(1) (2009), 68–78; Max Saunders and Brian Hurwitz, 'The To-Day and To-Morrow series and the popularization of science: an introduction' *Interdisciplinary Science Reviews* 34(1) (2009), 3–8.

[218] OP, Box 37, Malinowski to Ogden, 25 October 1923. Ogden's series was always advertised in the 1920s as 'Edited by C.K. Ogden, M.A., of Magdalene College, Cambridge'. Ogden had no post in the University or at his College; but at that time the two key features of Cambridge corporate life were membership of a College (and one becomes a member by matriculating – i.e. starting a course of studies) and membership of the pre-1926 Senate, conferred simply by being an MA – awarded automatically three years after the award of a BA (Cantab.). The 1926 reforms had introduced a new and crucial status in order to stop, as had happened with the vote over the admission of women to degrees in 1897 and again in 1920 and 1921, the MAs of the University, scattered across the globe but prepared to come to Cambridge for a crucial vote on a controversial policy matter, preventing University reform. The key democratic institution from 1926 on became the Regent House, and qualification to vote belonged solely to 'Resident M.A.s', and therefore was heavily skewed towards those in the employment of University or College. (See Brooke, *A History of the University of Cambridge*, Vol. IV, p. 351.) Even then, with his multiple residences in Cambridge, Ogden qualified as a member of the Regent House.

[219] Cf. Keith Hart's reflection: 'Certainly the 1920s were a fruitful period to examine the relationship between the new ethnography and psychoanalysis. Malinowski was actively engaged with Freudian ideas at this time, until the exchange went the wrong way from his point of view. There was support

commentators have noted, it followed Rivers's critique in staying on the psychoanalytic terrain Freud had already mapped.[220] Nonetheless the effect of Malinowski's attack, including his 'refutation' of the universality of the Oedipus complex, meant that for many any substantive link between anthropology and psychology, in particular psychoanalysis, was 'taboo' for the next fifty years.[221]

By 1923, as Rapp has documented meticulously, the general post-war excitement over psychoanalysis was mostly over.[222] Malinowski, Ogden and the publishers at Kegan Paul, Trench, Trubner began to feel the climate change in 1926.[223] Their judgement is accurate. By 1926 Ogden

from Seligman. But the trend, both in anthropology and psychology, was towards divorce, not marriage' (Keith Hart, 'The place of the 1898 Cambridge Anthropological Expedition to the Torres Straits in the history of British social anthropology', Lecture given in the opening session of a conference held at St John's College, Cambridge, 'Anthropology and psychology: the legacy of the Torres Strait expedition, 1898–1998', 10–12 August 1998). Adam Kuper uses even stronger language in his opening sentence: 'Like many social anthropologists of my generation, I was imbued very early with the understanding that psychology was taboo to a [British] social anthropologist' (Adam Kuper, 'Psychology and anthropology: the British experience', *History of the Human Sciences* 3 (1990), 397–413 at 397). The divorce between psychoanalysis and anthropology is undoubtedly linked to another feature of the history of anthropology which we cannot explore but is undoubtedly of importance for our topic: 'There is a common consensus that sex retreated from the center stage of anthropology sometime during the 1930s. It was more than 30 years before it reemerged as a major concern' (Andrew P. Lyons and Harriet D. Lyons, *Irregular Connections: A History of Anthropology and Sexuality*, Lincoln: University of Nebraska Press, 2004, p. 217).

[220] Malinowski's relation to psychoanalysis has been subtly analysed by George W. Stocking, Jr., 'Anthropology and the science of the irrational: Malinowski's encounter with Freudian psychoanalysis', in George Stocking, Jr. (ed.), *Malinowski, Rivers, Benedict and Others: Essays on Culture and Personality*, History of Anthropology 4, Madison: University of Wisconsin Press, 1986, pp. 13–49, and also in a series of excellent papers by Bertrand Pulman, 'Aux origines du débat ethnologie/psychanalyse: W.H.R. Rivers (1864–1922)', *L'Homme* 26(100) (1986), 119–42; 'Anthropologie et psychanalyse: "paix et guerre" entre les herméneutiques?', *Connexions* 44 (1984), 81–97; 'Le débat anthropologie/psychanalyse et la référence au "terrain"', *Cahiers internationaux de sociologie* 80 (1986), 5–26; 'Malinowski and ignorance of physiological paternity', *Revue française de sociologie* 45, Supplement: An Annual English Selection (2004), 125–46.

[221] Meyer Fortes (later William Wyse Professor of Social Anthropology at Cambridge and Fellow of King's College), as former psychologist, as Freudian colleague working with Emanuel Miller at the first Child Guidance Clinic in London, and later as faithful student of Malinowski (they met at the dinner table of the London psychoanalyst J.C. Flügel in 1931), was ideally placed to reflect upon this unexpected and, from his later point of view, untoward history; see M. Fortes, 'Malinowski and Freud', *Psychoanalytic Review* 45A (1958), 127–45 and more particularly Fortes, 'Custom and conscience in anthropological perspective', *International Review of Psycho-Analysis* 4 (1977), 127–54.

[222] Dean Rapp, 'The early discovery of Freud by the British general educated public, 1912–1919', *Social History of Medicine* 3 (1990), 217–43; Rapp, 'The reception of Freud by the British press: general interest and literary magazines, 1920–1925', *Journal of the History of the Behavioral Sciences* 24 (1988), 191–201. Stocking, following a variety of other sources, also remarks that the feverish interest in psychoanalysis had trailed off by 1924.

[223] James Strachey tried to cement a different alliance between psychoanalysis and anthropology by hosting a dinner for Ogden and Róheim in September 1925 'to try and induce O. to publish a book of Róheim's in one of his series' (BF, 17 September 1925, p. 297; see also OP, Box 69, F.11, JS to

had diversified his operations: J.B.S. Haldane's talk to the Heretics on 4
February 1923 was such a scintillating performance that Ogden published it
as a short book, *Daedalus, or Science and the Future*, which sold like
wildfire.[224] Needing a quick follow-up to build a new series, Ogden
persuaded Bertrand Russell to convert a lecture on science and the future
of civilization into a response – *Icarus, or the Future of Science* – completed
in a week.[225] With these two pamphlets, another of Ogden's series, To-Day
and To-Morrow, was born – all short, snappy, little essays, on scientific
and cultural themes and their future. Over a hundred of these pocket-sized
books were published in the next six years. The success of the series
persuaded him to utilize another of his forums as a jumping-off point for
a further publishing venture. He conceived the idea of Psyche Miniatures,
small books that were developments of papers published in *Psyche*, or of
other ideas being developed by his circle of regular contributors. The first
six were published in 1926, starting with I.A. Richards's *Science and Poetry*,
followed by Philip Sargant Florence's *Over-population: Theory and
Statistics*, while the sixth title was Malinowski's *Myth in Primitive
Psychology*.[226] The philosopher John Wisdom's first book, *Interpretation
and Analysis in Relation to Bentham's Theory of Definition*, published in the
series in 1931, includes the first usage of the phrase 'analytic philosophers'
and develops a novel definition of the distinctive, Cambridge-led approach
of 'philosophical analysis'.[227]

Ogden thus emerges both as a bellwether for the various sub-cultures in
and by which psychoanalysis was introduced and transmitted in
Cambridge and more broadly in England; and also as a principal patron,
the spider at the centre of a very large and complex web of correspondences
and projects. Before the war he created a resolutely intellectual,
Germanophile, secularist society and magazine. During the war, he turned

CKO, 24 September 1925). Róheim's *Animism, Magic, and the Divine King* was eventually
published by Ogden in the series in 1930.

[224] Marjory Todd, in Sargant Florence and Anderson (eds.), *C.K. Ogden*, p. 113.

[225] *Ibid.* and MBR II, pp. 28–30.

[226] At the same time, Ogden commissioned titles for Psyche Miniatures, Medical Series; three
appeared in 1926: *Migraine and other Common Neuroses* by Francis Graham Crookshank,
Ogden's close medical friend and theorist of language and diagnosis, who had earlier contributed
an Appendix to *The Meaning of Meaning* in 1923 and from 1931 on would be the leader of medical
Adlerians in Britain; *Aphasia* by Kinnier Wilson, first occupant in Britain of a Professorship of
Neurology, created in the wake of the war (see Macdonald Critchley, 'Remembering Kinnier
Wilson', *Movement Disorders* 3(1) (1988), 2–6); and *Types of Mind and Body* by Emanuel Miller,
Cambridge moral sciences graduate in the same cohort as Richards.

[227] Aaron Preston, *Analytic Philosophy: The History of an Illusion*, London: Continuum, 2007, pp.
69–71.

the *Cambridge Magazine* into the sole national organ for disseminating internationalist, and by implication, pacifist views. After the war, he hopped from directing an undergraduate magazine to producing the sole journal in Britain for the promotion of general psychological, predominantly psychoanalytic, ideas. Extending the ethos of the Heretics and *Psyche*, he founded a number of book series as diverse outlets for his circle – the Stephens, the Russells, the Haldanes, Wittgenstein, Bernal. In particular, the weightiest of his publishing ventures, the International Library, really does capture the intellectual heart of the psychological 1920s, just as Penguin and the Left Book Club were to capture the heart of the political 1930s. In the beginning, his Library was resolutely Cambridge-based: the first eight titles were all by dons who taught Moral Sciences (or, like Wittgenstein, were Cambridge-connected).[228] Its impact was national and international. In the early 1920s, *Psyche* vied with the *International Journal of Psychoanalysis* and the *British Journal of Medical Psychology* as a forum while Ogden's Library vied with the International Library of Psycho-Analysis for publication of the latest in psychological writing. Ogden may not have had access to translations of Freud's work or the Freudian faithful, but Jung, Adler and many other 'psychoanalytic' writers of the time would be published in his series.

As the 1920s went on, the gulf grew more and more gaping between Ogden's publications and the experimental psychology which Bartlett's restrictive idea of psychology entrenched in the University proper. Psychoanalysis might have lost out in the disciplinary scramble, but because of work Ogden published, its formal absence at the University was a visible lack. Margaret Gardiner later recalled of her year's study as an undergraduate (1924–25): 'I had imagined that psychology would be about behaviour and particularly about the theories of Freud, which were new and much discussed in undergraduate circles. I had already heard a little about psychoanalysis from Lionel [Penrose], who had told me that its aim

---

[228] These were Moore, Stephen, Wittgenstein, Whateley Smith, Rivers (3), Broad. The first non-Cambridge author was Jung (*Psychological Types*, 1923) – even this book owed its presence in the series to its translator, Baynes, being a long-standing Cambridge friend of Ogden's. From 1925 the series became less dependent on Cambridge authors but was still heavily dependent on Cambridge translators (e.g. for volumes by Vaihinger, Koffka, Koehler); even the oddity of Paul Radin (an eminent early anthropologist most usually associated with Berkeley and American anthropology and a figure roughly comparable in the US context to John Layard in England) being the translator of Adler, *The Practice and Theory of Individual Psychology* (1924), was due to his spending time in Cambridge in the early 1920s working with Rivers and at the same time developing a life-long interest in psychoanalysis – though his greatest psychologial interest, virtually from then to the end of life in 1959, was always in Jung.

was the acquisition of a quiet effrontery.' 'Moral Science wasn't turning out to be at all that I had expected.'[229]

Psychoanalysis and Anthropology

What was missing in Bartlett's post-Rivers and post-Myers Department of Experimental Psychology is well captured by the destiny of Reo Fortune. Born in Coromandel, New Zealand in 1903, he took a first class honours degree at Victoria University College in Wellington in 'advanced philosophy', which for him meant principally psychology. His BA dissertation of 1925 was on sleep and dreams; he was awarded a National Research Scholarship to Emmanuel College, Cambridge, the first to a psychologist. Before he left for England he sent off and had accepted for publication an intervention in the energetic debates about psychoanalysis: on the sexual symbolism of the snake. His target was Adolf Wohlgemuth's criticism of the universal phallic symbolism of the snake: Wohlgemuth claimed the snake was principally a symbol for immortality. Fortune's contribution was unambiguous evidence from Maori legends, linked in to a parallel chain of myths in Hindu mythology, that the snake or eel is a phallic symbol: 'the serpent is not obviously a symbol of the male organ in *Genesis*; but the eel obviously is in the Maori legend. And by the Maori legend we know the serpent for what he really is.'[230] Fortune drew upon Rivers's authority and methodology in his argument, relying heavily on the cultural diffusionist Rivers of the last ten years and on his interrogation of Freud's universal and sexual symbolism (which Rivers had, however, increasingly attacked in his last years). Fortune was more supportive of Freud's ubiquitous sexual symbolism, partly because it could be married neatly to a strong diffusionist account of myths and symbols. The best evidence, he asserts, leads us to believe that the Maoris derive their myths from Ur of the Chaldees via India: 'Maori and Jew and Gentile all have the legend of the creation and fall of man from the one source.'[231]

Fortune's time on the boat from New Zealand to England in the summer of 1926 changed his life:[232] he fell in love with his chance companion, Margaret Mead, returning from her first ethnographic field trip in Samoa. In her autobiography, Mead captures clearly the intellectual ideals

---

[229] Margaret Gardiner, *A Scatter of Memories*, London: Free Association Books, 1988, p. 69.
[230] R.F. Fortune, 'The symbolism of the serpent', *IJP* 7 (1926), 237–43 at 242.   [231] *Ibid.*
[232] Roger Ivar Lohmann, 'Dreams of Fortune: Reo Fortune's psychological theory of cultural ambivalence', *Pacific Studies* 32 (2009), 273–98 at 274.

and projects he was bringing to the Department of Experimental Psychology in Cambridge:

> to make up for the isolation in which New Zealanders lived in the days before modern communications, he had read deeply and with delight, ranging through the whole of English literature, and he had eagerly taken hold of whatever he could find on psychoanalysis. It was like meeting a stranger from another planet, but a stranger with whom I had a great deal in common.
>
> Reo had saturated himself in the work of W.H.R. Rivers, the Cambridge don whose work in physiology, psychoanalysis, and ethnology had excited people right around the world. I had never met Rivers and, of course, neither had Reo. But we both knew he was the man under whom we would like to have studied – shared an impossible daydream, because he had died in 1922 . . . He had been fascinated by Freud, but was critical of his theories. With the kind of insight that was so characteristic of him, Reo had pointed out in his prize winning essay that what Rivers had done was to stand Freud on his head, without changing the premises, by making fear, instead of libido, the driving force in man.

Despite Reo's entreaties to go on with him to Cambridge, she chose to travel on with her first husband Luther Cressman, who met her off the boat in Marseilles in September 1926.

The formal topic of Fortune's psychological research in Cambridge was 'Diagnosis of retardation'. He recorded his impressions in letters to Mead: how MacCurdy, despite his fearsome reputation, was affability itself – but inconsequential; how his own work on dreams, balanced between Rivers and Freud, was clearly inadequately developed:

> Again, I should be analysed. There is nothing below the conscious level in my analyses, they are the upper level mechanisms e.g. my dislike of Germany and of the training college is possibly due to some underlying factor which makes such feeling come to me as an individual. Such feeling need not be universal. To ignore childhood experience is scientifically discreditable.[233]

Fortune summarized his discussions with MacCurdy as focusing on critical comments directed at the eminent psychologist, rival instinct theorist to Freud, now Professor at Harvard, William McDougall. And then:

---

[233] Letter from Reo Fortune to Margaret Mead, 18 October 1926 (Library of Congress, Margaret Mead Papers, MMP R4); we would like to thank Roger Lohmann for sharing drafts of his research and Caroline Thomas for making available to us some of her research on Mead and Fortune and this manuscript letter.

> criticism in a very kindly manner – with just a quiet certainty of more sex underneath than I knew – without analysis, for that must be the root of the third factor underneath conscious factors – and maybe he is right – as to early childhood experience, or unconscious experience.

Seeing MacCurdy once a month did not enthuse Fortune, nor did the research supervision codes: 'non-interference unless desired'; there was no room for him in the laboratory, forcing him to work in his room in College. He concluded that psychology was distinctly 'unfriendly'. He had already come into contact with the anthropologists in Cambridge, including Gregory Bateson, who had achieved a First in Part II of Anthropology in June 1926. Within three months Fortune had applied to shift from psychology to taking the Diploma in Anthropology, for which he was the only candidate. After Cambridge, he toured Germany with Margaret Mead (whose husband told her and Reo where to acquire contraceptives in England) and then returned to Australia to take up a position working under Radcliffe-Brown;[234] in 1928 he did fieldwork in Dobu in Eastern Papua producing a classic of the new ethnographic monograph, *Sorcerers of Dobu* (1932), which embodied, Malinowski pronounced, 'the ethnographer's supreme gift: he can integrate the infinitely small imponderable facts of daily life into convincing sociological generalizations'.[235] Fortune was not far from Rivers's ideal of a confluence of psychoanalysis and anthropology around the methodology of the interpretation of 'hidden sources' (see Chapter 3). Indeed, in Radcliffe-Brown's eyes in 1927, Fortune was much too much of a psychologist to be altogether trusted as an ethnographer;[236] by 1931, after Fortune's fieldwork in New Guinea and in Omaha, Radcliffe-Brown changed his mind, regarding him as 'one of the very few first-class anthropologists round the world'.[237] But from then on, as Fortune bitterly complained to Malinowski, it was his fate to be kicked around by the fledgling discipline of anthropology throughout the 1930s and 40s[238]. It was also his fate to be sucked into a marriage with Mead only for her and Gregory Bateson to fall in love in front of his eyes when the three met up, under the proverbial ethnographic tent, at Christmas 1932 on the Sepik River.

[234] Jane Howard, *Margaret Mead*, New York: Fawcett Columbine, Ballantine, 1984, p. 102.
[235] *Ibid.*, p. 113.
[236] George W. Stocking, Jr., *After Tylor: British Social Anthropology, 1888–1951*, Madison: University of Wisconsin Press, 1995, p. 341, citing letter from Radcliffe-Brown to Haddon, dated 7 September 1927.
[237] Quoted Howard, *Mead*, p. 138.
[238] Howard, *Mead*, p. 267, Fortune to Malinowski, 7 July 1936, Malinowski Archive.

Fig. 5.4 Gregory Bateson, Margaret Mead and Reo Fortune, 1933.

After short-term positions in China, Canada and the USA, Fortune eventually returned in 1947 to Cambridge as a Lecturer in Anthropology, remaining there till his death in 1979.[239]

Before gravitating from psychology to anthropology, he produced a small book, *The Mind in Sleep*, its Preface dated January 1927, based on

[239] Peter Gathercole, obituary, Reo Fortune, *RAIN* 37 (April 1980), 9.

his New Zealand dissertation with supplementary discussion. The book was almost entirely based on detailed analysis of his own dreams – dreams dreamt in 1923 after he had read Freud but before he had read Rivers, together with some additional dream material, including the dreams Mead and he had dreamt for this purpose on the ship travelling from New Zealand to Europe in the summer of 1926. Initially opposed to Freud's theories, in particular those concerning hidden sexual wishes, he recounted how his own dreams had refuted his own theoretical position (thus mirroring the path Rivers himself had charted in *Conflict and Dream* (1923)). So the resulting book navigated between Freud and Rivers, decidedly on Freud's side, but still eager to assert its own discoveries of mechanisms involving symbolism and affect of which Freud, so Fortune maintained, was ignorant. The book received a haughty, somewhat patronizing review from Ernest Jones.[240]

Fortune's story reveals just how uninspiring, for young psychoanalytic enthusiasts, Cambridge psychology had become under Bartlett's and MacCurdy's leadership. Future eminent and wide-ranging psychologist Ralph Pickford, taking the Part II in Moral Sciences in 1927 while Fortune jumped ship to Anthropology, and having to find his education in psychoanalysis for himself in Glasgow in the 1930s while pursuing a varied and successful career in academic psychology, summed it up:

> At Cambridge [in the late 1920s], psychoanalysis was very much under a cloud. MacCurdy spoke about it but always in a critical way, and Bartlett (and other people) in his discussions used to spend a lot of time raising objections to Freudian psychology. In fact I suppose in a sense it was the only real gap in Bartlett's understanding of psychology, from my point of view at present, that he had no use for psychoanalysis and no insight whatever into its importance.[241]

Mead and Fortune were by no means the only young researchers enthused by psychoanalysis, braced to develop the new ethnographic methods and produce a psychological anthropology.[242] But certainly in Cambridge and overwhelmingly elsewhere, the psychologists were antipathetic to psychoanalysis; while the functionalist anthropologists (with the notable exception of Malinowski) were adamantly opposed to psychology. The idea of a psychoanalytic anthropology – the fusion of the methodology of fieldwork with the project of universal psychological understanding offered by

[240] J., E. 'The Mind in Sleep: By R.F. Fortune', IJP 8 (1927), 548–9.
[241] BPS, Interview with Ralph Pickford, 4 June 1973, Manchester, conducted by Mr Kana.
[242] Stocking, 'Anthropology and the science of the irrational'.

Freudianism – was repudiated, save by Malinowski, who fulfilled the programme, persuaded few of his colleagues, and in the process drove anthropology away from psychology. The greatest irony is that Malinowski is remembered as the anthropologist who decisively refuted psychoanalytic pretensions in anthropology[243] while he himself was adamant that a psychoanalytically informed psychology was foundational for anthropology and delivered as much in a series of important books.

However, another important line of potential rapprochement between psychoanalysis and anthropology, the American line associated with cultural anthropology as it was formed by Ruth Benedict and Margaret Mead, also had its Cambridge connection through Mead's connections with Fortune and Bateson, and her life-long collaborative friendship with Geoffrey Gorer. From a well-to-do Jewish family, Gorer was educated at the Sorbonne, then at Jesus College, Cambridge, where he took Classics and Modern Languages, graduating in 1927, and then spending a year at the University of Berlin. Successful through a 1934 study of Sade, Gorer travelled in Africa, the Far East and then to America, where he met Ruth Benedict and Margaret Mead – Mead became a life-long intimate friend. From then on, he became a freelance anthropological writer studying cross-cultural variation, often infused with psychoanalytically inspired speculations or analytic frames of reference. A study of 'Japanese character structure and propaganda' was followed by a study of opinion poll data to produce *The Americans* (1948). In 1951, he collaborated with John Rickman, drawing on Rickman's pamphlets from his time in Russia during World War I, to produce *The People of Great Russia*, in which he put forward his 'swaddling hypotheses': that the national character of the Russians was decisively influenced by the practice of tightly swaddling infants in blankets, producing cold and distant personalities and the highly concentrated, episodic 'maximum total gratifications'; the causal relation could be explicated using psychoanalytic accounts of early infant development.[244] This approach, in line with Benedict's and Mead's wartime

---

[243] In the debate over the 'universality' of the Oedipus complex in the 1920s.

[244] Margaret Mead's presentation of the hypotheses in 1954 was: 'Gorer's hypothesis is that an unusually long swaddling experience is a significant aspect of the educational process by which human infants, born to and reared by Russian parents, become Russians. By analyzing the way in which Russian swaddling differs from the swaddling in surrounding areas, the special features of Great Russian swaddling are identified and related to other aspects of Russian child rearing. Gorer then invokes developmental theory from another field, that of psychoanalytically oriented studies of character formation in children in our own society, to provide a theoretical basis for hypothesizing the intrapsychic mechanisms involved in the process of the formation of Great Russian character, one expression of which can be found in the traditional attitudes of Great Russians

studies of the relationship between 'culture and personality' (Japan, America, Britain)[245] was briefly endorsed, in particular by US funding organizations and military sponsors; but Gorer's swaddling hypothesis was severely criticized and its implausibility led, in large part, to the general demise of the project to link individual personality with cultural variation. Married to Mead in the key period of the development of 'culture and personality', Bateson also played his part in the development of Freudian-inspired cross-cultural hypotheses. As he declared in 1941:

> We may joke about the way misplaced concreteness abounds in every word of psycho-analytic writing – but in spite of all the muddled thinking that Freud started, psychoanalysis remains as *the* outstanding contribution, almost the only contribution to our understanding of the family – a monument to the importance and value of loose thinking.[246]

So these male figures, Cambridge-educated in the late 1920s, each an essential collaborator in developing the Mead empire in anthropology, sustained a dialogue, uneven, controversial and even sometimes infamous, between psychoanalysis and anthropology for many years. Two lively pieces of gossip that circulated for years give a clear picture of the bristling tension between the two nascent fields. Here is Gregory Bateson in a retrospective letter dated Christmas, 1973: 'I was gently dropped from Harvard because a rumor got around, "Bateson says anthropologists ought to be psychoanalyzed." I did not say this, and I don't think I even believed it, but if they thought this was a good reason for dropping me, then I was probably lucky to be dropped.'[247] And Margaret Mead's biographer picked up the following: 'It was rumored Fortune got fired [from the University of Toronto] for suggesting, in class, that the unique human feature of face-to-face sexual intercourse might have influenced human development. I once asked [the Head of Department] McIlwraith if this story was true: in a hushed voice, he confided, yes.'[248]

---

toward external authority as being both hateful and essential' (Margaret Mead, 'The swaddling hypothesis: its reception', *American Anthropologist* 56 (1954), 395–409 at 395).

[245] Peter Mandler, *Return from the Natives: How Margaret Mead Won the Second World War and Lost the Cold War*, New Haven: Yale University Press, 2013; see also John S. Gilkeson, *Anthropologists and the Rediscovery of America, 1886–1965*, Cambridge: Cambridge University Press, 2014, pp. 144–9.

[246] Gregory Bateson, 'Experiments in thinking about observed ethnological material', *Philosophy of Science* 8(1) (1941), 53–68 at 65.

[247] http://edge.org/conversation/gregory-bateson-the-centennial, accessed 30 July 2015.

[248] Edmund Carpenter, in Howard, *Mead*, p. 267.

Sociology and Criminology: Contrasting Failure and Success

As all histories of British sociology agree, the subject hardly arrived in Britain in a recognizable European or American form until well past 1945 – and Cambridge was very late in reluctantly admitting it as a subject into the University (1969). Instead, as we have seen, Cambridge had its own peculiar configuration of the moral sciences, with political economy the English equivalent, roughly, of sociology – and it nurtured anthropology from the beginning of the century. However, the story of the non-arrival of sociology in Cambridge is more telling than simply its late arrival.

Just at the moment Myers returned from war to implement his plans for the expansion of psychology in early 1919, the Board of Moral Science set up a two-man Committee to consider what steps 'can be taken to bring about a more adequate provision for the teaching of sociology in this University'.[249] Whatever the source of this initiative (Rivers?), it had little backing: when, a year later, the central University authorities asked the Board for its post-war priority list for posts and facilities, a Professorship of Psychology and one for Logic were at the top of the list, but a Professorship of Sociology was only 'desirable rather than urgent'.[250] Then in May 1925, a Cambridge conference with anthropology, economics, history and moral sciences all participating discussed a proposal to make Social and Political Theory a fourth part in the Part II for Moral Sciences Tripos.[251] This conference was clearly organized in full awareness of an offer that had recently been made to the University by the Laura Spelman Rockefeller Memorial Fund to establish two chairs, one in Political Science, one in Sociology, in the University. Despite all these signs of serious consideration of introducing new forms of teaching and support, it turned out that the University could not overcome the reluctance of the Board of Moral Science to accept this offer and restore to 'moral science' its full meaning and scope as found J.S. Mill's *Logic*.[252] On 23 January 1926, Bartlett moved

---

[249] CUL, Board of Moral Science, Minute Book, 5 March 1919.

[250] *Ibid.*, 28 February 1920. As anyone who has sat on a University committee knows, such language as 'desirable but not urgent' means that this can be safely forgotten for now.

[251] In addition to the three specialist options that had been operating for some twenty years, since the departure of political economy into the Economics Tripos: '(a) Metaphysical and Ethical Philosophy, together with the History of Modern Philosophy; (b) Logic; (c) Psychology' (HR 1910, p. 733). CU, Board of Moral Science, Minute Book, 18 May 1925.

[252] Martin Bulmer, 'The development of sociology and of empirical social research in Britain', in Bulmer (ed.), *Essays on the History of British Sociological Research*, Cambridge: Cambridge University Press, 1985, pp. 3–36 at 23–4. The moral scientists in Cambridge must have been aware of developments at Oxford, with the establishing of 'Modern Greats' (Philosophy, Politics and Economics) as a powerful pedagogic innovation, but their response to invitations and gifthorses remained unswervingly impervious.

that 'the proposal for an option in Social & Political Theory for Part II be dropped for present'.[253] Within a month the Board was being asked to consider a proposal by the Council of the Senate 'to set up Chairs of Political Science and of Sociology' and duly appointed two members to the Committee to consider the proposal – but it was already clear what the attitude of Moral Science, led by Bartlett, was: a strictly ostrich attitude. Moral Science wished to have nothing to do with either Political Science or Sociology.[254] As a result, history successfully scrambled to annex the gifted Chair in Political Science from moral sciences,[255] but the University was obliged to make clear to the Foundation that accepting the Chair in Sociology would be controversial and fraught with difficulty. Writing in 1926, with a radically new set of Statutes and mechanisms of governance, the University replied: 'The Council is necessarily in a difficult position in writing on the subject. Under the constitution of the University the Council has power of initiative only, while decisions rest with the Senate, comprising the whole body of teachers, some hundreds in all.'[256] In effect, the University administration turned down the gift because the relevant teaching staff (most specifically in moral sciences) would not welcome sociology. We can conclude that the small group of moral scientists repudiated innovations not only in psychology, including psychoanalysis, but also in political science and in sociology. The University administration, not to mention the faculties of history, anthropology and economics, expected that these new initiatives would find their natural

[253] CUL, Board of Moral Science, Minute Book, 23 January 1926.
[254] Bartlett's strategy in the period 1922 onwards was to bind the new Department of Psychology closer and closer in to the Natural Sciences Tripos and the new Faculty Board of Biology 'B', to reduce links with the Medical Sciences, to maintain the autonomy of Psychology within the Moral Sciences Tripos and to ensure through such committee manoeuvring as indicated here that no momentum for the development of the social sciences more generally within Cambridge could gather pace, which might threaten the now-established position of experimental psychology.
[255] The background to the importance of the history of political thought in the Faculty of History and its position within the Victorian and early twentieth-century Board of Moral Science is excellently expounded and analysed by Stefan Collini, Donald Winch and John Burrow, *That Noble Science of Politics: A Study in Nineteenth-Century Intellectual History*, Cambridge: Cambridge University Press, 1983, chapter 11, 'A place in the syllabus: political science at Cambridge', pp. 339–63.
[256] Vice-Chancellor to Laura Spelman Rockefeller Fund, spring 1926, quoted in Bulmer, 'The development of sociology', p. 24. Even with an untried and untested structure of governance, this was a radically conservative interpretation of the relationship between University and Faculties. Bulmer concludes that the University was implying that 'Senate was not thought likely to be willing to countenance the establishment of a chair in the subject', but it is more likely that the real sticking-point was resistance emanating from the natural locus within the University for the subjects, moral sciences. If there had been leadership in moral sciences of the dynamic Myers variety, history, anthropology, economics – subjects which had already been convened to confer with moral sciences on the question – and even English, classics and MML, might have offered the sort of general support that wins Senate votes.

home in the Whewellian dispensation of the moral sciences.[257] They were proved wrong. Later initiatives sputtered out in the 1930s; even when post-1945 the British government injected funds into Cambridge for the support of the social sciences, the University could not bring itself to create a Chair in Sociology (for want of suitable candidates, it seems: the two candidates, of course both Cambridge graduates, deemed suitable in private (T.H. Marshall and Gregory Bateson) having turned down the offer), and deployed the money for a Visiting Professorship in Social Theory.[258]

This sorry and telling story is to be compared with the administrative facility with which Myers, despite so much opposition to his other plans, prepared psychology for the post-war era: creating a new post in psycho-pathology to cement the bridge between psychology and medicine; rallying and bringing together the civic forces of Cambridge City and the University (previously tied to the problem of mental deficiency with a eugenic backdrop), for an initiative which would bring Maghull to Cambridge. This was a vision which, in 1932, would bring the psychother-apeutic expertise of the war neuroses to peacetime Cambridge, thus giving institutional ballast to the original pre-war plan to make Cambridge the national centre for psychological medicine. Myers envisaged a newly expanded Department of Psychology in Cambridge mirroring the recon-structed British Psychological Society in London, with educational, indus-trial and medical psychology sections. This was a new psychology that reached out to the enthusiastic interest in these branches of psychology. In Myers's plan, the University was to encourage and reflect the 'wild upsurge' that so alarmed the Regius Professor of Physic. But, precisely because of the opposition that then manifested itself in Cambridge, Myers resigned and departed in disgust to a new and non-academic institution of psychology – the NIIP. And Rivers died at that very moment. Instead of a transformed psychology based on industrial, educational and medical psychology, opening the University to schools, magistrates' courts, out-patient clinics, businesses and factories, Bartlett made an about-face and decided to concentrate psychology on its original experimental foundation. 'When the War was over psychology had, in a way, to be begun afresh at Cambridge', he recalled in 1937.[259] What he didn't recall in public was

---

[257] But we should be properly agnostic as to how welcoming economics or anthropology might have been of sociology, given 'Veblen's Law of Interdisciplinary Relations: to the extent that disciplines overlap in subject matter, they are unlikely to cooperate' (Marshall Sahlins, 'The conflicts of the Faculty', *Critical Inquiry* 35 (Summer 2009), 997–1017 at 1013).

[258] Bulmer, 'The development of sociology', p. 25.

[259] Bartlett, 'Cambridge, England, 1887–1937', p. 108.

that in 1919 Myers had begun to build psychology afresh in a revolutionary new configuration that by 1922 had run into the sand, so that Bartlett's 'rebuilding' in 1923 and after was in fact a second rebuilding, one with an exactly opposite character to Myers's.

Myers had created the permanent resources for a sustained interest in psychoanalysis and related fields without having to take this initiative to a vote of the Senate, simply by getting what he wanted from a University Review Committee. Sociology could have been built into the new Cambridge order that began in October 1926, with the new Statutes, if only a political fixer like Myers had been at work on its behalf. Cambridge moral sciences would have looked quite different if instead of two Chairs (the Knightbridge (1683) and Philosophy (1896)), there had been four, two in Philosophy, one in Political Thought, one in Sociology, together with the Readership in Psychology. Instead, Bartlett, also a political fixer, though not of the same order as Myers, worked in exactly the opposite direction, to keep sociology out. It is also salutary to reflect that, if the Professorship of Sociology had been created and if Rivers had survived, he would have been a perfectly suitable candidate for the post, just as he would have been a perfect candidate for a Chair in Psychology or Anthropology. Disciplinary boundaries were fluid and Rivers's range of interests were wide. The Cambridge man solicited two decades later for the Chair of Sociology, Gregory Bateson, had the same breadth: he too could have been a Professor of Anthropology or of Psychology.[260]

So how did one successfully create a new disciplinary formation in twentieth-century Cambridge? Nothing should be taken for granted or treated as inevitable, especially in Cambridge. Even history, a subject one might suppose would be an essential discipline in any modern university, was first bounced around within the Moral Sciences Tripos, then hived off in a short-lived alliance with Law before a separate Tripos was created in 1875. Its rapid expansion in numbers – from 36 (28 men, 8 women) in 1888 to 115 in Part I in 1910 (of whom 92 were men) and 101 in Part II in that year (of whom 83 were men) – still did not guarantee it a strong position in the Colleges, reluctant as ever to create College positions in such new and possibly only fashionable (as opposed to serious) subjects. It was only in the inter-war years that history established itself firmly in the Colleges, to the point where in the twenty-first century it is now one of those subjects

---

[260] As we have seen, Bateson had a no-nonsense respect for psychoanalysis similar to that of Rivers. When, while teaching at Harvard in 1947–48, he was approached to return to Cambridge as Professor of Sociology, Bateson was about to embark on his influential researches on communications, cybernetics and psychiatry, which led to the double-bind hypothesis.

excessively dependent upon College financial support – and thus seen as a venerable and conservative subject, as if the august antiquity of the College buildings rubs off on the teaching they promote. We also saw how key subjects in the sciences, notably physics, physiology and biochemistry, relied upon College (or external) funding to become established in the University, particularly from Trinity College. The configuration that permitted psychology to become established was support from both Natural Sciences and Moral Sciences Tripos (and Special Boards), but it only became firmly established with external funding from a private patron (C.S. Myers's family). The disciplinary entrepreneur Myers, acting both within Cambridge and on a national scale, attempted to create disciplinary alliances – with education, economics, anthropology and not least medicine; the creation of the Diploma in Psychological Medicine was designed to supplement experimental psychology with abnormal psychology and make Cambridge a national centre – another efficient way of establishing its credentials within Cambridge. In contrast, the creation of the English Tripos and the establishing of a 'discipline' of English was a strikingly contingent coup, building on the pre-war external endowment of a Chair, but then also dependent on the free market created by student demand, with most lecturers, coming from a range of neighbouring disciplines (classics, history, MML), dependent for their living on the fees they charged for attendance at lectures (Richards charged one guinea for his course on 'Modern novels' in 1919–20). Even when Richards was appointed to a College Lectureship in 1922, his position was jointly in English and in moral sciences. But the rapid expansion of student numbers ensured that English would eventually become well entrenched, both in the University and, like history at this time, in the Colleges.

An entirely different path to the consolidation of a discipline with an institutional grounding is presented by criminology – contrasting here strikingly with sociology, with which it has considerable links, but also a reminder of the 'what might have been' of psychoanalysis, since psychoanalysts were in the forefront of the first major attempt to introduce criminology into Britain in the 1930s.[261] Lacking a home-grown criminological tradition comparable to Italy, France or Belgium and relatively impervious to the criminal anthropological tradition associated with Lombroso, criminology effectively began in Britain with the influence of the 'new psychology' in a medical mode after the Great War: the roots of

---

[261] Michal Shapira, *The War Inside: Psychoanalysis, Total War, and the Making of the Democratic Self in Postwar Britain*, Cambridge: Cambridge University Press, 2013, chapters 5 and 6, pp. 138–97.

criminal behaviour being latent (unconscious) in all, the abnormality of the criminal was to be diagnosed and treated. Maurice Hamblin Smith,[262] a Birmingham prison doctor, is often accorded the title of first criminologist, and his inspiration was Freudian, while the penal context in which he practised was entirely in keeping with the practical and policy-oriented long-term character of British criminology. It was through Grace Pailthorpe's work with Hamblin Smith in 1922–23, assisting Smith in the psychoanalytic investigation of female offenders at Birmingham, followed by a five-year study, funded by the MRC, at Holloway Prison, that she became convinced of the importance of the psychoanalytic approach to and treatment of delinquency. This led her to convene a Committee composed of an eclectic range of medical psychologists and neurologists, which founded the Association for the Scientific Treatment of Criminals (1931), later becoming, in 1932, the Institute for the Scientific Treatment of Delinquency (ISTD).[263] The aim, she spelled out, was 'to eradicate crime by curing through psychological treatment and other measures the underlying psychological maladjustments and defects'. Being both preventative and therapeutic in its scope, the ISTD's programme was radically different from the conventional English policy orientation of initiatives associated solely with the prisons and courts. The ISTD's network overlapped considerably with the other inter-war initiatives associated with psychoanalysis and the new psychology: the out-patient Clinics; the Child Guidance Clinics (indeed a key figure in both child guidance and the ISTD was Emanuel Miller); but close links with magistrate courts, in particular the Juvenile Courts established by the Children Act (1908), and with the probationary system (Probation Act 1907) established before the Great War principally for juvenile offenders, allowed the preventative and therapeutic orientation of the ISTD to flourish with increasing numbers of cases. The ISTD was neither based in universities nor closely linked with the Home Office and prison system, but was located in the interstitial spaces, linked with magistrates' courts, child guidance and out-patient clinics (Tavistock) in the novel inter-war arena Nik Rose aptly calls 'the psychological complex'. In this arena, there was one supremely gifted political operator, Cyril Burt, author of *The Young Delinquent* (1925), the most influential work in the field in the inter-war years, which incorporated psychoanalysis and its projects into his wide

---

[262] P. Bowden, 'Pioneers in forensic psychiatry. Hamblin Smith: The psychoanalytic panacea', *Journal of Forensic Psychiatry* 1 (1991), 103–13.

[263] David Garland, 'British criminology before 1935', *British Journal of Criminology* 28(2) (1988), 1–17 at 7.

survey. However, Burt was thoroughly eclectic in what he took from psychoanalysis: while he was a formal Patron of the ISTD from its inception and was a full member of the BPaS from its earliest days, when Ernest Jones had been so proud of the public lectures he gave on 'Psychoanalysis and education' in the autumn of 1920, until 1939, he could be rebarbative in public, especially when it came to the psychoanalytic preoccupation with sex. Burt was not going to be the operator who would create an Institute of Psychoanalysis or even an Institute of Psychoanalytic Criminology in the University of London.

After the Second World War, the ISTD's Clinic was absorbed into the NHS (as the Portman Clinic) and the ISTD's activities were extended to establishing a journal, *British Journal of Delinquency* (1950), edited by Miller, Edward Glover and Herbert Mannheim, brother of Karl, a European-style criminologist who had emigrated to Britain in the early 1930s and been welcomed by the LSE, the sole major academic centre for the social sciences in inter-war Britain. But the development of criminology within academia was not to come from the ISTD; indeed, there is sufficient evidence that the ISTD was very deliberately bypassed in this process. The principal reason was that the impetus for the establishment of criminology came in the immediate post-Second World War years from the Home Office.

The key figure in the establishing of the Institute for Criminology in Cambridge in 1960 was Leon Radzinowicz. Wealthy and originally from Poland, in the inter-war years he had established a reputation throughout Europe, particularly through his initial legal training in Paris, Geneva and Italy, the home of criminology. Then after working in Belgium and advising the Belgian government, he taught in his home country of Poland, publishing in several languages a series of influential works (e.g. *Le crime passionnel* (1929); *La crise et l'avenir du droit pénal* (1929); *Nowy ustroj penitencjarny we woszech* (1935)). Operating usually as a private scholar of independent means, Radzinowicz was a master at cultivating connections with the highest authorities and best-placed figures in his field across many countries. In 1938 he visited England to study the English system of justice for the Polish government, with introductions to the Home Office, where he was welcomed, and to Cambridge, where he quickly cemented a crucial alliance with Cecil Turner, the only person in a distinctly conservative Cambridge Law Faculty interested in the area of penal reform. Radzinowicz rapidly gathered friends and allies; with Turner, he edited a book series under the auspices of the Cambridge Law Faculty, English Studies of Criminal Science and a series of Criminal

Science Pamphlets, avowedly broad-minded, including amongst its early pamphlets S.H. Foulkes's 'Psychoanalysis and crime' (with an introduction by Cyril Burt). His impressive research, writing and creation of a network during the war resulted in a special post being created for him in the University in 1946, as Assistant Director of Research; in 1948 he was elected into a special Fellowship at Trinity (shades of the discipline-fostering that Trinity had embarked on for Sidgwick, Foster and Hopkins) and in 1949 a Department of Criminal Science was created within the Law Faculty. Radzinowicz did intend to bring the social sciences more broadly into his vision of 'criminal science', but the fundamental orientation of his disciplinary vision was as advisory to government and parliament, to judges and lawyers; so his most important links were with Cambridge Law and the Home Office. With the Home Office initiating the commitment of research funding to issues concerning crime and penal policy in the late 1940s, this governmental research orientation came to sudden fruition with Home Secretary Rab Butler's push in 1957 to found a formal institution outside government which would research and advise on criminal matters. As Radzinowicz later succinctly described the unfolding process by which the Institute of Criminology came into existence in 1960: 'the initiative was taken by a national political figure of Rab Butler's stature; ... from the start it had enthusiastic support from the senior staff of the Home Office; ... it gained financial backing of considerable magnitude from a highly respected foundation [Wolfson]; ... a university as prominent as Cambridge had expressed its willingness to give criminology a permanent home'.[264] News of the proposed initiative was first announced by Butler in the House of Commons.

The Wolfson Foundation's generosity permanently funded two senior posts; the Home Office supplied other funds; the University (with a crucial ally in Trinity, Edgar Adrian,[265] both in the late 1940s and late 1950s) gave

---

[264] Leon Radzinowicz, *Adventures in Criminology*, London: Routledge, 1998, p. 192.

[265] Adrian had long-standing interests in criminology. On 3 February 1926, he gave a talk on 'Crime and mental defect' to the Emmanuel Law Society, where he surveyed the contributions of Lombroso, Goring, Healy and 'an extremely able & readable book just appeared' by Cyril Burt called *The Young Delinquent*. He concluded: 'Being a doctor I shall naturally come down on the side of criminals as patients.' The long-standing concern with mental deficiency and its broad social connections was substantially bolstered through his marriage – his mother-in-law, Dame Ellen Frances Pinsent, was a national figure in mental health and its care, and his wife Hester followed in her footsteps. However, when discussing with Radzinowicz in 1948 the setting up of the Department of Criminal Science in the Faculty of Law and the implausibility of the term 'criminal science', he had murmured: 'There is no such a thing as the science of crime, unless one is concerned with the most effective way of committing it' (Radzinowicz, *Adventures in Criminology*, p. 202).

support; the Faculty of Law supported Radzinowicz's elevation to the Wolfson Professorship. Virtually single-handedly, through networks of scholarship, the government and legal insiders, and the great and the good, Radzinowicz had settled a new social science institution within Cambridge – and the first of its kind in Britain. Essential to the success in Cambridge was the fact that the University of London, in particular LSE, had refused to consider creating an Institute of Criminology on the grounds that it would tread on too many well-established toes. The Home Office, at an early stage of discussions, in a Minute of 7 May 1958, had also made it clear that the ISTD was not an approved organization:

> The ISTD has been concerned mainly with the medical aspects of the subject, with a pronounced psychoanalytic bias, and although it has done excellent work on the treatment of offenders, notably in establishing the Portman Clinic, it is unlikely that the expansion of this body would lead to the establishment of high academic standards or a balanced approach to the different aspects of criminology, or that it would command general support.[266]

Radzinowicz shared this view: 'it would have been fatal to establish the first Institute of Criminology under the umbrella of the ISTD, an organization ... so committed to the psychoanalytic approach'.[267] So the establishment of the Institute of Criminology in Cambridge had the advantage from the point of view of the Home Office that, by the late 1950s, Cambridge was an entirely psychoanalysis-free zone. These negative factors – LSE's incapacity, the necessary detour around psychoanalysis – were supplemented by the remarkable positive achievement of Radzinowicz of persuading a very conservative Faculty of Law and a science-focused and conservative Trinity College to back him and his links to government (facilitated, it should be said, by the reforming Home Secretary being a Trinity man whose father had been Master of Pembroke College, Cambridge, who was himself High Steward of Cambridge University (1958–66) and who would himself be elected Master of Trinity in 1965 to follow Lord Adrian). Keeping criminology close to law – to penal policy, to research benefiting proposed government reform proposals – was crucial to his success. Given Cambridge's track-record of the previous forty years, any mention of 'sociology' could have been fatal to the project. Radzinowicz in addition made it quite clear that the entire

---

[266] Leon Radzinowicz, *The Cambridge Institute of Criminology: Its Background and Scope*, London: HMSO, 1988, p. 9n14.
[267] Radzinowicz, *Adventures in Criminology*, p. 180n24.

topic of 'the general problem of the criminal in society, its causes and its solution' originally ambitiously envisaged by the Home Office – the core problematic as the psychoanalysts saw it, seeking out causes so as to find cures through treatment – would not be addressed by the new Institute.[268] Instead, the Institute eschewed psychology, eschewed the medical model of the motives for crime and their 'treatment', and committed itself to methodologically rigorous statistical studies, very much part of the new sociologically oriented research into crime ('deviance and control'), following the long tradition, stretching back to France in the 1820s, in which 'criminal statistics' are the first and most important element in the knowledge and understanding of crime.[269] Radzinowicz himself spent his entire Cambridge career preparing and completing his monumental five-volume *History of English Criminal Law and its Administration from 1750*.

The Creation of Cambridge English

A curious unintended consequence of the Cambridge University Reforms following the report of the Oxford and Cambridge Commission in 1922 was to freeze disciplinary boundaries and make the University considerably less responsive to the initiatives of young teachers and researchers. Those who were let in under the 1926 bar effectively closed off opportunities for others for years to come. The thirty years prior to the reforms saw a spate of new Triposes: Mechanical Sciences (1894, renamed Engineering in 1970); Oriental Languages (1895, renamed Oriental Studies in 1958); Economics (1905); English (1919); Geography (1920);[270] Anthropology (1921, becoming Archaeology and Anthropology in 1928). In the thirty years after, only Music (1948) and Chemical Engineering (1950) were created. The distinctively 'modernist' projects of economics, English and anthropology got under the bar in Cambridge before 1926; the 'social sciences' failed to do so and had to await the next wave of University reform, during the 1960s, to be established in Cambridge.

With the fluidity of disciplinary boundaries before the mid-1920s coinciding with the upsurge of interest in psychoanalysis, some intriguing and

---

[268] *Ibid.*, p. 194.

[269] See Ian Hacking, *The Taming of Chance*, Cambridge: Cambridge University Press, 1990; Alain Desrosières, *The Politics of Large Numbers: A History of Statistical Reasoning*, trans. Camille Naish, Cambridge, MA: Harvard University Press, 1998.

[270] For connections between the study of geography and psychoanalysis at Cambridge, see Laura Cameron and John Forrester, 'Freud in the field: psychoanalysis, fieldwork and geographical imaginations in interwar Cambridge', in P. Kingsbury and S. Pile (eds.), *Psychoanalytic Geographies*, Farnham: Ashgate, 2014.

unexpected alliances were in the offing, to be cemented or passed up. The link between psychology and anthropology is well known, not least because of the mythical status that the Torres Straits Expedition of 1898 has acquired in the early history of both psychology and anthropology – Rivers's position in the development of both disciplines being crucial. A less well-known link is between the 'moral sciences', with its inclusion of psychology, philosophy and psychoanalysis, and the origins of 'English literature'.

Before considering these institutional developments, it is worth reflecting on the more general schema of the transformations of these disciplines in this period. In an essay on 'Company histories', Stefan Collini considers the official myths of the founding of a set of disciplines – those cited in Perry Anderson's classic account of English intellectual life as the sole viable habitats for a potentially radical critical vision: social anthropology and English literature. The received versions of these disciplines' histories share the same founding *annus mirabilis* of 1922, which saw the publication of T.S. Eliot's *The Waste Land* and Bronislaw Malinowski's *Argonauts of the Western Pacific*.[271] The role of discipline founder passed from Eliot, non-academic man of letters, to Richards in Cambridge, but the basic model provided by Malinowski seems to cover both cases: a revolution was engineered by the innovative work of a key figure or figures,

> displacing the eclectic and promiscuous inclusiveness of their Edwardian predecessors with the rigorously close analysis of a relatively restricted body of material. A small group of graduate students and young researchers, conscious of their 'outsider' status, gather round an acknowledged leader of charismatic power and legendary difficulty, and after 1945 this generation moves into a position of intellectual hegemony within an expanding discipline marked by fierce doctrinal and personal disputes.[272]

The core group of structural-functionalists in anthropology would eventually be cemented around the figure of Radcliffe-Brown (not Malinowski), just as F.R. Leavis (not I.A. Richards), despite his long-running feud with the Cambridge English Faculty, is easily recognizable as the charismatic leader who would, after long struggle, acquire a position of intellectual hegemony within an expanding and self-consciously embattled discipline. To this template, we could add the discipline of philosophy, with Ludwig Wittgenstein as the charismatic outsider who

---

[271] See the remarkable book by Michael North, *Reading 1922: A Return to the Scene of the Modern*, Oxford: Oxford University Press, 1999.
[272] Collini, 'Company histories', p. 280.

published *Tractatus Logico-Philosophicus* in 1922 and whose intellectual shadow, if not leadership, would be cast long and dark for decades.

As the critic Collini astutely points out, it is as much the break-up of the previous organizing framework of knowledge as it is the innovations of modernism that gives plausibility to this picture:

> The broadest context would be provided by the long-drawn-out disintegration of the evolutionary and historicist paradigm that had furnished the guiding assumptions for so many intellectual enterprises in Victorian and Edwardian culture. Assumptions of socio-cultural evolution provided a common frame for topics as subsequently distant as the origins of religion or the nature of the human mind, while the canons of Germanic historicism governed scholarly practices across the study of the classics, of philology, of literature, of jurisprudence, and so on. The rejection of this common framework was, in different ways, a founding moment for the modern form of disciplines like philosophy, literary criticism, and anthropology.

It is the indubitable existence of this overarching evolutionist and historicist framework that alerts us to the possibility that its rebellious progeny may also have participated in a similar replacement matrix, beyond their invention of individually distinctive disciplinary methods and shibboleths. Indeed, the vehicle that promotes – or, more conservatively, that presages – these inventions may be of as much historical moment as the disciplinary births that it facilitated, precisely because it retained a unique founding characteristic that could be transmuted into something quite different by each of those disciplines. To be such a transitional catalyst may well have been the destiny of psychoanalysis.

The distinctive peculiarity of Cambridge was the fluid and overlapping set of institutions into which psychoanalysis entered in the crucial period of disciplinary gestation. What was the distinctive and crucial ingredient that psychoanalysis brought to the disciplinary ferment? It was the injunction to a symmetrical self-reflection, an injunction to unqualified reflexivity. The revolutionary effect of psychoanalysis would always be first and foremost the requirement, indeed the anguish and the satisfactions, of self-analysis. The distinctive mark of an enthusiastic reception of Freud's work would always be self-analysis, most easily identifiable in the interpretation, with all its naïve stumblings and nervous jockeying with Freud, of one's own dreams. Turning from one's own dreams to the rigours of a new discipline, with its always initially uncertain conception of a new subject of knowledge, would engender very different results. Within the germinating field of psychoanalysis, its reflexivity spinning out into a larger social field could take the following form, as John Rickman recorded:

> In 1921 a proposal was made to the Governing Body of a world-famous Mental Hospital by a candidate for the post of Medical Superintendent that the patients should be put in a 'climate' or 'social field' where everyone had been analysed, i.e. all the persons in contact with a patient (not only the doctor in charge of his analytical therapy) would be in a position to recognize and not be unduly reactive to the deeper sources of conflict which heavily influence the behaviour of the patient; thus the expression of the patient's troubles would be less hindered by lack of understanding and he would not feel so socially rejected.[273]

The three figures most uncontroversially identifiable as the key Cambridge innovators display this mutation. For Malinowski, following the ghost of Rivers, the novel method of participant-observation required the observer to pass through the phases of participation and return, the going out and the coming back, capable of rendering that transmutation of his very being which he had undergone not only into objective data but also into a body of theory. For I.A. Richards, who established English as a modern field in Cambridge, the practice of criticism puts the entire personality of the reader to the test: it is only through the experience of one's own failures and misreadings, of exposing one's own stock responses and inhibitions, that the achievement of an adequate reading is possible. For Wittgenstein it will, finally, only be through dispelling one's own grammatical mistakes, or through the realization of the limits of one's own language, that the aim of philosophy, which is its own dissolution, can be achieved. These quintessentially modernist projects start with the demo-cratic confrontation of the charismatic leader with his own errors and the realization that new knowledge begins at home. After all, it is not as if Western men of science had not been going out and coming back again for centuries. The imperial side of the project was not new and nor was it given up by these new practitioners. What they added was an obverse, a require-ment that the interior and the exterior be mapped onto each other. This is the lesson they learnt from Freud; it is an irony of intellectual history that no one would ever judge Malinowski, Wittgenstein or Richards to be humble, spontaneously egalitarian members of their respective commu-nities. But then, one might add, neither was Freud.

---

[273] J. Rickman, 'Reflections on the function and organization of a Psycho-Analytical Society', *IJP* 32 (1951), 218–37 at 222.

The English Tripos was created in a flurry of opportunistic compromise while most of Cambridge was away fighting the war.[274] Two politically astute College duos collaborated to engineer this triumph: in Clare College, Hector Munro Chadwick, Professor of Anglo-Saxon, and the brilliant and eccentric young historian Mansfield Forbes; and in Jesus College, the King Edward VII Professorship of English Literature, Sir Arthur Quiller-Couch, and the young Greek ceramicist, E.M.W. Tillyard. Taking advantage of the anti-German sentiment of the war, they put in place a new English Tripos, designed for those who had already studied moral science, the classics or languages, which would, in the telling phrase of Basil Willey, an early teacher for the Tripos, be entirely free of 'the alien yoke of Teutonic Philology'.[275]

Recruiting lecturers for the new Tripos in the most informal way imaginable, Forbes corralled the twenty-six-year-old tubercular would-be mountain guide I.A. Richards, who had taken Part I of the Moral Sciences Tripos at Magdalene College from 1911 to 1915, falling equally under the spell of G.E. Moore and of C.K. Ogden, his ubiquitous fellow Magdalene graduate.[276] Back in 1911, when Richards first repudiated history, he needed help finding another subject; he was introduced to a third-year undergraduate, Ogden, on the verge of founding the *Cambridge Magazine*, who told him in encyclopaedic detail – his unique, life-long stamp – what all the other subjects in the University consisted in and what each lecturer would tell him. Richards chose to study moral science. But one might equally say that in that conversation in December 1911 he chose to study with Ogden. Certainly the next thirty years of his life were dominated by their collaboration. Richards's first two books, including the celebrated and inevitably eccentric *The Meaning of Meaning*, were written with Ogden. His next two books, his most influential, both published with Ogden's publishers, incorporated many of the themes from psychology,

---

[274] The fullest account is given by an insider, Tillyard, *The Muse Unchained*, here supplemented with the useful, stimulating and careful discussion in Chris Baldick, *The Social Mission of English Criticism, 1848–1932*, Oxford: Clarendon Press, 1983.

[275] Basil Willey, *Cambridge and Other Memories, 1920–1953*, London: Chatto & Windus, 1968, p. 14. Tillyard argued that the anti-German sentiment prevalent in the University during the war prevented the two key figures who were German by extraction, Breul and Braunholtz, bearers of the philological flame, who sat on the Special Board of Medieval and Modern Languages, from making their opposition felt, thus allowing the proposals to go forward unchallenged.

[276] Mountaineering remained a passion for Richards for many decades; he set out climbing at Easter 1925 or 1926 with Lionel Penrose, Margaret Gardiner and Phyllis Leon, but had a premonition of disaster, so they left him at Victoria Station 'sitting forlorn on his rucksack' (Gardiner, *A Scatter of Memories*, p. 78). A year earlier, Gardiner, Blackett and Richards were climbing in North Wales; after a swim in Lake Ogwen, Gardiner introduced Richards to Gerald Manley Hopkins (*ibid.*, p. 71).

philosophy and aesthetics that he shared with Ogden. Ogden fixed him up with jobs when they embarked for America in 1926, Richards turning the journey into an eighteen-month voyage round the world, in the course of which he married Dorothy Pilley, whom he had met in 1917 when climbing in Wales. From 1929 on, Richards committed much of his energy, first in China, then in the United States with returns to China, to Ogden's Basic English.

Like so many others, Richards – 'mischievous, ardent, curious, hospitable, stern'[277] – had become enthused by Freud and psychoanalysis in the middle of the war; and he had begun medical studies in a fitful fashion at the Cavendish in October 1918, in physiology and chemistry, in order to become a psychoanalyst, only, like James Strachey at roughly the same time, to abandon them forthwith.[278] The partnership of Ogden and Richards was already giving birth, following their seminal two-hour conversation on a staircase on Armistice Night, to their extravagant *The Meaning of Meaning*, embodying the quintessence of Moore's incessant interrogation of meaning and a first stab at characterizing 'meaning' not only in relation to the referent, but also necessarily to a context of use. But Richards was footloose in Cambridge until Forbes signed him up in the summer of 1919 to lecture for the new English Tripos. In 1922, he was appointed to a College Lectureship at Magdalene in both English and moral sciences – subjects which he and Ogden, working on theories of meaning and on aesthetics, melded effortlessly, adding in psychology and psychoanalysis in the tradition of the moral sciences at Cambridge.[279]

By the end of the 1920s, while teaching both moral sciences and English, Richards had laid the foundation for an autonomous and confident practice of the academic study of 'English' with two keys works: *Principles of Literary Criticism* of 1924 and the innovative and extraordinarily influential

---

[277] Helen Vendler, 'The Explorer. Review of J.P. Russo, *I.A. Richards: His Life and Work*', *New York Review of Books* 36(7), 27 April 1989.

[278] John Paul Russo, *I.A. Richards: His Life and Work*, London: Routledge, 1989, p. 48.

[279] Edward Bullough was the Cambridge University figure most closely linked to aesthetics in the pre-war period; he taught German, Russian and Italian as well as researching and lecturing on the psychology and physiology of aesthetics. When translating Freud's case-histories in 1924, Alix Strachey suspected she had picked up 'empathy' ('a vile word, elephantine, for a subtle process') as an appropriate translation for 'Einfühlung' from attending his lectures, which she had then used for translating Freud (BF, 2 January 1925, pp. 170–1). There is a more complicated history of the introduction of the word into the English language than this, but Strachey's memory is backed up by an examination question set in the Moral Sciences Tripos Part II in June 1911: '2. What is meant by empathy (*Einfühlung*)? Give some account of its supposed influence on aesthetic and other judgments.'

Fig. 5.5 Ivor Armstrong Richards, 1924.

Practical Criticism: A Study of Literary Judgment (1929). The preface to
these books was his work with Ogden on semantics in *The Meaning of
Meaning*, which was not only an aggressively iconoclastic attack on historical linguistics and a cascade of metaphysical verbiage, but attempted to
yoke together neurophysiology, ideas lifted from behaviourism and the
scepticism of G.E. Moore in a model of the 'general sign'. Along the way,
in the late summer of 1920, Ogden and Richards teamed up with their
Cambridge friend James Wood, painter and novelist, to produce a slim yet
dashing survey of the philosophy of aesthetics, examining in detail all
contemporary contending definitions of the aesthetic and advancing the
argument that the aesthetic is to be found in *synaesthesis*, a blend of

equilibrium and harmony, whereby a balance of competing psychological impulses is achieved.[280]

The main threads of these preliminary works were very much the product of Cambridge, most especially the moral sciences. The most secure and venerated points of reference are G.E. Moore's, Bertrand Russell's and James Ward's accounts of the philosophy of language and logic and of philosophical psychology. Richards proved the most dutiful of students of his Cambridge masters, transferring their coolly analytic arguments into the broader sphere of aesthetics, semantics and a general theory of interpretation. Or was it more general? Richards's interests not only reflect remarkably faithfully the contemporary currents of moral sciences in Cambridge, through Ward, Russell, Moore and Edward Bullough, but had intellectual ambitions that reached back to the more grandiose and expansive era of the founders, to Whewell and Mill. When he declares, at the end of his most accomplished work, *Practical Criticism*, that 'The methods I have here tried to apply to critical questions, have to be applied to questions of morals, political theory, logic, economics, metaphysics, religion and psychology',[281] he was not staking out new ground for a new discipline, but returning to the original broad conception of the moral sciences. This seems to be an ironic last throw for a project that failed, advanced by the theoretician whose own practical innovations would give a foundation to the newly specialised discipline that, dispensing with such a large vision, would take over the high cultural ground once aimed at by the moral scientists: literary criticism or 'Eng. Lit.' as it came to be known.

Richards became the theoretician behind the invention of 'Cambridge English', which, as Stefan Collini notes, was characterized by a vigilant and exclusive close verbal criticism, by applying these skills to recent and contemporary literature and by the conviction, so fateful to its ideological future, that literary criticism involves one's whole being and is thus an exemplary 'moral' activity.[282] Yet it was not this last but a pedagogic invention which was the making of Richards's vision. First conceiving it as a test for students, Richards in class asked them to comment upon copies of poems distributed without title, author or date of composition. The scientific model was obvious: this was neither a lecture, nor a Teutonic seminar, but an 'experiment'; he called the exercise 'practical criticism'.

---

[280] C.K. Ogden, I.A. Richards and James Wood, *Foundations of Aesthetics*, London: Allen & Unwin, 1922; see Russo, *I.A. Richards*, pp. 98–109.

[281] PC, p. 341.

[282] Stefan Collini, 'The study of English', in Sarah J. Ormrod (ed.), *Cambridge Contributions*, Cambridge: Cambridge University Press, 1998, pp. 42–64 at 62.

These were Practicals for literature students, Practicals without a laboratory bench but Practicals nonetheless, and Richards cast himself in the role
of Demonstrator rather than Lecturer.[283] There are moments when
Richards finds himself re-enacting with a different matter the experiments
that another Cambridge scientist had performed two hundred and fifty
years previously:

> It is sometimes convenient to regard a poem as a mental prism, capable of
> separating the mingled stream of its readers into a number of distinguish-
> able types. Some poems . . . merely scatter or throw back a large propor-
> tion of the intellectual-moral light that is applied to them. Others . . . are
> transparent; and, since they have, as it were, a high refractive index, they
> perform their analytic function to perfection. They split up the minds
> which encounter them into groups whose differences may be clearly
> discerned.[284]

Each of these mental prisms gave rise to a mass of data, which Richards
called 'protocols'. At times, then, Richards modelled his procedure on
the experiments he had conducted as an undergraduate taking the
Moral Sciences Tripos. But equally persistent was his espousal of the
model of the 'psychological clinic' and the understanding of pathological states of mind. This was the importing of not simply an external
framework into the field of understanding poetry, but a framework that
arose from the particular characteristics of the experiment he was
mounting.

The confrontation of readers with a poem that had been cut loose from
all the familiar accoutrements – name of author, date of composition, place
in the lineages and hierarchies of literary oeuvres – had a series of major
consequences, both negative (through elimination) and positive (through
induction or provocation). The negative consequences were to eliminate
all questions relating to authorship and historical context in the experience
of understanding the poem. Projects for understanding literature that were

---

[283] In the list of formal University offices in Cambridge, a Demonstrator was, as the OED puts it, 'A
person who demonstrates how something works or is done; *spec.* one who teaches by demonstration in a laboratory.' Demonstrators were junior to Lecturers and could often be promoted to
Lecturer. Collini glosses Richards's work as 'an experiment in applied psychology that Richards
conducted with his lecture-audiences in the 1920s' (Collini, 'The study of English', p. 54) Cf.
Helen Vendler: 'In his single quasi-scientific experiment in the 1920s, which made him
notorious in university circles, Richards asked his students in "practical criticism" at
Cambridge to write comments on poems (both good and bad) to which no author's name
had been attached.'
[284] PC, p. 53.

predicated upon knowledge of the past, knowledge of social context, knowledge of authorial biography were immediately eliminated as irrelevant. Richards's project looked as if it were aimed at producing the pure text as object of study – the text and nothing but the text. But this is where the positive consequences asserted their force: the experiment was designed to *induce* errors, *provoke* mistakes, *force* the reader to reveal the vice of his reading – his obstinacy and conceit, the 'stock responses' – a famous phrase that became a trademark of this tradition. Instead of a confrontation with a purified text, this was an experiment in which the text was simply an organized system of mental stimuli designed to produce reader-effects, effects which were above all marked by failure:

> The most disturbing and impressive fact brought out by this experiment is that a large proportion of average-to-good (and in some cases, certainly, devoted) readers of poetry frequently and repeatedly fail to understand it . . . They fail to make out its prose sense . . . And equally, they misapprehend its feeling, its tone, and its intention. They would travesty it in a paraphrase.[285]

Richards immediately asserted that this was not a result of immaturity or lack of knowledge or familiarity; every reader is vulnerable to all of these failures: 'no immunity is possessed on any occasion, not by the most reputable scholar, from this or any of these critical dangers'. This is not the prologue to a sweeping indictment of the parlous state of present culture (although the task of cultural diagnosis, prognosis and therapy is dear to Richards's heart), but rather the prelude to acts of individual critical therapy. Richards's model is the physician, specifically the neurological psychologist, and his knowledge of psychology, 'the only good legacy left by the War'.[286] Reading and engaging in the interpretation of the prepared texts is an exercise in mental hygiene: 'the critic is as closely occupied with the health of the mind as any doctor with the health of the body'[287] and Richards self-confessedly approaches the task with 'a few touches of the clinical manner'.[288] He is intent on eliciting mistakes which can be gathered together to form a psychopathology of everyday reading.

---

[285] PC, p. 14.
[286] I.A. Richards, *Principles of Literary Criticism*, London: Kegan Paul, Trench, Trubner & Co., 1924, p. 63. This odd use of hyperbole – or is it meant to be taken literally? – is almost identical to a phrase that Walter Langdon-Brown (see Chapter 8) used in a letter to Tansley (quoted in Chapter 2) in 1920: 'I sometimes feel that the only permanent advantage that will be derived from the war is that we know far more how man's mind really works.' 'Freud' is thus the only silver lining, the only sign of hope, to be seen when looking back at the Great War.
[287] Richards, *Principles of Literary Criticism*, p. 25.   [288] PC, p. 7n1.

So Richards was aware that he was eliciting the pathological phenomena from his experimental subjects, and that therefore it was appropriate to adopt some of the attitudes and techniques developed by 'alienists'. Our understanding of the protocols will be aided if we adopt 'the alienist's attitude, his direction of attention, his order or plan of interpretation':

> A few touches of the clinical manner will, however, be not out of place in these pages, if only to counteract the indecent tendencies of the scene. For here are our friends and neighbours – nay our very brothers and sisters – caught at a moment of abandon giving themselves and their literary reputations away with an unexampled freedom. It is indeed a sobering spectacle, but like some sights of the hospital-ward very serviceable to restore proportions and recall to us what humanity, behind all its lendings and pretences, is like.[289]

The model is clearly that of the mind-doctor[290] familiar to readers of the New Psychology, that is, the Freudian-inspired large vision of the immediate post-war period: the crafted uncovering, through these moments of abandon when everyone gives themselves away, of their 'latent' common humanity, in normal social or intellectual life embarrassing, shameful, certainly immoral, and even animal.

Was the new Faculty of English, fumbling for a raison d'être and a distinctive new project, thus founded on the importing of the psychoanalytic method into literary studies? The effect on many of Richards's readers might not have been immediate identification with and sympathy for his hapless experimental subjects.

> The scandal Richards produced by printing in his book *Practical Criticism* (1929) the mostly fatuous and mistaken 'protocols' of his students – who loftily patronized Hardy, Lawrence, and Hopkins, misread the texts, registered enthusiastic appreciation for drivel, and in every way showed themselves utterly unequipped to read and judge poems from their own culture – led to a revolution in university teaching.[291]

But the 'revolution' – which the most varied of scholars agree was one – was not transmitted as Richards had intended it, neither in Cambridge, nor elsewhere in Britain, nor even in the rise of the 'New Criticism' in the USA,

---

[289] PC, p. 7n1.

[290] Somewhat conservatively, Richards steers clear of the word 'psychiatrist', never using it in the 1920s; rather, he employs the terms 'alienist' or 'psychologist'. Similarly, the term 'hospital' is never used ('mental hospital' was to become the dominant term only after the Mental Treatment Act (1930)), but rather 'asylums' or that interestingly novel term 'psychological clinic', e.g. 'We cannot reasonably expect diagnosis here to be simpler than it is with a troublesome wireless set, or, to take an even closer parallel, than it is in a psychological clinic' (PC, p. 189).

[291] Vendler, 'The Explorer. Review of J.P. Russo, *I.A. Richards*'.

where Richards's allies and like-minded academics have most readily been identified.[292]

In Cambridge, there was hot dispute about the terms 'literary criticism' and 'criticism' itself – a local version of the decades-long expansion and floating of the term 'criticism' in both Britain and America. Was it a term of description or evaluation? Was it to be contrasted with 'theory' or was it a type of theory? Was 'criticism' to be contrasted with 'historical'? The practice of asking students to 'comment' on 'passages' in their examinations began in Cambridge while Richards was still there, in 1928 (although it only became a core stand-alone part of the curriculum and examinations after the Second World War).[293] But when eventually 'Practical Criticism', as an officially sanctioned term and a title for an examination paper, was introduced, Richards's aim of inducing each student's critical vices – whether perversely personal or dispiritingly commonplace – was no longer part of the pedagogical practice. This was inevitable, since from an examiner's point of view it was a nightmare: the more successful the practice, the greater the number of 'errors' and 'prejudices' induced and revealed. Would this mean a candidate who produced more errors was to be classed higher than one who produced a bland, error-free response?

Most historians of literature would not immediately recognize Richards's methods and their revolutionary outcome as inspired by psychoanalysis, principally because the idea of what literary psychoanalytic criticism looks like has been so dominated by other strands. Basil Willey remembered these glory days of the 1920s, the age of innocence, as prior to any such influence:

---

[292] See Wallace Martin, 'Criticism and the academy', in A. Walton Litz, Louis Menand and Lawrence Rainey (eds.), *The Cambridge History of Literary Criticism*, Vol. VII, *Modernism and the New Criticism*, Cambridge: Cambridge University Press, 2000, pp. 269–321. See also Chris Baldick, *The Oxford English Literary History*, Vol. X, *1910–1940. The Modern Movement*, Oxford: Clarendon Press, 2004, p. 70: 'Richards helped in the promotion and refinement of Basic English, but his more important work as a lecturer in the newly established school of English at Cambridge was the development of a "scientific" modern theory of poetry in *Principles of Literary Criticism* (1924) and its application in *Practical Criticism* (1929). In these works, which laid the foundations of the influential "Cambridge School" of modern critical analysis, he called for and demonstrated a new kind of detailed attention in the reading of poetry to the "words on the page" rather than to a poem's paraphrasable doctrine. As the Cambridge emphasis upon "close reading" of poems evolved in the Thirties, in the work of Richards's student William Empson and of other contributors to F.R. Leavis's journal *Scrutiny*, an inherited "Victorian" conception of poetry as the melodizing of noble sentiments was discarded as a typically moralistic fallacy. The key to poetry was no longer its moral message but its linguistic medium.' Tillyard thought it likely that the introduction of 'Practical Criticism' into Cambridge teaching and its dissemination into schools and worldwide was the 'greatest single achievement' of the Cambridge Faculty and its Tripos.

[293] Thanks to Pete de Bolla, for sharing his research on the early history of courses and examination papers in the English Tripos (email 19 January 2012), on which this is partly based.

of the major corrosives of the century, of Marx and Freud for example, hardly more than a *soupçon* had as yet trickled through into our fool's paradise ... I remember what a joke we all thought it when Ernest Jones psycho-analysed Hamlet, and diagnosed him as a sufferer from something called an Oedipus-complex; I can hear 'Q's' [Quiller-Couch] snort to this day. We none of us looked instinctively, and primarily, for Freudian or sociological explanations of writers or their works.[294]

But such memories were beside the point; and those who look for the influence of psychoanalysis often look for the wrong signs and symptoms. Psychoanalysing artists in the style of Freud's *Leonardo* or of his 'Dostoevsky and parricide' was to be a great temptation to analysts and their epigones. Psychoanalysing Hamlet as if he were a real 'character', a potential patient on the couch, raised equally the hackles of belles-lettrists such as Quiller-Couch and of modernists such as L.C. Knights, an early product, in 1928, of the Richards regime in the new Cambridge Tripos and a right-hand man of F.R. Leavis. In his jauntily entitled essay 'How many children had Lady Macbeth?'[295] Knights attacked the analysis of character. Knights's overt target was A.C. Bradley and a venerable tradition of home-grown Shakespeare criticism, but his covert targets were Freud's essays on *Macbeth* and *Rosmersholm*. Both genres would be endlessly anathematized by literary critics, who could certainly subsume the efforts of Freudians within their victorious critique of the Romantic cult of inspiration and the individual genius.[296] Richards's deployment of psychoanalysis eschewed concepts such as the Oedipus complex or repression, since it never occurred to him that literary criticism should focus on writers in the misery of their lives or their imaginary creations in the glory of theirs; at first glance, practical criticism cuts off and isolates the text itself. But it only does so in order to open up the *reader* and his misreadings for analysis. Like so many of the first uses of psychoanalysis in the early 1920s, its applications were first and foremost *self*-analytic. And this self-analytic mode, far from being moralising or leading to confessional *mea culpa*s, chimed very well with the playful and jocular tone to be found in Richards and Forbes, 'giving the impression that some kind of propriety or convention is being enjoyably flouted ... a characteristic of Cambridge English in the late Twenties'.[297]

---

[294] Willey, *Cambridge and Other Memories*, p. 23.
[295] L.C. Knights, *How Many Children Had Lady Macbeth: An Essay on Theory and Practice of Shakespeare Criticism*, Cambridge: The Minority Press, 1933.
[296] See F.R. Leavis, *The Common Pursuit* (1952), Harmondsworth: Penguin, 1962, p. 183.
[297] E.E. Duncan-Jones, 'The Wizard of Finella. Review, Hugh Carey, *Mansfield Forbes and his Cambridge*, Cambridge: Cambridge University Press, 1984', *London Review of Books* 7(1) (24 January 1985), p. 9.

The sign of the Freudian on whom a veritable mark had been left is in the first instance the enthusiasm with which they analysed their *own* dreams. The next step in bringing psychoanalysis into a wider world was to analyse their professional function in the same spirit. Teachers were to examine their relationship to their pupils, judges were to examine their fantasies about guilt, righteousness and punishment. And literary critics were to examine their own fantasies about texts for signs of the emergence of the perennial professional vices, such as a 'disorder of the self-regarding sentiment – a belated Narcissism, perhaps'[298] or the shadows of the 'Puritans and perverts' which beset a balanced reading.

When Richards approaches theoretical questions concerning interpretation, he is simultaneously wary of the help to be expected from psychology, in particular both psychoanalysis and behaviourism, the extreme and contrasted wings of modern psychology to which, at different times, he had himself been doctrinally drawn, and also certain that the future belongs to the new science: 'it is as certain as anything can be that in time psychology will overhaul most of our ideas about ourselves, and will give us a very detailed account of our mental activities'.[299] Richards's espousal of the psychological is in deed, not word, since the result of his experiments is a veritable catalogue of the many paths by which readers go astray, in each of the principal functions of language: 'a reader garbles the sense, distorts the feeling, mistakes the tone and disregards the intention'.[300] Yet he is well aware that his project bears a close kinship to a psychoanalytic investigation, as the following caveat indicates:

> The indispensable instrument for this inquiry is psychology. I am anxious to meet as far as may be the objection that may be brought by some psychologists, and these the best, that the protocols do not supply enough evidence for us really to be able to make out the motives of the writers and that therefore the whole investigation is superficial. But the *beginning* of every research ought to be superficial, and to find something to investigate that is accessible and detachable is one of the chief difficulties of psychology. I believe the chief merit of the experiment here made is that it gives us this. Had I wished to plumb the depths of these writers' Unconscious, where I am quite willing to agree the real motives of their likings and dislikings would be found, I should have devised something like a branch of psychoanalytic technique for the purpose. But it was clear that little progress would be made if we attempted to drag too deep a plough. However, even as it is, enough strange material is turned up.[301]

---

[298] PC, p. 251.   [299] PC, p. 323.   [300] PC, p. 183.   [301] PC, p. 10.

Without quite declaring it, Richards did indeed devise something rather
like 'a branch of psychoanalytic technique': his defence of these superficial
protocols reveals that he knew it too. It is the fluent ear of the analyst,
listening for the complexities that can give rise to absurdities, from which
we profit throughout his book. The fact that his project is an interpretation
of linguistic *failure*, and that out of this catalogue of failure is required a
new philosophy of interpretation, indicates a clear kinship with the psychoanalytic project, not least since chronic and repeated misinterpretation
is the lot of us all:

> The wild interpretations of others must not be regarded as the antics of
> incompetents, but as dangers that we ourselves only narrowly escape, if,
> indeed, we do. We must see in the misreadings of others the actualisation of
> possibilities threatened in the early stages of our own readings. The only
> proper attitude is to look upon a successful interpretation, a correct under-
> standing, as a triumph against odds. We must cease to regard a misinter-
> pretation as a mere unlucky accident. We must treat it as the normal and
> probable event.[302]

The profit to be had from the experiment is in the unravelling of the
complex mental processes by which readers go astray. To give some idea of
Richards's methods, take his discussion of sentimentality.

> Most, if not all, sentimental fixations and distortions of feeling are the result
> of inhibitions ... If a man can only think of his childhood as a lost heaven it
> is probably because he is afraid to think of its other aspects. And those who
> contrive to look back to the War as 'a good time' are probably busy dodging
> certain other memories.[303]

But Richards is not going to advocate complete release of inhibitions,
which are necessary for any kind of order in experience: 'the opinion
sometimes emitted that all inhibition (or repression) is bad, is at the least
an overstatement'; nonetheless, our 'mental health' requires us to envisage
aspects of experience which inhibitions too frequently 'blank out'.

> As a rule the source of such inhibitions is some painfulness attaching to the
> aspect of life that we refuse to contemplate ... The man who, in reaction to
> the commoner naïve forms of sentimentality, prides himself upon his hard-
> headedness and hard-heartedness, his hard-boiledness generally, and seeks
> out or invents aspects with a bitter or squalid character, for no better reason
> than this, is only displaying a more sophisticated form of sentimentality.[304]

[302] PC, p. 336.   [303] PC, p. 267.   [304] PC, p. 268.

The internal dynamic by which sentimentality mutates into cynicism takes many forms: here is one remarked upon in an individual protocol: 'As so often happens the reader's own revulsion at his own devious excesses is counted against the poet.'[305] A later psychoanalytic reading would be equally accurate, if perhaps less informative, in designating this process 'projective identification'. What is not in doubt is that Richards is firmly on the terrain mapped out, if not yet fully charted, by psychoanalysis.

It was the quasi-psychoanalytic innovation of practical criticism that was Richards's longest-lasting contribution to the field of literary studies that he had happened into.[306] He was perhaps one of the last of the moral scientists as well as the first principal of Cambridge 'Eng. Lit.' or 'the Cambridge School of English'. His colleague Tillyard attributed to Richards 'leadership and a policy' for the new discipline. He also judged that Richards's background and persistent preoccupation with Cambridge psychology had a twofold consequence:

> First through Freudian insistence on underlying motives it reinforced the influence of Moore in seeking multiple meanings in literature. And, second, it led Richards to attempt a change parallel to the change he attempted in the criticism of texts: a change from the old *a priori* type of aesthetics to the scientific examination of what actually went on in the mind and also in the body under the impact of a piece of literature.[307]

Whether or not Richards would have assented at the time to this characterization, it mirrored Tillyard's own approach in this field, which was as new to him as to Richards, as he remembered the glory days of the 1920s looking back from the 1950s:

> In sum what I then represented was an enthusiasm for the new academic liberty, an instinctive reaching after practical criticism, a dislike of the expansive and the florid in critical method, and a distrust of unmitigatedly mystical theories of literature. When it came to the type of practical criticism to which I tended, I expect I was influenced by the sort of Freudianism that was then in the air. As there is a mind below the surface mind, so, if you begin to dwell intently on the richer kind of poetry, something more than a surface meaning is likely to be revealed. And that other meaning was likely to be the more important.[308]

---

[305] PC, p. 88. Baldick characterizes this passage as 'amateur-psychoanalytic conjecture' (*The Oxford English Literary History*, p. 153).

[306] 'the practical criticism that is the most noteworthy innovation of academic criticism in the first half of the century' – Martin, 'Criticism and the academy', p. 269.

[307] Tillyard, *The Muse Unchained*, p. 90.   [308] *Ibid.*, p. 88.

Richards may have been the brains and the hands behind Cambridge English, but by the end of the 1920s he was widening his horizons beyond Cambridge; he left Tillyard and others to run the show and write its, which was also his, history.[309]

Richards had come to Cambridge in 1911 to read history, but very quickly realized he hated history – he told his history tutor that he 'didn't think History ought to have happened'.[310] Richards's dislike of history – only compounded by the insane slaughter of the war – was twinned with an optimistic vision of the future. Even *Practical Criticism*, for all its exclusion of all historical markers in order to produce a blank Rorschach test for the diagnosis of 'stock responses', for all its revelations of the ubiquitous pathology of reading, included a redemptive strand, one derived less from arguments concerning the inherent worth of 'great literature', or more specifically 'poetry', than from the purification process that the study of literary criticism could promote in readers. This purification was individual, in response to the diagnosis of a reader's failure; it was a matter of mental hygiene. In Richards's later work, this redemptive strand would find much freer exercise through the development of his preoccupation with a general theory of meaning in the purification of language and all its dialects, starting with Basic English.

In 1934, Richards was commissioned to provide a follow-up to *Practical Criticism*, dealing with prose instead of poetry. He conducted the 'research' for this study in Cambridge, in 1935, in a course entitled 'Practical Criticism – Prose', deploying a set of protocols consisting of excerpts from past critical writings.[311] He completed the book based on this research in 1936; it was published as *Interpretation in Teaching* in 1938. The second edition of 1973 gave Richards an opportunity to write a prefatory 'Retrospect' on the project whose fruit was *Practical Criticism* and *Interpretation in Teaching*, and whose aim had been the '[b]etter diagnosis of the causes of our manifold defects in comprehension, more efficient design in forfendings and remedies':

> this is the beginning of what should be a vast collective *clinical* study of the aberrations of average intelligence, and is duty bound to adduce its evidence as much as, say, any psychoanalyst's case-book. It is an extensive piece of

---

[309] Tillyard's 'gut' Freudianism must have been well known in the 1950s, since his name was mentioned (along with those of James Strachey, Cyril Connolly and Erwin Stengel) in a letter in early 1956 from the *British Medical Journal* consulting Ernest Jones about a proposed article on the Freud Centenary; see BPaS, CBC/F12/05, 17 January 1956. In the event, the *BMJ* commissioned Jones and Stengel to contribute, alongside a judiciously enthusiastic editorial leader.
[310] Russo, *I.A. Richards*, p. 35.   [311] *Ibid.*, p. 412.

natural history, collecting, arranging, displaying and hazarding observations upon its selected specimens.

So, fifty-five years after Richards had started training as a psychoanalyst, he would still align the project of diagnosing and curing the defects of understanding evident amongst readers and teachers of poetry and prose with the work of the psychoanalyst. Richards would always remain, unusually for his generation of students of literature, open to the ideal and model of the sciences; for him, the original model of a science oriented towards literature had been Cambridge moral science. Thus, when Collini characterizes one of the principal features of the Cambridge School of English from the 1920s to the 1960s as the conviction that literary criticism involves the whole being, that it is a 'moral' activity,[312] Richards's contribution to this underlying assumption is that this 'moral' activity derives from the 'moral' scientific project – and therefore in the starkest contrast to the later Leavisian understanding, in which 'moral' slides ineluctably towards 'moralizing'. Richards's great innovation of setting up 'practicals' for literature students was crucially inflected by the model of the psychoanalyst, seeking to find remedies in the exposure of aberration, of mistakes, of the ordinary and ubiquitous human error – not, make no mistake, just plain error, but, in the spirit of psychoanalysis, *motivated* error.

The Moral Sciences, Philosophy and Psychoanalysis after the Great War

The separation of philosophy and psychology is one of the fundamental givens, accepted by all historians, as a necessary condition for the emergence of a distinct discipline of psychology. But the story is usually told as the separation of psychology *from* philosophy – as if 'philosophy' were already a defined and stable entity. The processes by which philosophy disentangled itself from 'psychology' were complex, involving the fate of German idealist philosophy in Britain and the rise of the antagonistic school of analytic philosophy (the new realism, logical empiricism); nor should one underestimate the long-term effect of the hostility to most things German engendered by the First World War.[313]

There is one figure who embodies all these movements in philosophy: the abrupt turn away from idealist philosophy; the turn to logic and

---

[312] Collini, 'The study of English', p. 62.
[313] Stuart Wallace, *War and the Image of Germany: British Academics 1914–1918*, London: John Donald, 1988.

mathematics as the key problems – and solutions – to philosophy; the entanglement with pacifism and the sacrifice (willing and unregretted) of an academic philosophical career to violent chauvinistic currents of opinion during the Great War: Bertrand Russell. What is perhaps less well known is that Russell also had his encounter with Freudian psychoanalysis, just as he was detaching himself from the Cambridge world of his early maturity.

Bertrand Russell

Russell came up to Cambridge in 1890 to study mathematics. Disappointed by the overly technique-oriented teaching, after a First in Mathematics Part I he turned to the MST, gaining a starred First in Part II in 1894. He was elected to a Fellowship at Trinity College on the basis of a dissertation on the philosophical foundations of geometry. As an undergraduate, his most important discovery was companionship, through the Apostles but also in a close friendship with A.G. Tansley. A man of supreme intelligence, powerful passions and a capacity for unrelenting work, Russell put himself at the service of one of the most ambitious and unworldly of philosophical projects: the attempt to ground the entirety of mathematics upon philosophical logic. From his earliest years in search of certain foundations of mathematics, he came under the influence, successively, of Kant, Hegel and Leibniz until shaken out of 'idealist' system-building by the influence of the young G.E. Moore and his encounter with the new logical symbolism of Peano. From 1900 to 1910 he devoted himself, together with his former teacher A.N. Whitehead, to the project of showing that logic and mathematics are identical, first in *The Principles of Mathematics* (1903) and then in the *Principia Mathematica* (1911–13).

This period of his career, from 1898 to 1913, has been seen by many as the moment of the creation of 'analytic philosophy', the dominant form of philosophy within English-speaking universities throughout the twentieth century – professionalized philosophy.[314] Yet, after his prodigious labours of the first decade of the century, Russell was to leave academic philosophy: partly through intellectual exhaustion; partly when his failed marriage to Alys Pearsall Smith of 1894 – a cold, empty husk (Russell) after 1902 – finally collapsed in 1911 when he started an intense love affair with Ottoline

---

[314] Historians point back to assumptions such as Moore's 'A thing becomes intelligible first when it is analysed into its constituent concepts' (1898) as initiating the programme. The term 'analytic philosophy' began to come into use in the early 1930s (see the end of this section for further discussion).

Morrell; partly because Russell had always wished not only to establish truth with absolute certainty but also to change the world through political writing and action. With the start of the war in August 1914, Russell threw himself into action to change the world – most immediately to campaign against the war and resist the encroachments on freedom that the Warfare State implemented. Russell emerged as a prominent critic and opponent of the Government, acknowledged as such by the attention they paid him – restricting his freedom of movement, censoring his writings – and through two successful prosecutions, one in July 1916 under the Defence of the Realm Act for 'impeding recruiting and discipline' by publishing a pamphlet criticizing the controversial conscription laws introduced by the War Government; the second in 1918, for an inflammatory editorial about the British and American conduct of the war, which landed him with a six-month prison sentence. As a result of his first conviction, Trinity College's Council immediately voted to deprive him of his post, much to the horror of some of them, such as his close friend the mathematician G.H. Hardy.

Yet Russell was already out of love with the donnish world of Cambridge, as a result of his tempestuous and passionate affair with Ottoline, as a result of his drive to change the much larger world of war and injustice, but also in 1915 through an intense short-lived friendship with D.H. Lawrence. Lawrence ended up battering Russell for his dry intellectuality and personal hypocrisy, for the life-starved political views he advanced, for his overwhelming desire to 'jab and strike, like the soldier with the bayonet, only you are sublimated into words . . . You are simply *full* of repressed desires, which have become savage and anti-social [a]nd they come out in this sheep's clothing of peace propaganda . . . too full . . . to be anything but lustful and cruel.'[315] Keynes, over twenty years later, imagined how Cambridge then appeared in Lawrence's eyes, 'this thin rationalism skipping on the crust of the lava, ignoring both the reality and the value of the vulgar passions'.[316] Lawrence's exhortations to 'break his hard shell' meant, for Russell, a repudiation of Cambridge's rationalism and cynicism. He did break with Cambridge, he did recognize in his own way the volcano of the vulgar passions, but he refused to become a disciple to Lawrence's *Lebensphilosophie*. Years later, he diagnosed Lawrence as a visionary proto-fascist, under the spell of a woman infused with psychoanalysis:

[315] DHL to BR, 14 September 1915, quoted in MBR I, p. 426.
[316] John Maynard Keynes, 'My early beliefs', in *The Collected Writings*, Vol. X, p. 434, quoted in MBR I, p. 408.

Lawrence, though most people did not realize it, was his wife's mouth-piece. He had the eloquence, but she had the ideas. She used to spend part of every summer in a colony of Austrian Freudians at a time when psycho-analysis was little known in England. Somehow, she imbibed prematurely the ideas afterwards developed by Mussolini and Hitler, and these ideas she transmitted to Lawrence, shall we say, by blood-consciousness.[317]

Lawrence's utopian schemes – to establish a new community on a South Sea Island or in Florida – fleetingly captured Russell's imagination, and left him with a deepening conviction that the problems of politics and society were psychological in nature.[318] So, via Lawrence, via the necessity for rethinking his own understanding of human nature, Russell was moving towards a fundamental reappraisal of the significance of 'psychology'.

Russell cut most of his ties to Cambridge just as the interest in psycho-analysis began to change from a significant academic and medical influence to something more like a psychic epidemic. Nonetheless, it is worth considering Russell's response to Freud. It turns out he was a bellwether: the very broadening of his activities, into politics, social reform and education, parallels closely some of the many areas in which psychoanalysis was to have significant influence. After the intense friendship with Lawrence abated, 1916 was a year of whirlwind romantic activity as well as public martyrdom by the criminal courts. The beginning of 1917 was perhaps the low point of the war for him as for many; and then the February Revolution in Russia brought him back to political life and frenzied activity. In May 1917 he wrote to his mistress Colette:

> I am reading Freud on dreams, most exciting – I see in my mind's eye a great work on how people come to have the opinions they have – interesting scientifically, & undermining *ferocity* at the base (*unmasking*, I ought to have said) – because it is always hidden behind a veil of morality. The psychology

---

[317] BRA II, p. 23. Lawrence scholars agree that Frieda provided one, perhaps the main, source for Lawrence's knowledge of Freud, which appears to have been always at second hand; see Mark Kinkead-Weekes, *D.H. Lawrence: Triumph to Exile 1912–1922*, Cambridge: Cambridge University Press, 1996. Russell's relationship with Lawrence spanned 1915; in 1916, Lawrence and Frieda (née Frieda von Richthofen) were close to Ernest Jones, involving him in their turbulent marriage. Jones knew Frieda from his time in Munich when he first encountered psychoanalysis in the flesh, as it were, with Frieda's lover Otto Gross, and was having an affair with Gross's wife, née Frieda Schloffer; no doubt Jones gave them his authoritative version of Freud and Freudian theory. See Maddox, *Freud's Wizard*, pp. 121–3 and Ken Robinson, 'A portrait of the psychoanalyst as a Bohemian: Ernest Jones and the "lady from Styria"', *Psychoanalysis and History* 15(2) (2013), 165–89.

[318] MBR I, p. 401.

of opinion, especially political opinion, is an almost untouched field – & there is room for really great work in it – I am quite excited about it.[319]

The slip in this letter is informative: Russell had quickly taken the step from the 'unmasking' method of Freud and psychoanalysis to its hoped-for effect – the 'undermining' of the 'vulgar passion' of 'ferocity'. Tear away the veil of morality, unmask the ferocity that underpins it, and civilization can be restored. Russell had a new vision, a new project: a psychology of public opinion.

But what psychology would serve Russell's purpose? In fact, Russell did not develop his new project of a psychology of public opinion or even a psychology of political motives. Under the influence of Freud, many of those writing in the late 1910s and 1920s, amongst them J.D. Bernal and Kingsley Martin, viewed such a project as a matter of urgency, but none fully delivered on it. One complicating factor in assessing its scope in Russell's case stems from the fact that his programme in philosophical logic – the programme on which analytic philosophy would be based – had explicitly excluded 'psychology' from having any relevance to philosophy.[320] But the twists and turns of Russell's philosophical development in the period 1910–18 – trying to address the problems of the paradoxes, of the relation of propositions to 'objects', trying to escape from the searching criticism of Wittgenstein – had led him to reconsider the relation of psychology and philosophy. Under some interpretations, the question of 'meaning' could not be excluded from logic. Could an account of 'meaning' be given that was not psychological? How did logic 'attach' itself to the world? Russell's initial impetus had been a profoundly Platonic one: it is even plausible to describe his long philosophical career as starting with a new, a modernist, Platonism – which in the early years of the century he and others called the 'new realism' ('realism' in the medieval sense, contrasted with nominalism, referring to the reality of 'universals') – followed, through internal contradictions and the pressure from Wittgenstein, by a

---

[319] MBR I, pp. 495–6. The editors of BRCP (15, p. 472) note that Russell's library contains Freud's *The Interpretation of Dreams* in the Allen & Unwin edition (2nd edition, printed in 1915, unchanged reprint of the Brill translation first published in 1913). The editor of Russell's Letters (Nicolas Griffin, *The Selected Letters of Bertrand Russell*, Vol. II, *The Public Years 1914–1970*, Routledge: 2002, p. 108) reckons that this was Russell's first reading of Freud.

[320] See Martin Kusch, *Psychologism: A Case Study in the Sociology of Philosophical Knowledge*, London: Routledge, 1995. The exclusion of psychological considerations is regarded by many philosophers as the distinctive foundational position of modern analytic philosophy (pointing to Frege as setting out the necessity and the arguments for this exclusion). G.E. Moore's critical deployment of the term 'naturalistic fallacy' also had psychological accounts of the mind as one target. But 'psychologism' has kept on returning, just as has naturalism, much to the horror of many.

slow yet inexorable 'retreat from Pythagoras'.[321] Beginning to take psychology seriously, around 1913, provided Russell with one more possible solution to the problems of logic. When he was in prison in the summer of 1918, making full and fruitful use of all this 'free time' (placed in the luxury of the First Division, courtesy of Arthur Balfour[322] – aristocratic and Cambridge connections were still intact), he returned to philosophy proper, completed a book (*Introduction to Mathematical Philosophy*, writing 10,000 words a day for one week) and set out the skeleton of a major new project, based on a stupendous amount of reading of recent psychology and philosophy: a theory of language, logic and psychology that built upon both his own work in mathematical logic and that of behaviourists in psychology.[323] In the process, he rejected outright the behaviourist embargo on any mental contents, insisting on the reality of 'images' and in the process developing something remarkably like Wittgenstein's later (*Tractarian*) 'picture theory of meaning'.

Despite Russell's new enthusiasm for American philosophy and psychology – William James, John Dewey and elements of pragmatism, J.B. Watson's behaviourism – he was also delving into psychoanalytic writings: 'James's attack on "consciousness", the study of animal behaviour, and Freud, all naturally belong together.'[324] While in prison he drafted a manuscript entitled 'Morality and oppressive impulses', a response to reading Stanley Hall's 'The Freudian methods applied to anger' (1915) – eventually partly published by Ogden in the *Cambridge Magazine*.[325] A companion piece, 'Dreams and facts', published in the *Athenaeum* in April 1919, gives a full picture of Russell's engagement with Freud at this time:

> Man is essentially a dreamer, wakened sometimes for a moment by some
> peculiarly obtrusive element in the outer world, but lapsing again quickly
> into the happy somnolence of imagination. Freud has shown how largely
> our dreams at night are the pictured fulfilment of our wishes; he has, with an
> equal measure of truth, said the same of day-dreams; and he might have
> included the day-dreams which we call beliefs.

[321] 'My philosophical development, since the early years of the present century, may be broadly described as a gradual retreat from Pythagoras' – the opening sentence of *My Philosophical Development* (1959). In 1912, C.K. Ogden orchestrated a Heretics debate on the 'new realism' between C.S. Myers in the role of opponent, G.E. Moore and Bertrand Russell as commentators, and G.H. Hardy responding ('On the new, new realism'). Russell made clear the relation between the 'new realism' and Platonism in the 'Introduction' added to the second edition (1937) of *The Principles of Mathematics* (1903): 'Broadly, the result [of developments in logic since 1903] is an outlook which is less Platonic, or less realist in the mediaeval sense of the word. How far it is possible to go in the direction of nominalism remains, to my mind, an unsolved question' (p. xiv).

[322] BRA I, p. 242.   [323] MBR I, p. 534.

[324] BRCP 8, p. 276, Russell to Gladys Rinder, 17 June 1918.   [325] *CM*, 1 February 1919, p. 359.

There are three ways by which this non-rational origin of our convictions can be demonstrated: there is the way of psycho-analysis, which, starting from an understanding of the insane and the hysterical, gradually makes it plain how little, in essence, these victims of malady differ from ordinary healthy people; then there is the way of the sceptical philosopher, showing how feeble is the rational evidence for even our most cherished beliefs; and finally there is the way of common observation of men. It is only the last of these three that I propose to consider.[326]

In fact, rather than being a distinct science, theory or body of facts, Russell's appreciation of Freud was seamlessly folded into his own common observation of men. This serious reading of psychology was to have important implications for his philosophical programme. He brought his concerns about the status of logic and his readings in psychology together in a series of lectures at Dr Williams's Library, 14 Gordon Square, in May and June 1919 which became the must-show event of the Bloomsbury season:[327] these he published, after some revision, as *The Analysis of Mind* in 1921. The book clearly spoke to the general intellectual excitement over psychology in the immediate post-war period; it also was his most extended discussion of his final (so many historians claim) important contribution to philosophy – the theory of neutral monism.

This theory, inspired by William James, rejected both materialism and idealism for a mid-way position: 'the "stuff" of the world is neither mental nor material, but a "neutral stuff," out of which both are constructed. I have endeavoured in this work to develop this view in some detail as regards the phenomena with which psychology is concerned.'[328] A prominent theme of the book was an attack on the importance of consciousness, mounted on a number of fronts: on the one hand deploying 'analysis' (by which Russell meant philosophical analysis); on the other, exploring the findings of the psychologists, principally the psychoanalytical findings, which Russell was prone to re-interpret along behaviourist lines,[329] and the behaviourists' findings concerning animal behaviour. Both analysis and

---

[326] Russell, 'Dreams and facts' (1919), *The Athenaeum* 4,642–3 (18 and 25 April 1919), BRCP 15, pp. 13–19 at 14.

[327] 'The touchstone of virtue with (the select Norton, Alix & James) now is whether you attend Bertie's lectures or not' (VWD I, 16 May 1919, p. 273).

[328] Bertrand Russell, *The Analysis of Mind*, London: George Allen & Unwin, 1921, p. 6.

[329] Russell's critical comments about the anthropomorphic tendencies of psychoanalytic theory and vocabulary were eagerly taken up by psychoanalytic critics, e.g. in an 'Editorial: The nature of desire', *Journal of Neurology, Neurosurgery, and Psychiatry* 3(11) (1922), 274–6. Ernest Jones felt called upon to rebut the Russellian view by arguing that the psychoanalytic theory of unconscious desire was entirely in conformity with Russell's views (Ernest Jones, 'The nature of desire', *Journal of Neurology and Psychopathology* 3 (1922), 338–41).

empirical findings put in doubt common-sense views of the mind; the
findings of the psychoanalysts are brought to bear in particular upon the
relations of desire to consciousness.

> The first set of facts to be adduced against the common sense view of desire are
> those studied by psycho-analysis. In all human beings, but most markedly in
> those suffering from hysteria and certain forms of insanity, we find what are
> called 'unconscious' desires, which are commonly regarded as showing self-
> deception. Most psycho-analysts pay little attention to the analysis of desire,
> being interested in discovering by observation what it is that people desire,
> rather than in discovering what actually constitutes desire . . . But the deeper
> the Freudians delve into the underground regions of instinct, the further they
> travel from anything resembling conscious desire, and the less possible it
> becomes to believe that only positive self-deception conceals from us that we
> really wish for things which are abhorrent to our explicit life.[330]

And elsewhere in the book he could blithely lay down as certain new
knowledge what psychoanalysis had brought to science: 'dreams, as Freud
has shown, are just as much subject to laws as are the motions of the
planets'.[331] Overall, psychoanalysis played a supportive but subordinate
role in the philosophy of mind that Russell was developing, as he moved
out from logic to other traditional areas of philosophy.

More generally, though, Russell imbibed the resources that Freud
offered to 1920s intellectuals. His visit to Bolshevik Russia in 1920 horrified
him: instead of a realization of his ideals of Guild Socialism, he found a true
dictatorship of the Communist Party and returned convinced for good of
the fundamental virtues of freedom embodied in classic liberalism, while
now certain, like so many others in the 1920s, that socialism was the only
viable economic system. He had become convinced that Bolshevism was,
like liberalism and capitalism, premised on an erroneous psychology, the
psychology of rationalism: 'There is need of a treatment of political
motives by the methods of psychoanalysis. In politics, as in private life,
men invent myths to rationalise their conduct.'[332] Despite his repudiation
of Bolshevik Russia, so out of line with the views of so many of his friends,
including Dora Black whom he would marry in 1921, Russell could still
reach out for the formula, which would soon become quite standard, of
combining Freud and Marx in order to address a whole host of political,
social and ethical questions; in a book review of 1922, he wrote:

---

[330] Russell, *The Analysis of Mind*, pp. 59–60.   [331] *Ibid.*, p. 138.
[332] Bertrand Russell, *The Practice and Theory of Bolshevism*, London: George Allen & Unwin, 1920, p.
84, quoted in MBR I, pp. 586–7.

Is there, in any objective sense, such a thing as morality or immorality? Or is the conception of morality merely part of the police force by which dominant groups seek to enforce their authority? The combination of psychoanalysis with Marxian political theory has forced this question insistently upon many people. Psychoanalysis shows that the basis of a passion is by no means always, or even usually, what the patient thinks it is; and Marx suggests that all morality is derived from class-interest.[333]

Here Russell is reaching for psychoanalysis as a method and theory which documents the gulf between consciousness and mental reality, between the manifest and the latent, and, like Marxism, postulates the underlying reality which can be accessed, can be given its explanatory weight and perhaps even, as a consequence, can be transformed. Russell is in the grips of the explanatory machine that Paul Ricoeur would later describe as the 'hermeneutics of suspicion' derived from Marx, Freud and Nietzsche and which would be so tempting to intellectuals and theorists in the period from 1920 on. This is nowhere better expressed than in his defence of his own theories against the criticism that they were atavistic and pre-Kantian:

> I dislike the heart as an inspirer of beliefs; I much prefer the spleen. I take comfort in Freud's work, because it shows what we are to think of the heart, which, he says, makes us desire the death of our parents, and therefore dream that they are dead, with a hypocritical sorrow in our very dreams. The heart is the cause of the anti-rational philosophy that begins with Kant and leads up to the 'will to believe'. The heart is the inspirer of atrocities against negroes, the late war, and the starvation of Russia.[334]

Russell followed his disillusioning visit to Bolshevik Russia with a long visit to China, accompanied by Dora Black. He was enchanted by the traditional civilization of pre-industrial China, and this marked a turning-point in his life: while he was on what everyone took to be his deathbed with double pneumonia, Dora discovered she was pregnant. He recovered; they returned via Canada to Britain, he divorced, they married and in November 1921, nearly fifty years old, he became a father for the first time. '[D]uring the next ten years my main purposes were parental.'[335]

During the First World War, Russell had established what would eventually be a very long-standing affiliation with the publisher Stanley Unwin. However, in the crucial period that saw Russell definitively decide

---

[333] Russell, 'What is morality? Review of B.M. Laing, *A study of moral problems*, London: George Allen & Unwin, 1922', *The Nation and Athenæum* 32 (11 November 1922), 254–5, in BRCP 9, p. 348.
[334] Bertrand Russell, 'Dr. Schiller's analysis of the *Analysis of Mind*', *Journal of Philosophy* 19(24) (1922), 645–51 at 645.
[335] BRA II, p. 150.

Fig. 5.6 Bertrand Russell with pram, *c.* 1921.

to make his living out of his writing and not consider any appointments at a university – 'writing potboilers' as he called it[336] – it was C.K. Ogden who acted as his agent and publisher, thereby significantly affecting the direction of Russell's work. Ogden had been a long-standing friend of both Bertrand and Dora, who had been Secretary of the Heretics in 1918 and 1919;[337] in 1922 he persuaded Russell to write a popular science book, *The ABC of Atoms* (1923), based on a series of articles Russell had written in early 1923, thus inaugurating a series of 'ABC' books published by Kegan Paul, Trench, Trubner.[338] Due to its success Russell later wrote another for the series, *The ABC of Relativity* (1926). At the same time, Ogden inaugurated the influential To-Day and To-Morrow series, with J.B.S. Haldane's *Daedalus* (1924) and persuaded Russell to deliver a rapid response based in part on a Fabian Society lecture Russell had given in October 1923 entitled 'The effect of science on social institutions'. The new version, finished in a week, was entitled *Icarus*.[339] The following Christmas (1924), both Dora and Bertrand Russell wrote books for Ogden's series, hers, *Hypatia*, a fiery defence of socialist feminism and impassioned advocacy of birth control and sexual freedom, his *What I Believe* (one of his most often reprinted texts), where he wrote: 'The good life is one inspired by love and guided by knowledge.' By knowledge Russell meant science; and throughout the 1920s, psychology, or the promise of psychology, was a crucial element of the scientific knowledge which Russell believed should be the guide through life. In 1923, Ogden also persuaded Russell to commit himself to delivering the second half of his diptych on analysis, though it took him till 1926 to deliver: *The Analysis of Matter*, published in Ogden's International Library. Ogden was therefore the principal midwife for Russell's transformation from esoteric logician and fiery political campaigner to essayist, popular science writer and social critic.

However, the two books written in the late 1920s which sold best and secured Russell's finances for some time were not written for Ogden: *On Education: Especially in Early Childhood* (1926) and *Marriage and Morals* (1929), explicitly cited when he was awarded the Nobel Prize for Literature in 1950. Both books were devoted to topics for which psychoanalytic thought was relevant.[340] Russell wrote *On Education* over the summer of

---

[336] BRA II, p. 190.

[337] T.E.B. Howarth, *Cambridge between Two Wars*, London: Collins, 1978, p. 52.

[338] Each book in the series was entitled 'The ABC of . . .'; the eventual list of titles, published between 1923 and 1932, was: Atoms, Relativity, Jung's Psychology, Adler's Psychology, Nerves, Psychology, Biology, Basic English. Ogden himself wrote *Psychology* (1929) and *Basic English* (1932).

[339] MBR II, p. 28.   [340] BRA III, p. 502.

1926, a summer when life with Dora and his two young children, John (nearly 4¾) and Kate (2½) was at its most idyllic and care-free.[341] Indeed, nearly all of Russell's writings on 'education' are transparently rooted in his own personal preoccupations of the time: child psychology and the nature of the modern family.

On Education is about the making of character in early childhood. Russell relied, as he had done ever since his voracious prison reading in 1918, on an eclectic blend of Watson's behaviourism and psychoanalysis to provide the underpinnings for his optimistic view of how a scientific approach to child-rearing could bring in a new utopia, in which could flourish the 'four characteristics which seem to me jointly to form the basis of an ideal character: vitality, courage, sensitiveness, and intelligence'.[342] Russell's persistent high evaluation of the importance of modern scientific psychology both for human knowledge and for social reform had already been expounded in a lecture to the 1917 Club on 16 October 1925:

> Given a science equally definite, and capable of altering man directly, physics would be put into the shade. This is what psychology may become. Until recent times, psychology was unimportant philosophical verbiage – the academic stuff that I learnt in youth was not worth learning. But now there are two ways of approaching psychology which are obviously impor- tant: one that of the physiologists, the other that of the psycho-analysts. As the results in these two directions become more definite and more certain, it is clear that psychology will increasingly dominate men's outlook.[343]

Again taking up the contrast and conjunction of Marx and Freud, Russell declares his inclinations towards the latter:

> As against orthodoxy and Marxianism, the psycho-analysts say that the one fundamental human impulse is sex. Acquisitiveness, they say, is a morbid development of a certain sexual perversion ... If people think wealth constitutes happiness, they will not act as they will if they think sex the essential thing. I do not think either view quite true, but I certainly think the latter the less harmful ... In such important acts as choosing a career, a man is greatly influenced by theory. If a wrong theory prevails, successful men will be unhappy, but will not know why. This fills them with rage, which leads them to desire the slaughter of younger men, whom they envy unconsciously. Most modern politics, while nominally based on economics,

---

[341] MBR II, p. 57.
[342] Bertrand Russell, *On Education, Especially in Early Childhood* (1926), London: George Allen & Unwin, 1966, p. 41.
[343] BRCP 9, pp. 356–7.

is really due to rage caused by lack of instinctive satisfaction; and this lack, in turn, is largely due to false popular psychology.

Russell is here continuing his analysis of the war, using Freudian psycho-analysis in his usual comic and vulgarly direct way, as one more weapon against his opponents, one more tool to understand the downfall of civilization. However, he both promotes this sex-based theory of politics, wheeling out sex-starved spinsters for his purpose, and then downplays it:

> I do not think that sex covers the ground. In politics, especially, sex is chiefly important when thwarted. In the war, elderly spinsters developed a ferocity partly attributable to their indignation with young men for having neglected them. They are still abnormally bellicose. I remember soon after the Armistice crossing Saltash Bridge in the train, and seeing many battleships anchored below. Two elderly spinsters in the carriage turned to each other and murmured: 'Isn't it sad to see them all lying idle?' But sex satisfied ceases to influence politics much. I should say that both hunger and thirst count for more politically. Parenthood is immensely important, because of the importance of the family; Rivers even suggested that it is the source of private property. But parenthood must not be confounded with sex.[344]

Yet Russell now does link parenthood and sex closely together as the future foundations of universal happiness: the psychological understanding of the primordial importance of these instincts will lead to a transformed politics and a revolutionized society:

> If people were genuinely happy, they would not be filled with envy, rage and destructiveness. Apart from the necessities of life, freedom for sex and parenthood is what is most needed – at least as much in the middle class as among wage-earners. It would be easy, with our present knowledge, to make instinctive happiness almost universal, if we were not thwarted by the malevolent passions of those who have missed happiness and do not want any one else to get it.[345]

These themes are continued in his book *On Education*:

> A community of men and women possessing vitality, courage, sensitiveness and intelligence, in the highest degree that education can produce, would be very different from anything that has hitherto existed. Very few people would be unhappy. The main causes of unhappiness at present are: ill health, poverty and an unsatisfactory sex life. All of these would become very rare.[346]

[344] BRCP 9, p. 359.   [345] BRCP 9, p. 362.   [346] Russell, *On Education*, p. 46.

Absolutely committed to the enormous influence of early childhood on character development, jointly attributed to behaviourism and psycho-analysis, he blithely extracted the sensible parts of Freud's work while dismissing others:

> Some psycho-analysts have tried to see a sexual symbolism in children's play. This, I am convinced, is utter moonshine. The main instinctive urge of childhood is not sex, but the desire to become adult, or, perhaps more correctly, the will to power ... In [my son's] play, there was a sanguinary outbreak of cutting off ladies' heads. Sadism, Freudians would say; but he enjoyed just as much being a giant who ate little boys, or an engine that could pull a heavy load. Power, not sex, was the common element in these pretences.[347]

So, at times discounting sex as the source of any troubles, he alighted on fear as the greatest obstacle to trouble-free character development; it was here that the influence of Watson's behavioural conditioning was greatest. As so often throughout his work, Russell was both an enthusiastic advocate for the enormous powers unleashed by science for command of nature, here human nature, while on other occasions warning (most clearly and apocalyptically at the end of his life in relation to nuclear weapons) of the power of science and technology to destroy civilization and even human life. Science enthusiast on the one hand, technological dystopian on the other.

With their glittering wit, with his ready aptitude to shock by deploying his razor-sharp logic to arrive at apparently – and often truly – absurd conclusions, Russell's writings delight and appal. Take his eccentric adap-tation of psychoanalytic technique for the 'treatment' of a boy's interest in indecent pictures: some boys may have such an interest even though they have been instructed openly and decently about sex; if so, the boy should be 'treated by a doctor skilled in these matters. The treatment should begin by encouraging him to utter freely even his most shocking thoughts, and should continue with a flood of further information, growing gradually more technical and scientific, until the whole matter bored him to extinc-tion. When he felt that there was nothing more to know, and that what he did know was uninteresting, he would be cured.'[348] Russell takes the idea of free association, of allowing the most shocking the freedom to be expressed, but then slips without warning into conceiving of psychoana-lytic treatment as a cure through boredom: a didactic lecture, endless and

---

[347] *Ibid.*, pp. 68–9.   [348] *Ibid.*, p. 132.

ever more technical, bulldozing the boy out of his preoccupation with the sexually transgressive.

At Christmas 1926, after Russell had completed *On Education*, Dora and he decided to open a school, because they wished to create the perfect environment for the schooling of their own two children through putting into practice their educational ideals. John had just turned five and the question of his schooling was becoming pressing. Renting from his brother Frank Telegraph House, a grand country house on the South Downs in West Sussex, they opened Beacon Hill School in autumn 1927. Part of the movement for progressive education which was so active in the 1920s, whose best-known example was A.S. Neill's Summerhill and whose closest-to-hand example was the Malting House School in Cambridge, the Russell school would, over its fifteen years of existence, depend very much on the Russells' own family and friends for its children and its survival.[349] Dora and Bertrand's marriage came to pieces only three or four years after the opening of the school, with Bertrand withdrawing entirely from the school's running by 1932. It was Bertrand's series of money-making American lecture tours that supported the school in the 1920s and early 1930s, together with his book royalties and incessant journalism.

The aim of the school was to protect their and other children from the distorting effects of conventional education: 'I am every day more convinced that people who have the sort of ideas that we have ought not to expose their children to obscurantist influence, more especially during their early years when these influences can operate upon what will be their unconscious in adult life', he wrote to H.G. Wells in 1928.[350] But what made the Russells decide that no other progressive school would be appropriate for their children? '[W]e could not agree with most "modern" educationists on thinking scholastic instruction unimportant, or in advocating a *complete* absence of discipline.'[351] The most obvious choice, which would have provided sufficient scientific education and discipline, would have been the Malting House School in Cambridge – though children approaching university age were not catered for there when the Russells were making their decisions. Despite their lack of background in education and teaching, one cannot assume the Russells were not aware of the experiments and initiatives taking place across the country – quite the opposite: they knew about Freud, Adler, Piaget, Pestalozzi, Froebel,

---

[349] The Pritchards (see below) were faithful supporters of the school, sending their two boys there throughout the 1930s.
[350] Russell to Wells, 24 May 1928, BRA II, p. 180.   [351] BRA II, p. 152.

Montessori and Margaret McMillan, whose school they visited.[352] Of course they were aware of Summerhill, which Russell did not approve of because it neglected intellectual development. 'Geoffrey Pyke, in Cambridge', Dora later wrote, 'had opened the Malting House where a group of children were presided over and studied by Susan Isaacs, in a small community that could hardly be called a school, since its principle was to see just what children would do and find out in freedom on their own'.[353]

> With regard to Freudian and other psychiatric theories, we thought that knowledge of these might be applied in education in a preventive sense, rather than, as they then were, merely curatively. For instance, provision of good nursery schools might well be a better means to children's mental health than the Child Guidance Clinics which, by treating a young child as a patient, could induce hypochondria or self-importance.[354]

Given the formula – navigating between the excessive freedom of Summerhill and the laboratory culture of the Malting House School in Cambridge, aiming to implement Freudian theory in a prophylactic rather than therapeutic mode – their choice for the principal teacher at the inception of the School was Beatrix Tudor-Hart. How they came into contact with her is unclear; quite possibly she answered the advertisement they placed in *The Nation and Athenæum* in March 1927. But she came from a well-established radical family, well known to the Stracheys,[355] part of the Cambridge networks. With an older brother, Alex, preceding her to Cambridge (he studying with Keynes before becoming a doctor and a Communist), she took a Second in Part I of Modern and Medieval Languages at Newnham in 1924, then a First, specializing in 'Psychology', in the Moral Sciences Tripos in 1925 – the same year that John Wisdom got a Second in the speciality of 'Metaphysical and moral philosophy'. On graduating, she, like many others, gravitated to Vienna to work with Charlotte Bühler on child psychology, publishing a classic paper

---

[352] Dora Russell, *The Tamarisk Tree*, Vol, I, London: Elek Pemberton, 1975, p. 199.
[353] *Ibid.*, pp. 199–200.   [354] *Ibid.*
[355] Obviously well known to the Stracheys as well as the Russells: in a letter to his wife Alix dated 20 May 1925 reporting on his visit to Cambridge, James described a party in Richard Braithwaite's rooms 'in honour of Marjorie [Strachey] (who was reading a paper next day at the Heretics): a large party – 30 people, I should say, or more. You've no notion how many young ladies from Newnham I'm now acquainted with – though their names for the most part escape me. On the whole I prefer La Tudor-Hart (not, alas, sexually), who's less of a stick than most.' Partly in BF, p. 269; the final sentence quoted here is omitted from the published version. See BLSP 60715, Vol. LXI, March 1925–October 1925. Others report that Beatrix Tudor-Hart was striking, tall, vivacious and uninhibited.

on the responsiveness of new-born children to sound and voice.[356] From there, as Sebastian Sprott's favourite student,[357] she moved to Giessen to work with Karl Koffka[358] and then, sponsored by Bühler in Vienna and Thorndike in New York, to Columbia to compare the different moral coding of lying amongst Austrian and American schoolchildren.[359] Back in England in 1927 and preparing to open Beacon Hill in September 1927, she made clear the rivalry that she and the Russells felt with the Malting House School in her reaction to Pyke's advertisement for a science teacher in *The Nation and Athenæum* in July 1927. She sent a cutting of the advertisement to Russell with the following commentary:

> an amusing, and (for Cambridge) obviously naive envy & jealousy of Telegraph House which they are immediately going to try to imitate. It appeals to my sense of humour, which, I admit, may be perverse. I wish them luck from the bottom of my heart.[360]

However, as Dora reported in early 1928: 'We found that we did not see eye to eye with Beatrix Tudor Hart about methods, and we parted soon, but quite amicably. This was, to some extent, a disappointment, because we had looked on this association as the start of a partnership.'[361] In 2004, Beatrix's grand-daughter told a differently nuanced story: 'Beatrix told me she was invited by the Russells to set up a school. At the time it only consisted of their kids. She set the school up, got no credit for it, did not respond to Bertie's advances and left to set up her own school in Fitzjohns Avenue Hampstead London.'[362]

---

[356] H. Hetzer and B.H. Tudor-Hart, 'Die frühesten Reaktionen auf die menschliche Stimme', *Quellen und Studien zur Jugendkunde* 5 (1927), cited by John Bowlby, who took the Moral Sciences Tripos Part II in Psychology three years later than Tudor-Hart, in J. Bowlby, 'The nature of the child's tie to his mother', *IJP* 39 (1958), 350–73 at 358.

[357] And also, completely unwittingly, an obstacle to a love affair, as Frank Ramsey recorded in his diary on 2 January 1924: 'when Seb came into his [Stephen Morland's] room he [Stephen] said "A dreadful thing has happened; I have fallen in love with Tudor-Hart's sister and don't want to go to bed with you, though I should very much like to."'

[358] SSP, Karl Koffka to Sebastian Sprott, 22 December 1925.

[359] B.E. Tudor-Hart, 'Are there cases in which lies are necessary?', *Pedagogical Seminary and Journal of Genetic Psychology* 33 (1926), 586–641.

[360] BR 710.059094, Beatrix Tudor-Hart to Bertrand Russell, 19 July 1927; the Russell Archive also holds a copy of the preliminary announcement of the opening of the Malting House School on 6 October 1924, which may have been acquired at the time or sent by Susan Isaacs (the correspondent) in 1926–27 when the Russells were actively engaged in considering the education of their children.

[361] Dora Russell, *The Tamarisk Tree*, Vol. I, pp. 205–6.

[362] Deborah Gorham, 'Dora and Bertrand Russell and Beacon Hill School', *Russell: The Journal of Bertrand Russell Studies* N.S. 25 (summer 2005), 39–76 at 61n53. Gorham notes that Beatrix's daughter Jennifer was fathered by Jack Pritchard, engineer and designer, husband of Molly Pritchard, biochemist and psychotherapist, a couple who met at Cambridge in the early 1920s;

Tellingly, when Dora recalled the ethos of Beacon Hill, she did not mention Watson and behaviourism, which had earlier been Bertrand's enthusiasm. John's early years may, Ray Monk surmises, have been significantly affected by Russell's blithe implementation of Watson's views about the dangers of maternal affection and the use of conditioning techniques to correct and discipline babies; but as a result of the experience of living with and teaching a crowd of children at the school, by 1930 Russell's advice was: 'Read John B. Watson on mothers. I used to think him mad; now I only think him American; that is to say, the mothers that he has known have been American mothers. The result of this physical aloofness is that the child grows up filled with hatred against the world.'[363] His view of Freud also changed. Until the Beacon Hill years, Russell had deployed Freud in search of a psychological underpinning for an understanding of politics, he had marshalled the psychoanalysts in his attack on introspectionism (the view that only introspection can furnish knowledge of the mind and that this is a privileged and distinct form of knowledge) and materialism in philosophy of mind,[364] and he had picked out, magpie-like, the eye-catching bits of psychoanalytic theorizing that suited his libertarian views on sex. But he had also been eager to find fault with the Freudians, eager to find good cause to join in the general ridicule of their more extreme statements. He had drawn upon Adolf Wohlgemuth's psychological expertise in preparing *The Analysis of Mind* and, when Wohlgemuth's excoriating critique of psychoanalysis was published in 1923 (see Chapter 7), had written a congratulatory letter to him to express his pleasure that the 'Freudian bubble had at last been pricked'.[365]

---

she adds: 'the Pritchards, unlike the Russells, successfully navigated an unconventional, "open" marriage'. The original plans for a house Jack and Molly aimed to build on their plot in Lawn Road, Hampstead included a nursery to be run by Beatrix. In the course of discussions between Molly and her lover and architect Wells Coates in the course of 1930, a new conception emerged, which became the famous Lawn Road Flats – the first modernist building in England – consisting of flats for 'minimal living' for young professionals. (See Elizabeth Darling, *Re-forming Britain: Narratives of Modernity before Reconstruction*, London: Routledge, 2007, esp. pp. 88–9, and Boris Jardine, 'Scientific moderns', PhD dissertation, University of Cambridge, 2012, 'Chapter 2. Constructivism 1: From the Bauhaus to Lawn Road', pp. 91–9.) As local historian and family friend Jack Whitehead notes, Beatrix 'told Jack Pritchard, the designer, that she wanted his child. He was happily married but it happened and Jennifer was brought up openly as their child, so Jennifer had two amicable homes' (email 7 September 2012). Jennifer was the architect of the Pritchards' country home, built in 1961 (Jack Pritchard, *View from a Long Chair: The Memoirs of Jack Pritchard*, London: Routledge and Kegan Paul, 1984, p. 174).

[363] Russell to Rachel Gleason Brooks, 5 May 1930, BRA II, p. 179.

[364] 'The whole tendency of psycho-analysis is to trust the outside observer rather than the testimony of introspection' (*The Analysis of Mind*, p. 60).

[365] A. Wohlgemuth, 'The Freudian psychology', *The Literary Guide and Rationalist Review* 393 (March 1929), 60.

What changed Russell's mind about Freud's excesses was one year of actual experience of children; now, in late 1928, he leapt to the defence of the Freudians: 'I am myself by no means an out-and-out Freudian, and I have in the past criticized Freud somewhat severely; but experience with young children has led me to think that much of what appears fantastic in his psychology is nevertheless largely true.'[366] In 1932, when addressing the vexed issue of children's freedom, Russell maintained this new position:

> When children are left free as regards their language, they say from time to time such things as Freudian text-books assert that they must be thinking, but being able to express their thoughts freely, they are not obliged to give them some fantastic form and become to that extent out of touch with reality.[367]

In *Marriage and Morals*, originally titled *Sex Freedom*, Russell advanced his vision of freedom in sexual matters and the raison d'être of marriage as children. Throughout the book he attacks conventional ideals of sexual morality ('the lack of instinctive satisfaction has turned to cruelty masquerading as morality') and, in Freudian mode, derives much of his discussion of morality from the course of childhood development. Never concerned to be quite accurate in his depiction of Freudian theory, it is nonetheless clear that certain passages closely follow Freud, embellished by his own sharp wit and unstoppable logic:

> In addition to this intellectual damage [caused by the conventional refusal to satisfy natural curiosity about sex], there is in most cases a very grave moral damage. As Freud first showed, and as everyone intimate with children soon discovers, the fables about the stork and the gooseberry-bush are usually disbelieved. The child thus comes to the conclusion that parents are apt to lie to him. If they lie in one matter, they may lie in another, so that their moral and intellectual authority is destroyed. Moreover, since parents lie where sex is concerned, the children conclude that they also may lie on such topics. They talk with each other about them, and very likely they practise masturbation in secret. In this way they learn to acquire habits of deceit and concealment, while, owing to their parents' threats, their lives become clouded with fear. The threats of parents and nurses as to the bad consequences of masturbation have been shown by psycho-analysis to be a very frequent cause of nervous disorders, not only in childhood but in adult life also.[368]

---

[366] Bertrand Russell, 'Freudianism', *The Literary Guide and Rationalist Review* 392 (February 1929), 44.
[367] Bertrand Russell, 'Free speech in childhood', *New Statesman and Nation* 1 (30 May 1931), p. 486, quoted in David Harley, 'Beacon Hill School', *Russell: The Journal of Bertrand Russell Studies* 35–36 (autumn/winter 1979–80), 5–16 at 12.
[368] Bertrand Russell, 'The taboo on sex knowledge', chapter 8 in *Marriage and Morals*, New York: Liveright, 1929, pp. 93–117.

The result? To 'make people stupid, deceitful, and timorous, and to drive a not inconsiderable percentage over the border-line into insanity'. What is more, the law resolutely fails to recognize these facts, requiring a sensible person 'to choose whether he will break the law or whether he will cause the children under his charge irreparable moral and intellectual damage'. Changing the law requires persuading 'the elderly men', a hopeless task, since they 'are so perverted that their pleasure in sex depends upon the belief that sex is wicked and nasty. I am afraid no reform can be hoped for until those who are now old or middle-aged have died.'[369]

Sexual enlightenment – the freedom to explore and above all to *know* about sex, about bodies, anatomy, birth – was at the heart of the educational ideals that Dora and Bertrand promoted in their school, in common with many others in the progressive school movement. Thus, while Bertrand may have been attracted to the behaviourists' promise of seemingly unlimited power over the moulding of character in their early years, his emphasis on sex aligned him much more closely with Freud than Watson. And that emphasis on sex was strongly confirmed both by the feminist and egalitarian ideals promoted by Dora (always, in part as she recognized, because she came from the younger generation, more of a Freudian than him) and by his own interpretation of the liberal ideal of freedom and outspokenness for adults in their love relationships. So the eventual endorsement of Freudianism was made natural and easy, while still conforming to his enthusiasm for 'scientific psychology'.

In her 1926 review of *On Education*, Susan Isaacs had excoriated Russell for his naïve yet 'delightful belief in the goodwill of the parent and the reasonableness of the child'; these beliefs 'spring from the fact that Mr. Russell, face to face with his children, is unable to tolerate the possibility of hostility on either side'. Yet even on this point, by 1929, as a result of his experience with young children, maybe as a result of Isaacs's perspicacious review, Russell began to give way:

> such is the influence of Christian asceticism that people have been more shocked by Freud's insistence upon sex than by his picture of infantile hatreds. We, however, must try to make up our minds without prejudice as to the truth or falsehood of Freud's opinions concerning the passions of children. I will confess, to begin with, that a considerable experience of young children during recent years has led me to the view that there is much more truth in Freud's theories than I had formerly supposed. Nevertheless, I still think that they represent only one side of the truth, and a side which can

[369] *Ibid.*

easily, with a little good sense on the part of parents, be rendered very unimportant.[370]

How are parents to neutralize the child's dark passions? Here Russell combines vestiges of his Watsonian wresting of power from mothers with a perennial theme in his writings: the pathological effects on women and thence on their children of a lack of sexual satisfaction:

> Let us begin with the Oedipus complex. Infantile sexuality is undoubtedly stronger than anybody thought before Freud. I think, even, that heterosexuality is stronger in early childhood than one would gather from Freud's writings. It is not difficult for an unwise mother quite unintentionally to centre the heterosexual feelings of a young son upon herself, and it is true that, if this is done, the evil consequences pointed out by Freud will probably ensue.

Yet again, Russell slips from considering the passions of the child to the passions of the mother, thus in effect denying the very fact of autonomous 'heterosexual' desires in childhood that he has just affirmed: it is the mother's capacity unwittingly to make her child into her sexual partner that has evil consequences (and rouses Russell's immediate opprobrium). The solution? Keep the children away from their mothers, by allowing them to express their sexuality with other children, not with adults: 'The heterosexual emotions of young children can find a natural, wholesome and innocent outlet with other children; in this form they are a part of play, and, like all play, they afford a preparation for adult activities.'

In the section of his *Autobiography* devoted to Beacon Hill School, drafted in the 1950s, Russell deprecated his own educational efforts, blighted, so he thought, by foolishly wilful blindness to children's need for order and routine, by the hypocrisy about how much freedom was actually allowed to the children, but above all by the immense difficulties for all concerned – Dora, Bertie, their children and then all the other children living and learning at the school – by their being both teacher and parent.[371] But in 1932 he had given a more judicious report revealing the aims of the school: 'I firmly believe that our methods enable a child to acquire knowledge without losing the joy of life and to become scientific without ceasing to be spontaneous.'[372] Science and spontaneous love of life: these were the core of Russell's philosophy of education.

---

[370] *Ibid.*, chapter 14, 'The family in individual psychology'.
[371] MBR II, pp. 95–6 puts a clear case for the hurt the school did to the Russells' children.
[372] Russell, 'Free speech in childhood', pp. 486–8, quoted in Harley, 'Beacon Hill School', p. 16.

In an interview published in the *New York Post* in 1957, Russell declared his complete faith in his own variety of self-help therapy: the conquest of fear and anxiety by imagining the very best that could happen. No professional therapists for him. 'I believe in the discoveries of Freud, I accept the theories, but I've never known a human being who has been helped by psychoanalysis – except for very little things like writer's cramp.' The interviewer asked him how people he knew appeared after psychoanalysis: 'Why, they're exactly the same afterwards! But they are not aware of this. They think they are better, but nobody else does.'[373] Dogmatically certain of his own views, always liberally sprinkled with his glee in uncovering paradox, especially when it came to a judgement on self-deception, most especially that of others, and most especially where psychoanalysis entered the picture: the view of this formidably intelligent eighty-five-year-old who had been through so much pain and distress, much, seemingly, self-inflicted, bears the distinctive Russell hallmark.

'Bertie in particular sustained simultaneously a pair of opinions ludicrously incompatible. He held that in fact human affairs were carried on after a most irrational fashion, but that the remedy was quite simple and easy, since all we had to do was to carry them on rationally.'[374] Thus Keynes summed up the style and substance of Russell. Psychoanalysis was a means both of explaining the irrationality of the present and of offering a model of how to move from such irrationality to a brighter, better future. Russell was swayed by the promise of this rational study of the irrational but, being so convinced of his own rational powers, despite the overwhelming evidence to the contrary supplied by his own very messy life and passions, he never was to see its darker side, glimpsed by others through the perpetual irony and pessimism of Freud's style.

Ludwig Wittgenstein

'Freud delved into the world of our dreams and subconscious in an attempt to understand our aberrations. Bertie, though he intellectually "took in" Freud, was, I think, too well barricaded within the intellect really to comprehend Freud's meaning.' So wrote Dora Russell in her autobiography. All the evidence confirms the accuracy of her judgement. Despite his

---

[373] Russell, 'Appendix VII. Voice of the Sages [1957]', interview with Irwin Ross, early September 1957, for the *New York Post*; other 'Sages' included Ernest Jones, Somerset Maugham and Clement Attlee. The only version is the text in the *New York Post*, BRCP 29, pp. 385–8 at 385–6.

[374] Keynes, *Two Memoirs*, London: Rupert. Hart-Davis, 1949, p. 102, quoted in A.J. Ayer, *Bertrand Russell*, Chicago: University of Chicago Press, 1974, p. 150.

enthusiasm for Freud, his many references to him and the lip-service he paid, Freud did not get under Russell's skin. Nobody could say the same of Russell's one-time pupil and privileged logical interlocutor, Ludwig Wittgenstein. 'It will take a long time before we lose our subservience'[375] to psychoanalysis, he told Rush Rhees in the early 1940s. By 'we', he meant, first and foremost, himself.[376]

When Wittgenstein first read Freud is not certain. Growing up in Vienna around 1900, he would certainly have heard a great deal about Freud, not least through the veritable ding-dong fights that went on around Karl Kraus's satirical magazine *Die Fackel*, to which Wittgenstein was most probably introduced by his sister and intellectual mentor, Gretl (Margarethe): he surely must have been acquainted with Kraus's famous aphorisms concerning psychoanalysis, so resonant with later Wittgensteinian themes – 'Psychoanalysis is the disease for which it claims to be the cure.'[377] Studying engineering in Berlin in 1906–8, then in Manchester, his first time in Cambridge in 1911–13 was also the period when any student associated with the Moral Sciences Tripos would likely

---

[375] Ludwig Wittgenstein, *Lectures and Conversations on Aesthetics, Psychology and Religious Belief*, compiled from notes taken by Yorick Smythies, Rush Rhees and James Taylor; ed. Cyril Barrett, Oxford: Basil Blackwell, 1966, p. 41.

[376] In commenting on this passage, Jacques Bouveresse concludes: 'I do not know to what extent he really includes himself in the "we"' (*Wittgenstein Reads Freud: The Myth of the Unconscious* (1991), trans. Carol Cosman, Princeton: Princeton University Press, 1995, p. 21). Given that Bouveresse has devoted a whole book to the complex and delicate question of Wittgenstein's relationship to Freud, this sentence is inexplicable.

[377] However, it appears unlikely that Gretl provided some sort of privileged access to psychoanalysis for Ludwig. Some writers have assumed that Gretl, being his intellectual mentor and having been born in 1882 (so seven years older than Ludwig), was closely acquainted with Freud and psycho-analysis in the period before the Great War; we have not come across evidence of this. Her biographer, Ursula Prokop, quotes several diary entries from the late 1920s indicating Gretl's strong scepticism about psychoanalysis (in this respect like her brother) – which, in itself, would not be incompatible with a friendship with Freud, but would undermine a view of her as a 'psychoanalytic enthusiast'; Prokop infers that Gretl's friendship with Freud developed as a result of her friendship with Marie Bonaparte, whom she met in early 1919 when they were both staying at the Hotel National in Lucerne. If Gretl's friendship with Freud was mediated through Bonaparte, it would almost certainly have been subsequent to June 1925, when Bonaparte first met Freud, and would probably date from the period when Bonaparte was resident in Vienna and having daily analysis with Freud from October 1925 on. (See Lisa Appignanesi and John Forrester, *Freud's Women*, London: Weidenfeld & Nicolson, 1992; Ursula Prokop, *Margaret Stonborough-Wittgenstein: Bauherrin, Intellektuelle, Mäzenin*, Vienna: Böhlau, 2nd edition, 2005, pp. 113–14, 222; also Alexander Waugh, *The House of Wittgenstein: A Family at War*, London: Bloomsbury, 2008, p. 135.) Gretl's own analysis with Freud took place in early 1937, though it may have been a continuation of an earlier analysis; it apparently continued in summer 1937. There are no extant letters exchanged between them prior to that time (see Freud to Margaret Stonborough-Wittgenstein, LoCSF, letters of 21 April 1937 and 3 May 1937); there are five from Freud in the period 1937–39. She visited Freud in London in late July 1938. (Email from Michael Molnar to John Forrester, 27 August 2012; source: Ms. Freud's Blue Book.)

have been aware of the new psychoanalytic ideas. But the young engineer
turned logician may well not have read Freud seriously before the war.
There is no record of Freud having any impact on him at this time. As is
clear from the experiences of Tansley, Rivers and others, reading Freud
during or immediately after the war was a far more affecting experience than
before the war.

Rush Rhees remembered Wittgenstein telling him that 'when he was in
Cambridge before the War he had thought psychology a waste of time'.
Registered as an 'Advanced Student' in 1911–13, Wittgenstein did not sit
any exams and did not proceed to a degree. He nonetheless threw himself
into the moral sciences with his work on logic, under Russell's supervision,
and in the experiments he conducted in collaboration with C.S. Myers on
musical rhythm in experimental psychology – what he called
'Experimental aesthetics'.[378] It was through asking David Pinsent to be
his experimental subject that they became friends.[379] Wittgenstein was part
of the team of students who performed demonstrations of psychological
experiments for distinguished guests on the opening day of the new
Psychological Laboratory in May 1913. So when, later in life,
Wittgenstein declared this type of experimental psychology a 'waste of
time', he spoke from experience. He was hardly alone: Myers and Rivers,
the founders of experimental psychology in Cambridge, had come to the
same conclusion as a result of their wartime experience. And, like them,
Wittgenstein's eyes were opened following the war:

> 'Then some years later I happened to read something by Freud, and I sat up
> in surprise. Here was someone who had something to say.' I think this was
> soon after 1919. And for the rest of his life Freud was one of the few authors
> he thought worth reading. He would speak of himself – at the period of
> these discussions [1942–46] – as 'a disciple of Freud' and 'a follower of
> Freud'.[380]

'Soon after 1919' places Wittgenstein's reading of Freud in the darkest and
most turbulent period of his life. A much-decorated volunteer in the
Austrian Army from 1914 to 1918, serving in the East and in the South, a
prisoner of war in Italy till August 1919, Wittgenstein had completed the
project he had embarked upon in Cambridge and Norway before the war:

[378] Brian McGuinness (ed.), *Wittgenstein in Cambridge: Letters and Documents 1911–1951*, Oxford: Wiley-Blackwell, 2012, p. 31n.
[379] G.H. Von Wright (ed.), *A Portrait of Wittgenstein as a Young Man: From the Diary of David Hume Pinsent 1912–1914*, with an introduction by Anne Keynes, Oxford: Blackwell, 1990, pp. 3–7.
[380] *Lectures and Conversations*, ed. Barrett, p. 41.

the manuscript of what would be published, eventually, in 1921/22, as *Tractatus Logico-Philosophicus*. His personal post-war turmoil was, as always with Wittgenstein, the result of an internal spiritual struggle, but it did have its external counterpart: having completed his philosophical work, what should he do with the life he had tested to the limit in facing death? His awakening to religion during the war gave him the possibility of becoming a priest or monk. Certainly he prepared himself for the ascetic life he would ever after lead, divesting himself of his large fortune by assigning it to his siblings. In the winter of 1919–20, he made his decision: to become a teacher in remote, provincial schools. It was in this period of turmoil and decision that he read Freud: it was *The Interpretation of Dreams* that preoccupied him – as it did others like Tansley, Rivers and Penrose – and would continue to do so over the next thirty years. Putting himself and his own dreams to the test in Freud's crucible was key. Two dreams from this period, reproduced in his controversial study by William Bartley,[381] together with some of Wittgenstein's associations and a record of his discussion of the dreams with his friends, indicate that he was attempting to follow Freud's method of interpretation.[382] The minimal records show he had only some success. Nor is there much evidence of Wittgenstein's intellectual interests in the early to mid-1920s – a bit of autobiography and exchanges of dreams with Gretl, but little else.[383]

The interest in Freud would, however, last the rest of his life: it goes hand in hand with the records he kept of his own dreams, his preoccupation with their meaning or lack of it, and with what standing such interpretations have. It is this preoccupation with his own dreams, seen in a Freudian light – or taken as an occasion for criticizing Freud – that marks Wittgenstein's engagement with Freud as far more profound than the opportunistic dallying with Freudian ideas found in Russell's work and

[381] William Warren Bartley III, *Wittgenstein*, Philadelphia: Lippincott, 1973.
[382] In an Appendix to his biography of Wittgenstein, Monk weighs the evidence for Bartley's fiercely contested claim that Wittgenstein passed through a phase in 1919–20 of compulsive seeking out of 'rough, young men' in the Prater in Vienna; the best evidence he can find as to whether Bartley's claims are based not on hearsay but on sound documentary evidence – i.e. a document, though one not available to other scholars – is Bartley's quotation of Wittgenstein's dreams from this period, together with Wittgenstein's own interpretations. Monk argues: 'If it strains credulity to think that Wittgenstein's friends supplied Bartley with accounts of his trips to the Prater, it positively defies belief that they gave him reports of Wittgenstein's dreams, told in the first person … And the discussion of the dream that Bartley quotes, though more elaborate than anything else that has been preserved, is entirely consistent with the interest that Wittgenstein showed at various times in Freud's techniques of interpreting dreams' (MLW, pp. 583–4).
[383] Brian McGuinness, 'Wittgenstein and Ramsey', in Maria Carla Galavotti (ed.), *Cambridge and Vienna: Frank P. Ramsey and the Vienna Circle*, Vienna Circle Institute Yearbook 12, Dordrecht: Springer, 2004, pp. 19–28 at 21.

life. For Russell, Freud was either one resource amongst many for advancing anti-mentalist or anti-introspectionist views in the philosophy of mind or an important resource for advancing arguments about marriage, children and morals within a broader context of political and social reform. For Wittgenstein, Freud represented a privileged figure for self-reflection not only as a dreamer, as a passionate and tortured moral being and as an unceasingly vigilant inquirer into the human condition, but also as a model, and therefore a threat – and perhaps nemesis – for Wittgenstein's own fate as thinker and teacher. Freud had gripping and seductive charm, yes; he was an example of a rich and creative thinker making illuminating mistakes – and illuminating not only for any possible project of scientific psychology – yes; but also, more insidiously and more profoundly, Freud's fate stood as an ever-present warning as to the potential fate amongst its 'followers' of critical philosophy as undertaken by and embodied in Wittgenstein. Wittgenstein's fate would be, he feared, like Freud's: to be so misunderstood, precisely by his most faithful followers, that everything that had been potentially good in his teaching would be mischievous, if not worse, in its consequences. Who would prevent Wittgenstein's critique of the mythology of 'meaning' not being displaced by a new mythology of 'use'? Nor, most alarmingly, was it just a coincidence that Freud and Wittgenstein might share this fate; there were very good reasons, namely the similarities of their projects, why this should be so.[384]

Kinship with Freud: Critical Philosophy and Psychoanalysis in Parallel

Freud had something to say to Wittgenstein. His writings on jokes and dreams figured prominently in Wittgenstein's work from his first lectures in Cambridge in 1930 till the end of his life.[385] The older Viennese functioned for Wittgenstein in many different ways: as a warning about

---

[384] Cf. 'all his years of teaching had done more bad than good. And he compared it to Freud's teachings. The teachings, like wine, had made people drunk. They did not know how to use the teaching soberly. Did I understand? Oh yes, they had found a formula. Exactly' (O.K. Bouwsma, *Wittgenstein: Conversations, 1949–1951*, ed. J.L. Craft and Ronald E. Hustwit, Indianapolis, IN: Hackett, 1986, pp. 11–12).

[385] That Wittgenstein found Freud a continued resource is indicated by numerous examples, for example a passage in the *Blue Book* (1933–34), where he distinguishes 'two kinds of discussions as to whether a word is used in one way or in two ways', taking as his examples the Latin 'altus', 'standing for both "deep" and "high"', and whether the English word 'cleave' is 'only used for chopping up something or also for joining things together' (L. Wittgenstein, *Preliminary Studies for the 'Philosophical Investigations' Generally Known as The Blue and Brown Books*, Oxford: Blackwell, 1958, p. 58). These examples are taken directly from Freud's *Introductory Lectures* (*SE* XV, 179)

confusions concerning the limits of science and illegitimate claims concerning explanation; as an external standard against which Wittgenstein could judge his own capacities and limitations. Like Freud, he was 'geistrich' (clever); like Freud he might be a Jew whose creativity is 'reproductive' rather than genuinely creative; like Freud, his work's power came from the immense store of metaphors and analogies through which he developed his ideas, analogies which so often took over the imagination of his audience and created the danger of systematically misleading them, even of blinding them with their seductive charm. One such seductive psychoanalytic argument had already been singled out by Russell in *The Analysis of Mind* as a reason why Freud had become so popular:

> Freud and his followers, though they have demonstrated beyond dispute the immense importance of 'unconscious' desires in determining our actions and beliefs, have not attempted the task of telling us what an 'unconscious' desire actually is, and have thus invested their doctrine with an air of mystery and mythology which forms a large part of its popular attractiveness. They speak always as though it were more normal for a desire to be conscious, and as though a positive cause had to be assigned for its being unconscious. Thus 'the unconscious' becomes a sort of underground prisoner, living in a dungeon, breaking in at long intervals upon our daylight respectability with dark groans and maledictions and strange atavistic lusts. The ordinary reader almost inevitably thinks of this underground person as another consciousness, prevented by what Freud calls the 'censor' from making his voice heard in company, except on rare and dreadful occasions when he shouts so loud that everyone hears him and there is a scandal. Most of us like the idea that we could be desperately wicked if only we let ourselves go. For this reason, the Freudian 'unconscious' has been a consolation to many quiet and well-behaved persons.[386]

In a letter in December 1945, Wittgenstein seized on the same feature of psychoanalysis: purporting to portray us as worse than we hope we are, it in fact appeals to a desire to be wicked or simply to have a secret:

> I, too, was greatly impressed when I first read Freud. He's extraordinary. – Of course, he is full of fishy thinking & his charm & the charm of his subject is so great that you may easily be fooled. He always stresses what great forces in the mind, what strong prejudices work against the idea of psychoanalysis. But he never says what an enormous charm that idea has for people, just as it has for Freud himself. There may be strong prejudices

---

(although Wittgenstein's editors fail to note their source), where Freud discusses the way in which the dream-work treats contraries that occur in the latent dream.
[386] Russell, *The Analysis of Mind*, pp. 37–8.

against uncovering something nasty, but sometimes it is infinitely more attractive than it is repulsive ... All this, of course, doesn't detract from Freud's extraordinary scientific achievement. Only, extraordinary scientific achievements have a way these days of being used for the destruction of human beings ... So hold on to your brains.[387]

Often there is something of a contest in Wittgenstein's dealings with Freud. Trying not to be bamboozled by the attractiveness of Freud's topic and his striking analogies, Wittgenstein could, almost in relief, turn the tables on him, showing how philosophical mileage – useful clarification – could be derived from following the twists and turns of his suspect procedures:

> Freud's book on [Jokes] was a very good book for looking for philosophical mistakes ... the same was true of his writings in general, because there are so many cases in which one can ask how far what he says is a 'hypothesis' and how far merely a good way of representing a fact – a question as to which he said Freud himself is constantly unclear.[388]

Wittgenstein then developed what would become one of his major criticisms of Freud:

> this confusion between cause and reason had led to the disciples of Freud making 'an abominable mess' ... psycho-analysis does not enable you to discover the cause but only the reason of, e.g. laughter. In support of this statement [Wittgenstein] asserted that a psycho-analysis is successful only if the patient agrees to the explanation offered by the analyst. He said there is nothing analogous to this in Physics; and that what a patient agrees to can't be a hypothesis as to the cause of his laughter, but only that so-and-so was the reason why he laughed.[389]

The distinction between reasons and causes that Wittgenstein drew became immensely influential in the 1940s and 1950s, both for philosophers and psychoanalysts:[390] it went hand in hand with Wittgenstein's worry that Freud appeared to offer explanations of, but was actually only

---

[387] Letter to Norman Malcolm, 6 December 1945, reproduced in Norman Malcolm, *Ludwig Wittgenstein: A Memoir, with a Biographical Sketch by G.H. von Wright*, Oxford: Clarendon Press, 2nd edition, 2001, pp. 100–1.

[388] G.E. Moore, 'Wittgenstein's lectures in 1930–33', *Mind* N.S. 64(253) (1955), 1–27 at 20–1.

[389] Moore, 'Wittgenstein's lectures in 1930–33', pp. 20–1.

[390] E.g. Stephen Toulmin, 'The logical status of psycho-analysis', *Analysis* 9(2) (1948), 23–9. Its force was much lessened by Donald Davidson's seminal – and deeply Wittgensteinian – paper 'Actions, reasons and causes' (*Journal of Philosophy* 60(23) (1963), 685–700); since then, many philosophers have become less persuaded by Wittgenstein's argument and therefore less persuaded that Freud's construal of reasons as causes is illegitimate. Bouveresse's nuanced account ends somewhat in confusion on this point – but precisely for that reason is a good cartographic calendar of philosophical debate over a number of decades.

redescribing, mental events. Here Wittgenstein contrasted scientific explanations, based on experiment, with psychoanalytic mythology. And again,
there was a reflexive element to this worry, pertaining to what it was that
philosophy could offer. Can philosophy 'explain' something in the world?
No, that is reserved to the sciences. Yet philosophy, like psychoanalysis,
can offer something: it can make a problem disappear. But it does so by
showing how a philosophical problem arises not from the world, but from
a misuse of language – it offers a solution to a 'grammatical' problem. If
psychoanalysis truly were part of the sciences, it would be of *less* interest to
the philosopher. Wittgenstein would then have treated Freud with the
simple contempt he showed for scientists of the 1930s, like Sir Arthur
Eddington and Sir James Jeans, who tried to extract philosophical problems and solutions from their scientific work – this really was language
going on holiday, when scientists stopped doing hard work and took it
easy, dabbling in what they called 'philosophy'. Freud was no Eddington or
Jeans; to Wittgenstein, he was more like Spengler and Kraus, imparting a
deep vision, through example, and through penetrating and/or satirical
Kulturkritik.

So in surveying the possibilities – is psychoanalysis a science; a fashion or
fad; a seismic transformation of culture? – Wittgenstein patrolled the same
demarcation line as would Karl Popper in his famous lecture 'Conjectures
and refutations' (1953): Freud's revolution is not a scientific revolution. Yet
Wittgenstein was not particularly concerned with this issue of scientificity:
science for him was relatively unproblematic and definitely in need of
neither defence nor justification (contrast Popper). Unsurprisingly, given
his aptitude for mechanical ingenuity and his education, he had an
engineer's, not a theoretician's, view of science: a useful tool-kit, made
up of can-do and know-how. If it wasn't experimental – sophisticated and
ordered tinkering – it wasn't science, that's all that needs to be said; science
was *uninteresting* from a philosophical point of view – hence his furious
contempt for the philosopher-physicists, who relished spinning pseudo-
metaphysical conundrums, or even pseudo-rational religious apologetics,
out of the latest strange twist and turn of physics. No, in separating
psychoanalysis from science, Wittgenstein was more interested in determining more how 'philosophy' and 'mythology' worked. Take away
explanation, 'scientific' explanation, and then ask: what is really going on
in psychoanalysis? The answer: the creation of a mythology and a redescription. But this redescription has extensive consequences, amounting to
a new mythology for our time, alongside the similar 'end of philosophy'
that Wittgenstein's own redescriptions (or reconfigurations) embodied. As

with philosophical problems, one can see more clearly what psychoanalysis is by taking away its scientific heraldic costume: it is still standing – and is even more interesting.[391] One does not stop interpreting one's own dreams in the style of Freud just because one is not persuaded of the scientific rigour of Freud's theories, at least not if you are Ludwig Wittgenstein.

In the end, Wittgenstein's Freud is much more Wittgensteinian than Freudian. He touches on Freud so often because he can be so usefully used (as in the first Cambridge lectures, discussing *Jokes*) to advance the difficult philosophical ideas Wittgenstein was developing – about meaning as use, about family resemblances, about language-games, about rule-following, about the private language argument. But this use of Freud as useful example shades over into what is more recognizable as a sustained self-analysis through dreams. Retelling one dream to Rush Rhees in 1943, he said: 'I was interested by the dream; I wrote it down; and, as one influenced by Freud, I looked for an interpretation of it in its connections with events in my experience.'[392] The dream – constructed out of elements drawn from Hollywood films combining a western with a musical comedy designed to satirize his sister's charitable endeavours – prompted him to ask 'whether one can reasonably speak of the dream as a thought about my sister, a ridiculing of her attempts to help save people'. His interpretation of the dream was embedded in a conversation with Rhees devoted to the following topics:

> Whether a dream is a thought. The sense of the question 'What made you hallucinate such a picture at all?'. The idea of a dream as a kind of game: not everything in this game need have an allegorical significance, or call for an interpretation. Opposed to the idea that a hallucination requires a tremendous mental force, and that therefore only the most deep and fundamental and hidden wishes could be responsible for it.
>
> This compatible with regarding dreams in practice as a convenient means of discovering hidden wishes in a patient, and thus as important in curing neurosis. Viewing dreams in this way for a particular end.

The Freudian theses about the nature of dreams, together with the Freudian method of interpreting both dreams and neuroses is here embedded in the larger issues which so preoccupied Wittgenstein. Freud's example of dream analysis could provoke philosophical worries in a number of different ways.

---

[391] See Vincent Descombes, 'Foreword', in Bouveresse, *Wittgenstein Reads Freud*, pp. vii–xiii.
[392] Gabriel Citron, email to John Forrester dated 14 June 2011.

If there is anything in the Freudian theory of dream interpretation; then it shows how *complicated* is the way the human mind makes pictures of the facts.

So complicated, so irregular is the mode of representation that it can *barely* be called representation any more.[393]

And then: is dream-interpretation – or the very fact that we have 'dreams' – a 'kind of game' (a language-game)? What can we make of the claim that a dream is a thought or a form of thinking? In particular, how do we evaluate the claim that the elements in a dream are a language? Not just symbols, infamously the Freudian symbols: as Wittgenstein remarked, 'it is the most natural thing in the world that a table should be that sort of symbol [for a woman]'.[394] Rather a language on the periphery (a suburb) of what we might mean by 'language' ('it can *barely* be called representation any more'). Hence the questions: is dreaming a form of thinking? If it is, dreaming is also a language. But if it is a language, why can we only translate from a dream into waking language and not in the reverse direction?[395] And, if dreaming is a language-game in its own right, why should there be an *essence* of dreaming – modelled, as Wittgenstein thought Freud's idea was, on the *Urphänomen* of Goethe?[396] Just as there are games without balls – solitaire, chess – so there may be dreams without wish-fulfilment, and we may still want to call them, unproblematically, dreams – *pace* Freud. Similarly, the argument turns to another aspect of Freud's theory: can there be dreams which are not thoughts – or dreams in a language that one can only translate one way – and will we *still* be content to interpret them? The argument cuts both ways, as usual: Wittgenstein is testing the limits of Freud's theory, testing if it is truly a scientific theory; if not, what kind of activity is it (a scientific mythology?) and what might be the lessons for philosophy?

Why shouldn't I apply words in ways that conflict with their original usage? Doesn't Freud, for example, do this when he calls even an anxiety dream a wish-fulfilment dream? Where is the difference? In a scientific perspective a

[393] Ludwig Wittgenstein, *Culture and Value*, ed. G.H. von Wright, in collaboration with Heikki Nyman, trans. Peter Winch, Chicago: University of Chicago Press, 1980, p. 44e; modified translation.

[394] *Lectures and Conversations*, ed. Barrett, p. 44.

[395] Wittgenstein here offers the analogy of translating Chinese – clearly the antecedent and inspiration for Searle's famous Chinese room argument, addressing a different but related question: can machines (computers) be conscious? (John Searle, 'Minds, brains and programs', *Behavioral and Brain Sciences* 3 (1980), 417–57).

[396] Ludwig Wittgenstein, *Remarks on Colour*, ed. G.E.M. Anscombe, Berkeley: University of California Press, 1977, Part 3, section 230.

new use is justified through a *theory*. And if this theory is false, then the new extended use has to be given up. But in philosophy the extended use does not rest on true or false beliefs about natural processes. No fact justifies it. None can undermine it.[397]

In testing the limits of Freud's theory in a variety of ways, he is also testing the limits of philosophy: 'In philosophy we are always in danger of giving a mythology of the symbolism, or of psychology: instead of simply saying what everyone knows and must admit.'[398] Psychoanalysis is a very good example of running this danger; Freud's dangerous project is a dry-run for the kind of philosophy that Wittgenstein is exposing as ill-considered, gone astray.

Freud was both a constant companion as interlocutor and a constant warning of paths not to be taken. Wittgenstein, especially in the early

Fig. 5.7 Ludwig Wittgenstein (*right*) and Frank Skinner, Trinity Street, Cambridge, 1935.

[397] Wittgenstein, *Culture and Value*, p. 44e; translation modified.
[398] Ludwig Wittgenstein, *Philosophical Grammar*, ed. Rush Rhees, trans. Anthony Kenny, Oxford: Basil Blackwell, 1974, p. 57.

1930s, insisted on the great achievement of Freud, as Moore recorded:
'Freud ... had genius and therefore might sometimes by psycho-analysis
find the *reason* of a certain dream, but ... what is most striking about him is
"the enormous field of psychical facts which he arranges".'[399] More pertinently for Wittgenstein's immediate audience of moral scientists, fighting
off the influence of Freud became exemplary of how philosophy itself
should be done: it is the attempt to cleanse ourselves of errors we are
deeply attached to. In just the same way, we become, as Russell also saw,
deeply attached to the ideas put forward by Freud – yet we must free
ourselves of them. In 1925 Keynes had come away from a conversation with
Wittgenstein convinced that Freud was 'a sort of devil' for our time. This
was exactly the sort of language Wittgenstein would have used to give
warning; in the sinuous prose of the 'immoralist' Keynes, it became a
ringing endorsement.

So Wittgenstein could give an analysis of the resistance to his work in
terms knowingly parallel to the famous – notorious – analysis of the
resistance to psychoanalysis:

> A mathematician is bound to be horrified by my mathematical comments,
> since he has always been trained to avoid indulging in thoughts and doubts
> of the kind I develop. He has learned to regard them as something con-
> temptible and, to use an analogy from psycho-analysis (this paragraph is
> reminiscent of Freud), he has acquired a revulsion from them as infantile.
> That is to say, I trot out all the problems that a child learning arithmetic,
> etc., finds difficult, the problems that education represses without solving. I
> say to those repressed doubts: you are quite correct, go on asking, demand
> clarification! (1932)[400]

Wittgenstein knew very well that his position was extremely reminiscent of
Freud's: he took up the position of the psychoanalyst – the child – in
relation to the community of mathematicians, treating their contempt for
and revulsion at his infantilisms as resistance. Knowingly – using the
language of psychoanalysis (repression, the infantile) – he places himself
exactly in the position of analyst. Most graphically in the following remark
from 1931: 'I ought to be no more than a mirror, in which my reader can see
his own thinking with all its deformities so that, helped in this way, he can
put it right.'[401] The Wittgensteinian philosopher maps exactly on to the

---

[399] Moore, 'Wittgenstein's lectures in 1930–33', p. 20.
[400] Wittgenstein, *Philosophical Grammar*, p. 381. The passage in brackets drawing attention to the
analogy with Freud is Wittgenstein's own interpolation.
[401] Wittgenstein, *Culture and Value*, p. 18e (passage from 1931).

Freudian psychoanalyst, both in the therapeutic process (transference), but also in how Wittgenstein could see in Freud's relationship to his followers and to public life in general what doleful fate awaited him.

Therapeutics and the Making of Analytic Philosophy[402]

One of the several reasons why Wittgenstein's response to Freud was so much more profound and troubling than Russell's is that Russell, by temperament and in accordance with his life-long philosophical preconceptions, never let doubts about the aims of philosophy undermine his practice of philosophy. In the final pages of his monumental and best-selling *History of Western Philosophy* (1946), Russell reaffirmed what his philosophical school, the 'philosophy of logical analysis', took as its aim and its achievement: to find 'definite answers, which have the quality of science rather than of philosophy' by 'its incorporation of mathematics and its development of a powerful logical technique'.[403] Russell's philosophy could only distinguish itself from what used to be called philosophy by copying the sciences, finding definite solutions to well-defined problems, and recognizing that questions of value and feeling lie outside of its domain. 'Science alone, for example, cannot prove that it is bad to enjoy the infliction of cruelty';[404] by implication, we should reluctantly admit that if science cannot find such a proof, then neither can philosophy, whose practice is determined solely by the virtuous 'habit of careful veracity'.[405] Russell's blithe refusal, indeed disdain, for anything consolatory or uplifting is reminiscent of one side of Freud's own orientation; after all, his science attempted to give a naturalistic account of how we come to enjoy the infliction of cruelty – and, what is more, even our moral codes may themselves be simply socially sanctioned cruelty. Russell always warmed to the fellow scientific naturalist that he could immediately perceive in Freud.

It was also a reason why Wittgenstein could not take Russell himself seriously after the break in their relations, after the intensity and intimacy of their early dialogue had lapsed. Wittgenstein knew that the Russellian

---

[402] For the most useful and thorough guide to the emergence of 'analytic philosophy', together with vigorous criticisms of the accepted historical account, see Preston, *Analytic Philosophy*, pp. 69–71, who describes the 'Traditional Conception' of this history as specifying five historical phases: 1. Moore and Russell's linguistic turn, 1900; 2. Russell's logical programme, 1910 leading to logical atomism, 1910–1930s; 3. logical positivism, 1930s–1945; 4. ordinary language philosophy, 1945–65; 5. eclecticism, 1965–.

[403] Bertrand Russell, *A History of Western Philosophy* (1946), London: George Allen & Unwin, new edition, 1961, p. 788.

[404] *Ibid.*   [405] *Ibid.*, p. 789.

project had failed and that its very aim, of a philosophy modelled on the sciences, was doomed to failure; he thought Russell guilty of bad faith in not recognizing this fact and in pretending to a larger public, over several decades, that nothing fundamental had undermined his original logicizing project. Wittgenstein's critique had made urgent the question: what could the aim of philosophy be, if the project of logical analysis (and of its cousins logical empiricism and logical positivism) was chimerical?

It was at this point that the model of Freud's psychoanalysis provided an answer. Philosophy, like psychoanalysis, consisted in identifying the cause of a set of symptoms (the philosophical problems) and *the very process* of identifying this cause would dissipate the symptoms. Wittgenstein himself did not explicitly declare this new psychoanalytic or cathartic model for philosophy, but his followers and critics did. One of the first to do so was, as one would expect given the limpid clarity of his thought, Frank Ramsey. Still fresh back from his analysis with Reik in Vienna and his conversations with Wittgenstein in his high mountain village of Puchberg, Ramsey gave a paper to the Apostles in early 1925 after hearing Russell's lecture 'What I believe' in which he declared:

> the conclusion of the greatest modern philosopher is that there is no such subject as philosophy; that it is an activity, not a doctrine; and that, instead of answering questions, it aims merely at curing headaches ... If I was to write a *Weltanschauung* I should call it not 'What I Believe' but 'What I feel'. This is connected with Wittgenstein's view that philosophy does not give us beliefs, but merely relieves feelings of intellectual discomfort[406]

By the early 1930s, with Wittgenstein back in Cambridge and the next generation of philosophers crystallizing the institutions of the post-Moore and Russell mode of philosophizing around the journal *Analysis*, founded in 1933, it had become pressing to characterize this new way of philosophizing. Attempting to characterize the legacy of his teachers, Moore and Russell, C.D. Broad had, in his *Scientific Thought* (1923), drawn a distinction between two types of philosophizing: 'critical' and 'speculative' philosophy. Critical philosophy consisted of 'the analysis and definition of our

---

[406] F.P. Ramsey, *Philosophical Papers*, Cambridge: Cambridge University Press, 1990, pp. 246–9. This paper is an informal response to Russell's lecture (published by Ogden in March 1925) 'What I believe', composed by Russell in the last week of 1924 and probably delivered in January or February 1925. See MBR II, pp. 53–4. It is possible that this paper is the one Ramsey alluded to in a letter to Lettice, dated by his father, possibly incorrectly, as 'May 1925': 'Whetnall [?] asked a few questions, and [Max] Newman, Lionel [Penrose] and [John] Wisdom made remarks, but they weren't tiresome. Towards 11 everyone departed except Moore, who stayed a little longer; I feel slightly in love with him' (FRP, Letters to Lettice Ramsey, Letter 11).

fundamental concepts, and the clear statement and resolute criticism of our fundamental beliefs',[407] whereas speculative philosophy aims 'to take over the results of the various sciences, to add to them the results of the religious and ethical experiences of mankind, and then to reflect upon the whole'.[408] Broad was somewhat wavering on the status of Critical Philosophy: was it a science like others, was it the foundation of the other sciences ('The other sciences use the concepts and assume the beliefs; Critical Philosophy tries to analyse the former and to criticise the latter') or did it come along afterwards, as it were, and tidy up the loose ends? Whichever it was, this new Critical Philosophy was very different from the practical, speculative philosophy which tried to give answers to the most important things and divine the meaning of the Universe in general. This was, quite transparently, a manifesto for the new kind of esoteric, professional philosophy now being practised in Cambridge – which would proudly, almost condescendingly, have no truck with talk of wisdom or providing answers to questions about the mystery of existence or how best to live one's life. Its surgical separation from 'what laymen generally understand by' the name 'Philosophy', and specifically from the 'elaborate systems which may quite fairly be described as moonshine',[409] was a key moment for the future of the new university-bound philosophy, which would receive its new name from John Wisdom in the early 1930s. In 1931, he deployed the term 'analytic philosophers' for the first time, then in 1934, he opened his *Problems of Mind and Matter* as follows:

> **Analysis and Speculative Philosophy.** It is to analytic philosophy that this book is intended to be an introduction. Consequently it is not concerned with certain questions which are properly called 'philosophical'. Philosophers have asked whether God exists, whether good or ill prevails, whether men are immortal, whether the world is in spite of appearances merely material, and whether in spite of appearances it is wholly spiritual and a unity. These speculative questions are clearly a great deal more important than questions of the analytic kind which have also been asked by philosophers, such questions as 'What is the ultimate nature of the soul?'; 'What is the ultimate nature of matter, time and space?'. The speculative questions are the more important because, if they could be answered, we should thereby obtain new information about matters which concern us very much, while the answering of analytic questions does not provide us with knowledge of new facts but only with clearer knowledge of facts already

---

[407] C.D. Broad, *Scientific Thought*, International Library of Psychology, Philosophy, and Scientific Method, London: Kegan Paul, Trench, Trubner & Co., 1923, p. 18.
[408] *Ibid.*, p. 20.   [409] *Ibid.*

known ... Speculating and analysing are operations which differ in kind; the object of the one is truth, the object of the other is clarity. It is with the latter that we shall be concerned.

Having shown how the new philosophy, analytic philosophy, steers away from truth towards clarity, he makes explicit that the model of truth-seeking is not actually speculative philosophy, but rather science, so that 'analytic philosophy' should be contrasted equally with the sciences and with speculative philosophy:

> An introduction to a science, such as chemistry, will contain a selection of the easier and more fundamental chemical truths. In my opinion there cannot be such an introduction to analytic philosophy. For there is no special set of analytic truths. Analytic philosophy has no special subject-matter. You can philosophise about Tuesday, the pound sterling, and lozenges and philosophy itself. This is because the analytic philosopher, unlike the scientist, is not one who learns new truths, but one who gains new insight into old truths. In a sense, philosophy cannot be taught any more than one can teach riding or dancing or musical appreciation. However, philosophers can be made.[410]

Wisdom's statement of intent coincided with the foundation of the journal *Analysis* in November 1933, 'mainly devoted to short discussions of questions of detail in philosophy ... with the elucidation or explanation of facts, or groups of facts, the general nature of which is, by common consent, already known; rather than with attempts to establish new kinds of fact about the world, of very wide scope, or on a very large scale'.[411]

This programme was a decidedly odd way to cement a philosophical revolution which, by all accounts, has increasingly dominated professional Anglo-American philosophy from the 1930s onwards. Its explicit ambitions – restricted problems examined in detail and with clarity – were the result of an implicit recognition that these, the students of Moore and Russell, worked with the detritus of a truly grand but failed project, the logicizing of

---

[410] John Wisdom, *Problems of Mind and Matter*, Cambridge: Cambridge University Press, 1934, pp. 1–2.

[411] 'Statement of policy', *Analysis* 1(1) (1933), 1. For further clarification of the origins of the term, see, for instance, Max Black, John Terence Wisdom and Maurice Cornforth, 'Symposium: Is analysis a useful method in philosophy?', *Proceedings of the Aristotelian Society, Supplementary Volumes*, Vol. XIII, *Modern Tendencies in Philosophy* (1934), pp. 53–118. The authors there indicate that one of their sources for the term is the opening chapter of C.D. Broad's *Scientific Thought* (1923), which distinguishes between 'analytic' and 'speculative' aims in philosophy; Broad referred the term 'analytic' to the work of his teachers and colleagues, Russell and Moore. John Wisdom opened his *Problems of Mind and Matter* (1934) deploying Broad's terminology, explicitly committing the book to 'analytic philosophy', which aims solely at 'clarity', in contrast to 'speculative philosophy', which aims to discover new things – and thus aims at 'truth'.

mathematics and of all of science.[412] What was left was a fundamentally destructive – critical, vigilant, hard-headed – project: to avoid or undermine all sorts of mistakes and misapprehensions that so often went under the name of 'philosophy'. These aims might often be stated in the supremely brisk and confident tone of the Oxford enthusiast for logical positivism, A.J. Ayer – 'Philosophising is an activity of analysis'[413] or 'The function of philosophy is wholly critical'[414] – but a moment's reflection (a moment which has lasted several decades) shows that these are dwindling and thoroughly restricted aims; the tone may be exultatory but this is a philosophical hangover.

One version, perhaps the most tenacious, of this new aim for philosophy was explicitly seen by contemporaries as having a medical model and as being most strongly associated with the behind-the-scenes influence of Wittgenstein in Cambridge. The young Ernest Nagel, later to write a classic statement of the logical positivist vision in *The Structure of Science* (1961), visited Europe from his base at Columbia University in New York in 1936 and attempted to characterize what had become of Cambridge philosophy under the influence of Wittgenstein:

> The tendency to generate 'profound' questions not amenable to empirical inquiry – the perennial problems of philosophy – he regards as a disease which it is the task of sound philosophy to cure. The reinstatement of the unsophisticated, untroubled view of the man in the street is apparently the goal of philosophic activity, although I suspect that Wittgenstein would regard the man cured in this way of the dreaded affliction as somehow better off than the one who has never succumbed to philosophizing in the traditional manner. This desired goal is to be achieved by showing that the alleged problems of philosophy are either empirical (in which case they fall within the province of a special science); or that they arise from mistaking a synthetic material proposition for an analytic one true *a priori*; or that they assign meanings to words which they have not, and could not have, in the specific contexts in which they occur. Wittgenstein's primary concern is therefore with the clarification of the meanings of expressions.[415]

Wittgenstein may have been acutely aware in 1932, when he made his comments about the horrified response of mathematicians to his

---

[412] In 1931, in his first book (*Interpretation and Analysis in Relation to Bentham's Theory of Definition*, London, Kegan Paul, Trench, Trubner & Co.), written for Ogden's Psyche Miniatures series, Wisdom used the term 'logico-analytic philosophers' (p. 13) which he then shortened to 'analytic philosophers'. See Preston, *Analytic Philosophy*, p. 69 ff. for a good account of Wisdom's usages and after.

[413] A.J. Ayer, *Language, Truth and Logic* (1936), London: Penguin, 1971, p. 10.   [414] *Ibid.*, p. 32.

[415] Ernest Nagel, 'Impressions and appraisals of analytic philosophy in Europe. I', *Journal of Philosophy* 33(1) (1936), 5–24 at 17.

arguments, of the parallel between the psychoanalytic dialectic of repression, regression and resistance and his own relationship with mathematicians and philosophers, but the later development of the programme of philosophy as therapy was based on vivid analogies and similes that were only implicitly psychoanalytic. 'What is your aim in philosophy? – To shew the fly the way out of the fly-bottle.'[416] Or on other occasions, the aim of philosophical analysis was to relieve us of a 'mental cramp',[417] a 'cramp of philosophical puzzlement'.[418] Or even: 'Lack of clarity in philosophy is tormenting. It is felt as shameful.'[419] Whereas Ramsey had almost unselfconsciously blurted out the new Wittgensteinian philosophical sensibility – it 'merely relieves feelings of intellectual discomfort' – and the philosophical tourist Nagel saw a chronic disease which modern philosophy could cure, other contemporaries began making the analogy explicit for a much larger audience than those who regularly attended Wittgenstein's lectures in the early 1930s. For those following Wittgenstein, Max Black argued in 1938, 'philosophy is "the struggle against the fascination exerted by forms of representation"':

> So conceived, philosophical activity appears more like a psycho-analytical treatment than a scientific investigation. The 'patient' suffers from certain mental 'cramps' induced by incomplete awareness of the structure of his language. When the source of his fixation is brought to light by analysis the patient loses the desire to talk 'nonsense' any longer and is free to turn to the more profitable activities of ordinary life. (As in other psychoanalytical treatment, willing submission by the patient is an important factor in his recovery.) It would be interesting to have a full investigation of the psychological presuppositions of this method.[420]

Clearly for Black, likening philosophical to psychoanalytical investigation was automatically to the detriment, not the increased standing, of philosophy.[421] What is striking is that there is a long pedigree for the likening of

---

[416] Wittgenstein, *Philosophical Investigations*, #309 (p. 103ᵉ). Another version of the fly-bottle analogy is to be found in the beautifully crafted argument concerning spatial puzzle-solving in Ludwig Wittgenstein, *Remarks on the Foundations of Mathematics*, ed. G.H. von Wright, R. Rhees and G.E.M. Anscombe, trans. G.E.M. Anscombe, Oxford: Blackwell, 3rd edition, 1978, I-44, p. 56.

[417] Wittgenstein, *Preliminary Studies for the 'Philosophical Investigations'*, pp. 1, 58.

[418] *Ibid.*, p. 61.   [419] Wittgenstein, *Remarks on Colour*, p. 21e.

[420] Max Black, 'Relations between logical positivism and the Cambridge school of analysis', *Erkenntnis* 8 (1938/9), 24–35 at 32–3. He added: 'It is too early to say whether the nihilist methods of Wittgenstein will establish themselves firmly in English philosophy' (p. 33).

[421] Others took a different view. Lewis Feuer, remembering his first encounters in 1931–35 with psychoanalysis as a graduate student in the psychoanalytically hostile environment of Harvard University Department of Philosophy, wrote: 'Wittgenstein's critique of philosophy seemed to me a linguistic echo of what Freud had done much more powerfully on a psychological level while

philosophy to therapy, particularly in the Hellenistic schools (Epicureans, Stoics, Skeptics),[422] but also in Plato. The response to Wittgenstein and the 'Cambridge school' bore a modernist stamp: they looked to the immediate contemporary school of psychoanalysis for the template of therapy, not to the history of philosophy itself.

However, it must be said that the analogy of the method of Wittgenstein's version of analytic philosophy with psychoanalysis – most specifically with the structure of Breuer's and Freud's cathartic cure, in which finding the causes (previously forgotten scenes) of the symptom simultaneously removes the symptom – is very close. Here is Wittgenstein, in a passage drafted in the late 1930s or early 1940s:

> The real discovery is the one that makes me capable of stopping doing philosophy when I want to. The one that gives philosophy peace, so that it is no longer tormented by questions which bring *itself* in question . . . There is not *a* philosophical method, though there are indeed methods, like different therapies.[423]

In 1946, when A.J. Ayer was asked to give an overview of British philosophy on the BBC, he was somewhat nonplussed as to how to deal with Wittgenstein, who had not published a significant piece of work since 1922. Dashing off a witty phrase (which he later regretted), he suggested that Wittgenstein's 'effect on his more articulate disciples has been that they tend to treat philosophy as a department of psychoanalysis'.[424] Wittgenstein was furious for a number of reasons: first and foremost because Ayer's overview implied that he kept his work secret, but also because he refused the analogy between his method and psychoanalysis: 'they are different techniques'.[425] Another comment from conversations with Rush Rhees around this time gives a sense of his irritation: 'When you

remaining free from the grandiose solipsist-sounding pronouncements that made the *Tractatus Logico-Philosophicus* read at times like a *Tractatus Logico-Neuroticus*' (Lewis S. Feuer, 'A narrative of personal events and ideas', in S. Hook, W. O'Neill and R. O'Toole (eds.), *Philosophy, History and Social Action: Essays in Honor of Lewis Feuer with an Autobiographic Essay by Lewis Feuer*, Boston Studies in the Philosophy and History of Science, Dordrecht: Springer, 1988, at pp. 13–14).

[422] See Jerome Neu, *Emotion, Thought and Therapy: A Study of Hume and Spinoza and the Relationship of Philosophical Theories of the Emotions to Psychological Theories of Therapy*, London: Routledge and Kegan Paul, 1977, and Martha Nussbaum, *The Therapy of Desire: Theory and Practice in Hellenistic Ethics*, Princeton: Princeton University Press, 1994.

[423] Wittgenstein, *Philosophical Investigations*, #133, p. 51^e. This is akin to Wittgenstein's response to Farrell's version of the Wittgensteinians' project as 'therapeutic'. Wittgenstein had responded that philosophy is like psychoanalytic therapy, but it is different.

[424] A.J. Ayer, 'Recollections of Ludwig Wittgenstein', in F.A. Flowers III (ed.), *Portraits of Wittgenstein*, Vol. III, Bristol: Thoemmes, 1999, pp. 124–8 at 127.

[425] Malcolm, *Ludwig Wittgenstein: A Memoir*, pp. 46–7.

think of the investigation of philosophical problems as a form of psycho-analysis, you think like a physician – "We'll soon put that right!"'[426] What irritates him most is that a too hasty and unreflective endorsement of the analogy of the removal of neurotic symptoms and philosophical problems leads one to expect the problem symptom to vanish quickly. Freud too faced this response: hence his own quizzical riposte – one does not stop dreaming because one has gained access to the unconscious through dream interpretation.

Ayer was not the only quick-witted philosopher to think that Wittgenstein was transforming philosophy into psychoanalysis. When in early 1939 Wittgenstein applied for the Chair of Philosophy at Cambridge vacated by G.E. Moore, his old patron and friend J.M. Keynes was on the Board of Electors who asked Wittgenstein to submit a manuscript.[427] 'Of 140 pages, 72 were devoted to the idea that philosophy is like psycho-analysis. A month later Keynes met him and said that he was much impressed with the idea that philosophy is psychoanalysis.' The American philosopher O.K. Bouwsma, whose recollections of conversations with Wittgenstein these are, added: 'And so it goes ... Freud, of course, also did incalculable harm, much as Wittgenstein himself has done.'[428]

Knowingly endorsed and violently repudiated: this was Wittgenstein's double response to the analogy of his critical method in philosophy and the technique of psychoanalysis. Equally two-sided was his assessment of the effects of the two methods – sociological and cultural. As Bouwsma recognized, Wittgenstein was convinced of the great potential for evil that his philosophical work had: exactly the same *kind* of evil that Freud had done. A vivid analogy could easily become the seed for a new set of philosophical misunderstandings – 'they had found a formula'[429] – they had got drunk on Wittgenstein's teachings in the same way they abused (and Freud encouraged them, despite himself, to abuse) psychoanalysis: 'Freud's fantastic pseudo-explanations (just because they are so brilliant) perform a disservice. (Now every ass has these pictures to hand for "explaining" symptoms of illness with their help.)'[430] Despite Wittgenstein's protestations, it is not at all implausible to claim that his

---

[426] Brian McGuinness, 'Freud and Wittgenstein', in McGuinness, *Approaches to Wittgenstein: Collected Papers*, London: Routledge, 2002, pp. 224–35 at 233.

[427] This manuscript consisted of the substantial first part (roughly first 188 sections) of what was subsequently revised and posthumously published in 1953 as *Philosophical Investigations*. See McGuinness, *Wittgenstein in Cambridge*, p. 291.

[428] Bouwsma, *Conversations with Wittgenstein*, p. 36.   [429] *Ibid.*, pp. 10–11.

[430] *Culture and Value*, p. 55e (translation modified).

'therapeutic' model of philosophy was derived very knowingly from psychoanalysis. The term 'therapeutic positivism' indeed came into use in the late 1930s.[431] Herbert Feigl, émigré veteran of the Vienna Circle's meetings with Wittgenstein in the late 1920s,[432] elided the new terms in his 1954 survey:

> the trend [in philosophy] characterized variously as 'analytic philosophy,' 'therapeutic positivism,' or 'casuistic logical analysis,' originally introduced by G.E. Moore in England and most strikingly developed and modified by Wittgenstein. Nowadays Cambridge and Oxford, the second even more strongly than the first, are among the chief centers of this type of philosophizing. Here we find the Socratic method applied with extreme subtlety to the peculiarities (ambiguities and vaguenesses, strata and open horizons, implicit rules) of natural languages.[433]

Sometimes commentators would group Oxford and Cambridge together, sometimes it was Vienna and Cambridge.[434] The most waspish vision came in 1946 from an Oxford philosopher sniping at the strange tribe that had arisen in Cambridge, known as the 'W–ns' – a two-part article by Farrell entitled 'An appraisal of therapeutic positivism':

> It seems reasonable to suppose that a psycho-therapist has to have considerable confidence in the efficacy of his own particular medicine before he can muster the courage to try to cure his patients at all. Particularly is this the case with therapeutic methods that have a very unsettling effect on the patient. And the scientific enquirer is quite within his rights in asking for the evidence that entitles the therapist to his self-assurance. The W–ns are, I think, in a somewhat analogous position. Their self-confidence is unbounded. Present them with a philosophical problem, and they smile in a superior way, immediately place you in the role of patient, and plunge into therapy with a gesturing, grimacing, and stammering self-assurance. Where is the justification for this attitude? ... In part, I suspect (though here I may be wholly at fault), the smallness, and the isolated and specialist character, of the Cambridge Moral Science Tripos. Like the Mediaeval monastic orders,

---

[431] A.J. Ayer used the phrase, perhaps coined it, in an article in *The Listener* in 1937, prompting a sharp rebuke from Wittgenstein (MLW, pp. 356–7).

[432] MLW, p. 243.

[433] Herbert Feigl, 'Some major issues and developments in the philosophy of science of logical empiricism' (1954), in Herbert Feigl and Michael Scriven (eds.), *The Foundations of Science and the Concepts of Psychology and Psychoanalysis*, Minnesota Studies in the Philosophy of Science 1, Minneapolis: University of Minnesota Press, 1956, p. 5.

[434] The surprisingly long delay in the term 'analytic philosophy' or 'analytical philosophy' stabilizing and being accepted is still evident in 1962, when Gilbert Ryle, one of the leaders of ordinary language philosophy, supposedly the 1950s embodiment of a major strand of 'analytic philosophy', coined the phrase 'the Cambridge transformation of the theory of concepts' in order to describe this historical achievement (Gilbert Ryle, *Collected Papers*, Vol. I, London: Hutchinson, 1971, p. 189).

the W–ns tend to have a vision of great clarity, great intensity and astonishing narrowness.[435]

Further along, Farrell argues:

> Psychological theory here has been insufficiently general; for it has tended to regard neuroses and anxiety states as caused by non-verbal factors alone. The W–n contribution consists in an extension of the notion of anxiety states and their treatment to cover cases where the condition is produced, in part, by the ordinary verbal habits of the person himself. It is these habits that in certain people produce the conflict states known as philosophical puzzlement.

Like Black's, Farrell's portrayal of the W–ns is of their version of philosophical activity as a sub-variant of psychoanalytic therapy. But Black and Farrell had bundled together the social behaviour of the Wittgensteinians, viewed as a small sect or tribe, with the claim that the aim of philosophy is therapeutic – it is to rid oneself of torment, lack of clarity, confusion – of being infested or infected with philosophical problems.

It is noteworthy that these contemporaries of the development of the Wittgensteinian programme jumped on the analogy between philosophy-as-therapy and psychoanalysis in order to pour scorn on, or satirize, or hold up to ridicule its style and content. It was obvious to the intended reader that it was an ignominious fate for philosophy to end up as a sub-department of psychoanalysis. The force of the analogy thus depended upon a prior, contemptuous attitude to psychoanalytic therapy. That Wittgenstein himself did not share such contempt is clear: be careful, yes, but don't underestimate or denigrate the psychoanalytic project itself. Ironically, in recent years, several philosophers have argued vociferously in favour of the 'therapeutic reading' of Wittgenstein, but have dropped the analogy with psychoanalysis:

> [Wittgenstein's aim was] to help us work ourselves out of confusions we become entangled in when philosophizing ... as tracing the sources of our philosophical confusions to our tendency, in the midst of philosophizing, to

---

[435] B.A. Farrell, 'An appraisal of therapeutic positivism (II.)', *Mind* N.S. 55(218) (1946), 133–50 at 136–7. The parallel of Farrell's strategy with that of Ernest Gellner in his famous polemic against Wittgenstein and Oxford philosophy in *Words and Things* (1959) is striking. Both Farrell and Gellner also went on to write spirited critical appraisals of psychoanalysis (*The Standing of Psychoanalysis* (1981) and *The Psychoanalytic Movement* (1985) respectively); we might even call these belated critiques the return of the repressed – the original target, psychoanalysis, had been masked in Farrell's and Gellner's works on 'analytic philosophy' in the period 1940–61, but was now explicit in the 1980s, as if they had finally realized that Wittgenstein had been only the stalking-horse for the real prey.

think that we need to survey language from an external point of view ... as wishing to get us to see that our need to grasp the essence of thought and language will be met – not, as we are inclined to think in philosophy, by metaphysical theories expounded from such a point of view, but – by attention to our everyday forms of expression and to the world those forms of expression serve to reveal.[436]

The more ambitious commentators claim this aim of Wittgenstein's can be extended not only to the later philosophy, in which, as we have seen, Wittgenstein himself writes explicitly of the therapeutic aim, but also to the *Tractatus*. These new philosophers do not draw on the early – and contemporaneous – therapeutic readings offered by Wisdom, Black, Farrell, Feigl, Ayer and others. The *locus classicus* for this new therapeutic reading is a paper published by Stanley Cavell in 1962.[437] As a result, the conspicuousness of the name 'Freud' is missing from these accounts – Cavell's Wittgenstein stands in for Wittgenstein's Freud. The term 'therapeutic' has been reduced down somewhat in its scope, perhaps to include all the various philosophers with their different readings of Wittgenstein. But in Cavell's original 1962 formulation of the modern 'therapeutic reading' of Wittgenstein, Freud is conspicuously present. Cavell frames Wittgenstein's writing as a project of self-knowledge, as a style of writing constructed of necessity in the confessional mode, since to advance it must display its antagonists – the voice of temptation and the voice of correctness. Sensitive to the charge that Wittgenstein has harmed philosophy through the creation of a sect of disciples, he justifies the necessity of running this danger. For Cavell, the strength of the forces that are arraigned against the project of self-knowledge – which is also the cure and liberation from the past – justify this risk. And the methods that must be used show how closely aligned are Wittgenstein and Freud:

> Such writing [i.e. Wittgenstein's] has its risks: not merely the familiar ones of inconsistency, unclarity, empirical falsehood, unwarranted generalization, but also of personal confusion, with its attendant dishonesties, and of the tyranny which subjects the world to one's personal problems.

---

[436] Crary, 'Introduction', in Alice Crary and Rupert Read (eds.), *The New Wittgenstein*, London: Routledge, 2000, p. 1.

[437] Crary, 'Introduction', p. 7 writes: '(Stanley Cavell's writings contain the earliest descriptions, in this basic vein, of the method of Wittgenstein's later writings.)', referring to his 1962 paper in the 1976 reprint in his famous 1976 collection, *Must We Mean What We Say?*, with no indication that it was first published earlier than 1976. It is perhaps surprising, given that Crary is concerned with the 'earliest' description, that she pays such scant attention to the date of publication of Cavell's essay. However, she is by no means alone amongst philosophers in ignoring a complex history.

The assessment of such failures will exact criticism at which we are unpracticed. In asking for more than belief it invites discipleship, which runs its own risks of dishonesty and hostility. But I do not see that the faults of explicit discipleship are more dangerous than the faults which come from subjection to modes of thought and sensibility whose origins are unseen or unremembered and which therefore create a different [93] blindness inaccessible in other ways to cure. Between control by the living and control by the dead there is nothing to choose.

Because the breaking of such control is a constant purpose of the later Wittgenstein, his writing is deeply practical and negative, the way Freud's is. And like Freud's therapy, it wishes to prevent understanding which is unaccompanied by inner change. Both of them are intent upon unmasking the defeat of our real need in the face of self-impositions which we have not assessed (§108), or fantasies ('pictures') which we cannot escape (§115). In both, such misfortune is betrayed in the incongruence between what is said and what is meant or expressed; for both, the self is concealed in assertion and action and revealed in temptation and wish. Both thought of their negative soundings as revolutionary extensions of our knowledge, and both were obsessed by the idea, or fact, that they would be misunderstood – partly, doubtless, because they knew the taste of self-knowledge, that it is bitter. It will be time to blame them for taking misunderstanding by their disciples as personal betrayal when we know that the ignorance of oneself is a refusal to know.[438]

Coda: John Wisdom

Graduate of the Moral Sciences Tripos with a First in Part I in 1924 and a Second in Part II in 1925, undergraduate Secretary of Ogden's Heretics, employee of Myers's National Institute of Industrial Psychology after graduating, John Wisdom was a lecturer at the University of St Andrew's until 1934 when he returned to Cambridge as Lecturer in Moral Science,[439]

---

[438] Stanley Cavell, 'The availability of Wittgenstein's later philosophy', *Philosophical Review* 71(1) (1962), 67–93 at 92–3; reprinted in *Must We Mean What We Say?*, where this passage is at pp. 70–2 (and is cited as the *locus classicus* by Crary, 'Introduction', for the 'therapeutic reading' of Wittgenstein).

[439] Wisdom was eventually elected to a Fellowship at Trinity (probably the slot vacated by the lapsing of Wittgenstein's Fellowship). It may be thought surprising that Wittgenstein did not apply for the Lectureship to which Wisdom was appointed, since his five-year Trinity Fellowship ran from January 1931 to December 1935. If he had done so, he would doubtless have been appointed (just as he was appointed in February 1939 to a Professorship, although he then feared that it would go to Wisdom, who was also a candidate). However, Monk's biography indicates that, at the time of the appointment in 1934, Wittgenstein was not at all interested in continuing as an academic philosopher; his attitude changed markedly in 1938 and 1939, following the turmoil produced by Hitler's annexation of Austria, so he became quite anxious to secure the Professorship.

coming increasingly under Wittgenstein's influence. Indeed, during the late 1930s and 1940s, when Wittgenstein published nothing and restricted access to his lectures and classes, Wisdom's published papers were turned to by outsiders as the next best thing – as some guide to Wittgenstein's developing thought. Yet there would always be a subtle difference between Wittgenstein and Wisdom. When Wittgenstein was at his most critical, philosophical puzzles were simply the result of mistakes or weaknesses – ones we are all prone to, but nonetheless with little redeeming value in themselves. Wisdom never lost his respect for the sources of philosophical puzzlement. If there was a key word in his philosophical lexicon, it was 'question'. And the work of philosophy would not necessarily dismantle the question. '(If I were asked to answer, in one sentence, the question "What was Wittgenstein's biggest contribution to philosophy?", I should answer "His asking of the question 'Can one play chess without the queen?'.")'[440]

At the same time, his interest and involvement in psychoanalysis steadily deepened;[441] he published reviews of psychoanalytic books in *Mind*, one of the principal philosophical journals of the time, and then two papers in 1944–46 explicitly addressing the relations between philosophy and psychoanalysis, giving the title to his collection of essays published in 1953, a few months after he succeeded Wittgenstein in the second, younger Chair in Philosophy (1896) at Cambridge.

In contrast to the satirizing, critical intent of Black's and Farrell's demonstration of the analogy between Wittgensteinian philosophy and psychoanalysis, Wisdom's essays on philosophy and psychoanalysis accept their kinship without a knowing wink and snobbish tone; instead he attempted to draw out both the likeness and the differences. Wisdom allows them to be much closer than most philosophers would feel comfortable with: 'True, philosophy has never been merely a psychogenic disorder nor is the new philosophical technique merely a therapy', he reflects, without baulking at the possibility that they might just be that. However, he goes on, 'there's a difference. Philosophers reason for and against their doctrines and in doing so show us not new things but old things anew.' Wisdom differs from the more radical Wittgensteinian positions by allowing that philosophical problems won't go away – just

---

[440] A.J.T.D. [John] Wisdom, 'Ludwig Wittgenstein 1934–1937', in Flowers (ed.), *Portraits of Wittgenstein*, Vol. III, p. 274. It is typical of Wisdom, in the style of Wittgenstein, to make the grandest claim in a passage in brackets.

[441] A file of notes (closed access) relating to his self-analysis is held in his papers deposited in the archives of Trinity College, Cambridge; they are dated 1939–59.

like dreams, however well analysed. There is a difference: but the similarities are striking and given more detail than the differences:

> Nevertheless, having recognized how different is philosophy from therapy it is worth noticing the connections: (a) how philosophical discussion is the bringing out of latent opposing forces like arriving at a decision and not like learning what is behind a closed door or whether 235 × 6 = 1420; (b) how, often, when the reasoning is done we find that besides the latent linguistic sources there are others non-linguistic and much more hidden which subtly co-operate with the features of language to produce philosophies; (c) how, in consequence, a purely linguistic treatment of philosophical conflicts is often inadequate; (d) how the non-linguistic sources are the same as those that trouble us elsewhere in our lives so that the riddles written on the veil of appearance are indeed riddles of the Sphinx.[442]

Philosophical problems are deep and persistent. Linguistic clarification will help bring us some ease, but they will not make the conflicts go away. Most importantly: philosophy is not cut off from the rest of life – it is not cut off in any straightforward sense from what may happen in psychoanalytic therapy.[443] Wisdom demonstrates the intertwining of metaphysics and psychoanalysis[444] – how they mirror one another, can exchange between them, where the exhausted metaphysician can recognize in his speculations 'those other struggles in which something is for ever sought and never found, struggles which in their turn are connected with an earlier time when there was something, namely the world of the grown-ups, knowledge of which we desperately desired and equally desperately dreaded'.[445] And, in the end, the best solution (if not the only one) to the endless metaphysical spiral may be the psychoanalytic conversation which, after his twists and turns, is recognizably also a philosophical conversation:

> In the labyrinth of metaphysics are the same whispers as one hears when climbing Kafka's staircases to the tribunal which is always one floor further up. Is it perhaps because of this that when in metaphysics we seem to have

---

[442] John Wisdom, 'Philosophy and psycho-analysis' (1946), in *Philosophy and Psycho-Analysis*, Oxford: Blackwell, 1953, p. 181.

[443] Thanks to Mike Brearley for sharing his memories of Wisdom and for an apt and accurate summary of Wisdom's attitude to the question of the kinship of philosophy and psychoanalysis: 'how close [analytic philosophy] is to therapy (he would also have been careful about that – "say what you like but be careful") by which he means having gone through the detailed comparisons in the end may be less important whether you say it is or it isn't!' (email 10 October 2014). Janet Sayers (email 21 April 2015) recalls Wisdom's first lecture in the Moral Sciences Tripos in 1964 as devoted to Dostoevsky, Freud, and other minds.

[444] 'Besides all these ways in which the procedure, the difficulties, and the aims, of psycho-analysis are reflected in metaphysics' (Wisdom, *Philosophy and Psycho-Analysis*, p. 280).

[445] *Ibid.*, p. 281.

arranged by a new technique a new dawn then we find ourselves again on Chirico's sad terraces, where those whom we can never know still sit and it is neither night nor day?

We may hurry away and drown the cries that follow from those silent places – drown them in endless talk, drown them in the whine of the saxophone or the roar from the stands. Or, more effective, we may quiet those phantasmal voices by doing something for people real and alive. But if we can't we must return, force the accusers to speak up, and insist on recognizing the featureless faces. We can hardly do this by ourselves. But there are those who will go with us and, however terrifying the way, not desert us.[446]

The Unnamed Discipline

Tracing the vagaries attending the foundation in Cambridge of the four modernist 'disciplines' – psychology, anthropology, philosophy and English – it thus becomes evident that psychoanalysis was a crucial catalytic element in the process by which they crystallized out of the flux of novel interests and inherited practices. Psychology's eventual form was quite different precisely because of its close proximity and mutual intermingling with psychoanalysis as it burst upon the academic, medical and literary scenes of Edwardian and then post-war England. The 'experimental psychology' that was created in Cambridge after 1922 was rather like a reaction-formation to the enormous interest in psychoanalysis in that early period. The character of psychology in the University for the rest of the twentieth century was in large part due to the choices made in the 1920s, not only in its exclusion of psychoanalysis but also through its resolute refusal of American behaviourism – that odd combination of 'Darwinian naturalism, techno-cratic utopianism, and stimulus-response technology' to use Dorothy Ross's apt characterization. However, even if its unyielding commitment to scientific (i.e. experimental) psychology owed much to the 1920s turmoil, psychology's later twentieth-century antagonisms and difficulties were only distantly related to the psychoanalytic 'upswing' of the 1920s.

In both anthropology and English, a project of interpretation which was at the least homologous with, if not founded upon, psychoanalytic inter-pretation (both as model of inquiry and as set of rules for deciphering

---

[446] *Ibid.*, p. 282: end of paper and of book. Wisdom's quiet influence in Cambridge as a channel of communication between philosophy and psychoanalysis is visible in the easy references of his student, Stephen Toulmin, in his first book, to Freud (*An Examination of the Place of Reason in Ethics*, Cambridge: Cambridge University Press, 1950, p. 116) and the advice Wisdom gave his student Michael Brearley in the 1960s when he first expressed interest in becoming a psychoanalyst.

meanings) was essayed, with differing results. In anthropology, in parallel with psychology, the promise of a fruitful joint venture quickly turned to suspicion and boundary-work of a rigorously exclusionary character. In English, something else happened: an experimental procedure for the production of error and self-knowledge became a model pedagogic practice. No one would claim that the practice of practical criticism as intentionally perpetuated for so many decades was a form of psychoanalytic literary criticism. But the mark of Freud is quite clearly detectable in the original framing of Richards's pedagogic innovation.

In the 1940s, the phrase 'therapeutic positivism' began to be consistently employed to describe the Wittgensteinians, although its life turned out to be short-lived. It seemed such an appropriate phrase for the task of philosophy following Wittgenstein: to let the fly out of the bottle. How far was this view of philosophy due to the influence of Freud? Such claims are nearly always disowned – No, it is not really the idea that philosophical difficulties are like neuroses, constructed like dreams, susceptible to treatment, even to transference-analysis. But it was Wittgenstein himself who said: 'It will take a long time before we lose our subservience to Freud.'[447]

What is most striking about these four quite different episodes is that the influence of psychoanalysis in its era of enthusiasm in the 1920s was both unpredictable and long-standing. It is the contingency of the outcomes that is most striking: a rigorously experimental tradition of psychology systematically disdaining the larger movements in psychology, of which psychoanalysis was the most incendiary and infamous; a promising dialogue between psychoanalysis and anthropology that flipped almost within a few months into untouchability; a method of self-examination that became programmatic for the new subject of English studies while its 'scientific' origins in the self-examination inherent in psychoanalysis was forgotten; and the example, again, of the rigours of self-analysis as the model for the self-cure of philosophy. Undoubtedly the fact that these were all new disciplines in formation permitted these unpredictable influences to be felt. But what is equally striking is that the four disciplines all, in the end, repudiated the influence of psychoanalysis. Anthropology was, in this respect, the most uncomplicated: all psychology, of any sort, was banned, in the name of a quasi-Durkheimian orthodox dogma of the autonomy of the social. Perhaps these episodes are instances of Veblen's Law of

---

[447] Wittgenstein, *Lectures and Conversations*, ed. Barrett, p. 41.

Interdisciplinary Relations: 'To the extent that disciplines overlap in sub-ject matter, they are unlikely to cooperate.'[448]

The peculiar receptivity of Cambridge intellectual life to psychoanalysis might be attributed to a number of factors. The first and most obvious is the tradition of psychical research established in the last two decades of the nineteenth century and continuing uninterrupted almost to this day. An inordinate number of significant figures in the SPR, even from the 1920s on, were linked to Cambridge; not only Whateley Smith but the philoso-pher C.D. Broad (Fellow of Trinity; President of the SPR 1935–36 and 1958–60), R.H. Thouless (psychologist of religion and education, President SPR 1942–44), Donald West (Professor of Criminology, President SPR 1963–65, 1984–88, 1998–99) and the intriguingly sceptical William Rushton (Professor of Physiology, Fellow of Trinity, President SPR 1969–71). But the most obvious yet less tangible factor may have been the unique Cambridge configuration of the moral sciences. There is no better way to explain what this tradition meant than to invoke two classic Cambridge 'moral scientists', though ironically neither of whom took moral sciences as undergraduates, but instead imbibed it through living their whole lives in the academic groves of Cambridge: John Maynard Keynes and Frank Ramsey. Both manifested consistent and long-lasting enthusiasm for psychoanalysis. If one calls Keynes just an economist; if one calls Ramsey just a philosopher, obeying disciplinary injunctions to closure and purification that were beginning to grip fast in the 1920s, one misses the interconnectedness of their wide range of interests and effortless expertise. Keynes the philosopher, the probabilist, the economist, the political fixer; Ramsey the mathematician, logician, philosopher, econo-mist (savings and taxation) – these are perfectly formed moral scientists, indubitably English in the mould of John Stuart Mill and self-consciously so. It is no accident that Ramsey's imagined dialogue with Mill (see Chapter 6) would be deeply personal and also hangs on the question whether Freud has anything to add to what Mill has already left us. Freud, of course, as Mill's translator and life-long admirer of English empiricism, would have been deeply flattered – if he had accepted Keynes's invitation and come to Cambridge.

---

[448] Marshall Sahlins, 'The conflicts of the faculty', *Critical Inquiry* 35 (Summer 2009), 997–1017 at 1013.

CHAPTER 6

The 1925 Group

Early in 1925, a psychoanalytic discussion group was organized in Cambridge, described many years later by John Rickman:

> Soon after the Malting House School was started [in 1924] there arose in the vicinity, but independently, a discussion society which made the experience of personal analysis its major qualification for membership. It was a small group composed of two members of the Royal Society, three others clearly heading in the same direction, one literary person occasionally came and there was one nondescript member, all were graduates of Cambridge. A topic was taken at each meeting, announced beforehand, memoranda were sometimes circulated; the theme was outlined before lunch, discussed casually in the after lunch walk, seriously tackled before, during and after tea and brought to a close usually before but sometimes after supper.[1]

The nondescript person was Rickman himself; the 'literary person' was James Strachey – a joke the latter would have approved of. The other members were Harold Jeffreys, Lionel Penrose, Frank Ramsey and Arthur Tansley.[2] Of these men, only Tansley was FRS at the time, though Jeffreys was to be elected in 1925 and Penrose in 1953, after Rickman's death – so it is possible there is one member missing from this list, but more probable that Rickman miscounted. The three destined-to-be FRSs would indeed be distinguished.[3]

How did this discussion group come to be formed? There was one sole strict condition: that each member had been analysed. But some members of the group knew each other well for other reasons: three of them were Apostles (Strachey, Ramsey and Penrose); three of them had seen each other often in Vienna the previous year, when they were each being

---

[1] J.R. [John Rickman], 'Susan Sutherland Isaacs, C.B.E., M.A., D.Sc. (Vict.), Hon. D.Sc. (Adelaide)', *IJP* 31 (1950), 279–85 at 281.
[2] BF, 25 February 1925, p. 219.
[3] Neither Rickman nor Strachey became FRS – they became psychoanalysts instead. Rickman's account gives the membership of the society as seven; we can only determine its membership as six.

analysed (Tansley, Ramsey and Penrose). Two of the members had exceptional organizational proclivities (Tansley and Rickman), so almost certainly the initiative for the group's formation came from one of them. Of these two, the more likely candidate is Tansley, who, having resigned the Lectureship in Botany he had held in the University from 1907 to 1923 in order to continue his analysis with Freud, had in the summer of 1924 moved back to the home he had left in Grantchester. It was also probably Tansley who had originally volunteered, at Easter 1924, to organize the International Psychoanalytical Association Congress in Cambridge in 1925. However, by the autumn he had become curiously reluctant to organize the Congress; perhaps the formation of the Discussion Group was some kind of substitute (or recompense).

The most striking feature of the group is, however, that it was created by invitation only, and those invited all had to have been analysed.[4] The BPaS itself had no such precondition and was havering about whether to be an open scientific society or to be more exclusive. In that sense, the 1925 Group, to give it a name, was more akin to the Apostles than other Cambridge discussion groups. But perhaps the simplest way to describe the getting-together of the group is to observe that all of the members had criss-crossing long-term friendships and affiliations with the others, with the possible exception of the more solitary Jeffreys; even he had known Tansley in scientific circles in Cambridge for years, and was a resolute supporter of the Heretics and the *Cambridge Magazine*, as were Rickman and Tansley. When it came to Cambridge progressive causes and societies, they all had long-term history. With the exception of Jeffreys, they had all, of course, had analysis in Vienna.

John Rickman

John Rickman: ebullient, irascible, impulsive, outgoing, strong as a horse, hard-working as an ox. Here is Lionel Penrose's description of the Rickman who in 1920 swept him up into the project of an analytic adventure in Vienna:

---

[4] This was not unprecedented; Jung reported to Freud concerning 'the founding of a lay organization [*Gesellschaft für psychoanalytische Bestrebungen* (Society for Psychoanalytic Endeavours)] for ΨA. It has about 20 members and only analysed persons are accepted. The organization was founded at the request of former patients. The rapport among its members is loudly acclaimed. I myself have not yet attended a meeting. The chairman is a member of the ΨA Society. The experiment seems to me interesting from the standpoint of the social application of ΨA to education' (C.G. Jung, 'Letter from C.G. Jung to Sigmund Freud, February 25, 1912', *The Freud/Jung Letters: The Correspondence between Sigmund Freud and C.G. Jung*, ed. William McGuire, Princeton: Princeton University Press, 1974, pp. 487–8 at 487).

Fig. 6.1 John Rickman, *c.* 1925, by J. Palmer Clarke.

Tall, with a black beard and moustache and large spectacles, his appearance
was most striking. No wonder, he said, that people believed him to be
a bolshevik. Indeed he had only recently returned from Russia but was now
working at Fulbourn Asylum. 'The difference between me and the patients,'
he explained to me, 'is that I have a key and they haven't.'[5]

John Rickman (1891–1951) was educated by the Quakers in Dorking at
Leighton Park and then at King's College, Cambridge, where in 1913 he
took a third class degree in Part I, Natural Sciences Tripos, and gained
a Blue at rowing. He befriended Adrian Stephen, Virginia Woolf's brother,
and was part of the large group gathered round Ogden's *Cambridge
Magazine*.[6] He undertook clinical training in medicine at St Thomas'

[5] PP, 20/2, p. 108/12.
[6] The fullest account of Rickman's life and work is the Editor's Introduction in Pearl King (ed.),
No Ordinary Psychoanalyst: The Exceptional Contributions of John Rickman, London: Karnac, 2003.

Hospital in London, completing in 1916. At the same time he developed a strong interest in the law. During the First World War, as a Quaker pacifist he volunteered for the Friends' War Victims Relief Unit in Russia. Notably, he declined to write for Ogden and the *Magazine* about conditions in Russia on the grounds that he was forbidden to.[7] In Russia he met an American social worker, Lydia Cooper Lewis, whom he married there in March 1918. Forced to take the long way back to England via Vladivostok on account of the war in Europe and the Civil War in Russia, he then wrote a series of articles on his experiences while practising medicine amongst the Russian peasants, which, characteristically, were published anonymously at the time.[8] Having been a conscientious objector, he had difficulty finding a job, but eventually was taken on as a psychiatrist at Fulbourn Hospital near Cambridge which he envisaged running on revolutionary Freudian lines. He took up residence at the Old Vicarage, Grantchester, Rupert Brooke's old haunt, across the road from Tansley's home at Grove Cottage. Here the seventeen-year-old Frank Ramsey visited him in January 1920.[9] Rivers, his psychoanalytic mentor at this time, told Rickman, 'If you are going to do anything in the field of psychiatry or psychology you must get analysed',[10] and suggested he approach Freud. Rickman did exactly that, was accepted at a fee of two guineas a session and began analysis with Freud in April 1920.

After six months in Vienna, on a trip back to London in October 1920, Rickman spent a fortnight discussing analytic questions with Ernest Jones, who formed a dislike for the 'aloof' young man. Freud jumped to Rickman's defence: 'I think your judgment is too severe. His peculiarities are not beyond the measure of any young man getting aware and not yet sure of his powers. He is not conceited and full of passion. I rather like him.'[11] As it was for James Strachey, starting analysis at this time with Freud, Rickman's analysis set a trend for the rest of his family. During the fortnight Rickman was discussing analysis with Jones in London, his wife conceived a child and herself entered analysis with Jones, starting around 4 October 1920.[12] When their daughter was born in the summer of 1921, Rickman's mother took her place on Jones's couch.[13] While Jones

[7] OP, Box III, File 5, Rickman to Ogden, 3 July 1916.
[8] They were later revised and republished in a joint book with Geoffrey Gorer, the psychoanalytic culturalist, *The Peoples of Great Russia*, London: The Cresset Press, 1949.
[9] Margaret Paul, *Frank Ramsey (1903–1930): A Sister's Memoir*, Huntingdon: Smith-Gordon, 2012, [p. 179].
[10] John Rickman, 'An editor retires', *BJMP* 22(1) (1949), 1.   [11] FJ, 12 October 1920, p. 393.
[12] JF, 1 October 1920, p. 391.   [13] JF, 30 November 1921, p. 444.

'managed' the Rickman family, Freud was taking care of various other members of the British psychoanalytic enterprise:

> [Rickman] is a nice, strong fellow, his narcissism, I expect, a phase on the track of libido from mother-wife to some new, sublimated object. His analysis is just now rather sterile. I make excellent progress with Alix Strachey and expect to succeed with James Strachey, both of whom may become highly valuable members of your staff.[14]

Rickman and the Stracheys returned to England in July 1922, Rickman to 8 Fitzwilliam Street in Cambridge.[15] Undecided about his future, he had been playing with the idea of setting up a private psychoanalytic practice in Cambridge, where, as his friend Géza Róheim suggested: 'It would of course be an advantage to the movement if you settled at Cambridge. There is always a chance of some Professor getting neurotic and converted when analysed by you.'[16] Róheim also dreamt of a position in Cambridge to escape from the difficulties of Budapest and be near the heart of anthropology: Fraser at Trinity and Rivers at St John's were, in his eyes, the key figures in the field.[17] However, Rickman decided over the summer of 1922 to move definitively from Cambridge to London and set up his practice there.

Rickman had been elected an Associate Member of the British Society on 11 October 1920, while he was still in analysis in Vienna; as had been planned by Freud and Jones, he was put to work as an indispensable administrator both for publications and in the running of the Society. It was Rickman who, for important legal reasons which were beyond Jones's understanding, set up the legal framework, the *Articles of Association* – an 'ingenious document', which he was proud became a model for other scientific bodies in England[18] – for the Institute of Psycho-Analysis in 1924 and then for the Clinic which opened in 1926. So it was Rickman who, just at the time the 1925 Group was meeting, created the structure separating the *professional* institutions of British psychoanalysis (the Institute) from the *scientific* organization, the Society. All Members of the Society became Members of the Institute, which owned property and had substantial funds at its disposal. The Society became solely a forum for scientific discussion, where Associate Members

---

[14] FJ, 6 November 1921, p. 443.
[15] Charles Darwin returned from his voyage on the *Beagle* to live in the same street.
[16] RPBPaS, CRR/FO8/25, Róheim to Rickman, 6 December 1921.
[17] RPBPaS, CRR/FO8/14, Róheim to Rickman, February 1921 (exact date unclear).
[18] RPBPaS, CRR/K14/13, Rickman to Dr Paul Dane, 18 September 1939.

and guests could participate fully alongside the Members. Rickman would repeat these feats of constitutional (re-)organization as President of the Society after World War II (1947–50), overhauling the rules of the Society and Institute, and the acquisition of a new building, Mansfield House, in 1950, just as he had chosen the first building in 1925 in Gloucester Place.[19] The campaign to raise money to endow the Institute and Clinic of 1951 was initiated under his Presidency and bears his stamp, not least in the telling fact that the six signatories to the Appeal were two other members of the 1925 Group, Tansley and Penrose, together with F.R. Winton (another Professorial ex-Cambridge scientist), J.C. Flügel, Jones and the incoming President, William Gillespie.[20] The existence of the 1925 Group left its mark years later on the BPaS, but invisibly, as simple names in the list of its Great and Good patrons.

In 1926 Rickman completed his MD thesis on the aetiology of prolapse of the uterus; his first psychoanalytic publication, apart from reviews and a piece on psychoanalysis and alcoholism for the *British Journal of Inebriety* in 1925, was on psychosexual aspects of prolapse of the uterus.[21] He was also a supremely hard-working and skilled editor, acting as Assistant Editor (1925–34) and then Editor (1934–48) of the *British Journal of Medical Psychology*, thus ensuring that, alongside the traditionalist asylum-journal, *Journal of Mental Science* (which in 1969 became *British Journal of Psychiatry*), there was a prominent British 'psychiatric' journal which had a predominantly psychoanalytic, or psychoanalysis-friendly, orientation. Back in 1920, Freud had encouraged Rickman to follow up the psychoanalytic approach to the psychoses. In his systematic and exhaustive style, Rickman produced in 1928 two working books of exposition for students and psychoanalytic researchers: the *Index Psychoanalyticus*, a record of all psychoanalytical work from 1893 to 1926, and *The Development of the Psycho-Analytical Theory of Psychoses, 1893–1926*. The *Index Psychoanalyticus* and a later Bibliography of Freud's writings that Rickman appended to his *A General Selection from the Works of Sigmund Freud* of 1937 formed the basis for Tyson and Strachey's 'A chronological hand-list of Freud's works' (1956) around which the *Standard Edition* was organized. Rickman's editorial vigour was expressed in the series 'Psycho-Analytical Epitomes', published by the Hogarth Press and the Institute of Psycho-Analysis from 1937 on. Cut short

---

[19] King, *No Ordinary Psychoanalyst*, p. 4. He performed the same functions for the Medical Section of the British Psychological Society and its publication, the *BJMP*.

[20] Anon. 'News, notes and comments', *IJP* 32 (1951), 269.

[21] J. Rickman, 'A psychological factor in the ætiology of *descensus uteri*, laceration of the perineum and vaginismus', *IJP* 7 (1926), 363.

by the war, the series nonetheless saw four volumes published – two collections of Freud's work, including the *General Selection* and *Civilization, War and Death* (1939); *Love, Hate and Reparation* (by Klein and Riviere), and Roger Money-Kyrle's 1937 lectures on *Superstition and Society* (1939).

Rickman's analysis with Freud failed to satisfy him. In February 1924 he requested further analysis from Freud. He went to Vienna after the April 1924 Salzburg Congress at which it was decided to hold the next International Congress in Cambridge.[22] In July 1928 he contacted Freud again, requesting further analysis; this time it was impossible for Freud to take him on and, on Freud's recommendation,[23] Rickman spent periods of time in Budapest in analysis with Ferenczi over the course of the next three years and then, from 1934 on, with Melanie Klein in London, continuing fitfully throughout the war. He became a firm supporter of her work, to the extent that she could, in a Memorandum submitted to the British Society in 1942, refer to him (along with Joan Riviere, Susan Isaacs and Donald Winnicott) as one of the four training analysts in unqualified agreement with her on matters of method and technique, i.e. as part of the core group of Kleinian senior analysts.[24] He never became a notable training analyst, but one of his analysands was indeed notable: W.R. Bion, who had an extensive analysis with Rickman in the 1930s and with whom Rickman wrote an influential paper on groups for *The Lancet* during World War II.

Rickman became the backroom psychoanalytic organizer par excellence. At crucial times, he himself financially underwrote activities of the British Psycho-Analytical Society and Library. He was a leader in organizing lectures and events; repeating the success of Melanie Klein's lectures in July 1925, he organized Géza Róheim's London debut in his drawing room in September 1925, to which double the number expected turned up. (Strachey reported to Alix in Berlin, 'All our old friends were there – including Lionel (accompanied by the pug-faced Margaret Gardiner). Also Tansley, very pleased with himself after his Scandinavian tour.'[25]) Particularly in the 1930s Rickman was adept at organizing events which mediated between the small psychoanalytic community and the medical

---

[22] See Chapter 2 above. LoCSF, Freud to Rickman, 26 February 1924, Falzeder transcription (English in the original).

[23] LoCSF, Freud to Rickman, 30 July 1928, Falzeder transcription (English in the original).

[24] FK, Melanie Klein, 'Memorandum relating to some proceedings in the Society on and before May 13, 1942', p. 199.

[25] BLSP 60715, Vol. LXI, March 1925–October 1925, 15 September 1925 (not included in BF).

profession, the psychologists and the general public, from his firm base as a 'medically qualified analyst with a very considerable consulting practice'.[26] From 1935 on, as a regular (and anonymous) leader-writer for *The Lancet* (he wrote twenty-three in total), he had considerable influence on medical and public opinion and was much quoted in the national press on general medical topics.[27] As war approached, he was active in the Peace Pledge Union, the Medical Peace Campaign and the Quaker Medical Society; he also was involved in the All London Aid Spain Council formed to bring food to the starving children of Spain.[28] On the day Poland was invaded, he joined the Emergency Medical Service at Haymeads Emergency Hospital, Bishop's Stortford; Susan Isaacs and Melanie Klein joined him in the town a few days later. Starting with the Haymeads Memorandum, written three days after war was declared, he was incessantly organizing and communicating, finding new solutions and new strategies for the organization of psychiatric services in both the military and civil life.[29] Given that he was by then a recognized Kleinian training analyst and one of the pillars of the Society, he played a smaller part than one might expect in the wartime Controversial Discussions which split the society into groups, one loyal to Anna Freud, another to Melanie Klein, and another independent; this was largely due to his military duties posting him outside London. However, the moment he lost his temper with the Officers of the Society at a meeting on 17 December 1941 (for which he later formally apologised) is as good a moment as any to designate as the official start of the hostilities. Though ill from August 1944 on, at the end of the war Rickman travelled to Germany as part of the German Personnel Research Branch, whose aim it was to discover suitable people who had been opposed to the Nazis, and would be able to aid in the 'rehabilitation' of Germany. Rickman visited those analysts who had remained in

---

[26] William Gillespie, 'Reminiscences', *IRP* 17 (1990), 11–22 at 13.

[27] Sylvia Payne, 'Dr. John Rickman', *IJP* 33 (1952), 54–60.

[28] P.H.M. King, 'Activities of British psychoanalysts during the Second World War and the influence of their inter-disciplinary collaboration on the development of psychoanalysis in Great Britain', *IRP* 16 (1989), 15–32 at 17.

[29] Starting with his move to Sheffield in January 1940, his other notable activities included the 'Wharncliffe Experiment', in which Rickman initiated what was later to become the therapeutic community; the 'Northfield Experiments' (regarding neurosis as treatable by the leaderless groups), conducted when Rickman and Bion, from early 1942 both in the Army, were simultaneously posted to Northfield Hospital, Birmingham – experiments which were to become extremely influential after the war via the Tavistock Institute, an institution held very much at arm's length by the BPaS before the war, and with which Rickman helped to engineer a rapprochement in 1946. As part of the War Office Selection Boards system, Rickman played the part of the psychiatrist in their training film. See Tom Harrison, *Bion, Rickman, Foulkes and the Northfield Experiments: Advancing on a Different Front*, London: Jessica Kingsley, 2000.

Germany and advised Ernest Jones and the IPA on which of them had 'survived' the war as adequate human beings and which had not – as Rickman put it: 'If one just man could save Sodom and Gomorrah, Berlin is trebly saved in Fräulein Dräger, Steinberg, and Dr. Kemper.'[30]

'I never heard of anyone who was able to persuade him to consider his own welfare',[31] Bion said of Rickman. President of the British Society, Editor of the *International Journal* and the *British Journal of Medical Psychology*, Rickman's final paper for the Society before his sudden death in 1951 was titled 'Reflections on the functions and organization of a Psycho-Analytical Society'. There could have been no better-informed author.

Lionel Penrose

Lionel Penrose (1898–1972) may have been destined to be an FRS, but at the age of thirty nobody could have guessed from his initial distinctive gifts – a playful, speculative bent and a dogged commitment to empirical exactitude – alongside his scientific education in mathematics and logic, theoretical biology, psychology and medicine, in what field he would, in his own due time, make his distinguished, indeed fundamental, contribution. Starting as a psychologist and logician who metamorphosed into a psychoanalyst, he eventually emerged as the founder of modern human genetics.

Born in 1898 to an old and wealthy Quaker family, Penrose served from 1916 to the end of 1918 in the Friends' Ambulance Train of the British Red Cross in France.[32] In an unpublished memoir, almost certainly written in the 1960s, he noted:

> I think that my interest in psychiatry began very suddenly when, during the First World War, one evening I heard a short lecture on Freud's theory of

---

[30] King, 'Activities of British psychoanalysts during the Second World War', p. 32.

[31] See also Dimitris Vonofakos and Bob Hinshelwood, 'Wilfred Bion's letters to John Rickman (1939–1951)', *Psychoanalysis and History* 14(1) (2012), 53–94.

[32] For general accounts of Penrose's life and work, see the very well researched and contextualized account by D. Kevles, *In the Name of Eugenics: Genetics and the Uses of Human Heredity*, Harmondsworth: Penguin, 1985, for whom Penrose is the central figure in the mid-twentieth-century turn against eugenics within science; see also E. Barkan, *The Retreat of Scientific Racism: Changing Concepts of Race in Britain and the United States between the World Wars*, Cambridge: Cambridge University Press, 1992, pp. 260–6; H. Harris, 'Lionel Sharples Penrose 1898–1972', *BMFRS* 19 (1973), 521–61, and O. Penrose, *Lionel S. Penrose FRS: Human Geneticist and Human Being*, ed. Peter Cave and Oliver Penrose, Peterborough: Effective Print, 1999. Kevles's book, to which this study is greatly indebted for its later account of Penrose's life and work, is marred by a singular blindspot concerning Penrose's interest in psychoanalysis.

dreams, given by a lecturer at Manchester University.[33] The occasion was an
informal one when there was a break in the routine on the Ambulance Train
in Northern France on which I was then working. I was astonished to hear
that some fairly reasonable explanation could be given of the apparently
disordered sequence of ideas in the nocturnal theatre with an audience of
one. And I decided then, if possible, to give up mathematics and to study
something more exciting. When the war ended and I went to Cambridge,
I tried to study in this new field but it had not penetrated into the University
curriculum. The nearest possibility was psychology and this was linked to
philosophy and mathematical logic in the cumbersome academic configuration known as the Moral Sciences Tripos.[34]

Coming up to St John's College, Cambridge in January 1919, he recalled at
some point in that year walking down a Cambridge street and running into
John Rickman, whom he knew through the Quaker school, Leighton Park,
they had both attended. Rickman had just started work as a psychiatrist at
Fulbourn Hospital and was full of stories about 'the new psychology' and
his plans to go to Vienna. Some four months after arriving in Cambridge,
on 6 May 1919, Freud's birthday and a date only a thoroughly committed
Freudian would choose, Penrose gave a talk to a society entitled 'Treatise
concerning dreams and Prof. S. Freud's theories'.[35] The talk was well
informed about Freudian theory, to the extent of defending the thesis
which wasn't Freud's, but was commonly understood by Rivers and others
to be his, that all dreams are sexual. Going through the motions of
a Cambridge degree in the logic option of moral sciences, and disappointed
by the absence of Russell and the tedious lectures of Whitehead who was
substituting for him and which were enlivened only by the presence of the
one woman in the dozen or so students in the lecture-hall,[36] Penrose did,
however, meet and spend time with W.H.R. Rivers. Though the two were
never close,[37] Penrose attended Rivers's lectures in 1920–21 that became the
posthumously published *Conflict and Dream*. Penrose's election to the
Apostles in 1921 prompted Lytton Strachey's affectionately tart description:
'a complete flibbertigibbet, but attractive in a childish way, and somehow,
in spite of an absence of brain, quite suitable to the Society'.[38] Penrose's

[33] This was T.H. Pear, later Professor of Psychology at the University of Manchester, author of
Remembering and Forgetting, 1922, the Manchester psychologist already sympathetic to psychoanalysis mentioned by Jones to Freud before World War I, former student of C.S. Myers (see
Chapter 5). See John Forrester, 'Remembering and forgetting Freud in early twentieth century
dreams', *Science in Context* 19(1) (2006), 65–85.
[34] PP, 20/2 'Memoirs – Lunacy', pp. 100/4–101/5.   [35] PP, 21.
[36] PP, 20/2 'Memoirs – Lunacy', p. 106/10.   [37] PP, 20/2 'Memoirs – Lunacy', p. 102/6.
[38] Holroyd, Lytton to James Strachey, 28 November 1921, p. 501.

memories of his formal undergraduate studies being entirely devoid of psychoanalysis were by no means altogether accurate: on 4 June 1920, in the Part I examinations, Question 8 read: 'In what precise sense, if at all, should reference to the Unconscious be admitted in psychological analysis?' And the next day, another examination proposed to him an essay on 'Instinct and the emotions'.[39]

Graduating in 1921, Penrose remained in Cambridge, engaged in research in the Psychological Laboratory. As he himself recounted his career, when elected FRS in 1953:[40]

> After taking my degree at Cambridge I spent one year of psychological researches with F.C. Bartlett and then went off to continue postgraduate studies at Vienna. I conducted some experiments on memory & perception in Prof. Bühler's laboratory. What was for me much more important was the opportunity to meet S. Freud and other psychiatrists including Wagner-Jauregg and Paul Schilder. I was at that time writing a thesis on the psychology of mathematics but it was a failure.[41]

The less prosaic version was:

> So it came about that, after learning nothing at Cambridge except a little mathematical logic of the kind expounded by Russell & Whitehead's *Principia Mathematica*, I set off ... to Vienna with the vague idea of following in Rickman's footsteps.[42]

In an interview in 1965, Penrose informed the historian Paul Roazen that he first went to Freud, who referred him to Siegfried Bernfeld for analysis – Bernfeld spoke in German, Penrose in English and the analysis lasted about a year.[43] He also attended meetings of the Vienna

[39] PP, 22. The examination asked candidates to answer up to five questions; Penrose ticked four questions and placed a question mark against the question on the unconscious.

[40] Like all newly elected Fellows, he was asked by the Society to give a confidential account of his early development and career; each Fellow knew that the material they submitted would later form the basis for the extensive obituaries of its deceased Fellows that the Royal Society publishes in its *Biographical Memoirs*.

[41] PP, 20/4. That this project was redolent of significant future research projects is clear to any reader of Penrose's notes from this era who remembers how Alan Turing hard-wired the first computer to represent specific arithmetical operations in base 32 (Andrew Hodges, *Alan Turing: The Enigma*, London: Burnett Books, 1983, p. 399; it is a nice irony that Hodges's apprenticeship and work in mathematical physics was in the field of twistor theory founded by Penrose's son Roger). A later reader of the notebooks, almost certainly Penrose himself, probably thought so as well: there are a number of comments in biro (certainly not invented in the 1920s ...!) in the passages concerned with logico-neurology, including '(later 1969!)'.

[42] PP, 20/2 p. 100/4.

[43] Bernfeld's account of how he became an analyst indicates that Penrose was quite likely his very first patient: 'in 1922 I discussed with Freud my intention of establishing myself in Vienna as

Psychoanalytic Society.[44] Penrose's notebooks for this period – from 1920 to 1925 – range widely, over neurology, logic, psychological experiments, the mathematics of growth, chess problems and psychoanalytic metapsychology. His heart may have been with psychoanalysis, but his inquiring mind was as ever in many other places.

Between 1923 and the summer of 1924, Penrose appears to have been commuting between Vienna and Cambridge; officially, he was working on a higher degree in the psychology of mathematics with Frederic Bartlett, but he also went to the Salzburg Psychoanalytical Congress in April 1924.[45] He was also, as ever, tinkering with electrical circuits, electro-mechanical models of neurones representing logical and mathematical operations. Extremely prolific in producing extended drafts of papers he never brought to a satisfactory final form, in July 1923 he completed one 'dissertation' – possibly an unsuccessful Fellowship dissertation entitled 'The fundamental nature of neurotic reasoning'.[46] His brother Bernard (Beacus) was also in Vienna for a period, where he had a quick, failed analysis with John Rickman to cure him of his seafaring fever.[47] In equally light mode, sometime in 1923, Penrose collaborated with a friend he would see frequently in Belgrade, A.J.C. Brown, on a pretty dreadful poem they called 'The neurotic's progress'. In December 1923 he wrote 'A formulation of the fallacy in neurotic reasoning',[48] an attempt, based on Freud, to formalize the logical structure of obsessional neurosis – teasing out the original errors and mistaken beliefs constituting the neurosis. In August 1924, Penrose was attempting to put these papers together under the title 'Three papers on psycho-analysis and formal logic'[49]

a practicing analyst. I had been told that our Berlin group encouraged psychoanalysts, especially beginners, to have a didactic analysis before starting their practice, and I asked Freud whether he thought this preparation desirable for me. His answer was: "Nonsense. Go right ahead. You certainly will have difficulties. When you get into trouble, we will see what we can do about it." Only a week later, he sent me my first didactic case, an English professor who wished to study psychoanalysis and planned to stay in Vienna about one month. Alarmed by the task and the conditions, I went back to Freud; but he only said: "You know more than he does. Show him as much as you can."' Siegfried Bernfeld, 'On psychoanalytic training', *Psychoanalytic Quarterly* 31 (1962), 453–82 at 461–2.

[44] When Paul Roazen interviewed Penrose in 1965, Penrose told him that his analyst was Bernfeld (personal communications to John Forrester, Paul Roazen, email, 5, 6 and 13 February 2000). He also informed Roazen that he bought from a bookseller the psychoanalytic library of Herbert Silberer, who had committed suicide on 12 January 1923, and whose widow had sold his library to the bookseller. Penrose's Mss. mention Bernfeld by name on a number of occasions when discussing associations to dreams and chess problems.

[45] BF, 3 November 1924, p. 107.   [46] PP, 24.   [47] Holroyd, pp. 623–4.   [48] PP, 24.

[49] PP, 26.

Fig. 6.2 Lionel Penrose, 1922.

1. A formulation of the fallacy in neurotic reasoning
2. The intellectual aspect of the psycho-analytic cure
3. The elements of the psychology of mathematics.

As he noted in the Introduction, the first two papers apply logic to psychoanalysis, whereas the last applies psychoanalysis to 'the study of formal logic and mathematics'. These were, he noted, written over a period of months and may show changes since 'during the whole time I was intensively studying the subject of analysis and improving my acquaintance with it'.[50] The major influences on him at this point are also clearer: the first paper was being entirely rewritten in the light of comments and

---

[50] PP, 26/1.

criticisms by readers, 'who include Dr. Ernest Jones, to whom my special thanks are due'.[51] The third paper, the heart of his failed doctoral dissertation, arose when it

> was suggested to me by several people some two years ago that the psychology of mathematics was an interesting and practically unexplored field of research. I thought then that it would be quite impossible to make such a study, as the mathematical activity seemed to be carried on at such a deep layer of the mind and yet to be so abstract and mysterious that the mental processes concerned were quite invisible. Two factors have helped to throw light on these processes though. One is a study of the work of Wittgenstein who, carrying Russell's ideas about mathematics to their logical conclusion, produces a metaphysical view of mathematics and formal logic which reveals to some extent its psychological structure. The other factor which, in connection with the one I have mentioned, has revealed a sufficient amount concerning the workings of the mind of one who studies mathematical logic, is the fairly thoroughgoing psycho-analysis of myself, to which I have been sufficiently inquisitive to be subjected.
>
> The important point in the whole paper, to my mind is just the explanation of how the obsessional interest in tautology develops.[52]

It is quite possible that Frank Ramsey was the person who suggested this project to Penrose. Certainly the growing influence of Wittgenstein is visible in Penrose's Notebooks at this time, and Ramsey, as translator of the *Tractatus*, was becoming closely associated with Wittgenstein.[53] In the early 1920s, the burgeoning Cambridge Wittgenstein circle overlapped almost completely with the burgeoning psychoanalysis circle. In another paper from April 1925, entitled 'On the applications of elementary logic to some problems in psychology',[54] Penrose described how his original project of constructing a mechanical or electrical machine that would do logical work in the same way a slide rule does arithmetic had grown less interesting to him since he found out that he was

---

[51] PP, 26/2.   [52] PP, 26, 'Introductory note' to 'Three papers', pp. 3–4.

[53] On Penrose's death, David Garnett, writer and member of the Bloomsbury Group, wrote a letter to *The Times* with his memories of the distinguished scientist, whom he had first met in 1923, noting: 'At Cambridge he discovered that Frank Ramsey, then a fellow undergraduate, had the most remarkable mind in the university and he was profoundly influenced by him' (*The Times*, 22 May 1972). Thus Garnett placed the beginnings of the friendship between Ramsey and Penrose in the period 1920–21, when they were both undergraduates.

[54] PP, 27, dated 9 April 1925.

anticipated in some of these inventions by no less an authority upon logic than Wittgenstein himself. My reason for discontinuing these researches lay in the discovery that an excellent machine had already been prepared on similar lines for solving these very problems in formal logic. I refer to the invention of God, the human organism. And in this paper I propose to discuss the way in which this particular apparatus for solving logical problems works.[55]

While in analysis in Vienna, Penrose's interests shifted to the abnormal mind – as he described in his notes for the use of the Royal Society:

> My interests swung to the problems of the abnormal mind which seemed to offer so much opportunity for the enquirer. I determined to start afresh and to take a medical degree so that the field of abnormal psychology would be open to me. [Adds with asterisk at foot of page:] *I was much encouraged in this aim by my discussions with A. G. Tansley.* So I returned to Cambridge and took the 1st and 2nd M.B. examinations.[56]

Another influence may also have been making itself felt. On the return journey from Vienna at the end of the summer of 1924, where he had been holidaying not very agreeably in the Tyrol with Philip and Lella Florence,[57] he was in the company of two more of the Cambridge party, one of them Margaret Leathes, medical student daughter of a Professor of Physiology and FRS, ex-Bedales, close friend of fellow moral sciences graduate Frances Marshall at Newnham just after the war. Margaret and Lionel became close as Lionel crammed three years of medical training into one, catching up with Margaret's more orthodox pace. They qualified within three months of each other and married four years later.[58]

Back in England at medical school, Penrose was clearly fully integrated, via the Apostles, via Tansley, into the English psychoanalytic community. He also had extensive experience of the Continental psychoanalytic community, as a comment recorded by James Strachey in January 1925, nicely representative of the puerile tone often adopted by Bloomsbury, indicates. Alix had written from Berlin that 'Mélanie is rather tiresome, as a person – a sort of ex-beauty & charmer – & she's unpopular with a certain section of the Ψas';[59]

[55] PP, 27 'On the applications of elementary logic to some problems in psychology', f. 1.
[56] PP, 20/4. James Strachey reported to Alix on 3 November 1924 (BF, p. 107) that Lionel was settled in Cambridge studying medicine.
[57] BLSP 60714, Vol. LX, October 1924–February 1925, James Strachey to Alix, 10 December 1924: 'They were also very bitter about my brother Lionel Penrose, and I altogether gathered that their Tyrol party had been rather a strain.' 'Brother' is used – half-jokingly – because Lionel was also an Apostle; the correct description of other Apostles by an Apostle was 'brother'.
[58] See Oliver Penrose, 'Lionel Penrose, FRS, human geneticist and human being', pp. 9–10.
[59] BF, Alix to James Strachey, 11 January 1925, p. 180.

James responded: 'Lionel Penrose says he has always had a theory that Melanie [Klein] is a retired prostitute. I thought she was too domineering for that; but it sounds as though he was right.'[60]

Penrose's first published paper was entitled 'A note on the relation of rate of growth to structure in plants'. It appeared in late 1925 in the *New Phytologist*, the journal Tansley had founded with his own money in 1902 and edited until 1931. It was a much truncated version of the paper Penrose delivered to the psychoanalytic group on 20 May 1925 and which he also delivered to the British Psychological Society, Medical Section (of which Rickman was then Secretary), on 27 May 1925. This paper, 'The relation of the pleasure–pain principle of Freud to the question of growth', was far more concerned with the mathematics of growth than with Freud's metapsychology. Penrose was attempting to establish equations regulating the rate of growth in a variety of organisms, and drew on data from Quetelet's statistics of human weights and heights, rates of growth of leaves and plants, and, finally, the eugenicist Henry H. Goddard's figures comparing the growth of normal and defective children – data supplied to Penrose by Rickman in a letter two days before he completed the paper.[61] The next paper Penrose published was 'Some experiments upon inhibition and suggestion', in 1926 in the *British Journal of Medical Psychology*, of which Rickman was Assistant Editor.

On 3 February 1926, Penrose gave a talk to the BPaS on 'Psycho-analytic notes on negation' and a revised version appeared in the *International Journal*.[62] The immediate stimulus for this paper was the publication of Freud's remarkable four pages on 'Negation' in September 1925, which started out from the recognition that 'the content of a repressed image or idea can make its way into consciousness, on condition that it is *negated*'[63] and on that basis developed metapsychological reflections around the dichotomies 'external/internal', 'affirmation (Eros)'/'negation (instinct of destruction)'. Penrose had been thinking about the relations between logic and psychology since he was an undergraduate imbued with Freud's theories. Just as Penrose had supplemented D'Arcy Thompson's account of the mathematics of plant growth, so he supplements Freud's

---

[60] BLSP 60714, Vol. LX, October 1924–February 1925, James Strachey to Alix, probably 13 January 1925; not included in BF.

[61] PP, 29, Rickman to Penrose, 18 May 1925: 'I enclose the book references that I spoke about yesterday, and a few silly little diagrams of the weights of lunatic children [drawn from Goddard's 1912 paper].'

[62] D. Bryan, 'British Psycho-Analytical Society', *BIPA* 7 (1926), 533–4; L. Penrose, 'Some psycho-analytical notes on negation', *IJP* 8 (1927), 47–52.

[63] Freud, 'Negation', *SE* XIX, 235.

characterization of the 'not' or the '∼' of negation as a sign of repression, its trademark. He argues that there is a comparable process of assertion as a sign of the unconscious, captured in the logical notation developed by Russell and Whitehead, '∃'. "It is only when there is no repression in connection with the subject matter of a thought stated that neither assertion nor negation are necessary: all that need be expressed then is acquiescence, and this is done by affirmation.'[64]

The sole published fruits of Penrose's attempt to develop a psychology of mathematics found their way into this paper: 'The laws of formal logic are in fact symbols for the processes which allow or prevent thoughts entering the conscious mind in its normal state (i.e. preconscious mechanisms).'[65] And he characterizes the essential abstinence of mathematical and logical thought from assertions of content (following Russell and Wittgenstein) as a loss of contact with the external world: 'it can equally well be used for building up a systematic paranoic illusion as to help in the understanding of reality. And it is based solely upon the interaction of the two ideas, assertion and negation, which we have seen represent pleasure and pain.' Thus he is led to a disputed metapsychological point: 'I do not think Alexander is correct in stating that the laws of logic are copied from the laws of nature, and that they represent a fragment of introjected reality. It would seem to be more accurate to say that they are projections of the mechanisms by which the mental apparatus functions.'[66]

The place of the erotic in reasoning, Penrose argues, is to be found in induction, not deduction – in risking *not* finding the world to be as thought assumes it to be: 'the Ego reasons deductively and the Id inductively ... while negation represents a complete withdrawal from reality, assertion is indicative of the aggressive, sadistic attitude towards external objects. Both states show a disharmony between the conscious and the unconscious, whereas affirmation is the mark of harmony between these two.'[67]

The significance of this paper was as a bridge between foundational problems in philosophy and the metapsychology being developed by psychoanalysis. Penrose would have been acutely aware, as a logic specialist from Cambridge and avid reader of Russell and Wittgenstein, of the deep significance of 'negation' for Cambridge philosophical logic. There is a strong case for regarding the problem of negation as being as important for Russell's and Wittgenstein's development as the logical paradoxes. For

---

[64] Penrose, 'Some psycho-analytical notes on negation', p. 49.   [65] *Ibid.*, p. 50.   [66] *Ibid.*, p. 51.
[67] *Ibid.*, p. 52.

Wittgenstein, the problem of negation could not be 'solved': 'Here is a deep mystery. It is the mystery of negation: This is not how things are, and yet we can say *how* things are *not*.' Or: 'If e.g. an affirmation can be generated by double negation, is negation in any sense contained in affirmation? Does "p" deny "not-p" or assert "p", or both?' Yet the quality of an attempted answer to the problem of negation might sound its death-knell, in the same kind of way as a *reductio ad absurdam* (which is, of course, a form of argument deploying negation, in the form of the principle of non-contradiction).[68] Penrose's bridge from logic to psycho-analysis attempted a fusion between psychology and logic which would very soon become increasingly deprecated; Freud's enigmatic paper on 'Negation' would have no immediate answer from either philosophy or psychology.[69] Russell occasionally flirted with the idea that logic must be founded in psychology, but most later philosophers would dismiss such a project as the crassest form of 'psychologism'.

After Cambridge, Penrose moved to St Thomas' Hospital, London for clinical work, taking up residence in Bloomsbury, both socially and geo-graphically (first in Mecklenburgh Square, then in Brunswick Square, then with Adrian Stephen in Gordon Square when Karin was away in America in 1927).[70] Some sense of how he initially conceived of combining objective scientific methods and psychoanalysis can be gained from his account of data-gathering at this time:

> When attending hospital out-patients as a medical student I spent a little time enquiring into the dreams patients experienced under nitrous oxide when some minor operation was performed. The idea was to discover whether there was any regularity or uniformity to be found. I observed a frequent tendency to identify the whole body with the part which was operated upon. For example, a dream, during dental extraction, that the patient was being pulled out of a ditch.[71]

---

[68] For the philosophical and linguistic problem of negation, see the comprehensive and remarkable Lawrence Horn, *A Natural History of Negation*, Chicago: University of Chicago Press, 1989.

[69] There are two later significant attempts to take up the project of a 'psychoanalytic logic': Matte Blanco, I, *The Unconscious as Infinite Sets*, London: Duckworth, 1975 and Jacques Lacan in his *Seminars* of the mid-1960s.

[70] He lived at 27 Brunswick Square, Frances Marshall's mother's house, which had become a 'communal house' where she had bed-sitting rooms for her daughters and their friends (FPM, pp. 71–2); in 1927 he lived with Adrian Stephen in Gordon Square (Rice interview with Margaret Newman, 16 June 1984, cited in Gillian Rice, 'The reception of Freud amongst the Bloomsbury Group', MPhil dissertation, University of Cambridge, 1984, p. 66n132).

[71] L. Penrose, 'Psycho-analysis and experimental science', *IJP* 34 (1953), 74–82 at 79.

Lionel was up for election as Associate Member of the BPaS on 17 February 1926. On the 18th Lionel wrote to 'brother' (fellow-Apostle) Lytton Strachey:

> It was yesterday that the Psycho-analysts had their meeting about whether to elect me or not, and as I have heard nothing I suppose that James has had his own way – in which case I shall be forced to begin a campaign for the exposure of quacks and the first two to be exposed will be J & A. But then – what if they have elected me? Then it will be worse as I shall be forced to defend quacks or resign or prove that they aren't quacks or better still be one myself.[72]

Whatever James's views, Lionel was elected. When the Institute of Psycho-Analysis opened on Freud's birthday on 6 May 1926 – and amidst rebuilding works, John Rickman ceremoniously listened to the first patient – Penrose, along with Adrian Stephen and Marjorie Brierley, was amongst the first to be appointed as a Clinical Assistant and permitted to conduct analyses under supervision.[73] This entailed devoting five hours a week to the analysis of one patient. Rickman acted as go-between for Ernest Jones and Penrose in this appointment.[74] Penrose was also supervised by Ernest Jones,[75] and it may well have been on these occasions that Jones told Penrose he had too many other interests besides psychoanalysis, a view with which Penrose agreed.

Penrose's initial experiences with his clinic patients were not entirely happy.[76] His first, a newly qualified doctor from Ireland with a severe stammer, soon failed to turn up to his sessions. (Penrose noted after the first session: 'He is protestant, but [sets] great store upon the Catholic "confuc-fuc-fuc- confessional".'[77]) His second patient, a married woman, came to one session and then refused to come to any more, on the grounds that she had caught a cold in the draughty consultation room: 'If it is really necessary for me to lie for an hour in a complete draught at every treatment, I am afraid I shall be unable to attend again.'[78] To the Treasurer of the Institute she reported: 'I mentioned the discomfort of the draught to Dr. Penrose, but he did not see his way to alter the conditions.'[79] Whether any other of Penrose's analyses were more successful we do not know.

---

[72] BLSP 60690 Vol. XXXVI, Penrose to Lytton Strachey, 18 February 1926.
[73] A. Freud, 'Report of the Tenth International Psycho-Analytical Congress', *BIPA* 9 (1928), 132–56 at 147.
[74] PP, 46, Jones to Penrose, 18 October 1926.   [75] PP, 46 Penrose to Jones, 30 December 1926.
[76] Penrose interview with Paul Roazen, 1965, personal communication with Paul Roazen, 5, 6 and 13 February 2000.
[77] PP, 46.   [78] PP, 46, Mrs R. to Penrose, 2 February 1927.
[79] PP, 46, Mrs R. to Treasurer, 5 February 1927.

Lionel and Margaret may have considered moving back to Cambridge: Margaret applied – unsuccessfully – for the position as a teacher at the Malting House School when it was advertised in July 1927.[80] Lionel completed his medical degree in 1928 and then, while retaining a London address, took a position at the City Mental Hospital in Cardiff, where he completed a thesis for his Cambridge MD (1930) centred on one elderly man diagnosed with schizophrenia. He published a paper concerning this man in the *British Journal of Medical Psychology* in 1931. Penrose did not pretend to 'treat' this old man psychoanalytically. His paper portrayed it as a sufficient triumph firstly that an eighty-year-old man who had been in mental institutions for fifty years could, given his complex delusional system, be understood; and secondly, that this successful act of comprehension demonstrated that the man's mental powers had in no sense degenerated. Penrose's paper analysed the complex delusional system, including its new language, calendar and system of measurements of time and distance. And, he noted, it was this delusional system that kept the patient's mind alive, much as another kind of system might a philosopher or abstract scientist.

When Penrose considered the causation of the patient's mental illness, he dismissed the factor of heredity, noting – and foreshadowing his later life-work – that if insanity were a recessive Mendelian trait, 'we should expect to find the taint on both sides of the family'.[81] (As we have seen, Penrose's analysis with Bernfeld had alerted him to his own preoccupation with the different legacies of the two sides – 'White' and 'Black' – of his own family.) Having shown, in classic psychoanalytic style, such impatience with the doctrines of heredity and degeneration, he located the precipitating cause of the illness in the patient's early sexual history: an affair with a previously childless landlady which led to a pregnancy and stillbirth, combined with an episode of gonorrhea, seemed adequate in Penrose's eyes to qualify as the precipitating cause. While concluding that this was not a *sufficient* specific cause of the illness, and that a set of predisposing causes was necessary, he pointed out that such an aetiology was entirely analogous to the onset of infectious diseases. What is more, he modelled his account of the case on Freud's reading of Schreber, even to the point of seeking to locate the onset of the delusions in the repression of homosexuality and the patient's regression to masturbatory fantasy, in

[80] Pyke Papers.
[81] L. Penrose, 'A case of schizophrenia of long duration', *BJMP* II (1931), 1–31 at 28.

which anal and oral erotisms were dominant under a regime of primitive magical thinking.

Perhaps proud of his remarkable achievement in his limpid and sympathetic understanding of a man locked away in an asylum since 1875, long before psychoanalysis was dreamt of, Penrose sent this paper to Freud, who replied courteously (and in German):[82]

> Dear Dr Penrose,
>
> Thank you for your offprint. The case is really quite educational and it is once again good to see how a life-history only becomes understandable once one investigates the sexual life.
>
> With respects,
> Freud

However, in early 1931 Penrose had published a psychoanalytic paper which returned to one of his initial interests: the theoretical status of the biological principles underlying psychoanalysis, in particular that of pleasure–pain. He was contributing to a debate sparked by the attempt of his own analyst, Bernfeld, together with the physiologist Sergei Feitelberg to integrate Freud's theory of instincts and particularly of the death instinct with the physical sciences, giving a physical interpretation of the notion of 'psychic energy' and attempting to equate the death instinct with the second law of thermodynamics (the tendency of entropy to increase).[83] The other English commentator on their paper, Reginald Kapp,[84] was scathing in his dismissal of the elementary errors in physics, chemistry and biology committed by the Viennese Freudians.[85] Penrose's verdict on their theory was equally negative: the rapprochement with physics was preposterous and their interpretation of biology was deeply flawed; in addition, their biology was considerably different from Freud's, which had maintained a more appropriate tentativeness. Penrose's undoubted respect for Freud had not hindered him concluding that:

---

[82] Postcard dated 18 May 1931 to: 35 Lexden Road, Colchester, England. 'Dear Dr. Penrose, Danke Ihnen sehr für den Sonderabdruck. Der Fall ist doch sehr lehrreich und es ist wieder einmal gut zu sehen wie eine Lebensgeschichte erst verständlich wird, wenn man das Sexualleben erforscht, Ihr ergebener, Freud' (LoCSF, B-17, Lionel Penrose).

[83] S. Bernfeld and S. Feitelberg, 'The principle of entropy and the death instinct', *IJP* 12 (1931), 61–81.

[84] For further details on Kapp, see Laura Cameron and John Forrester, 'Tansley's psychoanalytic network: an episode out of the early history of psychoanalysis in England', *Psychoanalysis and History* 2(2) (2000), 189–256.

[85] R. Kapp, 'Comments on Bernfeld and Feitelberg's "The Principle of Entropy and the Death Instinct"', *IJP* 12 (1931), 82–6.

the attempt to shew that Freud's death instinct is the same as the second law of thermodynamics is rejected. It is concluded that if Fechner's principles of stability are made the basis of a theory of pleasure and pain the death instinct becomes of theoretical interest only.[86]

Both Kapp's and Penrose's papers were much quoted in later psycho-analytic literature as sounding the death-knell firstly for any rapprochement between the notion of energy as employed in metapsychology and that employed in thermodynamics, and secondly for the biological coherence of the death instinct.

On 21 October 1931, Penrose gave a further paper to the BPaS, entitled 'Recent psycho-analytical research in the psychoses', which consisted in 'a review of work done in the period 1926–1931'[87] – quite clearly a Supplement in the most literal sense to John Rickman's magisterial *The Development of the Psycho-Analytical Theory of Psychoses, 1893–1926* published in 1928. One has the sense that Penrose was still following in Rickman's footsteps. But it was to be his last psychoanalytic foray for many years.

Penrose family stories tell of a vague and tragic love affair some time in the 1920s, which ended traumatically with the mental illness of the woman. As a result, Penrose is said to have moved more towards a biological view of abnormality. Although the evidence for this is by no means clear,[88] his personal and professional lives took a distinct shift in the second half of the 1920s. In 1928 he married Margaret Leathes, two months pregnant;[89] they were to have four children, the first born in 1929.[90] In 1931, Penrose took up

---

[86] L. Penrose, 'Freud's theory of instinct and other psycho-biological theories', *IJP* 12 (1931), 87–97 at 97.

[87] E. Glover, 'British Psycho-Analytical Society', *BIPA* 13 (1932), 265–6.

[88] See Kevles, *In the Name of Eugenics*, pp. 154, 340n20. As if anticipating such an account, the Ms. which Penrose entitled 'Memoirs 2 – Lunacy' addresses the place that lunacy had had in his life, starting with his experience at the Cardiff Hospital, then immediately asserting, following Freud, the comprehensibility of the symptoms of mental illness, and finally moving back in time to those relationships of his own which had been marred by mental illness: the brother of a close friend from the war who had had a breakdown; the wife of the same close friend who turned out to be schizophrenic; finally, the woman he befriended in Whitehead's lectures, Molly, who, in the period 1919–21, was clearly an intimate friend with whom Penrose kept in contact, and who succumbed to mental illness many years later (but no earlier than the late 1930s, it is reasonable to estimate on Penrose's account, which carefully disguises her identity). These distressing stories of lunacy fade into the shadows once the bright sunny figure of John Rickman enters into his story.

[89] FPM, 6 November 1928, p. 155; Oliver was born 6 June 1929.

[90] All of whom followed in their father's many footsteps: Oliver and Roger became distinguished mathematicians; Jonathan was ten times British Chess Champion and a university psychologist; the youngest, Shirley Victoria, became a genetic paediatrician. Kevles infers, in part from his interviews with Margaret Penrose Newman (as she became after Penrose's death, when she married Lionel's closest friend in Cambridge, Vienna and after, the distinguished mathematician and computer pioneer Max Newman, who was also a close working colleague of Harold Jeffreys, particularly on

a post that was to reorient his scientific interests and produce the distinguished work on mental deficiency, hereditary factors in disease and mental illness, and the study of human genetics for which he became famous. He was appointed Research Medical Officer, sponsored by the Pinsent-Darwin Trust of the University of Cambridge, at the Royal Eastern Counties Institution at Colchester, a large asylum devoted to the mentally defective with a long history of linkage to the Cambridgeshire Voluntary Association for Mental Welfare, whose driving force, Lady Ida Darwin, was also the organizing spirit behind the Pinsent-Darwin Trust.[91] Within months of his arrival he was publishing studies on the heredity of mental defects, in particular in 'mongolian imbeciles' – a term which his work would do much to make archaic. Employing Mendelian genetics, painstakingly gathering enormous amounts of information about the family backgrounds of the patients in the Institution, giving up a part of his salary so he could employ a full-time assistant to help in this mammoth task,[92] he constructed a vision of what a truly scientific account of the factors – hereditary, congenital, environmental – in mental deficiency, mongolism and rarer defects (such as phenylketonuria) could look like. Based in Colchester till 1939, when he became Director of Psychiatric Research in Ontario for the duration of World War II, he carved out a singular identity as a human geneticist, chromosomal researcher, statistician *extraordinaire* and severe (while never polemical) critic of the wild eugenics of the first third of the twentieth century. In Daniel Kevles's classic history, *In the Name of Eugenics*, Penrose is the principal scientific hero of the victory over the eugenics of the early twentieth century, with his painstaking amassing of hereditary data and rigorous statistical analyses severely qualifying arguments for viewing the transmissability of mental illness, whatever its incidence, as of any eugenic concern.

Yet even in this work there are hints of his Freudian past. In his 1933 book *Mental Defect*, when discussing the supposed abnormally strong sexual drives of the retarded, he argued:

Scientific Inference), that it was partly her influence that cooled Lionel's passion for psychoanalysis in the late 1920s; while she no doubt made disparaging comments about the portentousness of psychoanalysts (who doesn't?) and appeared to adopt a distinctly sceptical attitude (see Kevles, *In the Name of Eugenics*, esp. p. 340n19), she herself had analysis in the 1930s (Rice, 'The reception of Freud amongst the Bloomsbury Group', p. 66n132, citing an interview 16 June 1984), attended as a Guest (often without Lionel) seven meetings of the Society in 1930 alone (BPaS, FSA 8–18) and still in 1946 (BPaS, FSC/41) and was again seeing an analyst in 1971 (FPM, 12 June 1971, p. 665).
[91] John Forrester, '1919: psychology and psychoanalysis, Cambridge and London: Myers, Jones and MacCurdy', *Psychoanalysis and History* 10(1) (2008), 37–94 at 57–9.
[92] CUEP, Minutes, Pinsent-Darwin Trust, 1932.

It is a well-known psychological mechanism that hatred, which is repressed under normal circumstances, may become manifest in the presence of an object which is already discredited in some way ... The greatest psychiatrist of modern times, Sigmund Freud, has pointed out very clearly that some of the most serious troubles affecting civilization come from man's imperfect mastery over his aggressive impulses against his neighbor. An excuse for viewing mentally defective individuals with abhorrence is the idea that those at large enjoy themselves sexually in ways which are forbidden or difficult to accomplish in the higher strata of society. The association between the idea of the supposed fecundity of the feeble-minded and the need for their sterilization is apparently rational, but it may be emphasized by an unconscious desire to forbid these supposed sexual excesses. It has been pointed out that the advocates of sterilization never desire it to be applied to their own class, but always to some one else.[93]

But Penrose's activity was now lodged elsewhere, including developing (with his assistant J.C. Raven, later also funded by the Pinsent-Darwin Trust) a new set of psychological tests, eventually known as the Penrose-Raven Progressive Matrices Test, which were made standard in the British Army during World War II by John Rickman.[94] In 1940, Penrose was corresponding with Rickman about another set of tests for schizophrenia that he was formulating.[95]

While in Canada, Penrose corresponded with J.B.S. Haldane about his possible return to Britain; in 1945 he was elected Galton Professor of Eugenics at University College London, where he created the leading postwar centre for human genetics research. He never cared for the name 'eugenics' with its associations of racial purification – indeed, his inaugural lecture culminated in a cool assessment of the intellectual poverty of eugenic proposals.[96] Until he succeeded in having the name changed to the Galton Professorship of Human Genetics in 1963, he used notepaper titled simply *The Galton Laboratory, University College*.[97] Working on statistical studies of heredity, including sceptical studies of the efficacy of psychiatric treatments (e.g. shock therapy), on the biology of mental defect (the title of his 1949 book), an early supporter of and participant in the Society for Social Medicine, so influential in the use of medical statistics

[93] L. Penrose, *Mental Defect*, London: Sidgwick & Jackson, 1933, pp. 172–4; see Kevles, *In the Name of Eugenics*, p. 108 and R. Gosselin, '*Mental Defect*, by L.S. Penrose', *Psychoanalytic Quarterly* 4 (1935), 529–31.

[94] L. Penrose and J. Raven, 'A new series of perceptual tests: a preliminary communication', *BJMP* 16 (1936), 97; King, 'Activities of British psychoanalysts during the Second World War', p. 24.

[95] PP, 165/3, Penrose to Rickman, 11 February 1940.

[96] L. Penrose, 'Phenylketonuria. A problem in eugenics', *The Lancet* 247(6409) (29 June 1946), 949–53.

[97] Harris, 'Lionel Sharples Penrose 1898–1972', pp. 537–8.

and epidemiology for public health in Britain[98], he remained an Associate Member of the BPaS till his death in 1972, and accepted an invitation in 1947 to give a paper to it, sandwiched between Melanie Klein on schizoid mechanisms and Donald Winnicott on hate. His title was 'Psycho-analysis and experimental science'.

Voicing more clearly the same muted scepticism concerning the fundamental biological and physical principles of Freud's metapsychology that he had expressed in his 1931 paper, he nonetheless declared himself, once again, convinced of the fundamental revolutionary character of Freud's scientific work:

> Freud's greatest contribution to psychological medicine has practically nothing to do with theories of mental energy, mental philosophy and metapsychology. It was a revolution in thought comparable in its effects to the discoveries of Darwin or Copernicus . . . There should be no need for me to discuss the nature of this nuclear discovery from which the recognition of repression, infantile sexuality, and the Oedipus complex followed. Suffice to say that it was the realization (1) that the phenomena of hypnosis, suggestion and so on were fundamentally sexual, in a word, that the patient obeys because he is in love with the doctor, and (2) that this neurotic love is itself a morbid symptom, which can be subjected to psychological analysis in terms of conditioned associations and ultimately resolved by a process of re-education. The analysis of transference is the key to Freud's contribution to therapeutics. It is this which differentiated his method from that of his teachers, Charcot and Bernheim, and which should differentiate present psychoanalytical from other forms of psychotherapy. It is perhaps unfortunate that the discipline was named psycho-analysis and not transference analysis. Much misunderstanding would thereby have been avoided.[99]

Having delivered this scientific *credo*, he gave his audience a sense of how he had kept alive his connection to psychoanalysis while he became involved in biological research that was increasingly remote from it:

> For many years I have tried to keep a watch on the developments in fields related to psycho-analysis in order to see how new information would help to strengthen or modify the assumptions inherent in psychoanalytic theory. It is difficult to present these points in any very coherent order since the original compilation is of the nature of a mental scrapbook.[100]

Penrose then went on to survey a number of fields of investigation which touch on psychoanalysis, and which psychoanalysts should either take note

---

[98] J. Pemberton, 'Origins and early history of the Society for Social Medicine in the UK and Ireland', *Journal of Epidemiology and Community Health* 56 (2002), 342–6.
[99] Penrose, 'Psycho-analysis and experimental science', p. 75.   [100] *Ibid.*

of or contribute to: the artificial cathartic cures (narcoanalysis, drug-induced convulsions); conditioning experiments in animals and humans; the relation of hormone chemistry and genetic variations to sexuality and sexual types, including the differential distribution of characteristics associated with males or females to the neuroses and psychoses; the experimental study of dreaming; the study of smell in relation to the unconscious; the part played by genetics in determining mental illness, including such factors as the choice of mate; the possibility that the superego may be localized in the prefrontal lobes. He concluded with a rousing call to analysts not to rest on the laurels of their founders: 'In science, we cannot go wrong so long as we stick to the facts.'[101] His final suggestion was: 'I should like to see every patient given a battery of psychological and physical tests before beginning treatment, and perhaps afterwards as well – children not excepted. But then, I am not a practising psycho-analyst; if I were I might think differently.' One suspects that Penrose knew just how far he had come since starting out as a practising analyst. Yet he was quite happy to let his name be used on an Appeal for the new building acquired by Rickman for the BPaS and Institute in 1950, alongside those of Dr J.C. Flügel, Dr William Gillespie (its Chairman), Dr Ernest Jones, Sir Arthur Tansley and Professor F.R. Winton.

Penrose and Rickman maintained their close friendship – in much the same style as in the 1920s perhaps. Penrose's psychoanalytic career had been very much in Rickman's footsteps up until the early 1930s. In the 1930s, following the Einstein–Freud published correspondence *Why War?*, Rickman, Penrose and others (including Adrian Stephen, Norman Glaister, John Bowlby, J.C. Flügel and Charles Madge) formed the Psychologists' Peace Society, which also published a Medical Peace Campaign Bulletin in good Cambridge wartime fashion. (Penrose was instrumental in founding a similar post-war organization, the Medical Association for the Prevention of War, in 1951 at the time of the Korean War, to which he devoted much energy in the 1950s and 1960s.)[102] Rickman introduced Penrose's psychological tests into the British Army in the 1940s. In 1946 Rickman offered to have Penrose elected a Fellow of the British Psychological Society.[103] Penrose read and commented on the proofs of the last paper Rickman wrote before his sudden death in July 1951.

[101] *Ibid.*, p. 81.
[102] Norman Macdonald, 'Lionel Penrose', *Medical Association for the Prevention of War Proceedings* 2(5) (1972), 117–22.
[103] PP, 165/3.

In a footnote to his long and striking paper on the organization of psycho-analytical societies, Rickman's discussion of the work of committees referred to Penrose's observation,

> in respect to the Unwritten Agendas, that the total number of topics of discussion including the Written Agenda is two to the power n, where n is the number of Committee members who stay awake. The practical importance of this formulation is that it gives a hint, perhaps a strong hint, that the enlargement of a small committee even by one member produces a greater effect than arithmetical proportions would suggest.[104]

This typically whimsical yet serious note was closely related to other work Penrose was then engaged on, his *On the Objective Study of Crowd Behaviour*, completed in May 1951. In the Introduction to his book, Penrose referred to Rickman's recent study, 'The factor of number in individual and group dynamics', published in 1950 (itself a part of a series of papers concerning number and groups[105]), which viewed Freud's Oedipus complex as having the characteristics it does because it is a 'three-person relationship'. Here Rickman introduced the notion of two- and three-body psychology – to be taken up by others, such as Michael Balint.[106] It was this notion of the importance of number that led Jacques Lacan to call Rickman 'one of the rare souls to have had a modicum of theoretical originality in analytic circles since Freud's death'.[107] While it was Bion, Rickman's other close collaborator in thinking about group dynamics, who was to have most influence, it is clear that Rickman and Penrose were thinking on very similar lines and sharing their ideas with one another. Penrose's lucid and original little book still incorporated psychoanalytic ideas: in its final few pages he cited Freud's account of the repressed homosexuality that lies behind paranoid delusions, indicating that it is this force 'which binds together members of the same sex into groups which follow a paranoid leader'.[108]

[104] J. Rickman, 'Reflections on the function and organization of a psycho-analytical society', *IJP* 32 (1951), 218–37 at 230n4.

[105] Collected together after his death in J. Rickman, *Selected Contributions to Psycho-Analysis*, compiled by W.C.M. Scott, London: Hogarth Press and Institute of Psychoanalysis, 1957, in particular 'Methodology and research in psycho-pathology' (1951) and 'Number and the human sciences' (1951). A roneoed version of 'Number and the human sciences' which Rickman had made up for use of the BPaS members is also in the Penrose Papers (54/3), though there are no notations on it.

[106] M. Balint, 'On love and hate' (1951), in Balint, *Primary Love and Psycho-Analytic Technique*, London: Hogarth Press and the Institute of Psycho-Analysis, 1952, pp. 141–56.

[107] J. Lacan, *The Seminar of Jacques Lacan*, Book I, *Freud's Papers on Technique*, 1953–1954, ed. Jacques-Alain Miller, trans. with notes by John Forrester, Cambridge: Cambridge University Press, 1988, p. 11.

[108] L. Penrose, *On the Objective Study of Crowd Behaviour*, London: H.K. Lewis, 1952, p. 62.

Penrose remained an Associate Member of the BPaS to his death. However, from 1960 onwards a curious footnote was attached to his name in the roster – 'Not eligible for Associate Membership of the International Psycho-Analytical Association.'[109] The reason for the sudden appearance of this forbidding qualification would have amused John Rickman, while it also indicates the scarcely visible trace of the historical episode of the Cambridge Freudians narrated in these pages. In 1958, the then President of the IPA, Heinz Hartmann, announced that

> the relationship of the associate members of Component Societies to the International Association has never been clearly outlined. The practice has been inconsistent. For years, the associate members of some Component Societies have paid dues to the International and have been regarded as its members; in others, they are not charged dues to our Association by their Component Societies and not considered as its members. I shall submit to your vote a change of statutes that tends to regularize this situation.[110]

Under the guise of an item concerning the correct dues to be paid to retain membership in the organization, an item that would be sure to send a large proportion of any Committee to sleep, the IPA then proposed a new regulation governing the membership by Associates of the IPA – a proposal that was adopted and incorporated into its Statutes in 1959:

> The Association consists of ordinary and associate members. Its ordinary membership is composed of the honorary and ordinary members of the component societies, whose election is therefore decided by the conditions valid for the individual societies. Its associate membership consists of the associate members of the component societies, but only in so far as the standing of an associate member in an individual society implies graduation from a recognized psycho-analytic institute.

Penrose thus found himself in a 'terminated' historic category, along with three other associate members similarly excluded: Dr Josephine Stross (Freud's doctor, along with Max Schur, in the last years of his life[111]), Mr Prynce Hopkins of Santa Barbara, California (an Associate Member of the

---

[109] Pearl H.M. King, 'List of members of the Component and Affiliate Societies of the International Psycho-Analytical Association 1959–1960', *BIPA* 41 (1960), 215–73 at 239.

[110] Ruth S. Eissler, MD, General Secretary, '113th Bulletin of the International Psycho-Analytical Association – Report on the Twentieth International Psycho-Analytical Congress', *BIPA* 39 (1958), 276–96 at 290.

[111] Recent research has shown that it was Stross who managed Freud's final days; despite the promise he had made Freud, Schur was too agitated to take the final steps to put Freud into a morphine-induced state from which he would not emerge. Appropriately, then, there is a memorial to Stross in the garden of the Freud Museum in London.

British Society on account of the fact that he had donated in 1925 a substantial sum to found the Institute of Psycho-Analysis), and Dr F.R. Winton, Cambridge physiologist, who had been elected an Associate Member in 1923 and would remain one until his death in 1985.[112]

The condition that made these relics of an earlier idea of a psychoanalytic society appear so anomalous was the new clause that 'the standing of an associate member in an individual society implies graduation from a recognized psycho-analytic institute'. Stross, Hopkins, Winton and Penrose had close links to the psychoanalytic movement, but they had never 'graduated'. From its early conception as a sign of *scientific* interest in and affiliation with psychoanalysis – the early Statutes of the IPA had only mentioned Associate Members in order to assert that 'Associate Members of the Branch Societies have the right to be present at the Scientific Meetings of the I.P.A.',[113] with no mention of any required qualifications – Associate Membership gradually came to mean trained but still junior member of a professional organization.[114] Back in 1924, Jones had been asked by the central powers in Vienna to explain the function of the Associate Member in the British Society; he answered:

Associate Members are elected year by year and thus can be released if they are not satisfactory. It gives a probation period to test whether they deserve promotion and membership and it affords an opportunity to have associated with us a few workers in other sciences who are seriously interested in

[112] Born in 1894, Frank Winton (originally Weintraud until he left Oundle School) read Natural Sciences at Clare College, Cambridge, getting a First at Part I in 1915, and a Third in Part II in 1916. He then studied physiology at University College London, specializing in renal autoregulation. He began a 'medical' analysis with Ernest Jones in the autumn of 1918, as he reported to James Strachey, completed by the end of 1920 (BPaS, Jones Papers, CWA/F03/01, Winton to Jones, 9 January 1921). After he had finished his analysis, Jones asked him to translate a paper from the German (Winton being bilingual) for the new *Journal* (Michael Joseph Eisler, 'A man's unconscious phantasy of pregnancy in the guise of traumatic hysteria', *IJP* 2 (1921), 255–86); this 'test' was not a success and Winton did no more translation. He was elected an associate member of the BPaS at the same meeting as the Stracheys, on 4 October 1922, and was familiar with them in the 1920s. In 1931, he was appointed Lecturer in Physiology at Cambridge, then Reader in 1933, before returning to London in 1938 as Professor of Pharmacology at UCL, a post he held till 1961. Author with Leonard Bayliss of a standard textbook, *Human Physiology*, he was President of the Renal Association, 1956–62. After leaving UCL, he took a position with the pharmaceuticals company May & Baker. He remained an associate member of the BPaS until his death in 1985.

[113] 'Statutes of the International Psycho-Analytical Association', *BIPA* 9 (1928), 156–9, Statute 4, Branches and Membership.

[114] The only other major Psycho-Analytic Societies affiliated to the IPA with associate members who had this asterisk attached to their names were the Italian Society (one associate member) and the Indian Psycho-Analytical Society (all associate members).

psycho-analysis without wishing either to practise or to profess a thorough knowledge of it.[115]

In 1958, when all such trace of a non-professional 'serious' interest in psychoanalysis had disappeared or was to be excluded on principle, such an explanation carried no weight. It could even be concluded that psycho-analytic organizations no longer cared about 'workers in other sciences who are seriously interested in psycho-analysis without wishing either to prac-tise or to profess a thorough knowledge of it'. We do not have a record of Penrose's response to this ruling (if any); however, he did not resign his Membership.

If we are looking for signs of a residual ambivalence in relation to Penrose's long-standing affiliation to psychoanalytic organizations, we may find it in the confidential autobiographical account he prepared for the Royal Society:

> Scientific or partly scientific societies to which I belong include, British Psychological Society, Biometric Society, Pathological Society, American Psychiatric Association, Royal Medical-Psychological Association, Harveian Society, Cambridge Philosophical Society.[116]

In this document drawn up in 1953, he singularly failed to mention his long-standing membership of the BPaS. And, unlike over his debt to Tansley, he failed to add a note at the bottom of the page correcting his oversight.

Penrose, gloriously and in the end triumphantly, did have too many interests: if there is one guiding thread to them, we would call it 'pleasure in puzzles'. If he had continued with his electrical circuit building in logic, he would have produced the first computer to model the operations of the mind – he would have been Alan Turing (or his close friend Max Newman, who actually did build the world's first electronic stored-program digital computer based on Turing's ideas in Manchester).[117] He was always inter-ested in the formal properties of biological organisms, so when Watson and Crick published their double helix model of DNA in 1953, Penrose quickly produced an alternative model, built out of blocks of wood he was always fiddling with, showing how the basic building blocks of an organism could self-replicate. The pleasure in puzzles had begun early. As a boy, in his strict Quaker household, one of the few games allowed was chess; Lionel became

[115] JF, 18 December 1924, pp. 564–5.   [116] PP, 20/4.
[117] See Hodges, *Alan Turing* and B. Jack Copeland, *Colossus: The Secrets of Bletchley Park's Codebreaking Computers*, Oxford: Oxford University Press, 2010.

an internationally known chess problematist before he entered his teens. While an undergraduate, on 30 April 1922 he gave a talk to the Heretics on 'The chess problem: a neglected form of art'.[118] There is even speculation that he played chess while an undergraduate with that other fine undergraduate chess problematist Vladimir Nabokov, so two varieties of Freudian lifelong 'enthusiasm' – one deeply sceptical and always malicious, the other wildly enthusiastic and generous – may have taken up the black and white pieces together.[119] Chess problems, electrical circuits, family lineages statistically mapped out, self-replicating mechanical automata, the Penrose endless stair, soon taken up by M.C. Escher – all these were the variants on his penchant for puzzles. 'Lionel', his close friend Margaret Gardiner recalled of their time together in Vienna in 1923, 'with his impish look of surprised discovery, was infinitely ingenious and forever inventing things, frequently absurd things – a chess problem on a slab of pink and yellow chequered cake, the mechanism of a homing pigeon'.[120] If you go on the internet looking for traces of his work, you will find him alongside John von Neumann as the builder of the first realized example of stochastic mechanical self-replication: the creator of alternatives to biological replication, and thus a founder of the hardware solution to the problem of artificial intelligence.

In the 1930s Penrose had given £1,000 to endow a Clinical Essay Prize at the BPaS; in the last part of the twentieth century the Prize was only intermittently advertised and no longer survives.[121] In the Deed of Trust, Penrose specified:

> The essay shall consist of a clinical record of a case investigated by psychoanalytical methods. It shall clearly illustrate events and changes in the mental life of the patient and their relations to external environment. In awarding the prize the Judges will pay attention to acuity of observation and the clearness with which the facts are stated. If the writer wishes to draw

[118] Damon Franke, *Modernist Heresies: British Literary History, 1883–1924*, Columbus: Ohio State University Press, 2008, p. 229.
[119] See Catharine Theimer Nepomnyashchy, 'King, queen, sui-mate: Nabokov's defense against Freud's "Uncanny"', *Intertexts (Lubbock)* 12(1/2) (2008), 7–24, 80–1.
[120] Margaret Gardiner, *A Scatter of Memories*, London: Free Association Books, 1988, p. 61.
[121] £1,000 in 1932 would be roughly £62,000 in 2015 prices. The Prize was awarded occasionally in the period from the 1930s to the 1960s and re-advertised in the 1980s (Eric Brenman, 'The Clinical Essay Prize – of the British Psycho-Analytical Society', *IJP* 63 (1982), 90). In 1995, Clifford Yorke reported that: 'For several years, I acted as a scrutineer for papers submitted to the British Psychoanalytical Society for their annual clinical essay prize. In the last year I served, only three papers were submitted, none of them outstanding. There were no submissions in the following year, and the prize was discontinued' (C. Yorke, 'Freud's psychology: can it survive?' *Psychoanalytic Study of the Child* 50 (1995), 3–31 at 19).

theoretical conclusions he must bear in mind the necessity of making the evidence for such conclusions carry conviction. It is recommended the length of the essay should not exceed Twenty thousand words.[122]

Penrose also made clear that this clinical essay was not intended for publication; quite the opposite. Its terms of reference encouraged the recording of the sort of confidential material that could not, by definition, be published; Penrose even envisaged how future changes in the management of the Library might affect access to the accumulated Prize Essays in the Institute's Library: 'Should the Library at any time be thrown open for public use the series of prize essays shall be withdrawn from the open shelves and put into a category for special readers.'[123] This donation has something of the character of an *envoi* about it, as if it symbolized Penrose drawing a line under his psychoanalytic career: since 1917 psychoanalysis had been his central intellectual passion – although he always had more passions than he could ever nurture at any one time; he had now made the definitive decision to leave that world for good – no longer dabble with it, and with other areas, but make his decision to be not a psychoanalyst but a scientific researcher in the field of human genetics. Of course he could never do only one thing, but the central thing, he had now decided, would no longer be and never again be psychoanalysis.

Some sense of the enduring effect of psychoanalysis in his life can be gained from the following story. In 1968, his old friend Frances Partridge – Frances Marshall as she had been when they were attending moral sciences lectures together in 1920 – recorded in her diary a visit to Vienna with Lionel to hear him lecture:

> After his lecture we went to look in vain for Freud's house in the Berggasse. Later, an enormous walk in the dark to places connected with Lionel's early life in Vienna, all round the old town through streets of portentously tall houses, under bridges, up steps. Lionel was being psychoanalysed at that time for bed-wetting, so Margaret told me later. Having heard this I naturally saw the course on which he led us in symbolical terms. Down a deep sunken street called Tief Graben (Deep Trench) under a bridge and towards the Hohe Markt. 'I'm always dreaming about Vienna,' he said as we walked along, 'and in my dream there's always somewhere I'm trying to get to. Now I realize that it's here.' When we found a large fountain in full operation in the middle of the marketplace, it seemed too good to be true.[124]

[122] Clinical Essay Prize, 1932 Deed of Trust, British Psycho-Analytic Society Archives: G05/BJ/F06/03.
[123] 'Clinical Prize Essay', *IJP* 13 (1932), 492–3.   [124] FPD, p. 568.

Frank Ramsey

Having 'romanced so much about Cambridge that to find myself sitting there was an anti climax', an excited Virginia Woolf visited King's College in February 1923, where for the first time she met Frank Ramsey, aged twenty: 'Ramsay [*sic*], the unknown guest, was something like a Darwin, broad, thick, powerful, & a great mathematician, & clumsy to boot. Honest I should say, a true Apostle.'[125] The young man was also a prodigy and, in the eyes of many, a genius. He was born in 1903 and died at the far too youthful age of twenty-six in 1930, having already made major contributions to probability theory, economics, philosophy and the foundations of mathematics.[126] Like Keynes his patron, Ramsey came from a Cambridge family – his father was President (Vice-Master) of Magdalene College from 1897 to 1934 and his brother Michael in the 1960s became

Fig. 6.3 Frank Ramsey aged eighteen.

[125] VWD II, 7 February 1923, p. 231.
[126] D.H. Mellor, 'Frank Ramsey', *Philosophy* 70 (1995), 243–62.

Archbishop of Canterbury. He was the sort of child who would be caught
in bed learning the Latin dictionary.[127] Ogden had supervised Ramsey's
learning of German while at school: he had recommended reading German
originals and English translations in concert; Ramsey turned the advice
around and read English philosophy (Berkeley and Hume) in German.[128]
As a second year mathematics undergraduate in 1922, who had already
presented a paper in late 1921 to the Moral Sciences Club (on 'The nature
of propositions'[129]), Ogden needed 'to find something for F.P. Ramsey to
employ himself on between whiles',[130] so set him to work as the principal
translator of Wittgenstein's *Tractatus Logico-Philosophicus*; Ramsey
then wrote in *Mind* one of the most perspicacious expositions and criticisms of Wittgenstein's argument.[131] He was Secretary of the Heretics
for a while; Kingsley Martin remembered the manner in which he fulfilled
the task:

> It is Frank Ramsey who comes back most vividly to my mind when I speak
> of the Heretics. I am remembering an occasion when, as secretary of the
> society, he read from the minute book a very learned and complete summary
> of a philosophical lecture we had listened to the previous week. He passed
> the book for signature to Ogden, in the chair, who stared in astonishment at
> finding the pages completely blank.[132]

The most vivid description of him is the valedictory by Keynes:

> His bulky Johnsonian frame, his spontaneous gurgling laugh, the simplicity
> of his feelings and reactions, half-alarming sometimes and occasionally almost
> cruel in their directness and literalness, his honesty of mind and heart, his
> modesty, and the amazing, easy efficiency of the intellectual machine which
> ground away behind his wide temples and broad, smiling face.

By 1923, he was helping Bertrand Russell revise *Principia Mathematica* in
the light of criticisms made of it and was a friend of Keynes, whose *Treatise*

---

[127] FRP, 'Ramsey's father's memoir'.
[128] Gabriele Taylor, 'Frank Ramsey – a biographical sketch', in Maria Carla Galavotti (ed.), *Cambridge and Vienna: Frank P. Ramsey and the Vienna Circle*, Vienna Circle Institute Yearbook 12, Dordrecht: Springer, 2004, pp. 1–18 at 6.
[129] In Frank Plumpton Ramsey, *On Truth*, ed. Nicholas Rescher and Ulrich Majer, Dordrecht: Kluwer, 1991.
[130] Richards to Gellner, 7 March 1960, in J. Constable (ed.), *Selected Letters of I.A. Richards*, Oxford: Clarendon, 1990, p. 159.
[131] F.P. Ramsey, 'Critical notice. Ludwig Wittgenstein, *Tractatus Logico-Philosophicus*, with an Introduction by Bertrand Russell. (International Library of Psychology, Philosophy and Scientific Method.) London: Kegan Paul, Trench, Trubner & Co., 1922. Pp. 189. 10s. 6d.', *Mind* 32(128) (1923), 465–78.
[132] KMFF, p. 108.

on Probability he reviewed in the *Cambridge Magazine*, in effect destroying its central arguments, in the eyes of many, including Keynes himself.[133] Ramsey spent the summer of 1923 in Vienna, where he met Wittgenstein who was then teaching at a village school, and became one of the few philosophical friends Wittgenstein could engage with. In part, as Strachey reported to Alix, because of an unrequited passion for Mrs Geoffrey Pyke, wife of the founder of the Malting House School managed by Susan Isaacs from September 1924 on, he was considering psychoanalysis. As he put it to Wittgenstein, in a letter dated 12 November 1923:

> most of my energy has been absorbed since January by an unhappy passion for a married woman, which produced such psychological disorder, that I nearly resorted to psychoanalysis, and should probably have gone at Christmas to live in Vienna for 9 months and be analysed, had not I suddenly got better a fortnight ago.[134]

Some sense of Ramsey's broodings can be gained from a talk he gave on 26 January 1924 to the Apostles. He had been elected a member in 1921, probably around the time of Penrose's election. James Strachey had been a member since 1906, and he combined meetings of the two societies on his visits from London. Ramsey's paper was originally entitled 'Psyché: or will it bear the light'; its final title was 'An imaginary conversation with John Stuart Mill'.[135]

John Stuart Mill walks in the door while Ramsey is sitting beside the fire feeling bored and opens the conversation: 'Have you never read my Autobiography?' Ramsey replies that he has done, remembering particularly his 'remarkable education'; Mill replies that he specifically wants to talk about 'the account of my mental crisis . . . one to which all young men are liable, if they use in introspection analytic intellect' (p. 302). Quoting extensively from Mill's *Autobiography*, their discussion turns on the question whether Freud's discoveries represent a decisive advance – and thus a satisfactory account of and answer to mental depression – when

---

[133] *CM* 11(1) (1922), 3–5, reprinted as 'Mr Keynes on probability', *British Journal for the Philosophy of Science* 40 (1989), 219–22. The best guide to Ramsey's work in the period 1923–27 on Russell's logicizing project is I. Gratton-Guinness, *The Search for Mathematical Roots, 1870–1940: Logics, Set Theories and the Foundations of Mathematics from Cantor through Russell to Gödel*, Princeton: Princeton University Press, 2000, esp. pp. 443–8.

[134] L. Wittgenstein, *Letters to C.K. Ogden, with Comments on the English Translation of Tractatus Logico-Philosophicus*, ed. with intro. by G.H. von Wright, and an Appendix of Letters by Frank Plumpton Ramsey, Oxford: Blackwell/London: Routledge and Kegan Paul, 1973, pp. 81–2.

[135] F.P. Ramsey, 'An imaginary conversation with John Stuart Mill' (1924), in Frank Plumpton Ramsey, *Notes on Philosophy, Probability and Mathematics*, ed. Maria Carla Galavotti, Naples: Bibliopolis, 1991, pp. 302–12.

compared with Mill's own associationist psychology ruled by the pleasure–pain principle.

RAMSEY: But you know psychology has advanced since your day, yours is very out of date.

MILL: Has it? I don't think so. You have advanced in philosophy in a way that excites my profound admiration but in psychology hardly at all. Perhaps you are thinking of the followers of Freud, who seem to regard the analysis of the mind as a panacea.

RAMSEY: Yes, I am thinking of them; you are probably put off by their absurd metaphysics, and forget that they are also scientists describing observed facts and inventing theories to fit them.

MILL: I doubt that; they alone observe these facts, and the different schools do not agree as to their nature still less in their interpretations . . .

RAMSEY: But of course [Freud] would dispute your psychology; he would say, that the most important associations in determining your desires were those formed very early in life and no longer accessible to consciousness . . .

MILL: I don't agree . . .[136]

And thus the argument swings back and forward, until Ramsey 'concedes' with the very Cambridge phrase 'You may be right . . .' so as to move on to another topic: obliging Mill to recount the details of his depression and his mode of self-cure, only to play the omniscient analyst to Mill by demonstrating that his own account of that cure and his new goals in life is inadequate. Mill rebuffs the suggestion and Ramsey concedes:

RAMSEY: But I do heartily agree that we must not too frequently ask ourselves or discuss with one another whether we are happy, which perhaps is your chief contention.

MILL: Yes that is what I came to tell you.
And walked out of the room.[137]

Ramsey was clearly trying out an identification with another English philosopher who had been, like him, an extraordinarily precocious thinker, and who had also confronted a major mental collapse: Ramsey feared this was what awaited him and wondered if the major difference between 'then' and 'now' – Freud – could make a difference to their respective destinies. Central to Ramsey was the question of happiness.

By February 1924, Ramsey had decisively revised his proposal to Wittgenstein:

[136] Ramsey, 'An imaginary conversation', pp. 306–7.   [137] Ibid., p. 312.

if I live in Vienna I can learn German, and come and see you often (unless you object) and *discuss my work with you*, which would be most helpful. Also I *have been very depressed* and done little work, and have *symptoms so closely resembling some of those described by Freud* that I shall probably *try to be psychoanalysed, for which Vienna would be very convenient*, and which would make me stay there the whole six months. *But I'm afraid you won't agree with this. Keynes still means to write to you;* it really is a disease – his procrastination; but he doesn't (unlike me) take such disabilities so seriously as to go to Freud![138]

In March 1924 Ramsey left Cambridge for Vienna and, after a brief meeting with Freud, he began analysis with Theodor Reik. He spent much time with Wittgenstein's sister, Margarete (Gretl) Stonborough and her circle, seeing Wittgenstein himself only on four occasions. Living with Penrose, he probably spent time with Rickman and Tansley as well.[139] Reporting to his mother:

> Everything is going very well. I like my analyst though he is a Jew (but all the good ones are). But being analysed is different from what I expected in being at any rate at first much more exhausting and unpleasant. But yesterday it didn't seem so bad.
>
> I have settled in with Lionel. He is quite industrious and out a good deal and so I ought to be able to work without him stopping me. He has found an absorbing vocation; he is being analysed, goes to lectures, classes in psychiatry (lunacy), and experiments in a lab. But he won't ever do for an analyst as he has no critical capacity or common sense.[140]

A week later, he had somewhat sharper things to say about Lionel: 'I have the idea of not going on living with Lionel, as he is impossible to talk to. Psychoanalysis has destroyed his brain altogether.'[141]

Writing to Sprott about his analysis, he was more revealing than to his mother:

> It is surprisingly exhausting and unpleasant ... For about 2 times I said what came into my head, but then it appeared that I was avoiding talking about Margaret, so that was stopped and I was made to give an orderly account of my relations with her ... I rather like him, but he annoyed me by asking me

---

[138] Ludwig Wittgenstein, *Letters to C.K. Ogden*, Ramsey to Wittgenstein, 20 February 1924, p. 84.
[139] When Penrose was interviewed by Paul Roazen in 1965 about his years in Vienna and his contact with psychoanalysis, he spontaneously informed Roazen, who knew nothing of Ramsey, that Ramsey had been in analysis with Theodor Reik. As we have seen, Penrose specifically mentioned Tansley as an important influence in his decision to seek medical training, clearly while they were both in analysis in Vienna in the first half of 1924, so it is plausible that Tansley also supported Ramsey's psychoanalytic enthusiasms while they were all in Vienna.
[140] FRP, letter to mother, probably 23 March 1924.   [141] FRP, letter to mother, 30 March 1924.

to lend him Wittgenstein's book and saying, when he returned it, that it was an intelligent book but the author must have some compulsive neurosis.[142]

During his analysis, he would make the odd request of his mother – concerning the identity of early figures from his childhood, or the book from his childhood, *Peter Pan*, that his analyst wanted to physically handle. And he would report on the importance of psychoanalysis, aware that he might be criticized for devoting himself to psychoanalysis rather than mathematics:

> Psycho-analysis is very important even I think to one's work. You see obscure unconscious things may decide your attitude about certain things, especially personal factors in a controversial subject. Lots of work on the Foundation of Mathematics is emotionally determined by such things as
>
> (1) love of mathematics and a desire to save it from those (villainous and silly) philosophers
> (2) whether your interest in mathematics is like that in a game, a science, or an art
> (3) General Bolshevism towards authority
> (4) The opposite, timidity
> (5) Laziness or the desire to get rid of difficulties by not mentioning them.
>
>   If you can see these in other people you must be careful and take stock of yourself.[143]

Always modest, never sure if he was working quite the right way, his speed and range are clear in this July survey of his intellectual interests:

> I haven't been working very hard but I've solved some things I thought almost impossible. I just can't keep on thinking about it more than a few hours a day it is so immensely difficult. I read a good deal of psychoanalytic literature, but am thinking of going back to relativity. I'm becoming rather an enthusiast for psychoanalysis. I've been reading a book by Reik on the psychology of religion which is most awfully good. That is his special subject but he isn't a good writer, but rather heavy. We really live in a great time for thinking, with Einstein Freud and Wittgenstein all alive (and all in Germany or Austria, those foes of civilization!).[144]

Ramsey followed Reik to his summer resorts, taking off only one week until the beginning of October; over the summer he was joined by another of Reik's patients, Lewis Namier from Balliol, whose culture and erudition

---

[142] FRP, Letter to Sprott, 2 April 1924, also quoted in Taylor, 'Frank Ramsey – a biographical sketch', p. 8.
[143] FRP, letter to mother, 4 June 1924.   [144] FRP, letter to mother, 22 July 1924.

he appreciated.[145] Having seen Wittgenstein one last time, Ramsey returned at the last minute to take up his Fellowship at King's in the Michaelmas Term.[146]

As soon as his analysis was over, the gossip networks connecting Vienna and Berlin informed Alix Strachey of the progress of the *Wunderkind* and she naturally passed the news on to James in London: '[Reik] was enthusiastic about Frank Ramsay's [*sic*] beautiful character, & seemed to think, analytically, that all was for the best.'[147] A few weeks later, in early November, with Ramsey back in England, James delivered a further progress report, this time from brother Lytton's visit to Cambridge:

> Ramsey (who, before he went to Vienna didn't know that he wanted to fuck Mrs. Pyke) discovered there that he did, but thought himself cured of such wishes. On returning & meeting her, however, he was more bowled over than ever; but asked her to go to bed with him – which she declined.[148]

The London analytic gossip network was equally informative:

> Glover said Ramsey (so the story went) is under the impression that he's completely analysed, & that Reik said – 'if you'd had unlimited time & money, we couldn't have gone deeper'.[149]

James's view of Ramsey on bumping into him in the Bloomsbury bookshop in December was less sanguine:

> He struck me as having simply been hypnotized. He at once began a long & violent tirade on the subject of the 'active' technique [of Ferenczi]. 'Ho! Weren't *you* analysed actively? Awhawch! Okk!' There are no analysts in England who know their job. It's monstrous for an analysis to last more than 6 months, etc. etc. I formed the darkest view of the Herr Dr.[150]

---

[145] FRP, letter to mother, August 1924.
[146] For some of these details on Ramsey, see Ray Monk, *Ludwig Wittgenstein: The Duty of Genius*, London: Jonathan Cape, 1990, esp. pp. 215 ff.; personal communications from Ramsey's sister Margaret Paul, and Margaret Paul, *Frank Ramsey (1903–1930): A Sister's Memoir*, Huntingdon: Smith-Gordon, 2012. In 1926, with the drawing up, following the introduction of the new University Statutes, of lists of 'established' University Lecturers attached directly to the new Departments and Faculties, he was appointed a Lecturer in Mathematics in the University.
[147] BF, 13 October 1924, p. 86.   [148] BF, 3 November 1924, p. 10.
[149] BF, 6 November 1924, p. 108. The tone of James's remarks implied a criticism of Reik, which explains why Alix in Berlin jumped to his defence: 'I think all that about Reik is nonsense . . . He said to me that he'd done all he could to Frank [Ramsey] *in the short time* at his disposal – that the analysis had gone very well owing to Frank's crystal-clear mind & soul – was enthusiastic about him; and wound up by saying that there'd never been anything much wrong with him. All of which seems fairly reasonable' (BF, 7/8 November 1924, p. 112).
[150] BF, 20 December 1924, pp. 153–4.

Fig. 6.4 Lettice Ramsey, painting by Frances Baker, *c.* 1915.

But within weeks of his return from Vienna, Ramsey had become involved with Lettice Baker, a former Cambridge undergraduate a few years his senior, Treasurer of the Heretics in 1921 when Sprott was Secretary.[151] She had returned in 1924 to work in the Psychological Laboratory on 'mental imagery and motor skill' under Bartlett's supervision.[152] As Strachey reported to Alix:

> Incidentally, [Sprott] said that Ramsey *has* been cured. He's abandoned Mrs. P[yke]; has taken on a new lady with whom (though, before, the idea

---

[151] FRP, Diary, 22 January 1921. This seems to be the first occasion when Ramsey met both Baker and Sprott.

[152] *CUR*, 17 March 1926, p. 783. In 1926 Bartlett reported that she had produced for the Laboratory the first report in Cambridge on vocational guidance. After Frank's death, she worked in 1931 for the Cambridge Women's Welfare Association (Newnham College Register) and then in 1932, with Helen Muspratt, set up the best-known photographic studio in Cambridge, servicing the Colleges and University for the next fifty years – Ramsey & Muspratt (see www.loftyimages.co.uk/gallery_408953.html, accessed 13 April 2015).

had filled him with repulsion) he proceeds to the furthest limits . . . Perhaps we'd better all go on to Reik.'[153]

And the same day Frank wrote to Lettice: 'I wrote a long letter to my psychoanalyst saying how happy I was and how grateful I felt to him. Because he did make it possible though you may not see how. Darling it is very wonderful.'[154] The couple married in late August 1925.

The marriage appears to have been happy, if studiedly unconventional. Two daughters were born, Jane on 12 October 1926 and Sarah on 28 March 1929. Lettice was a decidedly independent woman, five years older and more sexually experienced than Frank: 'I think she broke him in as it were', Frances Partridge, like Lettice, Bedales and Newnham, close friend to both, reflected in 2001.[155] Frank began an extended affair with Elizabeth Denby, a friend of Lettice's, a pioneer housing consultant.[156] This was with Lettice's knowledge and blessing, but it took him away from the family (for instance, Christmas with Denby at Le Lavandou on the Côte d'Azur) and inevitably caused tensions; however it was Lettice's brief but intense affair with the Irish writer Liam O'Flaherty in 1927 that caused both of them most pain. The hardest time for them was after Frank's mother was killed in a motoring accident in 1927.

Three months after his marriage, on 24 November 1925, Ramsey again spoke to the Apostles, this time on 'Civilisation and happiness'. It was only ten months since his debate with J.S. Mill, but the tone was significantly weightier, more sombre but less stricken – not quite the tone one would expect of the newly married twenty-one-year-old he in fact was:

> I have only lately begun to feel that civilisation is opposed to happiness; I feel it as a burden which I am forced to carry and cannot throw off, and I should be interested to discover whether we all suffer under it or whether I am merely objectifying the heaviness of my heart.[157]

[153] BF, 22 December 1924, p. 157. Naturally others gossiped about the relationship. Richard Braithwaite, who had stayed with Lytton Strachey in early January 1925, wrote to him on 16 January 1925: 'I was horrified to have repeated to me the other day the remarks I was alleged to have made to you about Miss Lettice Baker. "Perfect whore" was the most picturesque of the expressions: though I agree with the description I am sure it is too neat to have originated from me' (BLSP 60660, Vol. VI, Braithwaite to Lytton Strachey, 16 January 1925).

[154] FRP, Ramsey to Baker, headed King's College, Cambridge, Monday 22–24 December [1924].

[155] Interview with Frances Partridge by John Forrester, 10 September 2001.

[156] Elizabeth Darling, '"The star in the profession she invented for herself": a brief biography of Elizabeth Denby, housing consultant', *Planning Perspectives* 20 (July 2005), 271–300.

[157] F.P. Ramsey, 'Civilisation and happiness' (24 November 1925), in Ramsey, *Notes on Philosophy, Probability and Mathematics*, pp. 320–4 at 320.

In this inquiry, so akin to Freud's *Civilization and its Discontents*, which had yet to be written, Ramsey was at his most Freudian: happiness comes from the satisfaction of instincts, but civilization, which induces the sublimation of those instincts, deprives us of happiness: 'I think that it is just because they are the products of sublimated and not of primitive instincts, that our pursuits so often seem not really worth while.'[158] The pursuit of truth does not bring happiness, because we are really interested in it not for its own sake, but rather on account of the 'diversion' of other instincts, such as infantile sexual curiosity and a desire to triumph over our parents. 'And it is not the truth which will make us happy, but the satisfaction of those other repressed desires which our conscience will not allow us.' Then Ramsey hit a new note, a more personal note, clearly the fruit of his time spent in analysis:

> In my own case I think that my interest in philosophy and all kinds of criticism, which is much greater than my interest in constructive thought, is derived from a fairly well repressed infantile rivalry with my father and my wish to kill him.
>
> This means that I can never get any great satisfaction from philosophising, never anything like the pleasure I should have got from killing my father, which my conscience or rather my love for him forbade me to do when I was small.
>
> This has incidentally another unfortunate consequence, namely that my philosophical criticisms should always be regarded with suspicion, as I am probably identifying the man I am criticising with my father, generally in his hostile aspect, so that I am biased against the philosopher who in my unconscious mind represents my father. I am also liable to identify someone like Wittgenstein with my beloved father and attach a most exaggerated importance to his every word.[159]

At this point, Ramsey's argument switched to the position of women in civilization and to the feminist movement's influence on general happiness. Recognizing a historical inevitability in the success of women's demand for education and the same position as men, he could not but see this as both loss and gain:

> Not merely is feminism bad for the race but it is unfortunate for the women also, who are forced away from the kind of life which they are fitted by

---

[158] *Ibid.*, p. 321.

[159] *Ibid.*, pp. 321–2. As Penrose's unpublished papers and notebooks make clear, he also was extremely impressed by Wittgenstein at this time; but, with a different character and less sunnily murderous and idealizing relation to his father, the consequences were different, perhaps playing a part in his turn away from logic, philosophy and psychoanalysis.

nature to enjoy. But it seems to me bound to happen, and also to some extent excites my admiration. They are taking upon themselves the burden of civilisation and turning from sexual to intellectual activities which though less satisfying seem to me more excellent. We have here again in opposition culture and happiness, and what I really feel about that business is that I should like myself to be happy and other people to be cultured.[160]

Ramsey spoke on another occasion to the discussion society of Apostles in 1925; his topic was whether there was any discussable subject and was a response to Russell's recent lecture 'What I believe'.[161] He had come to observe that what usually passed for discussion was in point of fact 'comparing notes';[162] there used to be discussion, but the rise of modern science and the decline of religion have

> resulted in all the old general questions becoming either technical or ridiculous. This process in the development of civilization we have each of us to repeat in ourselves. I, for instance, came up as a freshman enjoying conversation and argument more than anything else in the world; but I have gradually come to regard it as of less and less importance, because there never seems to be anything to talk about except shop and people's private lives, neither of which is suited for general conversation. Also, since I was analysed, I feel that people know far less about themselves than they imagine, and am not nearly so anxious to talk about myself as I used to be, having had enough of it to get bored.[163]

Responding again to Russell's recent lecture – such an odd piece to be asserting there is no discussion possible, when it is manifestly a continuous dialogue with his mentor Russell – Ramsey concludes:

> If I was to write a *Weltanschauung* I should call it not 'What I Believe' but 'What I feel'. This is connected with Wittgenstein's view that philosophy

---

[160] *Ibid.*, pp. 323–4.

[161] This paper (without title) was given to the Apostles in 1925 and is an informal response to Russell's lecture (published by Ogden in March 1925) 'What I believe', composed in the last week of 1924, probably delivered in January or February 1925 and published by March 1925 (MBR II, p. 53). If the book had been published, Ramsey would have referred to it as a book rather than a lecture, which he probably heard Russell deliver in London. Alister Watson, five years younger than Ramsey and like him an Apostle and a mathematical product of Winchester and Trinity, later recalled: 'Frank's attitude to his work was affected by his having been psychoanalysed. Having spent so much time talking about himself Frank felt bored with the subject, and this had cured him of part of what Wittgenstein would have called "the disease of philosophy"' (Paul, *Frank Ramsey*, p. 268). This 'recollection' tracks rather closely this passage from Ramsey's paper to the Apostles in 1925; although Watson was too young to attend this meeting, he may well have read it in the first collection of Ramsey's papers (edited by Richard Braithwaite and published in 1931) and unconsciously reproduced it.

[162] Ramsey, *Philosophical Papers*, p. 247.   [163] *Ibid.*, p. 248.

does not give us beliefs, but merely *relieves feelings of intellectual discomfort.*[164] (Italics added for emphasis)

Here, in dialogue with Russell, Ramsey calls on Wittgenstein to voice what many would later associate with the influence of psychoanalysis: the therapeutic cleansing of problems that dissolve when viewed right. Ramsey's debt to psychoanalysis for curing him of his own interest in himself and Wittgenstein's preoccupation with the martyrdom of becoming a leader with disciples that he predicted for himself, both lie inescapably in Freud's aura. With this parentage – Russell, Wittgenstein, Freud – Ramsey might have become the philosopher of psychoanalysis he had enthusiastically promised to be. He died, not yet twenty-seven, in January 1930, of complications following jaundice.[165] He had turned his fine mind so quickly and profoundly to so many fields that the implications of his work are still being pursued.

Harold Jeffreys

Harold Jeffreys (1891–1989) was another brilliant young polymath, elected into a Fellowship at St John's College, Cambridge in 1914 which he held until his death, a tenure longer than the entire duration of the Soviet Union.[166] It is unclear exactly how his interest in psychoanalysis began, but despite his early background he moved in circles where Freud was as omnipresent as mathematics. The son of County Durham mining village schoolteachers, Jeffreys was educated at Rutherford College and then Armstrong College, Newcastle (later the University of Newcastle), receiving a Major Scholarship to enter St John's College, Cambridge to study mathematics (1910–13). His ever wide interests ranged from mathematics, via mathematical physics (he and his wife, Bertha Swirles, wrote *Methods of*

---

[164] *Ibid.*, pp. 248–9.

[165] FRP, File Ramsey to Sprott, letter from Braithwaite to Sprott: 'Sunday' [19 January 1930]: 'Frank Ramsey has died. When they operated for his jaundice (on Tuesday) they found some horrible condition of the liver. He died early this morning in London at Guys. Richard Braithwaite.'

[166] On Jeffreys, see Alan Cook, 'Sir Harold Jeffreys. 2 April 1891–18 March 1989', *Biographical Memoirs of the Royal Society* 36 (December 1990), 302–33; D. Howie, 'Interpretations of probability 1919–1939: Harold Jeffreys, R.A. Fisher, and the Bayesian Controversy', PhD dissertation, University of Pennsylvania, 1999; an excellent overview of his contributions to geophysics is Bruce A. Bolt, 'Jeffreys and the Earth', in Keitti Aki and Renata Dmowska (eds.), *Relating Geophysical Structures and Processes: The Jeffreys Volume*, Geophysical Monograph 76, IUGG 16, Washington, DC: International Union of Geodesy and Geophysics and the American Geophysical Union, 1993, pp. 1–10. We benefited greatly from a generous interview with Bertha Jeffreys (7 January 1999) and an interview with Dan McKenzie (9 May 2002) who has worked in Cambridge on geophysics in the early 1960s and since.

Fig. 6.5 Harold Jeffreys in his St John's College rooms. Possible self-portrait.

Mathematical Physics in 1946, which went through three editions to 1962 and has been endlessly reprinted, most recently in 1999), to geophysics, which established his world-wide reputation (the earthquake travel time-tables he produced with K.E. Bullen in 1940 are still the basis for the modern revised versions), theorist of solar system formation, as well as probability theory, econometrics and botany. Jeffreys taught alongside Tansley and Haddon in the new Geographical Tripos in the early 1920s. His book *Scientific Inference* (1931) is still a classic in the philosophy of science, and included a robust and original defence of the findings and methods of psychoanalysis, quoted at length by his former analyst Ernest Jones with almost paternal pride in his review for the *International Journal*.[167] *Theory of Probability* (1st edition, 1939) by Harold Jeffreys, 'this giant of Bayesian Statistics', is 'rightly considered as the principal reference in modern Bayesian statistics'[168] and led to a notable dispute with the statistician and geneticist R.A. Fisher.

A Wrangler and Smith's Prize Winner, Jeffreys worked in the Cavendish Laboratory from 1915 to 1917 and was part of the Ogden circle: he took photographs of Ogden's bookshops for publications in the *Cambridge Magazine* on 16 June 1917. At the suggestion of Professor Newall, he moved to the Meteorological Office in London as Senior Professional Assistant, to work on hydrodynamics for war-related work, staying there until 1922.[169] During this period, he was close friends with the extraordinary Dorothy Wrinch,[170] a mathematician who became amanuensis to Bertrand Russell and functioned as an intellectual inspiration to Jeffreys, as well as being the force behind an ever reluctant Wittgenstein publishing the *Tractatus*. James Strachey, when he finally met Jeffreys at the first meeting of the 1925 Group, attributed – almost certainly falsely – Jeffreys's turn to psychoanalysis to his being ditched by Dorothy.

> The ΨA meeting itself was very gloomy. Mr Jeffreys, in whose rooms we met, turned out to be … Miss Wrinch's fiancé. Do you remember? – a rather rugged figure, like a wire haired terrier, whom one constantly met on the stairs. I was so aghast at the recollection that I was on the *point* of saying: 'I used to know your wife, Miss Wrinch', but luckily restrained myself. I suppose his jilting was what brought him into Jones's hands. He's incredibly dull-minded, and comes so much from the North as to be almost incomprehensible.[171]

---

[167] E. Jones, 'Scientific inference', *IJP* 12 (1931), 381.
[168] Christian P. Robert, Nicolas Chopin and Judith Rousseau, 'Harold Jeffreys's theory of probability revisited', *Statistical Science* 24(2) (2009), 141–72 at 141.
[169] HJPC A.15, Records for Royal Society.
[170] On Wrinch (1894–1976), P. Abir-Am, 'Synergy or clash: disciplinary and marital strategies in the career of mathematical biologist Dorothy Wrinch', in Pnina G. Abir-Am and Dorinda Outram (eds.), *Uneasy Careers and Intimate Lives: Women in Science 1789–1979*, with a foreword by Margaret W. Rossiter, New Brunswick: Rutgers University Press, 1987, pp. 239–80, 342–54, and P. Abir-Am, 'Dorothy Maud Wrinch (1894–1976)', in L.S. Grinstein et al. (eds.), *Women in Chemistry and Physics*, New York: Greenwood, 1993.
[171] BF, 2 March 1925, p. 223.

Close friends with Ogden and Dora Black in organizing the Heretics in 1916,[172] she introduced Russell and Dora, who would become Russell's wife in 1921. Lecturing at University College London in the period 1917–20, Wrinch shuttled back and forth between UCL and Girton, and, in 1919, began a collaborative research project in the mathematical and philosophical theory of probability with Jeffreys which would result in three substantial joint papers in those areas in the period 1919–23. Their collaboration extended into other areas of physics: firstly a series of joint responses to Einstein's relativity theory[173] – Jeffreys had been present at the 'historic' meeting of the élite mathematical physics $\nabla^2 V$ Club meeting in St John's College on 22 October 1919 when Eddington announced that the light deflection confirmed Einstein's general relativity theory.[174] Wrinch and Jeffreys also collaborated on a joint paper concerning the possibility of using the seismographic records of the gigantic explosion at a chemical works on 21 September 1921 in the village of Oppau in Bavaria to produce a standard for establishing more accurate measurements of the travel times, position and size of earthquakes, from which more sophisticated models of the internal structure of the Earth could be developed.[175] Jeffreys, who had long been interested in articulating geophysics with theories of the origin of the solar system, would go on to make this one of his principal areas of expertise, particularly with the publication of his book *The Earth* in 1924. This book went through five influential editions. The final edition of 1970 remained adamantly opposed to the continental drift theory he had definitively refuted in 1924 (in Wegener's version). His grounds were the impossibility, according to the accurate and well-confirmed quantitative model he had put forward, of their being convection currents in the earth's mantle – an argument bypassed with the rise of plate tectonic theory in the 1960s.[176]

[172] OP, Box 112, F.12, Wrinch to Ogden, 8 July 1916, where she suggests as speakers Russell, Karin Stephen, '– even Mr. Adrian Stephen?' In the event, Adrian Stephen did speak to the Heretics on 29 April 1917 on the topic 'In defence of understanding' (see *CM* VII, 28 April 1917, p. 551, 'Calendar of Events').

[173] D. Wrinch and H. Jeffreys, 'On certain aspects of the theory of probability', *Philosophical Magazine* 38 (1919), 715–31; 'On certain fundamental principles of scientific inquiry', *Philosophical Magazine* 42 (1921), 369–90; 'The relation of geometry to Einstein's theory of gravitation', *Nature* 106 (1921), 806–9; 'On certain fundamental principles of scientific inquiry', *Philosophical Magazine* 45 (1923), 368–74; 'The theory of mensuration', *Philosophical Magazine* 46 (1923), 1–22.

[174] Andrew Warwick, *Masters of Theory: Cambridge and the Rise of Mathematical Physics*, Chicago: University of Chicago Press, 2003, p. 480n100.

[175] D. Wrinch and H. Jeffreys, 'On the seismic waves from the Oppau explosion of 1921', *Royal Astronomical Society Monthly Notices*, Geophysics Supplement, January 1923, pp. 15–22.

[176] See J.A. Stewart, *Drifting Continents and Colliding Paradigms: Perspectives on the Geoscience Revolution*, Bloomington: Indiana University Press, 1990, esp. pp. 116 and 34–5. To give some sense of the style and argument of Jeffreys's continued opposition, we quote the introduction to his six-volume *Papers* (*Collected Papers of Sir Harold Jeffreys on Geophysics and Other Sciences*, ed. Sir

Before Jeffreys established a major reputation in geophysics with this work, his collaboration with Dorothy Wrinch had come to an end, their ways parting with dramatic changes for both of them. Jeffreys, discontented with the Met Office in early 1922, had resigned and returned full-time to Cambridge. But his academic future was by no means secure, and in both 1923 and 1924 he journeyed to North America, partly in response to overtures from Harvard and Yale. In the event, unimpressed by the transatlantic atmosphere, as he informed Ernest Jones, he decided to remain in Cambridge. Together with his earlier studies of fluid dynamics, dynamical meteorology and celestial mechanics, the success of *The Earth* secured him Fellowship of the Royal Society in 1925.[177] He, like Frank Ramsey, benefited from the introduction of the new University Statutes in 1926, being made a University Lecturer, and he was promoted to Reader in Geophysics in 1931.[178] Whether he was upset, as Strachey immediately assumed, by Wrinch's quick marriage to an older Cambridge mathematician is unclear and remained unclear to the woman he married in 1940, Bertha Swirles. In 1999 she expressed a distinctly agnostic position on Jeffreys's feelings about the termination of his collaboration with Wrinch: 'I don't know that Harold minded all that much; he was never really clear about that – as one wouldn't be.'[179] Certainly Wrinch had marked his life – he repeatedly emphasized the importance of the joint papers they wrote in 1919–23.[180] And it should not be forgotten that he later co-wrote with his wife a major

---

Harold Jeffreys and Bertha Swirles (Lady Jeffreys), London: Gordon & Breach, 1977, Vol. VI, p. viii): 'Taylor's result on tidal friction in the Irish Sea is extended to other shallow seas. The damping of the 14-monthly variation of latitude, together with the fact that S pulses [one of the two principal reverberations of earthquakes through the mantle] are reasonably sharp at a distance of 80°, leads to a law of creep under which the creep under constant shear stress is like t^α ($0 < \alpha < 1$). For a free vibration the decrease of amplitude in a period would vary as $(period)^\alpha$; for a wave travelling a given distance, like $(period)^{-1+\alpha}$. The limiting case $\alpha \to 0$ leads to Lomnitz's law, but the two data together indicate that α is about ¼. S. Crampin and I have shown this to account for the Moon's rotation, the persistence of its ellipticities, and the undetectability of its free oscillations. ($\alpha = 1$ is the elastiviscous law, and leads to hopeless contradictions) .... [T]he law with $\alpha < 1$ forbids convection and continental drift.'

[177] Jeffreys obviously did not keep up with every field he worked in. In 1984, his former student, an American meteorologist, informed him that his specific mathematical approach to atmospheric eddies in the 1920s had recently been vindicated; the ninety-three-year-old Jeffreys responded: 'I don't think I have anything new to say, but I should like to know about later work. Who decided that I was right and when?' (Jeffreys to Hide, 6 April 1984, HJPC A.26).

[178] Jeffreys and Ramsey served together as Examiners in the Mathematical Tripos Part II in 1928 and 1929.

[179] Lady Jeffreys, Interview with Laura Cameron, 7 January 1999, Cambridge. Lady Jeffreys recalled that she and Harold 'lost touch' with Wrinch: 'last time we saw her was in New York about 1951'.

[180] 'Harold always stressed the importance of those papers which were done while she was a fellow of Girton. First I saw of him was going up the stairs to her room I think': Lady Jeffreys, Interview with Laura Cameron, 7 January 1999, Cambridge.

textbook in mathematical physics: evidently he was a man who thrived on scientific collaboration with women. Hence it is not altogether surprising that in 1986, sixty-one years after his election, the ninety-five-year-old Jeffreys wrote a letter to the Executive Secretary of the Royal Society in response to an administrative request to update the personal details lodged with the Society:

> I notice that in the Status codes /W is added for a woman Fellow. This is a form of sex discrimination unless /M is added for a man Fellow, since all are elected in the same way ... I make this comment as the Senior Fellow of the Society without consultation with the women Fellows. I have amended my form accordingly.[181]

Yet, mysteriously, when it came to publishing his *Collected Papers*, Jeffreys omitted his joint papers with Wrinch, explaining that: 'These papers contain the first statements of several principles, but only summaries are given, since all the subject matter is further developed in my *Scientific Inference* and *Theory of Probability*.'[182] For the historian and philosopher of science, this judgement is actively misleading: David Howie's careful reconstruction of the Jeffreys–Fisher debate over the status of probability devotes more space to these joint papers with Wrinch than to any other of Jeffreys's writings. Which leads one to conclude that the place of Dorothy Wrinch in his life and work remained somewhat enigmatic.

Where did Jeffreys's interest in psychoanalysis belong, both conceptually and chronologically? In August 1918, Dorothy Wrinch was already corresponding about psychoanalysis with Russell, then in prison and at the height of his voracious reading binge in psychology – 'extremely interesting ... but not quite scientific enough'.[183] Jeffreys also became friends with W.H.R. Rivers on his return to Cambridge in 1919; they corresponded on Rivers's book on *Instinct and the Unconscious*, and on Jeffreys and Wrinch's theory of probability in relation to the origins of belief. Whether she, Rivers, Tansley, Farrow or anyone else was the specific source for Jeffreys's psychoanalytic interest or whether she knew him by the time that interest was aroused is unclear. Long-standing his interest did remain – it was Jeffreys who, in 1936, was instrumental in Freud being elected a Foreign Member of the Royal Society and he obviously took pleasure in writing personally to Freud to inform him of the fact.[184]

---

[181] Jeffreys to Executive Secretary, Royal Society, 7 August 1986, Jeffreys Papers, A.17. The first two women Fellows of the Society were elected in 1945, twenty years after Jeffreys.
[182] Jeffreys, *Collected Papers*, Vol. VI, p. 251.   [183] Abir-Am, 'Synergy or clash', p. 246.
[184] Tansley, with Bertrand Russell's support, unsuccessfully attempted to get Jones elected to the Royal Society (BPaS, CTA/Fo2/o3, Tansley to Jones, 12 April 1953).

In replying to Freud's inquiry as to the identity of his English scientific patron, Jones gave a quick description of him:

> Harold Jeffreys is a Cambridge professor – I think of geo-physics.[185] He was in analysis with me for a couple of years, and later with Miss Sharpe. The remarkable feature of his case was a combination of the highest intellectual capacities with a very low degree of ordinary social capacities, such as *savoir faire*, common sense, etc. I think the success of the analysis was only moderate. He has an astounding facility for rapidly acquiring the profoundest knowledge of any subject, e.g. botany, higher mathematics, physics, etc., and has made himself world famous through his mathematical researches into physics and the structure and movements of the earth, on which he has written some heavy books.[186]

It seems highly probable that the first of Jeffreys's analyses, that with Ernest Jones, took place while he was living part-time in London working at the Met Office from 1917 to 1922. Jones recorded the end of Jeffreys's analysis in early 1923[187] and a letter from Jones to Jeffreys, dated 21 August 1923, displays that strange quality of impersonal and guarded informality so often found in letters between analyst and former patient:

> Dear Mr. Jeffreys,
>
> How quickly time goes! So you are not greatly impressed by the Transatlantic atmosphere. I was shocked to hear of your liking Toronto, though I have often heard the same from other people, who have not had the misfortune to live there.
>
> Many thanks for your fierce essay, with which I intend to have a serious wrestle.[188]

A year later, when Jeffreys may have been still undecided about taking up a post at Harvard, he had also made contact with leading psychoanalysts on the East Coast, Smith Ely Jelliffe and William Alanson White, joint editors of the *Psychoanalytic Review*, in which in October 1924 they published Jeffreys's forty-page analysis of Ibsen's *Peer Gynt* (which dealt with Ibsen's use of alliteration and the urinary complex amongst other themes). Jeffreys tried to arrange to visit Jelliffe and White en route from Canada to

---

[185] Jeffreys became a Reader in Geophysics in 1931; in 1946 he was elected Plumian Professor of Astronomy and Experimental Philosophy; he retired in 1958, at the statutory retirement age of sixty-seven.

[186] JF, 6 July 1936, pp. 753–4; Freud had asked Jones specifically who was his correspondent 'Jeffreys'.

[187] Jones to Kathleen Jones, 25 April 1923, BPaS, CJB/F02/27.

[188] HJPC, Jones to Jeffreys, 21 August 1923.

New York in late August.[189] (Strangely enough, when the group of psycho-analytic 'insiders' met in Cambridge in early 1925, the geophysicist Jeffreys was the only one to have published a substantial paper in a psychoanalytic journal.[190]) While it is certain that Jeffreys had analysis with Jones well before 1924, the dates of his analysis with Ella Freeman Sharpe, the doyenne of English analysts in the first half of the twentieth century,[191] are even less certain. But the analysis left a trace in a paper of hers published in 1930:

> I have registered during one week a number of things which, had I personally known more about them, would have enabled me to reach more quickly the unconscious themes that were being given to me in a representative way. In one analysis I needed an intimate knowledge of Peer Gynt, and a swift recognition of the rôles that Asa, Ingrid and Solveig were playing at that moment in terms of the patient's own identifications. In another an immediate recall of a Dutch picture would have given me the link I needed between an actual scene and an unconscious phantasy. The knowledge of the exact duties of a trustee; the differences between two ways of calculating commission on sales; a knowledge of the differences between two makes of motor cars; the appearance of a cider-press and the way it works; the precise meaning of football terms; an understanding of the processes of etching – all these would have enabled me to grasp more quickly than I did the unconscious significances that were being represented.[192]

That *Peer Gynt* was a leitmotif in Jeffreys's relation to psychoanalysis is clear from the fact that he published two papers on the play, the one noted above in 1924, and a further short note in the journal *Scandinavica* in 1964.

Jeffreys's public connection with analysis was to be restricted to his philosophical defences of it and a small number of published papers. His paper of 1924 on Ibsen's *Peer Gynt* proposes a psychobiography of the character Peer Gynt principally through the abundant free associations that the text, so baroque and redolent of fantasy, offers. Another remarkable paper was 'The unconscious significance of numbers',

[189] HJPC, Jelliffe to Jeffreys, 25 August 1924. Jelliffe gives Jeffreys's address as 436 Gilman Street, Ottawa.
[190] James Strachey's first psychoanalytic publication under his own name (as opposed to translation) was a critical review of two psychoanalytic studies in German of August Strindberg, in the *International Journal* for 1923. Rickman's psychoanalytic publications in 1924 also comprised a series of reviews.
[191] More candidates successfully completed training with Sharpe than any other English analyst in the period up to 1944 (see FK, TC Minutes); she attracted near universal respect and liking across the Society in this period. Her teaching for the Society (as partly published in *Dream Analysis* (1937), which quickly became a psychoanalytic classic) was also much respected.
[192] E.F. Sharpe, 'The technique of psycho-analysis', *IJP* 11 (1930), 251–77 at 255.

published in 1936.[193] Some sense of the argument of this latter paper, in which he inquired into the superstitions and significance given to certain numbers, particularly odd ones – 3 (representing the male genitals), 5 (whose Pythagorean magical status is due to the pentacle representing the 'phantasy of the father's penis preventing the angry child's return to the mother') – can be gained from the following hypothesis he put forward:

> The notion of a 'prime' number, one possessing no factor other than unity and itself, is clearly a development of that of an odd number, which merely does not possess the factor 2. Thus a prime (other than 2), while carrying the attribute of maleness [the '1'] like other odd numbers, is specially resistant to separation into parts. The theory of primes is therefore essentially a play mechanism designed to provide a defence against the fear of destruction by tearing apart into pieces of comparable size, and avoids consideration of the danger popularly associated with odd numbers, that they may lose a unit and become even ... To sum up, numbers in language, folklore and superstition, appear to carry affects derived from pre-genital situations, mainly oral and urethral. The interest in odd numbers, and especially in primes, is originally phallic, while even numbers and especially those with a large number of factors are associated with ambivalent attitudes to the mother.[194]

Jeffreys considered even numbers to be less freighted with fantasy. Thinking about numbers in a particular way is sexualized, and Jeffreys went on to speculate that opposition to a decimal system of weights and measures is affected by the fear of danger associated with too much preoccupation with the number 10, the base of the decimal system which is intimately associated with the ten fingers of the hands. Referring explicitly to the cosmos of language and its exotic system of measurement expounded in Penrose's case of schizophrenia discussed below, he noted that 'extreme decimilization was associated with habitual masturbation'.

Both Jeffreys and Penrose were preoccupied with the psychology of mathematics, and may well have exchanged psychoanalytic views in that area. Whereas Penrose had given a paper in 1925 on the psychology of chess

---

[193] H. Jeffreys, 'The unconscious significance of numbers', *IJP* 17 (April 1936), 217–23. While undoubtedly eccentric, the paper was later used for a further psychoanalytic exploration by another mathematical meteorologist (Raymond Hide, 'On the unconscious significance of the number thirty-one', *IRP* 11 (1984), 119–20), who had been a student of Jeffreys in Cambridge in the early 1950s. Jeffreys published a further paper linking the problem of inference with psychoanalysis by using material from Melanie Klein's book on the psychoanalysis of children to show how children and scientists use the same – reliable – principles of inference. See Harold Jeffreys, 'The problem of inference', *Mind* N.S. 45(179) (1936), 324–33.

[194] Jeffreys, 'The unconscious significance of numbers', p. 223.

drawn from the findings of his own analysis, in his paper of 1936 Jeffreys revealed some of the elements in his personal relations with numbers:

> In my own analysis the number 4 appeared with strong female attributes. This was traced partly to the four teats of the cow, and partly to the diamond on the label of one of Bass's beers often seen in childhood. This diamond was thought of as two equilateral triangles on a common base, and represented the breasts. Seven was thought of as 4 + 3, representing a hermaphrodite figure with the external organs of both sexes. It had associations with irresistibility and perfection; we have seen above that these are just what are found in the folklore of the number.[195]

It is possible that Tansley was another catalyst for Jeffreys's curiosity about psychoanalysis. Certainly Jeffreys's botanical interests had early on brought him into contact with Tansley. His three papers on plant ecology, two of which were local studies of the area of north-east England where he grew up, the third being an investigation of the comparative viability of certain grasses in Durham and Suffolk, all date from the period 1916–18 and were published in Tansley's *Journal of Ecology*.[196] Their theoretical framework is indebted to Tansley and to Farrow; he footnotes both men extensively. It is thus possible that it was Tansley who introduced Jeffreys to psychoanalysis, as well as serving as his professional patron in ecology.

In 1974, the eighty-year-old Jeffreys, by then knighted, opened a short note concerning the mathematical geneticist R.A. Fisher and inverse probability – a topic on which the two men had clashed formidably at the Royal Society in 1932–34 – with a statement of his origins:

> I think that I should begin with my first use of probability theory, which was in a paper with Dorothy Wrinch in the *Phil. Mag.* for 1919. My interest started through Dr E.P. Farrow, a plant ecologist, who introduced me to Karl Pearson's *Grammar of Science*, still the best general work on scientific method.[197]

---

[195] *Ibid.*, p. 222.

[196] H. Jeffreys, 'On the vegetation of four Durham coal-measure fells, I and II', *Journal of Ecology* 4 (1916), 174–95. 'On the vegetation of four Durham coal-measure fells, III and IV', *Journal of Ecology* 5 (1917), 129–54; 'On the rarity of certain heath plants in Breckland', *Journal of Ecology* 6 (1918), 226–9, to which Jeffreys added a note in 1970.

[197] H. Jeffreys, 'Fisher and inverse probability', *International Statistical Review* 42 (1974), 1–3 at 1. Jeffreys made an even stronger claim in an article published in the St John's College magazine, *The Eagle*, in 1984 (HJPC A.20): 'Many people have commented on the wide range of subjects treated in my works. In fact there is a connecting thread between all. It goes back to Dr E.P. Farrow, a plant ecologist at Trinity, who had been at University College London under Prof. F.W. Oliver. He introduced me to Karl Pearson's *Grammar of Science*, first published in 1892, and still the best general account of scientific method, though unknown to philosophers of science. He introduced the method of direct experiment in the field.'

Farrow also taught Jeffreys how to ride a bike – he would take his Sunday exercise by biking to south London and back to Cambridge, a round trip of 140 miles.[198] The influence of Pearson's positivist primer is clear throughout Jeffreys's writing on science, particularly in his *Scientific Inference* of 1931.[199] Farrow may even have been a further catalyst for Jeffreys's interest in psychoanalysis. Certainly this friendship with Farrow lasted longer than that with Wrinch – Lady Jeffreys remembered Farrow as 'an odd fish – rather!'[200] She also had decided views on her late husband's relationship to psychoanalysis: 'After Harold was married, he didn't need psychoanalysis anymore.'[201]

We thus have a fragmentary picture of Jeffreys at the time of his involvement with the other Cambridge psychoanalytic insiders: mathematically magisterial, socially ill at ease, restlessly polymathic while only at that time finding, from amongst the many scientific fields he had become interested in, the lines of inquiry in geophysics which would make his name.[202]

Despite the acerbic wit of his gossipy letters to his wife Alix, the last member of the 1925 Group, James Strachey, was impressed by the extraordinary range of scientific interests in the group. When, during the Second World War, amid the heated debates over the future of psychoanalysis in Britain, he was asked to draft a Memorandum on Training for the confidential consideration of the Training Committee, he opened with reflections on the relations of psychoanalysis to other sciences. Out of the blue, he wrote this: 'There are a number of people who are interested in the phenomena of geophysics; but we shall not insist upon all Candidates mastering the truths about the surface-tension of the earth, and indeed we shall discourage those who conduct our seminars from expatiating too freely upon the laws determining the epicentres of

[198] Cook, 'Sir Harold Jeffreys', p. 308.

[199] Despite Ramsey and Jeffreys both being members of the 1925 Group, despite Ramsey's important review of Keynes's *A Treatise on Probability* in the *Cambridge Magazine* in 1921, despite both Jeffreys and Ramsey being appointed to two of the nineteen new University Lectureships in Mathematics in 1926, Jeffreys was not aware of Ramsey's interests in probability theory until after his death in 1930. (Communication from David Howie, 8 June 2000, referring to transcripts of several interviews with Jeffreys.)

[200] Lady Jeffreys, Interview with Laura Cameron, 7 January 1999, Cambridge.

[201] Lady Jeffreys, Interview with Laura Cameron, 7 January 1999, Cambridge.

[202] Jeffreys's 'very low degree of ordinary social capacities' were, it turns out, quite sufficient to establish his scientific networks and sufficient collegiality to be remembered by Gregory Bateson in 1973, when he recalled the handful of dons at St John's who were important to him as a young research fellow in the late 1930s: 'L.S.B. Leakey, Harold Jeffreys, Claude Guillebaud, Reginald Hall, Teulon Porter, Sir Frederick Bartlett ...' (Christmas 1973. See www.edge.org/3rd_culture/batesono4/batesono4_index.html).

earthquakes.'[203] This evocation of the relation of psychoanalysis to geo-physics would have been incomprehensible to the other members of the Committee – Melanie Klein, Anna Freud, Sylvia Payne, Edward Glover, John Rickman, Ella Sharpe and Marjorie Brierley, with the exception of two, Rickman and Sharpe, who would have known exactly of whom he was talking. By choosing geophysics as his exemplary science, he reminded the historically well informed (and himself) that, in the wild early days of English psychoanalysis, there had been a distinguished geophysicist whose interests in psychoanalysis were entirely continuous with his interests in mathematics, in ecology and in the philosophy of probability. In so doing, he was making a very sharp point about the historical transformation of the intellectual and professional world of psychoanalysis from the early 1920s to 1944.

The First Meeting of the Group

On Saturday, 28 February 1925 James Strachey made his way to Cambridge for the first meeting of this quintessentially Cambridge group, to be held on Sunday, following a Saturday night meeting of the Apostles – whose membership partially overlapped with the Psychoanalytic Group. He spent breakfast, lunch and dinner with Ramsey, Sprott, Braithwaite and Penrose, in addition to paying his respects to his sister Pernel, Principal of Newnham College, and to Lella Secor Florence, Alix's sister-in-law; he complained to Alix about having to talk about nothing but psychoanalysis.

> The worst of it was that the only topic of conversation throughout was psycho-analysis. All the non-analysts took the opportunity of my being there to ask the usual string of questions. And all the analysed ones seemed to me to show they hadn't really been analysed by talking incessantly of their own analyses.

But this incessant talk about analysis was different from the London ambience:

> I was crushed by the unaccustomed intellectual level – especially of Ramsey. And it was rather like the third Act of Siegfried to hear the tone that he adopted about poor old Dr Moore. He seemed on the whole to accept Ψα, but thought the theory very muddled. The theoretical work of the Prof.'s which he most admired was – Das Ich und das Es. He is thinking of

[203] James Strachey, 'Discussion Memorandum by James Strachey. Member of the Training Committee – February 24th, 1943', CD, 604.

devoting himself to laying down the foundations of Psychology. All I can say is that if he does *we* shan't understand 'em. He seems quite to contemplate, in his curious naif way, playing the Newton to Freud's Copernicus.[204]

The agenda of the meeting consisted of two papers: Jeffreys's 'On the psychological significance of death duties' and Penrose on 'Psychoanalysis and chess'. Although neither was ever published in any shape or form, copies of both these papers have survived, simply because both Penrose and Jeffreys kept meticulous archives of their own lives and work. They are graphic guides to the unique meld of philosophy, science, psychoanalysis, cultural critique and intimate personal lives which this Cambridge group was developing. Detailed consideration of them will reveal much of the character of the 'Cambridge' reception of psychoanalysis.

Jeffreys's paper was a seven-page formal argument, entirely lacking in the details of psychoanalytic treatment which was the starting-point for Penrose's paper. Writing in early 1925, a few weeks after the first Labour Government had left office, at a time when the fundamental doctrines both of socialism and of Labour Party policy – state ownership of the means of production (1918) – were at the centre of political discussion, when the justice of inheritance along with free love and egalitarianism were incessant preoccupations of the post-war generation of undergraduates,[205] Jeffreys posed the following question:

> In contemporary discussions concerning the desirability of state ownership of the means of production, much attention is naturally devoted to the possible methods of accepting the transfer of property from private owners to the state. Those usually favoured are forcible confiscation and purchase.[206]

The means of implementing this policy, despite the objections to these methods – hardship for those deprived and the cost of the former – is simple: 'to enact that the property in question shall accrue to the state on the death of the owner'. Here Jeffreys was taking up an observation by the then Prime Minister, Lloyd George, in 1919: 'Death is the most convenient time to tax rich people.'[207] The aim of Jeffreys's paper is to discover why this straightforward option is 'so seldom alluded to'. Jeffreys diagnoses 'a specially severe resistance ... shared by the great majority of socialists' against thinking about this policy, instancing the 'violent outcry in 1909' at

---

[204] BF, 2 March 1925, p. 223.   [205] FPM, pp. 59–60.
[206] All quotations from Jeffreys's paper are in HJPC C.6.
[207] Lord Riddell, *Lord Riddell's Intimate Diary of the Peace Conference and After*, 1918–1923, London: Victor Gollancz, 1933, entry for 23 April 1919.

the time of Lloyd George's 'People's Budget' against a proposed increase of death duties in many cases to 20 per cent:

> it was regarded as an act of robbery that the heirs should be deprived of a fraction of what had never been theirs. Thus the resistance in question appears to be common to persons of extremely different political opinions, and it may therefore be conjectured that it rests on a psychological basis common to a large fraction of mankind.

Jeffreys then outlines the basic structure of primogeniture as it is found in British law, with inheritance by the eldest son, from which he infers that 'the family even in the father's lifetime are cherishing a wish that they may acquire his possessions after his death'. Then he advances his principal hypothesis:

> The wish for certain benefits arising in the event of the father's death bears an obvious resemblance to the Oedipus wish of the infant boy for uninterrupted possession of his mother's favours, which can only be obtained by the removal of the competing father ... Let us then adopt the hypothesis that the wish for the father's property is the modified expression of the childish wish for the mother, and see how it fits the other facts.

Amongst these facts: 'no daughter has hopes of gain by her father's death' and for the younger sons the eldest son is an additional barrier to unimpeded enjoyment of the mother. 'Thus the eldest son is the only member of the family who could at any stage of his life gratify his Oedipus wish by the simple removal of his father, and primogeniture represents a regression to this stage.' Going over each instance – eldest son, eldest son with elder daughter, independence of heirs' wishes from the father's wish – Jeffreys concludes that 'the simple system of inheritance is completely explained on the hypothesis that it is designed to gratify the Oedipus wishes of the children ... the affect originally directed towards the mother having been displaced to the father's property'.

The extent of the resistance is now explained: socialists do not mention it since the primitive elements cannot be changed and capitalists resent any interference, as in partial death duties, 'for this amounts to the intervention of the state as a new obstructing father'. The same Oedipal wish is found in the act of burial itself, a process of 'manuring, a fertilization of Mother Earth' involving unpleasant smells and hence of faeces, overdetermined by the identification of faeces and semen; the identification of a landed estate with the mother is particularly strong. Jeffreys does note that this account only applies to one system of inheritance; in addition, the father's own wishes – for immortality and in regard to his own procreative power – have not been included in the analysis, but do in fact support a system of primogeniture. The paper closes with a reminder of the strong connotations of the term 'dismemberment', applied to landed estates and also as a fantasy of the murder and castration of the father.

Jeffreys was writing this paper as the Committee on National Debt and Taxation, also known as the Colwyn Committee, was preparing its 1927 report and when certain socialist circles, led by Hugh Dalton, were considering the novel proposals of the evolutionary positivist Italian philosopher Eugenio Rignano, to introduce a progressive version of death duties in which the 'age' of an estate (the number of times it had already been bequeathed) would determine the rate of taxation at death of its owner.[208] Rignano's *The Social Significance of Death Duties* was published in England in 1925.[209] Jeffreys's paper shows no explicit signs of having delved into these contemporary detailed discussions; however, it is clearly a response, almost from on high, to them.[210] It is a mathematician's hypothetico-deductive argument: take a psychoanalytic hypothesis and demonstrate, step by step, how distinctive features of British inheritance law map onto that hypothesis.[211] The clean mapping of socio-legal formation onto psychological complex is the only direction Jeffreys considers running the argument; he did not consider that the Oedipus complex might be an effect rather than the cause of features of patriarchal social organization. Jeffreys assumes that the Oedipus complex is the 'nuclear complex of mankind'. In point of fact, his way of viewing matters is not very far from the contemporaneous arguments of Malinowski, who employed the term 'nuclear complex' to refer to that psychological structure, whose first

---

[208] See also Martin Daunton, 'Equality and incentive: fiscal politics from Gladstone to Brown' (May 2002), www.historyandpolicy.org/archive/pol-paper-print-06.html, accessed 10 September 2006. Thomas Piketty, *Capital in the Twenty-First Century*, Cambridge, MA: Belknap Press, 2014, p. 432n29, refers to the debates over Rignano's tax proposals and gives as a principal contemporary reference: G. Erreygers and G. Di Bartolomeo, 'The debates on Eugenio Rignano's inheritance tax proposals', *History of Political Economy* 39(4) (2007), 605–38.

[209] It had been published in New York in the previous year under the title *The Social Significance of Inheritance Tax*.

[210] Note also Keynes's interest in this same year of 1925 in the question of inheritance taxes discussed in Chapter 8, p. 486 below. Critical as the background to these debates was the increase in top rate of inheritance tax in Britain from pre-1914 rates of 15% to 40% in 1919–25; the rate would continue to increase, reaching 80% in 1949, a level it maintained till it peaked at 82% in 1972, before the Thatcher government's budgets reduced it by 1987 once again to 40%, the level at which it has since remained.

[211] Jeffreys's later remarks about psychoanalysis in his classic *Scientific Inference* (Cambridge: Cambridge University Press, 1931) are more philosophical in character, e.g. they defend the appropriateness of psychoanalytic concepts with regard to the empirical domain of facts, even if one were to believe that, for instance, all psychological facts are eventually physiological (Jeffreys, *Scientific Inference*, pp. 203–5). See Howie, *Interpretations of Probability, passim.*

version is the Oedipus complex, but which may have different forms giving rise to – or being shaped by – a different social organization; in the case of the Trobriand Islanders, the matrilineal structure confers repressive authority on the mother's brother with incestuous desire focused on the sister rather than the mother.[212]

Perhaps even more consonant with Jeffreys's intentions was the psychology of capitalism expounded in Keynes's 1919 tract for his times and his generation, *The Economic Consequences of the Peace*, discussed more fully later. For the men of Jeffreys's generation, Keynes's words announcing the possible doom of the capitalist world-order were deeply resonant:

> The war has disclosed the possibility of consumption to all and the vanity of abstinence to many. Thus the bluff is discovered; the labouring classes may be no longer willing to forgo so largely, and the capitalist classes, no longer confident of the future, may seek to enjoy more fully their liberties of consumption so long as they last, and thus precipitate the hour of their confiscation.[213]

Equally convinced that there is a direct link between psychology and economic forces was that other psychoanalytically educated Cambridge moral scientist Roger Money-Kyrle, whose *Aspasia: The Future of Amorality* (1932) also proposed that unconscious processes stemming from Oedipal structures inhibit rational public discussion of economic and social questions, such as the 'castrating' response of the victorious Allies to Germany's difficulties in paying its enormous bill for reparations.[214] This readiness to link psychology to economics was a distinctive feature of Cambridge High Science of the 1920s, and was entirely in tune with the grand speculative style that Freud displayed in *Totem and Taboo* and even in *Civilization and its Discontents*.

The other paper offered to the first meeting of the Cambridge Psychoanalytic Group was very different. In his own personal analysis, Penrose had discovered a decidedly unconventional method of work, and its thinly disguised fruit was 'Psycho-analysis and chess'. At one point in his analysis, as he told it, he asked his analyst Bernfeld what could be the cause of his strong interest in the game of chess, to which Bernfeld had replied

[212] Malinowski's 'Psychoanalysis and anthropology', published in April 1924 in *Psyche* 4, 293–332, and also published, courtesy of his contacts with Otto Rank, as 'Mutterrechtliche Familie und Ödipus-Komplex', *Imago* 1924; see also George W. Stocking, Jr., *After Tylor: British Social Anthropology, 1888–1951*, Madison: University of Wisconsin Press, 1995.

[213] J.M. Keynes, *The Economic Consequences of the Peace*, London: Macmillan, 1919, p. 19.

[214] See Neil Vickers, 'Roger Money-Kyrle's *Aspasia: The Future of Amorality* (1932)', *Interdisciplinary Science Reviews* 34(1) (2009), 91–106 at 100.

that 'he probably projected his infantile family conflicts onto the chessboard'.[215] Soon after this exchange, Penrose had a dream 'whose manifest content is the attitude of the dreamer towards the position of certain pieces upon a chessboard'.[216]

> I see before me a certain chess position. It is a problem. White is to checkmate Black in two moves. The location of the pieces is not however quite settled, and I feel as though there may be a misprint in the case of the Black Queen. The White Queen ought to be 'pinning' her, I think, in order that the White Pieces should succeed. As it is, too many moves of Black are unprovided for. But in two cases the nature of the mating move is actually known to me. If the Black Pawn (on the square d6) takes the White Pawn (on c5) I know that the White Queen can then checkmate by moving down to the square b2. If the same Black Pawn simply chooses to move on to d5, the White Queen will mate somewhere else.

Penrose followed the classical Freudian technique of associating to each of the dream elements in turn. For instance:

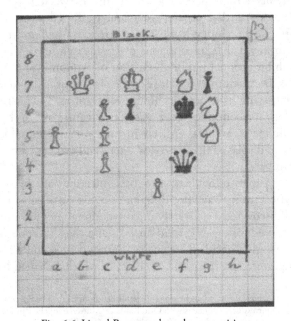

Fig. 6.6 Lionel Penrose, chess dream position.

[215] PP, 5/1 'Psycho-analysis and chess', March 1925, 18 f., f. 2.   [216] Ibid., f. 1.

White is to checkmate Black in two moves. 'White' means 'good' and 'clean' whereas 'Black' meant 'bad' and 'dirty'. This led to the discovery that 'to check' could mean 'to check the action of defaecating' and also 'to give money in the form of a cheque'. 'To check-mate' was translated into 'to kill the king'.

The location of the pieces is not however quite settled. This was associated with the dreamer's dissatisfaction with his position as analysand and his wish to alter the positions of persons in his family.

The position represented a dramatization or pictorialization of the dynamics of his family conflicts: on one side were the Black King (his father), the Black Queen (a governess who had taught him to read, a 'Miss Binney' – hence 'Mis-Print' and 'pinning'[217]) and himself (a pawn); on the other side were the White King (his grandfather, Lord Peckover), the White Queen (his mother) and a series of pawns (representing his brothers) and knights (representing himself and his brothers): 'the Black forces are composed of the unpunctual people [much criticized by the grandfather]: agreeing with the rule that White moves first'. The grandfather also held power over the Black King through his wealth: 'in the action of the problem as in real life, the White Queen was to come from the White King and give "cheque" to the Black King'.[218]

The most forcefully dynamic aspect of the dream's latent thoughts, however, attached to the dreamer's thought that '*in two cases the nature of the mating move is actually known to me*'. This referred to his knowledge of the sexual relations between his father and mother: his reaction against this knowledge was the most powerful source of the dream. The 'mating position' in the dream depended upon the actions of the Black Pawn, himself. It was thus, he concluded, a dream of omnipotence, a dream that he could affect the 'mating' of the King and Queen, whereas in reality he had no such power. As he remarked, his youngest brother Bernard (born in 1903) was not yet born at the time of the events associated with the

[217] The name and identity of Miss Binney were not revealed in the paper, but are clear elsewhere in the notebooks.

[218] PP, 5/1, 'Psycho-analysis and chess', f. 4. His grandfather Lord Peckover was indeed a man of wealth and influence; Lord Lieutenant of Cambridgeshire at the time of Lionel's birth, he paid for the new operating theatre at Addenbrooke's Hospital in 1898 and gave a series of supplementary cheques to the sorely pressed hospital in the next two decades up to about 1914 (see Arthur Rook, Margaret Carlton and W. Graham Cannon, *The History of Addenbrooke's Hospital,* Cambridge: Cambridge University Press, 2010, pp. 227, 263 and 289). In the early 1930s Lionel Penrose himself became a philanthropic patron: to the BPaS (its Clinical Prize) and to the Royal Eastern Counties Institution at Colchester through giving up a substantial portion of his own salary from the Pinsent-Darwin Fund in order to fund a full-time research assistant (CUEP, Minute-Book, Darwin-Pinsent Fund Committee of Management, 1931).

formation of the dream (though a younger brother, Roland, born 1900, was in the world).[219] He was clearly dating the impulse to master his family constellation through chess as arising from the era when his youngest brother was conceived, when he himself was aged five; the notebook page following the outline of the dream chess problem and his associations was the research for this inquiry: a chronology of the key events in his life from birth to 1922.

After this thinly disguised account of his own analysis, Penrose passed on to a more general consideration of psychoanalysis and chess, starting with a consideration of the symbolism of each piece and its connection with unconscious fantasy: murderous, urethral, sadistic, birth, oral and anal. He then turned to consider the infantile sexual activities which find satisfaction in fantasy in the game itself, e.g. the close relation of anal sadism and omnipotence of thoughts so manifest in blindfold chess; the considerable libidinal charge of 'sacrifice', with the 'Queen sacrifice fetish' belonging 'in England precisely to chess of the Victorian period'; considerations of homosexual fantasies related to some of the peculiar rules concerning movements of the King; and finally, following Freud's account of children's games in *Beyond the Pleasure Principle*, in repeating and mastering situations of anxiety.[220]

Throughout his Notebooks from this period are scattered chess diagrams and notes on problems. These would not be surprising, given his early interest in chess – he even had chess problems published when still a boy. His grandfather Lord Peckover was a keen chess player and the game was one of the few less than serious activities allowed on the Sabbath. Yet it

---

[219] In the autobiographical notes he prepared for the Royal Society, he recorded the date of marriage of his parents, his own date of birth followed by a list of the birth of his siblings and himself: 'Male stillbirth 1894/Male Alexander, 1896, Fellow of King's died 1950/Lionel/Roland, 1900, Art expert and critic/Bernard, 1903, Mariner, Lt Commander in Navy'.

[220] Penrose was not the first to consider chess from a psychoanalytical point of view. On 15 March 1922, a dentist by the name of Dr Fokschaner gave a paper 'On the game of chess' to the Vienna Psycho-Analytical Society, followed by a discussion led by Freud, Bernfeld, Federn, Kolnai and Schmiedeberg (*IJP* 3 (1922), 137). According to Willi Hoffer, Fokschaner drew a parallel between chess and obsessional neurosis and attempted to interpret symbolically the pieces and their movements on the chessboard; according to Kenneth Colby (a pioneer in the 1950s of artificial intelligence, computer chess and a computer program simulating psychotherapy), Bernfeld recalled how Fokschaner was heavily criticized by Freud for his simplifications (Norman Reider, 'Chess, Oedipus, and the Mater Dolorosa', *IJP* 40 (1959), 320–33 at 323). How Bernfeld's response to Penrose's demand for a psychoanalytic understanding of his own interest in chess might have been informed by Freud's comments is a matter of speculation; the next contribution to consider chess in the light of psychoanalysis – always regarded as the first major contribution in this area, since Penrose never published his paper – was written by Ernest Jones in 1930 (E. Jones, 'The problem of Paul Morphy – a contribution to the psycho-analysis of chess', *IJP* 12 (1931), 1–23).

is highly probable that Penrose was doing more than explore the source of his own interest in chess through dream analysis; he was conducting a substantial portion of his analysis in the form of this curious method of dream, chess problem and free association. The dream was only one in a series of notes on chess positions and problems which, as it were, were being formulated for their analytic utility. Thus in his notebooks[221] there are associations additional to those in the paper he delivered, specifically to the names of women currently preoccupying his erotic imagination. And there are a series of other chess problems to which Penrose adds comments like: '(I am composing the problem myself this time)/Dream problem in the making./I first compose with bishop on g2 and then change it to e4' – almost as if he never knew if his night-life would be concerned with constructing chess problems or dreaming in the language of chess for analytic purposes, or perhaps both.

Later Meetings of the Group

Another meeting of the Psychoanalytic Group two months later gives a further sense of the enthusiasms and wide differences of viewpoint in this curious Cambridge gathering:

> Tea at the Tansleys ... Dinner at the Union with Lionel, Rickman, & Tansley; and afterwards Lionel's paper on the Biological implications of the Pleasure Pain principle, with all his measurements of leaves & of the growth of the human body. It was rather amusing, because Frank began very hubristically, declaring that Lionel's mathematical formulae were dotty; but eventually it turned out that on the contrary Frank had made an absurd mistake in his calculations. He was thereafter so much upset at the revelation that he declared he was greatly impressed by the paper & its statistics & curves (though personally I think it's all my eye).[222]

A final meeting took place on 14 June 1925, when the Group met at Tansley's home in Grantchester and invited Susan Isaacs, then Director of the Malting House School, to give a talk about the school in relation to psychoanalysis.[223] The meeting indicates that the Group were attempting to extend their interests beyond their own circle towards the other initiatives involving psychoanalysis being launched in Cambridge.

---

[221] And there are at least two notebooks in which are found associations and analysis of the chess dream discussed in his paper.

[222] BF, 20 May 1925, p. 270.

[223] BLSP 60715, Vol. LXI, March 1925–October 1925, 15 June 1925 (not in BF).

In certain respects, the 1925 Group bears the hallmark of Tansley's gift for organizing informal research groups. Tansley was by far and away the eldest and most senior of the members – their ages were 53, 37, 33, 33, 26 and 22. He had much scientific organizing already under his belt: the Group's ethos may not have been unlike the Cambridge Ecology Club he convened in 1921. The psychoanalysis group had two striking features: firstly, the criterion that all members should have experienced a personal analysis; secondly, the fact that Rickman was the sole member with a medical degree or any extra-analytic interest in the 'profession' of psychopathology.[224] The traditional discipline best represented in the group was mathematics – Ramsey, Jeffreys and Penrose were all fine and creative mathematicians. A physicist, a philosopher of mathematics, a philosopher-psychologist, a botanist, a translator cum practising analyst and a 'nondescript' (a psychiatrist working at this time at the out-patient clinic at St Thomas' Hospital in London) do not make a conventional cohort of psychoanalytic researchers.

One possible reason for the Group's formation was Tansley's eagerness to establish a psychoanalytic group in Cambridge. Once the IPA congress held in Salzburg in 1924 had agreed to hold its next congress in Cambridge in 1925, the most probable host would have been Tansley. The plan faltered, for unknown reasons – quite possibly the effects of the hyperinflation of the early 1920s: England would have been exorbitantly expensive for central European analysts. Perhaps Tansley was left with the desire to substitute for the one-off international congress in Cambridge a more informal set of meetings. Hence the discussion group.

---

[224] Penrose was just beginning his medical training. A curious seeming contradiction in the historical evidence may be interpreted in this light as well. In 1925, Ramsey married Lettice Baker, a physiologist and psychologist who worked in the Cambridge Psychological Laboratory from 1924 on. Recalling her marriage some fifty years later, she made the following statement: 'All the philosophy and ethics and logic went in one ear and out at the other, I think. That was far above my head. It wasn't my subject, really, at all; and he wasn't a psychologist. So we didn't discuss psychology' (Mellor, 'Frank Ramsey'). This memory is on the face of it an odd one, given that James Strachey reported Frank Ramsey as ready to devote himself to the foundations of 'psychology' in March 1925. Part of the solution to this disparity of evidence may be found in the vast distance between 'psychology' as practised in the Cambridge Psychological Laboratory and 'Ψα' as discussed and practised by the Cambridge Group. Ramsey may indeed have had no interest in 'psychology' as it was to be found in the Psychological Laboratory, and this statement may have held true for all members of the Cambridge Ψα group, including the psychiatrist John Rickman. The Stracheys certainly had considerable contempt for Bartlett, the key figure in Cambridge University's psychology from 1922 on – 'that *mugwump*' (BF, Alix to James Strachey, 30 November 1924, p. 132); Penrose's private papers indicate that he too had a low opinion of Bartlett, his one-time research supervisor, judging that he was 'not refined enough' to teach in the Moral Sciences Tripos (PP, 20/2, 'Memoirs 2 Lunacy', p. 102/6).

A New Science

If one views psychoanalysis in the early 1920s as a new science, exciting great interest amongst many other kinds of scientists, the pattern represented by the 1925 Group takes on a telling cast. The several individual members of the Cambridge group who had ventured out to Vienna in the early 1920s to study psychoanalysis in its birthplace were part of a much larger cohort of young scientists going to Vienna for research and for other purposes.[225] Ramsey himself made contact not only with Wittgenstein but with Hans Hahn the mathematician-philosopher and Felix Ehrenhaft the experimental physicist;[226] Penrose worked in the psychologist Carl Buhler's laboratory while in Vienna, and made contact with psychiatrists Wagner-Jauregg and Paul Schilder as well as attending meetings of the Vienna Psychoanalytic Society. Tansley's sister-in-law Harriette Chick and his former student Margot Hume spent two years in Vienna researching dietary factors in rickets; Roger Money-Kyrle used the camouflage of a PhD in philosophy with Moritz Schlick to undertake his analysis with Freud. Vienna was a favoured destination for English scientists in these years, renewing contact after the war with one of the world scientific centres, where extreme suffering made it an urgent and fertile site for scientific and clinical research and where the sterling currency made English visitors rich. Some of these scientists were following in the charitable footsteps opened up by the Cambridge-based sisters Eglantyne Jebb and Dorothy Buxton when they founded the Save the Children Fund in spring of 1919 in order to save the victims of famine in Germany and the former Austro-Hungarian Empire.

Tansley's later decision not to devote himself full-time to psychoanalysis is both representative and emblematic of the paths taken by these informal

---

[225] In the late nineteenth and early twentieth century, patterns of travel to other universities and centres of research had developed gradually but with conspicuous success (e.g. Torres Straits) in Cambridge and were to become formalized and explicitly recognized with the new Statutes of 1926, which created the statutory right to sabbatical leave and a Research Fund to which all University Teaching Officers were entitled to apply. These practices were part of the process by which the collegiate teaching university was transformed into a modern 'German' research university. See Heike Jöns, 'Academic travel from Cambridge University and the formation of centres of knowledge, 1885–1954', *Journal of Historical Geography* 34(2) (2008), 338–62; Michael Heffernan and Heike Jöns, 'Research travel and disciplinary identities in the University of Cambridge, 1885–1955', *British Journal for the History of Science* 46(2) (2003), 255–86; and Peter J. Taylor, Michael Hoyler and David M. Evans, 'A geohistorical study of the rise of modern science: mapping scientific practice through urban networks, 1500–1900', *Minerva* 46 (2008), 391–410. Heike Jöns kindly made available to us her extensive databases on research travel by Cambridge academics in the period 1885–1954.

[226] FRP File 2.1.4, letter home, 13 April 1924.

colleagues. Certainly the existence of these psychoanalytic enthusiasts of the 1920s tells us about a moment in the history of psychoanalysis and the more general organization of 'scientific' inquiry in Britain in the early part of this century; but equally telling are the later destinies of Tansley and the others, which reveal that this particular and transient opening up to psychoanalytic enthusiasm would very soon be closed down, would soon become impossible.

Ernest Jones, of course, welcomed Tansley with open arms into the small BPaS – his Cambridge connections, his scientific standing and his FRS counted for much.[227] Jones was certainly more welcoming of Tansley initially than he was of the more literary Stracheys and the Stephens. Tansley's analysis with Freud and Freud's evident approval of him made his election as a Full Member of the Society a straightforward matter. Yet in the same circular letter in which Jones reported the election to the other European leaders of psychoanalysis was another item of news which would make psychoanalysis in the future less welcoming to the Tansleys of this world:

> At our annual [London] business meeting last week, the same officers were elected. Mr. Tansley was promoted to be a full member, the name of one associate member who had recently written an unsatisfactory book was omitted and two new associate members [Inman and Kapp] were elected. Following the instructions of the Congress, we elected an Education Committee of five, consisting of the four directors of the Institute (Bryan, Glover, Rickman and myself) with the addition of Flügel.[228]

A resolution passed at the Bad Homburg Congress in early September 1925 – the Congress which Jones and other English analysts had originally planned would take place in Cambridge under Tansley's aegis – required each national society to set up an Education and Training Committee. Combined with Max Eitingon's proposals for a formal system of psycho-analytic training, also submitted to the Bad Homburg Congress, this system of Education Committees spelt the beginning of a new development in the national psychoanalytic societies. To put it starkly, they ceased to be primarily 'scientific societies or colleges' and became 'training institutes'; they ceased to be relatively informal meetings of equals devoted to psychoanalytic ideas and instead became more strictly hierarchical institutions devoted to training professional psychoanalysts. As the fine British psychoanalyst Pearl

---

[227] We suspect that he has the distinction of being the only knighted full member of the Society in its history.

[228] Ernest Jones, circular letter, 15 October 1925, Library of Congress.

King has said, the founding of the Education Committees marked the beginning of the end of the era of the gentleman (and gentlewoman) psychoanalytic scholar.[229] We might put it another way: it marked the beginning of the end of the era when psychoanalysis was first and foremost a science and only secondly a profession.

Symbolically, then, Tansley became a full member of the British Society the same day as the international psychoanalytic movement changed direction, away from distinguished savants like him, towards a more restrictive conception of the responsibilities of organized psychoanalysis. From the 1920s on, there would be fewer Tansleys and Stracheys who could become practicing analysts via Freud's, or anyone else's, couch, without the 'appropriate' qualifications. And, equally importantly, there would be fewer Ramseys and Jeffreys, polymathic scientists who included psychoanalysis as one amongst their many interests. Tansley was welcomed into the analytic institutions just when they were becoming more vigilant both about external threats, whether vitriolic critiques or undisciplined and semi-ignorant popularity, and internal threats (who is permitted to be an analyst?). Hence, a month later, Jones is writing another circular letter, again mentioning Tansley, but in the context of a different but related debate – that on lay analysis:

> In regard to lay analysts, no one except Professor has supported them more consistently than I have. From the beginning no discrimination has been made in our Society between lay and medical analysts and at present nearly 40% of our members are laymen. Among these are several of the most competent such as Flügel, Tansley, the Stracheys and Mrs. Riviere.[230]

While it is true that Jones never objected to lay analysts, he was also amongst the keenest of the close Freudian followers to secure the respectability and acceptability of psychoanalysis and its practitioners, especially amongst the medical profession. He was to spend many hours in the late 1920s battling against the scepticism of the British Medical Association so as to win their approval – in the event, distant and qualified – for analysts trained by the Education and Training Committee under the aegis of the British Society. What is curious is how easily, once again, Tansley is made to fit into one of his schemes – alongside those other distinguished graduates of the Cambridge-and-Berggasse-19 ethos.[231] Within fourteen

[229] Pearl King and Alex Holder, 'Great Britain', in Peter Kutter (ed.), *Psychoanalysis International: A Guide to Psychoanalysis throughout the World*, Stuttgart-Bad Cannstatt: Frommann-Holzboorg, 1992, pp. 150–72 at 154.

[230] Ernest Jones, circular letter, 19 November 1925, Library of Congress.

[231] On Alix Strachey and Joan Riviere, see Lisa Appignanesi and John Forrester, *Freud's Women*, London: Weidenfeld & Nicolson, 1992, pp. 352–71.

months of this letter being posted, Tansley had finally decided against becoming a full-time analyst and accepted the Chair in Botany at Oxford.

So what audit can be drawn up of the relations of these disparate and creative individuals to psychoanalysis? Penrose's biological interests took him a long way from psychoanalysis, but he never renounced his conviction of the importance of Freud's discoveries. Ramsey's energetic and roving intelligence might have brought him back to psychoanalysis if he had lived. In the early history of psychoanalysis in Britain, beyond the medical, the psychological, the literary, there are the Jeffreys, the Tansleys and the Penroses – scientists, whose predisposition towards psychoanalysis stemmed in large part from their having, effortlessly and without ambiguity, subscribed to what Freud would call the 'scientific Weltanschauung' in his *Introductory Lecture* of 1932. But as the definition of 'official' psychoanalysis came into focus in the mid-1920s, with the rules governing training, qualification and the overall aim of psychoanalytic institutions, such larger than life natives of the 'scientific' world would find it less easy to include psychoanalysis in their public preoccupations – although many of them would, naturally, continue to end up on the couch. Did the rise of 'official' psychoanalysis put an end to such promiscuous pursuing of knowledge? Did its rise exclude the youthful and restless, fine and inquiring minds, so prominent in Tansley's network, so central to the ethos of Cambridge, from the circles of psychoanalysis?

The professionalization of psychoanalysis may well have had this effect; but in mitigation of the implication that the process of professionalization deprived psychoanalysis of exactly the openness that its technique seems to imply, one might, with some justice, point to developments wholly outside psychoanalysis which required scientists to specialize, display the necessary credentials, communicate more exclusively with immediate colleagues, buckle down to the demands of big and corporate science.[232] In other words, it may not have been only psychoanalysis that lost its mercurial Ramseys and its polymathic Jeffreys. After all, Freud himself shared their vision of science as open to all who had the wit and graft – hence his resolute defence of the Farrows and Faggs of this world. Resonating with this individualism, Cambridge of the 1920s balanced an unselfconscious

---

[232] For a discussion of such changes due to the professionalization of British plant ecology, see for instance P. Lowe, 'Amateurs and professionals: the institutional emergence of British plant ecology', *Journal of the Society for the Bibliography of Natural History* 7(4) (1976), 517–35. The literature on professionalization of the sciences in this period is enormous. See for instance, Mary Jo Nye, *Before Big Science: The Pursuit of Modern Chemistry and Physics, 1800–1940*, Cambridge, MA: Harvard University Press, 1999.

elitism with democratic, universalist cravings – one can think of no more representative examples of these twin values than Penrose and Ramsey. Nor should the defender of professionalization and the necessary 'maturing' of psychoanalysis neglect to point to the local and unrepeatable circumstances of this episode – to the opening up of English society in general following the First World War, with a chafing tolerance of new ways and paths, its new-found but possibly short-lived awareness of the pounding blood of fragile youth and the skull beneath the skin. Lenity towards eccentricity surely greatly benefited the Freudians. The paradoxical lesson of this episode is, thus, that professionalization may have been intended to create Freudians, but its effect may also have been to unmake them. This may appear an improbable conclusion, but stumbling upon the Cambridge Psychoanalysis Group of 1925 makes it less improbable than it used to be.

CHAPTER 7

The Malting House Garden School

In a little grassy bay between tall clumps of Mediterranean heather, two children, a little boy of about seven and a little girl who might have been a year older, were playing, very gravely and with all the focused attention of scientists intent on a labour of discovery, a rudimentary sexual game.

Aldous Huxley, Brave New World, 1932

An Unusual Opening

In March of 1924, a curious job advertisement was placed by a Mr Geoffrey Pyke of Cambridge in the *British Journal of Psychology*, the *New Statesman* and *Nature*:

> WANTED – An Educated Young Woman with honours degree – preferably first class – or the equivalent, to conduct the education of a small group of children aged 2½–7 as a piece of SCIENTIFIC WORK and RESEARCH.
>
> A LIBERAL SALARY – liberal as compared with either research work or teaching – will be paid to a suitable applicant who will live out, have fixed hours and opportunities for a pleasant independent existence. An assistant will be provided if the work increases.
>
> Previous educational experience is not considered a bar, but the advertisers hope to get in touch with a university graduate – or some one of equivalent intellectual standard – who has hitherto considered herself too good for teaching and who has probably already engaged in another occupation.
>
> The advertisers wish to obtain services of some one possessing certain personal qualifications for the work and a scientific attitude of mind towards it. Hence training and experience in any of the natural sciences is a distinct advantage.
>
> Preference will be given to those who do not hold any form of religious belief, but this is not by itself considered to be a substitute for other qualifications.

The applicant chosen would require to undergo a course of preliminary training, 6–8 months in London, in part at any rate, the expense of this being paid by the advertisers.

Communications are invited to: Box 168, 'NATURE,' c/o MACMILLAN & CO., LTD., St. Martin's Street, London, W.C.2.

The owner of *Nature* was unsettled by the wording of the advertisement and regretted its publication, suspecting that it was an agency associated with the white slave trade. His fears would be allayed by Pyke's subsequent reassurance that the grandson of Lord Rutherford, president of the Royal Society, was a pupil at the school.[1] Pyke's vision for his Malting House Garden School was, clearly, resolutely unconventional. What Pyke, speculator and inventor, desired was an extraordinary nursery teacher for a revolutionary kind of educational environment created in Cambridge properly to prepare his own son, along with other young children, for the complexities of the twentieth century. Inspired by psychoanalysis and a penchant for untrammelled thought, the guiding idea was that the children would be allowed to discover the natural world freely like little scientists, through the medium of fantasy and play. After gaining assurance from a mutual friend, the analyst James Glover, that Pyke was not a crank, and meeting the Pyke family in person, Susan Isaacs, psychologist, logician and a full member of the BPaS, accepted the offer to co-direct the school and moved with her husband Nathan to Cambridge. Informally Pyke declared himself the Crown of the Malting House School, and Susan Isaacs the Prime Minister. And when it opened on 6 October 1924, the pupils, to be understood as distinguished foreign visitors, began to explore the new realm.

Pyke's 'Preliminary Announcement' for the school boasted a prominent cast for its first Advisory Council, including Professor of Education at the University of Bristol, Helen Wodehouse,[2] and two members of the BPaS, T. Percy Nunn, Professor of Education at London University, and Pyke's first analyst, Dr James Glover.[3] Of the many educational experiments

---

[1] W. van der Eyken and B. Turner, *Adventures in Education*, London: Allen Lane, The Penguin Press, 1969, pp. 48, 50. Further documentation of this incident is in the Pyke Papers.

[2] Helen Wodehouse (1880–1964) accepted an undergraduate exhibition to Girton College to read Mathematics in 1898 but changed the following year to Moral Sciences. From 1919 to 1931 she held the Chair of Education at the University of Bristol, creating one of the leading education departments in the country. Her writings on psychoanalysis include: 'Natural selfishness, and its position in the doctrine in Freud', *British Journal of Medical Psychology* 9(1) (1929), 38–59 and 'Instincts of death and destruction: a comment on Freud', Presidential Address given at the Froebel Society Annual Meeting, 5 January 1932, published in *Child Life* 35(156) (1932), 5–14. She returned to Girton in 1931, staying on as Mistress until 1942.

[3] 'Preliminary Announcement. The Malting House Garden School', Cambridge: Allied Press, 1924. Pyke Papers.

Fig. 7.1 Malting House Garden School, 1927. The hall, which remains a Cambridge
landmark, is opposite Mill Pond; the words 'Malting House' are still prominent
above the front door. The image is a still from a film produced by Mary
Field of British Instructional Films Ltd.

initiated in the early decades of the twentieth century, such as Maria
Montessori's Casa dei Bambini in Rome (1907) and Rudolf Steiner's
Waldorf School in Stuttgart (1919), the Malting House School was one
of the first children's schools in the world to gather its inspiration from the
ideas of Freud.[4] The other psychoanalytically oriented school was the
Psychoanalytic Kindergarten founded by the Moscow State Psycho-
Neurological Institute in 1921, patronized by Leon Trotsky and, remarkably enough, attended by Joseph Stalin's son Vasily.[5] Another famously
pioneering school with a measure of psychoanalytic inspiration was also
founded in 1921: A.S. Neill, who three years earlier as a 2nd Lieutenant
suspected of neurasthenia had his dream (of a snake which he had killed

[4] Several 'laboratories-cum-nurseries' or 'nurseries-cum-laboratories', principally inspired by
Watsonian behaviourism, were opened in the 1920s in the USA (Teachers College, NY; Gesell at
Yale; Iowa Child Welfare Station and others); see Rebecca Lemov, *World as Laboratory: Experiments
with Mice, Mazes and Men*, New York: Hill and Wang, 2005, p. 61.
[5] C.J. Wharton and V. Ovcharenko, 'The history of Russian psychoanalysis and the problem of its
periodisation', *Journal of Analytical Psychology* 44 (1999), 341–52 at 345. Sabina Spielrein joined the
nursery in 1923: see Lisa Appignanesi and John Forrester, *Freud's Women*, London: Weidenfeld &
Nicolson, 2001, p. 225.

repeatedly coming back to life) analysed by Rivers,[6] started 'a school in which the emotions would be primary'.[7] It soon moved from Dresden to the village of Sonntagsberg, Austria and later, in 1924, to Lyme Regis, where it was known as Summerhill. Ernest Jones would remind Anna Freud of the Malting House School's pioneering efforts when she began her own, second nursery fifteen years later: 'The Kinderheim sounds most interesting and I should like to hear more about it. I suppose you know of a similar experiment Mrs Isaacs made here with the Malting House in Cambridge, but yours is probably better organised.'[8]

Although Susan Isaacs and Geoffrey Pyke would part bitterly in 1927, two years before the Pykes closed the doors of the school for good, the Malting House School figured centrally in the subsequent highly influential work of Isaacs, who became the leading educational psychologist of the inter-war years and a household name for those seeking to emulate Isaacs's new 'listen to understand' approach to children. In this sense, the Malting House School was the most important 'observational laboratory' for education in Britain between the wars. The experiment allowed Isaacs to fuse general child psychology with psychoanalysis rather seamlessly and her subsequent writing thus greatly increased the influence of psychoanalysis not least because it promoted a dissemination of psychoanalytic ideas in an acceptable form. Isaacs's classic books, based on three years of recorded data from the Malting House School (October 1924 to December 1927), include: *The Nursery Years* (1929), *Intellectual Growth in Young Children* (1930), *The Children We Teach* (1932) and *Social Development in Young Children* (1933). The latter book, drawing also on studies of children's play and emotional disorders by Melanie Klein, and containing cross-indexed detailed evidence supporting infantile sexuality in its oral, anal and genital phases, particularly unsettled the nursery education world. In *Intellectual Growth in Young Children*, Isaacs challenged the work of Swiss

---

[6] Jonathan Croall, *Neill of Summerhill: The Permanent Rebel*, London: Routledge and Kegan Paul, 1983, p. 79. 'Rivers concluded that if Neill was sent to France, he would either win the Victoria Cross or be shot for desertion' and recommended six months' convalescence. Homer Lane would introduce Neill to Freudian psychoanalysis as well. In Neill's 1920 *A Dominie in Doubt*, he expressed some unease regarding his own dedication to Freud but stated: 'I firmly believe that Freud's discovery will have a greater influence on the evolution of humanity than any discovery of the last ten centuries. Freud has begun the road that leads to superman.'

[7] A.S. Neill, *Talking of Summerhill*, London: Victor Gollancz, 1971, p. 118.

[8] 17 February 1937, Jones to Anna Freud, JPBPaS, CFA/F02/91. Jones was (conveniently?) forgetting that Anna Freud had already started a school in Vienna, very much comparable to the Malting House School: the Hietzing or 'Matchbox' school built in Eva Rosenfeld's backyard, using Dorothy Burlingham's money, in 1927, which had twenty pupils (half of them in analysis) and lasted five years; see Appignanesi and Forrester, *Freud's Women*, p. 292.

psychologist Jean Piaget[9] (who, like Klein, made a visit to the Malting House School), mainly his view that children developed in fixed stages which set limits on their ability to use logic. Isaacs's conviction that children could use reason from an early age and could question, hypothesize and make inferences about the natural world when the 'right' environment stimulated their interest would become an important argument again in the 1960s, tempering the warm embrace given to Piaget's theories in English primary school science reform.[10]

Besides Susan Isaacs and Geoffrey Pyke – whom crystallographer J.D. Bernal called 'one of the greatest and certainly the most unrecognised geniuses of the time'[11] – the other leading adult figures in the school were Geoffrey Pyke's wife Margaret and Susan's second husband, Nathan Isaacs. Evelyn Lawrence, holding an economics degree from the London School of Economics, arrived two years into the experiment and would become Nathan's second wife after Susan's death in 1948. In the concentrated world of Cambridge, a tangle of intimate relations sustained the school; beyond the two major disciplinary threads, psychoanalysis and natural science, these included the friends who brought their children to the school and the school's local connections to powerful scientific networks. A brewery of progressivism for both children and adults, the knowingly unconventional Malting House School helped to foster both an atmosphere of psychoanalytic experimentalism and a reforming zeal linked to women's rights, sex and access to birth control.

The Pykes and the Origins of the School

> The school was supposed to be Freudian in avoiding inhibitions in the scholars through interference by their teachers. Geoffrey also interpreted Freud as warning against anything interfering with his son eating old orange peel lying in the middle of any main road and I can still see him signalling buses, lorries and cars to stop while his son crouched in the puddles eating.[12]

[9] On Piaget's relation to psychoanalysis, see Eva M. Schepeler, 'Jean Piaget's experiences on the couch: some clues to a mystery', *IJP* 74 (1993), 255–73.

[10] J. Hall, 'Psychology and schooling: the impact of Susan Isaacs and Jean Piaget on 1960s science education reform', *History of Education* 29(2) (2000), 514–18.

[11] J.D. Bernal, 'Mr. Geoffrey Pyke: an appreciation', *Manchester Guardian*, 25 February 1948, p. 3.

[12] Philip Sargant Florence, hand-written draft of memoirs created for Barbara Moench Florence, editor of *Lella Secor: A Diary in Letters 1915–1922*, New York: Burt Franklin, 1978, located in the Birmingham City Library Archives, MS 1571.

How did the school happen to open its doors in Cambridge? Although in retrospect it might be placed in a Cambridge 'tradition' of progressive education following Caldwell Cook's pioneering 'play way'[13] curriculum at the Perse School, it had no formal links to any other Cambridge school, College, University laboratory or Department. The answer rather is entwined in the personal geography of the school's founder, Geoffrey Pyke, who maintained close connections to both Cambridge and Bloomsbury. A key figure for Pyke was the person who recollected the striking orange-peel-puddle-eating tableau above: Philip Sargant Florence, the brother of Alix Strachey. Philip and Geoffrey had met at Cambridge in 1910 where they were studying law and economics respectively. Geoffrey was the son of Lionel Edward Pyke, a prominent Jewish QC, and Mary Rachel, née Lucas. His father died when he was five with great consequences for the health, wealth and happiness of the family. When he did come up to Cambridge, he did not complete his degree, being too enthusiastically preoccupied with everything else that was going on, contributing to his friend C.K. Ogden's *Cambridge Magazine* and attending gatherings of the Heretics. In November 1913 the young Pyke penned a note of apology to G.E. Moore for monopolizing him in conversation during one of its meetings: 'I am that extremely silly young man who sat on the corner of the sofa tonight and told you all about the philosophy of Brandes. Please forgive my impertinence. I had no idea who you were, and I thoroughly deserved a good snubbing which I should have got. I shall grow older one day.'[14] Philip and Geoffrey remained friends and together took a walking holiday across Norway and Sweden in 1914, with Pyke, inspired by the Futurist Marinetti's recent talk to the Heretics, eager to improve the local farmers' machinery.[15] Soon after the start of the war, Geoffrey was captured in Berlin where he was working as a Special Correspondent for the *Chronicle* and placed in the Ruhleben internment camp near Charlottenburg. He made what he called a 'scientific escape' in daylight and wrote his tale *To Ruhleben – and Back; a Great Adventure in Three Phases* (1916); its publication made him something of a celebrity. Pyke then went to work for Ogden, as the London advertising manager for

---

[13] Caldwell Cook, *The Play Way: An Essay in Educational Method*, London: Heinemann, 1917. Cook emphasized action and experience in learning over reading and listening: spontaneous play was understood to be the natural mode of education. 'By Play I mean the doing of anything one knows with one's heart in it' (p. 17).

[14] Letter from Geoffrey Pyke to G.E. Moore, 24 November 1913, Papers of G.E. Moore, Cambridge University Archives, Add. Mss. 8330 8P/24/1.

[15] Henry Hemming, *Churchill's Iceman*, London: Arrow Books, 2014, p. 15.

Fig. 7.2 Geoffrey and Margaret Pyke on honeymoon, 1918.

the *Cambridge Magazine*. Through a variety of means, creative, tenacious and sometimes deceptive, Pyke managed to keep the magazine supplied with newsprint in a time of national shortage. It was around this time he met Margaret Chubb, who had been a student of history at Somerville College, Oxford, and they married three months later.

The Pykes began their life together in London, sub-leasing a flat in Bloomsbury from Alix Strachey, Philip Sargant Florence's sister, while she and James Strachey were in Vienna being analysed by Freud. J.D. Bernal met Geoffrey at a Bloomsbury party of intellectuals in the mid-1920s: Bernal, who would become his colleague at Combined Operations during World War II, teased him for wearing colourful spats which marked him for a member of the commercial class. Geoffrey protested that they 'obviate socks which I'd have to change much more often!'[16] Geoffrey had in fact become a speculator, encouraged by some of the bright economists at the Bloomsbury parties as well as, he claimed, the dullness of some of his own relatives who were involved in the Stock Exchange.[17] With his unconventional system and extreme secrecy, he did well – sometimes extremely well – becoming the largest private speculator on the London metal market.

A son David was born in May 1921, the same year that Geoffrey's friend Philip moved with his young family back to Cambridge. Geoffrey decided that David was to have a perfect childhood, unlike his own which had been darkened by his mother's erratic mood swings and decisions that impacted him harshly, such as her demand that he attend Wellington School as a practising Jew: he would suffer for two years as the only Jew in an intolerant atmosphere of 'Pyke Hunts' and ridicule. Geoffrey would keep his distance from his mother Mary all of his adult life and never allow his wife to meet her. In Geoffrey's blueprints for a better world, his son David would never be forced to endure an institution such as Wellington. He would grow up with no guilt and no neuroses. Reading voraciously anything available on early childhood education to prepare himself for non-neurotic fatherhood, Geoffrey was also himself psychoanalysed for a brief period by James Glover,[18] and later, it seems, by Edward Glover, who also analysed his wife Margaret and his younger brother Richard.[19] Gradually he began to hatch a plan to provide companions and an ideal learning environment for his son: he would create his own school with the help of leading educators in the place where his own educational experience had been happiest – Cambridge.

The Pykes relocated there in 1922, not long after his mother's death from cancer in August, living first in Chaucer Road and then at the Malting House on Newnham Road. Originally just that – a Malting House – it comprised an oast house, small brewery and warehouse sited beside the Mill Pond; beer had been brewed here since the 1830s, at a time

---

[16] David Lampe, *Pyke: The Unknown Genius*, London: Evans Brothers, 1959, p. 42.
[17] *Ibid.*, p. 35. [18] *Ibid.*, p. 36. [19] FRP, Diary of Frank Ramsey.

when beer was much safer to drink than the river water. In 1909, the Reverend H.F. Stewart, Dean of Trinity College Chapel, had part of the property converted into a home, and later another part was renovated for use as a hall for musical concerts and lectures. Settling with his family here near the city core, Geoffrey reacquainted himself with the Heretics (his younger sister Evelyn, a Newnhamite, had recently been closely involved)[20] and socialized with key society figures, amongst them Agnes Ramsey, a leader in the National Council of Women. Worried that only children could get lonely and become unbalanced, Geoffrey asked his friend Philip to allow one of his sons to move in with the Pyke family. Philip declined, telling Geoffrey that he too wanted his sons to have a well-balanced childhood and reminded him he had only two boys.[21]

Geoffrey's younger brother Richard was also living in Cambridge while nursing a strong crush on Geoffrey's wife, Margaret.[22] After serving in the war, he read economics at King's College (Part I, 1921; Part II, 1922) and then became involved in research in psychology at the Psychological Laboratory. In 1924 he undertook an investigation on 'The legibility of print' for the Medical Research Council which was published in 1926; in 1928 he published a novel, *The Lives and Deaths of Roland Greer*, which was praised for its power of psychological analysis.[23] Frank Ramsey became entwined with the Pyke family through Richard, who was his friend. Frank's mother, Agnes, was part of the circle, too.

The Pykes made Frank Ramsey godfather to David and they all travelled together in Italy in the 1923 Easter vacation. Frank developed a passion for Margaret, so all-absorbing that he was now, himself, considering psychoanalysis. Writing home to his parents in the spring of 1924, he said: 'It seems to me perfectly proper to spend a scholarship being analysed as it is likely to make me cleverer in the future, and discoveries of great importance are made by remark-

---

[20] OP, Box 110, F. 4, Letters from E. Pyke to Ogden, *c.* 1920.   [21] Lampe, *Pyke*, p. 37.
[22] This recollection came from Janet Pyke, wife of Geoffrey's son, David.
[23] Pyke's novel *The Lives and Deaths of Roland Greer* (London: R. Cobden-Sanderson, 1928. New York: A. and C. Boni, 1929) was loosely autobiographical and based on the analysis of family relations involving a cruel mother and Roland's fixations on his older brother and sister-in-law: an *Evening Standard* review commented that the book is 'charged with emotional quality that could scarcely be better ... [Pyke] is so preoccuped with mind-states that he forgets that mind-states ought to have some physical dress' (Arnold Bennett, 'Fiery sign on the horizon', 26 July 1928. Richard Pyke died suddenly due to illness in Shanghai, 2 March 1938. Times Obituary, 5 March 1938).

able people and not by remarkable diligence.'[24] While in Austria, Ramsey discovered progressive education at its most adventurous on a visit to A.S. Neill with Lionel Penrose.[25] In a letter home he wrote: 'It is a superb place on the top of a mountain where he has a school with 9 children mostly British who do what they like, nevertheless learn a certain amount. They are mostly children no other school would take, pathologically naughty or neurotic. He seems to me a most remarkable man in the way he deals with them rather based on psychoanalysis.'[26]

Ramsey's motivations in seeking out Neill were probably derived from the Pykes, who were now steaming ahead with their school project. Geoffrey had been devising teaching techniques for the education of David, following him and making copious notes on whatever David said or did. Pyke's Malting House formula seems to have blended the *Totem and Taboo*-inspired imperative to undo the relations between the generations with the credo that the child's natural curiosity could be freed through science. Both Susan and Nathan Isaacs, as Nathan recalled later, quickly identified themselves 'with the furtherance of his schemes and plans'.[27] Pyke, it was clear to Nathan, was an educational genius and had thought of everything. However, apart from the stream of increasingly detailed and creative job advertisements, Geoffrey Pyke published nothing else regarding his thoughts on the school. There is, though, an unpublished series of lectures he himself had planned to give to the BPaS: this offers up an idea of his scheme and what he thought it had to do with psychoanalysis.

In the draft lecture, written sometime after attending Klein's London lectures in July 1925, Pyke's concern was the transmission of culture. He believed he had discovered what he called 'important sociological applications of Freud's far-reaching discoveries'. He began:

> I have been asked to tell you this evening something about the school at Cambridge. This task, however, will, I hope, be fulfilled later by ~~one, who is a member of the Society and who is more fitted to it than I am.~~ Mrs. I. It occurred to me that you might nevertheless be interested to hear

---

[24] FRP, letter dated 13 April 1924, File 2/1/4.
[25] Penrose already knew Neill 'slightly' (FRP, File 2/1/4, 30 March 1924).
[26] FRP, letter dated 30 March 1924, File 2/1/4.
[27] N. Isaacs to M. Pyke, 1927, Institute of Education Archives, NI/D/2, 68 pages.

a little of the general ideas from which the school arose, which, lying like some ~~geographical~~ geological strata some way beneath the surface have, as it were, outcropped in places into notions that are more particular, and on which the school itself was founded.[28]

A point that had always puzzled Pyke was 'why the human race brings up its children at all'. And he speculated: it is 'our knowledge of the child's hatred'.

> Tremendous emotional forces meet at the frontier between the generations. The aims which are commonly assumed to lie behind the upbringing and education of our children are not the only aims which possess us.

First, it is fear which drives us to transmit our cultural inheritance, and second, it is the Oedipus situation which lies behind the drama. Sons are hostile to their fathers, and the father, identifying with the son, becomes aware of the son's hatred and guilt; unconsciously fearing the son and attempting to alleviate the pain of fear, the father exploits the guilt and applies the moral laws that were applied to him as a child. Fear and narcissism make it imperative 'that our children should be as like us as possible'.

> In the light of the awareness of our own hate impulses we despair of eradicating these in our children. The better we know our children the better we can defend ourselves. If by identification we *are* our children, we know the nature of every move before it is made and above all we know the limits of what they will dare to do. A hostile and mysterious son is most perturbing. Hence the fathers' moral laws must also be those of the sons. Behind the suppression of deviations lies the influence of fear. And it is in *this* way that God made man in his own image.

Education was the key mechanism of succession, for 'Not only is our education a culture transmission ... our protective device against the next generation, it is our over-determined neurotic protection against the production of the superman.' But to survive the future it would be imperative to 'breed a race who can survive the great changes' we are creating in our environment. In his paper, Pyke played with G.E. Moore's ideas on the limitations of external morality, J.B.S. Haldane's speculations on ectogenesis (the development of foetuses in artificial wombs)[29] and Lord Rutherford's laboratory aims and the possibilities of

---

[28] G. Pyke, *c.* 1925, draft lectures for the British Psycho-Analytical Society, Pyke Papers. Crossings out in the original.

[29] J.B.S. Haldane considered the idea in *Daedalus; or, Science and the Future*. Published in 1924, it was the written version of a lecture given to the Heretics Society on 4 February 1923.

atomic energy. Remarkably, Rutherford and Haldane would become two-thirds of the three-member final interviewing committee in Pyke's search for a new science teacher in 1927 – and G.E. Moore's two sons attended the school. Not only did these Cambridge networks provide Pyke with children and scientific advice, but their ideas inspired his self-mocking anticipation of 'super-Jews' and more serious prognostication:

> The future, seems to me, to be very full of chances and dangers. I see attempts, not very conscious attempts, being made to control our quality, probably with something very utilitarian in view, like the creation of better soldiers or smarter financiers – species of small human tanks immune from shell-shock, or super-Jews of telepathic sensitivity to market movements, or the robot Mr. Karel Kapek has put upon the stage. I believe these efforts may succeed in their immediate utilitarian purpose, but will nevertheless fail in their effect of showing men how, as such, to change themselves.

To show human beings how to change themselves and thus survive the future, one required a radically new educational method.

Pyke would later send his lecture drafts to Edward Glover (by then Ernest Jones's right-hand man in running the BPaS) who, in early August 1931, returned the annotated manuscripts with several pages of commentary and the encouraging words: 'Your niche is the exploitation of data in order to advance your ideas – or in other words, your slogan should be "the New Sociology" (rooted in individual psychology, of course) not another "New Psychology".'[30] Pyke had been impressed with Klein and seems to have agreed with her view – in contrast to Anna Freud who saw a possibility of analysts working closely with teachers – that education and analysis cannot be the same thing: both are separately necessary. Klein came to visit the Malting House School during this trip. According to family myth, David, at the age of three, was the first child in England to be analysed by Klein.[31]

The Isaacs

Once decided that they were a good match with the Pykes, the Isaacs moved to Cambridge, living initially in spacious quarters in Hoop Chambers off Bridge Street which they rented from Ogden, with

---

[30] Glover to Pyke, 9 August 1931, Pyke Papers.

[31] If Klein saw David in the summer of 1925, their meeting more likely would have constituted a consultation, not an analysis. With the use of twin studies, David Pyke later would become an expert in diabetes and for twenty-seven years was a consultant physician at King's College Hospital, London and Registrar of the Royal College of Physicians.

Geoffrey as constant guest.[32] For Susan, this was a return to a familiar town. After taking a first class honours degree in philosophy at the University of Manchester where she greatly impressed Professor S. Alexander and the newly appointed Mr T.H. Pear, C.S. Myers found support for her growing interests in 'the concrete and living field of psychology and the possibilities of knowledge-increasing research'[33]to come to Cambridge to work with him as an advanced student. She would spend the academic year of 1912–13 studying possible connections between visual imagery and good spelling and finding none. In an article written for the *Manchester University Magazine*, she reflected on her experiences of Cambridge and found much to criticize: 'The problems of the world outside fall into an unreal distance ... Women are essentially intruders there.'[34] In choosing to go back to Cambridge with Nathan ten years later, she would be revisiting the old haunts of male entitlement but also would be diving into the real 'turbulent streams' of modern child life and what earlier had been out of sight: the 'big primitive needs of existence'.[35]

Nathan, born in Germany to Russian Jewish parents, was a senior manager at the British metals dealer, Bessler, Waechter & Co. Ltd.[36] Beyond working at the firm, he was also a student of logic, psychology and epistemology. He encountered Susan in London in 1920 while attending her lectures on psychology for the Workers' Educational Association. Nathan, ten years Susan's junior, was an enthusiastic pupil writing 95-page essays and, over the course of the term, more intimate letters. At first giving a cold response, Susan reciprocated his feelings but was then married to the botanist William ('Willie') Brierley. At this time, Susan was finishing her first book, *An Introduction to Psychology*, which was pioneeringly and resolutely psychoanalytic, but at the same time eclectic.[37] Her interest in

---

[32] Eight rooms plus bathroom and water closet for £100 per annum. OP, Rental Agreement dated 20 October 1924, between Ogden and Susan Isaacs, 29 September 1924–24 June 1925. From 1925 on, they lived at 47 Hills Road.

[33] Nathan Isaacs, 'Notes by Nathan Isaacs', The Isaacs Collection, Institute of Education, S1/A/3, p. 2. Nathan notes that after obtaining a 'brilliant first' at Manchester, she immediately was employed there as a lecturer in the Philosophy Department.

[34] A Manchester Girl [Susan Fairhurst], 'Cambridge and Manchester', *Manchester University Magazine* 9(8) (1913), 173–5.

[35] *Ibid.*, p. 173.

[36] Philip Graham, *Susan Isaacs: A Life Freeing the Minds of Children*, London: Karnac, 2009, p. 75.

[37] The 1928 version of this book includes an appendix which augments further her earlier perspective in light of new developments in psychology – psychoanalysis being the most important. 'An introduction to psychology', original notes by Susan Isaacs for revisions, Susan Isaacs Papers, Institute of Education Archives, S1/B1, 1928.

psychoanalysis was growing steadily, reflecting perhaps the influx of the psychoanalytically inclined to the BPS meetings at the end of 1918 and the beginning of 1919.[38] Some time after 1916 she had become actively involved in the Medico-Psychological Clinic of London (also known as the Brunswick Square Clinic), then a key site for psychoanalytic education and discussion, where Susan became acquainted with May Sinclair, Sylvia Payne, John Carl Flügel, James Glover and a ferment of psychoanalytic ideas (not always strictly Freudian).[39] According to Laura Price, Susan was a student in analysis at the Clinic with Julia Turner and possibly Jessie Murray.[40] In September 1920, Isaacs began an analysis with Flügel which continued until July 1921. That same year, following a three-month analysis with Rank in Vienna (her final analysis would be with Joan Riviere starting in 1928 and ending in 1933), she began attending meetings of the BPaS and was elected an Associate Member in December. Her marriage to Willie Brierley was dissolved, and in 1922 she married Nathan and her younger friend Margery married Willie Brierley. Interestingly, both Margery and Susan were in analysis with Flügel: the two couples remained friends and Margery and Susan became key figures in the BPaS, both close to Melanie Klein.

During this period of personal upheaval, Susan was cutting herself off from old friends who might have objected to her being divorced and re-marrying.[41] Before the Malting House advertisement came her way, there were many directions in which she could have gone given her multi-faceted background and existing commitments. Following Cambridge, she had taught at Darlington Training College (1913–14), lectured in logic at Manchester University (1914–15) and from 1916 was a part-time tutor in psychology at London University, a post she held until 1933. From 1921 to 1929 she was a member of the Council of the National Institute of Industrial Psychology (NIIP) founded by Myers, her graduate supervisor;

---

[38] BPS Minutes, 1919. Nearly half of these new members were women; Susan Brierley was already a key force in the Society, nominating all the Brunswick Square people, along with Homer Lane and many others. Also relevant to Susan's interest in psychoanalysis may have been the stimulus of Education Professor and BPaS member Thomas Percy Nunn, who assisted her with the book.

[39] On the Clinic, see Theophilus E.M. Boll, 'May Sinclair and the Medico-Psychological Clinic of London', *Proceedings of the American Philosophical Society*, 106(4) (1962), 310–26 and Suzanne Raitt, 'Early British psychoanalysis and the Medico-Psychological Clinic', *History Workshop Journal* 58(1) (2004), 63–85.

[40] These details are from 1962 letters of Laura Price, part of Boll's research papers which Dr Ken Robinson had copied. Robinson also examined training records from the BPaS which indicate that Isaacs reported and dated her analyses with Flügel, Rank and Riviere, but does not, intriguingly, report analyses with Turner or Murray at the Brunswick Square Clinic.

[41] D.E.M. Gardner, *Susan Isaacs*, London: Methuen Educational, 1969, p. 50.

many Cambridge graduates in moral sciences found work in the NIIP, including John Wisdom, Lettice Baker and Margaret Gardiner. From 1923 to 1931 she served on the NIIP's advisory committee on vocational guidance headed by Cyril Burt (whom she had first encountered in the pre-war Cambridge Laboratory and who was her co-advisor to the Editor of the BJP).

Susan's analysis with Flügel was bringing up anxieties related to separation (she had lost her mother after the birth of a younger sister) and Gardner surmises that Isaacs, who wanted children of her own, was dealing with her own anxieties regarding childbirth. Neither William Brierley nor Nathan wanted children, and by the time Susan was mentally ready, she was feeling, at the age of thirty-eight, perhaps uncertain if it would be physically wise to begin. She had started medical training with the object of becoming a medical psychoanalyst. Despite passing her preliminary examinations she gave up the training because, according to Gardner, she did not want to put undue financial pressure on Nathan. She felt she had already saddled him with the purchase of their Bloomsbury home at 53 Hunter Street and did not want to burden him further. By 1923 she was a full member of the BPaS.[42] It would appear that, from the time of her marriage to Nathan to early 1924, her main occupation was her psychoanalytic practice working out of Hunter Street combined with WEA teaching;[43] in addition, throughout the 1920s she continued to assist both Myers (till 1924) and then Bartlett with the editing of the *British Journal of Psychology* and was the Secretary of the Committee for Research in Education of the Educational Section of the British Psychological Society, drafting its report in December 1923 on educational research across the UK. She was, quite clearly, the best-qualified person in the country to take Pyke's job.

---

[42] As was and is usual with 'scientific Societies', Membership of the Society did not require any qualifications; according to the rules formulated in May 1920 and adopted in October 1920, she had to be proposed and seconded, and then: 'All elections by ballot. One adverse vote in six shall exclude.' She was duly elected and became a full member on 3 October 1923. She had been an Associate Member since 1921. The structures of 'qualification' were introduced when the Institute of Psycho-Analysis (and the Training Committee) was established in 1925, but the Constitution of the Society was not changed to reflect this.

[43] Both Janet Sayers, 'Susan Isaacs: Children's Phantasies', chapter 3 in *Kleinians: Psychoanalysis Inside Out*, Cambridge: Polity, 2000, and van der Eyken and Turner, *Adventures in Education* (p. 21) suggest she could no longer work for the WEA as she had been banned after becoming involved with a student while married; however, in the brief notes on Susan Isaacs left by Nathan in the Institute of Education Archives, The Isaacs Collection S1/A/3, p. 3, he says, 'In 1922 she was married to me, and for the next two years combined a certain amount of psycho-analytic practice with W.E.A. teaching.'

Whether or not her decision to take the Malting House School job was symbolically equivalent to the definitive decision not to have her own children, Pyke was offering a good salary and his enthusiasm was infectious. The Isaacs even fitted his preference for those holding no religious beliefs: Susan had abandoned the non-conformist Methodist religion of her childhood and Nathan was an agnostic Jew. Throughout the period 1924–26, there is, in Susan's and Nathan's letters and works, a strong sense of shared purpose with Geoffrey Pyke. By 1926 Nathan was also placed on Pyke's payroll – and he was sent off to write a treatise on why children ask 'why?'[44] He produced several manuscripts – one called 'Psychology, education and science' – which envisioned a much expanded version of the Malting House: an Institute. In a chapter entitled 'The life-course as unit: psychology as ecology, and therefore as education'[45] he wrote that psychology could benefit from ecology's shared interest in the genetic method and the making of complete life-histories. To render the field of psychology more scientific, he argued that what was necessary was the 'adequate observation, description and comparison of a large number of behaviour-histories'. What was needed was no less than the 'complete education-study of a considerable number of children from birth'. In the planned Institute for which the Isaacs and the Pykes all hoped, ecological wisdom would resonate in the human nursery.

What was a 'natural' or 'normal' child in the Malting House? According to the editorial Nathan ghost-wrote for *Nature* in 1927, the public answer was the 'curious' child for whom an education in science provided 'natural continuity' to its 'natural roots'.[46] 'We should be endeavouring so to guide, reinforce, and develop this curiosity of the normal child in the world around him that it could pass continuously by its own activity into the same interest, informed and organised, in the world – not different but greater – of science.'[47] Although any 'moulds' were wrong, the scientist (the 'correlator' in Pyke's terminology) was the highest type, and, according to Nathan, reality would teach the child 'its own discipline, its own morality, which are the method, temper and spirit of science; and perhaps these are not so very much worse, after all, than those of two random

---

[44] Susan Isaacs, *Intellectual Growth in Young Children: With an Appendix on Children's 'Why' Questions by Nathan Isaacs*, London: Routledge and Kegan Paul, 1930, pp. 291–349.

[45] N. Isaacs, *c.* 1926–27, 'Psychology, education and science', Institute of Education Archives, NI/B/6.

[46] Nathan's article 'Education and science' is an unattributed editorial in *Nature*, 23 July 1927, but the authorship is revealed in reprinted form in Susan Isaacs's *Intellectual Growth in Young Children*, pp. 350–4.

[47] Isaacs, *Intellectual Growth in Young Children*, p. 351.

human beings haphazardly mixed together and compulsorily administered to an unwilling third'.[48]

In a long letter written in 1926 clarifying her sense of the school's purpose, Susan Isaacs explained: 'Science has become a way of life, one that was foreshadowed from the time when man began to stand erect on his feet and to look round upon the world as it is, and one which unless some unforeseeable cataclysm nullifies his history, will determine not only the material setting of his life, but more and more its inner conditions and most intimate values.'[49] Susan's very critical review the same year of Bertrand Russell's *On Education, Especially in Early Childhood*, in the *International Journal of Psychoanalysis*, tackled Russell for dismissing the so-called Oedipus complex as 'misleading or more airily as moonshine'. She asks: 'How is it possible for Mr. Russell to overlook the complexity of human motive in this relation of parent and child? We cannot avoid the conclusion that this delightful belief in the goodwill of the parent and the reasonableness of the child spring from the fact that Mr. Russell, face to face with his children, is unable to tolerate the possibility of hostility on either side; or, at any rate, of a hostility that will not yield to mild justice and a gentle logic.'[50]

> He speaks of 'the parental instinct in its purity' (p. 154), and of 'different kinds of natural affection' (p. 158). He has a similar quaint simplification of the sexual life in his discussion of sex education, where he taboos 'obscenity', and would seem to shut it out not merely from education and morality, but from psychology also. In fact, Mr. Russell is not letting his right hand know what his left hand is doing; he is at one and the same time saying that certain things are undesirable and that they do not occur. The tactics of his position was beautifully stated by the small boy who recently said to the reviewer, 'I do not like dreams, they are horrid things; and another thing, I don't have them.'[51]

While Russell advocated 'Science wielded by love',[52] the Malting House was (less naively in Isaacs's estimation) exploring how science should oblige us to recognize much darker feelings. Susan's trenchant remarks may have played a role in transmuting 'moonshine' into 'the truth': the next year

---

[48] Nathan Isaacs to unknown, most likely Nathan's father, 1924, Institute of Education Archives, NI/C/1.
[49] Susan Isaacs to unknown, most likely Nathan's father, 1926, Pyke Papers.
[50] Susan Isaacs, 'Review: *On Education, Especially in Early Childhood* by Bertrand Russell', *International Journal of Psychoanalysis* 7 (1926), 514–18 at 517.
[51] *Ibid.*, p. 517.
[52] Bertrand Russell, *On Education, Especially in Early Childhood* (1926), London: George Allen & Unwin, 1966, p. 127.

Russell and Dora Black would start their own school at Beacon Hill on the Sussex Downs. Russell would change his mind regarding Freud, writing in the *Literary Guide and Rationalist Review* in 1928 that 'experience with young children has led me to think that much of what appears fantastic in his psychology is nevertheless largely true'.[53]

The School: Inner and Outer Geographies

In their shared work at the school creating what Evelyn Lawrence called 'a new generation less nerve-ridden than the old',[54] the Malting House teachers recognized strong emotion in children and encouraged the expression of aggressive inner feelings. Indeed it underlay their understanding of scientific practice and curiosity in children. All strong passions including sexual interests were openly expressed but 'canalized by being turned into scientific channels'.[55] Pyke's technique had two entwined components, the first being the environmental design of the school itself, concerned with critical connections between the children's inner and outer geographies, and the second, his detailed methods. The latter were based on the imperative that we must not impose our cultural inheritance upon children but rather 'we must tentatively and most delicately, offer it to him ... He must discover the human world as an anthropologist discovers Torres Islanders [*sic*].'[56] Disporting his Cambridge view with the reference to the 1898 Torres Straits Expedition involving Haddon, Rivers, Myers et al., Pyke's technique was infused with Cambridge-specific geographical imaginations of science and discovery.

The interior of the school opened out to a large garden 'with plants (which may without taboo be dug up everyday to see how they are getting on, leading mainly to the discovery that this is a temptation best resisted if growth is desired)'.[57] The children were free to move from inside to outside or back again: if they saw an airplane, they might run inside to draw it or model what it might be looking at. The space was understood by Pyke and Isaacs to open the child to the facts of the external world, as opposed to conventional classrooms which created barriers between children and their natural living interests, discouraging rather than encouraging children to

---

[53] Bertrand Russell, 'Freudianism', *The Literary Guide and Rationalist Review* 392 (February 1929), 44.
[54] Van der Eyken and Turner, *Adventures in Education*, p. 42.
[55] *Ibid.*, p. 44. As above, the phrase is quoted from Evelyn Lawrence's written impressions of the school.
[56] Geoffrey Pyke, *c.* 1925, draft lectures for the British Psycho-Analytical Society, Pyke Papers.
[57] Geoffrey Pyke, advertisement in *The Nation and Athenæum*, 4 June 1927.

'find out' about the world around them. That freedom was qualified, however; and though a nursery, it is worth noting that this was simultaneously the home of freely roaming, erratically present King Pyke.

Subsequently, in *Intellectual Growth in Young Children*, Isaacs would indicate that, although their work with children was not as such a laboratory experiment, the 'real' world environment of the school was not to be seen as a design flaw or omission in the creation of significant data; on the contrary, 'Watching the spontaneous cognitive behaviour of a group of children under conditions designed to further free inquiry and free discussion may, therefore, reveal facts which would scarcely yield to the direct assault of test or experiment.'[58] Noting the 'vast bulk of barren experiment which has clogged the growth of psychology', she proposes that what escapes experiment '*may* be the more vital and genetically significant part'[59] and quotes F.C. Bartlett's own 1929 article on the dangers of simplifying the environment in studying complex response.[60] The Malting House building also had an observation gallery, rather akin to a bird-hide for ornithological fieldwork, from which visitors and teachers could watch the children without disturbing them; from this position it was observed that Geoffrey Pyke's recurring appearances in the school caused certain children to become hostile to his son David ('Dan' in Isaacs's published account of the school).[61]

The school's enclosed garden, which in previous years hosted Jane Harrison and friends,[62] provided an area for outdoor play. Susan provided a detailed description of the school spaces in the 1930 book:

> The school met in a large hall, from which easy steps ran to the garden, where there was plenty of room for running and climbing, for communal and individual gardening, and for various sheds and hutches for animals. The garden had two lawns, and plenty of trees, many of them bearing fruit. The large hall had a gallery with stairs at each end, and a low platform, on which the piano stood. The horizontal framework supporting the roof made excellent bars for the children to hang on or climb up to.
> Besides the large hall, there were four smaller rooms, and a cloakroom and lavatory. Part of the cloakroom was used as a kitchen by the children;

---

[58] Isaacs, *Intellectual Growth in Young Children*, p. 6.   [59] *Ibid.*, pp. 6–7.

[60] *Ibid.*, p. 3, from Bartlett, 'Experimental method in psychology', *Nature*, 31 August 1929, pp. 341–5.

[61] Susan Isaacs, *Social Development in Young Children: A Study of Beginnings* (1933), London: George Routledge and Sons, 1946, p. 53. Several clues exist in the published records, but by correlating the published Malting House notes with an unpublished list of the children with their mental ratios and real names found in the Pyke Papers, it is clear that David Pyke is 'Dan'.

[62] See photo in A. Robinson, *The Life and Works of Jane Ellen Harrison*, Oxford: Oxford University Press, 2002, p. 186. The previous owner of the house was Dr Hugh Frazer Stewart, Lecturer in Modern Languages and Dean of Trinity College, also pictured here in the Malting House Garden.

Fig. 7.3 Susan Isaacs with children at the Malting House School.

the gas cooker, and shelves and tables for crockery and cooking utensils were kept there. The large hall was used for general purposes, and for music and dancing. In the first year, one of the smaller rooms was used as a rest room, and another as a reading and writing room for the older children of the group.

Later on, one of the rooms became a quiet room for the older children, with shelves for the local school library, and the general reading and writing equipment. One large room was fitted up as a combined carpentry room and science laboratory. (The children at one stage called this the 'cutting-up room', as most of their biological work was done there.) The third was a handicraft room, with equipment for modelling, drawing and painting; and the fourth, a quiet room for the smaller children, in which reading and writing materials suitable for them was kept, and movable tables and chairs. The school was attached to a house, in which the children who were in residence lived.[63]

As the school expanded to require another residence, Susan became house-mistress at St Chad's on Grange Road. Beyond walking between its own buildings, Malting House activities were frequently extended into Cambridge excursions:

[63] Isaacs, *Intellectual Growth in Young Children*, pp. 14–15. Cited in van der Eyken and Turner, *Adventures in Education*, pp. 24–5.

Besides the usual walks and picnics on the fens and the river, and bus journeys into the country, the older children went to a ladder-maker's and wheelwright's to order ladders and see them being made; a sawmill to watch logs being cut up; the Round Church and some of the Colleges; the Botanic Garden; the Ethnological, Natural History and the Fitzwilliam Museums; to the Zoo in London; and up on the Gogmagog hills at dawn on the occasion of the June eclipse of the sun.[64]

The school buildings had multiple and overlapping uses: it was a private home for the Pykes yet a public sphere for Susan Isaacs and the other teachers in which to work. It was originally a day school but children quickly started living in residence. Few visual records of the school remain except for a number of stills from a film, now lost, produced in 1927 for Pyke by Mary Field (who worked for a company specializing in natural history films). The camera man E.W. Edwards said afterwards: 'In all our experience of photographing every kind of wild creature, not excepting cultures of bacilli, the problem of photographing children in their wild state proved the most difficult to tackle.'[65] By the time the film was made, there were twenty children at the school, ranging in age from 2½ to 8, four of whom were girls. The children were filmed trying to see if leaves would burn and how much water was necessary to put out a fire. Some listened to a gramophone which they worked themselves. They were also shown dissecting Susan's pet cat as well as working with a real drilling machine and a lathe.

There were few rules but the teachers restricted access to the key to turn on the bunsen burners. What these stills do not reveal are the 'shrieks and gurgles and jumpings for joy'[66] and the record-keeping technique (likely modelled after Pyke's earlier notations of his growing son's actions and conversations). Teachers would take turns making records of what the children were doing and saying inside and outside of school hours. From 1927–28, this task would be performed by stenographers hidden in the balconies. Lydia Smith draws a useful comparison which highlights the importance and novelty of this work:

> Gesell used one-way glass, Burt administered mental tests, Charlotte Bühler described the minute-by-minute behaviour of infants, Piaget talked with

---

[64] Isaacs, *Intellectual Growth in Young Children*, p. 290.

[65] Van der Eyken and Turner, *Adventures in Education*, p. 56.

[66] Lydia Smith, *To Understand and to Help: The Life and Work of Susan Isaacs, 1885–1948*, London: Fairleigh Dickinson University Press, 1985, p. 73, citing Gardner, *Susan Isaacs*, pp. 65–6 who quotes Evelyn Lawrence's retrospective thoughts on the school. This is also quoted in van der Eyken and Turner, *Adventures in Education*, p. 42.

Fig. 7.4 Malting House children with a gramophone.

Fig. 7.5 Children with an adult-sized drill press.

and asked questions of children one by one, and Bridges devised a developmental rating scale. But only at the Malting House were extended observations made of children interacting within an environment specifically designed to stimulate their powers of inquiry and imagination, to

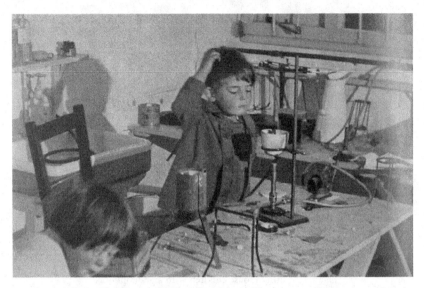

Fig. 7.6 Children working in the Malting House science laboratory.

provide considerable verbal freedom and the company of other children, and to include adults who dealt with them according to a deliberate planned technique.[67]

For Isaacs, reflecting in 1933, the school was not, as Russell had described it in *The Scientific Outlook*, an 'application of psycho-analytic theory to education'; she maintained that her more active inspiration was the educational philosophy of John Dewey.[68] However, her psychoanalytic experiences 'naturally lead' her to 'be interested in *all* the behaviour of the children' and she was 'just as ready to *record* and to *study* the less attractive aspects of their behaviour as the more pleasing', all observations then being ripe for psychoanalytic interpretation.[69]

[67] Smith, *To Understand and to Help*, pp. 75–6.
[68] Isaacs, *Social Development in Young Children*, p. 191n1. Russell does not name the Malting House in this book but he does write: 'The most important applications of psycho-analytic theory are to education. These applications are as yet in an experimental stage, and owing to the hostility of the authorities they can only be made on a very small scale.' He then gives a footnote: 'For experimental data on this subject see Susan Isaacs, *The Intellectual Growth in Young Children*, 1930' (Bertrand Russell, *The Scientific Outlook*, 2nd edition, 3rd impression, London: Allen & Unwin, 1954, p. 186).
[69] Isaacs, *Social Development in Young Children*, p. 19

8.10.24. Robert and George again half-chanted 'wee-wee thing' and 'do-do thing', when running in the garden. With this they tended to get very excited, and to make a half-screaming, half-wailing noise.[70]

16.10.24. Robert wanted to smash a plant pot, and when prevented by Mrs. I., he bit her wrist severely.[71]

19.6.25. The children had made a bonfire of garden rubbish, and Duncan (7:0) put an old tennis ball on to the fire, telling the others, 'It will make a smell and perhaps it'll explode'. But nothing happened, as it did not fall into the hot part of the fire.[72]

25.2.25 . . . Frank and Dan were climbing on the window-sill, and Frank said he would 'push Dan's foot up'. Paul (3:11) said, 'Yes, we'll push Dan up to God'. Frank (5:4): 'Yes, and perhaps He'll kill him'.[73]

The records also contain many examples of friendliness and cooperation as well as delightful insights such as six-year-old David Pyke's ('Dan's') spontaneous reflection that 'a thing can't be nowhere – it must be somewhere if it's a thing at all'.[74]

In a 9 July 1927 advertisement in *Nature*[75] for a new science teacher,[76] Pyke wrote that 'the directors believe that the learning of how to learn and a scientific scrutiny of familiar things, an attitude of critical curiosity and intellectual aggression to the unknown, require to be preceded by the discovery of the idea of discovery'.

> The method employed at Cambridge . . . is on the one hand to eliminate the arbitrary authority of the pedagogue and to substitute for it the attitude of the co-investigator ('Let's find out' and not on any verbal information is the answer given to most questions), and on the other hand to provide an environment with more than usual scope for activity, intellectual and social, including apparatus which shall both set problems and provide their solution.

The three-man Appointments Committee, which included Lord Rutherford and J.B.S. Haldane, appointed an American, Richard Slavson. He would write in four-year-old Peter Fowler's report for the

---

[70] *Ibid.*, p. 138.   [71] *Ibid.*, p. 41.   [72] *Ibid.*, p. 114.   [73] *Ibid.*, p. 114.   [74] *Ibid.*, p. 71.

[75] Macmillan had reversed his decision to forbid publication of Pyke's copy after Pyke wrote with reassurances that the woman hired in 1924 was both married and published and that the grandchild of Rutherford, the President of the Royal Society, was a pupil at the school (van der Eyken and Turner, *Adventures in Education*, p. 50). Further documentation concerning this incident is in the Pyke Papers.

[76] Margaret Leathes, who would marry Lionel Penrose the following year, also applied for the job along with over 200 others.

winter term of 1927, building on earlier records of the activities of Rutherford's grandson[77]:

> Peter's interest in fires still persists but we have succeeded both in extending that interest into other channels than merely bonfire and to divert it ... Peter is outstandingly of an experimental and investigatory nature, but, at the present time, these tendencies are focused around fire.[78]

Such 'instances of direct interest in the physical world and mechanical causality in children so young' would be interpreted by Susan 'as positive evidence to be weighed against Piaget's negative evidence that such interests are not shown'.

> They suggest strongly that the extent to which they do appear and are sustained must in large part be a function of the environment, and of the degree of response which they meet with in influential adults. If they win no help or attention, they will not be pursued and sustained. Little children are profoundly at the mercy of grown-ups and of the environment which grown-ups determine, and are always ready to draw in the sensitive feelers which they put out to test the world.[79]

On 4 March 1927, the same day he would address the Cambridge Education Society on 'La causalité chez l'enfant', Piaget visited the Malting House School.[80] 'Dan', Pyke's son David, then 5 years and 9 months with a measured IQ of 142, happened to be sitting on a tricycle back-pedalling just as Piaget was telling Isaacs that 'he had found that the appreciation of mechanical causality does not normally occur until eight or nine years of age, and that with regard to bicycles, for instance, children of

---

[77] Peter Fowler appears as 'Phineas' in the Malting House records. See Isaacs, *Intellectual Growth in Young Children*, pp. 82–3, for other instances of his early interest in fire. Isaacs notes that 'he bulks so largely in the notes' on organized interest in physical experiments 'partly because it happened that the staff had more leisure to write down their observations at that period; and the majority of the other children were just then engaging in other forms of play' (p. 82).

[78] Arnold Wolfendale, 'Peter Fowler', *Dictionary of National Biography*.

[79] Isaacs, *Intellectual Growth in Young Children*, p. 82.

[80] Thanks to Paul Harris for alerting us to Piaget's address to the Cambridge Education Society published in the *British Journal of Psychology* 18 (1928), 276–301. For his commentary on the address which suggests that Isaacs's criticisms of Piaget have been upheld in many respects, see Paul L. Harris, 'Piaget on causality: the Whig interpretation of cognitive development', *British Journal of Psychology* 100 (2009), 229–32. The Cambridge Education Society may have been nascent: few records have been located; the *Cambridge Review* (21 November 1930, p. 127) noted a meeting of the Cambridge University Education Society on 22 November 1930 to hear C.W. Valentine, Professor of Education at Birmingham University, speak on 'The psychology of the unconscious and its bearing upon discipline'. Bartlett was the editor of the BJP, Isaacs assisting him in this task throughout the 1920s and sitting on the Council of the BPS, so it is possible that they jointly engineered Piaget's visit to Cambridge, but it is more likely that Isaacs led in this.

less than that age rarely have any understanding of the function of the pedals'.[81] Isaacs went over to 'Dan' and said:

> 'You're not going forward, are you?' 'No, of course not, when I'm turning them the wrong way.' She asked, 'How does it go forward when it does? What makes it?' He replied, in a tone of great scorn for her ignorance, 'Well, of course, your feet push the pedals round, and the pedals make that thing go round (pointing to the hub of the cranks), and that makes the chain go round, and the chain makes that go round (pointing to the hub of the wheel), and the wheels go round, and there you are!'[82]

Although Piaget may have recognized David Pyke's clear understanding of the tricycle mechanism as a counter-example to the argument he would expound that evening, there is no mention of the incident in Piaget's published address.

In Pyke's draft lectures, he wrote that language would be the key to an improved technique of education. What was needed was 'amoral' men who would not associate emotions with the words 'Father and Mother' or 'Christmas' or the knowledge of religion. He also wrote of the need to 'produce folk who, in a specific direction, are better equipped, by the intensifying of the creation of conflict'. His advertisements highlighted the threats to 'the vigorous survival of an intelligent bourgeoisie'[83] and it would seem that Pyke's belief in the potential hostility of the environment was entangled with his perceived 'need' for aggressive intelligence. John Rickman would write in his obituary of Isaacs: 'It was characteristic of Susan Isaacs to take the risk of such an enterprise, to sober down the extravagance of her fellow-workers and by great common sense and intellectual grasp of the essential issues to lead the venture to novel, practical and constructive projects and finally achievements.'[84] Placed in Prime Minister Isaacs's hands, King Pyke's ambitions appeared less extreme in practice; however, situated in Cambridge, where the 'state of nature' *was* 'being a scientist', the Malting House process of turning pupils into little scientists could not have seemed more 'natural'.

The Cambridge Context

The University took no formal interest in the Malting House. However, its life and existence were closely entwined with the University and Colleges

---

[81] Isaacs, *Intellectual Growth in Young Children*, p. 44.

[82] *Ibid.*, p. 124; contrast with the slightly different account on p. 44.

[83] Advertisement in *The Nation and Athenæum*, 4 June 1927.

[84] John Rickman, 'Susan Sutherland Isaacs, C.B.E., M.A., D.Sc. (Vict.), Hon.D.Sc. (Adelaide)', *IJP* 31 (1950), 279–85 at 279.

through multi-stranded informal connections, as parent-academics responded to intellectually fashionable pedagogical ideas of the time, such as child-centred education, and to the unique blend of experimental science and psychoanalysis that the school embodied. Indeed, it makes sense to look upon the school as a continuation in a different way of a distinctive Cambridge style of 'making families' (Keynes, Darwin, Adrian, Hill, Trevelyan, Huxley, etc.),[85] which Maynard Keynes had high-lighted in observing the families formed through the close connection between Newnham College and moral sciences at the end of the previous century. Now thirty years on, with Pyke as the entrepreneur, the Malting House School became an institution for reforming Cambridge habits so that wives, families, children could be sewn into the social fabric of the reformed multi-monastery that was Cambridge.

Pyke's friend, the economist Philip Sargant Florence, who enrolled his son Tony in the school, was much interested in the interrelation of social, psychological and economic issues: he lectured on 'Psychology and indus-try' in Psychological Studies,[86] published in Ogden's Psyche Miniature series as *Economics and Human Behaviour: A Rejoinder to Social Psychologists* (1927) and took over the presidency of the Heretics in 1924 from Ogden. The Malting House provided the first formal schooling for the fire-loving future physicist Peter Fowler, eldest son of the mathematical physicist Ralph Fowler and physicist Elaine Rutherford and grandson of Lord Rutherford. When Elaine died in 1928, the Fowler children were cared for by Derek and Phyllida Cook, parents of another Malting House schoolchild, Lesley, and the two families (seven children in total) grew up together in Ralph's home, Cromwell House.[87] Philosopher G.E. Moore brought his two sons, Nicholas, later a poet associated with the New Apocalyptics,[88] and Timothy (Timmy), a future teacher of music at Dartington College, who at 2½ was the youngest student when the School began. Very likely 'Tommy' in the Malting House records, Timmy brought his mouth organ to school, which excited David Pyke:

[85] Noel Annan, 'The intellectual aristocracy', in J. Plumb (ed.), *Studies in Social History: A Tribute to G.M. Trevelyan*, London: Longman, 1955, reprinted with light revisions in Noel Annan, *The Dons*, London: HarperCollins, 1999, pp. 307, 315.

[86] *CUR*, 8 October 1924, p. 116. This he co-taught with F.C. Bartlett and would again the following year. The lectures for the Board of Psychological Studies were reorganized to come under the Board of the Faculty of Biology 'B' in 1927–28.

[87] William McCrea, 'Lecture: Sir Ralph Howard Fowler, 1889–1944: a centenary', *Notes and Records of the Royal Society London* 47 (1993), 61–78.

[88] Nicholas Moore was author of *The Glass Tower*, a selected poems collection from 1944, which appeared with illustrations by Lucian Freud.

'Shall we be a band and *I'll* go first?'[89] The physiologists Gleb Anrep and Edgar Adrian sent their children. Adrian's daughter Anne ('Alice' in the records, later becoming The Hon. Mrs Anne Keynes) recalled that she particularly enjoyed moving to the music they played on the gramophone.[90] Gordon Carey, Scholar of Caius and Educational Secretary of the Cambridge University Press from 1922 to 1929, enrolled both of his sons Adrian and Hugh ('Alfred' and 'Herbert'). Adrian was later in life the Canon Adrian Carey; brother Hugh, who grew up knowing the architect of the English Tripos as 'Uncle Manny', wrote the biography *Mansfield Forbes and his Cambridge* (1984).

Boarding students included Susan Foss (later the Kleinian child psychoanalyst Susanna Isaacs Elmhirst) daughter of 'Bohemians' Hubert Foss (founder of the music department at Oxford University Press) and wife Kate, who placed Susan for two terms in the Malting House.[91] Jack and David Polishuk were the sons of Joe and Phoebe Polishuk, the Isaacs's closest friends. Joe Polishuk, imprisoned as a conscientious objector in the Great War, would later change his name to Peter Pole. Jack came in 1925, age three, and boarded at the school residence, St Chad's, joined later by his younger brother. Little Jack Polishuk, later Professor Jack Pole, would return to Cambridge in 1963 as Reader in American History and Government. Interviewed by Philip Graham nearly eighty years after first being sent to the Malting House, he said he 'had never forgiven his parents for sending him away and, as he saw it, rejecting and abandoning him';[92] he interpreted his recorded behaviour of throwing Geoffrey Pyke's jacket into the Cam as aggression towards those who had taken him away from his mother and father.

Talk of the scandalous goings-on at the school fed the gossip mills of Cambridge. Susan Isaacs kept a collection of her favourite fabrications: in one of them, Malting House children were told to climb trees with ladders and once they were up, the ladders were removed to study how the children would react.[93] Actual incidents no doubt fuelled the chatter, such as the time five-year-old David Pyke was having tea at Newnham College: 'tucking into the cake, [he] announced to the startled ladies present: "You have no idea of the smell inside a dogfish!"'[94] News of the Cambridge school

---

[89] Isaacs, *Intellectual Growth in Young Children*, p. 31.

[90] Personal communication to Laura Cameron.

[91] Leslie Baruch Brent, 'Susanna Isaacs Elmhirst obituary: child psychiatrist adept at observing disorders from play', *The Guardian*, 29 April 2010.

[92] Graham, *Susan Isaacs*, p. 121.

[93] Van der Eyken and Turner, *Adventures in Education*, p. 36; see also Gardner, *Susan Isaacs*, p. 69.

[94] Van der Eyken and Turner, *Adventures in Education*, p. 36.

travelled quickly through the complementary networks as well, including those of Bloomsbury. In February 1925, Alix Strachey was in Berlin partaking in a whirl of balls, analysis with Karl Abraham and a scheme to bring Melanie Klein to England for the first time. Her husband James remained in England and, in his letters to her, reported on his travels that month which took him from London to Cambridge three weekends in a row. Keeping her attuned to all the Cambridge gossip, James wrote with regard to their young nephew, Tony Sargant Florence, and his new and unorthodox nursery, the 'Pyke-Isaacs school'.

> I must say I can't make out the point of it. There seem to be about 8–10 children, of ages from 3 to 5½. And all that appears to happen is that they're 'allowed to do whatever they like'. But as what they like doing is killing one another, Mrs. Isaacs is obliged from time to time to intervene in a sweetly reasonable voice: 'Timmy, please do not insert that stick in Stanley's eye.' There's one particular boy (age 5) who domineers, and bullies the whole set. His chief enjoyment is spitting. He spat one morning onto Mrs Isaacs's face. So she said: 'I shall not play with you, Philip,' – for Philip is typically his name – 'until you have wiped my face.' As Philip didn't want Mrs Isaacs to play with him, that lady was obliged to go about the whole morning with the crachat upon her. Immediately Tony appeared Philip spat at him, and in general cowed and terrified him as had never happened to him before. That may be a good thing; but it doesn't precisely seem to be the absence of all repressive influences. However I suppose all these accounts come from a highly resistant source. And why on earth does Lella send the child there?[95]

Alix responded with interest, reflecting on Klein's views concerning the impossibility of psychoanalytical education and the potential moderating value of her lectures on pedagogy for the British audience: 'In fact, Mrs. Isaacs sounds positively mischievous.'[96] The Cambridge psychoanalytic discussion group would meet at Tansley's Grove Cottage on 13 June 1925 to discuss the work at the Malting House School.[97] As John Rickman, recalled:

[95] BF, 17 February 1925, p. 205.
[96] BF, 19 February 1925, p. 209. Alix wrote: 'Your description of the Pyke-Isaacs institution interests me very much – I suppose there's some truth in it? It all goes to confirm die Klein's view, which is that (1) Psycho-analytical education is an impossibility since it involves 2 contradictory processes a.) necessary repression – of killing one's little friends & pumping over the porridge, etc. This is education; & b.) the 'Abbau' [demolition] of what is too violently repressed – from internal causes mostly, so that it is unavoidable – & the substitution of conscious condemnation for unc. repression. This is analysis. (2) Ψa can only be applied to education in a very limited way, such as not to frighten the little dears out of their wits, etc. what every good educator already knows. (3) What she *really* thinks is the function of analytical knowledge is to provide trained teachers to spot early signs of inhibitions in learning & in behaving socially, & packing the child off to a regular analysis.'
[97] BF, 13 June 1925, pp. 280–1. Regarding Susan Isaacs's talk, James remarked, 'I hope to be very rude to her.'

[Isaacs] spoke of the insight into the development of the ego there displayed and the very important part played by freedom for the children to speak of, to think of and to experience as part of their daily life the erotically tinged excitements and interests in their excretory and on occasion genital functions; the role of aggression in the social relations of children was given a greater place than would have been usual among psychoanalysts at that time. Someone referred to the view that boys and girls in other, warmer climates, proved to be ineducable (in the scholastic sense) after puberty and that this was perhaps due to the fullness of their sexual freedom, so the question arose whether the freedom in the school might not have some of the quality of a 'pre-genital brothel' and so hinder the development of the cultural gains which are bought at the cost of erotic deprivation. Where would the energy for the sublimation necessary for cultural achievement come from if erotic satisfactions were not denied?[98]

Though firmly convinced that the school 'in fact was no pregenital brothel', Rickman recalled that the question on sublimation irked Isaacs and, with its 'half truth', remained 'stuck, barbed like a fish-hook, in her memory'.[99] Whether or not it actually caused Isaacs to refuse further invitations to the meetings as Rickman claimed,[100] it was an issue that Isaacs had already addressed outside the realm of critical academic interrogation. Two weeks previously, she had recorded:

26.5.25 While a visitor who was sceptical was talking to Mrs. I. about the concept of 'sublimations', Mrs. I. had instanced the child's play with mud and water as an example; just then Harold came out of the schoolroom to where the children were playing with water, carrying in one hand a can of water, and in the other the roll of toilet paper from the lavatory, which he put down in the mud, and poured water over it.[101]

If in many respects Cambridge was a vanguard city for progressive movements in this period, the Malting House was its headquarters. Two months after the school opened, the following notice was placed in the *Cambridge Review*: 'All those who are interested in the education of children are invited to join the Cambridge Child Study Society and to send their names to Hon. Sec., Margaret Pyke, The Malting House, as soon as possible.'[102] Margaret Pyke, managing the daily affairs of home and school, also acted as Secretary for the new Child Study Society, which met every Friday night at the Malting House to hear talks by experts on the latest educational ideas. Eva Hartree, serving Mayor of Cambridge (its first Lady

---

[98] Rickman, 'Susan Sutherland Isaacs', p. 281. [99] *Ibid.*, p. 282.
[100] Gardner, *Susan Isaacs*, p. 69. [101] Isaacs, *Social Development in Young Children*, p. 124.
[102] *CR*, 3 December 1924, p. i.

Mayor), was its President. Speakers included James Glover on 'The child in Utopia',[103] Mrs Meredith on 'Play as an educational method', Mrs Mackenzie on 'Some problems in freedom', Susan Isaacs on 'Authority and freedom in school and home', Jessie White on 'Scientific pedagogy', Professor Cyril Burt on 'The young delinquent' and in December 1924, Miss Ella Freeman Sharpe.[104] One meeting on 13 March 1925 was held in a larger venue (St Andrew's Hall opposite the New Theatre) to accommodate the audience for Professor Percy Nunn who lectured on 'First things – in home and school'.[105]

Three days prior to Nunn's Friday lecture, another progressive group was formed, the Cambridge Women's Welfare Association, whose express purpose was to open a clinic where 'instruction in methods of contraception' would be given to 'married women in poor circumstances'.[106] Mayor Hartree, also active with Mrs Ramsey (Frank's mother), Mrs Keynes (Maynard's mother) and Lady Darwin in the Cambridge Branch of the National Council of Women which approved the scheme[107] (Hartree would serve as National President in the 1930s), would be elected chairman of the Executive Committee. Tight-knit social connections between these distinguished upper-middle-class women, most of them wives of University dons, facilitated the creation of Cambridge's first birth control clinic in the summer of 1925.[108] The accompanying injection of support from young Malting House parents and other of their backers was also key to the clinic's success. Lella Sargant Florence was its honorary secretary and the motive force of the Association; Edgar Adrian was treasurer. General Council members included the Pykes, the Moores, the Fowlers, the Tansleys, the Ramseys, Adrian's wife Mrs Hester Adrian (Hon. Sec. of the Cambridge Voluntary Association for Mental Welfare[109]), Lady Rutherford, and the Sargant Florences, including

[103] Ernest Jones, 'James Glover 1882–1926', *IJP* 8 (1927), 1–9 at 8.   [104] *CR*, 7 May 1926, p. 380.
[105] *Cambridge Daily News*, 6 March 1925.
[106] Cambridge Women's Welfare Association, *Annual Reports 1925–74*, Cambridgeshire Collection, Cambridge Central Library, First Annual Report.
[107] Minutes, National Council of Women, Cambridge Branch, 8 March 1926, Cambridgeshire Record Office.
[108] See Marin Katherine Levy, 'Birth control on trial (1930): Lella Secor Florence and the Cambridge Women's Welfare Clinic', unpublished paper, 2003.
[109] Hester Adrian was the sister of Wittgenstein's friend David Pinsent and the daughter of Ellen Hume Pinsent, an active figure in the care for the mental deficient before and after the Great War; the original 'Association for the Care of the Feeble-Minded', founded in 1909, had changed its title to the Cambridge Voluntary Association for Mental Welfare in 1921, when its remit had broadened to include the activities of the Psychological Clinic at Addenbrooke's (see Chapter 5 and John Forrester, '1919: psychology and psychoanalysis, Cambridge and London: Myers, Jones and MacCurdy', *Psychoanalysis and History* 10(1) (2008), 57–63).

Philip's mother Mary. Akin to Eglantyne Jebb's studies earlier in the century,[110] the activities of the CWWA involved upper-middle-class left-leaning women (and men) organizing facilities solely for working-class women. The birth control clinic would not thus be centrally located but placed in the east of the city, in a core working-class neighbourhood, in Fitzroy Hall off East Road. Geographically, one can imagine another class separation: books about birth control being sold in Ogden's shops on King's Parade and the actual consultations and dispensing of materials taking place on the other side of town.

News spread quickly, even to India: 'A former undergraduate, writing from his native land, felt that the opening of a birth control clinic in a town so old and respected, must be sufficient guarantee of its scientific character' and was encouraged to open his own clinic.[111] Professor Sir Humphrey Rolleston, Regius Professor of Physic at the University, was the CWWA's first President followed by his successor as Professor of Physic, Mrs Keynes's brother, Walter Langdon-Brown, in 1933. However, no matter how important the circle of modern-minded dons and their wives, pro-gressive attitudes had only so much grip: in matters of sex and morality, the University was still digesting the recent failure of its moral tribunal, the Sex Viri – 'meaning the six men not the Sex Weary', as J.B.S. Haldane put it[112] – to deprive Haldane of his Readership for committing an act of adultery with Charlotte Burghes, a married woman – the 'divorce proceedings against Cambridge University',[113] in Ogden's version. When the CWWA attempted to enlist support for laboratory research on contraception, the absence of departmental support from the University ensured that the initiative inevitably ran into the sand, even though the author of the foundational text, *The Physiology of Reproduction* (1910, 2nd edition 1922), F.H.A. Marshall, was Reader in Agricultural Physiology in the University's School of Agriculture and was a member of the CWWA's Executive Committee and later its president.[114]

Nevertheless, informal data collection had already begun. The CWWA's second annual report noted that the work of the clinic

[110] Eglantyne Jebb, *Cambridge: A Brief Study in Social Questions*, Cambridge: Macmillan & Bowes, 1906. Eglantyne was the sister of Dorothy Buxton, who provided the Foreign Section of the *Cambridge Magazine* (see Chapter 4).

[111] CWWA, First Annual Report.

[112] Ronald Clark, *J.B.S.: The Life and Work of J.B.S. Haldane*, London: Bloomsbury, 2011, p. 83.

[113] C.K. Ogden, 'Review of *Encyclopaedia Britannica's* Supplementary Volumes, 13th edition 1926', in P. Sargant Florence and J.R.L. Anderson (eds.), *C.K. Ogden: A Collective Memoir*, London: Elek Pemberton, 1977, p. 201.

[114] Levy, 'Birth control on trial'.

contradicted statements made by medical opponents of birth control 'that the use of contraceptives causes neurosis'. Patients 'invariably replied that their nerves and their general health have improved since the fear and dread of unwanted pregnancies has been removed'.[115] In the absence of a thorough study, Lella decided then to conduct her own research, personally interviewing the first 300 women who had visited the clinic between 5 August 1925 and 24 May 1927. She also went to Vienna to gather information on the success of the silver pessary.[116] Frank Ramsey's wife Lettice was appointed as the investigator to follow up the work of Lella when she and Philip moved to Birmingham in 1929.[117] The result of Lella's study was *Birth Control on Trial*, a remarkable treatise on *failure*: the failure of science to develop any contraceptive that was as yet reliable and the failure of clinics to distribute non-contradictory information. Of 247 Cambridge women for whom data was available, 78 experienced contraceptive failures after receiving advice: even after visiting the clinic and obtaining the recommended contraceptive devices, such as pessaries, women found them difficult to use; for many of them *coitus interruptus* remained the main method of contraception.[118]

Lella and Philip withdrew their son Tony from the School when, as Philip described it, Lella's fascination with watching her son from the balcony 'turned to horror' as his tower of bricks was repeatedly destroyed without intervention.[119] The Pykes remained staunch supporters of the CWWA though undoubtedly Geoffrey's friendship with Philip suffered.[120] James Strachey reported to Alix that amongst the Sargant Florences 'the

[115] CWWA, Second Annual Report.

[116] Lella visited several Viennese doctors including Johann Fersch, author of *Birth Control*, trans. Christian Roland and ed. Miss A. Maude Royden, which declared that in Vienna the silver pessary was being 'practised with complete success'; Lella came away convinced that there is no 'adequate statistical evidence to be found in Vienna'. Although the condom was not generally distributed in clinics, Cambridge was an exception, its use always urged 'in cases when men can be relied upon'. Lella Secor Florence, *Birth Control on Trial*, London: George Allen & Unwin, 1930, pp. 36–40.

[117] Philip and Lella lived the rest of their lives in Birmingham where he was Professor of Commerce. They kept in close touch with James and Alix Strachey; in her will Lella appointed Alix as her executor. See www.psychoanalysis.org.uk/P17/P17-G-A.htm, accessed 5 December 2014, File P17-G-A-01.

[118] Florence, *Birth Control on Trial*, pp. 58–9; Levy, 'Birth control on trial'; Kate Fisher and Simon Szreter, '"They prefer withdrawal": the choice of birth control in Britain, 1918–1950', *Journal of Interdisciplinary History* 34(2) (2003), 263–91; Kate Fisher, *Birth Control, Sex and Marriage in Britain, 1918–1960*, Oxford: Oxford University Press, 2006; Kate Fisher and Simon Szreter, *Sex before the Sexual Revolution: Intimate Life in England 1918–1963*, Cambridge: Cambridge University Press, 2010.

[119] Sargant Florence, handwritten draft of memoirs created for Barbara Moench Florence, MS 1571.

[120] Tansley also remained a staunch supporter of the CWWA, including giving regular financial support, until his death in 1955.

Fig. 7.7 Philip Sargant Florence and family, possibly by James Strachey. Philip, mother Mary, Lella, and the boys with Alix Strachey, late 1920s.

emotional tenseness on the subject of the Pyke school is extraordinary. Their voices shake with feeling. Even your mother observed it and said to me in private that "she wasn't surprised that in some respects Mrs. Isaacs took Lella for an ordinary mother".'[121] By this time, however, more general conditions of 'emotional tenseness' were beginning to threaten the very foundations of Malting House life.

The Tangle of Relations

So far from being in a position to be cool and sensible and detached and judicial, we were in fact only one degree less involved, less pulled and pushed, less worked up, than Geoff and Susie themselves. It was bound to be so in the tangle of all our relations.

> Nathan Isaacs to Margaret Pyke, October 1927

[121] BF, 13 September 1925, p. 293.

Besides sharing a strong sense of shared purpose with Geoffrey Pyke, the two couples were also having to face their own strong feelings for one another. Geoffrey and Susan were in 'full and open love with one another' by Easter 1925, with Margaret's 'blessing and active encouragement',[122] and Nathan was informed. 'Actual intercourse became the natural symbol for going on, as against leaving off', but, as far as Nathan knew initially, Susan had 'decided to go no further'. The situation, infused with modernist experimentalism and Pykian sensibilities, was both strained and, for a period anyway, stable.

According to Nathan in his typed account of events addressed to Margaret in the autumn of 1927 (a month after finding out that Susan had consented to sex with Geoffrey but 'withheld the unnecessary and absurd pain of knowing' from him), the sexual relations between Susan and Geoff had ended badly with both parties feeling rejected, Geoff physically and Susan mentally. Geoff had pressed her for sex on 'the very ground of the School' and their deteriorating personal relations began to affect their work, Geoff becoming increasingly domineering and possessive. The 'explosion' occurred in February 1927:

> She had referred to some compliment of Nunn's in connection with the School. It had been pointed out that that had been applied to her. She answered: 'I am the School.' An unfortunate form certainly, but in the context as in her mind it could only mean that through her that compliment extended to the School, that she stood there for the School.

Geoff became enraged: 'I told Margie long ago that you wanted to push me out of the School. The School is mine, the child of my brain, I created it . . . As Margie agrees with me, all the technique in the School is mine.' Susan collapsed with what Nathan called 'the worst psychological shock of her life'.

Nathan's letter, which does not mention his own recently begun affair with Evelyn Lawrence,[123] lists incident after disturbing incident, some of which he attempts to analyse and understand: that the 'breaking up of the personal love relation released these elements together with other much deeper ones that had helped in that break-up itself (Geoffrey's ambivalency to women, looking for the mother he never had, and when he found her finding in her the mother he never had)'. There was Geoffrey's desire to be the only paternal figure in the school; his insistence she remove her name

[122] N. Isaacs to M. Pyke, 1927, The Isaacs Collection, Institute of Education Archives, NI/D/2, 68 pages.
[123] Graham, *Susan Isaacs*, p. 144.

from the letter-head: 'I will have no scrambling for credit.' Geoffrey's increasingly verbose advertisements were becoming another of many intensifying sources of friction and hostility. Nathan was alarmed by the 'open megalomania' of the first one to appear and encouraged Pyke to seek the opinions of Nunn and their weekend guest, Willie Brierley, who condemned it with 'bell, hook, and candle'. Nathan had a 'very long and late debate about the term "bourgeoisie" which' he tried to get Pyke to remove, 'but he all the more insisted upon as the very thing he wanted to say.' There was the upset, vague on details, during Jean Piaget's visit to the school:

> I won't go into the details of the ominous Piaget incident. It was only another example of Geoff's native method of tearing a single fact out of its context, glowering at it until he had worked himself up into a furious rage, and then rushing.

Nathan also recalled Geoffrey's anger after Susan, upon learning with surprise of the existence of Russell's school, had drafted a letter offering cooperation. And then there was Geoffrey's decree to no longer take Cambridge children so that, presumably, all of the children would live in residence and the parents would have less influence in their lives. Nathan's friend Joseph Polishuk had offered his own son Jack to this system early on. But the differences between Susan and Geoffrey became clearer: for Isaacs, the family remained the primary relation for the child; you could not in fact break the circle which 'binds one generation to the next'; mothers and fathers were a fact; one could not ignore their influence and importance – nor was it particularly wise to try. But by now Geoffrey was not open to objection: in dealings with Susan, he was 'brilliantly insane, with hate'. With all their collective attention to the aggressive forces alive in children, it did not occur to them that, as Nathan put it, 'they themselves might be driven by forces unknown to them and much stronger than anything they knew'.

The Malting House at the British Psycho-Analytical Society

On 16 March 1927, one month after 'Geoff exploded his mine under her', Susan Isaacs would explain the work of the school at a meeting of the BPaS attended by visitors Geoffrey Pyke and Nathan Isaacs.[124] Geoff's extant 1925 drafts for his own lecture to the Society suggest he had wanted to give this first statement regarding the school but

---

[124] BPaS Minutes, FAA/194. Isaacs was elected an associate member in 1921 and a member in 1923.

also that he had envisaged Susan as the one to provide a more detailed account. Nathan wrote that she cancelled once and 'waited for him for two years to go in first!' Susan's 1927 paper for the BPaS, entitled 'The reaction of a group of children to unusual social freedom' would begin by crediting Pyke for originally suggesting 'much of the detailed technique that has been tried out in the school'.

> Many of the particular methods suggested by Mr. Pyke had been worked out by him with his own son from earliest infancy, with such promising results as made it desirable to test them out further with a group of children.

Here she placed great emphasis on the reality that the children were hardly free to do as they pleased. Indeed, with respect to the Malting House experiment, the concept of freedom was 'psychologically worthless, since any parent figure is of necessity a powerful psychological factor'.[125] One is never dealing with 'The Child', she asserted, but only with the child in a particular relation. As shaped through the fantasies of the child, the parent figure 'had a prestige' he or she could not escape. Each child was marked by genetic histories of past care and disturbance and could not be understood as an isolated subject: even if it appeared to onlookers like Strachey that the children were 'running wild', the notion of the innocent wild child or the passive teacher had no place in the Malting House.[126] This is what Isaacs here referred to as 'the ecological point of view', correcting in part the 'natural history' point of view she associated with Montessori and her emphasis on the necessity for passivity on the part of the educator.

> This point of view of the natural phases of development of the child's growth, immensely valuable as it has been, had, of course, in its turn to be corrected by the ecological point of view in psychology, by the recognition that since the human child is a social animal, and is shown by psychoanalysis to be so at a far earlier age than mere external observation can reveal, we have at every point to see the child in a particular social relation. We now know that the babe at the mother's breast is not merely a young animal drinking in nourishment, but is a social human being in his first and fundamental social relationship.

[125] Susan Isaacs, 'The reaction of a group of children to unusual social freedom', delivered to the British Psycho-Analytical Society, 16 March 1927, Pyke Papers.

[126] One former student, who attended the Malting House School from 1927 to 1929 (and remembered it as Margaret Pyke's School), recalled he never was aware he was free there to do as he wanted. Margaret Pyke told him years later (in answer to his query about why his memory was at variance with popular legend) that 'You were such a good little boy . . . and you always wanted to please. You had to do as your mother told you at home and it would never have occurred to you that you would do as you liked at school' ('Random memories of the Malting House, 1927–9', 24 January 2001, Pyke Papers).

What concerned the Malting House staff was not what the child was permitted to do when left alone but rather, 'in view of the fact we act upon the child in any case', what 'is in detail the most valuable form of action'.

The lecture set out to clarify how the attitude of the child's adult co-investigator would respond to the child's hostility and curiosity. The child's dramatic play was to arise spontaneously from the child, and he or she was to work out any inner conflict with the physical world alongside the co-investigator. As the teacher became the hated (or loved) parent figure, the transference formed the first element of the technique. This was followed by the symbolic gratification of the children's unconscious wishes 'through our sharing in the children's play, the resultant lessening pressure of the ego-ideal, and the consistency and intelligibility of the real environment'. Intimately connected to the superego and the Oedipus complex, for Freud the ego ideal was the source of the perfection children attribute to their parents. Isaacs suggested that the lessening of inner tension would make it easier for children to accept the limitations of the real physical world and help them control their emotions and behaviour. 'The demands made upon the children, and above all the way in which the demands were made, involved the minimal stirring up of guilt, and a minimal use of penalties – penalties which themselves belonged to a clear and simple reality, which were within the grasp of the child's understanding, and, there, less likely to create anxiety and aggression.' The most troublesome boys, such as Philip/ 'Frank', who regularly spat on his teacher and knocked down the blocks of other children, saw marked improvement in behaviour over time: a marked lessening of the ambivalence, overt oral and excretory interests, hostility to the mother-figure (accompanied by a decrease in signs of feminine identification) and maliciousness towards the younger children. Science, the dramatic play with reality, including flowers, fire and heavy machinery, would permit the working through of hostility, providing the technique alluded to in Pyke's earlier draft lecture that could produce his 'superman', supplanting the old educational method of 'culture transmission' that would protect us from him.

Here Isaacs explains rather than disavows her psychoanalytic interpretations made while working with the children, and describes their adjustments to the technique with regard to spitting (the act is 'too strongly libidinised ... to yield to affectionate identification with the adult, who does not spit') and active excremental interests:

> At first I was inclined to regard the marked hostility to the parent-figure which always accompanied any outbreak of these interests as a simple

expression of the primal sadism of the anal level. It gradually became clear, however, that the situation was much more complicated than this – that the hostility to adults, the frequent references to 'Dirty old giant,' 'Horrid, dirty Mrs. I.' etc., was in part projected guilt, and in part a definite anger with us because we appeared to be tolerating these interests: because, in fact, we did not give the child's super-ego the necessary help but appeared to ally ourselves with the hated impulses. Passivity was, in fact, interpreted as approval. Here also, therefore, we began to interfere more actively, with sharper and earlier withdrawal of social co-operation, following on a definite request that these things should not be talked about: and this development in technique has been justified by the lessening of the signs of conflict in the children and the increase of social constructiveness.

Isaacs ended her account of the Malting House School and its possible significance for psychology by conceding full epistemic priority to psychoanalysis:

> Most of the questions upon which we have attempted to throw some light in our work at this school cannot, however, be adequately solved until the methods adopted have been tried out with a group of children who have been fully analysed at an early age, and in which the strictly educational problem is thus not obscured by the problems of neurosis.

Nikolas Rose has argued that Isaacs's published work on the Malting House would help problematize, indeed 'dissolve' the inter-war project of the New Psychology to normalize the English family.[127] The ideal for the New Psychology was a 'science of social contentment' to create norms of family relations likely to produce the normal, well-adjusted child. This was possible because it was believed that the child's character was created in real interactions with the family unit. Rose contends that Isaacs challenged that relation of the real and the possibility of a 'normalising practice of inter-vention and therapy'. For her, the child clearly exists in a relationship to a parent figure but Isaacs stressed that 'Our real behaviour [to children], and the actual conditions we create, are always *for them* set in the matrix of their own phantasies.'[128] This denegation of the real, Rose argues, made it impossible to prescribe general norms of right conduct for parents to help them raise well-adjusted children.

Isaacs indeed would increasingly follow a Kleinian route, which her work at the Malting House prefigured, to suggest that along with parental figures, mud, fire, engines and animals in the physical world have

[127] Nikolas Rose, *The Psychological Complex: Psychology, Politics and Society in England, 1869–1939*, London: Routledge and Kegan Paul, 1985, pp. 189–90.
[128] Isaacs, *Intellectual Growth in Young Children*, p. 8.

profoundly symbolic meanings for the child, rooted in infantile fantasy. Robert Young remarks that 'Kleinians have consistently written in a language which eschews analogies drawn from natural science'[129] and indeed Isaacs would increasingly drop the reverential tone for science along with terminology such as 'the ecological point of view'.[130] In tension with this, though, Isaacs had vigorously criticized Jean Piaget's excessive stress on the social world of the child referring to human relations only and had, following Pyke, laid more emphasis on the key importance of the environment as a sort of super-parent and the first measure of reality (a critique that Rose overlooks). In *Intellectual Growth in Young Children* she writes, 'one cannot shut one's eyes to the influence of direct contact with the physical world ... The burnt child dreads the fire even in the stage of ego-centrism.'[131]

> While it is certainly true that the first value which the physical world has for the child is as a canvas upon which to project his personal wishes and anxieties, and this first form of interest in it is one of dramatic representation, yet, as I have already urged, this does not prevent him from getting direct actual experience of physical processes. Physical events become, in fact, the test and measure of reality. There is no wheedling or cajoling or bullying or deceiving them. Their answer is yes or no, and remains the same to-day as yesterday. It is surely they that wean the child from personal schemas, and give content to objectivity.

This physical environment was hardly benign (as a parent) but potentially violent, biting, painful and explosive. It was also the counterpart of the aggressive intellect Pyke so desired for his child subjects. Significantly, Isaacs would stress that these physical events were consistent and intelligible to the child. This insistence on consistent real nature curiously echoes the New Psychology's hope for a project of normalization. However, for Isaacs the arena of normalization was no longer the human family but more importantly the non-human world: the practices of normalization visible to her in the Malting House School flowed from curiosity, play, scientific method, and laboratory and field technique.

---

[129] Robert Young, 'Melanie Klein 1', lecture in series 'Psychoanalytic pioneers', given at the Tavistock Centre in London, p. 12: http://human-nature.com/rmyoung/papers/pap128.html.
[130] *Intellectual Growth in Young Children* has a section, pp. 6–10, which closely follows her 1927 lecture to the BPaS but the 'ecological point of view' and other analogies with natural science are largely expurgated.
[131] Isaacs, *Intellectual Growth in Young Children*, p. 79.

The Final Break-up

Susan and Nathan Isaacs remained another seven months at the Malting House after Susan's address to the BPaS. Finally, however, 'Susie got out.' In Nathan's analysis, the school had become Geoffrey's child and Susan found herself 'in the middle of a struggle for a "child"':

> Geoff carried the personal relation over into the school, accused Susie of doing so (the less conscious she was of it, the more clearly her Unconscious was doing it) attacked her over and over again when she least expected it in the school field and forced her into such a defencelessness against the savagery and incessancy of his attacks that she had to abandon her job, rights, her merely decent dues.

In 1928, Pyke lost all his money within hours due to a bad deal, with liabilities for over £73,000. Edgar Adrian and his family moved to St Chad's to offset Pyke's cost of maintaining the residence. An appeal to the Laura Spelman Rockefeller Trust in New York to save the school, in part for its 'copious and careful record of phenomenon',[132] signed by seventeen people (altogether eight Fellows of the Royal Society) including Jean Piaget, A.G. Tansley and Sir Charles Sherrington failed. Margaret Pyke closed the doors of the school on her own as Geoffrey began to descend into deep depression. Margaret left Geoffrey, forming a partnership with Lady Denman, the wealthy wife of the former Governor-General of Australia, with whom she would found the National Birth Control Council, later the Family Planning Association.[133] In World War II Geoffrey would become a key member of Mountbatten's backroom advisory team along with Bernal, in which capacity he designed an unsinkable ship made of 'pykrete' – a frozen mixture of water and wood pulp. A successful prototype was

---

[132] Van der Eyken and Turner, *Adventures in Education*, p. 64.

[133] In 1930, Lady Denman appointed Margaret secretary to the NBCC, a group of birth control clinics and associated societies which had previously been independent. The NBCC became the NBCA in 1931 (changing 'Council' to 'Association') and, in 1939, the name changed to the Family Planning Association (FPA), of which Margaret Pyke became honorary secretary. She became chairman after Lady Denman's death in 1954, a post she held until her own death in 1966. She was also one of the founding officers of the International Planned Parenthood Federation and was awarded the OBE in 1965 for her work in family planning. She maintained connections with Lella Secor Florence, reading and commenting on the draft manuscript for Lella's 1956 book, *Progress Report on Birth Control* (London: William Heinemann Medical Books). The FPA in the 1950s was still 'strongly dominated by the founding generation of women' which included Lella who had created the Birmingham FPA after leaving Cambridge and remained involved into the 1960s. (Hera Cook, *The Long Sexual Revolution: English Women, Sex, and Contraception, 1800–1975*, Oxford: Oxford University Press, 2004, p. 276 and note 17.) Information on Margaret Pyke comes from unpublished notes in the Pyke Papers; Ann Dally, 'Margaret Pyke', *ODNB*, www.oxforddnb.com/index/54/101054693/.

constructed on Lake Patricia near Jasper, Alberta, but the project was eventually judged unnecessary after the Normandy Landings. Rusting refrigeration pipes and a timber skeleton that once supported the 'Habbakuk' model now rest on the bottom of a cold Canadian lake. Susan Isaacs and Geoffrey Pyke both died in 1948 – first Geoffrey, a suicide; Susan, a few months later, of cancer.

The zoologist Solly Zuckerman wrote that Pyke 'was constantly being overwhelmed by the fantasies he created, and by his prejudices against the conventional'.[134] That he was able to achieve the support he did speaks to the wide appeal of his vision at the time, resonating through such popular works of the period as H.G. Wells and Julian Huxley's *The Science of Life*. There it was argued that ecology and psychoanalysis could be enlisted together in the understanding and moulding of 'Life'.[135] As in Aldous Huxley's ambiguous parody of erotic play at the State Conditioning Centre garden nursery in *Brave New World*,[136] contemporaries also articulated the potential dangers of such sweeping control. Russell warned in *On Education* that 'The power of moulding young minds which science is placing in our possession is a very terrible power, capable of deadly misuse; if it falls into the wrong hands, it may produce a world even more ruthless and cruel than the haphazard world of nature.'[137]

Haphazard or consistent, enemy or resource; what counted as 'nature' for scientists in 1920s England was shaped by their ambitions and fears for the future, but also by historically and geographically specific practices through which they understood, engaged and produced 'nature'. Psychoanalysis was one of those practices. The practices of the Malting House were marked by its place on the margins of an institution dedicated to the natural sciences, by Pyke's concern to promote the survival of his class, and perhaps by the raw proximity of the trauma of the Great War. Through new educational techniques that recognized powerful emotional tensions between generations, Pyke sought to help create an aggressively intelligent race that would be able to survive the great changes which he expected science to create in our environment.

---

[134] S. Zuckerman, 'Patents and precedents: review of *Pyke, the Unknown Genius*', *The Spectator*, 30 October 1959.

[135] H.G. Wells and J. Huxley, *The Science of Life*, London: Hazell, Watson & Viney, 1930, pp. 1502–3.

[136] Both Julian Huxley and Pyke had been captivated by Haldane's speculations in *Daedalus*. The Malting House School is the likely inspiration for the scene in Aldous Huxley's *Brave New World* placed as epigraph to this chapter (A. Huxley, *Brave New World* (1932), New York: Harper Perennial, 2004, p. 38).

[137] Russell, *On Education*, p. 127.

Isaacs and Pyke parted company during the experiment. Winnicott briefly points to a more complex understanding of Isaacs: not just tolerant and understanding but a figure who found a valued place for the aggressive intellect. 'When I heard Nathan ruthlessly criticize her ideas and formulations I felt maddened, but I found that she valued exactly this from him and that she made positive use of his ruthlessness, as of his terrific intellect.'[138] This account would support his insight, but perhaps most interestingly our interpretation of the Malting House work would argue that Isaacs's connection with the eccentric Pyke was hardly an eccentric cul-de-sac in her life. Far from it: it provides an alternative genealogy for her pre-Kleinian psychoanalytic development and suggests how her subsequent and widely read works are marked with psychoanalytic ideas and passions she had earlier explored with Pyke in the Malting House.

In relation to Cambridge, the Malting House School emerges not only as the brewery for progressive education in this period, but as a centre of progressivism and experimentalism of all sorts, even in terms of the socio-sexual relations between the members of staff. In the appeal to the Rockefeller Trust, the illustrious group of signatories wrote that 'Science owes much to pioneers who have worked independently in the No Man's Land between the different sciences until they have reached a stage at which they could command the support of established institutions.'[139] Pyke in fact had done more than struggle in the 'No Man's Land': he had created the Malting House in which the sciences mingled playfully and riskily in both theory and practice.

[138] Gardner, *Susan Isaacs*, p. 5.   [139] *Ibid.*, p. 65.

CHAPTER 8

A Psychoanalytic Debate in 1925

In the 13 June 1925 issue of *The Nation and Athenæum*, Tansley published a complimentary review of the third volume of Freud's *Collected Papers*, the 'Case histories', translated by James and Alix Strachey for the Hogarth Press, alongside a review of a translation by James S. Van Teslaar of Wilhelm Stekel's *Peculiarities of Behaviour*. Tansley extolled the virtues of this form of the scientific foundations of psychoanalysis: 'Case histories are of special value to students because every psycho-analysis is, or should be, an organic whole, a definite development of the patient's mind in relation to the analyst.'[1] A wide-ranging controversy, whose echoes were transmitted by Ernest Jones to Freud in Vienna,[2] erupted in response to his review. On 4 July 1925 the magazine's letters column featured an 'emphatic protest' from E.C. Allmond, B.Sc. (Lond.) against the tone of Tansley's review, on the grounds that this tone would appear to make of Freud's doctrines a contribution 'comparable with the Copernican theory in Astronomy and the Einstein theory in Physics'.[3] The letter-writer cited a number of recently published criticisms of Freud's works and insisted that, until these criticisms had been satisfactorily answered by the exponents of Freud's teaching, no review as favourable as Tansley's should be published, on the grounds that it is 'entirely misleading'.

Despite the fact that Klein's first London lectures, attended by Rickman, James and Alix Strachey and many others, were then taking place at the Stephens' house in Gordon Square,[4] Tansley's response to Allmond came on 8 August 1925 from Sweden, where he was attending the second post-

---

[1] A.G. Tansley, 'Review of Sigmund Freud, *Collected Papers*, Vol. III, Authorized translation by Alix and James Strachey; Wilhelm Stekel, *Peculiarities of Behaviour*, trans. James S. Van Teslaar', *The Nation and Athenæum*, Supplement, 13 June 1925, p. 348.
[2] 'Tansley, who is the last man to welcome polemics, has nevertheless been drawn into a controversy in a weekly magazine where he had been extensively attacked.' JF, 19 September 1925, p. 581.
[3] E.C. Allmond, 'Psycho-analysis', *The Nation and Athenæum*, 4 July 1925, p. 425.
[4] Phyllis Grosskurth, *Melanie Klein: Her World and Her Work*, New York: Knopf, 1986, pp. 137–8.

war International Phytogeographic Excursion.[5] Tansley noted that it was true that many psychologists take no notice of psychoanalysis, but that was because they were ignorant of it. Many others condemn it and Tansley presumes that the reason for this is that:

> Freud's teaching is undoubtedly very astonishing, and his theories certainly give a first impression of being bizarre and grotesque to an extreme degree – and this apart from the disgust and general repugnance they arouse in many people ... Freud probes far more deeply and painfully [even than Darwin], and is even more bitterly attacked.[6]

Tansley went on to concede that psychoanalysts and their supporters had not replied sufficiently to a major critical study, cited by Allmond, that had caused considerable stir in England, A. Wohlgemuth's *A Critical Examination of Psycho-Analysis* (1923).[7] Nonetheless, Tansley went on, 'I have devoted what intelligence and critical judgment I may possess to a first-hand as well as to a literary study of Freudian analysis'; he found no alternative to accepting Freud's hypotheses because they give explanations for hitherto neglected or inadequately explained fundamental phenomena of the human mind.

The issue of 22 August 1925 contained three letters responding critically to Tansley: from Allmond again, from Dr A. Wohlgemuth himself, and from Sir Bryan Donkin. Donkin argued that the silence of psychologists on the question of psychoanalysis demonstrated their refusal to accept its doctrines; the silence of the analysts in replying to Wohlgemuth gets to the '*basic* question as to the soundness of the principles and the value of the practice of the Freudian cult ... [Namely] there are many who deem that the doctrine contains the seeds of its own dissolution, and therefore are disinclined to discuss it.' Wohlgemuth's letter referred readers to the debates in the specialist journals and emphasized that psychoanalysis displayed a 'flagrant and persistent disregard of scientific method'. And he added: 'I feel quite convinced that Professor Tansley's observance of the [scientific method] in his own domain of botany is more rigorous than in his *New Psychology* or he would not enjoy the great reputation he does.'[8]

---

[5] William Cooper, 'Sir Arthur Tansley and the science of ecology', *Ecology* 38(4) (1957), 658–9.

[6] A.G. Tansley, 'Psycho-analysis', *The Nation and Athenæum*, 8 August 1925, pp. 566–7.

[7] A. Wohlgemuth, *A Critical Examination of Psycho-Analysis*, London: George Allen & Unwin; New York: Macmillan, 1923.

[8] A. Wohlgemuth, 'Freudian psycho-analysis', *The Nation and Athenæum*, 22 August 1925, p. 619. In July 1924, shortly before he intervened in the debate with Tansley, he penned a long defence of his own position culminating in: 'Not only did I tear [Freud's] "schema of Ψ systems", upon which the *Traumdeutung* depends, to tatters, but I showed also that it is quite unintelligible. In fact, it has

Miss Allmond's letter introduced a new note:

> the doctrine of psycho-analysis is not an objective explanation of certain phenomena of the human mind at all, but a subjective reaction to these phenomena; that the result of dream analyses carried out by Freud are revelations of the mind of the latter even more than of the early history of the patient. [This charge of subjectivity] could only be completely refuted by the carrying out of experiments, carefully controlled in accordance with the ordinary laws of scientific method.[9]

The controversy rumbled on in the next number of the magazine, with two letters, one by P. McBride, whose depiction of the issue was stark and simple: 'Does the theory of psycho-analysis rest upon a scientific basis, or does it depend upon imagination?' And he addressed the following question to Tansley: 'Has he found [in Freud's *The Interpretation of Dreams*] any data which have been verified or which are capable of being verified on lines such as would meet the demands of a critic accustomed to weigh scientific arguments? ... Repeated investigation, however, convinced me that the large volume contained no such proof.'[10]

The other letter – by far the most interesting intervention in the debate – published on 29 August 1925 was signed by Siela, a pseudonym for John Maynard Keynes.[11]

Here is the entire text of Keynes's letter:

> Sir, – I venture, as an outsider, to suggest that the truth about the importance to be attached to the ideas of Professor Freud lies somewhere between the views expressed by your learned correspondents.
>
> Professor Freud seems to me to be endowed, to the degree of genius, with the scientific imagination which can body forth an abundance of innovating ideas, shattering possibilities, working hypotheses, which have sufficient foundation in intuition and common experience to deserve the most patient

---

occurred to me that Freud could never have written it himself, and that it might be the composition of one of his paranoiac patients' (A. Wohlgemuth, 'Wohlgemuth and his reviewers', *Journal of Mental Science* 70 (1924), 495–503 at 502).

[9] Miss Allmond, 'Freudian psycho-analysis', *The Nation and Athenæum*, 22 August 1925, p. 619.

[10] P. McBride, 'Freudian psycho-analysis', *The Nation and Athenæum*, 29 August 1925, p. 644.

[11] For information concerning Keynes in this period, the principal resources employed are Skid I or Skid II. Also indispensable are E.G. Winslow, 'Keynes and Freud: psychoanalysis and Keynes' account of the "animal spirits of capitalism"', *Social Research* 53 (1986), 549–78 and Ted Winslow, 'Bloomsbury, Freud, and the vulgar passions', *Social Research* 57 (1990), 786–819.

What significance does the pseudonym 'Siela' have? Skidelsky gives no clue. Perhaps it is a deliberately inaccurate anagram of 'alias', with a possible play on the ambiguous pronunciation of the word 'Keynes', which should be pronounced as 'canes', not 'keens'. The coded message is: spell the 'e' in 'Siela' as you pronounce the 'ey' in 'Keynes', i.e. as an 'a'. Perhaps only the 'sort of devil' who enjoyed the plays on words Freud found in dreams and in slips would come up with such an 'alias'.

Fig. 8.1 John Maynard Keynes (1883–1946) (oil on canvas),
by Roger Eliot Fry (1866–1934).

and unprejudiced examination, and which contain, in all probability, both theories which will have to be discarded or altered out of recognition and also theories of great and permanent significance.

But when it comes to the empirical or inductive proof of his theories, it is obvious that what we are offered in print is hopelessly inadequate to the case – that is to say, a very small number of instances carried out in conditions not subject to objective control. Freudian practitioners tell us that they are personally acquainted with a much greater number of instances than those which have been published. But they must not complain if others base their criticisms merely on what is before them.

I venture to say that at the present stage the argument in favour of Freudian theories would be very little weakened if it were to be admitted that every case published hitherto had been wholly invented by Professor Freud in order to illustrate his ideas and to make them more vivid in the minds of his readers. That is to say, the case for considering them seriously mainly depends at present on the appeal which they make to our own intuitions as containing something new and true about the way in which human psychology works, and very little indeed upon the so-called inductive verifications, so far as the latter have been published up-to-date.

I suggest that Freud's partisans might do well to admit this, and, on the other hand, his critics should, without abating their criticism, allow that he deserves exceptionally serious and entirely unpartisan consideration, if only because he does seem to present himself to us, whether we like him or not, as one of the great disturbing, innovating geniuses of our age, that is to say as a sort of devil. – Yours, &c., SIELA.[12]

The form of Keynes's five-paragraph argument is worth spelling out. The first paragraph announces that he will position himself between the combatants, a man of objectivity without *parti pris*. The second paragraph acknowledges Freud's extraordinary fertility, yet hints that some of his striking ideas may not stand the test of time – the jury is not yet even out. In the third paragraph, he castigates Freudians for not making available to others the evidence they affirm they possess and, in consequence, they should respect the scepticism and misgivings of their opponents. All of these arguments are indeed those of an even-handed intelligent layman sympathetic both to the bold ideas of Freud and to the rational misgivings of sceptics.

It is in the final two paragraphs that Keynes introduces an entirely different note, which renders his previous arguments almost besides the point. He imagines the effect on the impartial layman of learning that Freud's cases were pure invention, and concludes that this would not change one's attitude one jot. So he draws the implication: Freud's impact does not lie in evidence as commonly understood – objective results, carefully recorded and published – but depends instead on 'the appeal which they make to our intuitions as containing something new and true about the way in which human psychology works'. In the final paragraph, Keynes turns to address the warring factions, much as he was used in these years to addressing the financiers and statesmen of Europe as they struggled to maintain in place the Versailles agreement, the reparations imposed on Germany and the banking system of Europe and America. Both sides should admit that Freud speaks directly to our intuitions and, now speaking as the pre-eminent theorist of probability, hardly at all to our judgement of the weight of 'inductive verifications'; in consequence, both sides should cease to blame the other. The upshot is not entirely neutral: the onus of Keynes's argument falls on Freud's critics, rather than his partisans.

---

[12] 'Siela' [John Maynard Keynes], 'Freudian psycho-analysis', *The Nation and Athenæum*, 29 August 1925, pp. 643–4; see also Winslow, 'Keynes and Freud' and Skid II, 414.

It is they who have to make room for something new in their view of Freud – this appeal he makes to our intuitions, which Keynes then expands upon: 'he does seem to present himself to us, whether we like him or not, as one of the great disturbing, innovating geniuses of our age, that is to say as a sort of devil'.

Thus Keynes has inserted a different kind of figure between the two invoked by Freud's partisans and critics: instead of the great scientist, discoverer of new truths to place alongside those of Copernicus and Darwin, and instead of the unscientific purveyor of fantasies that are the product of his own feverish imagination, Keynes's Freud is a hybrid of the two, and thus something beyond both. Yes, Freud is a great scientist akin to Darwin (or, to give a hint of an argument yet to come, to Keynes himself); yes, Freud is a man of unmatched fantasy and great speculative leaps. The little phrase 'whether we like him or not' introduces the notion of some kind of objective measure of Freud's cultural standing. Freud, Keynes intimates, stands above personal likes and dislikes, since he is a genius of the age, perhaps its very own *Zeitgeist*. What difference would it make if one liked or disliked Freud?

Let us leave for the moment Keynes's two significant contributions to the debate over psychoanalysis in 1925 – the fundamental importance of Freud as the principal figure of our *Zeitgeist*, and the fact that Freud calls principally on our intuitions, not on the scientific requirement to be shown evidence – to proceed to show that our understanding of this debate can be deepened by placing it in some context. Who are the protagonists in the debate? Where and when is the debate taking place? What are the implicit criteria guiding and funnelling the debate?

The obvious first question concerns the forum of the debate. Keynes's contribution appears as an ordinary letter in *The Nation and Athenæum*. Why, then, was it written pseudonymously and signed Siela? Most probably, the principal reason is that Keynes at the time effectively owned and controlled the magazine, having taken it over in early 1923 as an organ for the dissemination of his brand of New Liberal political and economic views.[13] Installing a colleague from Cambridge as Editor, bringing in Leonard Woolf, founder of the Hogarth Press (publisher of the Freud translations from 1924 on) as Literary Editor when T.S. Eliot turned the job down, in the period 1923–25 Keynes oversaw every detail of the magazine, and in the entire life of the magazine from May 1923 to February 1931, when it merged with the *New Statesman* and he

---

[13] Skid II, 134–9.

appointed Kingsley Martin as Editor, Keynes contributed 155 pieces, including fifty articles on domestic policy, forty on debts and war reparations, five book reviews, twelve anonymous contributions and fourteen letters – a journalist on another newspaper described him as an 'ungovernable soda-water syphon'.[14] This superabundance of contributions may in part explain his bashful pseudonymity. And the control he exerted, alongside that of Leonard Woolf, who brought on board all his Bloomsbury and Hogarth Press friends and authors, meant that it was clearly a house journal for Cambridge economics and for Bloomsbury's literary, aesthetic and moral views.

Keynes's position as controller of the public medium meant that he may have cared to keep the debate going. As Ernest Jones was to remark thirty years later, this was a 'heated polemical discussion between [Tansley] and three very bitter opponents ... Some of the language used by the latter rivalled the German outbursts before the war.'[15] Keynes would no doubt have liked a partisan of Freud's to have stepped into the breach. With Tansley botanizing out of England in August, he may have been casting around for someone to make good copy for a lively debate; as an interventionist proprietor he eventually had no compunction in playing that role himself. It is not certain that this played a factor, but it is plausible. Of such conjunctions may heated polemics be on occasion wrought.

August 1925, when Keynes wrote his letter for the magazine, was a busy month for a man who was never anything but busy. Having married early in the month, he was forced to accept Ludwig Wittgenstein as a houseguest a few days later. Wittgenstein arrived penniless on 20 August and stayed for six days at the Keynes's summer cottage.[16] At the end of the month, Keynes and his new wife, the ballerina Lydia Lopokova, left for the Soviet Union, so he could meet his new relations. Do we have any evidence as to what Keynes and Wittgenstein talked about in those days?

On 20 August 1925, the day before the philosopher Frank Ramsey was himself to be married, he visited Keynes's cottage expressly to see Wittgenstein. He wrote to his fiancée:

> I got here at tea time yesterday, and went for a long walk with Keynes and Wittgenstein and had a very good dinner ... Keynes and Wittgenstein are awfully nice together but I can't get a word in, they both talk such a lot. I got slightly heated because W said that Freud was morally deficient though very

---

[14] Skid II, 136; the syphon quote is from Skid II, 139.
[15] Ernest Jones, *Sigmund Freud: Life and Work*, 3 vols., London: Hogarth Press, 1953–57, Vol. III, p. 119.
[16] Skid II, 208.

clever.[17] To-day K has gone up to town on business and it is pouring with rain; how I shall amuse W I can't think because he doesn't much like ~~almost incapable of~~ any but the most serious conversation, which tends to lead to such violent disagreement as to make it impossible.[18]

Thus, leaving his house-guests Wittgenstein and Ramsey on that August day to work in London, Keynes may well have written his intervention in the debate about Freud with their heated debate in the forefront of his mind: as the exemplary Freudian enthusiast, alongside all the others he knew, and there were many, he might have been thinking of young Frank Ramsey, not long back from his analysis in Vienna; as critic of Freud, he might have had not only Wohlgemuth and his acolytes in mind, but also the redoubtable Wittgenstein.

And what might Wittgenstein's views on Freud at that time have been? Whatever the complexities and ambiguities of that position as parsed by his commentators,[19] Wittgenstein never ceased to recognize Freud's extraordinary power over him and others – indeed, he might well have been a source for Keynes's view that Freud's hold stems from his speaking to our *intuitions* rather than our *assessment* of inductive inferences. He was not that far off from that period, in 1935, when he was considering training as a Freudian psychiatrist in order to make the best use of his gifts.[20]

---

[17] It is probable that this heated dispute over Freud is the quarrel that led to Wittgenstein severing relations with Ramsey for the next two years (MLW, p. 231, describing Wittgenstein's visit to England in August and September 1925: 'He also met with Ramsey, with whom, however, it appears he quarrelled so fiercely that the two did not resume communication until two years later'). It is possible that the argument over Freud was a contributing factor in Wittgenstein's further delay in 'returning' to Cambridge: on 22 July 1927 Ramsey wrote to Moritz Schlick: 'last time [LW and I met] we didn't part on very friendly terms, at least I thought he was very annoyed with me (for reasons not connected to logic)' (quoted in Friedrich Stadler, 'Editorial', in Maria Carla Galavotti (ed.), *Cambridge and Vienna: Frank P. Ramsey and the Vienna Circle*, Vienna Circle Institute Yearbook 12, Dordrecht: Springer, 2006, p. vii). The return eventually took place on 18 January 1929, when Keynes announced to Lydia Lopokova: 'Well, God has arrived. I met him on the 5.15 train' (MLW, p. 255). McGuinness's version is close to ours: 'Wittgenstein's long walks with Ramsey, which gave them an opportunity to quarrel about psychoanalysis. (The quarrel is attested, the subject probable.) At all events a return to Cambridge and philosophy was deferred' (Brian McGuinness (ed.), *Wittgenstein in Cambridge: Letters and Documents 1911–1951*, Oxford: Wiley-Blackwell, 2012, p. 6). Gabriele Taylor, 'Frank Ramsey – A biographical sketch', in Galavotti (ed.), *Frank P. Ramsey and the Vienna Circle*, pp. 1–18, also records the important quarrel between Ramsey and Wittgenstein in August 1925, but has no information about its topic (probably, we surmise, because she relied upon the biography by Margaret Paul (Ramsey's sister), rather than on the correspondence between Ramsey and Lettice Baker).

[18] FRP, Ramsey to Lettice Baker, 20 August 1925; deletions in the original.

[19] The most substantial overview remains Jacques Bouveresse, *Wittgenstein Reads Freud: The Myth of the Unconscious*, trans. Carol Cosman, foreword by Vincent Descombes, Princeton: Princeton University Press, 1995.

[20] MLW, p. 356.

As Wittgenstein in 1936 put it when he gave one of his close students, Maurice Drury, *The Interpretation of Dreams* for a birthday present: 'Here at last is a psychologist who has something to say' – an attitude not far off, perhaps, Keynes's sense of Freud speaking closely to our intuitions. Ramsey may have realized too late how passionate Wittgenstein's response to Freud was – in part because of the sense of self-recognition he experienced with Freud – and therefore how intense a discussion of Freud might be. A few years after their argument about Freud, Wittgenstein wrote in his diary:

> Freud certainly is mistaken very often and in what pertains to his character so he is really a pig or something similar but in what he says there is an awful lot. And that is true of me. There is a LOT in what I say.[21]

Moving beyond a depiction of Keynes's Freudian circles, it is useful to consider the general attraction Freudian ideas had for him within his own special areas of interest: the economics and politics of post-war Europe.[22] Keynes had become an international figure at exactly the same time that Einstein and Freud had, in the period immediately following the war, with, in his case, the publication of *The Economic Consequences of the Peace* in December 1919. That polemical indictment of the Versailles Treaty included a number of different elements, from an audit of the ruinous state of world capitalism to his excoriation of the blindnesses and character defects of Wilson, Clemenceau and Lloyd George. The savage dissection of their characters already included, in 1919, a Freudian rationale for the blatant inconsistencies of Wilson's positions:

> In spite of everything, I believe that [Wilson's] temperament allowed him to leave Paris a really sincere man ... The reply of Brockdorff-Rantzau inevitably took the line that Germany had laid down her arms on the basis of certain assurances, and that the treaty in many particulars was not consistent with these assurances. But this was exactly what the President could not admit; in the sweat of solitary contemplation and with prayers to God he had done nothing that was not just and right; for the President to admit that the German reply had force in it was to destroy his self-respect and to

[21] Ludwig Wittgenstein, *Denkbewegungen: Tagebücher 1930–1932, 1936–1937*, hrsg. von Ilse Somavilla, Frankfurt: Fischer, 1999, Diary entry 27 April 1930 (our translation), p. 21; the final sentence is in English in the original; emphasis in the original. Thanks to Martin Kusch for drawing this passage to our attention.

[22] Edward G. Winslow ('Bloomsbury, Freud, and the vulgar passions', *Social Research* 57 (1990), 786–819; 'Keynes and Freud: psychoanalysis and Keynes' account of the "animal spirits of capitalism"', *Social Research* 53 (1986), 549–78) has argued in convincing detail the extent of Keynes's reliance on a psychology underpinning or entangled with more purely economic arguments.

disrupt the inner equipoise of his soul; and every instinct of his stubborn nature rose in self-protection. In the language of medical psychology, to suggest to the President that the treaty was an abandonment of his professions was to touch on the raw a Freudian complex. It was a subject intolerable to discuss, and every subconscious instinct plotted to defeat its further exploration.[23]

Throughout Keynes's writings – from his early work on probability (demolished, he recognized, without any rancour, by Ramsey in the 1920s), through the *Economic Consequences* of 1919 into his busy political manoeuvring and polemicizing of the 1920s, when his opposition to Britain's return to the gold standard and his attempts to stave off the economic, political and eventually military consequences of the reparations exacted on Germany were his principal concerns, then into the 1930s with the publication of his magnum opus *The General Theory of Employment, Interest and Money* in 1936 – there is an elemental vision of economics as grounded on psychology. Like his other Bloomsbury Group members, Keynes was fundamentally critical of a foundational Victorian value: that of saving and fear of the future. His account of the Great Depression, then, would point to the underperformance of the economies of the West as due to too great account being given to uncertainty and fear, and too little to the present desires to consume. From a certain point of view, saving was a rational means to secure a more prosperous future; Keynes, however, emphasized that the excessive desire to save stemmed from general anxiety about the future and an inability to enjoy the present. Keynes was the 'sort of devil' that would undermine the Victorian virtues of thrift, hoarding and miserliness with any means he had to hand.

Freudianism thus could help supply Keynes with a general psychology of the cultural unconscious – that is, the basic psychological dispositions underlying the class attitudes of successive eras. In 1919, he had described capitalism as

> a double bluff or deception. On the one hand the labouring classes accepted from ignorance or powerlessness, or were compelled, persuaded or cajoled by custom, convention, authority and the well-established order of society into accepting, a situation in which they could call their own very little of the cake that they and nature and the capitalists were co-operating to produce. And on the other hand the capitalist classes were allowed to call the best part of the cake theirs and were theoretically free to consume it, on the tacit underlying condition that they consumed very little of it in

---

[23] J.M. Keynes, *The Economic Consequences of the Peace*, London: Macmillan, 1919, pp. 49–50.

practice. The duty of 'saving' became nine-tenths of virtue and the growth of the cake the object of true religion. There grew round the non-consumption of the cake all those instincts of puritanism which in other ages has withdrawn itself from the world and has neglected the arts of production as well as those of enjoyment.[24]

The 'psychology' of the capitalist classes was thus of crucial importance to the functioning of the system.

Keynes insisted on the importance of the psychological – at times, in classic Cambridge fashion, he thought it could equally be called 'moral' – in relation to economic argument. Capitalism is based on 'the money-making and money-loving instincts'; 'love of money' is a phrase that Keynes repeats and gives great weight to. It is also possible to envisage future economic arrangements without such love of money. Indeed, in his essay 'The economic possibilities of our grandchildren' (1928/30), he predicts that the future era of abundance will present our grandchildren with the problem of how to use the freedom, 'the leisure, which science and compound interest will have won for him, to live wisely and agreeably and well':

> The love of money as a possession – as distinguished from the love of money as a means to the enjoyments and realities of life – will be recognised for what it is, a somewhat disgusting morbidity, one of those semi-criminal, semi-pathological propensities which one hands over with a shudder to the specialists in mental disease.[25]

Closely linked on this and other occasions with the pathology of excessive love of money is a pathological relation to time and the future. '*In the long run* we are all dead' – this most famous of Keynes's many aphorisms is not directly about the love of money, but in fact comes from his *A Tract on Monetary Reform* (1923) and is concerned with how the changes in the speed with which people spend money may change prices in the short term, whatever the long-term trends.[26] But it does indicate Keynes's willingness to launch satirical and conceptually profound challenges concerning the limits of economic rationality in relation to time. The indefinite postponing of consumption, consummation or satisfaction is a pathology – excessive 'purposiveness' is one of the names Keynes gives it.

---

[24] *Ibid.*, p. 17.
[25] J.M. Keynes, *Essays in Persuasion*, New York: W.W. Norton & Co., 1963, pp. 358–73; Keynes, *The Collected Writings of John Maynard Keynes*, Vol. IX, *Essays in Persuasion*, London: Palgrave Macmillan for the Royal Economic Society, 1972.
[26] See Skid II, 156.

For purposiveness means that we are more concerned with the remote future results of our actions than with their own quality or their immediate effects on our own environment. The 'purposive' man is always trying to secure a spurious and delusive immortality for his acts by pushing his interest in them forward into time. He does not love his cat, but his cat's kittens; nor, in truth, the kittens, but only the kittens' kittens, and so on forward forever to the end of cat-dom. For him jam is not jam unless it is a case of jam to-morrow and never jam to-day. Thus by pushing his jam always forward into the future, he strives to secure for his act of boiling it an immortality.[27]

In the long run, even all the cats are dead. Keynes would have thoroughly agreed with D.H. Lawrence's pithy observation: 'Life is ours to be spent, not to be saved.' An unpublished note on 'love of money' from 23 December 1925 shows the full span of this argument, by linking it to the seductive abstractness of money, an abstractness that always outweighs the concrete good: this argument stretches from Keynes's early work on probability to what he judges should be our fundamentally primitive and corporeal attitude to the future:

It is not right to sacrifice the present to the future unless we can conceive the probabilities of the future in sufficiently concrete terms, in terms approximately as concrete as the present sacrifice, to be sure that the exchange was worth while ... Abolition of inheritance would help by making it more difficult to shirk the problem of concrete comparisons.[28]

In a later review, Keynes recognizes that abstractness can obliterate all concreteness in those who have no concrete desires in the present any longer, but are possessed only by 'the grand substitute motive, the perfect ersatz, the anodyne for those who, in fact, want nothing at all – money'.[29]

In his *Treatise on Money* (1930), a diatribe against those who would re-introduce the gold standard, freely employing the language of pleasure postponed or indulged in, a language that stemmed from its utilitarian and now Freudian versions of the reality principle, he depicted those conservative forces who saw in gold the 'sole prophylactic against the plague of fiat moneys' as throwing over themselves 'a furtive Freudian cloak' – the unconscious attachment to gold that Freud's essay on anal erotism had described.[30]

---

[27] 'Economic possibilities for our grandchildren', in *The Collected Writings of John Maynard Keynes*, Vol. IX, pp. 329–30.

[28] Skid II, 241.

[29] *The Collected Writings of John Maynard Keynes*, Vol. IX, p. 320, quoted in Winslow, 'Kleinian psycho-analysis and Keynes's treatment of capitalist "purposiveness"', mimeo, York University, 1999.

[30] *The Collected Writings of John Maynard Keynes*, Vol. VI, *Treatise on Money*, Vol. II, *The Applied Theory of Money*, London: Macmillan, 1971, p. 258.

Money Keynes described as a 'subtle device for linking the present to the future'.[31] But if money were held for long out of circulation, it ceased to be money, it de-monetizes – in the Freudian dialect that Keynes appreciated, gold turned back into faeces. Excessive anxiety about the future based on an inability to enjoy the present provoked regression back to a past fixation, that of the anal stage, in which pleasure was gained in hoarding faeces; money is thus a device 'through which the fear of the future takes its revenge on the hopes of the present'.[32]

In 1929, Keynes had asked James Strachey to help him research the psychoanalytic theories of money to which these arguments were linked. Strachey had passed him the relevant papers, including 'The theory of symbolism', in which Jones, writing in 1916, had linked, rarely for him, a socialist vision of the economy with psychoanalysis:

> the idea of wealth means simply 'a lien on future labour,' and that any counters on earth could be used as a convenient emblem for it just as well as a 'gold standard'. Metal coins, however, and particularly gold, are unconscious symbols for excrement. The ideas of possession and wealth, therefore, obstinately adhere to the idea of 'money' and gold for definite psychological reasons and people simply will not give up the 'economist's fallacy' of confounding money with wealth. This superstitious attitude will cost England in particular many sacrifices after the War, when efforts will probably be made at all costs to reintroduce a gold standard.[33]

In his 1919 review of Jones's *Papers*, the Cambridge philosopher C.D. Broad, in a sympathetic review that concluded that 'no psychologist can safely neglect the Freudian school, whether he likes their conclusions or not', had singled out this passage for specific ridicule: the 'contention that a warming-pan[34] is an erotic symbol is certainly not in the least further fetched than Dr. Jones's *obiter dictum* that people cling to a gold-standard because gold is a well-known symbol for excrement, "the material from which most of our sense of possession in infantile times was derived"'.[35] The contrast with Keynes's view in 1929 is enlightening: 'Jones's forecast, written in 1916, as to the troubles in which the passion to return to gold

---

[31] J.M. Keynes, *The General Theory of Employment, Interest and Money*, London: Macmillan, 1936, p. 293. See Thomas Crump, *The Phenomenon of Money*, London: Routledge and Kegan Paul, 1981, p. 3 ff.

[32] Skid II, 543.

[33] Ernest Jones, 'The theory of symbolism', in *Papers on Psycho-Analysis*, 2nd edition, London: Bailliere, Tindall & Cox, 1918, p. 172; 5th edition, p. 129.

[34] A reference to a rascally barrister, Serjeant Buzfuz, in *The Pickwick Papers*.

[35] C.D. Broad, 'Review, Ernest Jones, *Papers on Psycho-Analysis*, 1918', *Mind*, N.S. 28(III) (1919), 340–7 at 340.

would involve this country must be reckoned one of the triumphs of psycho-analysis.'[36]

There are, to be sure, many sources for Keynes's preoccupation with 'psychological' or 'moral' theories in relation to economics, in particular from very early on in the long gestating thesis on probability. Keynes was the principal voice (but not the only one – yet again Jeffreys was thinking along similar lines) arguing that probability was not an extension of statistics or a purely mathematical theory of frequency but was part of a general theory of logic: what is the connection between the truth of one statement and a second, when there is no syllogistic relation between the two, but nonetheless the first statement does provide some grounds or provide real evidence for believing the second? Probability is the name for this relation between propositions. And Keynes thus shifted the grounds of the theory of probability from objective measures of frequency to the subjective grounds of degrees of belief. By 1910 Keynes could transfer this view of probability to an account of the conditions of successful investment, conditioned not by objective risks but by the subjective risk, which depends upon 'purely irrational waves of optimism or depression' and upon the 'degree of his ignorance'.[37] Keynes was a pure product of the Moral Sciences Tripos in that he would always combine into a systemic unity economics and psychology within an overarching general logic.[38]

Thus Keynes's economics *required* a psychological underpinning for its portrayal of the economic virtues which, under changed circumstances, would mutate into vices leading to the disaster of the Great Depression. Keynes was a psychologist of economics before he became a Freudian; but Freud was ideally suited to the kind of portrait of the bourgeoisie and its unconscious character traits that Keynes's economics required. When he spoke in 1925 of 'the appeal which [Freud's theories] make to our own intuitions as containing something new and true about the way in which human psychology works', he meant not only our intuitions about, for example, why he himself was bisexual, or why his friends' character traits were the way they were, but also intuitions about what are the principal motors of world economic history: is it the entrepreneurs or the savers, the buccaneers or the bankers, who have created wealth? This economic-historical question was also, for Keynes, a question about psychology.

---

[36] JPPBaS, CSD/F03/07, Keynes to James Strachey, 12 July 1929.   [37] Skid I, 208.

[38] The contrast here would be with those economists who, rather like Durkheim in sociology, conceived of a science modelled on thermodynamics: pure concepts, defined through mathematical interrelations, able to pick out items in the real world through the good luck that Platonizing physicists have always relied on.

No wonder that, in October 1919, when he was completing the *Economic Consequences* with its quasi-Freudian account of capitalism and of diplomatic manoeuvrings, he met up with the new King's College undergraduates and spent an hour talking about Freud with one of them, Richard Braithwaite, remarking in a letter afterwards: 'Thank God, there's an intelligent man in College.'[39] Keynes's serious interest in psychoanalysis would be remarkably sustained, even aside from his psychological approach to economics: he was the initiator in 1922 of Sprott's expedition to the Austrian mountains to invite Freud to Cambridge; he orchestrated the debate over Freud in his magazine in 1925; and in 1939 he was even keen on what he took to be Wittgenstein's assimilation of philosophy to psychoanalysis.[40]

Let us return to the 1925 debate about psychoanalysis in *The Nation and Athenæum*. Having sketched in what a cynic might call the Freudian coterie, and given some account of why Keynes was so sympathetic to Freud, we can ask: what of the critics? Miss Allmond was a teacher, a member of the British Psychological Society who collaborated with Wohlgemuth in 1921 in testing children to establish whether, as Freud averred, unpleasant experiences are remembered less easily than pleasant ones.[41] Sir Bryan Donkin was an eminent alienist, the first Medical Commissioner of Prisons, and a public authority on heredity and criminality, much concerned with public health and mental deficiency, retired and knighted in 1911, and one of the lecturers at the course founded in 1920 to teach the Diploma in Psychological Medicine at the new Maudsley Institute; he wrote an introduction to a small book of 1924 *Psycho-Analysts*

---

[39] Skid II, 4. In his obituary of Keynes, Braithwaite wrote: 'For Keynes economics was always one of the "moral sciences": though he used mathematical techniques when helpful, he always remembered that mathematics is a good servant but a bad master, and that economics is not an abstract deductive system, but a science concerned with one aspect of human welfare. He liked to consider himself as standing in the tradition of what he has called "the High Intelligentsia of England" (which included Locke, Malthus, Mill, Sidgwick, Alfred Marshall), and his own ethic was essentially that of Moore's *Principia Ethica*. Never has there been a more humane Utilitarian. Keynes was born and bred a moral scientist in the Cambridge atmosphere' (Richard Braithwaite, 'John Maynard Keynes, First Baron Keynes of Tilton (1883–1946)', *Mind* 55 (1946), 283).

[40] Keynes was a great collector of art and books (his acquisition of Newton's alchemical papers and his coup during the First World War in persuading the Treasury to spend money at auction on post-impressionists for the National Gallery are just two out of his many famous collector's triumphs). He acquired a rare first edition of Freud's *Die Traumdeutung*, from the collection of Dr Hugo Schwerdtner, an early follower of Freud (see *Elke Mühlleitner, Biographical Lexicon of Psychoanalysis*, Tubingen: Diskord, 1992, p. 297). See also Catalogue of King's College, Cambridge, Keynes Bequest.

[41] A. Wohlgemuth, 'The influence of feeling on memory', *British Journal of Psychology*. General Section 13(4) (1923), 405–16 at 413.

Analysed by a retired Edinburgh opthalmologist, Peter McBride. He was a life-long sceptic of all 'psychic phenomena', his public opposition dating back fifty years.[42] So his opposition to psychoanalysis brought the fresh air of Huxley and the evangelical materialists of the 1870s to the debate orchestrated by Keynes.

The most substantial voice amongst these critics was that of Wohlgemuth. In his book of 1923, he examined all the principal claims of psychoanalysis chapter by chapter. His tone is varied, often contemptuous, often angry, expounding his disagreements in very great pedantic detail: railing against the way Freud writes German, or against the complex interpretation of serpents in dreams and mythology developed by Herbert Silberer and Ernest Jones.[43] His principal argument is the baselessness and

[42] The BMA Committee on Psycho-Analysis interviewed McBride, having accepted his extremely critical submission to its proceedings; under cross-examination, it turned out he had been retired since 1910 and had never read any of Freud's work (Richard Overy, *The Morbid Age: Britain and the Crisis of Civilization, 1919–1939*, London: Allen Lane, 2009, p. 148). For Horatio Bryan Donkin (1845–1927), knighted 1911, see 'Crime and responsibility', *Journal of Mental Science* 67 (July 1921), 324 and W. Norwood East, 'Obituary. Horatio Bryan Donkin', *Journal of Mental Science* 74 (1928), 1–12. His Majesty's Commissioner of Prisons in 1910, a member of the Royal Commission on the Care of the Feeble-Minded and Director of Convict Prisons, Donkin was sceptical of claims concerning the heredity of crime and alcoholism, entering into vigorous disputes with Karl Pearson. In 1909, Donkin acted for the Home Office in the forcible feeding of suffragettes in prison, leading to a vigorous dispute in the correspondence columns of *The Times* with Sir Victor Horsley, who regarded Donkin's forcible feeding as a breach of medical ethics amounting to common assault (*The Times*, 21 December 1909, p. 10). Donkin was also engaged in the first half of 1925 in a vigorous debate in the correspondence columns of *The Times* contra Lord Balfour, core member of the Society for Psychical Research, amongst others over the issue of thought transference, in which he teamed up once again with his old friend Ray Lankester, with whom he had waged a similar war against psychic research in the 1870s (Bryan Donkin, Ray Lankester et al., 'A spirit-medium', *The Times*, 16 September 1876, p. 7). This earlier spat was remembered in 1924 by the physicist and co-founder of the Society for Psychical Research Sir William Barrett: 'For weeks a great controversy ensued in the London *Times*, which, like all the other newspapers (with the exception of the *Spectator*), poured ridicule upon my daring to bring such a contemptible subject before the British Association. Among other of my vigorous opponents in *The Times* were Professors Lankester and Donkin (now Sir Ray and Sir Bryan); both of these distinguished men are still living and have not abated their hostility to the subject' ('Some reminiscences of fifty years' psychical research' read at a private meeting of the Society on 17 June 1924). In the course of the correspondence, Mr W.R. Bousfield, KC FRS, recent author of *A Neglected Complex and Its Relation to Freudian Psychology* (1924), declared on 7 May 1925 that Sir Bryan Donkin's state of mind resulted from 'a complex [which] renders him impervious to evidence or argument in matters which are opposed to his mental constellation ... The complex closed his mind against even the consideration of the evidence' ('Telepathy and science', *The Times*, 7 May 1925, p. 10). Donkin responded in kind: 'please allow me to inform him that he is probably suffering from ill-developed or disordered neurones' (*The Times*, 9 May 1925, p. 10). Having been accused of having a complex in May, Donkin was no doubt raring for a set-to in July. Note that W.R. Bousfield, Cambridge Wrangler, engineer, senior patent lawyer, physical chemist and FRS (1916), conservative MP 1892–1906, became a Freudian enthusiast and 'psychologist' in the 1920s, just when his son, Dr Paul Bousfield (see Chapter 4), was establishing himself as a Harley Street psychoanalyst (though never a member of the BPaS).

[43] A second debate emanates from this concern with the universality of snake or serpent symbolism. In response to Wohlgemuth's denial of the universal sexual significance of the snake, Reo Fortune,

illogicality of Freud's principal arguments for the existence of the unconscious, for repression, for censorship. Yet Wohlgemuth makes no attempt
to conceal his emotional reactions, common to many of Freud's readers, in
the early years of this century; indeed he clearly prizes his emotional
reactions and his accurate record of them:

> When I read for the first time Freud's exposition of the 'Oedipus-Complex'
> I passed, as probably most people have done on like occasions, through
> a series of emotional states . . . there was first a violent moral shock, followed
> by extreme disgust, outraged self-respect, and bitter resentment turning to
> rage. This gave place to a transitory contempt for Freud, turning soon to
> sadness, pessimism, and melancholy. Quickly I reacted against this. I said
> to myself, if Freud's view is true, all our outcry, gesticulation, and denial will
> not alter it a whit, for truth is truth to the end of reckoning. We have simply
> to face it.[44]

student dream-researcher at Cambridge (see Chapter 5), in transition from Riversian psychology to
Riversian ethnology, future husband of Margaret Mead, published an article on snake symbolism,
where he refuted Wohlgemuth in detail and endorsed the Freudian sexual symbolism: 'the serpent is
not obviously a symbol of the male organ in *Genesis*; but the eel obviously is in the Maori legend.
And by the Maori legend we know the serpent for what he really is' (R.F. Fortune, 'The symbolism
of the serpent', *IJP* 7(1926), 237–43 at 242). The Rivers that Fortune deployed was the Freudian-
diffusionist Rivers. Simultaneously, in *Psyche* (the general journal of psychology run by C.K. Ogden,
which was preponderantly psychoanalytic in orientation for the period 1921–27, hosting
Malinowski's psychoanalytic and anti-psychoanalytic writings before their publication in book
form) there was the publication in January 1926 of 'The significance of the snake in dreams'
(*Psyche* 6 (January 1926), 12–21), by Ursula H. McConnel, sister-in-law of Elton Mayo (formerly
Professor of Philosophy at Brisbane and psychoanalytic enthusiast who by the mid-1920s was
performing the function that Myers performed in Britain, of founding industrial psychology in
the USA). She had, with advice from Mayo's friend Malinowski, arrived in London to work with
Elliot Smith on a doctoral dissertation. Taking up the Elliot Smith cudgels against universalist
theories of symbolism in her article, she took psychoanalysis to task for advancing such a theory.
McConnel confronted Freud's theory of the universal significance of the snake as a symbol of
'repressed sex desire' with Rivers's criticisms of such universalist interpretations – his example,
published shortly before his death, was that of water, which he denied was always a symbol of rebirth
(see Rivers, 'Presidential address: the symbolism of rebirth', *Folklore* 33(1) (1922), 14–33). McConnel
argued that snakes appeared as symbols of sex because of the religious tradition which made the
snake into a symbol of sin. The Rivers McConnel deployed was the Elliot Smith anti-Freudian sceptic. Ernest
Jones could not resist responding to McConnel in the next number, April 1926, defending the
psychoanalytic theory against the Riversian (see Ernest Jones, 'Snake symbolism in dreams:
a rejoinder', *Psyche* 6 (April 1926), 89). Oddly enough, Freud never discussed directly whether the
snake is a universal symbol of the penis; the closest he came was in the *Introductory Lectures*: 'Among
the less easily understandable male sexual symbols are certain reptiles and fishes, and above all the
famous symbol of the snake' (Freud (1916), *Introductory Lectures on Psycho-Analysis, SE* XV, 155). Did
Freud include the snake as one of the 'less easily understandable' male symbols of the penis? And
what did Freud mean by 'famous'? One wonders: where did the idea that the snake symbolizes
a penis originate?

[44] Wohlgemuth, *A Critical Examination*, p. 146. We might contrast his response with the Mafia boss
played by Robert de Niro in *Analyze This* (1999), also confronting Freud's theories in a prelapsarian
state of innocence: 'Fuck Freud! After what you told me [about Oedipus] I'm afraid to call my
mother on the telephone.'

Wohlgemuth also confessed that he was strongly attracted to Freud's theory of dreams until he read Freud's analysis of Jensen's *Gradiva*, and realized with a shock that Freud applied his method to both real and fictional dreams. It was at that moment that he became a virulent anti-Freudian.[45] This moment – when the distinction between fiction and the real is elided, when he realizes that the new psychologist is clearly as at ease with Dostoevsky as he is with Wundt, combines the virtues and tasks of both William and Henry James – had also, as we have seen, been recognized by Keynes, who was much less perturbed by the porous boundary between psychoanalysis and the imagination, the world of fiction. Then, as now, for many of Freud's critics the permeability of the boundary between fiction and 'reality' is what alarms them and leads them to an outright condemnation of psychoanalysis in the name of science.

Besides the analysis of concepts and the refutation of the coherence of the unconscious and the censor, Wohlgemuth's principal means of refuting Freud is 'experimental', by which he means trusting in his skill as an introspectionist psychologist in detecting the psychological contents Freud posited. With the shift in the meaning of 'objectivity', we are ill-placed to recognize that a principal aim of the discipline of laboratory psychology until behaviourism came on the American scene, to which it would remain confined until the second half of the twentieth-century, was the experimental investigation of the subjective by employing a disciplined subjectivity.[46] What else could psychology be, the first experimentalists argued, than the trained observation of the 'subjective'? The doctrines of psychoanalysis put this ideal under considerable pressure: precisely the most interesting features of subjectivity, those which are unconscious or susceptible to the influence of unconscious forces, will by definition not be within the grasp of the subjective observer, whose discipline will fail him the closer he gets to the phenomenon. It is as if an astronomer were to discover – as they often have done – that the higher the power of his observational instrument, the greater the artefactual effects were; yet in astronomy, there is still the ideal of an observation purified of its artefactual effects, whereas in psychology it is precisely the artefactual effects that may become the object of interest.

Born in Berlin, Adolf Wohlgemuth (1868–1942) moved to London as a young man and established an import and export business in sausage casings, which he ran throughout his life. In 1902 he began to study for

---

[45] Wohlgemuth, *A Critical Examination*, 'Preface', pp. 7–8 and the final page of the book, p. 246.
[46] See Lorraine Daston and Peter Galison, *Objectivity*, New York: Zone Books, 2010.

a psychology degree at University College London, being awarded a third class degree in 1905; he then became one of the first doctoral students in the UCL Psychological Laboratory, gaining a DSc in 1909 from the classic experimental study he eventually published in 1911, *On the After-effect of Seen Movement.*[47] A further study in the experimental introspectionist style followed in 1919, *Pleasure–Unpleasure: An Experimental Investigation of the Feeling Elements.* For over twenty-five years he was affiliated to the Department at UCL, without having a position, well known for his considerable abilities in the construction and use of apparatus and for keen induction of students into the rigorous methods of experimental introspection.[48] This older experimental psychology, centred on the disciplined self-observer, the field in which Rivers had been the supreme expert, was increasingly under attack; by 1919 it had been explicitly repudiated by Rivers as having 'proved unfruitful as a direct means of increasing our knowledge', especially when compared with the psychoanalytic methods of 'observation' (deliberately contrasted with 'experiment').[49] However, Wohlgemuth was immovable in his fidelity to this discipline. Thus, in order to refute Freud's claim that human beings oscillate between homosexual and heterosexual feeling, Wohlgemuth subjected himself to the rigours of examining every man he met,

> in railway-carriages and omnibuses, in the park, at theatres and concerts, at lunch in the restaurants, at dinners, meetings, and so on, I chose as objects for these experiments youthful men of varying types: the martial figure of the dashing soldier or the brainy and intellectual countenance of the thinker, the athlete, or the delicate and dreamy artist – all men who would probably please and whose exterior decidedly attracted me. I contemplated them and dwelt upon their personal advantages, having constantly in mind the purpose of the experiment. I imagined the preliminary period of a sexual approach; but I think it unnecessary to enter here into further details of this process, and will state at once the result of these experiments. In no single case have I ever been

[47] The fullest biographical account of Wohlgemuth is N.J. Wade, P. Thompson and M. Morgan, 'Guest editorial: The after-effect of Adolf Wohlgemuth's seen motion', *Perception* 43 (2014), 229–34; see also E. Valentine, 'The other woman', *The Psychologist* 21 (2008), 86–7, which gives details of Wohlgemuth's first marriage, his relationship with the psychologist Nellie Carrey and their eventual marriage.

[48] Carl A. Murchison, *Psychological Register*, Vol. II, Worcester, MA: Clark University Press, 1929, and J.C. Flügel, 'A hundred years or so of psychology at University College London', *Bulletin of the British Psychological Society* 23 (1954), 21–30 at 25.

[49] This was also the variety of scientific psychology Russell attacked with no holds barred in *The Analysis of Mind* (1921), with weapons supplied to him by a correspondence with Wohlgemuth, whom he thanked in the Preface to the book, deploying his combination of behaviourism and psychoanalysis to undercut the pretensions of consciousness.

able to discover the slightest trace of libido, whilst I discovered invariably decided repulsion and disgust.[50]

When he conducted a similar control experiment studying his heterosexual feelings, he deliberately chose

> women whose exterior was decidedly repulsive. Old and decrepit women, and such as were afflicted with some nauseating complaint, women of varying degrees of cleanliness, or rather uncleanliness. In all these cases I have invariably been able to discover sexual conative tendencies and unmistakeable libido. It was vanishingly small as compared with the colossal repulsion felt, yet it was unmistakeably there. I thus satisfied myself that my introspection in the experiments on men was accurate and reliable, as I could not have missed in the one case what I was able to discover in the other.[51]

Wohlgemuth was not only intrepid in scrutinizing his own libidinal ('sexual conative tendencies') responses in this experimental fashion; his own life had been indelibly marked by the vagaries of the libido. In 1913 he had married his housekeeper, Clemence Morrelet, a French widow; their marriage was riven with incessant quarrels. That same year he had rented a flat in St Pancras to be near the British Museum; there from 1914 on he was visited twice a week by a 'Miss X', with whom he had scientific discussions. His wife became jealous; in addition, she later reported that their quarrels took on a chauvinistic character, with Wohlgemuth crowing over the German victories during the war at the expense of the French. In June 1918, unable to stand the quarrels any longer, he moved out to a hotel; on returning, a new row broke out and Clemence picked up her revolver and shot him in the back. She was remanded in custody.

At Clemence's trial in September 1918, the defence attempted to portray Wohlgemuth as a German fifth-columnist, depicting his expertise with scientific instrumentation as cover for the use of 'wireless telegraphy' to communicate secrets to the German authorities. His relationship with 'Miss X' was probed, including a will he had drawn up following the shooting, leaving all his possessions to her. But it was the extent of his support for the German war aims that most preoccupied the defence in court. However, the jury found Clemence not guilty on the more serious charges of intent to murder and grievous bodily harm, but guilty on the charge of unlawful wounding. The judge in his concluding remarks dismissed the flurry of insinuations about Wohlgemuth's patriotism and commended the jury for 'a very proper verdict', endorsing their view that

---

[50] Wohlgemuth, *A Critical Examination*, pp. 157–8.　　[51] *Ibid.*, pp. 158–9.

the firing of the pistol was an accident that occurred while Clemence was in a 'state of nervous excitement'.[52]

'Miss X' was Nellie Carey, daughter of a carpenter and a silk-weaver in Hornsey, London, who entered UCL in 1905 and took a degree in psychology in 1908.[53] Working as a teacher in an elementary school, she went on to research work in psychology, earning a DSc and being awarded the prestigious Carpenter Medal. During the war, when she and Wohlgemuth were meeting in his St Pancras flat, she also published major papers in experimental psychology, on mental processes in schoolchildren. Wohlgemuth's shooting and his wife's trial did not disrupt Carey's relationship with him; in 1921, having changed her name by deed-poll to Wohlgemuth, she gave birth to their daughter; a baby boy was to follow in 1929. In 1936, Clemence Wohlgemuth died and the couple were finally able to marry. In his book on psychoanalysis, Wohlgemuth gave the obliquest of hints of these dramatic events in his life, even quoting Freud to the effect that the relevant 'details are of too intimate a nature to allow of publication'.[54]

When on 19 September 1925 Wohlgemuth entered the controversy in the letter columns of *The Nation and Athenæum*, after Tansley had responded for a second time on 12 September, one of his two principal arguments concerned the possibility of experimental examination of Freud's scientific claims. Tansley had written that Wohlgemuth's '"experimental evidence" was quite valueless for the use to which Dr. Wohlgemuth put it'[55] and went on to describe Miss Allmond and Dr Wohlgemuth's accusation of Freud's 'flagrant and persistent disregard of scientific method' and 'careful control' as 'mere pseudo-scientific bombast'; he added:

> We all know that controlled experiment is by far the most satisfactory method of establishing any scientific conclusion. But the method of controlled experiment is simply not available in many spheres of scientific investigation, and no one denies them the name of science or refuses to give credence to results based on converging lines of evidence.[56]

There spoke the field botanist, ecologist and psychoanalyst. Wohlgemuth was not going to allow this argument; he regarded his own introspective experiments as controls on psychoanalytic findings – he had retraced

---

[52] 'Wohlgemuth Trial', *The Times*, 13 September 1918, p. 2.
[53] Valentine, 'The other woman', pp. 86–7.
[54] Wohlgemuth, *A Critical Examination*, pp. 214–15.
[55] A.G. Tansley, 'Freudian psycho-analysis', *The Nation and Athenæum* 12 (September 1925), 699–700 at 700.
[56] *Ibid.*, p. 700.

Freud's steps and come up with 'better and less far-fetched'[57] Freud-analyses than Freud had managed in his own analysis. He repudiated Tansley's claim that there were fields of science in which controlled experiment was not available, asserting that he himself had made and published such experiments. 'Such controls will', he added, 'suggest themselves easily enough to the trained psychologist, though they may worry the dilettante.'[58] He continued: 'The critic [i.e. himself] has also *practised* psycho-analysis upon himself and others, as Freud did, and has repeated Freud's experiments upon which Freud rests his doctrine, and this with the result that he could not accept it.'[59] Indeed, betraying a remarkable capacity for overlooking the uniquely individual character of the method of free associations, Wohlgemuth then claimed that he had taken an example of self-analysis of a seemingly random number, 426718, that Freud had published in *The Psychopathology of Everyday Life*,[60] and '"analyzed" it with the greatest ease, and my published "analysis" has been adjudged better and less far-fetched than Freud's, whose "Unconscious" was concerned in the prompting, and not mine.'[61] Wohlgemuth clearly regarded his experiments with Freud's numbers and his own self-examination, out strolling on station platforms and peering at diners in restaurants, as practising psychoanalysis.

The reviewer of Wohlgemuth's book for the *International Journal*, James Strachey, who would have been flattered to be called a dilettante, saw nothing to worry about; perhaps forgetting that he was not reviewing for the new 'house magazine'[62] of his set, *The Nation and Athenæum*, but for a 'scientific' journal, he opened his review in nonchalant style: 'This volume is chiefly remarkable for its dust-cover, and we therefore propose in this instance to review the dust-cover instead of the book which it contains.'[63]

---

[57] A. Wohlgemuth, 'Freudian psycho-analysis', *The Nation and Athenæum* 19 (September 1925), 729.
[58] Wohlgemuth, 'Freudian psycho-analysis', p. 729.   [59] *Ibid.*
[60] Freud, *The Psychopathology of Everyday Life: Forgetting, Slips of the Tongue, Bungled Actions, Superstitions and Errors* (1901), *SE* VI, vii–296 at 247. The analysis of the number is presented as the achievement of the patient in conversation with Freud.
[61] Wohlgemuth, 'Freudian psycho-analysis', p. 729.
[62] Almost literally: in 1922, Keynes took over the lease of 41 Gordon Square from Geoffrey Pyke, founder of the Malting House School in Cambridge – Pyke used it as a London base, including for his and his wife's analysis with James Glover; Pyke himself was leasing the house from Alix and James Strachey, who were in Vienna in analysis with Freud. In 1925, the configurations of inhabitants of Gordon Square were a little different. The Stracheys were installed at 41; at No. 50 were Adrian and Karin Stephen, who had taken over the part rented by Bertrand and Dora Russell, both sharing some of the time with the Bells (with Adrian moving out to 51 during periods of marital disharmony in the late 1920s); C.K. Ogden was at 45 Gordon Square, Homer Lane was at 18.
[63] James Strachey, '*A Critical Examination of Psycho-Analysis*. By A. Wohlgemuth, D.Sc. (Lond.) (London: Allen & Unwin, 1923. Pp. 250. Price 10s. 6d. net.)', *IJP* 5 (1924), 222–5.

Strachey's peroration reveals his argument: "'Unsavoury phantasies" (p. 160). "Defiles the child's mind" (p. 187). "Utter absurdity" (p. 237). "Simply nonsense" (p. 241). "Shallow", "faulty", "spurious" (p. 164). "Superficial and cavalier" (p. 238). "Museum specimens" (p. 239). "Threadbare confidence-trick" (p. 241). "Dead as a doornail" (p. 246). "Cagliostro" (p. 183). "Cagliostro" (p. 216). "Cagliostro" (p. 246). . . . The book closes; but as it slips from the reader's trembling grasp his eyes fall once more upon the publisher's pronouncement: "A sober and dispassionate examination of Freud's teachings." . . . This volume is chiefly remarkable for its dust-cover.'[64]

The dust-cover struck Ernest Jones as well: it had also advertised Wohlgemuth's conclusion that his book had abolished 'the right of the science to exist'. Jones was so incensed with the publishers Allen & Unwin, who also published Brill's translation of *The Interpretation of Dreams* and the Bernays edition of Freud's *Introductory Lectures* alongside the first volumes in the International Psycho-Analytical Library, that in late 1923 he broke off relations with them, leading to the crisis in psychoanalytic publishing from which James Strachey's brokerage of a deal with the Hogarth Press rescued him.[65]

Wohlgemuth captures well two quite different tones in which stringent criticisms of Freud were couched in this period: the tone of moral outrage, and the tone of professional censure. His entire critical project and his last dig at Tansley – who, he implied, being a distinguished botanist, was not sufficiently aware of the sophisticated experimental protocols which come second-nature to the well-trained psychologist – were based on the view that psychoanalysis represented a threat not to this well-trained psychologist (for whom, he wrote at the end of his book, 'in general, psychoanalysis was still-born, and has ever been as dead as a door-nail'[66]), but to the general public, to medical men and to educationists, who he observed were increasingly taken in by 'the psycho-analytic confidence trick'.[67] The tone of moral outrage is also candidly avowed, as befits Wohlgemuth's committed disciplined introspectionist methodology; yet he puts this to one side, he claims, as only a disciplined observer like him can, since he is explicitly and only committed to the truth, whatever its consequences.[68]

---

[64] The ellipses are in the original.   [65] Overy, *The Morbid Age*, pp. 144 and 416n23.
[66] Wohlgemuth, *A Critical Examination*, p. 246.   [67] *Ibid.*, p. 246.
[68] Wohlgemuth's book and its arguments are presented by Borch-Jacobsen and Shamdasani as a major contribution to the critical examination of psychoanalysis conducted by psychiatrists and psychologists in the period from the early 1900s to the 1920s; their presentation of Wohlgemuth's work is highly selective, omitting his empirical research (discussed and quoted here) almost entirely and presenting solely his arguments and conclusions.

Set against such critics, Tansley noted, were others, also outside psycho-analysis, who 'find its doctrines in harmony with their independent observation of human life'.[69] Yet Tansley immediately shifts from this general observation of life to the specific business of overcoming one's scepticism with regard to Freud:

> I confess that at first I was sceptical of very many of the Freudian theses, and even now there are interpretations which strike me as far-fetched. But I have become very chary of downright unbelief, for in so many cases I have been forced by accumulating evidence to accept interpretations which at first I rejected as overstrained. Freud generally turns out to be right.[70]

This was not, however, to be the last word in the debate. Well into October, McBride and Wohlgemuth were attacking Tansley and Freud. Then a different voice was heard, that of the anonymous ex-patient:

> in the interests of thousands of others who have been sufferers, through no fault of their own, I may perhaps be permitted to protest against the foolish and ignorant manner in which such correspondents as Miss Allmond and Dr. McBride have attempted to discredit a system of medical treatment whose powers to strengthen and to heal the mentally afflicted have already been vindicated in numerous instances.[71]

The ex-patient then described his own case, his years of 'unimaginable mental torture', during which time 'various eminent specialists of the older schools ... [which] included several famous names ... enriched themselves at the expense of my parents, and later of myself', his reluctant decision to try psychoanalysis as a last resort, solely because he came into 'contact with three men who had been cured by psycho-analysis of so-called "shell-shock"', and the happy outcome of his three-year analysis:

> by the end of that period, every trace of the shadow had faded into lucid daylight. Since then it has never returned, and I have the best of reasons for knowing it can never return – this reason being that, in the course of my slow recovery, nearly every aspect of my mental life has been completely and permanently transformed.[72]

Giving an account of the process of the analysis in which his long-held doubts and scepticism were finally overcome by the weight of evidence and his entire transformation – 'physical, mental and moral' – the ex-patient took special pains to refute Wohlgemuth's 'pretentious treatise' on three

---

[69] Tansley, 'Freudian psycho-analysis', p. 700.   [70] *Ibid.*, p. 700.
[71] 'Ex-patient', 'Freudian psycho-analysis', *The Nation and Athenæum* 10 (October 1925), 46.
[72] *Ibid.*, p. 46.

grounds: his refusal to accept the concept of subliminal mind in general, entirely independently of Freudian theories; his refusal to take account of the confirmatory experiences of patients themselves; and, lastly, 'the whole body of evidence adduced by psychical research'. For this ex-patient, the experiments of these researchers can be explained in either one of two ways: through the 'spiritualistic alternative' or through the hypothesis of an 'active sub-conscious'. For this reason, the critics of psychoanalysis should 'explain why they have not adopted the spiritualistic alternative'.

In his biography of Freud, Ernest Jones revealed the identity of the author of this anonymous letter: Dr Ivor Lloyd Tuckett.[73] Tuckett was a physiologist, an exact contemporary of Tansley's studying Natural Sciences at Trinity College, Cambridge – their names sat side by side in the published list of the First Class of the Natural Sciences Tripos Part I in 1893 and Part II in 1894, though Tuckett got a Starred First in both years. Tuckett became a Fellow at Trinity in 1895 and later University Demonstrator in Physiology (1899 and then again in 1905), having also trained as a doctor at University College London and the Royal London Opthalmic Hospital (qualifying in 1898). At UCH, he was the medical student Ernest Jones's much admired house-physician, 'a very sceptical and highly scientific thinker, demanding close evidence for all opinions; I would almost call him cynical in his attitude to official teachers'.[74] Jones and Tuckett kept in contact: Tuckett referred patients to Jones in 1906.[75] At some point in these years, Tuckett tried his hand as a farmer in New Zealand, but eventually returned to be an opthalmic surgeon. He had a long-term interest, like many Cambridge scientists, in spiritualism, but in his case it was a passion driven by scepticism; author of *Evidence for the Supernatural* (1912), a scathing attack on the accumulated evidence provided by the Cambridge psychical researchers,[76] he became for Ogden's *Cambridge Magazine* a regular heretical voice of ridicule of and disbelief in all manner of hauntings, rappings and telecommunications.[77] Ogden could always rely on Tuckett to produce good copy, such as his article in January 1915 'The theory and practice of Christians. The need for experiment';[78] while preparing this he reminded

---

[73] Jones III, 119. For further information on Tuckett, thanks to David Palfrey (email 19 June 2006).

[74] Ernest Jones, *Free Associations: Memories of a Psycho-analyst*, London: Hogarth, 1959, pp. 79–80.

[75] JPBPaS, CTA/Fo4/o1, 3 July 1906.

[76] See Alan Gauld, *The Founders of Psychical Research*, London: Routledge and Kegan Paul, 1968, p. 185; Gauld also notes the strength of Tuckett's 'black case' against Mrs Piper (p. 361).

[77] See *CM* II, 1 February 1913, p. 263; 22 February 1913, and 1 March 1913, p. 383, on the experiments conducted by Mrs Verrall and Mrs Piper for the Society for Psychical Research, whose results he attributed to 'natural association of ideas in minds occupied with the same themes' (p. 383).

[78] *CM* IV, 23 January 1915, pp. 194–5.

Ogden to read Hart's *The Psychology of Insanity*, which 'explains so clearly the phenomena of religious, material and all kinds of bias illustrated so well during this war'.[79]

It was very probably his former student Ernest Jones who referred Ivor Tuckett to James Glover for analysis in the early 1920s. In the summer, Tuckett would follow his analyst to the holiday resorts that Glover's own analyst, Karl Abraham, frequented, with sessions under the olive trees in Gardone Riviera[80] or beside the mountain lake at Seefeld in the Salzkammergut; Roger Money-Kyrle, another Trinity man in analysis, in his case with Freud, a year or so later, and related to Tuckett via marriage,[81] thought of these summer analyses as 'Shades of the Peripatetic School of Athens in the third century B.C.';[82] Ferenczi described them as 'a chronic psychoanalytic Congress'.[83]

Tuckett's participation in 1925 in this concerted defence of Freud is emblematic of the argument of this book. A Cambridge scientist, Fellow of Trinity College, which was and remains the powerhouse of British science ever since the 1850s, embroiled with psychical research in the early years of the century, although as an out-and-out sceptic rather than an enthusiast, a spirited participant in the Cambridge critique of organized religion and 'Authority', drawn against all the advice of his expert advisors into psychoanalysis by the success of psychoanalysis in treating shell shock during the war and then more than happy to go public as 'Satisfied Ex-Neurotic' in a spirited defence of Freud which was equally a swingeing attack on what, with hindsight, we know was his very own profession – his trajectory maps perfectly the manner in which Cambridge figured importantly, though in unseen or forgotten ways, in the reception of psychoanalysis in Britain. All orchestrated in a magazine run by the Cambridge School of Economics in the front pages and the Cambridge Apostles in the back.[84]

---

[79] OP, Box 112, F.3, Tuckett to Ogden, 2 January 1915. This strand of satirical criticism of Christianity can be seen emerging in Tripos Examinations; in 1921 in the Moral Sciences Tripos, candidates were asked to answer: 'Is religious conversion rightly regarded as psychopathological?'

[80] L.S. Kubie, 'Edward Glover: a biographical sketch', *IJP* 54 (1973), 85–94 at 88.

[81] Money-Kyrle's wife, Helen Juliet Rachel Fox, was the daughter of Samuel Middleton Fox (1856–1941), son of Samuel Lindoe Fox who in 1854 had married Rachel Elizabeth Fox (1833–1923); Samuel Lindoe Fox died in 1862 and Rachel Fox married Philip Debell Tuckett in 1867. We will give this genealogy because it forms the sole evidence of deducing reliably that Tuckett was the relative Money-Kyrle was referring to, and thus provides evidence of the date of Tuckett's analysis and the identity of his analyst.

[82] R. Money-Kyrle, 'Looking backwards – and forwards', *International Review of Psycho-Analysis* 6 (1979), 265–72, this passage from p. 266.

[83] S. Ferenczi, 'Letter from Sándor Ferenczi to Sigmund Freud, August 17, 1922', in *The Correspondence of Sigmund Freud and Sándor Ferenczi*, Vol. III, *1920–1933*, ed. Ernst Falzeder and Eva Brabant, Cambridge, MA: Belknap Press, 2000, pp. 85–6.

[84] Skid II, 138 makes this claim explicit.

So, with Tansley, Keynes (pseudonymously) and Tuckett (anonymously) making up the trio of defenders of Freud against the intemperate attacks of Wohlgemuth and his 'flying-squad'[85] of co-conspirators, it looks very much as if, once the true identities are revealed, the Freud enthusiasts in the 1925 debate consisted solely of eminent Cambridge scientists, one a botanist-analyst, one a physiologist-analysand, one a mathematical economist-'sort of devil'. Keynes ordered the debate closed in late October 1925 and then published a two-part article by James Glover to put the case for the scientific credentials of psychoanalysis in respectable – and unanswered – prose.

A public spat, it turns out, may be a rather well-orchestrated ritual dance. The three principal Freudian critics of 1925, Donkin, McBride and Wohlgemuth, referred to each other's writings, wrote Prefaces for each other's books and adopted something of a united front. On the opposing side, Keynes probably intervened to keep the debate going because he regarded it of first-rank importance; it is also fair to say that, on a normal day, his magazine was stuffed full of his close friends, amongst whom there were numbers of staunch Freudians. Those defending Freud turn out, surely not solely by coincidence, to be Cambridge dons, from Trinity and King's. However, this public spat in a weekly magazine was a dry-run for a more weighty set of public deliberations on the standing of psychoanalysis. The English 'Freud craze' of the early 1920s has been well documented;[86] Gordon Leff recalled of his parents how they were 'part of

---

[85] James Glover, 'Freud and his critics', *The Nation and Athenæum*, 14 November 1925, p. 259.

[86] See Rapp's ground-breaking papers. Despite his wide coverage of British magazines, Rapp consistently underestimates the growing body of general approval and advocacy for psychoanalysis both in medical circles and more widely in the early 1920s. He and others do highlight the development of a widespread British tendency to take only the 'jam' and not the 'powder' (to employ Edgar Adrian's 1919 simile), where the powder was very generally regarded as the excessive Freudian emphasis on sex (both adult and infantile); but there was equally widespread and important advocacy of a more full-blown Freudianism in some quarters where two of the standard accounts (Stone, Rapp) would lead one to expect unremitting hostility. To single out one: at the AGM of the Royal Medico-Psychological Association held in August 1920 (*The Lancet* 196(5060) (21 August 1920), pp. 403–4), a string of speakers, including Stoddart, Stanford Read and Dr Rees-Thomas, backed up by William Brown (publicly declaring that he 'had submitted himself to analysis, and he had been surprised at the way in which unconscious trends of thought sprang to the surface'), voiced overwhelmingly approving views on Freudian psychoanalysis; the only unreservedly hostile views were voiced by Sir Robert Armstrong-Jones. Prof. George Robertson, Physician-Superintendent of the Royal Asylum, Morningside, Edinburgh, 'speaking as an asylum physician', commented that 'anyone who studied Freud on dreams and tried to analyse his own dreamstates could not fail to be convinced of the truth of Freud's contentions. He referred to a number of very instructive cases in support of his views.' Finally, 'Dr. R. G. M. Ladell agreed that on first becoming acquainted with Freudism it seemed repulsive, but on proceeding to study it and on submitting oneself to analysis one was forced to the conclusion that Freud was justified in his contention that these tendencies did exist. Children were admittedly innocent, and

the bright young things of the 1920s, when psychoanalysis was in vogue' – his mother was analysed as a wedding present.[87] Widespread enthusiasm, increasingly to be found in medical circles as well as general interest or avant-gardiste magazines, alternated with expressions of alarm and high-minded dudgeon. The small psychoanalytic Society had certainly been appalled by *The Times*'s headline on 25 March 1925: 'Psycho-analyst convicted', detailing Homer Lane's appearance before a magistrate as an alien who had failed to register with the police. The main scandal turned round rumours recounted to the magistrate that Lane was living the high life off money extorted from patients and credulous believers in his psycho-analytic talents. Four months later, in July 1925, just as the debate in *The Nation and Athenæum* was getting going, the British Medical Association's Sussex Branch tabled a motion at the AGM that 'the Representative Body instruct the Council to consider certain practices alleged to be prevalent among some medical men practising psycho-analysis and to report'.[88] The 'certain practices' that the Sussex doctors found so appalling were free discussions about sexuality with child patients and a certain school that recommended that children of both sexes bathe in the nude together. The Ethics Centre, chaired by Dr Reginald Langdon-Down, took no action, since if these were valid complaints, they were a matter for the police not for the Medical Association. But in December 1925, a barrister who committed suicide, said to have been driven to distraction by the revelations of his psychoanalysis, prompted *The Times* to publish a leader on psychoanalysis. While respectful, sympathetic and balanced in its view of the new science, the Thunderer called for an inquiry:

> This weapon, whether truly forged or not, is capable of inflicting terrible injuries, a point made with urgency at the last annual meeting of the British Medical Association. There is no doubt that an inquiry by competent persons into the whole subject of psychological treatment is now overdue. The present position, which is giving rise to a great deal of anxiety, is one of the consequences of the neglect of psychology by the medical schools of the country.[89]

that was the reason they were "indecent"; the indecencies of childhood were natural.' None of these members of the alienists' professional organization were principally psychoanalysts (save Stoddart, one of the first to advocate a psychoanalytic psychiatry in Britain early in the war); they were mainstream asylum physicians – Ladell, for one, was advocating liberal views concerning sexuality in the correspondence columns of the *BMJ* up until 1958.

[87] Jonathan Croall, *Neill of Summerhill: The Permanent Rebel*, London: Routledge and Kegan Paul, 1983, p. 171.

[88] 'British Medical Association. Psycho-analysts' sex teaching. Call for inquiry', *The Times*, 18 July 1925, p. 14.

[89] 'Psycho-analysis', *The Times*, 31 December 1925, p. 11.

Sir Bryan Donkin endorsed this proposal vigorously, convinced as he was that Freudism was an 'unsubstantiated fiction';[90] the National Council for Mental Hygiene volunteered itself as a suitably competent body to conduct such an inquiry. At the AGM of the BMA held in July 1926, a further debate was held on a new motion calling for an inquiry concerning therapeutic aspects of psychoanalysis, putting improper practices to one side. The motion was carried with considerable support and the announcement 'was received with cheers'.[91]

Starting its formal deliberations in 1927, after twenty meetings over two years, the *Report of the Psycho-Analysis Committee* of the BMA, chaired by Langdon-Down, was published on 29 June 1929. As the most hard-working and energetic member of the Committee, Ernest Jones was instrumental in drafting the Report and hence secured his aim: the Committee withheld from adjudicating either on the questions of theory and method or on the therapeutic claims of psychoanalysis, leaving these to be decided by the test of time. However, it ruled that two senses of 'psychoanalysis' should be firmly distinguished: 'a loose, popular sense' incapable of further definition; and 'the strict sense of the technique devised by Freud, who first used the term, and the theory which he has built upon his work'.[92] For many decades after, this BMA Report provided both the shelter of respectability that Jones craved and a guarantee that, in official eyes, the moniker 'psychoanalysis' was a trademark of the IPA.

In the course of its deliberations, the BMA Committee commissioned thirteen reports which were prepared by members of the Committee; it also received five memoranda from non-members: Peter McBride, Adolf Wohlgemuth, J. C. Flügel, Edward Glover and Walter Langdon-Brown.[93] These five experts are very familiar names: Flügel and Glover, stalwarts of the BPaS; McBride and Wohlgemuth, the two critics of psychoanalysis who had already participated in the debate in the *Nation and Athenaeum*. It is in this sense that we can see the 1925 debate as a dry-run for the BMA's Psycho-Analysis Committee's deliberations. Tansley, not being a physician, would not be called to participate. But there was a fifth memorandum, from Walter Langdon-Brown – cousin of the

---

[90] 'Letter: Inquiry by competent persons', *The Times*, 5 January 1926, p. 8.
[91] 'Doctors' views on psycho-analysis', *The Times*, 19 July 1926, p. 9.
[92] 'Supplementary Report of Council', *Supplement to the British Medical Journal*, 29 June 1929, p. 270.
[93] A further communication on medical psychology was received from Dr E.H. Connell, a practising analyst in Edinburgh who had no contact with the BPaS, but had already trained W.D. Fairbairn. Connell had contributed a paper, based on a talk to the Society in Cambridge, entitled 'the use of the method of psychoanalysis in medicine', *Cambridge University Medical Society Magazine* 1 (Easter Term 1923), 97–105.

Chairman, Reginald Langdon-Down. Walter was a member in the 1890s of St John's College, Cambridge, where he was a friend of Rivers; pioneer endocrinologist; soon to be Regius Professor of Physic at Cambridge (1932–35); enthusiast for psychodynamic psychotherapy, in particular Adler's views, and thus elected President of the Medical Society of Individual Psychology on Adler's death.[94] To the chief orchestrator of the 1925 debate, his nephew John Maynard Keynes, he was 'Uncle Walrus, huge of frame and moustache and endlessly anecdotal'.[95] The BMA Report of 1929 marked the closing down of the swirl of debate over psychoanalysis in British medical circles and reinforced the conversion of psychoanalysis from a field for wide debate and discussion in academic and more general intellectual circles to an esoteric corpus of knowledge and technique reserved to the professionally qualified. Once again, that close web of institutional, class and family affiliation which characterized public debate over psychoanalysis, here even within the medical profession, is evident. Some of the same cast of characters reappear in that more famous Cambridge–Bloomsbury axis, known as the Bloomsbury Group. What was the significance of the Bloomsbury Group for the development of psychoanalysis in Britain in the early twentieth century?

---

[94] For a good example of the medical territory he covered, see Walter Langdon-Brown, 'The influence of the endocrines in the psychoneuroses', *British Journal of Medical Psychology* 2(1) (1921), 1–12. In his account of his own approach to Freud's and Adler's views, Langdon-Brown wrote: 'Then [around 1910] came a stormy dawn, and amid thunderous protests from the orthodox, Freud began to loom on the horizon of our vision. I remember one of my senior colleagues, the late Dr. J.A. Ormerod, an able, learned and conspicuously fair-minded physician, in his criticism of Freud instanced the case of a nursemaid who micturated into the shoes of the children, as conclusive evidence against a sexual basis for neuroses! All of which indicates the gulf between pre- and post-war medicine' (Walter Langdon-Brown, 'Adler's contribution to general medicine', in Philip Mairet and Drs H.C. Squires, Cuthbert Dukes, O.H. Woodcock, Sir Walter Langdon-Brown and others, *The Contribution of Alfred Adler to Psychological Medicine*, London: C.W. Daniel, April 1938, pp. 47–60 at 50). Even in its obituary, the Royal College of Physicians managed effectively to dismiss both Freud and Langdon-Brown with one laughable historical claim: 'For some years he was the only English physician to accept and apply Freud's teachings' (*Munk's Roll*, Vol. IV, p. 491); but this does at least provide further evidence that he was memorably and publicly a supporter of Freud. And, after all, if it *were* true, it would mean that the sole English physician who accepted and applied Freud's teachings was the Regius Professor of Physic in the University of Cambridge.

[95] Skid II, 286.

CHAPTER 9

Bloomsbury Analysts

Freud has a wider appeal than simply to psychoanalysts.[1]

Leonard Woolf to John Rickman, 1926

As cause and consequence of our general state of mind we completely misunderstood human nature, including our own. The rationality which we attributed to it led to a superficiality, not only of judgment, but also of feeling. It was not only that intellectually we were pre-Freudian, but we had lost something which our predecessors had without replacing it.[2]

John Maynard Keynes, 1937

The rationalism and liberalism mark [Bloomsbury] off very sharply from the common or garden esthetic bohemians of modern times, and also very sharply from the Russian modernism of the opening of the century. Being theoreticians of the passive, dividend-drawing and consuming section of the bourgeoisie, they are extremely intrigued by their own minutest inner experiences, and count them an inexhaustible treasure store of further more minutious inner experiences. They have a high opinion of Dostoievsky and of Freud.

But even these writers are taken by them without any trace of common bohemian gluttony. They are agile-minded, and of Freudian concepts they make a very special kind of mental discipline; I may say that a prominent bloomsburian once told me how he had trained himself, every time he wakes in the night, be it only for a single minute, immediately to take up his pencil and record all dreams experienced up to that point.

---

[1] Letter concerning the translation of the title of Freud's book *Das Ich und das Es*, 13 August 1926, unpublished, Hogarth Press Papers, Reading University, quoted in R. Steiner, 'Bloomsbury/Freud. The letters of James and Alix Strachey, 1924–1925', *IRP* 15 (1988), 404–7 at 407.

[2] John Maynard Keynes, 'My early beliefs' (1938), in *The Collected Writings of John Maynard Keynes*, Vol. X, *Essays in Biography*, London and Cambridge: Macmillan and Cambridge University Press for the Royal Economic Society, 1972.

Yet the interest these people show in Dostoievsky and Freud is quite equalled by their interest in Voltaire and Spinoza. They believe (or should I say, they hope?) that reason plus education will some day bring an age in which people will be enlightened ladies and gentlemen much like themselves, and there will be no more wars or revolutions.[3]

D.S. Mirsky, 1935

Bloomsbury was not a 'school' in any literary sense ... It included novelists, critics, painters, college dons but, curiously, no important poet (if one counts Virginia Woolf as a novelist) or composer. Nearly all its members had been to Cambridge and came from distinguished upper-middle-class families; i.e., without being aristocrats or large landowners, they were accustomed to efficient servants, first-rate meals, good silver and linen, and weekends in country houses. In rebellion against the rhetoric and conventional responses of their Victorian parents, hating dogma, ritual, and hypocritical expressions of unreal feelings, they, nevertheless, inherited from the Victorians a self-discipline and fastidiousness that made bohemian disorder impossible.[4]

W.H. Auden, 1954

'Old Bloomsbury' had been formed before the First World War with the cementing of friendships between the two Stephen sisters, Vanessa and Virginia, and a group of young men from Cambridge, linked by membership of the Apostles. The name came following the Stephen siblings' move to Bloomsbury in London after the death of their eminent father, Leslie Stephen, in 1904. This group continued and metamorphosed during and after the war; a later definition takes membership of the 'Memoir Club', formed in 1920, as the condition.[5] Core Bloomsbury Group members numbered perhaps twenty, including figures crucial to the cultural, intellectual and artistic life of England of the first half of the twentieth century: E.M. Forster novelist,

---

[3] Prince Mirsky, quoted in Robin Majumdar and Allen McLaurin (eds.), *Virginia Woolf: The Critical Heritage*, London and New York: Routledge, 1975, p. 348.

[4] W.H. Auden, 'A consciousness of reality', review of *A Writer's Diary*, by Virginia Woolf. *The New Yorker*, 6 March 1954, reprinted in Auden, *Forewords and Afterwords*, selected by Edward Mendelson, London: Faber & Faber, 1973, pp. 411–12.

[5] Quentin Bell, *Bloomsbury*, London: Weidenfeld & Nicolson, 1968, pp. 14–15. The Memoir Club was formed in 1920 by Molly MacCarthy, and included Leonard and Virginia Woolf, Vanessa and Clive Bell, E.M. Forster, Roger Fry, Duncan Grant, Maynard Keynes, Desmond and Molly MacCarthy, Adrian Stephen, Lytton Strachey and Saxon Sydney-Turner. See also Hermione Lee, *Virginia Woolf*, London: Vintage Books, 1997, pp. 263–5 for the question of definition.

Fig. 9.1 Bertrand Arthur William Russell, 3rd Earl Russell, John Maynard Keynes, Baron Keynes, and Lytton Strachey, by Lady Ottoline Morrell, 1915.

J.M. Keynes economist, Virginia Woolf novelist and publisher, Leonard Woolf political writer and publisher, Lytton Strachey writer and biographer, Vanessa Bell and Duncan Grant painters. Vanessa Bell thought Bloomsbury had died in 1914, with so much else; but it is more accurate and more helpful to point to the deaths of Lytton Strachey and Dora Carrington in early 1932 and to Virginia Woolf's suicide in 1941 as the end of their cultural moment.

It was the publication of Strachey's *Eminent Victorians* and the establishment of the Hogarth Press by the Woolfs, both in 1918, followed in 1919 by Keynes's *The Economic Consequences of the Peace*, that gave the group a distinctive and original cultural presence. They became arbiters of taste, facilitating the reception of Proust, Dostoevsky and Chekhov, of Cezanne and Matisse, and eventually, through the Hogarth Press, of psychoanalysis into British high culture.[6] As Virginia Woolf reflected in 1924 when the calls on her writing had swelled: 'very likely this time next year I shall be one of those people who are, so father said, in the little circle of London

---

[6] Skid II, 17.

Fig. 9.2 Lytton Strachey, Virginia Woolf and Goldsworthy Lowes Dickinson, by
Lady Ottoline Morrell, 1923.

Society which represents the Apostles, I think, on a larger scale. Or does
this no longer exist? To know everyone worth knowing . . . just imagine
being in that position – if women can be. Lytton is: Maynard; Ld Balfour.'[7]
When the BBC was created, key members of the group became frequent
wireless voices. But it was not only their productions as individuals – their
books, their paintings, their political articles – that gave the Bloomsbury
Group cultural significance. The Group fashioned a way of life seamlessly
intertwined with a set of aesthetic and moral ideals. The wing of the group
who had been to Cambridge – the overwhelming majority, save for the
Stephen sisters and the painter Duncan Grant – never forgot the powerful
effect of G.E. Moore's ethical philosophy and his attitude to argument, a
rational scepticism which blew away the cobwebs of Victorian religious
doubt and moral aspiration. 'States of mind': these were the basic and sole
elements of Mooreian ethical examination. And the principal good states of
mind would be promoted by the cultivation of friendship and
the contemplation of beauty. The supremacy of immediate personal relations was the discovery that the turn of the century generation of Apostles
made: it grew from their recognition of the failure of authority of all

---

[7] VWD II, 319, 17 October 1924. 'Ld Balfour' is Arthur Balfour, philosopher, psychical researcher and
former Prime Minister, who figured in Chapter 5.

institutions – family, state, codes of honour, rules of thumb.[8] Moore's philosophical grounding was the legitimization of this shift in values, a revolutionary doctrine for the hyper-rational young men and women who already espoused in their lives and actions this new orientation away from duty, away from the higher, away from the obligations of the social. Keynes, the only core Bloomsbury member to live the whole of his life in Cambridge, reminded them of this decades later:

> We repudiated entirely customary morals, conventional wisdom. We were, that is to say, in the strict sense of the word, immoralists. The consequences of being found out had, of course, to be considered for what they were worth. But we recognised no moral obligation on us, no inner sanction, to conform or to obey. Before heaven we claimed to be our own judge in our own case ... It resulted in a general, widespread, though partly covert, suspicion affecting ourselves, our motives and our behaviour. This suspicion still persists to a certain extent, and it always will. It has deeply coloured the course of our lives in relation to the outside world. It is, I now think, a justifiable suspicion. Yet so far as I am concerned, it is too late to change. I remain, and always will remain, an immoralist.[9]

Morality was replaced by the hegemony of the personal and the aesthetic, 'classifying as aesthetic experiences', Keynes wrote, 'what is really human experience and somehow sterilising it by this mis-classification'.[10] From this personal aestheticism flowed the requirement of absolute honesty and a boundary-less confrontation with human passion, including, very centrally, the sexual.

> It was a spring evening. Vanessa and I were sitting in the drawing room ... At any moment Clive might come in ... Suddenly the door opened and the long and sinister figure of Mr. Lytton Strachey stood on the threshold. He pointed a finger at a stain on Vanessa's white dress.
> "Semen?" he said.
> Can one really say it? I thought & we burst out laughing. With that one word all barriers of reticence and reserve went down. A flood of the sacred fluid seemed to overwhelm us. Sex permeated our conversation. The word bugger was never far from our lips. We discussed copulation with the same excitement and openness that we had discussed the nature of good. It is strange to think how reticent, how reserved we had been and for how long.[11]

---

[8] MacCarthy, paper read to the Apostles, December 1900, cited in Skid I, 134–5.
[9] Keynes, 'My early beliefs', pp. 447–8.   [10] *Ibid.*, p. 450.
[11] Virginia Woolf, 'Old Bloomsbury', in *Moments of Being*, ed. Jeanne Schulkind, St Albans: Triad/ Panther, 1978, pp. 195–6.

The ensuing 'libertinism and comprehensive irreverence' coupled with the
constant display of 'intellectual *chic*' transmuted the values of the Apostles
into the characteristic Bloomsbury mode: 'unadorned honesty in conversation and personal relationships, a passion for literature and the visual arts.
Politics, together with bourgeois sexual conventions, were damned as
prime examples of cant; but there was profound insight into the niceties
of intellectual and cultural rank, and a constant concern with the problems
of domestic servants.'[12]

Ascendancy of personal relations – sexual variety and tolerance – absolute honesty: these ingredients made the Apostolic Bloomsbury reception
of psychoanalysis a meeting of ideas whose kinship is obvious – a revea-
lingly partial union. Yet the rationalism of Bloomsbury was later to be
contrasted by Keynes himself with psychoanalysis, with its altogether darker
vision of human nature.

> It seems to me looking back, that this religion of ours was a very good one to
> grow up under. It remains nearer the truth than any other I know, with less
> extraneous matter and nothing to be ashamed of . . . It was a purer, sweeter
> air than Freud cum Marx. It is still my religion under the surface.[13]

Keynes was here contrasting Bloomsbury's ethos with the altogether darker
vision of the 1920s and 1930s, dominated, in turn, by Freud and then by
Marx. But in between the Edwardians and the political 1930s came the
modernist moment which was also the Bloomsbury moment. And the
Freud moment.

On 18 May 1924, Virginia Woolf gave a talk to the Heretics in Cambridge
entitled 'Character in fiction'. In the final published version of the talk,
which T.S. Eliot had already sought from her and which appeared in *The
Criterion* in July, Woolf declared:

> My first assertion is one that I think you will grant – that every one in this
> room is a judge of character. Indeed it would be impossible to live for a year
> without disaster unless one practised character-reading and had some skill in
> the art. Our marriages, our friendships depend on it; our business largely
> depends on it; every day questions arise which can only be solved by its help.
> And now I hazard a second assertion, which is more disputable perhaps, to
> the effect that on or about December 1910 human character changed.[14]

---

[12] Skid I, 244; note that this accurate but nonetheless flip reference to the class statuses of the members
  of the Bloomsbury Group and their servants is rather like characteristic Bloomsbury humour and the
  incessant self-deprecation that is an integral element in its supremely self-confident tone.
[13] Keynes, 'My early beliefs', p. 442.   [14] VWE III, 421.

This dramatic claim – 'on or about December 1910 human character changed', now so famous there are a book and many articles that take it as their title[15] – is only to be found in the third version of this essay. In the talk in May to the Heretics, she had instead said:

> No generation since the world began has known quite so much about character as our generation ... The average man or woman today thinks more about character than his or her grandparents; character interests them more; they get closer, they dive deeper in to the real emotions and motives of their fellow creatures. There are scientific reasons why this should be so. If you read Freud you know in ten minutes some facts – or at least some possibilities – which our parents could not have guessed for themselves. And then there is a ... vaguer force at work – a force which is sometimes called the Spirit of the Age or the Tendency of the age. This mysterious power is taking us by the hand, I think, and making us look much more closely into the reasons why people do and say and think things.[16]

As Woolf refined her argument in the three versions of this piece, Freud flared up – temporarily – as a principal driving force behind the claim that her generation not only is more concerned with character but also is in a position to *know* more about human character. Then, displacing talk of Freud's science or the Spirit of the Age, Woolf alighted upon her now famous phrase, 'on or about December 1910'. In her final formulation the actual *cause* of the change in human character is left completely vague. Instead, the fact of change stands in its full assertiveness – and implausibility – thereby stirring the reader's attention into speculation and excited association, a pleasurable oscillation between credulity and scepticism.

Woolf had been stung into her reflections on character by Arnold Bennett's review in March 1923 of her novel *Jacob's Room*: 'the characters do not vitally survive in the mind', Bennett had written. She had responded to Bennett's charge that the 'Georgian' writers – James Joyce, D.H. Lawrence, E.M. Forster, Lytton Strachey, T.S. Eliot, Edith Sitwell, Dorothy Richardson as she named them – failed to create realistic and therefore memorable characters. Her rebuttal had firstly accused Bennett's 'Edwardian' generation of themselves failing to generate truly memorable – i.e. 'Victorian' – characters, because they were in flight, justifiably, from the superficial abundance of the Victorians. The Edwardians' attempt to find a heightened literary voice for moral aspiration had been undercut by

---

[15] Peter Stansky, *On or about December 1910: Early Bloomsbury and its Intimate World*, Cambridge, MA: Harvard University Press, 1996.
[16] VWE III, 504.

two new forces: the effect of the exposure, expressed clearly in Butler's *The Way of All Flesh*, of the 'family secrets' of the Victorians: 'It appeared that the basement was really in an appalling state . . . The social state was a mass of corruption.' But it was the impact of Dostoevsky that militated most strongly against the Victorian idea of character, creating 'characters without any features at all. We go down into them as we descend into some enormous cavern. Lights swing about; we hear the boom of the sea; it is all dark, terrible, and uncharted.'[17] The task the Georgians faced was:

> to bring back character from the shapelessness into which it has lapsed, to sharpen its edges, deepen its compass, and so make possible those conflicts between human beings which alone rouse our strongest emotions – such was [the Georgians'] problem. It was the consciousness of this problem and not the accession of King George [in May 1910], which produced, as it always produces, the break between one generation and the next.[18]

Having delineated the problem, Virginia Woolf then performs a little of the novelist's magic: she allows to steal on stage besides Mr Wells, Mr Galsworthy and Mr Bennett a 'character – shall we say? – [a] Mrs Brown' and concludes: 'The capture of Mrs Brown is the title of the next chapter in the history of literature.'[19]

Who is this Mrs Brown sidling on to the stage of world literature? In her talk to the Heretics in Cambridge, Woolf introduces a version of Mrs Brown – a woman sitting in a railway carriage from Richmond to Waterloo – in order to pose the problem of describing character in fiction. She throws down the gauntlet to Mr Bennett, ventriloquizing his version of Mrs Brown with all its details: her father's shop, the rent and the price of calico, so that she then finds herself crying out – 'Oh I cried, oh, stop. Stop.' Mr Bennett and the Edwardians laid enormous stress on the fabric of things, their houses, and forgot the people who live in them. The task of the novelists who began writing around 1910 is to invent other tools to find Mrs Brown.

To the student of Virginia Woolf's biography, Mrs Brown is Mrs Dalloway: Woolf had written the doctor chapter of this novel in April and she was set on finishing the book over the summer. To the historian of modernism, Mrs Brown is Molly Bloom – Woolf trails her coat by deliberately making a Freudian slip when depicting Joyce as the foremost contemporary destroyer of the past tools of the novel – '"I will break up the language, the grammar; I will bring the whole building down about my

---

[17] *Ibid.*, p. 386.   [18] *Ibid.*, p. 387, 'Mr Bennett and Mrs Brown'.   [19] *Ibid.*, p. 388.

head" he seemed to say "if by so doing I can keep absolutely close to my idea of Mrs Brown – Mrs Bloom, I mean."'[20] To the historian of Bloomsbury's in-jokes and the historian of proto-celebrity culture, Mrs Brown is Queen Victoria – 'we do not even know whether her villa was called Albert or Balmoral',[21] Woolf has the 'British public' complain, and then wheels in Lytton Strachey's *Eminent Victorians* and *Queen Victoria* (1921) – evoking the fantasy, common to all, based on the great Royal 'family secret' known to all, that Queen Victoria may have married her dearest servant, Mr Brown: 'Ulysses, Queen Victoria, Mr Prufrock – to give Mrs Brown some of the names she has made famous lately'.[22] But most seriously and importantly, and strictly in apposition to Queen Victoria, Mrs Brown is Mrs Everywoman, thereby giving voice to the clear red thread of defiant and sober feminism that ran through so much of Woolf's work, including her most famous pamphlet, *A Room of One's Own*, which unfolds out of the experience of giving another Cambridge talk, this time in October 1928. Finally, of course, Mrs Brown is the involuntary burden of the novelist: 'a little figure rose before me . . . "My name is Brown. Catch me if you can."'[23] Every novelist is in pursuit of this phantom: 'they are lured on to create some character which has thus imposed itself upon them'.[24]

And why did Freud steal on to Mrs Brown's stage only to be rapidly ushered off? The weekend of 17–19 May 1924, when Virginia was in Cambridge to talk to the Heretics, was the most intense and delicate moment in the negotiations Leonard and Virginia Woolf were conducting with James Glover (the analyst at that time of both her brother Adrian and her sister-in-law Karin), representing the Institute of Psycho-Analysis, to become the publishers of Freud's *Collected Papers*, his books and the International Psycho-Analytic Library.[25] So Virginia Woolf's enthusiastic recommendation of Freud – 'if you read Freud you know in ten minutes' – may well have owed something to her having Freud on her table as a business proposition. By the end of May, the Woolfs had agreed to the deal,

---

[20] *Ibid.*, Appendix III, p. 515, 'Character in fiction' Cambridge version.

[21] *Ibid.*, p. 433. The Cambridge version named the houses as 'Balmoral or Albert Edward *Stratford*' (p. 514, italics added for emphasis). In going the whole hog in the final version (Balmoral AND Albert!), Woolf may have been pleased with the nice trick her unconscious had played on her – clearly Shakespeare's Stratford represented a loss of nerve. Queen Victoria and Prince Albert bought Balmoral estate in 1852; it was their private residence and is not part of the Crown Estate.

[22] *Ibid.*, p. 435. Lytton had written: 'With an absence of reticence remarkable in royal persons, Victoria seemed to demand, in this private and delicate matter [of the rituals of mourning for John Brown], the sympathy of the whole nation; and yet – such is the world! – there were those who actually treated the relations between their Sovereign and her servant as a theme for ribald jests' (*Queen Victoria* (1921), London: Penguin, 1971, pp. 218–19).

[23] VWE III, 420. [24] VWE III, 421. [25] By 1931, they had twenty volumes in the Library.

paying £800 for the existing stock of the Psycho-Analytical Library;
undoubtedly it was the biggest financial gamble that the Press had undertaken since its foundation in 1917 and a major new commitment for a
highbrow literary publishing house. The contract was signed on 3 July 1924
and stock arrived the same month, 'dumped in a fortress the size of
Windsor castle in ruins upon the floor' in the basement at Tavistock
Square.[26]

But the more immediate, internal reasons for Freud appearing in
Virginia's Heretics talk were, firstly, her sense of her audience – she
knew very well she would be speaking to the younger generation, who
had already been infected by the epidemic of psychoanalysis that had
passed through the University (and her own family) immediately post-
war, so it was deliberately 'elementary and loquacious, meant for undergraduates', as she explained to Eliot.[27] 'Cambridge youths', as she called
them, so often triggered her sharp-edged curiosity and sense of 'the repetition of certain old scenes from my own past: the obvious excitement, &
sense of being the latest & best (though not outwardly the most lovely) of
God's works, of having things to say for the first time in history; there was
all this; & the young men so wonderful in the eyes of the young women, &
young women so desirable in the eyes of the young men.'[28]

Secondly, her argument was groping towards an account of why it was
that the 'British public' clearly saw that 'the Edwardian version of human
nature was ... very superficial, conventional, all mixed up with clay and
stucco brick and mortar, empire and commerce, conditions and circumstances'.[29] She was reaching outside the novel for some explanation of the
change of *Zeitgeist*, and 'Freud' came easily to hand, the most recognizable
delegate of this 'mysterious power'. So many of her contemporaries (Lord
Keynes, Lord Adrian) would make the same gesture. (But we might still
ask: why did she (and those others) not mention the war?) Yet she withdrew from her own gesture when she came to revise her text, noting in the
margin of her manuscript beside the excised Freud passage: 'That is a very
debatable point: how much we can learn & make our own from science.'[30]
'Science' (Freud) was insufficiently powerful to account for this change in

---

[26] VWL III, 133; 119; 134–5 – letters to Roger Fry in July and to Molly MacCarthy in September (gull-
like imbecility). On date of contract, see Richard Overy, *The Morbid Age: Britain and the Crisis of
Civilization, 1919–1939*, London: Allen Lane, 2009, p. 416n23.

[27] VWL III, 106, Woolf to Eliot, 5 May 1924.

[28] VWD I, 103, 9 January 1918, describing 'a large semi-circle of Cambridge youths' at the 1917 Club,
home of the anti-war circles, grouped around Lancelot Hogben: 'We tried still to overhear young
Cambridge, & Lytton finally decamped, on my daring him, to that party.'

[29] VWE III, Appendix III, 514–16, 'Character in fiction'.   [30] *Ibid.*, p. 504.

our demand for a description of human character. So in her final version, she leaves to one side the question of the *motor* of this change and simply *asserts*: 'human character changed'. She gives examples, all of women: her cook now emerges from the 'lower depths' and seeks advice about a hat or borrows the *Daily Herald*. And, she continues:

> Do you ask for more solemn instances of the power of the human race to change? Read the *Agamemnon*, and see whether, in process of time, your sympathies are not almost entirely with Clytemnestra ... All human relations have shifted – those between masters and servants, husbands and wives, parents and children. And when human relations change there is at the same time a change in religion, conduct, politics and literature. Let us agree to place one of these changes about the year 1910.[31]

Why is our transformed appraisal of Clytemnestra the index of the power of the human race to change? Could it be because of an incident she recorded in February 1917?

> As to Aeschylus ... I've been reading him in French which is better than English ... Aeschylus however excited my spirits to such an extent that, hearing my husband snore in the night, I woke him to light his torch and look for zeppelins. He then applied the Freud system to my mind, and analyzed it down to Clytemnestra and the watch fires, which so pleased him that he forgave me.[32]

Although the passage points more to the famous passage where Clytemnestra describes the beacons that brought news of the taking of Troy back to Greece – and reveals Leonard Woolf to have been a proud owner of the newly marketed pocket torch – it is singular and curious that she suppresses Freud from her text only to substitute a feminist reading of Clytemnestra which for her, personally, maybe unconsciously, stands for the Freud method or 'system' (the method of free association) as the marker of the change in human character since 1910.

Five months after her talk in Cambridge, in October 1924, Virginia Woolf glanced at the proofs of Freud's *Collected Papers*, Vol. II and recorded her reaction in a letter:

> we are publishing all Dr Freud, and I glance at the proof and read how Mr A. B. threw a bottle of red ink in the sheets of his marriage bed to excuse his impotence to the housemaid, but threw it in the wrong place, which unhinged his wife's mind, – and to this day she pours claret on the dinner

---

[31] *Ibid.*, p. 422. [32] VWL II, 141, Woolf to Sydney-Turner, 3 February 1917.

table. We could all go on like that for hours; and yet these Germans think it proves something – besides their own gull-like imbecility[33]

Her tone is withering – as it so often was – a reaction of intense dislike. Yet she did not *dispute* these 'facts'. Quite the reverse: just like Keynes, Virginia perceived that 'we could all go on like that for hours', as if such versions of human motivation were *now* the simple bread and butter for us all – 'the appeal which they make to our own intuitions', as Maynard put it ten months later. What caused her to spit fire, like Freud's critics, was the idea that something substantial – something scientific and proven – can be built on such facts. To quote Freud, 'the case histories I write . . . read like short stories and . . . lack the serious stamp of science . . . [yet] a detailed description of mental processes such as we are accustomed to find in the works of imaginative writers enables me, with the use of a few psychological formulas, to obtain at least some kind of insight into the course of that affection'.[34] Woolf immediately recognized Freud as working on *her* terrain and was appalled, then, in late 1924, at what he did with her everyday toolbox in the name of science.

The Reception of Freud by Bloomsbury

Virginia Woolf had every reason for being interested in Freud: her husband was a keen enthusiast; she became Freud's English publisher; her brother, sister-in-law and two more of her close circle were amongst the small group of early analysts; she herself had suffered since the death of her mother from

---

[33] VWL III, 134–5, Woolf to Macarthy, 10 October 1924. Elizabeth Abel, *Virginia Woolf and the Fictions of Psychoanalysis*, Chicago: University of Chicago Press, 1989, p. 18, mistakenly assumes that the proofs were of Lecture 17 of *Introductory Lectures*, where there was an extended discussion of the 'table-cloth lady'; however the *Introductory Lectures*, in Joan Riviere's translation, were not published by Hogarth Press but by George Allen & Unwin – and in 1922. The first edition of the *Introductory Lectures* to be produced by the Hogarth Press was that in the *Standard Edition* (1961). Abel's interpretation, built on this assumption, is inadvertently taken up by Lee, *Virginia Woolf*, p. 496 and n62. The proofs that Virginia Woolf actually saw were of 'Obsessive acts and religious practices', translated by R.C. McWatters (a friend of Ernest Jones, one of three British members of the Indian Psycho-Analytical Society, then serving as a Lieutenant Colonel in the Indian Medical Corps in Saharanpur); the translation of this paper appeared in *Collected Papers*, Vol. II and includes a much shorter paragraph-long version of the symptom and its analysis. However, in neither of Freud's versions does the patient herself produce a stain on the table-cloth; and 'claret' is a nice addition, very Virginia (Freud's text does not specify the source of the stain) and is, unsurprisingly, not a word in Freud's lexicon; in an earlier part of the letter Woolf had joked about not having yet reached the menopause. The person – the audience – whom Virginia excludes from the scene (and without whom the symptom has no meaning) is the maid: 'she always arranged the cloth in such a way that the housemaid was bound to see the stain' (*SE* IX, 121).

[34] Freud, *Studies on Hysteria, SE* II, 160–1.

acute bouts of insanity and her doctors – though much ink has been spilt in debate over this – were perfectly well positioned in 1912 to consider psychoanalytically oriented psychotherapy as a possible course of treatment.[35] But she steered clear of reading Freud seriously until 1939 – although she felt she knew quite enough to pass judgement upon him in the meantime. There was no 'party line' in the reception of Freud by Bloomsbury.

The publication of Brill's American translations of *The Interpretation of Dreams* (1913) and *The Psychopathology of Everyday Life* (1914) supplemented the steady uptake of psychoanalysis into scientific and medical circles from the late 1890s on. The medical journals had kept their readers well informed on the development of Freud's ideas, both for enthusiastic supporters and for sceptical critics.[36] Inevitably interest gathered pace

[35] The much-discussed question of Virginia Woolf's contacts with psychoanalysis, both as literature and as therapeutic practice, is best addressed in two papers: Jan Ellen Goldstein, 'The Woolfs' response to Freud – water-spiders, singing canaries, and the second apple', *Psychoanalytic Quarterly* 43 (1974), 438–76; B. Hinshelwood, 'Virginia Woolf and psychoanalysis', *International Review of Psycho-Analysis* 17 (1990), 367–71; there is a useful website devoted to the question:www.malcolmin gram.com/doctors.htm (accessible via 'Wayback' at: web.archive.org/web/20080221104420/http:// www.malcolmingram.com/vwframe.htm). One of Virginia Woolf's doctors, Dr Maurice Wright, whom she consulted not long after her engagement to Leonard and some months before her marriage on 9 March 1912, calling him a 'psychologist' (the term 'psychiatrist' being rarely in use in English English before the First World War), was a founding member of the London Psycho-Analytical Society on 30 October 1913 (*Zeitschrift* 2 (1914), 411) and was listed by Ernest Jones, along with three others ('Berkeley-Hill, Forsyth and Eder' – note the striking omission of Hart) amongst 'the earliest' physicians 'interested' in psychoanalysis in the period while Jones himself was in Canada (i.e. 1908–13) (E. Jones, 'Reminiscent notes on the early history of psycho-analysis in English-speaking countries', *IJP* 26 (1945), 8–10 at 10); Wright gave a paper on 'The psychology of Freud and its relations to the psychoneuroses' in February 1914 at St George's Hospital Medical School, and was elected an associate member (proposed by Joan Riviere) of the BPaS on 6 November 1919, and a full member on 4 October 1922; he remained a member till the late 1940s and was a key figure in the Tavistock Clinic throughout the 1920s and 1930s (see H.V. Dicks, *Fifty Years of the Tavistock Clinic*, London: Routledge and Kegan Paul, 1970, *passim*). In 1929, Wright was described in George Pitt-Rivers's letter to Bertrand Russell as 'a psychologist, gen.[eral] prac.[titioner], + also treats s.[exual] repression cases according to the Gospel of St. Freud; he is intelligent, reputable, but includes among his friends Ernest Jones and Joynson-Hix [*sic*] – Home Sec.[retary]!' (Ivan Crozier, '"All the world's a stage": Dora Russell, Norman Haire, and the 1929 London World League for Sexual Reform Congress', *Journal of the History of Sexuality* 1(1) (2003), 16–37 at 27). A doctor who moves in the best circles, including those of St Freud, would be precisely the sort of doctor Leonard and Virginia Woolf would call upon in 1912 – and they did; all the endless debates about Woolf's (lack of) knowledge of psychoanalysis – and whether she was deliberately kept by her entourage from coming into contact with psychoanalytically informed professional advice – are redundant given these overlooked facts about Dr Wright.

[36] In England, the *BMJ*, *Journal of Mental Science*, *Brain* and *The Lancet* regularly featured articles addressing or employing Freud's latest papers or books, or reviewing them. A review of Freud's *Selected Papers*, 2nd edition, 1912, could refer casually to 'the well-known case of hysterical pains in the legs, "Elizabeth v. R."' (*Brain* 35 (1913), 325). See Philip Kuhn, 'Subterranean histories: the dissemination of Freud's works into the British discourses on psychological medicine, 1904–1911',

with Brill's first translations of Freud's *Selected Papers* in 1909 and Freud's American Clark lectures published in 1910 in the *Journal of Abnormal Psychology*.[37] Leonard Woolf was later proud of having recognized Freud's importance relatively early on. He reviewed the *Psychopathology of Everyday Life* in 1914, praising Freud's analytic grasp, noting 'his sweeping imagination more characteristic of the poet than the scientist'.[38] His chance first encounter with Walter Lippmann on a train back from the Fabian Summer School in July 1914 led to an immediate intimate conversation as soon as they discovered their mutual enthusiasm for Freud.[39] Lippmann had been part of the small team helping Brill translate *The Interpretation of Dreams* in the summer of 1912 and had been much impressed. If there was little altogether distinctive or 'avant-gardiste' in these early literary and intellectual responses of Bloomsbury to Freud, what was distinctive and unusual was the decision of four core Bloomsbury figures to become psychoanalysts. What was unique was that these four constituted two married couples, closely bonded with one another for many years.

James and Alix Strachey

James Strachey and Adrian Stephen were both the youngest children of large families; both had lives seemingly overshadowed by supremely well-endowed and dominating older siblings – Lytton Strachey; Vanessa and Virginia Stephen. Both were swept away by intense homosexual loves in their early manhood and then turned, without much of a nostalgic look back, to women for the rest of their adult amorous lives. In the course of that turn from men to women, they both fell in love with Noel Olivier.[40] Young and overshadowed by their older siblings, both cultivated the critical faculty – Adrian with 'the right sneer for every occasion' and James with an overwhelming desire to '"stab humbug dead"'. Sympathetically but firmly written off by their nearest and dearest, both men finally achieved much and did so late in life.

Psychoanalysis and History 16(2) (2014), 153–214, for a detailed evaluation of the reception of psychoanalysis by the English medical profession in the period up to 1911.

[37] Dean Rapp, 'The early discovery of Freud by the British general educated public, 1912–1919', *Social History of Medicine* 3 (1990), 217–43 at 219.

[38] Leonard Woolf, 'Everyday life', *New Weekly* 1 (1914), 412.

[39] Ronald Steel, *Walter Lippmann and the American Century* (1980), New York: Little, Brown, 2008, pp. 70–1.

[40] One of the four beautiful Olivier girls, whose father was one of the founders of the Fabian Society; the actor Laurence Olivier was their first cousin.

Fig. 9.3 Alix Strachey (née Sargant Florence) and James Beaumont Strachey by
unknown photographer, mid-1930s.

Lytton's Cambridge years certainly had a great influence on James,
always known as 'Little Strachey' at Trinity, where, after St Paul's
School, which he'd attended in the company of his cousin Duncan
Grant, in 1905 he followed his older brother. Elected an Apostle immediately, simply on his family connections, James slid effortlessly into Lytton's
witty and blasé social groups; but his most passionate friendship was with

Rupert Brooke, with whom he had been at school when they were aged eleven. During these years James shared Lytton's sexual tendencies as well; he became infatuated with Brooke, who failed to reciprocate his feelings. The situation was reversed when George Mallory, another friend of these years, fell in love with James.[41]

Footloose as an undergraduate, taken up into the London literary world through his job in the years 1910–15 on the *Spectator*, edited by his uncle John St Loe Strachey, James was still without a vocation, although his reviews displayed his deep preoccupations with the borderline between philosophy, psychical research and psychology.[42] Coinciding with a seemingly definitive break with Noel Olivier,[43] he asked Loe for several months off in the first six months of 1914: 'The fact is that I have for some time been getting into a very depressed state, which, among other things, makes me inert and incompetent, and which I suppose may be the beginning of what people call "a nervous break-down", (though I have always had my doubts as to the objective reality of that condition).'[44] As a cure, he travelled to Berlin and Moscow, returning via Vienna, Budapest and Venice.[45] Despite his perennial torpor, James had some political fire in his belly; before leaving England at the end of 1913, he resigned from the National Liberal Club: he was no longer able to think of himself a 'liberal in politics' because 'of the attitude adopted by the leaders of the Liberal Party towards the movement for the Enfranchisement of Women'.[46] By implication, James still thought of himself as a liberal – and indeed this is

[41] Mallory would later die in 1924 attempting the ascent of Mount Everest. His passion for James Strachey is passed over in silence in his most recent biography, but is a major topic in the Strachey–Brooke correspondence; this erasure of Mallory's homoerotic relationships thus repeats the long erasure (orchestrated by Geoffrey Keynes) of Rupert Brooke's complicated sexuality. However, Wade Davis, *Into the Silence: The Great War, Mallory and the Conquest of Everest*, London: Bodley Head, 2011, does narrate Mallory's early homosexual passion.

[42] See in particular [Strachey], 'Memory and the individual', *The Spectator*, 16 April 1910, p. 618, which addresses the question whether individuality (e.g. of a communicant from beyond the grave) depends essentially on memory (it doesn't) and how thought-experiments concerning replaceable memories (derived from McTaggart) can resolve questions about our equanimity in the face of memory-loss/death.

[43] Noel wrote to Rupert Brooke (3 January 1914): 'I started well. After Scotland, James & I became very sane & very courageous & said goodbye. He wrote no more notes & I never called. When we met by chance it was taken fairly calmly & quite silently, & one of us went away at once. I found it very dreary, never being able to talk to him. And he found it worse. (In a few days he's going abroad for six months.)': Pippa Harris (ed.), *Song of Love: The letters of Rupert Brooke and Noel Olivier, 1909–1915*, London: Bloomsbury, 1991, p. 260.

[44] James Strachey to St Loe Strachey, 5 December 1913, BL Strachey Papers, 60713, Vol. LIX, 1903–October 1924.

[45] 'Bloomsbury/Freud', p. 21.

[46] BL Strachey Papers, 60713, Vol. LIX, 1903–October 1924, 31 December 1913.

Fig. 9.4 James Strachey, 1910, painting by Duncan Grant (1885–1978).

the best term for him: he held immoveably to 'liberal' attitudes, especially towards sex, morals and religion, for the rest of his life.

James's political positions during the war would, through a knock-on effect, eventually lead to his new choice of profession. In November 1915, he was forced to resign from the *Spectator* 'since the Editor wished me to attest under Lord Derby's scheme and my convictions made it impossible for me to do so'.[47] He sought work from the more liberal *Manchester*

---

[47] James Strachey to C.P. Scott, 24 January 1916, BL Strachey Papers, 60713, Vol. LIX, 1903–October 1924. Lord Derby's scheme was introduced in late 1915 as a method intermediary between volunteer and full conscription for securing forces for the Army. 'Disappointed at the results of the Derby Scheme, the Government introduced the National Military Service Act on 27 January 1916. All voluntary enlistment was stopped. All British males were now deemed to have enlisted – that is, they were conscripted – if they were aged 18 to 41 and resided in Great Britain (excluding Ireland) and unmarried or a widower on 2 November 1915. Conscripted men were no longer given a choice of which service, regiment or unit they joined, although if a man preferred the navy it got priority to take him. This act was extended to married men on 25 May 1916.' www.1914–1918.net/recruit ment.htm.

Guardian, only to encounter the same difficulties created by his refusal to support the Government's wartime policies. He became active in the National Council against Conscription, of which Adrian Stephen was the Treasurer, and suffered the general panic prevalent amongst his fellow-refuseniks when obliged to run the gauntlet of the Conscription Board Tribunal. In March 1916, he enlisted the help of Keynes, who 'testified before the wicked leering faces of the Hampstead Tribunal to the genuineness of James's conscientious objections'.[48] From then on, James was active in the No Conscription Fellowship and with the Quaker-organized Emergency Committee for the Assistance of Germans, Austrians and Hungarians in distress, where his work was admirable – tactful, kind and, apparently, superbly well organized.[49] He also offered his services to that other defiantly anti-war organ, Ogden's *Cambridge Magazine*, suggesting he write a weekly London Letter focused on politics and art, 'all of it viewed from a rather Cambridge standpoint';[50] as a taster he suggested an exposé of a newly self-appointed Commission on the Cinema intent on implementing a general policy of cinema censorship. He also approached Ogden in 1918 proposing a corrective and critical review of a recent bland de-sexualized biographical memoir of his close friend Rupert Brooke; the idea eventually came to nothing.[51]

It was about the time of his tribunal appearance that, in parallel with his love affair with Noel Olivier, Alix Sargant Florence set her sights firmly on James. He had known both women from the close-knit circles of the Apostles, Neo-Pagans, and Fabian camps of the pre-war period. In fact, James Strachey had first singled Alix out to Lytton in September 1910 as 'a delightful Bedalian ... an absolute boy'.[52] True harbinger of the life and work in common they were eventually to share, on 15 October 1915 Alix had borrowed from James his copy of Brill's translation of *The Interpretation of Dreams*.[53] The notoriety of his simultaneous dalliance over some years with Noel and Alix is probably the background for Keynes's uncharacteristic refusal to invite James to his end-of-ballet-season party in July 1919, on the grounds that he had 'ruined two young ladies' – relenting only when

[48] Keynes to his parents, 26 March 1916, Skid I, 325–6.
[49] Barbara Caine, *From Bombay to Bloomsbury: A Biography of the Strachey Family*, Oxford: Oxford University Press, 2005, pp. 345–6.
[50] OP, Box III, F.12, Strachey to Ogden, n.d. (probably 1916 or 1917); on the Commission on the Cinema, 12 October 1917.
[51] Keith Hale (ed.), *Friends and Apostles: The Correspondence of Rupert Brooke and James Strachey, 1905–1914*, New Haven: Yale University Press, 1998, p. xi.
[52] BF, 12 September 1910, p. 23.   [53] BF, p. 27.

Duncan Grant (whom Keynes had seduced away from Lytton back in 1908) attacked him for moral hypocrisy.[54]

Alix Sargant Florence – born in 1892, so five years younger than James – came to Newnham College, Cambridge in 1911 via Bedales and the Slade School of Art to study Spanish and French. Falling in with the Heretics (along with her brother Philip and mother Mary) and the *Cambridge Magazine* while an undergraduate, already reading Freud, she concocted in June 1914 a plan to work in psychology with C.S. Myers after graduating.[55] However, the war intervened: in Finland in August 1914, she ended up stranded in Petrograd, where she sent regular foreign correspondent pieces to the *Cambridge Magazine*, learned some Russian and learned to drive. On her return, she settled into the Bloomsbury world. By September 1915, she was driving James to a country weekend at Lytton's, despite James's continuing preoccupation with Noel. Characteristically, it was Alix who was to be the persistent wooer: falling in love with James in early 1916 when staying with Lytton, according to Holroyd, she wooed him intellectually, with her knowledge of Freud and her practicality. She organized their joint rental of a house in Hampstead in 1918 and then, in January 1919, she took out a lease on 41 Gordon Square 'chiefly in order to live with James'.[56] While this nest-building was going on, she conducted a therapeutic affair with Harry Norton, mathematician at Trinity College and one of James's oldest and closest friends: 'Copulation every 10 days in order to free his suppressed instincts!'[57]

James Strachey dated his first acquaintance with Freud to the activities of the Society for Psychical Research (SPR) while he was an undergraduate in 1905–9, and the first Freud he actually read was the article on the unconscious Freud published in the *Proceedings of the Society for Psychical Research* in 1912.[58] However, in a paper to the Apostles in 1907 Strachey had

---

[54] Quote from Skid II, 12; on the Lytton–Grant–Keynes triangle, and James's part in it, see Skid I, 188–205.

[55] BLSP Add. Mss, 60701, Alix Sargant Florence to Mary Sargant Florence, 7 June 1914.

[56] VWD I, 237, 30 January 1919; James Strachey to Holroyd, cited Holroyd, p. 733 n61. The Stracheys occupied all or part of 41 Gordon Square from 1919 to 1956.

[57] VWL II, 319, Woolf to Vanessa Bell, 22 January 1919.

[58] BPaS, CSD/F03/08, Strachey to Jones, 18 July 1945. That Strachey's first contact with Freud came from the SPR also highlights an obvious fact: becoming acquainted with Freud would take very different paths for a German-speaker and for a non-German-speaker. Any Cambridge scientist of the period inevitably spoke and read German, the world-language of science from 1870 to 1920. James Strachey didn't.

    BF accepts Winnicott's assertion in his obituary of Strachey that 'I understand that he was positively influenced by a quotation from Freud in a book by C.G.S. [*sic*] Meyer. From here he developed the idea of becoming a psychoanalyst' (*IJP* 50 (1969), 129). Winnicott's obituary of Strachey is so full of inaccuracies that this statement has to be treated with liberal scepticism

drawn on James's *Principles of Psychology* and flirted with SPR themes of survival; so he was already known for his 'psychological' interests.[59] In 1912, with his characteristic diffidence, he accepted an invitation to a party thrown by his sometime lover, John Maynard Keynes, with the words: 'D'you really think I'd better come, though? It might make your party psychological – which is always a thing to be avoided.'[60]

In 1945, when Jones wrote a typically tendentious version of the 'early history of psycho-analysis in English-speaking countries', he sent a draft to Strachey who commented that Jones had been 'slightly unjust to the shade of Fred Myers'; he also indicated that for many English readers, 'Freud', at this time, came packaged with other continental and American writers:

> I have a personal feeling about this S.P.R. episode, as that was actually my own road of approach to Ψα. The S.P.R. was still very lively (and not at all exclusively spiritualistic) at Cambridge when I was an undergraduate (1905–9), and, though I was never spooky, I was very much interested. I read a lot of the current literature on abnormal psychology – Janet, Prince, Flournoy etc. But I remember quite well the impression made on me by Freud's paper in 1912 – which was the first thing of his I ever read.[61]

The year 1917, as we have seen, was crucial for the upsurge of enthusiasm for psychoanalysis in England, and not simply because the growing numbers of doctors employing psychoanalytic methods became evident in the

(Winnicott clearly confused C.S. Myers and F.W.H. Myers, amongst many other imprecisions). In his letter to Jones correcting his historical oversights and highlighting the SPR's reception of Freud, he wrote: 'On March 12th 1897 (85th General Meeting) Mr F.W.H. Myers delivered an address on "Hysteria and Genius". This was briefly summarized in the *Journal* of the S.P.R. Vol. 8, April 1897. Pp. 55–6 of this gave an account of *Studien über Hysterie*. A very much fuller account of it is of course contained in *Human Personality*, Vol. I, pp. 50–6, published posthumously in Feb. 1903.' The 'of course' in this passage, written expressly for Jones, is a clear pointer to Strachey assuming that Jones was 'of course' familiar with Myers's book, in the same way that Strachey was. (And if he weren't, he should be ashamed of himself.)

59 Erik Linstrum, 'The making of a translator: James Strachey and the origins of British psycho-analysis', *Journal of British Studies* 53(3) (2014), 685–704 at 691.

60 BLSP 60713, Vol. LIX, 1903–October 1924, Strachey to Keynes, 17 July 1912.

61 JPBPaS, CSD/Fo3/o8, Strachey to Jones, 18 July 1945. Strachey added: 'It was because of that that I got hold of your *Papers on Ψα*' – which was, in fact, published in the same month, November 1912, as Freud's paper. He gave a similar account in his memorial tribute to Joan Riviere: 'I used to read [the SPR's] published proceedings, not because I was much interested in the question of survival, but because that was almost the only place (apart from Janet's works) where I could read anything about abnormal psychology' ([James Strachey], 'Joan Riviere (1883–1962)', *IJP* 44 (1963), 228–35 at 229). Like many others 'out of the same middle-class, professional, cultured, later Victorian, box', James would of course read French, but German with more difficulty. For a careful examination of other eccentric claims by Jones concerning the early history of psychoanalysis in England, see Philip Kuhn, 'Subterranean histories. The dissemination of Freud's works into the British discourse on psychological medicine: 1904–1911', *Psychoanalysis and History* 16(2) (2014), 153–214.

medical journals. Key figures in English intellectual life – Rivers, Russell, Tansley – plunged into dream analysis in that year.

Exactly when James – and Alix – turned their joint interest in Freud, psychoanalysis and psychology into something more serious is not absolutely certain. 'It was just before and during that war that my own interest in psycho-analysis began to become crystallized', James recalled in 1963, and he added, 'I soon learnt through common acquaintances that Joan Riviere was by way of being an authority on the subject.'[62] At some point in the middle of the war, probably in 1917, James contacted Ernest Jones, already Riviere's analyst, and was advised by him to study medicine if he wished to become an analyst. He began a medical course, probably at University College Hospital and probably in October 1917 – only to abandon it after three weeks. By February 1918 he was working at the *Athenaeum*.[63] But Sylvia Payne recorded that James undertook the two-minute walk from Gordon Square to the Brunswick Square Clinic to investigate the possibility of becoming an analyst with them and may even, like her, have had analysis there.[64] Strachey volunteered to work in one of the special neurological hospitals which were deploying a variety of psychotherapeutic techniques, including those based on Freud's psycho-analysis; however, his services were refused on the grounds that his anti-war politics constituted an insult to the soldiers under care.[65] Virginia Woolf reported in November that 'Poor James Strachey was soft as moss, lethargic as an earthworm. James, billed at the 17 Club to lecture on "Onanism", proposes to earn his living as an exponent of Freud in Harley Street. For one thing, you can dispense with a degree.'[66] So, from 1917 on, James Strachey had the intention of becoming a psychoanalyst, but had yet to decide on the right course of action to make it become a reality.

As it had been for years, sex remained a staple of Bloomsbury interest and gossip. Virginia records Lytton's report to her of a meeting of the British Society for the Study of Sex Psychology in 1918:

> Among other things he gave us an amazing account of the British Sex Society which meets at Hampstead. The sound would suggest a third variety of human being, & it seems that the audience had that appearance. Notwithstanding, they were surprisingly frank; & 50 people of both sexes and various ages discussed without shame such questions as the deformity of

---

[62] [Strachey], 'Joan Riviere', p. 229.   [63] Holroyd, p. 414.
[64] Sylvia Payne, 'Notes on James Strachey's death', BPaS Payne Papers, Miscellaneous MS/SP/03/F02.
[65] BLSP, Add. 60711, James Strachey to Lytton Strachey, 24 April 1918. See Linstrum, 'The making of a translator', p. 694.
[66] VWD I, 221, 21 November 1918.

Dean Swift's penis; whether cats use the w.c.; self-abuse; incest – incest between parent and child when they are both unconscious of it, was their main theme, derived from Freud. I think of becoming a member. It's unfortunate that civilisation always lights up the dwarfs, cripples, & sexless people first. And Hampstead alone provides them. Lytton at different points exclaimed *Penis*: his contribution to the openness of the debate. We also discussed the future of the world; how we should like the professions to exist no longer; Keats; old age, politics, Bloomsbury hypnotism – a great many subjects.[67]

In Bloomsbury circles, James appeared to be the leader of the serious interest in psychoanalysis, even though Lytton was reviewing Freud for the *New Statesman* in early 1917.[68] James eventually acquired the post of drama critic for the *Athenaeum*, the literary magazine then coming increasingly under the control of Keynes and his Cambridge circles, in March 1919.[69] In April, Karin and Adrian Stephen were consulting him and Alix about psycho-analysis: out of these conversations their own parallel project to become analysts was kindled. In May 1919, the two couples and many others attended the much-talked-about lectures by Bertrand Russell in Bloomsbury, lectures which were to become *The Analysis of Mind* and in which Russell drew upon psychoanalytic findings for crucial starting-points for his argument: 'the touchstone of virtue with [Norton, Alix and James] is whether you attend Bertie's lectures or not'.[70] The couples also had further long discussions about the possibility of becoming psychoanalysts, which led to Karin and Adrian starting medical training in the autumn of 1919. It took James and Alix a little longer to get to the point, but when they did, it was in style.

In May 1920, James was staying with Harry Norton at Merton House, at the back of St John's College, Cambridge.[71] Two summers previously, André Gide, also Norton's guest, provided a vivid portrait of the ambience of the house:

> Never in my life have I lived in such comfort . . . I have two servants waiting on me. I write in front of a Picasso in the latest style and an admirable

[67] VWD I, III, 21 January 1918.
[68] Lytton Strachey to Dora Carrington, 31 January 1917, LLS, p. 339.
[69] VWD I, 254, 15 March 1919. See Chapter 8 for more details.   [70] VWD I, 273, 16 May 1919.
[71] Norton was a Fellow of Trinity College. The house, which sits at the back of both St John's and Trinity Colleges, was owned, together with the School of Pythagoras and Merton Hall, by Merton College, Oxford between 1270 and 1959 (see Peter Linehan (ed.), *St John's College, Cambridge: A History*, Woodbridge: Boydell Press, 2011, pp. 581–3, esp. n93 on its sale to St John's College in order to make possible the building of the Cripps Building in the early 1960s). The same house had been used by Henry Sidgwick, also a Fellow of Trinity, in 1872–75 to provide interim accommodation for a group of women students before they decamped to the first purpose-built buildings of Newnham Hall, later Newnham College.

Persian vase, turning my back as best I can on an incomparable library, where a first edition of [Apollonius'] *Treatise on Conic Sections* nestles besides first editions of the Elizabethans. The plonk I am served is a Mouton-Rothschild '78.[72]

This was the Cambridge world to which James was accustomed. But, aged thirty-two, what had he achieved? 'A discreditable academic career with the barest of B.A. degrees, no medical qualifications, no knowledge of physical sciences, no experience of anything except third-rate journalism' was his verdict forty years later.[73]

On 31 May, James drafted a letter to Freud on Merton House letter paper:

Dear Sir,

I believe that Dr Ernest Jones has mentioned to you that I am anxious if possible to arrange to go to Vienna to be analysed by you. He will, I expect, have explained that my object in doing so is to try to obtain the essential empirical basis for such theoretical knowledge of psychoanalysis[74] as I have been able to derive from reading. For this purpose I should be prepared to remain in Vienna for at least a year. I understand from Dr Jones that you will not in any case have a vacancy until the autumn; but I am venturing to write to you on his advice to enquire whether you will then be willing to take me. I fear however that the financial question may prove an obstacle. In face of the possibility and (from my point of view) the desirability of a prolonged analysis, allow me to afford a fee of more than one guinea [inserted above line: an hour] in English currency.

I do not know whether a satisfactory arrangement would in these circumstances be possible; but, must add the expression of a desire, which it would not be easy to exaggerate, that I may be fortunate enough to obtain the benefits of your personal teaching in psychoanalysis.

Yours faithfully,
James Strachey[75]

---

[72] André Gide, diary entry, 2 September 1918 and letter to André Ruyters, 21 August 1918, cited in David Steel, 'Gide à Cambridge, 1918', *Bulletin des Amis d'André Gide* 125 (janvier 2000), 11–74 at 43–4.

[73] Quoted in M.R. Khan, '*The Psychoanalytic Study of the Child*. Vols 34 and 35: Edited by Albert J. Solnit et al. New Haven: Yale University Press. 1979/1980', *IJP* 63 (1982), 98–100 at 99–100.

[74] Note: no hyphen!

[75] BLSP 60713, Vol. LIX, 1903–October 1924, Strachey to Freud (draft), 31 May 1920.

Strachey's family knew of this plan: three days later, somewhat prematurely, given that Freud had not yet sat down to reply to James's letter, his sister Dorothy told André Gide that James was going to be analysed by 'le grand Freud en personne'.[76] On the 4 June, Freud replied to James:

> The obstacle you mention so frankly is not an absolute one. In fact, it would not have existed at all before the war. But now you know, things have changed much for the worse. I have become very poor and must work hard to make the two ends still meet. So I would not accept a patient for the fee of one guinea, but the case of a man who wants to be a pupil and to become an analyst is above these considerations. As long as the English pound continues to be equivalent to 600 Kr or about so, I am ready to take you as you propose, and I am glad you are allowing yourself so spacious a term for the work as most people who want analysis do sin against this postulate.
>
> Pray do not forget to apply again at the end of Sept when I come back to Vienna.[77]

That same day James and Alix married in preparation for the summer travels through Europe that they had long planned, and which would now include staying on in Vienna. James started analysis with Freud on 4 October 1920. Soon after, following Alix's 'palpitation attacks' while at the opera, James requested, at her behest, that Freud analyse her as well. He agreed, although he had originally thought it a 'technical impossibility' to analyse husband and wife. While reporting the start of this analytic experiment to Lytton, James also gave him a stylish description of what it was like to be analysed by Freud:

> Each day except Sunday I spend an hour on the Prof.'s sofa (I've now spent 34 altogether), – and the 'analysis' seems to provide a complete undercurrent for life. As for what it's all about, I'm vaguer than ever; but at all events it's sometimes extremely exciting and sometimes extremely unpleasant – so I daresay there's *something* in it. The Prof himself is most affable and as an artistic performer dazzling. He has a good deal rather like Verrall in the way his mind works. Almost every hour is made into an organic aesthetic whole. Sometimes the dramatic effect is absolutely shattering. During the early part of the hour all is vague – a dark hint here, a mystery there –; then it gradually seems to get thicker; you feel dreadful things going on inside you, and can't make out what they can possibly be; then he begins to give you a slight lead; you suddenly get a clear glimpse of one thing; then you see another; at last a whole series of lights break in on you; he asks you one more question; you

---

[76] D.A. Steel, 'Escape and aftermath: Gide in Cambridge 1918', *Yearbook of English Studies* 15, Anglo-French Literary Relations Special Number (1985), 125–59 at 156.

[77] BLSP 60667, Vol. XIII, Freud to Strachey, 4 June 1920.

give a last reply – and as the whole truth dawns upon you the Professor rises, crosses the room to the electric bell, and shows you out at the door.

That's on favourable occasions. But there are others when you lie for the whole hour with a ton weight on your stomach simply unable to get out a single word. I think that makes one more inclined to believe it all than anything. When you positively feel the 'resistance' as something physical sitting on you, it fairly shakes you all the rest of the day.[78]

Deep in research on Queen Victoria, Lytton's response also had an inimitable blithe style:

Your account of the Doctor sounds odd – personally I have never been able to believe that *I* should suffer from *any* 'resistance' – but one never knows. Does he ever make jokes? Or only German ones? And in what language does it all go on? I wish to God he could have analysed Queen Victoria. It's quite clear, of course, that she was a martyr to analeroticism: but what else? What else?[79]

From the beginning of Strachey's analysis, both Jones and Freud envisaged making him a translator of psychoanalytic works into English. When Jones had written to Freud on his behalf, on 7 May 1920, he had noted: 'He is a man of 30, well educated and of a well-known literary family (I hope he may assist with translation of your works), I think a good fellow but weak and perhaps lacking in tenacity.' Within days of James's analysis beginning, Jones had suggestions for his translation projects: 'Will you or Rank suggest that he does Federn's *Vaterlose Gesellschaft*,[80] which is already partly translated.'[81] And by early November 1920 Strachey had clearly passed some tests of his translator's skills for Freud: 'Strachey seems to prove a good acquisition. His translations to me seem excellent, I am ready to give him the *Jenseits*.'[82] Between the Sunday when he wrote this and the Thursday when Freud and Rank penned a Rundbrief, Rank had probably informed Freud that *Jenseits* had already been assigned by Jones to one of his analysands, Miss C.J.M. Hubback.[83] So the Rundbrief reported that: 'In Mr. Strachey aus Cambridge, der sich jetzt in Wien aufhält, haben wir einen sehr guten und tüchtigen [very good and capable] Übersetzer gefunden; er soll jetzt eine Arbeit von Professor übernehmen, wahrscheinlich:

---

[78] BF, James Strachey to Lytton Strachey, 6 November 1920, pp. 31–2.

[79] LLS, Lytton to James, 24 November 1920, p. 476.

[80] Paul Federn, *Zur Psychologie der Revolution: die vaterlose Gesellschaft*, Vienna: Anzengruber, 1919.

[81] JF, 17 October 1920, pp. 394–5. Federn's famous essay has, as of 2015, yet to be translated into English, though its sub-title was made famous by Mitscherlich's *Society without the Father: A Contribution to Social Psychology* (1963), an influential essay for the men's movement.

[82] FJ, 7 November 1920, p. 398.

[83] JF, 12 November 1920, p. 399 (though Jones did not mention Hubback by name).

ein Kind wird geschlagen.'[84] And so it was that 'Mr Strachey aus Cambridge' was assigned 'A child is being beaten'; by February, he had completed the task, and Freud judged him 'excellent but apt to fall into laziness if not admonished'.[85] Freud was now planning to give him the Five Case-Histories; as things turned out, Freud started him immediately on the newly written *Group Psychology and the Analysis of the Ego*, 'as he is near me and I can collaborate with him';[86] but once this task was complete, the plan for both Stracheys to translate the case-histories was hatched. As the analysis drew to the conclusion of its first year, Freud clearly came to feel affection and respect for James, taking him very much under his own wing and politely insisting to Jones that he recognize his value:

> I found [Strachey's translation of *Massen*] absolutely correct, free of all misunderstandings and I hope the rest will prove the same. I am no judge of the style, it seemed to me plain and easy, your claims for elegance may be stronger than mine. In any case, don't be too hard on him, it is not easy for us to get efficient translators ...
>
> Strachey and his wife might become very useful to you. They are exceptionally nice and cultured people though somewhat queer and after having gone through their analysis (what is not yet the case) they may become serious analysts. Perhaps he is the man to assist you in editing the *Journal* and both would be fit to take an active part in the management of the Brunswick place. This is left to their further development, for the moment their conviction is not yet completely assured. I have to warn you that they are rather sensitive and critical. But they are not to be rebuked, they are good stuff.[87]

---

[84] CFC/FO5/11, Vienna, 11 November 1920. 'In Mr. Strachey of Cambridge, who is already resident in Vienna, we have discovered a very good and capable translator; he should be ready to take on some work from the Professor, probably: "a child is being beaten".'

[85] FJ, 7 February 1921, p. 409.

[86] FJ, 12 April 1921, p. 419. It should be noted that Strachey introduced the neologism 'cathexis' in his translation of *Group Psychology*; Freud's interest in collaboration on the translation and his eventual verdict (quoted) are therefore of interest if one thinks it of importance whether or not Freud approved of 'cathexis' as the translation for 'Besetzung' before the translation was finalized. Barbara Caine claims that the Stracheys began to translate Freud's work before they even arrived in Vienna; 'indeed, they arrived with their first translation of *Group Psychology* almost complete' (Barbara Caine, 'The Stracheys and psychoanalysis', *History Workshop Journal* 45 (1998), 145–70 at 151; the same claim is made in Caine, *From Bombay to Bloomsbury*, p. 258). This claim is certainly erroneous: Freud's writing of *Massenpsychologie* was not *completed* until February 1921, six months after James's analysis began, and was published in German only in June–July 1921. The mistake arises probably from her misreading of a letter from Strachey to Jones, dated 28 April 1920 concerning a translation – but the text in question to be translated was not one of Freud's, but the *Tagebuch eines halbwüchsigen Mädchens*, published by the Psychoanalytische Verlag in 1919; Strachey wished to know if it was being translated already and, if not, to whom the copyright belonged (see CSD/F03/01). A translation was in fact under way, done by Eden and Cedar Paul; the book was published in London by Allen & Unwin in 1921.

[87] FJ, 14 July 1921, p. 431. Freud wrote this letter in English.

With such signs of Freud's approval, Jones also warmed to them – 'both very attractive and cultured people'[88] – and Freud used their interchange of views on the singular and eccentric Stracheys to make a crucial point about the future of psychoanalysis: 'I am glad you liked the Stracheys, you will see we can get the best people for ΨA if we only drop the professional condition soon enough, for it will show unavoidable in the long run.'[89] Freud was implicitly admonishing Jones for his persistent twin desires to attempt to restrict psychoanalysis to medically qualified men and to achieving respectability. Core members of the Establishment ranged along the Cambridge–London axis, literate and distinctly unmedical, but entirely lacking in respectability – the Stracheys were precisely the sort of followers Freud wanted and was confident psychoanalysis would get, a confidence that would prove misplaced as the 1920s drive towards professionalization and respectability intensified.

Freud was fitting out Jones with the wherewithal to run an efficient psychoanalytic organization in London: 'I am glad you like Rickman and intend to make him the helpmate you need so much ... I make excellent progress with Alix Strachey and expect to succeed with James Str[achey], both of whom may become highly valuable members of your staff.'[90] Jones now proposed to make Rickman co-director of the Psychoanalytic Press.[91] The Stracheys were already committing themselves to the work of the 'Glossary of Psycho-analytic Terms', a project Ernest Jones had envisaged in 1918 when he added elements of such a glossary to the second expanded edition of his own *Papers on Psycho-Analysis*.[92]

Over the summer of 1921, James and Alix had substantially added to Jones's glossary[93] and were beginning to confront the task of producing a considered set of standard English terms for translating Freud's German – as is evident from James's lengthy disquisition, in a November 1921 letter to Jones, concerning the problem of 'Besetzung'.[94] Jones hammered home to Freud the importance of writing:

---

[88] JF, 22 July 1921, p. 432.   [89] FJ, 27 July 1921, p. 434.   [90] FJ, 6 November 1921, p. 443.
[91] JF, 15 June 1921, p. 430.
[92] See the invaluable and comprehensive treatment of the early history of this project in R. Steiner, 'To explain our point of view to English readers in English words', *IRP* 18 (1991), 351–92. Perhaps behind Jones's emphasis on English to the exclusion of German there was a strategy that took account of the general chauvinism of English culture at this point during the war. And many of the terms, perhaps most, were not strictly psychoanalytic terms but rather general psychiatric and sexual terms (e.g. Jones defines 'masturbation', 'coitus interruptus', 'voyeurism').
[93] JF, 10 August 1921, p. 436.
[94] See in particular the letter Strachey wrote to Jones dated 27 November 1921, reprinted and discussed at illuminating and controversial length in D.G. Ornston, 'The invention of "cathexis" and Strachey's strategy', *IRP* 12 (1985), 391–8 at 393 and after; see also D.G. Ornston, 'Strachey's

> A knowledge of good English is almost unbelievably rare here, and of course rarer still in America, and is valued correspondingly highly. It is difficult to convey to anyone not English how completely a man is estimated here by his speech and his writing. The average doctor writes worse English than a poor tradesman writes German in Austria. Last week, for instance, I had the occasion to read for the first time Brill's translation of your Leonardo, and I was deeply shocked time and again to see punctuation as illiterate as that of a servant girl's, with expressions of a similar order. Men of sensitive feeling, taste and education like Rickman and Strachey rightly shudder at such things.[95]

The first English translations of Freud's work were done by A.A. Brill in the United States following his agreement in 1908 to be Freud's translator; so it was imported books, *Selected Papers on Hysteria and Psychoneuroses* (1910/12), *Three Contributions to the Sexual Theory* (1910), *The Interpretation of Dreams* (1913), *The Psychopathology of Everyday Life* (1914) and *Wit and its Relation to the Unconscious* (1916), that Strachey and others were reading before and during the First World War. Jones's important first book of psychoanalytic papers was published by a conventional publisher with strengths in medicine, Baillière, Tindall & Cox, in December 1912. After the war, Freud, Sachs and Rank founded the Internationaler Psychoanalytischer Verlag, and Freud fully expected Jones to bring English-language psychoanalysis, of whose increasing importance he was only too aware, under the umbrella of the Verlag (as it was always known).

Jones was willing as ever to agree to Freud's policy and to bring his great organizational skills to bear in implementing it. So, in the same spirit of independent intellectual entrepreneurship, he opened a shop – in New Cavendish Street in London, close to the medical world of Harley Street – to sell psychoanalytic books published by the Verlag and related material. In order to facilitate the coordination between the Verlag based in Vienna and the publishing activities in Britain and other English-speaking countries, Jones sent Eric Hiller, a Mancunian non-medical analytic acolyte of his, a founder member of the BPaS in February 1919, to Vienna to coordinate, translate and mediate on his behalf between England and the psychoanalytic centre. In the very early 1920s, Hiller was joined in this task by the steady stream of English analysands – the Stracheys, Joan Riviere and John Rickman were the most important. Jones's letters to Freud, and the Rundbriefe of the early 1920s, indicate how much time and energy were

influence: a preliminary report', *IJP* 63 (1982), 409–26 and 'Freud's conception is different from Strachey's', *Journal of the American Psychoanalytic Association* 33 (1985), 379–412.
[95] JF, 15 December 1921, p. 448.

devoted to this coordination of London and Vienna and how conflicts arose in which different parties were periodically blamed for the difficulties – Hiller, the Stracheys with their slow, careful methods, even Jones, with his abrasiveness and his compulsion to control every stage of the translation and publishing process.[96] In these years, Jones regarded himself as an 'amateur publisher',[97] principally preoccupied with the dissemination of psychoanalysis in Britain through the Psychoanalytic Press and the translations he and his growing group of reliable translators and proof-readers were engaged on – Riviere, the Stracheys, Hubback, Flügel, Stoddart and Low, with 'no fewer than thirteen people translating the *Sammlungen*, scattered from India to America'.[98] The hope was to produce what he already called, in 1920 and 1921, a 'Standard Edition' of Freud's writings[99] and in the process wrest control of translations of Freud into English from the American Brill. From 1922 on, Joan Riviere, in Freud's clear opinion obviously 'the strongest personality among them',[100] was given the official position of Translation Editor for the *International Journal*, though it was James Strachey's translations Freud singled out as 'done with the utmost care for style and truthfulness'.[101] Dora Carrington would not have been surprised by Freud's praise of the virtues of James's translations: in the terrible days in early January 1932 when Lytton was dying of cancer, she expressed her admiration for him: 'James is such a truthful *exact* person that I believe everything he says and looks.'[102]

James had been targeted by Jones as a potential translator even before he had met Freud.[103] But it was Freud, James and Alix together, independently of Jones, who arranged for Alix to join him in this task in early 1921; by 9 March the plan had become for James and Alix together to translate the case-histories.[104] By 19 May, James was translating *Massenpsychologie* and the 'Wolfman', Alix 'Dora' and the 'Ratman'; these three case-histories were almost complete by the summer, when they had also completed a draft of the *Glossary of Psycho-analytical Terms for the Use of Translators* published in 1924. But the Stracheys were not only translators. By January 1922, James could write to Lytton: 'we've been passed as fit to practise by

---

[96] See FJ, 6 April 1922, p. 468 for Freud's brusque diagnosis that the 'wheel in the machinery' causing all the trouble was Jones himself.
[97] Jones to Rank, 1 April 1921 and 14 February 1922, quoted in Steiner, 'To explain our point of view', p. 387.
[98] JF, 10 April 1922, p. 472.
[99] Steiner, 'To explain our point of view', p. 388; see also Jones to Rank, 26 February 1920, CRA/FO6/09: 'our Standard Edition of collected works which I hope one day the Press will publish'.
[100] FJ, 16 April 1922, p. 475.   [101] FJ, 4 June 1922, p. 486.   [102] Holroyd, p. 672.
[103] JF, 7 May 20, p. 378.   [104] Holroyd, p. 736n10, James to Lady Strachey, 9 March 1921.

the Prof.'.[105] When the Stracheys returned to England in the summer of 1922 with Freud's full imprimatur, he issued Jones with his orders:

> As regards Rickman and the Stracheys I send them back to you within a week. Both will prove of great help to you if you treat them generously. I propose the Stracheys should become members (full) of the Society as they have gone through 1½ years of serious analysis, are theoretically well informed and people of a high order. To be sure their conflicts have not been decided, but we need not wait so long, we can only instigate the processus which has to be fed by the factors of life. Becoming full members – as well as Rickman – would bind them to the interests of the Society. Stracheys are likely to remain in England next winter. Do not put back her for him, she is very valuable.[106]

At the Annual General Meeting of the British Society in October 1922, Jones did indeed try and get both Stracheys elected as full members, but, as he had anticipated,[107] the proposal was voted down by the Society, which declined to accept them as members but duly elected them associate members.[108] They became full members the following October. But the Stracheys did not get as involved in the work of the British Society as Jones might have hoped. Alix's serious illness in early 1922 had provoked a real scare, with Bloomsbury (Carrington) and family (Sargant Florences) rushing from England to Vienna to nurse her and look after James.[109] So, in view of Alix's still fragile health, they spent much of the spring of 1923 travelling in the south – in Algeria with Lytton, Carrington and Ralph Partridge, then on to Tunis, Palermo, Naples and Rome.[110] Jones reported to Freud in February 1924 on their standing in the psychoanalytic London world:

> The Stracheys are harder people to get close to, and have also been away from London a great deal on account of her health, but behave quite correctly. They have been to dinner a few times, he oftener, and also for a visit to Elsted, I have sent him a couple of patients and also some books for review, and recently got him to assist in our company meetings. He is of course chiefly engrossed in his translation work, and makes no other contribution, e.g. to the work of the Society.[111]

---

[105] Holroyd, p. 736n10, 22 January 1922. [106] FJ, 25 June 1922, pp. 491–2.
[107] 'The Stracheys will also certainly be made associate members, and they shall be made full members if I can over-ride the rule to the effect that members must have been associate members first for at least a year' (JF, 19 July 1922, p. 494).
[108] BPaS, Minute Book, 4 October 1922. [109] Holroyd, p. 505; LLS, p. 505.
[110] Holroyd, pp. 521–2. [111] JF, 9 February 1924, p. 538.

But it was just at this time that James would intervene in Jones's imperial plans with considerable consequence.

The post-war plan for the Verlag in Vienna to print English-language psychoanalytic works had foundered on personal misunderstandings and business difficulties, not least because of the severe economic uncertainties and hyperinflation in both Germany and Austria. By the summer of 1923, it was clear to Jones that English-language psychoanalysis would have to find another organizational form, independent of Vienna or Berlin – an organizational secession negotiated by Rickman and Jones acting together. Rank had begun to behave more and more mysteriously, unpredictably, unreliably, as he and Ferenczi put forward their technical novelties in *The Development of Psycho-Analysis*, as he prepared his innovative theory of *The Trauma of Birth*. He and his fellow Committee members were also responding to the shattering news of Freud's first operation in the spring of 1923, the dire prognosis emerging over that summer leading to the serious operations for jaw cancer he underwent in the autumn. By the Salzburg conference of April 1924 – at which it was decided on the final day that the next IPA Conference would be held in Cambridge – Jones farsightedly saw that London and Berlin would be the new centres of psychoanalysis.[112] Or perhaps he was determined that this was how it should be. Jones may have said he would wait on Rank's enquiries about finding an American publisher for international publishing in English, but he had long been at war with Rank and now seized the opportunity James Strachey offered him to separate definitively from Rank's Viennese operations and establish England as the centre for psychoanalytic publishing in English. The new format for the English institutions, put in place in 1924 through the timely interventions of James Strachey and John Rickman, would have a long psychoanalytic life.

In 1917, Leonard and Virginia Woolf had established the Hogarth Press in their house in Richmond, partly as recreation from reading and writing, partly as a therapeutic regime for the troubled Virginia, whose mental balance Leonard was always to supervise and tend in the most remarkably loving and attentive fashion. Afternoon hours could now be devoted to typesetting and printing, more efficacious than any of the nostrums of the alienists, specifically Savage and Craig, under whose care she had been placed. In mid-March 1924, they moved to the centre of London, to 52 Tavistock Square, close once more to their friends. They brought the Press

[112] Phyllis Grosskurth, *The Secret Ring: Freud's Inner Circle and the Politics of Psychoanalysis*, London: Cape, 1991, p. 159.

with them to the lower part of the house. Alix Strachey had been in on their foundation of the Press, acting as their very first assistant – for all of the three hours Virginia recorded it took her to decide this job was not for her:

> Alix solemnly & slowly explained that she was bored, & also worried by her 2 hours composing, & wished to give it up. A sort of morbid scrutiny of values & of motives, joined with crass laziness, leads her to this decision; as I expect it will lead her to many more. She has a good brain, but not enough vitality to keep it working. The idea weighed upon her, & I assured her there was no need for it to weigh.[113]

By 1924, the Hogarth Press had become an important and respected small house for the publication of 'advanced' literature and ideas, often of the Woolfs' friends – amongst them Katherine Mansfield, T.S. Eliot, Roger Fry, E.M. Forster, J.M. Keynes, Bertrand Russell and Robert Graves. Jones's efforts to find an English alternative to the Viennese Verlag for publishing psychoanalysis in Britain were foundering: he had tried a whole slew of publishers, starting with Cape in the autumn of 1923; by April 1924 he was in negotiation, as his final hope, with Cambridge University Press. So with Jones's plans in disarray, James Strachey provided the bridge between Jones and Leonard Woolf, who noted: 'Some time early in 1924 James asked me whether I thought the Hogarth Press could publish for the London Institute.'[114] The date must have been in very late April or early May; on 26 May 1924, eight days after Virginia had given her talk to the Heretics in Cambridge, James Glover, deputizing for Jones and Rickman,[115] came to Tavistock Square to discuss with Leonard and Virginia the proposal to take on the Psychoanalytic Series.[116] On 30 May, Jones reported to Freud: 'After a good deal of bargaining we have come to a definite agreement with the Hogarth Press here.'[117] The chief selling point for the Woolfs, Leonard later remembered, was the prospect of publishing

---

[113] VWD I, 61, 16 October 1917.
[114] Leonard Woolf, *Downhill All the Way: An Autobiography of the Years 1919–1939*, London: Hogarth, 1967, p. 164.
[115] Jones was in Italy.   [116] VWD II, 302.
[117] JF, 30 May 1924, p. 545. Jones indicated that Rickman was the natural negotiator with the Woolfs, but at this time was out of the country: 'When the Institute of Psycho-Analysis was formed, of which he was a member of the Board of Directors, Glover was entrusted, during Dr. Rickman's absence abroad, with the delicate negotiations concerning the transfer of the International Psycho-Analytical Library to the present publishers, the Hogarth Press, and he skilfully brought these to a successful issue' (E. Jones, 'James Glover 1882–1926', *IJP* 8 (1927), 1–9 at 4). As Jones's co-director at the Press, Rickman would have usually played the key role in the negotiations, as he had been doing increasingly in legal and business arrangements since his return from Vienna in 1922; see for example Rickman to Jones (JPBPaS, CRA/F14/04 22/10/10) concerning American law governing journals sent through the post.

the two volumes by Freud already printed by the Verlag in English (*Beyond the Pleasure Principle* and *Group Psychology*) and the agreement to publish the four-volume *Collected Papers*, on which James, Alix and Joan Riviere were already hard at work.

The contract with Hogarth promised delivery of the manuscripts (or unbound pages) of four volumes of Freud's *Collected Papers* by 1 September 1924.[118] Now, instead of the Verlag printing and binding in Vienna and using a large English publishing house for distribution, they agreed for Hogarth to take over all functions.[119] And so psychoanalytic publishing in Britain became annexed to literary Bloomsbury rather than to the medical publishers, the university presses or the larger commercial publishers. The alliance marked what was perhaps the inevitable failure of Freud's vision of a genuinely international publishing house for psychoanalysis, one in line with the internationalist wartime hopes he had shared with Einstein and Romain Rolland; one, too, as one of the American leaders of psycho-analysis William Alanson White put it, that could be under the centralized control of the 'Pope in Vienna'. The new arrangements for publishing Freud and other psychoanalytic writings were part of the larger institu-tional arrangements being put in place for British psychoanalysis: it became necessary to create the Institute of Psycho-Analysis, so that, as a company (unlike the Society), it could legally enter into the financial arrangements necessary for managing relations with the Hogarth Press. John Rickman, the architect of these legal and administrative arrange-ments, took the opportunity to create an institutional shell within which the new clinic could also be established.[120]

As a core member of Bloomsbury, James Strachey played a sustained and central part, perhaps the main part, in the distinctively *English* literary culture of psychoanalysis: first as one of the key translators of Freud and other psychoanalytic writers into English, forming with Jones, Riviere and Alix the Glossary Committee of the British Society; then with his key intervention at a timely moment in 1924 in establishing a long-term link between British psychoanalysis and the Hogarth Press. One of Jones's

---

[118] JF, 29 September 1924, p. 554.
[119] However, Hogarth did not take on the publication of the journal, for which Jones found a parallel stable arrangement with Baillière, Tindall & Cox as of September 1924.
[120] Pearl King, 'Background and development of the Freud–Klein controversies in the British Psycho-Analytical Society', in FK, p. 11: 'The Institute of Psycho-Analysis was set up as a company in British law in 1924, largely through the initiative and energy of John Rickman, in order to deal with financial and other matters concerning book publication, and especially to facilitate the publication of books in the International Psycho-Analytical Library series, with the Hogarth Press, which in 1924 thus became joint publishers with the Institute.'

highest priorities had been to seize back an *English* Freud from the Americans, in particular from the fast-working Brill whose translations appalled so many.[121] James's role in securing what now promised to be an ideal arrangement for publishing Freud and psychoanalytic books in England may have contributed to Jones's assessment in August 1924 of his team: 'Strachey also, though terribly slow, has improved in his work so much that he now ranks as easily the best translator here or in America, much better than either Mrs. Riviere or myself.'[122] Yet, once he and Alix had completed the task of translating the *Five Case-Histories*, the persistently self-effacing Strachey never put himself forward as the pre-eminent translator, not even of Freud. He would always have to be persuaded to take on any such new tasks. None the less, he became a much sought after path of access to Freud and the truths of psychoanalysis, and indeed something of a psychoanalytic celebrity. If one could find the right way of approaching him, however, James would facilitate contact between Freud and the literary world, Bloomsbury and beyond, in which he had moved for so many years. André Gide had quite probably heard of Freud's writings from James in the summer of 1918, when he had been taken in by the Strachey family, lodging in Cambridge for the summer at Merton House and also in Grantchester, communing with Rupert Brooke's absence. He thus met James's older sister, Dorothy Bussy, who became first his English teacher then his sole English translator and intimate friend for the rest of his life. In early 1921, Gide read Freud for the first time, the 'Five Lectures', published in French in the *Revue de Genève*; he asked Dorothy to request James for advice on contacting Freud about translating his work into French under the auspices of *La nouvelle revue française*. James's lengthy response then paved the way for Gide's own greater interest in psychoanalysis and the long history of the NRF's publication of Freud's work in French.

It was clearly James who led the way in the decision to seek analysis with Freud in 1920; it was never clear if Alix was interested in an analysis or in becoming an analyst. But once she was in analysis Freud certainly treated her as an equal to her husband, had great respect for her qualities, and by July 1921 was urging Jones to do likewise. As we have seen, Alix slid into analysis with Freud and then into translating Freud, starting with 'Dora' and the 'Ratman'. There is every indication that, allergic to dull work

---

[121] And still do: 'an appallingly dull and inaccurate translation' (Richard D. Chessick, *The Future of Psychoanalysis*, Albany: State University of New York Press, 2007, p. 51).
[122] JF, 12 August 1924, p. 550.

though she was, her philological drive was highly developed – in 1918, the odd visitor to Lytton and Carrington's Tidmarsh might find her 'puzzling over Rabelais with the aid of six dictionaries'.[123] Alix was the first of the Stracheys to go into print as a psychoanalytic author, with two papers, both very personal and auto-analytic, neither of which has been referred to by biographers or other commentators.[124]

In June 1922, A.S. Strachey published a paper entitled 'Analysis of a dream of doubt and conflict' in the *International Journal of Psycho-Analysis*.[125] While the patient and analyst depicted in the paper are not named, it is clear that it is a record of an episode from her own analysis with Freud. The paper recounted a series of three interconnected dreams, dreamed in one night, and their subsequent interpretation; all three dreams were preoccupied with pregnancy and childbirth. Hard at work translating Freud's case-history of 'little Hans',[126] she had identified with little Hans's concern about the blood plainly visible in the bedroom after his little sister's birth. She was also discovering, in parallel with little Hans, her own idiosyncratic childhood theory of conception – in her case, the belief that eating cake or taking pills would lead to pregnancy. The pressure of these concerns had, it turned out, been provoked by Freud's active intervention: believing that there was a psychogenic element to her chronic constipation for which she regularly took laxatives, he had suggested she cease taking the pills, for the purpose of releasing material into the analysis. The dream was 'a reaction to this abstinence.'[127]

The dream analysis was working on two distinct levels: firstly, there was her scepticism about psychoanalysis, specifically with regard to the existence of infantile fantasies concerning childbirth; secondly, there was her own emotional conflict over the prospect of pregnancy and birth.

One of the most elegant features of the analysis was the shift from questions of translation which occupied the manifest dream-content to the underlying preoccupation with her fear of the pains of childbirth. The text of the second dream was:

---

[123] Holroyd, p. 414.

[124] But see John Forrester, 'Remembering and forgetting Freud in early twentieth century dreams', *Science in Context* 19(1) (2006), 65–85.

[125] A.S. Strachey, 'Analysis of a dream of doubt and conflict', *IJP* 3 (1922), 154–62.

[126] Since Alix had been occupied translating 'Dora' and the 'Ratman' in the spring and summer of 1921, it is reasonable to infer that the events in her analysis of which she writes took place in the period between October 1921 and February 1922, when she fell ill.

[127] Strachey, 'Analysis of a dream of doubt and conflict', 159.

I was correcting some manuscripts and asked my husband, concerning some word, whether it would 'wo' in italics, or whether it was 'wo' in italics – I can't exactly remember the words I used. I think I asked him more than once, and did not use exactly the same words each time.

The term 'wo' was connected to 'woe' and thus to the cognate German term '*Weh*' ('pain'); she added 'that in that translation there had been a description of childbirth and that the word "Wehen" had been used'. There is only one passage in Freud's work that fits this description: a graphic description of the birth of little Hans's sister and his reaction to the event.[128] Thus behind the manifest textual preoccupations with German translation and typographical accuracy there was a latent level of emotional preoccupation: she was asking herself 'whether the pains of childbirth are very excessive in reality, or are only supposed to be so'. The bridge from translating Freud's account of little Hans to her own preoccupations was accomplished with a clever verbal play: the manifest '"*wo*" in italics' conceals and allows into consciousness the latent '"wo" in it, Alix'.

'Is this "wo" (i.e. childbirth) in it, Alix?', meaning 'Are your thoughts engaged upon the subject of giving birth to a child?' ... We therefore see that in its more general aspect – taken, that is, in relation to the patient's analysis – the dream is of great importance as marking the step from a repudiation of an unconscious phantasy to an acceptance of it.[129]

In the final dream, the figure of her own mother emerges, together with her own scepticism about the reality of her mother's illnesses – and the reality of the pain of her mother's childbirth.

It is quite likely that these dreams, interesting and momentous enough to publish in a concealed form, prompted the following scene in her analysis which she later recounted to Masud Khan:

It had been a critical week in her analysis, which resulted in her having a significant dream. She recounted her dream to the Professor and they worked around it. Then the Professor gave an interpretation, at the end of which he got up to fetch a cigar for himself, saying: 'Such insights need celebrating.' Alix Strachey mildly protested that she had not yet told the

---

[128] Freud, *GW* VII, 247; *SE* X, 10: '"At five in the morning," he writes, "labour began [*mit dem Beginne der Wehen*], and Hans's bed was moved into the next room. He woke up there at seven, and, hearing his mother groaning, asked: "Why's Mummy coughing?" Then, after a pause, "The stork's coming to-day for certain." ... After the baby's delivery ... he was called into the bedroom. He did not look at his mother, however, but at the basins and other vessels, filled with blood and water, that were still standing about the room. Pointing to the blood-stained bed-pan, he observed in a surprised voice: "But blood doesn't come out of my widdler."'

[129] Strachey, 'Analysis of a dream of doubt and conflict', 161.

whole dream, to which the Professor replied: 'Don't be greedy, that is enough insight for a week.'[130]

As so often with Freud, the memory of his off-hand remark might have summed up all the complex interpretations they had been engaged upon: 'Don't be greedy!' The first dream was a surrealistic case of the missing cake:

> *I had eaten a slice of cake that had been put by in a tin. My husband commented on the fact. I replied that he would still find the slice there; that it was not eaten. He again pointed out that I had eaten it. I wanted to tell him that I thought I had only eaten it in my dream; but all I could say was that I had somehow not really eaten it, and that he would still find it in the tin.*[131]

The night before the dream James had suggested she not eat cake, since it might interfere with her digestion, so she had put the cake aside. The 'unconscious' question being posed by the dream was whether or not she could get pregnant by eating cake. The dream affirmed that she had *in reality* a fantasy that she could.

Alix's serious illness in early 1922 had caused the breaking off of her analysis and then extended into a protracted convalescence in 1923. By 1924, she was adamant about her need for further analysis; in September she moved to Berlin to have analysis with Karl Abraham, partly because she had viewed Freud as an unsatisfactory analyst. While there she threw herself into the vigorous life of the Berlin Psychoanalytic Society and its associated Policlinic. Abraham reported to Freud on how this second analysis was proceeding:

> It struck me from the very beginning that the long years of work with you are like extinguished. We have to discover everything afresh, as all the facts elicited by the first analysis have disappeared, while the general knowledge of psychoanalysis is intact. You will remember that the patient lost her father in the first weeks of her life, and has no memories of her own of him. Besides other reasons for the amnesia there is a complete identification of your person with her father she does not remember either of you. On the other hand she has directed to you the same rescue phantasies as to her father.[132]

Curiously enough, another repetition took place – while in analysis with Abraham, Alix published a psychoanalytic paper: a two-paragraph observation in German, entitled 'Eine Zeugungstheorie'.[133] A fifteen-year-old

[130] M.R.R. Khan, 'Mrs Alix Strachey (1892–1973)', *IJP* 54 (1973), 370.
[131] Strachey, 'Analysis of a dream of doubt and conflict', 154.   [132] FA, 20 October 1924, p. 519.
[133] *International Zeitschrift für Psychoanalyze* XI.H.1–2 (1925), 87. It was published by 4 March 1925 when James alerted her to its appearance (BLSP 60715, Vol. LXI, March 1925–October 1925).

girl tells a younger girl that, as she has actually observed happening, when a woman is about to give birth to a baby, her husband's genitals swell up in sympathy with the swelling of his pregnant wife. Strachey attributes this theory to a repression of knowledge of the swelling up of the man's genitals during *copulation* and its displacement on to the act of *childbirth*: the connection between erection and the birth of the child is recognized, but at the expense of any recognition of sexual intercourse, so that the man in fact makes no contribution to the development of the child.

One cannot but conclude that, not only was the topic of copulation, pregnancy and childbirth of central importance to Alix, but on two occasions she managed, while in analysis, to give birth to two psycho-analytic papers on the topic – as if by herself, without the contribution of the analysts. The analysts are, as it were, erased, just as her memory of her dead father (and of her husband, we can surmise) were. Abraham's bad luck – to repeat the cutting short of Alix's analysis in the autumn of 1925 by himself falling ill, fatally as it turned out – was also Alix's bad luck: father, Freud, Abraham. All were to be erased.

Hovering behind her dreams while in analysis with Freud were husband and her analyst, unproblematically assimilated one with the other (i.e. in relation to getting pregnant).[134] James prohibits cake-eating, Freud prohibits taking laxatives. Throughout her life, a preoccupation with food recurred again and again. As an undergraduate, she had had an anorexia-like episode, which led to her transformation from a 'beefy' young woman into that tall, gaunt, mannish and troubled figure in the photographs. She even relished drawing up character types according to their oral proclivities: 'But we want so much to make the party a success', she wrote to her mother in 1917, '& every member is a glutton – Lytton (fussy), Maynard (voracious), James (systematic), Norton (hypochondriac), Carrington (healthy) & myself (greedy).'[135] Track forward to World War II, when Frances Partridge observes her dear friend Alix guzzling mustard because it's become so expensive.[136] In the 1960s, when Paul Roazen visited the Stracheys, lunch was taken seated outside in ramshackle deck-chairs and served from tins. When Michael Holroyd visited in 1962, spam was the *pièce de résistance*, everything 'swatched in protective cellophane'.[137] And reflecting on the last weeks before Alix's death in 1973, Angela Richards – Noel Olivier's eldest daughter, who completed the final volume 'Indexes

[134] Cf. 'The dispute between her and her husband in the first dream represents this difference of opinion between her and the analyst. The substitution of her husband for the analyst needs no explanation' (159n3).

[135] Holroyd, 3 April 1917, p. 384.   [136] FPD, 3 June 1941, p. 55.   [137] Holroyd, p. xiv.

and Bibliographies' of the *Standard Edition* with Alix after James's death – wrote: 'She ate terribly little: if one was there, one had a job to persuade a few extra mouthfuls into her, but once one's back was turned I suspect she scarcely ate at all. And there was another thing: I have had the impression the last two or three times I have visited her, before this last illness, that she was really not interested in living much longer.'[138]

James was to suffer the same fate as Alix, losing his analyst to death. On 15 September 1925, he had begun a second analysis with James Glover,[139] who, a sufferer from diabetes, was to die suddenly a year later. On her return from Berlin, Alix began a third analysis, with Edward Glover (and subsequently she was to have analysis for many years with Sylvia Payne). She did not become an office-holding stalwart of the psychoanalytic institutions as James did; nor did she have a substantial psychoanalytic practice, though she and James reorganized their menage at Gordon Square in 1929, evicting the Partridges in the process, so that they both had consulting rooms on the first floor.[140] Certainly her encounter with Melanie Klein in Berlin in 1924–25 did have important consequences for British psychoanalysis; in effect, it was Alix's stubborn insistence, to James and to Jones, that Melanie was a psycho-analytic force to be reckoned with (whatever Berlin back-stabbing might say, and despite Anna Freud's hostility), that led to Klein's lectures in London in the summer of 1925, and to her eventual relocation to London some months later. Alix gave Klein continual support, translating *The Psycho-Analysis of Children* in 1932, following her work on Abraham's papers in 1927. She also translated Freud's *Hemmung, Symptome, Angst* in 1935.[141] Alix was a con-siderably more active translator than James in the 1920s and 1930s. And, in the 1930s, she continued to demonstrate her flair for partying, particularly with her lover Nancy Morris.

Adrian and Karin Stephen

> at a New Year's fancy-dress party, she [Karin] had gone as Medusa with snakes in her hair, and reappeared on the first stroke of midnight wearing lilies, having been analysed.[142]

---

[138] Letter to Ilse Grubrich-Simitis, 7 May 1973, Freud, *GW. Nachtragsband*, S. 20–1.
[139] BLSP 60715, Vol. LXI, March 1925–October 1925.   [140] FPM, p. 167.
[141] Jones reported to Freud that the Stracheys were translating this jointly; in the end the translation bore only Alix's name. In the same letter, Jones wrote: 'He and his wife are doing very useful work at present in an extensive revision of our Glossary' (JF, 13 July 1935, p. 746). Again, the eventual completed *Glossary* (1943) bore only Alix's name.
[142] Story told by Karin Stephen near the end of her life to Jean MacGibbon, in *Lighthouse*, p. 169.

Fig. 9.5 Adrian and Karin Stephen by unknown photographer, 1914.

Who was the woman whom Bertrand Russell described to Ottoline
Morrell in 1911 as having 'more philosophical capacity than I have ever
seen before in a woman, and now will probably give up philosophy'[143] –
the woman who would marry Virginia Woolf's youngest brother and
turn herself into one of the first psychoanalysts to have emerged from
Cambridge?

Karin Costelloe came from a family as rooted in the upper-middle-class elites as the Stephens. Her mother, Mary Pearsall Smith

[143] Barbara Strachey, *Remarkable Relations: The Story of the Pearsall Smith Family*, London: Victor
Gollancz, 1980, pp. 263–4.

(1864–1945), from a distinguished American Quaker family, was educated at Smith in Gertrude Stein's cohort, and married Frank Costelloe, an Irish Catholic barrister in 1888; the Pearsall Smiths moved with the marrying daughter to England, where Mary's sister Alys married Bertrand Russell in late 1894. Her uncle, Logan Pearsall Smith, became an influential essayist and critic. Mary and Frank had two daughters, Ray, the elder, and Karin, born in 1889. When her daughters were still tiny Mary met Bernard Berenson at a Pearsall Smith family occasion in 1890; three years later she and Frank drew up a formal separation agreement. Mary lived with Berenson and agreed not to interfere in the raising of her daughters, being restricted from seeing them to four weeks a year. In the late 1890s, Frank discovered he was seriously ill and attempted to prevent his wife or her family gaining control at his death over his daughters. Mary's mother, Hannah Whitall Smith, intervened to have the girls made wards of court. In 1899, when Karin was ten, Frank died – 'a stroke of great good luck', as Russell put it – and Mary was able to marry Berenson; in later years they lived at the magnificent Villa I Tatti (now the Harvard Center for Renaissance Studies).[144] Throughout much of Karin's later life (and indeed of Ray's), particularly in the 1910s and 20s, she was often calling on her mother's financial support[145] and submitting her annual accounts to Berenson, who had become extremely wealthy through his astute capitalizing on his reputedly unrivalled expertise in the attribution of Renaissance paintings.[146]

Through the networks of family friends, from their teens the Costelloes were linked closely to the circle out of which the Bloomsbury Group would develop. The newly graduated John Maynard Keynes spent a week with Mary Berenson and the Costelloe girls in 1906, enjoying the usual japes and transvestite entertainments.[147] Ray went up to Newnham in 1905 to read mathematics; her closest friend was Ellie Rendell, daughter of the eldest Strachey daughter and later Virginia Woolf's doctor,[148] through whom she met the entire Strachey family and adopted them as a family superior in

[144] See E. Samuels, *Bernard Berenson: The Making of a Connoisseur*, Cambridge, MA: Belknap Press.

[145] VWD I, 119, 5 February 1918; on accounts prepared for Berenson, see Karin Stephen to Mary Berenson, BPaS, KS/14, 29 January 1924.

[146] See Strachey, *Remarkable Relations* and Barbara Strachey and Jayne Samuels (eds.), *Mary Berenson: A Self-portrait from her Letters and Diary*, London: Hamish Hamilton, 1983, pp. 20–65.

[147] Skid I, 170 ff.   [148] VWD III, 46n1, 27 June 1925.

every way to her own. In 1911, following the death of her beloved grand-
mother, she officially entered that family by marrying Oliver Strachey after
the break-up of his first marriage.[149] Her sister Karin was to make a similar
union with the other principal Bloomsbury family, the Stephens. Karin
was a hearty sportsplayer at Cheltenham Ladies' College and followed her
sister to Newnham, taking the Moral Sciences Tripos. Her deafness had
already begun to be noticeable even in her teens; it was one reason why she
intermitted from Cambridge at Bryn Mawr for a year. But she did return
and achieved a double First, with a very rare starred First in Part II. Her
uncle Bertrand Russell supervised her and thought her philosophical
capacity very considerable. Consorting with the Neo-Pagans, Virginia
Stephen's name for the group around Rupert Brooke,[150] she travelled in
France with her close friend Hope Mirlees after graduating in the summer
of 1911[151] and she took to the wild Bloomsbury parties with an enthusiasm
she would never lose. Lytton Strachey and the Woolfs turned up to offer
moral support when she read a paper, 'What Bergson means by "interpenetration"', to the Aristotelian Society in 1913 and described the scene:

> an odd affair – Karin rather too feminine and boring; Bertrand Russell
> very brilliant; Moore supreme; Ethyl Sands, having come for the sake of
> Karin's beaux yeux, silent and watchful . . . There was a strange collection of
> people . . . sitting round a long table with Bertie in the middle, presiding,
> like some Inquisitor, and Moore opposite him, bursting with fat and heat,
> and me next to Moore, and Waterlow next to me, and Woolf and Virginia
> crouching, and a strange crew of old cranky Metaphysicians ranged along
> like half-melted wax dolls in a shop window . . . Miss Sands! . . . the
> incorrigible old Sapphist – and Karin herself, next to Bertie, exaggeratedly
> the woman, with a mouth forty feet long and lascivious in proportion. All
> the interstices were filled with antique faded spinsters, taking notes.[152]

Encouraged by Uncle Bertie, she completed a Fellowship dissertation on
Bergson in May 1914 and was elected a Fellow of Newnham. But Karin
already shared the Bloomsbury curiosity about the varieties of sexual
activity, 'abnormal psychology' as it was known, asking Duncan Grant
for French books on spanking, whipping and homosexuality.

---

[149] See Barbara Caine, 'Mothering feminism/mothering feminists: Ray Strachey and "The Cause"',
Women's History Review 8(2) (1999), 295–310 at 300.
[150] Skid I, 242; see Paul Delany, *The Neo-Pagans: Friendship and Love in the Rupert Brooke Circle*,
London: Macmillan, 1987.
[151] Mary Beard, *The Invention of Jane Harrison*, Cambridge, MA: Harvard University Press, 2000.
[152] Holroyd, p. 279.

Even though she was sure she did not love him, Karin married Adrian Stephen in late 1914 after a hurried and stuttering courtship. Adrian was the younger son of Leslie and Julia Stephen, born 27 October 1883. His father was a former Cambridge don, examiner for the moral sciences, who had given up his Tutorship at Trinity Hall, Cambridge for religious reasons and then resigned his Fellowship, as one was obliged to at that time, to marry in 1867; Leslie had remarried with one child, Laura, from the first marriage and four children, Vanessa, Thoby, Virginia and Adrian, from his second to Julia Duckworth. At the time of Adrian's birth, he was embarking on a great and ultimately triumphant adventure in scholarship as founding Editor and 'considerate autocrat' of the *Dictionary of National Biography.*[153]

During the childhood of the Stephen children, the family bonds were intense, intimate and emotionally exhausting. Leslie Stephen's health began to fail when Adrian was eight. His mother died when he was eleven, to be followed by his half-sister Stella, who had taken over managing the household, when he was thirteen; eventually so did his father, in 1904, when he was twenty and up at Trinity College, Cambridge, following in his much admired and loved older brother Thoby's footsteps. With the death of their father, the Stephens children moved to 46 Gordon Square in Bloomsbury, where Thoby began to host his 'Thursday evenings' for his Cambridge friends, mainly Apostles. In 1905, Adrian took his Ordinary degree, a Third, having studied history and law. In 1906, the siblings journeyed through the Balkans to Greece and Turkey, Thoby and Vanessa both returning ill to be nursed by Virginia. Thoby died of his illness, typhoid fever. Two days later, Vanessa accepted Clive Bell's proposal; they were married in February 1907. The two remaining siblings – 'poor Virginia! And Adrian!', wrote Lytton Strachey – then set up house together in Fitzroy Square.[154]

Adrian seemed always to live in and with the shadows of their dead. The poignancy of inexplicable gloom pervades portrayals of the 'ever cadaverous'[155] Adrian Stephen, who seemed never quite to find the right path to exercise the easy talents he, like his siblings, inherited from their distinguished family. Throughout her life, Virginia would periodically muse in her diary on his failure to become someone substantial – until at last, as a psychoanalyst, he did.

---

[153] Alan Bell, 'Stephen, Sir Leslie (1832–1904)', *ODNB*, www.oxforddnb.com/view/article/36271, accessed 19 September 2006.

[154] Hermione Lee, *Virginia Woolf*, London: Vintage Books, 1997 p. 232.   [155] FPM, p. 177.

Living in London with his sister Virginia, the two forever sparring, she had taken on the role of protector without having her heart in it. His circle of friends were those of Virginia, of Bloomsbury, of the Neo-Pagans associated with his old childhood friend Rupert Brooke and the Costelloe sisters. He spent four years in an intense erotic relationship with Duncan Grant. He was called to the Bar in 1907, but in January 1909 Brooke reported to James Strachey that Adrian was giving up the Bar and going on the stage.[156] One fruit of his passion for acting was the famous Dreadnought Hoax he orchestrated in February 1910, when a group of Adrian's friends, including Duncan and Virginia, masqueraded as the Abyssinian Emperor and entourage to inspect the flagship HMS *Dreadnought* at Weymouth.

For the next few years, Adrian worked under Sir Paul Vinogradoff, the Professor of Roman Law at Oxford, deciphering and translating medieval manuscripts. In late 1911, recognizing that they were not happy living as a twosome, he and Virginia moved into a larger menage in Brunswick Square with Keynes, Grant and, in December, Leonard Woolf. How might the menage be sustained? 'I should like Lytton as a brother in law better than anyone I know, but the only way I can perceive of bringing that to pass would be if he were to fall in love with Adrian – & even then Adrian would probably reject him.'[157] Thus Vanessa Bell commented when the one-day-old engagement between Virginia and Lytton was ended, much to their mutual relief. In August 1912, Adrian was in Munich happily learning German with James and Marjorie Strachey – missing his sister's wedding to Leonard Woolf in the process.[158] He was also following in Rupert Brooke's earlier amorous path, as was James Strachey at much the same time,[159] by falling unhappily in love with Noel Olivier, who repudiated him with passionate ardour.[160] Then, in August 1914, there came his rushed court-ship of Karin Costelloe and their marriage in October.[161]

They set up home in Hoop Chambers in Cambridge as Karin took up her Newnham Fellowship in Moral Sciences. She, however, was already moving in a new direction towards 'abnormal psychology as throwing light

---

[156] Hale (ed.), *Friends and Apostles*, p. 55.   [157] VWD I, p. 129.

[158] They were married 10 August 1912 (Lee, *Virginia Woolf*, p. 322).

[159] Hale (ed.), *Friends and Apostles*, p. 207, citing letter from James to Lytton in July 1912.

[160] Despite, according to Brooke, Virginia 'buttering her up in order to get her for Adrian' (Brooke to James Strachey, 8 August 1912, in *ibid.*, p. 261).

[161] Skidelsky's view was that 'it was the war, in fact, which enabled Bloomsbury to catch up with heterosexuality. The "buggers" belatedly discovered the joys of domesticity. Adrian Stephen had led the way by marrying Karin Costelloe' (Skid I, 328). This account lacks some plausibility: James and Adrian had turned the other way some time before the war.

on memory and personality'.[162] The topic suggests that her shift towards psychoanalysis was in process as early as the end of 1914, by which time James Strachey had also read and been struck by Freud's essay 'A note on the unconscious in psycho-analysis'. But Adrian likewise: 'Morton Prince's *Dissociation of a Personality* led him to the conclusion that the study of human personality was the most interesting and important task for anyone in our generation, it was the growing edge of knowledge.'[163] War politicized Adrian, as it did many of the other members of Bloomsbury. Early in the war, Adrian gave up his work on medieval history for the Shelden Society and became involved in the Union of Democratic Control in Cambridge and then Honorary Treasurer of the National Council against Conscription,[164] where he was aided by Karin and James Strachey.[165] When conscription came in early 1916, Adrian was called before the Hampstead Tribunal in July and was given twenty-one days to find suitable work as a conscientious objector. Adrian, Karin and their newborn daughter Ann moved to a farm near Cheltenham, where Adrian milked cows, and then to friends with an estate in Hertfordshire, well placed for London and Cambridge; he gave a talk, 'In defence of understanding', to the Heretics in Cambridge on 28 April 1917.[166] As the war dragged on, Karin became seized with socialist enthusiasm[167] and entered a world more familiar to her sister Ray, who was working closely with Lord Robert Cecil on the arguments in favour of establishing Leonard Woolf's new pet idea, the League of Nations. Karin wrote a pamphlet entitled 'Arbitration in history' for the League of Nations Society.[168]

---

[162] *Lighthouse*, p. 103.

[163] Anonymous, N., 'Adrian Leslie Stephen 1883–1948', *IJP* 29 (1948), 4–6.

[164] In 1916, the NCAC changed its name to the National Council for Civil Liberties, which disbanded in 1918; an organization with the same name was founded in 1934, with E.M. Forster as president and Kingsley Martin as vice-president.

[165] James Strachey to Keynes, prob. March 1916, BLSP 60713, Vol. LIX, 1903–October 1924; see also *Lighthouse*, p. 105.

[166] *CM* VII, 28 April 1917, p. 551, 'Calendar of events'. [167] VWD I, 119, 5 February 1918.

[168] Karin's sister Ray was writing her classic history of the British feminist movement, *The Cause*, just when Karin was beginning her practice as an analyst; later feminists have noted that Ray's history is completely silent on the relations of feminism to the issues of brutality within marriage or to the sexual double standard – issues, particularly the latter, which lay at the centre of the contribution that psychoanalysis made to the politics of the relations of men and women. This was even noted at the time: Mary Stocks contrasted the 'Freudian view' of Mary Wollstonecraft, barely mentioned by Ray Strachey, with the stance of J.S. Mill, for Strachey the prime instigator of the modern women's movement, who saw only an unjust 'external imposition' (Mary D. Stocks, 'The spirit of feminism', *The Woman's Leader*, 13 September 1929, p. 239, quoted in Caine, 'Mothering feminism/mothering feminists', p. 306).

Adrian's health suffered from farming and by the end of 1917 he was allowed to undertake desk work for the estate. After this, he was discharged from his obligations to do war work. On their return to Hampstead in February 1918, they conceived their second child, Karin Judith – 'not welcomed, but made the best of'[169] – born on 7 November 1918. Adrian still had no clear idea of a profession or way to live with fulfilment; Karin was active as ever, but no longer committed to philosophy.[170]

'Why did Adrian marry her?' Virginia Woolf recorded the question in her diary in August 1918 as being the frequent starting-point of conversations with Leonard.

> First & foremost she makes him like other people. He has always, I believe, a kind of suspicion that whereas other people are professionals, he remains an amateur. She provides him with a household, children, bills, daily life, so that to all appearances he is just like other people. I believe he needs constant reassurance on this point; & takes constant delight in her substantiality . . . He is very proud of her vitality. I suppose it provides him with a good deal of the stuff of life, which he does not provide for himself.[171]

'She intends him [Adrian] to have a career. At any rate she is going to ask for that, too; for certainly Adrian will never ask for anything for himself.'[172]

In conversations both with James and Alix Strachey in April 1919 and with W.H.R. Rivers, the couple now broached the project of psychoanalysis as a career.[173] In June their decision was recorded by Virginia:

> I went off, as I now remember, to call on Adrian, as I was early for Ray; & found that strange couple just decided to become medical students. After 5 years' training they will, being aged 35 & 41 or so, set up together in practice as psycho-analysts. This is the surface bait that has drawn them. The more profound cause is, I suppose, the old question which used to weigh so heavy on Adrian, what to do? Here is another chance; visions of success & a busy, crowded, interesting life beguile him. Halfway through, I suppose, something will make it all impossible; & then, having forgotten his law, he will take up what – farming or editing a newspaper, or keeping bees perhaps.[174]

---

[169] VWD I, 221, 30 November 1918.

[170] However, Karin did contribute to a symposium at the Joint Session of the Aristotelian Society, the Mind Association and the British Psychological Society on 11–14 July 1919, on the topic of 'Time, space and material, are they, and if so in what sense, the ultimate data of science?', alongside A.N. Whitehead, Sir Oliver Lodge, J.W. Nicholson, Henry Head and H. Wildon Carr.

[171] VWD I, 187, 27 August 1918.   [172] VWD I, 185, 24 August 1918.

[173] *Lighthouse*, pp. 110–11.

[174] VWD I, 282, 18 June 1919. See also letter from Virginia to Vanessa Bell in June 1919.

Virginia Woolf was wrong, at least about the surface; Karin and Adrian stuck at their plan, through considerable difficulties and setbacks. Their choice of James Glover as analyst is a possible indication that they had already taken the Brunswick Square route (in contrast with the 'Ernest Jones' route) to psychoanalysis, just as the Stracheys had initially,[175] since Glover was a principal force behind the Clinic at this time.[176] Adrian and Karin also kept up their attachment to Cambridge, seeking relevant education and advice there from Rivers; for part of each of the next few summers, they rented a house in Cambridge in order to attend Rivers's lectures.[177] The reason was obvious: his lectures were probably the most directly psychoanalytic teaching to be had in Britain at that time. Where else would a course of nineteen lectures be devoted to Freud's ideas, their development and their criticism? As she wrote to her mother in August 1919:

> I dropped in to see Goldie[178] and found Roger Fry and Paula with him. We had a jolly evening talking about psychoanalysis. Next morning I went to Dr Rivers to explain why I had missed his first lecture – he lent it to me. He dined here on Friday with Roger and Paula and they talked on all sorts of subjects, hypnotism, art. Ethnology is his principal interest.[179]

At the end of 1919, Karin bought the lease on 50 Gordon Square, which was to be their base for the next twenty years – they moved to No. 59 during the Second World War. Over the next few years their shared analyst was the object of tussles which mirrored the ups and downs of their

---

[175] See Suzanne Raitt, 'Early British psychoanalysis and the Medico-Psychological Clinic', *History Workshop Journal* 58(1) (2004), 63–85, and Philippa Martindale, '"Against all hushing up and stamping down": the Medico-Psychological Clinic of London and the novelist May Sinclair', *Psychoanalysis and History* 6(2) (2004), 177–200. The Strachey contact with the Brunswick Square Clinic prior to any discussion with Ernest Jones is recorded by Sylvia Payne, 'Note on the death of James Strachey' (BPaS Payne Papers, Miscellaneous MS/SP/03/F02).

[176] See the account Sylvia Payne gave of her own introduction to psychoanalysis, while she was practising as a physician in Torquay: 'I attended Clinics at the R.A.M.C. Nerve Hospital opened at Seal Hayne on Dartmoor. It was here that I heard of Ps. An. which was decried, but aroused my interest & I wrote to a medical woman in London working at the Brunswick Square Clinic where psycho-analysis was said to be studied.

  When the Torquay hospital was closed in 1919, & my husband sold his practice and bought another in Eastbourne, I visited the Brunswick Square Clinic to enquire about the study of Psycho-Analysis, & saw James Glover who was also interested in studying Freud's work – He told me that psychoanalysis could only be learnt reliably by the personal experience of the technique. I therefore arranged to study the method with him' (BPaS, Sylvia Payne Papers, Misc. Papers).

[177] 'Tremendous talk with Rivers about Psychoanalysis. He thinks it is very dangerous though a few people might come out of it all right!' (BPaS, KS/27, Karin Stephen to Jones, 14 June 1921).

[178] Goldsworthy Lowes Dickinson, Fellow of King's College, Apostle, member of the SPR, who, along with Moore, was the academic father-figure for Bloomsbury and many others from Cambridge. The original idea of the League of Nations is often said to have been Dickinson's.

[179] Karin to her mother and Aunt Loe, 9 August 1919, BPaS, KS/40.

marriage, its strains reflected in a separation in late 1922. They both also failed parts of their medical exams and had to retake them.

Adrian's path through medical school and personal analysis was slow and difficult. 'Adrian is altogether broken up by psycho analysis', Virginia reported implacably in 1923.

> His soul rent in pieces with a view to reconstruction. The doctor says he is a tragedy: & this tragedy consists in the fact that he can't enjoy life with zest. I am probably responsible. I should have paired with him, instead of hanging on to the elders. So he wilted, pale, under a stone of vivacious brothers & sisters. Karin says we shall see a great change in 3 months. But Noel would have done what none of these doctors can do. The truth is that Karin, being deaf, & as she honestly says, 'Your sister-in-law lacks humanity, as perhaps you've noticed', the truth is she does not fertilise the sunk places in Adrian. Neither did I. Had mother lived, or father been screened off – well, it puts it too high to call it a tragedy ... For my part, I doubt if family life has all the power of evil attributed to it, or psycho-analysis of good.[180]

In November 1923, Karin and Adrian agreed to separate, and then got back together – a pattern repeated regularly across the next fifteen years. 'Incompatible, is what they say: and this they've realised for 8 years, and ground their teeth over, while appearing in public the most love-locked of couples', so Virginia reflected, not being able to help feeling this was a whole family failing, by adding: 'We Stephen's are difficult, especially as the race tapers out, towards its finish – such cold fingers, so fastidious, so critical, such taste. My madness saved me; but Adrian is sane.'[181] On one of the periodic occasions when he separated from Karin (in late 1924), Adrian was living in Gordon Square with James Strachey. James found Karin and Adrian both 'very reduced'[182] and tried to avoid, unsuccessfully, spending too much time with Adrian. Always, one suspects, as Virginia had noted back in 1921, between Adrian and James, 'There's Noel, too, in the background. Noel, for James, as for Adrian, the unattainable romance.'[183] Though they kept failing their medical exams and were required to re-sit, though their analyses went on and on, Adrian finally qualified in 1926, Karin a little later. Even his acerbic sister, having converted Adrian's childhood miseries into her latest novel ('Master Adrian Stephen was much disappointed at not being allowed to go' to the lighthouse, the Stephens' childhood newspaper had reported in 1892[184]),

---

[180] VWD II, 242, 12 May 1923.   [181] VWL III, 92–3, Woolf to Jacques Raverat, 8 March 1924.

[182] BLSP 60713, Vol. LIX, 1903–October 1924, James Strachey to Alix, 2 (?) October 1924.

[183] VWD II, 135–6, 12 September 1921.

[184] As recorded on 8 September 1892 in *Hyde Park Gate News*, a newspaper the Stephen children produced. See *Lighthouse*.

began to see light at the end of his tunnel in June 1927: 'Adrian came to tea on Sunday, & fairly sparkled. At last I think he has emerged. Even his analysis will be over this year. At the age of 43 he will be educated & ready to start life.'[185] This didn't stop the painful game of musical houses and changing partners. In the late 1920s, Adrian rented a room in No. 51; in June 1932, he moved out when he was having an affair.[186] How true was Kingsley Martin's description of Bloomsbury: 'demographically a place where the couples were triangles who lived in squares'.[187] But Adrian and Karin's drawing-room was sufficiently welcoming and capacious to become Melanie Klein's lecture-theatre in July 1925.[188]

Karin left more ample traces of the progress of her analysis with Glover than Adrian. When in May 1922 her book *The Misuse of Mind: A Study of Bergson's Attack on Intellectualism* appeared, the first in Ogden's new International Library, she sent a copy to her uncle Bertrand Russell and commented:

> What I should like to do would be to take walks together sometimes & talk as we used to do in Cambridge ... I have just been reading Freud's new book.[189] Adrian has been writing a review of it for the *New Statesman*. I wonder how it would strike someone not already corrupted by having been analysed. I am surprised to find how true it all seems to me, particularly all the 3rd part about the neuroses. Also my own analysis isn't done yet and I find it insufferably difficult to get back my repressed memories. I must have done something shocking as a child! But I believe the mind really does work as Freud says, and it would be the greatest work of mercy in the world to rid people of their conflicts which sap their energy & shock their joy in life. The new technique for getting at the Unconscious *must* be elaborated (by Adrian & me!)[190]

One wonders whether Karin uncovered any childhood memories relating to the flirtation Bertie had with Karin's mother, his bride-to-be's sister, in the run up to his marriage in December 1894 (Karin was five at the time, although not in Paris with her mother).[191] In June 1922, she reported to her mother:

> Attacking your unconscious is like walking with a ghost, you can't get to grips with it at all! However I will stick to it for I know it is worth while and I have never sat down [?word uncertain] under a failure yet. Glover is very

[185] VWD III, 141, 23 June 1927.   [186] VWD IV, 109, 13 June 1932; 115, 8 July 1932.
[187] C.H. Rolph, *Kingsley: The Life, Letters and Diaries of Kingsley Martin*, London: Victor Gollancz, 1973, p. 115.
[188] BLSP 60715, Vol. LXI, James Strachey to Alix, 20 June 1925.
[189] No doubt this was the new translation by Joan Riviere of the *Introductory Lectures*, previously available in a poor American translation rushed out by Freud's nephew Edward Bernays under the title *A General Introduction to Psycho-Analysis* (1920).
[190] BR 0075408, 1,027, Box 1, letter dated 14 May 1922.   [191] MBR I, pp. 94–8, 120–1.

helpful though he says I am the most obstinate case he has ever had to handle! ... We had another lovely cruise again last week but getting back today I hear of Dr Rivers' sudden death on Saturday. I feel upset as if I had lost a friend I had known a long time.[192]

Karin's deafness required her to use an ear-trumpet for certain functions, such as attending her medical lectures. In 1925 she underwent surgery to try and alleviate the problem; the surgeon cut a nerve during the operation and she was left with a half-paralysed face; however, it was said to have greatly improved her hearing.[193]

Virginia, relentlessly acute, painted several unforgettable portraits of Karin's purportedly difficult character: 'There's an unhappy woman if you like. But what is happiness? I define it to be a glow in the eye. Her eyes are like polished pavements – wet pavements. There's no firelit cavern within.'[194] Karin's growing deafness, which was already noticeable when she was seventeen, was an important contributory factor to that unhappiness.

> I can't help being reminded by her of one of our lost dogs – Tinker most of all. She fairly races round a room, snuffs the corners of the chairs & tables, wags her tail as hard as she can, & snatches at any scrap of talk as if she were sharp set; & she eats a great deal of food too, like a dog. This extreme energy may be connected with deafness ... she sits over the fire, & I have to shout. But I see that Adrian must find her energy, her not fastidious or critical but generous & warm blooded mind, her honesty & stability a great standby.[195]

On the sudden death of James Glover in September 1926, Karin went into analysis with Sylvia Payne and Adrian with Ella Freeman Sharpe. Karin and Adrian were both elected associate members of the British Society in June 1927.[196] In the late 1920s, Karin spent time in America, first in New York and then in Baltimore at the Sheppard and Enoch Pratt Hospital, where she worked with Clara Thompson in analysis.[197] As ever, her old

---

[192] BPaS, Karin Stephen Papers, KS/16, 6 June 1922.
[193] VWD III, 45–6, 30 September 1925. Karin found a method in later years for ensuring clear communication with her patients. Noel Bradley wrote of his training analysis with her in the period 1948–52: 'There was no impediment to mutual communication both at the time and during the following four years of analysis. On the couch I held, similar to a present-day hand-held small microphone, one end of a flexible tube about an inch in diameter and a few feet long that attached to the device she wore as she sat behind the couch. The set-up soon ceased to seem unusual and later became an aspect of transference interpretation according to the context of my associations' (Noel Bradley, 'Response: Karin Stephen, the superego and internal objects', *Psychoanalysis and History* 4(2) (2002), 225).
[194] VWD II, 286, 12 January 1924.   [195] VWD I, 118–19, 5 February 1918.
[196] M. Eitingon, 'British Psycho-Analytical Society', *BIPA* 8 (1927), 558–9.
[197] M. Milner, 'Karin Stephen 1889–1953', *IJP* 35 (1954), 432–4.

Fig. 9.6 Adrian Stephen by Rachel Pearsall Conn ('Ray') Strachey (née Costelloe).

friend C.K. Ogden proved the most reliable source of contacts, giving her introductions to Abraham Flexner and Smith Ely Jelliffe.[198] By 1930 she and Adrian were both involved in the running of the London Psycho-Analytical Clinic.[199] Finally, Adrian, in 1930, and Karin, in 1931, were elected full members of the British Society.[200]

Karin preserved her links to Cambridge, her 'spiritual home'.[201] When she returned from the USA and established her private practice in London, she was formally invited to give lectures to students taking both the Moral Sciences Tripos and the new Part II in 'Psychology and Physiology' in the Natural Sciences Tripos: 'Mrs Adrian Stephen' offered a course in 'Psycho-analysis' in Lent Term 1931. In anticipation of delivering the lectures, she had already been in correspondence with Ogden about their possible publication in his International Library of Psychology at Routledge.[202]

[198] OP, Box 69, F.7.   [199] A. Freud, 'Business meeting', *BIPA* 10 (1929), 510–26.
[200] 'British Psycho-Analytical Society', *BIPA* 11 (1930), 514–16 at 511 and 'IV. Reports of Proceedings of Societies', *BIPA* 12 (1931), 509–23 at 514.
[201] *Lighthouse*, p. 114.
[202] OP, Box 69, F.7, Karin Stephen to Ogden, 16 March 1930(?) and 17 March 1932(?).

However, the lectures were eventually published by Cambridge University Press in 1933 as *Psychoanalysis and Medicine: A Study of the Wish to Fall Ill*; Karin's exposition of psychoanalysis was very much oriented towards its clinical practice. The early stages of the development of the libido as dominated by the pleasure principle – 'IV. Infantile Pleasure-seeking by the Mouth'; 'V. Excretory Pleasure-Seeking and Creation'; 'VI. Phallic Pleasure-Seeking. The Oedipus Complex and Castration Fears' – gave Karin her expository framework. She interwove the theoretical exposition with vivid clinical material drawn from her own analytic cases. The most remarkable feature of the book was its faithful reflection of the lectures: there was no other work, no external authority, referred to, not even any of Freud's work – no footnotes, no bibliography. A masterly performance, lucid, confident and modest in its claims.[203]

She delivered lectures on 'Clinical aspects of psycho-analysis' in the Lent terms of 1931–34; in 1935 the title changed to 'Psycho-analytic studies of neurotic patients' and her course continued to 1940 at least, when courses were no longer listed in detail again until 1944.[204] Her lectures had a

---

[203] A sign of the diffidence with which psychoanalysis was increasingly treated by Cambridge libraries in the 1930s is the fact that only two copies of her book were acquired (by the University Library and Trinity College) – her former College, Newnham, where she had been a Fellow, did not acquire the book; the Library of Experimental Psychology does not hold the book, even though she supposedly offered the lectures as part of the Department's teaching for nearly ten years. When reissued in 1960, Cambridge University Press dropped the word 'psychoanalysis' from the title, publishing it under its subtitle: *The Wish to Fall Ill*. On this occasion, the University Library (a copyright library) declined to accept the book, even from Cambridge University Press; the sole copy to be found in Cambridge is in the Department of Experimental Psychology – which did acquire the 1960 edition. In contrast, Harry Banister, Lecturer in the Department of Experimental Psychology from 1926 on, had given lectures in 1930 on 'Psychology for medical students' and published in 1935 (with Cambridge University Press) a slim volume entitled *Psychology and Health*, which surveyed the views of Janet, Freud, Jung, Adler, with an anti-Freudian bias (according to an acerbically dismissive review by Louis Minski, in *Journal of Mental Science* 81 (1935), 694); copies of this book are to be found in the libraries of Newnham, Pembroke and St John's colleges as well as the Psychology Department and the University Library.

[204] It was these lectures that Marion Milner described in her obituary of Karin Stephen as 'the first course of lectures on psycho-analysis ever given at Cambridge University' (Milner, 'Karin Stephen', p. 433). As is clear from the whole of this book, this claim is highly misleading. It is possible that these were the first lectures on psychoanalysis with 'psychoanalysis' in the title given by a *full* member of the BPaS in Cambridge, but all other claims to priority are empty: MacCurdy, a member of the American Psychoanalytical Association, gave lectures with the title 'Psycho-analysis' in 1927; Rivers's lectures on dreams, first given in 1919, were essentially an examination of Freud's theory of the dream (and Rivers was an associate, though not full, member of the BPaS); Needham testified to the psychoanalytic content of some of Tansley's lectures in the Natural Sciences Tripos in the post-war period, and Tansley (who did become a full member) was certainly discussing psychoanalysis in his lectures before the war, probably around 1912, when he brought the proofs of Hart's book on *Insanity* into the lecture-room to discuss with his students. It is even possible that Cyril Burt (also a full member from 1920 on), who had already given lectures on psychoanalysis at Liverpool University in 1908, gave lectures dealing with psychoanalysis in the Moral Sciences

significant effect on at least one member of her audience, Heinz Wolff (1916–89), whose schoolboy interest in Freud was rekindled by her as he turned away from mathematics to enter medical training and simultaneously had a personal analysis in the late 1930s before becoming a psychiatrist in India during the war. He eventually took over Willi Hoffer's position as consultant psychoanalytic psychotherapist at the Maudsley Hospital from 1961–81 – even though he had no formal training in either psychoanalysis or psychiatry; he was also one of the founders of the Psychosomatic Research Society.[205] Karin also gave papers to the Cambridge Moral Sciences Club in the 1930s ('Is the unconscious in touch with reality?' (1938))[206] and another Cambridge connection led to Gregory Bateson adding a footnote to his classic anthropological study, *Naven*, of 1936, recording his conversations with Karin and wrestling with integrating her psychoanalytic terms of description of 'transvesticism' with his ethnographic framework of analysis of the Iatmul.[207]

In 1933, as she, Adrian and Rickman managed the public lectures the BPaS offered, she gave lectures, together with Susan Isaacs, on 'Unconscious wishes in daily life'. Her commitment to securing psychoanalytic influence within medicine found expression in her sitting, with Edward Glover, on the (Walter) Langdon-Brown committee on postgraduate training in psychological medicine which reported in 1943; the committee was self-consciously divided between neurological and psychological orientations and was self-consciously concerned with future training under the new National Health Service already envisaged for the post-war period.[208]

In 1934, Virginia reported to her diary once more on the progress of her younger brother, now turned fifty, and his wife:

> Yesterday just as we had done tea, Adrian's gaunt form appeared; Karin's tousled shape, grown very thick & large. Her inferiority complex takes the form of praising Adrian. Clever old Adrian, she exclaims, if he bowls a good

---

Tripos in 1913–14 – which, as we saw in Chapter 5, already included psychoanalysis as an examinable topic before the Great War when Rivers was lecturing on psychopathology, including psychoanalysis.

[205] Heinz Wolff, in Wilkinson Greg (ed.), *Talking about Psychiatry*, London: Gaskell, 1993, pp. 178–91, esp. 181–2; see also 'Heinz Hermann Otto Wolff', *Psychiatric Bulletin* 13 (1989), 584–5.

[206] Milner, 'Karin Stephen', pp. 432–4.

[207] Gregory Bateson, *Naven: A Survey of Problems Suggested by a Composite Picture of the Culture of a New Guinea Tribe Drawn from Three Points of View*, Cambridge: Cambridge University Press, 1936, p. 188n1.

[208] D.V. Hubble, 'Brown, Sir Walter Langdon- (1870–1946)', rev. Michael Bevan, *ODNB*, www.oxforddnb.com/view/article/34401, accessed 12 October 2015; see also FK, pp. 492–3.

bowl. This is by way of saying – what? My marriage was not so bad after all? He remains perfectly unmoved, quiet, sensible; I suppose curiously immature, though able to go through all the actions correctly of a grown man, father, husband.[209]

Their 'house full of lodgers and patients',[210] an ever active and busy Karin also became in 1935 the spiritual patron of a 'Society of Creative Psychology' established by her analysand Basil Rakoczi, a mature medical student at Cambridge, and Herbrand (Billy) Ingouville-Williams. She was also a patron of the White Stag Group of artists in Fitzrovia, who fled to Dublin with the outbreak of the war.[211] Another of her protégées, this time from earlier but also with a strong Cambridge connection, was Marion Blackett, later Milner, with whom she discussed psychology and went sailing from the mid-1920s on.[212] It was also a conversation with Karin that persuaded Charles Rycroft, a callow undergraduate at Trinity College, more huntin' 'n' fishin' than Bloomsbury, but more Marxist than liberal in politics, to apply to train as a psychoanalyst in the mid-1930s.[213] And in the early 1950s Karin invited to live in her house Marie Singer (1910–85), an African American graduate of Smith College, US Army psychiatric social worker, drawn to London by the training course at the Anna Freud Centre. It was through Karin that Marie met her husband, James Burns Singer, the poet, literary critic and marine biologist; they soon moved to Cambridge, where she became one of the two psychoanalysts actively at work in Cambridge in the 1960s and 1970s. Active she was: 'she liked to complain that she could hardly go anywhere in Cambridge without meeting people who had been or were in treatment with her', her obituarist recorded.[214]

By 1927, the four Bloomsbury analysts were embedded in their profession. Other Bloomsberries made use of their interest in or familiarity with psychoanalysis in different ways. That year Lytton recorded a conversation with old close friends: 'there was much talk about the old subjects – Freud

---

[209] VWD IV, 234, 4 August 1934.   [210] VWD IV, 276, 2 February 1935.
[211] S.B. Kennedy, 'The White Stag Group', www.modernart.ie/en/downloads/whitestagbk.doc, accessed 29 January 2006.
[212] BPaS, Karin Stephen Papers, KS/18. Milner, Patrick Blackett's sister, was to write Karin Stephen's obituary for the *IJP*. It was her brother who introduced her to Freud: 'Patrick sent me for my 21st birthday [in 1924] Freud's *Introductory Lectures*, which he must have got onto through Rivers. Although he was caught up in physics, he read it a bit himself' (Andrew Brown interview with Marion Milner, 22 February 1997). Blackett was the Secretary of the Memorial Fund for Rivers (*CR*, 13 October 1922, p. 4).
[213] Charles Rycroft, 'Where I come from', in Rycroft, *Psychoanalysis and Beyond*, London: Chatto & Windus/The Hogarth Press, 1985, pp. 204–5, and Jenny Pearson, *Analyst of the Imagination: The Life and Work of Charles Rycroft*, London: Karnac, 2004, p. 203.
[214] Pauline Cohen, 'Marie Battle Singer', *Bulletin of the Anna Freud Centre* 8 (1985), 213–15.

and love, Sainte-Beuve and love, Cambridge and love'.[215] Lytton was completing an explicitly psychoanalytic interpretation of *Elizabeth and Essex* (1928).[216] Two years earlier, as we have seen, Keynes intervened in a heated debate over its validity (Chapter 7) and was ruminating throughout the late 1920s on psychoanalytic theories of money. Even Virginia Woolf, always suspicious and resolute that Freud should remain amongst her Great Unread, found herself reflecting in 1939 on her great achievement of 1927 and the troubled familial ghosts she had transmuted into a masterpiece, *To the Lighthouse*:

> Until I was in my forties ... the presence of my mother obsessed me. I could hear her voice, see her, imagine what she would do or say as I went about my day's doings ... she obsessed me, in spite of the fact that she died when I was thirteen, until I was forty-four. Then one day walking around Tavistock Square I made up, as I sometimes make up my books, *To the Lighthouse*; in a great, apparently involuntary rush. One thing burst into another. Blowing bubbles out of a pipe gives the feeling of the rapid crowd of ideas and scenes which blew out of my mind, so that my lips seemed syllabling of their own accord as I walked. What blew the bubbles? Why then? I have no notion. But I wrote the book very quickly; and when it was written, I ceased to be obsessed with my mother. I no longer heard her voice; I do not see her. I suppose I did for myself what psycho-analysts do for their patients. I expressed some very long felt and deeply felt emotion. And in expressing it I explained it and then laid it to rest.[217]

She had first recorded a reflection on her achievement in *To the Lighthouse* in 1928. But back then she had not taken the next step, which was to view it as her personal *psychoanalytic* labour. The walk in Tavistock Square may have taken place in October 1924,[218] just when she was dismissing Freud for his gull-like imbecility; but she would only allow herself to see what she *might* have been doing back then much later in life, after she had met Freud, in early 1939. Later that year, after his death, she began to read Freud. But already in 1924, at the time when the *Lighthouse* bubbles and the syllables burst out of her, the form that psychoanalysis would take in Britain was becoming established. And Virginia, her family and her close circle had much to do with the making of that form.

---

[215] Holroyd, p. 571.

[216] Holroyd, pp. 609–16; see also D.W. Orr, 'Virginia Woolf and psychoanalysis', *IRP* 16 (1989), 151–61.

[217] Woolf, *Moments of Being*, p. 81.   [218] Lee, *Virginia Woolf*, p. 475.

Bloomsbury Analysts

In the 1930s, the four Bloomsbury analysts, Adrian and Karin Stephen, James and Alix Strachey, eased themselves, in their idiosyncratically distinctive ways, into their professional lives. James, Adrian and Karin all had full-time private psychoanalytic practices and they each contributed significantly by writing and in other ways. Adrian Stephen made use of the German he had learnt, good German, by helping Alix Strachey translate Klein's *The Psycho-Analysis of Children*, finally published in 1932, and in 1932–34 for the *International Journal* wrote brief abstracts of psychoanalytic papers published in German. His 'membership paper', as it would later be called – the *rite de passage* for progressing to full membership of the Society – given on 15 October 1930, and published in 1936, showed the stamp of the Bloomsbury literary culture: '"Hateful", "Awful", "Dreadful"'. It opened: 'I was casting about in my mind the other day for a good translation of the German "*gehässig*", when the English "hateful" occurred to me. The obvious objection was that while the German word is used to qualify a person who feels hatred himself, the English word qualifies a person who causes others to feel hatred.' Adrian contrasted 'beautiful' with 'hateful': the person described as 'beautiful' is full of beauty or 'graceful', but the person who is 'hateful' is not full of hate. 'Perhaps it relieves us of our sense of guilt in expressing hatred if we project our feeling into the hated person. It is more comfortable to hate one who is full of hatred himself.' Analogies from French and Latin words that have the same peculiarity clarify; but he then returned to English:

> I wrote down a list of all the words I could think of that were made on the same model. Obviously such a list cannot pretend to be complete, but excluding words like 'unfaithful' and 'distasteful', which are made from others by the mere addition of a negative prefix, I found that I could think of eighty-seven, a number which I felt sufficient for my purpose.[219]

This paper combines the kind of literary parlour-game Bloomsbury spent many happy hours engaging in with a translator's problem, producing psychoanalytic insights out of this work on the nuances and implicit claims of the English language. There is a cultural continuity with the work of his father, Leslie, editor of the *Dictionary of National Biography* (that Victorian sister of the *Oxford English Dictionary*).

Another side of Adrian's character, his fundamental contrariness – and courage – prompted him to give a second paper to the Society on 20

---

[219] A. Stephen, '"Hateful", "Awful", "Dreadful"', *IJP* 17 (1936), 109–12 at 109. The 1930 paper was subtitled 'A study of the factors determining the choice of these and other expressions.'

January 1932 entitled 'On the repetition compulsion', in which he vigorously contested the utility of Freud's concept.[220] A more reflective, equally ambitious paper, was published in 1935 in the *British Journal of Medical Psychology*: 'On defining psychoanalysis':

> The object of every intervention on the analyst's part is simply to allow more freedom to the patient's mind, to allow fresh phantasies and memories to come into his consciousness, and to trace down to their origin as many as possible of his stereotyped reactions.

While 'freedom' is undoubtedly a fundamental feature of Freud's conception of the achievement of analysis, it is striking that Adrian thought to bring this value, so resonant for the Bloomsbury culture which he inhabited, to the fore in his account of the task of the analyst.

While Karin inclined more to public engagement, Adrian felt impelled to engage in reform of the BPaS. Elected to its Council in 1935, a proto-Oedipal signal of his future professional intentions appeared in a brief book review he published in 1937: 'One small but definite mistake may be noted: the address of the British Psycho-Analytical Society is not identical with that of its President's consulting room.'[221]

When war was declared, in contrast to his CO status in the Great War, Adrian was quickly into uniform (though the Army found great difficulty in covering his huge, gaunt frame) as an Army psychiatrist, work he found he relished; Karin found work at Shenley Mental Hospital. And in another theatre of conflict, the BPaS, Adrian and Karin now found a mission which would have considerable long-term consequences.

Compared to Adrian's and Karin's long uphill struggle to qualify in both medicine and psychoanalysis, the Stracheys had had a straightforward entry to the profession. The magnitude of their eventual achievement in producing the *Standard Edition* has misled commentators (including the admirable 'Introduction' and 'Epilogue' of Meisel and Kendrick's *Bloomsbury Freud*[222]) into casting the Stracheys as devoting most of their psychoanalytic lives to the task of translation. However, it is not the case that James Strachey settled down in the 1920s to his life-work of translating Freud. Indeed, he left translation to one side for most of the years from 1925 to 1940, save for his preferred position of

---

[220] 'II. Reports of Proceedings of Societies', *BIPA* 13 (1932), 389–99 at 389.

[221] A. Stephen 'Practical aspects of psycho-analysis', *IJP* 18 (1937), 66.

[222] Meisel and Kendrick write of the 1930s: 'the Stracheys' ever-increasing responsibility for the translating and editing of psychoanalytic works . . . left them little time for original composition'. Caine writes: 'Having taken up [psychoanalysis] in the early 1920s, why did they choose translation, rather than clinical work or the development of their own ideas?' (Barbara Caine, 'The Stracheys and psychoanalysis', *History Workshop Journal* 45 (1998), 145–70 at 145).

second reader. Alix did substantially more translation than James in these years. None of Freud's major writings of the late 1920s or 1930s and barely a handful of the brief writings were translated by James. Instead he devoted himself to the practice of psychoanalysis.

By the end of 1927, James Strachey was analysing eight patients a day.[223] Amongst his first patients was D.W. Winnicott (who started on 7 October 1924[224]). How was this intriguing encounter initiated? James was very much dependent on the more established practising analysts to get his own practice started; the referral of Winnicott to Strachey came from Jones. But why did Jones refer the young Winnicott to the analytically inexperienced Strachey? After all, Winnicott had an elite educational and medical pedigree of Cambridge and Bart's; James Strachey, unlike Winnicott, did not even take the Tripos at Cambridge and had no professional qualifications at all. The answer is probably to be found in Jones's astute assessment of the ramifications of class and its partial corollary, the Cambridge background: Winnicott came from a 'merchant' background, his father Mayor of Plymouth in 1906–7 and again in 1921–22, knighted in the New Year's Honour Lists of 1924, just when Winnicott married and sought analysis. But it was most probably the fact that Winnicott and Strachey shared a Cambridge history – Winnicott for seven years, from 1910, when he entered the Leys School, to 1917, when he completed Part I in the Natural Sciences Tripos at Jesus College – that prompted Jones to match them up.[225]

In the 1920s Strachey also analysed Alec Penrose (1896–1950), older brother of Lionel, who had read English at King's College – 'Gentleman

---

[223] JF, 5 December 1927, p. 638.

[224] BLSP 60713, Vol. LIX, 1903–October 1924, James Strachey to Alix, 'Saturday', internal evidence allows dating this letter to 4 October 1924: 'You'll be relieved to hear that there was a letter last night from Winnie melding himself an. So I shall start the two of them next Tuesday.' ('Anmelden' means 'to announce, register'.) Most writers state that Winnicott began analysis in 1923, without giving an exact source; F.R. Rodman, *Winnicott: His Life and Work*, Boston, MA: Da Capo Press, 2004, p. 70, indicates that Winnicott first consulted Jones in 1923, was given a list of possible analysts, from which he could not choose, and then was assigned to Strachey; again, there is no source for this information. There is one specific source from Winnicott that gives this date, but Winnicott was so often inexact about the objective details of his own autobiography that this evidence from Strachey's contemporary letter is more reliable than Winnicott's published memories. See D.W. Winnicott, 'A personal view of the Kleinian contribution' (1962), in D.W. Winnicott, *The Maturational Processes and the Facilitating Environment: Studies in the Theory of Emotional Development*, London: The Hogarth Press and the Institute of Psycho-Analysis, 1965: 'it was Jones to whom I went when I found I needed help in 1923. He put me in touch with James Strachey, to whom I went for analysis for ten years' (p. 171).

[225] Winnicott wrote the obituary of Strachey, 'a good sample of an Englishman', for the *IJP* in 1969: 'He will always be my favourite example of a psychoanalyst' (D.W. Winnicott and Anna Freud, 'James Strachey – 1887–1967', *IJP* 50 (1969), 129–32 at 131).

Penrose' as a friend called him, later a gentleman farmer.[226] This must have been an odd analysis, since both patient and analyst were members of the Apostles. In 1927, Strachey served with Jones and Rickman on a Society committee devoted to lay analysis – this was at the time of the controversies sparked by Theodor Reik being taken to an Austrian court for illegal practice of medicine, with Freud's pamphlet *The Question of Lay Analysis* leading to extensive public discussions and publications in psychoanalytic societies and journals in Europe and America.[227] In Britain, the committee canvassed the views of the Society by issuing a questionnaire.[228] In 1928, Strachey also sat (with Barbara Low and Stoddart – a man with considerable financial resources) on a three-man committee managing the James Glover Memorial Fund with a view to building up a specialist psycho-analytic library[229] – already James was the analyst to turn to for bookish advice. By 1929 Strachey was a training analyst.[230] Indeed James was, with Ella Freeman Sharpe, the most successful training analyst of the entire inter-war period: he had three candidates successfully qualify (Sharpe had five, Klein had one, Riviere two, Glover two, Jones one).[231] Over the long haul, he had even managed to impress Jones: 'he is proving distinctly the best of the three helpers [Rickman, Riviere and Strachey] you sent me some ten years ago'.[232]

---

[226] Gillian Rice interview with Frances Partridge, 10 July 1984, in Gillian Rice, 'The reception of Freud amongst the Bloomsbury Group', MPhil dissertation, University of Cambridge, 1984, p. 66n133. Rickman had earlier analysed another of Lionel's brothers, Bernard, in Vienna in 1924 (FRP, 30[?] April 1924): 'his brother Bernard is here being analysed by Rickman. It is rather a bore for me because though nice Bernard is terribly stupid; he is being analysed because it seemed that otherwise he would never pass the little go. It has already got him through 2 parts!' The 'Little Go' (or 'Previous Exam') was the first part of the three parts of examinations for those taking an Ordinary degree at Cambridge.

[227] See Robert S. Wallerstein, *Lay Analysis: Life in the Controversy*, Hillsdale, NJ: Analytic, 1998, pp. 8–20 and K.R. Eissler, *Medical Orthodoxy and the Future of Psychoanalysis*, New York: International Universities Press, 1965 for a very full account of the debates in the 1920s and later.

[228] 'Abbreviated Report of the Sub-Committee on Lay Analysis', *BIPA* 8 (1927), 559–60 at 559.

[229] D. Bryan, 'British Psycho-Analytical Society', *BIPA* 9 (1928), 276–7 at 277.

[230] A. Freud, 'Business Meeting', *BIPA* 10 (1929), 510–26 at 521. How did Strachey become a training analyst? One important factor may have been that Winnicott's analysis with Strachey began before the formal system of 'training analyses' was introduced, although there was already in 1924 the expectation that the most important element of any training to become a practising analyst was a personal analysis. By the late 1920s, rules had been introduced requiring personal analysis with a designated 'training analyst'. Thus, for Winnicott to qualify as an analyst, he would either have to change analysts or James Strachey would have to be designated a training analyst (and Winnicott's earlier years of analysis to be retrospectively designated a 'training analysis'). These considerations may have had some weight in the decision to appoint James as a training analyst. Rodman, *Winnicott*, p. 74 states that Winnicott began his analytic training as a candidate in 1927 (but gives no source) – meaning, presumably, that he took on his own analytic cases at that point.

[231] FK, pp. 192–5.   [232] JF, 5 April 1930, p. 670.

One unwavering theme the Bloomsberries brought to psychoanalysis
was their deep attachment to the interweaving of sharp wit and transgressive humour. Taking something seriously their way nearly always ends up
in laughter amidst ribald and bawdy banter and jokes at the expense of
psychoanalysis. Psychoanalysis crossed with Bloomsbury produced the
following opening sentence to a letter sent by James to Alix Strachey in
Berlin in late 1924:

> Dearest Alix
>
> The long awaited child was successfully born to-day. It was a fine
> turd, weighing a hundred and fifty pounds. In accordance with the
> usual practice it was immediately circumcised, and I enclose the
> foreskin.[233]

Undergraduate scatology meets psychoanalytic symbolism in an intimate
correspondence. Earlier that year James's sister Marjorie threw a party at 50
Gordon Square, when she produced her version of Schnitzler's *Reigen* (*La
Ronde*), complete with a copulation scene in the dark and rear of stage,
horrifying some and embarrassing others. Vanessa Bell commented to Roger
Fry: 'It was a great relief when Marjorie sang hymns.'[234] So when James
turned his hand to psychoanalytic writing, just as with Adrian's musings on
'awful' and 'hateful', humour bubbled beneath the surface.[235] Writing in
1930 on the quintessential Bloomsbury topic, 'Some unconscious factors in
reading', he made some waspish observations: 'though it may be impossible
to suck down the works of Mr. Bertrand Russell, or to chew up those of Miss
Ethel M. Dell, it is surprising what successful efforts may be made in both
these directions'.[236] Reading, Strachey observed, in contrast to speaking, is 'a
way of eating another person's words'; but his sharpest wit and powers of
observation were reserved for reading habits centred on defecation, leading
him to offer the following hypothesis: 'a coprophagic tendency lies at the
root of all reading. The author excretes his thoughts and embodies them in
the printed book; the reader takes them, and, after chewing them over,
incorporates them into himself'.[237] Despite this, Strachey reserves the climax
of his reading of reading for the full Oedipal drama:

---

[233] BLSP 60714, Vol. LX, October 1924–February 1925, James Strachey to Alix, n.d. [1924].
[234] VWD II, 304n2.
[235] For an early example, see Strachey's review in 1924 of Wohlgemuth's attack on psychoanalysis, discussed in Chapter 8.
[236] James Strachey, 'Some unconscious factors in reading', *IJP* 11 (1930), 322–31 at 326.
[237] *Ibid.*, p. 329.

if the book symbolizes the mother, its author must be the father; and the printed words, the author's thoughts, fertilizing and precious, yet defiling the virgin page, must be the father's penis or fæces within the mother. And now comes the reader, the son, hungry, voracious, destructive and defiling in his turn, eager to force his way into his mother, to find out what is inside her, to tear his father's traces out of her, to devour them, to make them his own, and to be fertilized by them himself.[238]

Then, in 1934, James wrote a classic paper on the technique of analysis and its metapsychological underpinnings, 'The nature of the therapeutic action of psycho-analysis' – one of only three papers cited more than 600 times in the psychoanalytic journals.[239]

His starting-point was the absence of methodical discussion of the means by which psychoanalytic therapy achieves its goals. Much had been found out about mental processes and constructing scientific laws concerning the considerable quantities of data gathered.

> But there has been a remarkable hesitation in applying these findings in any great detail to the therapeutic process itself. I cannot help feeling that this hesitation has been responsible for the fact that so many discussions upon the practical details of analytic technique seem to leave us at cross-purposes and at an inconclusive end.[240]

What he proposed was to develop the *theory* of analytic technique, using the tools supplied by Freud, in order to help *conclude* discussions, and give clarity and certainty to matters relating to technique (and thence to training).

> I have set up a hypothesis which endeavours to explain more or less coherently why these particular procedures [agreed upon by practitioners] bring about these particular effects; and I have tried to show that, if my hypothesis about the nature of the therapeutic action of psycho-analysis is valid, certain implications follow from it which might perhaps serve as criteria in forming a judgment of the probable effectiveness of any particular type of procedure.[241]

The highly influential concepts James introduced were the 'mutative interpretation' and the 'auxiliary super-ego'. The second of these was derived from Sandor Rado's account of the hypnotist functioning as a 'parasitic

---

[238] *Ibid.*, p. 331.
[239] PEP-Web statistics, 15 March 1914. In the year September 2011 to September 2012 it was read by 580, and therefore was still one of the most frequently read papers in the psychoanalytic literature.
[240] J. Strachey, 'The nature of the therapeutic action of psycho-analysis', *IJP* 15 (1934), 127–59 at 127.
[241] *Ibid.*, p. 128.

super-ego'; James extended this theory so as to describe the psychoanalyst's function as that of the 'auxiliary super-ego': 'The most important character-istic of the auxiliary super-ego is that its advice to the ego is consistently based upon *real* and *contemporary* considerations and this in itself serves to differ-entiate it from the greater part of the original super-ego'.[242] The 'mutative interpretation' is one that allows the patient to become aware of the differ-ence between the 'real nature of the analyst' and 'the patient's "good" or "bad" archaic objects', with all their greater power of cruelty, vindictiveness and seduction. The interpretation was effective in so far as it breached the positive feedback circuit ('the neurotic vicious circle'[243]) and induced a negative feedback circuit ('a *benign* circle') in its stead.

Many readers of Strachey assumed that, because it causes change, the mutative interpretation causes a significant, large-scale shift. He empha-sized the opposite feature:

> For the mutative interpretation is inevitably governed by the principle of minimal doses. It is, I think, a commonly agreed clinical fact that alterations in a patient under analysis appear almost always to be extremely gradual: we are inclined to suspect sudden and large changes as an indication that suggestive rather than psycho-analytic processes are at work. The gradual nature of the changes brought about in psycho-analysis will be explained if, as I am suggesting, those changes are the result of the summation of an immense number of minute steps, each of which corresponds to a mutative interpretation.[244]

What is curious about these features of Strachey's account of the mechan-ism of psychoanalytic interpretation is that it draws on two contempora-neous technological and scientific developments. The mechanisms, while called vicious and benign circles, were clearly those of the electrical circuits becoming familiar to anyone who tinkered with radio (as Strachey did[245]); he did not employ the term 'feedback' (a term being developed at precisely this time but restricted in the late 1920s and 30s to describing the function-ing of radio valves), nor 'amplification' or 'inhibition', but the system he described had precisely those properties which would become so important in the development of cybernetics in the 1940s and 50s.

To become aware of the second scientific field on which Strachey may have drawn, one needs to ask the question: why did he use the rare term 'mutative'? This term had a quiet and obscure life in the history of the language until the 1910s, when it began to be used by the laboratory

[242] *Ibid.*, p. 140.   [243] *Ibid.*, pp. 137–8.   [244] *Ibid.*, p. 144.   [245] See BF, *passim.*

scientists, the lords of the fly,[246] associated with the development of Mendelism in connection with genetic mutations. Intriguingly enough, James Strachey had direct access to the mathematical side of this scientific development through his friendship with brother Apostle Harry Norton, the Trinity mathematician.[247] In 1915, the Cambridge entomologist R.C. Punnett had, for the second time, called on G.H. Hardy's mathematical skills for help with a problem in genetics. The first time, in 1908, Hardy had produced a mathematical demonstration of what is now known as the Hardy-Weinberg Law: simple enough (allele frequencies are constant between generations), but fundamental to population genetics.[248] Hardy saw that the second question involved some lengthy calculations: Punnett, developing his theory of mimicry, wanted to know the effects of selection at a single Mendelian diallelic locus under random mating – the relative efficacy of positive and negative selection when transmitted by Mendelian mechanisms across a number of generations. So, in one version of the story, Hardy, also a Fellow of Trinity, passed the question to Norton, who produced the statistical table that Punnett then published. It was this table that stimulated a number of biologists, most notably J.B.S. Haldane in 1922, to start work on developing the mathematical theory in which Darwinian natural selection and Mendelian genetics were eventually satisfactorily brought together: the modern evolutionary synthesis. In fact, according to Haldane in another version of the story,[249] Norton had already started work as early as 1910 on these mathematical questions, had completed his work by 1922 and eventually published a fundamental paper 'Natural selection and Mendelian variation'[250] in 1928 (although the paper was often passed over until it was revived as the foundation-stone for specific questions in mathematical demography in the 1970s,[251] possibly in

[246] Robert E. Kohler, *Lords of the Fly: Drosophila Genetics and the Experimental Life*, Chicago: University of Chicago Press, 1994.

[247] *Eminent Victorians* was dedicated to Norton, who had given Lytton financial support at crucial moments in his life. James said of Norton that he was 'one of the only three or four people I have ever known in the same intellectual category as Russell – with whom he was perfectly able to argue on equal terms'. He later suffered, Levy writes, from 'hypomania which later became severe depression' (LLS, pp. 80–1).

[248] A.W.F. Edwards, 'G.H. Hardy (1908) and Hardy-Weinberg equilibrium', *Genetics* 179 (July 2008), 1143–50.

[249] J.B.S. Haldane, 'A mathematical theory of natural and artificial selection' (Part IV), *Mathematical Proceedings of the Cambridge Philosophical Society* 23 (1927), 607–15.

[250] H.T.J. Norton, 'Natural selection and Mendelian variation', *Proceedings of the London Mathematical Society* 28 (1928), 1–45.

[251] See B. Charlesworth, *Evolution in Age-Structured Populations*, Cambridge: Cambridge University Press, 1980, 2nd edition 1994, esp. pp. 73–4, and W.B. Provine, *The Origins of Theoretical Population Genetics*, Chicago: University of Chicago Press, 1971.

large part because Haldane had already published, with due reference to Norton's work, the main mathematical results in a series of papers from 1924 to 1927).[252] While James Strachey was mathematically illiterate, he would have had ample opportunity to discuss Norton's work on population genetics with him. The terms 'mutation' and 'mutative' would have been familiar to him, as would also the notion that the cumulative effect of many small 'mutations' may, when acted upon by natural selection, give rise to species change. Lurking behind Strachey's use of 'mutative', then, may have been a suspicion that the processes of transformation of the mind are analogous to the changes wrought by mutation and selection in organic populations.

Absolutely characteristic of Strachey's choice of problem and method of approach was his disdain for theoretical dispute and group politics – 'my non-committal attitude to questions of psycho-analytic theory',[253] as he called it late in life. Strachey tackled the question of interpretation, in other words of psychoanalytic *practice*, in order to help psychoanalysis find a way of avoiding its self-destructive theoretical battles, already perennial by the early 1930s, the most present of which was taking place in 1934, with Glover's first attacks on Klein. The resultant desire to find practical consensus separate from – and as protection from – theoretical disputes would underpin Strachey's interventions in the Training Committee during the wartime battle between psychoanalytic groupings within the Society, which goes by the name Controversial Discussions. Elected to the Training Committee in 1939, it was not surprising that the Committee called upon him to draft its Report. Not only was he perceived to be above the battle of the competing factions, independent but with sympathetic links across the board, to Melanie Klein, Anna Freud, Jones, Glover and others, and not only had he made a major contribution to psychoanalytic theories of technique, but he had also become one of the most experienced and successful analysts in the Society.

The Controversial Discussions

With the external and internal upheaval caused by war, with the profound inner confusion and grief caused by the death of Freud in September 1939,

---

[252] Haldane in 1936 wrote: 'In the last twenty years a considerable body of evolution theory based on Mendelism has arisen. The pioneer in mathematical theory was Norton, of Trinity College, whose work has been continued by Haldane, Fisher, and Wright' (J.B.S. Haldane, 'Forty years of genetics', in Joseph Needham and Walter Pagel (eds.), *Background to Modern Science*, Cambridge: Cambridge University Press, 1938, p. 241).

[253] [Strachey] 'Joan Riviere', p. 230.

with the consequent scattering of many analysts from London and the economic uncertainties and belt-tightening, the British Society went into a military campaign of its own over three interconnected issues in the period 1942–44, with three different theatres of operation. Disagreements concerning the relationship between psychoanalysis and its wider public led to a series of Extraordinary Business Meetings between February and June 1942: these expressed concern at the Constitution of the Society, which had allowed a few members – Ernest Jones and Edward Glover in particular – to hold powerful offices for many years. The movement against this autocratic exercise of power and the consequent corruption of economic and scientific allegiances eventually resulted by the end of the war in reforms of the Society's rules so that members could not hold offices for longer than two or three years.

However, well before this issue could be resolved, the disputes had already overflowed into the second theatre: the scientific debates about the character of the Society's dominant theoretical principles – the conflict between those who regarded the ideas developed by Melanie Klein, particularly in the years since 1934, as inimical to the fundamental Freudian teachings on which the Society, in its very Constitution, was founded, and those who regarded Klein's theories as legitimate extensions of, as new scientific discoveries building upon, Freud's ideas.

The principal theoretical disagreements concerned the following questions: What do we know of the early life of the infant? Are psychoanalysts to revise their view of the pre-eminence of libido in the light of Klein's findings about the importance of the destructive impulses? How should analysts manage or conceive of the axiomatic but permanently unresolved question of the relations between biology and psychoanalysis, between a theory of instinctual development and the findings of clinical work with adult and child patients? Finally, and most irresolvable: Who is the legitimate heir of Freud – Anna Freud, who had recently arrived with her father from Vienna, or Melanie Klein, whom the British Society had embraced since the 1920s, creating the 'English School' of psychoanalysis? Is psychoanalysis the exposition and filling in of Freudian doctrine or is it a developing theory, ripe for expansion using new and persuasive ideas and techniques? The Kleinians in particular demonstrated their orthodoxy by displaying themselves as hermeneutic virtuosi when it came to Freud's texts, forcing the Viennese Freudians, together with Glover, to display barely controlled outrage at having their settled canonical readings exposed as narrowly conceived textbook versions of Freud's writings.

The 'Series of Scientific Discussions on Controversial Issues existing within the Society' began on 27 January 1943. In the course of seven meetings in 1943 and 1944, leading Kleinians presented a series of position papers, followed by extensive discussions sometimes stretching over four months per paper: first, Isaacs on the nature of fantasy, then Heimann on introjection and projection in early development, then, in early 1944 Heimann and Isaacs on regression, and lastly Klein herself on the emotional life of the infant. However, by the time Klein came to give her paper in early 1944, the battle had already been won and lost. A less visible, but finally more destructive, closed battle was being waged in the third and most secret theatre of operations, the Training Committee (TC), which had been asked to come up with a means of preventing the Society's training of analysts from dissolving into factional preferment. The fall-out from the deliberations of that committee were to transform the Society's governance and undermine any resolution of scientific differences within the Society.

One army was constituted by an alliance of convenience between the Viennese (Anna Freud and her group of Viennese refugees defending the Freudian orthodoxy they had brought with them) and a group willing to attack Mrs Klein and her group with any weapon that came to hand – Edward Glover, Melitta Schmideberg (Klein's daughter) and her husband Walter Schmideberg.[254] The opposing army in these scientific discussions was made up of Klein's group: Susan Isaacs, Joan Riviere, Paula Heimann, Donald Winnicott and John Rickman. The four Bloomsbury analysts were ranged on neither side, joining with Sylvia Payne, Ella Sharpe, Marjorie Brierley, William Gillespie, John Bowlby and Michael Balint in trying to find common ground for the whole Society.

However, on the other fields of battle, the protagonists lined up rather differently. The first front was opened over the issue that Adrian and Karin cared much about: the public face of psychoanalysis. On 5 November 1941, Barbara Low gave a paper on 'The Psycho-Analytic Society and the public', discussed at three subsequent meetings, at which complaints were voiced about how the Society was run, its poor relations with other professional bodies, as well as their disquiet over the scientific differences; at a discussion on 17 December 1941, Rickman lost

---

[254] It is important to note that Melitta Schmideberg and Edward Glover had both been supporters of Melanie Klein in the early 1930s; indeed it was Melitta who introduced the concept of 'internal objects' which Klein made her own and which was to be the source of considerable debate in the late 1930s in the British Society. See Melitta Schmideberg, 'The rôle of psychotic mechanisms in cultural development', *IJP* 11 (1930), 387–418 and R.D. Hinshelwood, 'The elusive concept of "internal objects" (1934–1943). Its role in the formation of the Klein group', *IJP* 78 (1997), 877–97.

his temper, attacking the officers of the Society for their rudeness to the public and their lack of capacity to respond to the needs of the wider community.[255] Low and the Stephens had been active members of the Lecture Committee dedicated to giving psychoanalytic lectures to the general public, so it was not surprising that, 'following these Scientific Meetings, four members (Barbara Low, Melitta Schmideberg, and Adrian and Karin Stephen) demanded that the Council call an Extraordinary Business Meeting to discuss the state of affairs in the Society'.[256] Now battle would be joined. Resolutions were invited. Adrian submitted the following: 'That it is desirable that the rules of the Society be so altered that the continual re-election to office (other than the office of Honorary Treasurer) is rendered impossible.'[257] The decisive issue for Adrian was the constitution of the Society and, behind it, the power that Ernest Jones and Edward Glover had exerted for nearly twenty years. In this struggle and debate, the émigré analysts held back from involvement, sensing that the Constitution of the Society was a matter for the 'indigenous' English members.[258] At this First Extraordinary Meeting, it was Karin who made explicit the other principal issue: not only, agreeing with Adrian, the Constitution, but also the question of training. She introduced a key term – 'autocratic' – to describe the Society's mode of governance in its first twenty years of existence[259] and gave a socio-psychological analysis of the patterns of intimidation to which this gave rise:

> Those who become entrenched in power are apt to grow contemptuous of and condescending to the rest of the members and this always arouses resentment. Resentment leads either to intimidation or else to non-cooperation. In the end it usually results in open revolt, with ill-will on both sides. Our Society has suffered for years under the first phase, intimidation and non-cooperation; we are now witnessing the second phase, open revolt and ill-will, which requires courage, independence and a free exchange of ideas. Both are fatal to creative work. By redistributing and reducing the concentration of power the proposed changes might end the existing causes of friction and create a better atmosphere for real scientific work. But resentment in a state of helplessness has not been the only reason for intimidation; there have also been more practical reasons for it: I mean economic dependence, which is

---

[255] FK, pp. 33–4.   [256] FK, p. 34.   [257] FK, p. 41.
[258] Mitchell Ash, 'Central European émigré psychologists and psychoanalysts in the United Kingdom', in Werner E. Mosse et al. (eds.), *Second Chance: Two Centuries of German-speaking Jews in the United Kingdom*, Schriftenreihe wissenschaftlicher Abdhandlungen des Leo Baeck Instituts 48, Tübingen: J.C.B. Mohr (Paul Siebeck), 1991, pp. 101–20.
[259] FK, p. 48.

another bad effect of the concentration of power in one or two hands. The inevitable result of this is that those who wield the power come to be regarded by the public as the sole official representatives of the Society and therefore have in their hands the distribution of the vast majority of the cases who apply for psychoanalytic treatment. Members who depend on having cases passed on to them cannot afford to make themselves unpopular with those on whom their livelihood depends, either by criticism or even by vigorous independent thinking.[260]

On her next topic, the problem of the training analysis, she spoke some hard words:

> It is unsatisfactory in the extreme that so many members should be in the patient–analyst relation with one another over periods of years and years. Straightforward adult equality relations, such as should hold between fellow scientists, are hardly possible in these circumstances, and if it is true, as was stated at one of our previous meetings, that members and candidates who are being analysed cannot always be sure that their analysts will respect their confidence, the position is even more impossible still.[261]

At the second Extraordinary Business Meeting on 10 June, Adrian spoke forcibly to the issue of institutional power:

> by allowing the prestige and the professional patronage of the Society and Institute to be permanently concentrated in the hands of a few men we have given these few men the power, should they wish to exercise it, to make or to break a large proportion of our Members . . . and that this power, whether actually exercised or not, is disastrous to freedom of speech.[262]

Karin, in her more forceful style, with the energy that her sister-in-law Virginia Woolf could not help admiring, put the issue even more starkly: members of the Society should vote 'in favour of altering our methods of election in such a way as to make it impossible in future for any individual or group to capture power over the Society or to retain power for more than a limited period'.[263] The targets were clear: Jones and Glover, who had, she implied, exercised year after year a greater and greater pressure of domination over all other members, by occupying all the central official functions for two decades, interviewing candidates, speaking on behalf of the Society and referring patients to those they favoured.

> It is not good for human beings to wield unchecked power: they become dictatorial and arrogant and there is too much temptation for them to favour supporters and penalize opponents or rivals. And it is humiliating

---

[260] FK, p. 48.   [261] FK, p. 49.   [262] FK, pp. 178–9.   [263] FK, p. 182.

for those who allow themselves to be ruled in this way: either they become subservient, or they become aggressive, or they become depressed and apathetic.[264]

All this debate was conducted with Ernest Jones in the chair. The meeting agreed 'that a committee be appointed to investigate the questions of tenure of office and the holding of multiple official positions'. By secret postal ballot, Anna Freud, Edward Glover, Barbara Low, Sylvia Payne and Adrian Stephen were elected to the Committee: not a Kleinian in sight, but with a nice balance of critics and defenders of the status quo.[265] With the proposal to limit the tenure of the Society's officers now under serious discussion, the scene of conflict shifted to the Scientific Meetings and the confidential proceedings of the TC. As Adrian Stephen declared in a letter to his sister Vanessa after the June 1942 meeting:

> I believe I have won the first round against the opposition of the President and most of the committee to vote in favour of a committee to review our rules and make it compulsory for officers to retire from office periodically ... I find that in my old age I have become quite a good fighter ... It is odd how pleased I feel, but I have had this row brewing for years and have always been persuaded against really starting it because the time was not right.[266]

Prior to the organization of the Scientific Discussions, a so-called 'Armistice Resolution' was passed in response to some of the inflammatory attacks, often personal, often by Edward Glover, that had been heard in meetings: 'That the Society require all Members to refrain from personal attack or innuendo in discussion, but also, strongly affirm the right of all Members to complete freedom of speech within the limits of common courtesy.'[267]

At the Annual General Meeting in July 1942, the next two theatres of conflict were opened up within the Society: following Marjorie Brierley's cool-headed call, quickly endorsed by the Society, for this 'temporary armistice', she drew up a programme for a set of meetings to examine the scientific differences in the Society. These Scientific Discussions began in the autumn of 1942, under the auspices of an organizing committee consisting of Glover, Brierley and James Strachey – in the event, Strachey played no part in their organization (principally managed by Brierley) and did not attend any of the discussions. At the same AGM, the TC was asked to 'discuss the effect of the present divergences on the training situation'.[268] The next round was prepared. Throughout 1943 the

---

[264] FK, p. 184. [265] FK, p. 210n1. [266] *Lighthouse*, pp. 158–9. [267] FK, p. 174.
[268] FK, p. 206, AGM, 29 July 1942.

Scientific Discussions continued. But, in the event, when the smoke of battle cleared, no resolution had been found, no scientific truth emerged victorious, just an exhausted stalemate.

Strachey and the Training Committee

It was events on the TC which decided the Society's future; this time the instigator, witting or unwitting, of mayhem and uproar was James Strachey. It was Strachey's standing in the Society, secured gradually over many years, that put him in a position to achieve this end. Allocated by the belligerent Glover to the 'English Freudians' (or 'Middle Groupers') as opposed to the 'Klein Party' and the 'Viennese Group',[269] he very pointedly took no part at all in the scientific controversies, attending none of the meetings; instead, he guided the Society's *International Journal* safely through the war. He made no enemies and was universally respected as clinician, administrator and training analyst.[270]

Above the battle he may have wished to be, but in private he had exceedingly strong peace-making views. Just at the moment when the Phoney War was about to turn into the Fall of France, on 23 April 1940, he wrote to Edward Glover to apologize for being unable to attend a meeting of the TC and allowed himself to let loose about the bickering and politicking between the Viennese and Klein factions:

> if it comes to a show-down – I'm very strongly in favour of compromise at all costs. The trouble seems to me to be with extremism, on both sides. My own view is that Mrs K. has made some highly important contributions to PA, but that it's absurd to make out (a) that they cover the whole subject or (b) that their validity is axiomatic. On the other hand I think it's equally

---

[269] FK, p. 136, 13 May 1942. In the discussions, the term 'English analysts' referred to what would later be called the Middle Group – it had a smidgeon of a reaction-formation to Strachey's complaints about 'bloody foreigners' about it; the 'English analysts' were referred to as those who were born and bred English men and women. On the other hand, during the 1930s, the term 'English School' was a term (as Anna Freud and others averred) used by those outside Britain to refer to Melanie Klein and her (English) followers. King and Steiner variously use the term 'indigenous' to refer to English analysts, born and bred, who engaged with Melanie Klein in the Second Series of Scientific Discussions (1944), or to those analysts who represented the 'indigenous culture' (256) of British psychoanalysis – the representative names are Ernest Jones, Joan Riviere, the Stracheys and John Rickman (i.e. those established as analysts and as key figures in the mobilization of organized psychoanalysis in Britain – through Society and institutional activity and through translation – before the arrival of Melanie Klein). One might say that the very term 'English' was part of the contest in the Controversial Discussions.

[270] Two other comparable figures, equally admired and respected by different 'parties', Marjorie Brierley and Sylvia Payne, also played crucial roles in keeping the Society from falling apart during the Controversies.

ludicrous for Miss F. to maintain that PA is a Game Reserve belonging to the F. family and that Mrs K's ideas are totally subversive.

These attitudes on both sides are of course purely religious and the very antithesis of science. They are also (on both sides) infused by, I believe, a desire to dominate the situation and in particular the future – which is why both sides lay so much stress on the training of candidates; actually, of course, it's a megalomaniac mirage to suppose that you can control the opinions of people you analyse beyond a very limited point. But in any case it ought naturally to be the aim of a training analysis to put the trainee into a position to arrive at his own decisions upon moot points – not to stuff him with your own private dogmas.

In fact I feel like Mercutio about it. Why should these wretched fascists and (bloody foreigners) communists invade our peaceful compromising island? – But I see I'm more feverish than I'd thought. Anyhow, I feel that any suggestion of a 'split' in the society ought to be condemned and resisted to the utmost.[271]

Ironies of prose there may be, but Strachey's forceful views condemned all warring factions as religious, power-hungry, megalomaniacal. Evoking the beleaguered position of Britain, beset by warmongering fascists and communists, a month before Churchill's famous speeches were mobilizing patriotic sentiment ('we shall prove ourselves once more able to defend our island home, to ride out the storm of war, and to outlive the menace of tyranny, if necessary for years, if necessary alone'), he was evoking 'our peaceful compromising island'. Strachey's initial resolve under these circumstances was to withdraw. He hardly attended the Society's business meetings and did not attend any scientific meeting. He took on the editorship of the *International Journal*, moved out of London to Alix's mother's country house at Lord's Wood; even there his dogmatic attachment to compromise at all costs was visible to the pacifist Frances Partridge:

> [Alix thinks] we should have imported everything we wanted for the war right from the start, tinned it and buried it, thus leaving all available manpower free for the services and munition-making. In spite of our stupidity in *not* doing this she thinks we may well gain a crushing victory. James thought a compromise peace might be better.[272]

Strachey's means for reforming training and bringing the conflicts in the Society over training to an end were a refined version of his outburst to Glover: 'Why should these wretched fascists and (bloody foreigners) communists invade our peaceful compromising island?' Strachey

[271] FK, pp. 32–3.   [272] FPD, 3 June 1941, p. 55.

recommended the elimination of all extreme views ('bloody foreigners') and the English partiality for compromise *at all costs* as an essential element of psychoanalytic training. But he had to manage the situation adroitly to achieve this end. Once he had, there were extraordinary consequences which he undoubtedly had not foreseen.

Having been assigned the task of responding to the influence of the 'present divergences' on the training in the Society, Edward Glover, as Chair of the TC, opened the discussion with an inflammatory and dogmatic 'Introductory Memorandum' (21 September 1942), which set down that the central question that the TC should address was 'teaching'. What should be taught? And who should be authorized to teach it? The answer to the first question was simple: 'the main body of psychoanalytic knowledge, in a word mainly Freud'.[273] The answer to the second question flowed straightforwardly from the answer to the first: 'Only analysts whose views do not contradict in any important respect the main body of Freud's teaching. Should any appointed teacher find himself developing an important disagreement with Freud's teaching he should inform the Training Committee of the fact.'[274] The entire thrust of Glover's remarks indicated that the Klein group could not conform to his dogmatic criteria; hence they should be excluded from teaching.

However, the TC pointedly did not endorse Glover's Memorandum. This is not altogether surprising: the forces ranged generally within the Society were all reasonably evenly represented on the Committee: Glover (Chair), Sylvia Payne, Anna Freud, Ella Sharpe, Melanie Klein, James Strachey, Marjorie Brierley and John Rickman. To neutralize Glover's confrontational approach, they asked Strachey to provide another Memorandum, which was discussed at a meeting on 24 February 1943. Strachey opened his draft with an inimitable Cambridge flourish:

> this problem which has been confided to us is ultimately a *political* and *administrative* problem and not a scientific one. A student of psychology, having arrived at what seemed to him the truth about certain phenomena, might quite well, like Henry Cavendish in the field of physics, be satisfied with leaving things at that – and decide to keep his conclusions to himself. The Members of the Psycho-Analytical Society, for various reasons, have decided on the contrary that they wish to disseminate their conclusions as widely as possible, and the business of this Committee is to consider how best to carry that decision into effect. We are thus concerned with problems not of theory but of practical expediency.

[273] FK, p. 597.   [274] FK, p. 597.

Henry Cavendish was one of the most remarkable natural philosophers of the eighteenth century, devoting his entirely solitary life to experimental labour, making major discoveries in chemistry, magnetism, astronomy, geology and electricity, deploying instruments he refined himself to a remarkable degree of sensitivity and accuracy. He published very little. Like Gauss, his unpublished manuscripts proved, when examined in the course of the nineteenth and early twentieth centuries, to be full of anticipations – of heat theory, electrical theory and many other areas.[275] Strachey was therefore deploying, highly contentiously, the mythical figure of the lone discoverer of scientific truth to make his first point against Glover. Science and its truth, he declares, is nothing to do with its dissemination – witness Henry Cavendish. (And, he implies, and this was very probably his view, truth will never be arrived at by public debates, with incessant communication, of the sort the Society was indulging in.) So what the TC is concerned with is politics and administration – purely practical problems, the privileged field of compromise.

> Scientific problems must be faced with ruthless logic and clear-cut consistency: there is no half-way house between truth and falsehood. Administrative problems, with their considerations of expediency, their constant balancings of probabilities, call for flexibility and compromise. Yet there is no contradiction here. Political adaptability is not in the least incompatible with the strictest regard for scientific truth; nor for the matter of that,[276] is rigidity in the application of a belief any evidence that the belief so applied is a true one.[277]

But before turning to these practical, administrative issues, Strachey could not leave the question of truth for good and took a swipe at Glover in citing Freud's view (the only time he mentioned him, in contrast to Glover) that, unlike deductivist and dogmatic philosophy, psychoanalysis 'keeps close to the facts in its field of study, seeks to solve the immediate problems of observation, gropes its way forward by the help of experience, is always

---

[275] The editors of Vol. II of *The Scientific Papers of the Honourable Henry Cavendish, F.R.S.*, Cambridge: Cambridge University Press, 1921, noted that the publication of Cavendish's manuscripts was felt to be an obligation on the University of Cambridge 'on account of the intimate connection of the House of Cavendish with the University, maintained now through many generations' (p. v). Volume I, covering Cavendish's electrical and magnetic researches, had been edited and published by James Clerk Maxwell, first Director of the Cavendish Laboratory. By citing this famous son of a famous family with intimate connections to Cambridge through many generations, especially through the foundation of the Cavendish Laboratory, Strachey was reaffirming a particular ideal of a scientist (drawn from a Cambridge source) in order to contrast this with the social and professional obligations of the BPaS.

[276] This phrase is probably a mistranscription of 'nor for that matter'.   [277] FK, p. 603.

incomplete and always ready to correct or modify its theories'.[278] And again, Strachey could not avoid taking yet another detour on the question of truth before getting down to practicalities, pointing out that, even when it comes to scientific truth, it is entirely a question of expediency whether candidates are educated in 'psychology, certain departments of medicine and physiology, education, anthropology', let alone 'the phenomena of geophysics ... the surface-tension of the earth, the laws determining the epicentres of earthquakes'.[279] As we noted in Chapter 6, Strachey was here expressing an oddly formed nostalgia – repelled and yet admiring – for the Cambridge polymath Harold Jeffreys, who had thought it perfectly sensible to inquire into psychoanalysis, geophysics, applied mathematics, statistical reasoning and ecological modelling in the same spirit. At least two of the TC would have known of whom he was talking: Ella Sharpe, Jeffreys's second analyst, and John Rickman, also a member of the 1925 Group.

So, Strachey argues, given that psychoanalytic truth will always be, according to Freud, partial, tentative and changing, and that there will be disputes over the truth of different theories, even disputes over what counts as an important difference, 'To what extent and in what respect do false or *defective* views about the findings or theories of psychoanalysis imply incompetence to carry out a training analysis, to do control work, to conduct a seminar or to give a course of lectures?'[280] Strachey's next step was to argue that there is a considerable gap between theory and technique:

> I am prepared to insist that upon the whole a valid technique is not necessarily or even chiefly the *product* of our scientific findings and theories, but that it is rather the most efficient instrument by which they can be reached. And I suggest that the essential criterion of whether a person is fit to conduct a training analysis is not whether his views on aetiology or theory are true, but whether his technique is valid.[281]

Strachey thus shifts the grounds entirely towards a problem which no one has an adequate answer for: 'what are the essentials of this valid technique. That must be our first and principal task ... At all events, if there are to be divisions in this Committee, it is surely at that point that they must be focused and not in the realm of inference and theory.'[282]

---

[278] FK, p. 603, quoting 'Two encyclopaedia articles' (1923), whose first English translation Strachey had in the previous few weeks prepared and published in the *IJP* for 1942. See *SE* XVIII, 263 and 254.
[279] FK, p. 604.   [280] FK, p. 605.   [281] FK, p. 607.   [282] FK, p. 608.

Tenacious convictions about the errors of the opposing clique were, in his diagnosis, the heart of the problem, so Strachey, with dry humour, evoked 'the extraordinary and wildly unlikely situation that would arise if I were *not* absolutely and positively certain of these [findings and theories] but merely considered them on the whole probable or was even, horror of horrors, in complete doubt as to where the truth lay?'[283] Strachey concluded that the only way to get clear about the impact of the Society's differences on the issue of training was to establish 'what are the essentials of a valid psychoanalytic technique', and then decide if 'the parties to our current controversies diverge from those essentials'. If so, the TC can address 'the purely political question of whether it is expedient to inhibit any of those concerned from functioning as training analysts or in any other educational capacity'.[284]

Not surprisingly, Glover did not agree: technique in his view was much more closely tied to theory than Strachey argued. And the Society's problems were insoluble because theoretical differences had become inextricably bound up with training as a result of the unresolved webs of transference and countertransference relations which made up the various groups or factions within the Society. Glover seemed intent on producing a confrontational stalemate between these warring factions. But the TC, in particular Payne and Brierley, were not to be cornered like this; they took up Strachey's recommendations that the Committee investigate the individual techniques of members of the Committee in order to try and throw light on what Strachey had called the 'unholy mystery'[285] of each analyst's technique. So, in the course of the autumn of 1943, first Brierley, then Freud, followed by Klein, Sharpe and Payne, delivered memoranda on their personal way of conducting analysis.[286] Despite their differences, the English analysts (including Klein, who expressed her 'full agreement'[287] with Strachey's paper) indicated that when it came to interpretation as part

[283] FK, p. 609.   [284] FK, p. 609.   [285] FK, p. 608.
[286] The three male members of the TC, Glover, Strachey and Rickman, never got to deliver their memoranda. It should also be noted that Anna Freud did not disclose much of her technique, speaking in generalities and establishing what she took to be the orthodox view of the development of Freudian technique – in contrast to Brierley, Klein, Sharpe and Payne, who wrote riveting personal accounts. And we can note how Payne, for example, gave firm support to the open-minded, 'compromise-at-all-costs' approach of Strachey, e.g. when she stated: 'no analytical technique is sound however applied if the analyst regards it as the only method of saving the patient and as an exact method depending on exactitude for its success' (FK, p. 648). She may have been firing a covert shot across Edward Glover's bow, since he was the author of a well-known paper entitled 'The therapeutic effect of inexact interpretation: a contribution to the theory of suggestion', *IJP* 12 (1931), 397–411.
[287] FK, p. 638n1.

of technique, Strachey's classic paper of 1934 had made him the authority.
Yet this unprecedented, leisurely close examination of the details of analytic technique was cut short by the draft report Strachey had been asked to
prepare. Strachey could now be bolder in this second version, the Draft
Report, since his first Discussion Memorandum had succeeded in shifting
the focus from the question of orthodoxy to the question of a 'good-
enough' technique.

Strachey's argument in the Draft Report of November 1943 was carefully
constructed to undermine Glover's attempt to exclude from analysing,
supervising or even giving seminars to trainees anyone who departed from
the accepted teachings of Freud. Rather Strachey proposes to exclude
anyone, including a so-called follower of Freud, who displays excessive
investment in theoretical questions and demonstrates dogmatism in relation to others: 'persons who have a strong emotional interest in the
prevalence of their own theoretical views among the present and future
generations of psychoanalysts – however eminent their services to the
science of psychoanalysis may have been – are not likely to be best fitted
to carry out the particular tasks of a training analyst'.[288] Turning to the
control analyst, who supervises the cases of trainee analysts, he emphasized
the 'quite peculiar qualities of judgement, balance and tact' required –
qualities very probably absent in theoretical grandees. At every turn,
Strachey argues against dogma and authoritarianism, neatly turning the
tables on Glover when it came to lectures and seminars:

> the Committee believes it to be important that no attempt should be made
> to shield the Candidate from the impact of what may be regarded in one
> quarter or another as heterodox beliefs. It should be an essential part of every
> Candidate's training to attend seminars and lectures given by partisans, and
> even by extreme partisans of any conflicting trends of thought that may be
> active in the Society.[289]

Strachey's now relentless argument about the noxiousness of dogmatic
theoretical views even extended to the election of members of the TC:

> it may be asked whether those who have taken or are taking an active part in
> controversies and are both bitter and personal can reasonably be chosen to
> superintend the education of young practitioners. There is even a risk of
> training analyses being generally regarded as a means of diffusing particular
> sets of opinions so that the Training Committee may cease to be an organ of
> education and become instead a battle-ground for warring parties and a

---

[288] FK, pp. 655–6.   [289] FK, p. 656.

political key-point, the control of which will decide the fate of this or that faction in the Society.[290]

If you have strong and publicly held views on theoretical controversies, Strachey implies, you are disqualified from being a training analyst or member of the TC. Voicing such views in the middle of the Controversial Discussions, when Anna Freud and the Viennese group were locked in debate with Melanie Klein and her 'followers' (as they were always called), targeted key members of the TC: Freud, Klein and Glover. Strachey's superb summary of his own argument was as unambiguous as his resentment of bloody foreigners had been back in 1940:

> If training analyses are to be conducted with complete concentration upon the one essential of improving the Candidate's mental functioning, if control analyses are to give him sympathetic but non-dictatorial help in his early efforts, if seminars and lectures are to offer him a truly impartial field of controversy – then the Training Committee itself must be possessed of such qualities as patience, moderation, tolerance, open-mindedness, and (in practical affairs) a taste for compromise. And it must be added that there seems to be no place in it for such ingredients as over-enthusiasm for innovation or over-rigidity in opposing it, to say nothing of personal enmity and resentment.[291]

English virtues – 'patience, moderation, tolerance, open-mindedness, and (in practical affairs) a taste for compromise' – were James Strachey's response to the battles threatening to tear his Society apart. So anodyne a recommendation in other circumstances – who could disagree that these quiet virtues were a Good Thing? Yet in these circumstances, advocating peace, love and fraternity would also have been inflammatory – Strachey's virtues proved to be incendiary devices.

His proposed recommendations flowed directly from his analysis, even adding some new subtleties to the imperturbably radical argument:

1. That as soon as practicable after the end of the war in Europe, the Society should proceed to the election of a new Training Committee.

Mentioning the 'war in Europe' at this moment, November 1943, the day when the Opera House in Berlin was destroyed by Allied bombing, when the campaigns in Italy and in Ukraine were turning the war decidedly in the Allies' favour, was also a subtle way of underlining how the period of hostilities in the British Society would also reach its inevitable end and

[290] FK, pp. 657–8.   [291] FK, p. 658.

enter a new era of peace, needing renewed institutions in an atmosphere of hope. His recommendations concerned a TC which would be a peace-time TC.

2. That, in choosing the members of the new Training Committee, the Society should deliberately bear in mind the undesirability of appointing persons who are prominently involved in acute scientific or personal controversies.

3. That it should be a definite instruction to the new Training Committee that it should so far as possible avoid selecting for the functions of training or control analysts persons whose desire to enforce their own extreme or rigid views shows signs of impairing the correctness of their technical procedure or of interfering with the impartiality of their judgement.

Strachey's recommendations were deliberately and explicitly targeted at both the TC and the very idea of a training analyst. Correct technical procedure and ability to stand back from any controversy: these were the essential qualities; in addition, if these were compromised by extremism or dogmatism, an automatic disqualification from being a training analyst followed.

4. That, on the other hand, it should be a definite instruction to the new Training Committee that every Candidate should, by means of attending Seminars and Lectures, be given an opportunity of obtaining the closest and most extensive knowledge of all sections of opinion in the Society, including the most extreme.[292]

So extremism in lectures and seminars was quite another matter, almost to be welcomed. Strachey's arguments had deployed the specific logic of psychoanalytic transmission. Unlike Glover, for whom all components of psychoanalytic training – analysis, supervision, lectures and seminars – could be subsumed under the rubric 'education' (and education was conceived of extremely dogmatically), Strachey distinguished sharply between on the one hand analysis and supervision, whose aim was entirely the same as therapeutic analysis, and on the other lectures and seminars.

Having written an excoriating response to James Strachey's Draft Report, immediately rebutted by Sharpe, Edward Glover resigned from

---

[292] FK, p. 659.

the Society with effect from 25 January 1944.[293] He chose that date so that his riposte to the Draft Report to the TC could be taken into account at its meeting on the 24 January. After that meeting, Melanie Klein immediately wrote, in strictest confidence (!), to her followers, that the reason Glover had resigned was because the 'majority of the members of the Training Committee, not consulting me at all in this matter, had united against him and expressed their distrust of his partisanship'.[294] But just as important as Glover's resignation and the purportedly united distrust of Glover, one might have thought, was the fact that Strachey's recommendations threatened Klein with exclusion from the TC.[295] Anna Freud certainly interpreted Strachey's Second Report in this way: 'Miss Freud asked if she was one of the parties alluded to in Mr Strachey's draft report and said that if so she considered it to be an insult, and she wished to resign from the Training Committee.'[296] Even though the Committee asked her to reconsider her decision, she did resign. Anna Freud's group immediately withdrew from the Scientific Discussions. The fire went out of all debate on each of the three fronts.

At the meeting where Glover's and Freud's resignations were announced, James Strachey suggested discussion of his report be postponed. At a meeting two weeks later, after hearing Klein's response, Strachey was asked by the TC to 'reconsider some items in his report and submit a draft of the revised version to the members of the Training Committee for criticism and comments'.[297] This revised version, now public to all members of the Society, was submitted to a business meeting of the whole society on 8 March 1944, discussed and accepted. For once, Strachey attended and spoke – but only about matters relating to the Report.

On that day, what members of the Society who were not members of the TC wanted to know was which parts of the confidential Draft Report had so enraged Glover that he left the Society and so offended Anna Freud that she resigned forthwith from the TC. Sylvia Payne and James Strachey did a very good job of ducking and weaving on this issue. James's third go at drafting the Report was subtly altered: the Final Report no longer

---

[293] FK, p. 853. Years after these events, Glover wrote: 'More than any other factor the compromise organized by the training committee in 1944 determined my ultimate resignation' (L.S. Kubie, 'Edward Glover: a biographical sketch', *IJP* 54 (1973), 85–94 at 93). It was Strachey's insistence on 'compromise' ('our peaceful compromising island') that would always be impossible for Glover to accept.

[294] FK, pp. 667–8.

[295] She did write a response to Strachey's report, read at a meeting a fortnight after the news of Glover's and Freud's resignations; unfortunately this response has not survived.

[296] FK, p. 665.   [297] FK, p. 667, 9 February 1944.

recommended excluding from the TC 'persons who are prominently involved in acute scientific or personal controversies', nor did it recommend explicitly that the new TC 'so far as possible avoid selecting for the functions of training or control analysts persons whose desire to enforce their own extreme or rigid views shows signs of impairing the correctness of their technical procedure or of interfering with the impartiality of their judgement'. But it retained Strachey's crucial distinctions between science and administrative politics and between theoretical views and analytic technique, so that all the weight of the recommendations now fell on the Report's portrait of two principal contrasting disqualifying traits: 'an enthusiast seeking for converts to his own way of thinking, and on the other hand a timid sceptic, paralysed by his insistence upon impartiality, moderation and compromise'.[298] Was this new figure of the 'timid sceptic' paralysed by his insistence on compromise a piece of self-analytic self-portraiture drawn by James Strachey in the wake of the departure of Glover and Freud? Or was it a piece of collective self-analysis, most probably performed by Sharpe, Payne, Brierley and Strachey? The sources give no answer.

Having depicted the Scylla of enthusiasm and the Charybdis of timid scepticism, the Report's main weight for choice of members of the TC now fell on the positive character traits required – traits which, it has to be said, were more akin to those displayed by the sceptic than the enthusiast, very like Strachey's original virtues which had provoked Glover's and Freud's resignations: 'patience, moderation, tolerance of opposition and criticism, open-mindedness, width of outlook, good judgement of character, capacity for co-operation and (in practical, though not in theoretical, affairs) a taste for compromise'.[299]

To the Society, Strachey clarified the genesis of the Report: 'I was asked to draft the report and I did the wording, but the content represents the opinions of the Training Committee as far as possible. In fact the report was drafted and re-drafted three or four times and as it stands now it represents the views of those who signed it and in some parts also the views of those who did not sign it.'[300] And he then gave his view of the basic thrust of the Report:

> I made a fair attempt of dealing with the difficult question of relating theory
> and practice of training. The report is not attempting to deal with the
> differences and disputes in the Society which I hope are now finished, but

[298] FK, p. 675.   [299] FK, p. 678 – an expanded version of Strachey's original list.
[300] FK, p. 881.

tries to prevent such disputes coming up later. All our experience shows that there are splitting tendencies not only in our Society but also in others, for instance in the U.S.A. If we could avoid this happening again it would be a great advantage. There are always personal emotional factors involved with which we cannot deal. But we can try to revise the machinery of training and devise a machinery which will as far as possible prevent these splits. The report puts the stress in training on the technique of the training analyst rather than on the theoretical views he holds. It is possible to disagree with an analyst's theoretical views and at the same time this analyst can be a competent analyst. If this is generally adopted we can hope to be able to prevent splits.

I move that the recommendations of the Training Committee be adopted by the Society.[301]

After further discussion, these recommendations were accepted.

With the battle drawn to a close for want of opponents to Klein's group, the Society's Constitution was now changed: Sylvia Payne, the new Chair of the Society, presided over a meeting held on 26 June 1944, attended by the smallest of groups (Bowlby, Gillespie, Isaacs, Barbara Low, Helen Sheehan-Dare, Karin and Adrian Stephen, Winnicott), all supporters of a change in the rules that they now voted in:

> 6. No member who has served as President, Scientific Secretary, Training Secretary or Business Secretary, for three consecutive years shall be elected to the same office until two further years have elapsed.[302]

'Complete victory. I have never fought a battle like that in my life before', Adrian wrote to his sister Vanessa that same day, 'and I never dreamt it would be so successful' – it was 'great fun'.[303] The Cambridge chapter (Bowlby, Isaacs and Winnicott led by the Stephens) were present to confirm their victory.

Adrian Stephen had won his battle. He had sought confrontation to achieve a constitutional revolution. He had reasserted the liberal values of freedom of rational discussion as the foundation for any scientific Society, pushing to one side arguments derived: from the history of the Society (the founders – Ernest Jones, James Glover, Edward Glover – of necessity had to give firm leadership when psychoanalysis was as yet inchoate and a controversial medical practice); from the uniqueness of the development of psychoanalysis (it is more a professional school than a scientific forum); and from the internal logic of Freud's charismatic leadership, which required each fledgling Society to repeat the inevitably autocratic style of

---

[301] FK, pp. 881–2.   [302] FK, p. 897.   [303] *Lighthouse*, p. 173.

the Primal Founder (an argument Ernest Jones, defensively, but with powerful logic, himself rebutted by deflecting the charge of 'autocracy' with a defence of 'aristocracy').[304] During the First World War, Adrian Stephen and James Strachey had worked together on the National Council against Conscription. They and their Bloomsbury friends fought the political warmongery of militarists, the hypocrisy and cynicism of politicians who, through introducing conscription, had in their view sunk as low as the Germans they intended to crush. During the Second World War, within the BPaS, they found themselves engaged in another battle against the abuse of authority and blind dogmatism. Yet, in this battle, Adrian Stephen and James Strachey found themselves on opposite sides. And ironically it was James Strachey, more than any other single individual, who forced the most dramatic of the events: the resignations of Edward Glover from the Society and Anna Freud from the TC. How did this come about?

If Adrian Stephen came to relish the confrontation with the Society's authorities and to find the resources within himself to become an uncompromising leader of a palace revolution, James Strachey was virulently opposed to confrontation and fought for compromise at any cost. He opposed Adrian Stephen's call in June 1942 for a reform of the Society's election rules, denying that a battle for power between cliques was taking place and that there were both ideological (scientific) and economic (access to patients) causes fuelling this battle: 'He is creating a melodramatic situation. What are the two parties that struggle for power? They are in fact represented among the officers. So I do not see the point at all.'[305] Adrian clashed swords with James in public over the reality of the economic power exerted by politically powerful individuals, but made his strong views clear in a letter to his sister Vanessa about his fight in the Society: 'I find that in my old age I have become quite a good fighter. That miserable James Strachey, lickspittle that he is, was against me – I think because his practice depends on the goodwill of Jones and Glover.'[306] So, to Adrian, it looked as if he and James were in that rivalrous and oppositional position they had occupied so many times in their lives, probably

---

[304] Jones to Anna Freud, early 1942: 'I am very dissatisfied with the present unproductive activities of the Society, for which I hold Dr Glover partly, though by no means wholly, to blame. By nature I believe in aristocratic leadership, but I think there are occasions and I wonder if this is not one, where it is more successful to exert that leadership indirectly instead of covertly. Thus I am inclined to the solution of reducing the responsibilities of officials, making their policies or decisions more a matter of business meetings, and having the officials re-elected annually' (FK, p. 235).

[305] FK, p. 187.   [306] Letter, 5(?) June 1942, *Lighthouse*, pp. 158–9.

most passionately when they were both in love with Noel Olivier in their youth before the First World War. Perhaps long-held resentment at the different paths that he and James had had to follow to become analysts still rankled: James's so-easy passage via Freud's couch (surely one ingredient for the 'blood-curdling feelings of anxiety and remorse'[307] he felt in 1963 when considering the professional qualifications required of those now entering the psychoanalytic profession compared with his own risible qualifications in 1920) to full membership of the British Society compared with Adrian's and Karin's decade of slog and misery. But beneath this rivalry, there is a more interesting underlying unity in their political activities during the Controversial Discussions.

To understand James's opposition to extremist positions in the psycho-analytic world we must turn back for a moment to his long-term develop-ment. When he and Adrian Stephen fought against conscription they were both fighting for freedom (not pacifism) – freedom from state control. As we have seen, Adrian viewed psychoanalysis as a continuation of the struggle for freedom – the freedom of internal mental life that is the outcome, when successful, of the psychoanalytic process. James followed a similar path but, with greater originality, he made this the central problem of his professional intellectual life. Positively allergic to splits and scientific disputes, James in his classic paper worked to develop a theory of psychoanalytic practice that would *insulate* that practice from theoretical enthusiasms and dogmatisms. He adopted the same position ten years later in drafting the Report of the TC in the midst of the Controversial Discussions, but his efforts to take analytic practice out of the heated debate, by asking for members to agree on the basic technique of analysis and to eliminate enthusiasts ('bloody foreigners') from pedagogical power, then resulted in the plague of resignations and withdrawals of Glover and Freud, together with her group of refugee Viennese analysts. It was his implacable and unswerving search for compromise at all costs that had the exact opposite effect from his peace-loving intention. Although the possibility had been on the cards for years, it was Strachey who catalysed the eventual division (brokered in 1945 by Sylvia Payne, a true diplomat and peace-maker) between the Kleinian and the Anna-Freudian training programmes within the Society – a division that con-tinues, in less exclusive form, to this day. His drive for compromise was in the name of freedom for analytic practice, i.e. attempting to protect the space of analysis from external political forces. The description of the aim

---

[307] Quoted in Khan, '*The Psychoanalytic Study of the Child*', pp. 99–100.

of analysis written into the Final Report on Training in early 1944 indicates
this – and also shows close affinities with Adrian Stephen's therapeutic
ideal of internal freedom ('to allow more freedom to the patient's mind'):

> The primary purpose of a training analysis, like that of a therapeutic
> analysis, must be to facilitate normal mental functioning. It would thus
> include amongst its aims the liberation of the Candidate so far as possible
> from the influence of unconscious prejudice, including those arising from
> transferences, so that he might be free to make independent observations
> and unbiased judgements and to draw valid inferences from what he
> observes. This would afford the best guarantee against the inevitable limita-
> tions and errors to which the theoretical knowledge of even the best of
> training analysts must from the very nature of things be prone. For the
> Candidate, in proportion as he has been freed from his unconscious pre-
> judices, will be able to accept those of the analyst's views which can be
> confirmed, to supplement those which prove incomplete, and to correct or
> reject those which seem false.[308]

Adrian and James were thus resolutely united in their characterization of
psychoanalysis as aiming solely at mental freedom and such a characterization was entirely of a piece with their liberal political ideals. But the
political paths they then followed within psychoanalysis diverged considerably: Adrian wished to reform the political structure of the Society in
order to throw off the twin oppressions of oligarchic power and the
economic power that flowed from long-term domination by two individuals; James wished to protect at all costs analytic practice, entirely devoted
to freedom, from the surrounding pressures of scientific politics and the
formation of cliques. This antipathy to cliques they shared. In September
1944, when the palace revolution had achieved its goals and Adrian had
been appointed Scientific Secretary, he was called upon to explain himself
at the Society for his careless polemical views, expressed in a letter to an
analyst in training to whom he had written: 'I think we have really now an
opportunity of turning over a new leaf but we can only do it if we prevent
the Society falling into the hands of any "party", whether they are called
"Kleinians" or "Freudians".' Barbara Low, a long-time affiliate of Edward
Glover and the Viennese 'Freudians', 'asked how, since the Society is a
Freudian Society, there could be a Freudian *party* within it. She certainly
thought there should be no Kleinian party in a Freudian Society.'[309]
Adrian explained 'that he had not meant that the Society should not be a

---

[308] FK, p. 672; this description was taken over virtually unchanged from Strachey's Draft Report (FK,
p. 655).
[309] FK, p. 902, AGM, 4 October 1944.

Freudian Society. He had meant that he did not want it to get into the hands of any party.'[310] James and Adrian were united in wanting to rid the Society of parties, sects, those who were intolerant of others (e.g. on theoretical or 'dogmatic' grounds). Neither James nor Adrian cared deeply about the scientific disputes dividing the Society; they did care passionately about the freedom that was synonymous with the practice of analysis.

Oddly enough, the one Bloomsbury analyst who did care about the scientific controversies was Alix. She intervened in her inimitable, scholarly way by writing a paper, in early 1941 (which James published in the *Journal*), entitled 'A note on the use of the word "internal"'.[311] She distinguished three distinct senses of the term 'internal': mental, imaginary and 'inside'. The manner in which she illustrated the pitfalls attendant on the term 'internal' sparkled with Bloomsbury mordant wit at the expense of her psychoanalytic colleagues:

> When we say 'box' we usually know whether we mean a box on the ear or a box at the opera or merely a cardboard box. But when we say 'internal' we do not always know whether we mean 'mental', 'imaginary' or 'inside'.

The term 'internal', most specifically in the phrase 'internal objects', was the topic of intense debate within the BPaS from 1935 until the start of the Controversial Discussions.[312] So, very much in the style of the analytic philosophy that was now well ensconced at Cambridge and elsewhere, the style which Alix as a student at Cambridge and an avid reader of Bertrand Russell and G.E. Moore would have imbibed, she was warning that the confusions between the different senses, particularly between what is 'imaginary' and what is 'inside', might lead psychoanalysts to espouse 'a general theory that all imaginary phenomena are thought of by the subject as inside himself, or at any rate to proceed on that assumption' – which would obviously be erroneous. Similarly the confusion of 'mental' and 'imaginary' senses of 'internal' would lead us 'to confuse a *figment* of the mind with a *function* of the mind'.

---

[310] FK, pp. 901–2.

[311] Alix Strachey, 'A note on the use of the word "internal"' *IJP* 22 (1941), 37–43.

[312] Less diplomatic was a follow-up paper developing Alix's argument, by Karin Stephen, given to the Society in May 1945, 'Relations between the superego and the ego'. Like Adrian, Karin was not fearful of stepping on the most august of toes in this critical examination, part philosophical, part psychoanalytic, of two of the most important Freudian concepts from a point of view sceptical of the well-foundedness of the general concept of 'internal objects'. See Hinshelwood, 'The elusive concept of "internal objects"', pp. 877–97.

But Alix did not stop at this critical, 'analytic' argument – 'analytic' in the modern philosophical sense – castigating her colleagues for being fooled by verbal ambiguity; she took the next step, to a properly psycho-analytic (i.e. more 'genetic' or 'genealogical') argument:

> For obviously the idea of insideness is imbued with a great intensity and variety of feeling. It carries with it a sense of power, mystery and special truth, of guilt, knowledge and ignorance. Thus an inward conviction is more strong and true than just a conviction; inside knowledge is more knowing and often more guilty than knowledge alone.

Psychoanalysis, Alix claimed, has made considerable discoveries about the underlying phantasies about "inside things":

> They are, as we know, primarily occupied with ideas of discovering, destroy-ing or seizing objects inside the mother's body and of keeping and preserving from attack the objects in one's own body. They are very early phantasies and they are motivated by very strong desires and conflicting impulsions of rage, anxiety, jealousy and love. In consequence they have become very highly charged with feelings of guilt and have been deeply repressed.

The psychoanalyst is subject to the force of these fantasies; their distorting effect on the development of psychoanalytic theory can be very great.

> if he [the subject] has a special interest in psychology, he will be in danger of making some kind of compromise and of regarding other people's imagin-ary objects as objects not indeed inside their bodies but inside their minds in too concrete a sense.

Alix's target here is very clear: Melanie Klein's recent work. Alix is on Klein's side, unambiguously convinced of how much psychoanalysis owes her when it comes to knowledge of these 'inside things'; but the direction of her criticism is to both sides of the dispute developing in the British Society: against those who confused 'mental structure' (the theory of the 'inside') with 'mental contents' (the primacy of certain fantasies of 'inside-ness'). So, unfolding her conclusion, she gives a Kleinian account of why the Kleinians and the Freudians are at war:

> our misuse of the word 'internal' is a symptom rather than a cause of the difficulty we have in distinguishing between what is inside, what is mental and what is imaginary; and that this difficulty is inherent in the nature and status of phantasies about the things and events inside one's own and other people's bodies.

We see that neither side emerges unscathed from Alix's sceptical, philosophical-cum-analytic scrutiny.[313] Like her fellow Bloomsbury analysts, Alix would take neither side.

Strachey and the *Standard Edition*

By the late 1920s, James Strachey was analysing eight patients a day and doing little translation.[314] Freud often enough gave rights first to Americans, who therefore took the lead on translation. And in 1927 *The Future of an Illusion* was translated in England not by Strachey but by German scholar William Robson-Scott, graduate of Oxford where he was a friend of Christopher Isherwood,[315] Jones's analysand and an old school-friend of the art critic Adrian Stokes, from whom he had learned of Freud, and whom he in turn introduced to Jones and Klein.[316] James did translate 'Some psychological consequences of the anatomical distinction between the sexes' and *An Autobiographical Study* (for an American publisher) in 1927,[317] and he also was very willing, perhaps even preferred, to give assistance by going over the translations of others – most conspicuously his brother Apostle Sebastian Sprott's version of *New Introductory Lectures*. Eventually, in 1939, with the war, with Freud's death, with Ernest Jones evacuating to semi-retirement in his country cottage, Strachey took over editorship of the *International Journal of Psycho-Analysis*; he did the job for the duration of the war.

Obviously short of copy, the war editor drew on his own resources. The next three years of the journal were slimline, on rationing. James and Alix abstracted articles by the dozen (signing them A.S. and J.S.). A prominent part of each issue was 'Unpublished Freud', ten pieces in all between 1940 and 1943, starting in the first issue of 1940 with *An Outline of Psycho-Analysis*, a partly completed manuscript found amongst Freud's papers at

[313] K. Stephen, 'Relations between the superego and the ego', *Psychoanalysis and History* 2(1) (2000), 11–28; see also Noel Bradley, 'Response: Karin Stephen, the superego and internal objects', *Psychoanalysis and History* 4(2) (2002), 225–30.

[314] JF, 5 December 1927, p. 638.

[315] Peter Parker, *Isherwood: A Life*, London: Picador, 2004, p. xxx.

[316] W.M. McIntyre, 'The late William D. Robson-Scott', *IRP* 8 (1981), 119.

[317] See Editor's Note to 'Some psychical consequences of the anatomical distinction between the sexes', *SE* XIX, 243. However, in 1925 James informed Alix that he was translating Freud's 'Die Verneinung' ('It's by no means as easy as you might think' – 23/26 September 1925, BLSP 60715, Vol. LXI, March 1925–October 1925 (not in BF)), but he then dropped it when he was rapidly commissioned to translate Freud's entry 'Psychoanalysis' for the *Encyclopaedia Britannica* (BF, 30 September 1925, p. 301). The translation of 'Die Verneinung' by Joan Riviere appeared in the *IJP* in late 1925.

his death. It was as if he needed the editorial responsibility – and no doubt wartime shortage of copy as well as paper – to entice him back into serious translation, even of Freud. Perhaps Freud's death, and its effect on James, by adding to his sense of wanting to pull his weight in the life and death struggle of the war, may have played a part. Characteristically, Strachey never mentioned in these translated texts that he was the translator.[318]

Attending Freud's funeral on 26 September 1939, Leonard Woolf spoke to Ernest Jones of the need for a prompt biography of Freud. Jones was more exercised by his own plan for the 'Memorial Edition' of Freud's complete works in English, writing to James three days later of the urgent need to 'secure a definitive edition for generations to come: if it is done after our time, it can never be done so well ... I would suggest that we form an "ad hoc" committee, say we two, Mrs Riviere, Dr Payne and Rickman to thrash the matter out.'[319] Jones's conception of the 'Memorial Edition' emphasized this closeness of the translators and editors to Freud and his time, making possible, and only now, in this historical moment, the possibility of 'uniformly editing and annotating his writing in a way that would elucidate allusions and references the meaning of which would otherwise be lost for those after us'.[320] In the next few months, various forms of proposals were set out, consultations were entered into with American analysts, with Ernst Kris, with Anna Freud and others. On a proposal of 9 March 1940 from Jones, setting out the organizational structure that would be a permanent feature of the project, namely of a 'bottom level' of translators supervised by 'reviewers', James scrawled the following: 'Not inclined to be involved – anyway not unless paid. Journal takes up too much time.'[321] The Americans did go ahead with administrative organization in preparation for the edition; in May 1940 a committee, consisting of Franz Alexander, William Menninger and Sándor Radó, was appointed to advance the project.[322]

[318] We can only establish that he was the translator by examining the editorial notes in the *Standard Edition* published many years later.
[319] Riccardo Steiner, 'A world wide international trade mark of genuineness? – some observations on the history of the English translation of the work of Sigmund Freud, focusing mainly on his technical terms', *IRP* 14 (1987), 33–102 at 42–3.
[320] Jones to Sir William Bragg, President of the Royal Society, 13 December 1939, quoted in Steiner, 'A world wide international trade mark of genuineness?', p. 47.
[321] James Strachey, note appended 22 March 1940 to letter from Ernest Jones to Joan Riviere, Alix and James Strachey, concerning arrangements for translation and reviewing of proposed 'Memorial Edition', 9 March 1940; from BPaS website.
[322] 'The Forty-Second Meeting of the American Psychoanalytic Association, The Netherland Plaza, Cincinnati, Ohio May 19, 20, 21 and 22, 1940', *Bulletin of the American Psychoanalytic Association* 3 (1940), 16–65 at 39–40.

However, with the end of the Phoney War and the fall of France, even Jones let the plans for the Freud edition drop.

In July 1945, a new plan emerged from America, instigated by Otto Fenichel. Jones conferred again with Strachey: 'There is obviously one way in which a really proper Collected Edition translation could be done, namely by you two people [i.e. Riviere and James] and my self being indemnified for spending the rest of our lives in the task.'[323] In the event, with Riviere declining to be involved in any way, Jones and Strachey divided up the tasks which would indeed occupy the rest of their lives: Jones to the biography, Strachey to the *Standard Edition*. Handing over the editorship of the *Journal* to his old friend and sparring partner Adrian Stephen, Strachey was now more ready to take on the task. Jones worked hard to ensure that the project did not escape across the Atlantic to be run by the 'one American and six refugees'[324] on the Americans' Committee. Strachey's considered response to Fenichel spells out his clear vision – and preconditions – involved in a complete edition of Freud's work:

> Much of the work [i.e. the previous translations of Freud], it has always seemed to me, was done by people who understood neither psychoanalysis, German, nor English ... I do however feel that except in the very rarest cases, no one who has not been brought up from childhood in a language can possibly acquire an adequate grasp of its literary idiom. And that is where your team of editors and checkers seems to me to fall short ... failing a better alternative, I myself might possibly be employed usefully as some sort of stylistic checker. What I have in mind is that I might perhaps be given an opportunity of reading through each volume as it is completed and of tentatively noting for editorial consideration any suggestions of a stylistic kind that might occur to me. I may perhaps remark that I have in the past gone through a large number of the Hogarth Press translations in this way.[325]

'A sort of stylistic checker': this was to be James's self-assigned role. Very pointedly, his inflexible criteria excluded émigré analysts – most obviously Brill, but including his correspondent Fenichel – from the task of translating into English. (And he had immediately made a clarification of an obvious worry: 'what I am thinking of is nothing to do with differences of usage here and in America, which seem to be of trivial importance and easily adjustable.') Strachey's highest priority – which had been perceived and approved of from the start by Jones in the early 1920s – was for stylistically acceptable

[323] Steiner, 'A world wide international trade mark of genuineness?', p. 48.
[324] *Ibid.*, Jones to Strachey, 26 July 1945, p. 49.   [325] *Ibid.*, pp. 47–8.

English, attainable solely by those who had been swimming, linguistically speaking, in the bath of the English language from birth.

Otto Fenichel died suddenly in January 1946. Other projects were mooted, including a short-lived plan concocted by Leonard Woolf together with Anna Freud and James in 1946. The one constant in all the various projects was naming James Strachey as chief translator and editor. One crucial decision was taken: Strachey's proposal for the 'Memorial Edition' of eleven thematically organized volumes was productively criticized by Ernst Kris.[326] His insistence that 'only a chronological edition seems to me beyond criticism' (as in the *Gesammelte Werke* hastily produced in 1940) persuaded Anna and eventually others, including Strachey. Jones was not formally invited by the Freud family to write the biography until the late 1940s; progress on the 'Memorial Edition' of Freud's works was at a snail's pace in the late 1940s. So a crucial factor in getting both the projects moving was Ernst Freud, architect and Anna's brother, throwing himself into a role in 1950 he would increasingly adopt in the coming years: chief negotiator and trouble-shooter for the enormous administrative obstacles facing the project, in particular the chaotic copyrights Freud had negotiated so lackadaisically with his American and British publishers.

Faced with the problem of funding such a large project, Strachey now reversed his previous position and stated in a Memorandum on Remuneration for Translations that: 'In so far as the object is one of public interest, that is to say of providing improved editions of Freud, I am prepared to do what I can to help without any immediate or indeed any certain prospect of pay.'[327] But by the next year a firm funding model was in place, probably as a result of John Rickman, President of the British Society from 1947 to 1950, taking organizational matters in hand. A public announcement to the members of the American Psychoanalytic Association, who would be crucial to the financial underpinning of the project, was made. Members of the Association were asked to sign up to 400 sets out of the 700 to be printed, with the Commonwealth Fund (founded in 1918 by Anna Harkness and devoted to promoting health care) 'to be responsible for 200 sets to place in Educational Institutions throughout the U.S.A.'[328] Rickman grandiosely compared the edition to the '24 or 20 volume Oxford English Dictionary which came out letter by letter over about a dozen or more years'; he also encouraged subscribers to envisage

---

[326] Ernst Kris to Anna Freud, 13 November 1946, discussed in *ibid.*, pp. 40 and n8.
[327] *Ibid.*, p. 40n6.
[328] 'Committees at Work', *Bulletin of the American Psychoanalytic Association* 5B (1949), 24–32 at 28–32.

the eventual reader to be posterity: 'what in our short lives may seem to be a distant future' will see the copyright lapse.

> In that distant time, the reprinting of any of these 24 volumes will be at anybody's disposal (as with Shakespeare); what the Memorial Committee wants is to put before the public a set of translations made by persons who were in collaborative touch with the author that will be used by all English reading publics. A wide distribution of this accurate translation is therefore one of our long term aims so that this shall become the standard for all editions in the future.

He reiterated again that the English contribution was 'to put at the disposal of the community the unique attribute of some of our members, viz., that they have discussed Freud's translations with Freud himself'; by implication, the sole contribution of the Americans was to provide the money in a Marshall Plan for psychoanalysis – not only the rebuilding of psychoanalysis in Europe but also ensuring a Freud for eternity.[329]

Strachey was already in place as the principal translator and, as usual, kept a very low public profile. His name was not mentioned in this public announcement. But Jones and he were both setting to work in tandem; when Jones completed a draft of the first chapter of the biography, his first reader was Strachey, who, already involved in preparing an entirely new translation of *The Interpretation of Dreams*, in the variorum style the 'Memorial Edition' proposed, was able to pick up intricate contradictions and text-puzzle fragments which advanced the biographical reconstruction Jones was engaged on.[330] In 1950, Strachey had seen the final volume of Freud's *Collected Papers* into print. Marjorie Brierley declared in her review that 'the publication of this fifth volume of *Collected Papers* has made the whole of Freud's work available in English'.[331] This fifth volume, filled mainly with late works which he had already translated for the *Journal* during the war, together with some previously unpublished early papers, was part of the dry run for the *Standard Edition*, as was another entirely new translation by Strachey, of *Beyond the Pleasure Principle*, also published

---

[329] It is not certain that Rickman wrote the fund-raising letter to the Americans; certainly the person who did had consulted Strachey. To which of them we owe the comparison with the OED and Shakespeare is not clear – but it sounds more like the expansive Rickman than the self-effacing Strachey.

[330] Brenda Maddox, *Freud's Wizard*, London: John Murray, 2006, p. 261.

[331] Marjorie Brierley, Review: 'Collected Papers: Vol. V. By Sigmund Freud. Edited by James Strachey. No. 37, The International Psycho-Analytical Library, edited by Ernest Jones. (London: Hogarth Press and the Institute of Psycho-Analysis, 1950. Pp. 396. Price 25 s. net.)', *IJP* 32 (1951), 247.

in 1950. In addition, in 1949 Strachey revised his 1940 translation of *An Outline of Psycho-Analysis* for separate publication as a book.

The 1949 American advertisement for the 'Memorial Edition' had declared that 'about one-third of the whole material has already been adequately translated into English (subject to revision). Another third will require very considerable re-writing, while the remainder will have to be translated afresh.'[332] This sizing up of the task was a rephrasing of Jones's view in 1940 that 'one third need to be freshly translated, one third to be thoroughly revised and one third to be read through'.[333] In 1951–52, the BPaS reported that 'four books have been retranslated',[334] one of them *The Interpretation of Dreams*, which in 1953 was the first of the volumes of the *Standard Edition* to appear. With his prodigious pace of work, Jones completed his biography of Freud in less than the ten years he had anticipated. Strachey's pace was slower, averaging two volumes a year. Right from the start, he relied on a team: the indispensable Alix, but also Anna Freud, always ready to advise and generally hover, just as she did with Jones's biography. And then younger helpers: Charles Rycroft and Frances Partridge for the proof-reading, Frances for the indexes. The ample editorial introduction to *The Interpretation of Dreams* set out the case for his particular way of producing an edition of Freud's psychological writings in English: a variorum edition, especially for the book on dreams which had expanded and been changed so much from its first edition in 1900 to its eighth and final one in Freud's lifetime in 1930. Many cross-linking notes were inserted and Strachey envisaged many readers would be irritated by their number. So he broke cover and set out the explicit editorial principles on which he would base the whole of the *Standard Edition*:

> In any case, the fact must be faced that *The Interpretation of Dreams* is one of the major classics of scientific literature and that the time has come to treat it as such. It is the editor's hope and belief that actually the references, and more particularly the cross-references to other parts of the work itself, will make it easier for serious students to follow the intricacies of the material. Readers in search of mere entertainment – if there are any such – must steel themselves to disregard these parentheses.

Freud as the producer of scientific classics; an edition for serious students – a key term Strachey would repeat in 1966 when he completed the final

---

[332] 'Committees at work', pp. 28–32.
[333] Steiner, 'A world wide international trade mark of genuineness?', 9 March 1940, circular letter, pp. 45–6.
[334] R.S. Eissler, '107th Bulletin of the International Psycho-Analytical Association', *BIPA* 35 (1954), 379–400.

volume, Volume I, with its 'General Introduction': 'from first to last I have framed this edition with the "serious student" in mind'.[335]

Strachey's turn back to preoccupation with Freud's writings began even before Freud died in September 1939. His 'Preliminary notes upon the problem of Akhenaten' was published in the *Journal* immediately following the translation of 'If Moses was an Egyptian . . .' which eventually became Part II of Freud's *Moses and Monotheism*. Strachey's essay was something of a rebuke to Freud, who had failed to mention Karl Abraham's pioneering study of Akhnaten in 1912 in his *Moses*. This rebellious and critical strand of Strachey's relationship with Freud, while a minor note when heard alongside the major refrains of admiration and fidelity, would continue. It definitely found expression in his approach to the *Standard Edition* and is perhaps most clearly voiced in some critical comments he made to Ernest Jones in 1951, at a time when he was beginning the editorial labour on his variorum edition of Freud's complete works:

> Freud was quite extraordinarily inaccurate about details. He seemed to have had a delusion that he possessed a photographic memory . . . Actually a large proportion, I should say the majority of his literary quotations turn out to be slightly wrong if you verify them . . . I think he was probably quite right not to bother about such things, but it requires an effort on my part not to be irritated by them.[336]

Every time Strachey assiduously noted in Freud's books and papers the appearance of a seemingly new idea which was, in fact, a restatement or reshaping of an older idea, there was a knife that cut both ways: a reaffirmation of psychoanalytic findings or a covert recycling of old ideas. Yet this critical aspect of Strachey's labours is scarcely visible, since these continuities were always overshadowed by the sheer profligacy of Freud's new ideas and changes of mind.[337] One might even say there was a hint of obsessional defence about Strachey's 'fundamental rule': 'Freud, the whole of Freud, and nothing but Freud.'[338] This rule was erected in part, perhaps, to prevent Strachey himself doing too much tinkering. As he put it to Jones, who was negotiating with Anna over the Freud family's proposal to

---

[335] James Strachey, 'General Preface', *SE* I, xv.

[336] Strachey to Jones, 27 May 1951, quoted in Riccardo Steiner, '"Et in arcadia ego . . . ?" Some notes on methodological issues in the use of psychoanalytic documents and archives', *IJP* 76 (1995), 739–58 at 748.

[337] Freud's changes of mind – the developments of his theories – so much admired today were one reason for his contemporaries concluding that his work was 'unscientific'.

[338] *SE* I, xix.

do some 'rewriting' in better English of Freud's letters, about which Strachey was deeply shocked and angry: 'If one starts improving on his work I shall have quite a number of suggestions to make.'[339]

Monumental: this is a word that has often been used to describe the achievement of the *Standard Edition*. Strachey did emerge from the discussions of the 1940s to lead the project and he saw it through to the end, save for the final volume, *Indexes and Bibliography*, 'compiled' by Angela Richards, who had in fact been assisting the project virtually from the first volume in 1953 but was only finally granted a place on its title page in Volume I (published 1966) as Editorial Assistant. The official team, named on every title-page, was the General Editor, James Strachey, 'in collaboration with Anna Freud, assisted by Alix Strachey and Alan Tyson'. Tyson was a late addition to the 1953 team, alongside Frances Partridge bringing her years of experience editing the Greville diaries, who was commissioned to produce the indexes.[340]

---

[339] Steiner, ""Et in arcadia ego ... ?"", p. 751.

[340] From 1934 on, James had a hand in the organization of the founding and flourishing of the annual Glyndebourne Festival of opera – an hour's stroll across fine country from Fyrle (the Keyneses (at Tilton) and the Bells (at Charleston)) and two hours from the Woolfs at Monk's House in Rodmell. While James's enthusiasm for Mozart and Alan Tyson's later remarkable musicological discoveries through his ingenious study of Mozart and Beethoven manuscript sources, rooted in his collecting of rare editions starting in the very early 1950s, is the obvious tie between the two men, it appears that Tyson's enrollment on the team producing the *Standard Edition* was more the result of his own early enthusiasm for psychoanalysis and his subsequent Bloomsbury contacts. After serving in the Navy and taking a double First in Classics at Magdalen College, Oxford, Tyson (1926–2000) was reluctant to take the easy path of becoming a teacher like his father and was led to psychoanalysis in part by his intimate relationship with the African American psychiatric social worker Marie Battle (later Singer), an accomplished pianist with an interest in jazz who accompanied her own singing, sixteen years his elder, then (*c.* 1951) working in Hackney. Marie was training at the Anna Freud Clinic and living with Karin Stephen. Starting an analysis with Ilse Hellmann-Noach, one of Anna Freud's friends from Vienna who became her close collaborator in London, Tyson began training at the Institute in 1953 (qualifying in 1957 and practising until 1963). Around the same time he tutored in classics Burgo, the son of Ralph and Frances Partridge (FPD, 1 May 1953, p. 189), who just at that moment had agreed to provide James with an index for the *Standard Edition*. How Tyson made connections with the Bloomsbury analysts and their circle is not certain, but James Strachey quickly enrolled him in assisting in editing the *Standard Edition*. Despite his evident polymathic brilliance (and his recently awarded Prize Fellowship at All Souls'), Tyson barely knew German, but acquired it rapidly, so that his translation of *Leonardo* was ready by September 1955 and received Anna Freud's admiring approval. Another important achievement in collaboration with Strachey was 'A chronological hand-list of Freud's works', *IJP* 37 (1956),19–33, for which Tyson was lead author; this was clearly the basis for the detailed plan of the entire *Standard Edition*. In the event, Tyson contributed significantly and substantively to the *Standard Edition*: his entirely new translation of the daunting *The Psychopathology of Everyday Life* (1960) makes up Vol. VI; most of Vol. XI (1957), including *Leonardo da Vinci and a Memory of his Childhood* and the papers on the psychology of love, was his. He was an essential part of Strachey's team – to which was added around the same time Noel Olivier's daughter, Angela Harris – 'Miss A. M. O. Richards', as they called her in the 1956 'Handlist'. Tyson spent the years 1959–67 qualifying in medicine, before turning his attention in 1970 full-time for the next two decades to Mozart's and

It was undoubtedly the fruit of collaboration – but a collaborative project that had been long under way by the time of Freud's death in 1939. What Strachey's *Standard Edition* accomplished was to implement not his own personal vision but what Riccardo Steiner has aptly called 'the whole weight of the *élite* of the English psychoanalytical establishment within the international movement, and its own particular history'.[341] Included in that history was the crucial fact, for them, that they had known Freud: for Jones, as a long-time colleague, disciple and, in the end, intimate; for Rickman, Riviere and the Stracheys, as Freud's analysands and correspondents. The core translation team of Riviere and the two Stracheys had been put together by Jones in the early 1920s, with Rickman sometimes in, sometimes out; at that time, Riviere was the senior partner. But quite quickly it was the Stracheys who came to the fore, not in quantity of translation, since Riviere translated as much as they did in the 1920s, but as implementors of Jones's strategy for a standardized English Freud. *The Glossary for the Use of Translators of Psycho-analytical Works*, published for the Institute of Psycho-Analysis in 1924, was revised intermittently by both Stracheys from the mid-1930s on and eventually published under Alix's name in 1943. So, their work on a glossary for Freud's terminology started in 1921 and was completed with the publication of Volume I of the *Standard Edition* in 1966.

The extent of this terminological work was not made fully evident until Volume XXIV was published by Angela Harris Richards, in which she compiled an 'Index of terms and their uses', listing some 240 terms, around 50 of them German, to which Strachey had dedicated specific commentaries or footnotes. This Index was in addition to that of 'Editorial commentaries and annotations, given under key-words', numbering over 400 entries. It is these editorial commentaries, not so much the translation itself, that made the *Standard Edition* into the standard against which all other editions of Freud have been measured and from which they have borrowed much – notably in the Fischer edition in German, the *Studienausgabe* of the 1970s, a fine modern partial edition entirely reliant upon *SE* for its editorial annotations and scholarship. Indeed, James Strachey's blistering criticism of Alexander Mitscherlich's initial proposal in 1964 for a German edition modelled on the English *Collected Papers* ('an out-of-date relic which should

Beethoven's musical manuscripts. (See Oliver Neighbour, 'Alan Walker Tyson', *Proceedings of the British Academy* 115 (2002), 367–82.)

[341] Steiner, 'A world wide international trade mark of genuineness', p. 50. See also the pertinent comments in Steiner, 'Bloomsbury/Freud', p. 406.

be buried and forgotten as soon as possible'[342]) was then followed up by his offering, together with Angela Richards, to co-edit the *Studienausgabe* himself. After Strachey's death in 1967, Angela Richards became the English hand responsible, together with Mitscherlich and Fischer Verlag, for the production of the ten-volume *Studienausgabe*, whose editors were listed as 'Alexander Mitscherlich, James Strachey, Angela Richards'. With his habitual self-effacement, Strachey had written to Mitscherlich in 1965: 'My reason for wanting us [Richards and himself] to take back seats as editors of the paperback Ausgabe is merely that it will be absurd for a couple of unknown English people to set themselves up as editors of a German classic.'[343] His self-effacement was, this time, ineffective: the principal advertisement today on Fischer Verlag's website is that 'der Herausgeber der berühmten englischen Freud-Ausgabe, James Strachey, hat an der Konzeption der *Studienausgabe* noch selbst mitgewirkt' ('the editor of the famous English Freud edition, James Strachey, himself collaborated in the planning of the *Studienausgabe*').

If the presiding hand over the translations of Freud and of the psychoanalytical literature in general into English had been, since the Great War, Ernest Jones, who initiated a policy that had remarkable success once the team of Riviere, Rickman and the Stracheys had been put in place, it was the literary skills and sensibilities of this team that gave the English Freud its tone. When in 1921 Jones was in rivalrous competition with Brill over the Freud translations, he instructed Freud on how 'men of sensitive feeling, taste and education like Rickman and Strachey rightly shudder' to see Brill's 'punctuation as illiterate as that of a servant girl's, with expressions of a similar order'.[344] Jones had made the most of the stream of erudite Cambridge men – Tansley, Rickman, Strachey, Penrose – who formed the first cohort of analysts in the British Society of the early 1920s – and fussed, cajoled and flattered as needs be to keep them as core members. Alongside his surprisingly collegial long-term relationships with these men, did Jones feel intimidated, even envious, of the natural superiority that they exuded within British society of the inter-war years, with their Bloomsbury connections, their membership of the elite Societies, whether secret like the Apostles or open (but not to Jones) like the Royal? How

---

[342] Letter from James Strachey to Ilse Grubrich-Simitis, 16 March 1964, in Ilse Grubrick-Simitis, *Back to Freud's Texts: Making Silent Documents Speak*, trans. Philip Slotkin, New Haven: Yale University Press, 2006, p. 52.
[343] Letter from James Strachey to Alexander Mitscherlich, 25 January 1965, in Grubrick-Simitis, *Back to Freud's Texts*, p. 53.
[344] JF, 15 December 1921, p. 448.

daunted was he when he was obliged to recognize, as no doubt he did, that each of them was, as the anonymous obituarist for Adrian Stephen put it in 1948, 'a representative of a culture that has almost passed away. That generation of stern and brilliant thinkers among whom he lived did much to shape the course of England's destiny – their books we daily use, their formulations we still employ.'[345] Jones negotiated his own personal reaction to these representatives of the English cultural elite who had, like cuckoos, come to lay their eggs in his psychoanalytical nest: it was his achievement to have bound them to the task of realizing his vision of an English Freud.

So if it is clear that the language of the *Standard Edition* did not flow solely from a decision of one isolated individual, it nevertheless bears a particular stamp. Strachey willingly and knowingly reflects his class, his interests and his historical situation and turns it into editorial choices: 'The imaginary model which I have always kept before me is of the writings of some English man of science of wide education born in the middle of the nineteenth century. And I should like, in an explanatory and no patriotic spirit, to emphasize the word "English".'[346] This decision makes of Strachey's Freud a *historical* translation, in the language of an 'English man of science' of Freud's era – coming into his prime in the 1890s and 1900s, a man like Strachey's father, a notable engineer, geographer and meteorologist moving in the Huxley and Tyndall circles of Victorian positivism.[347] It is said to be a natural vice of translators to slip into a style that is just a bit out of date;[348] Strachey turned this natural temptation to his advantage by deliberately choosing an archaic voice for Freud. No writer writes outside of his own time, not even Chaucer, Shakespeare or Dickens – which is why we say 'Chaucerian English', 'Shakespearian English' or 'Dickensian'. Strachey could have attempted to render Freud in contemporary English – that is, English of the 1950s. (Though it would have been difficult for him, a man who came of age in Edwardian England, to have had his ear tuned to the Angry Young Men and Larkinian English, or the just out of date Eliot and Orwell.) In this sense, Ornston is accurate in his description of the *Standard Edition* as a 'deliberately elegant period

[345] Anonymous, N., 'Adrian Leslie Stephen 1883–1948', *IJP* 29 (1948), 4–6. It is most likely that Rickman wrote the obituary, but it is so well informed about Adrian's childhood and family that he was probably assisted by James Strachey and Karin.

[346] Strachey, 'General Preface,' *SE* I, xix.

[347] Linstrum, 'The making of a translator', pp. 685–704.

[348] D.G. Ornston, Jr., 'Alternatives to a Standard Edition', in Ornston (ed.), *Translating Freud*, New Haven: Yale University Press, 1992, pp. 97–113 at 103.

piece'.[349] Precisely: it was intended as such. Clearsightedly making a strategic decision, Strachey's Freud was never a Freud for our time (as Gay's biography wished itself to be) nor a Freud for Strachey's time, but an English Freud in *Freud's* time – the turn of the century, the time of Freud's generation of Cambridge men of science.

While the achievement of the *Standard Edition* has been widely recognized, the criticisms of its translations have also been fierce and wideranging. They are very telling – not so much for their scattergun accuracy, but for the misunderstandings and preconceptions they express concerning Strachey and, in his shadow, Freud. According to a range of critics, Strachey produced a stiff, overly scientific and medical Freud, standardizing terminology where the original German is always flexible. Many of these critics deploy questions of translation to advance covertly psychoanalytic theses against Freud, using Strachey as the stalking-horse; this is particularly true of Bruno Bettelheim, who wished to annex psychoanalysis to a humanistic psychology, idiographic rather than nomothetic, where there is palpably not a trace of this conception in Freud's work. At the end of his life, as throughout his psychoanalytic writings, Freud conceived of psychoanalysis as part of psychology: 'Psychology, too, is a natural science. What else can it be?'[350] Darius Ornston's heated polemics against Strachey's translation are equally attacks on Freud's scientistic self-understanding – Habermas's influential phrase, in his important book *Knowledge and Human Interests*, whose explicit agenda is also to annex Freud to a German tradition of philosophical hermeneutics. Every single one of Strachey's critics focuses considerable wrath on the translation of supposedly flexible German terms, immediately understandable by ordinary German speakers, into Latin and Greek terms – 'hellenisms' is the standard barbed epithet used by Freud's most recent translators, in the Penguin Classics translations of the early 2000s. These translators, who no doubt would have no compunction translating 'Fernsehen' as 'television' rather than 'far-seeing', 'Stickstoff' as 'nitrogen' rather than 'choking substance' or 'Kernspaltung' as 'nuclear fission' rather than 'kernel splitting', uniformly baulk at the translation of Freud's neologism 'Fehlleistung' as 'parapraxis' and 'Besetzung' as 'cathexis'. Exploring these antipathies to Strachey's terminological choices tells us much about the place of

[349] D.G. Ornston, 'Improving Strachey's Freud', in Ornston (ed.) *Translating Freud*, New Haven: Yale University Press, 1992, p. 222.
[350] *SE* XXIII, 282.

psychoanalysis in scientific and wider cultures – and also about the problems facing Strachey and the knowingly inadequate solutions he proposed.

Why would one assume that Freud's German terms have a non-hellenic and non-Greek or non-Latinate translation? Only if one assumed that Freud was a writer who deliberately avoided technical terms, 'that sort of technical *"esperanto"* based on Ancient Greek and Latin, which circulated through scientific and medical texts through Europe',[351] as Steiner fittingly describes it. Freud's career as a neurologist belies any view that, despite Strachey reporting that Freud 'disliked unnecessary technical terms',[352] he consistently avoided such technical terms: his monograph on aphasia and those on the hemiplegias and palsies of childhood are as replete with standard medical terminology as the work of any of his colleagues; in addition, he was an enthusiastic neologizer, employing both German and Greek terms: 'Agnosia' (1891), 'Angstneurose' (1894), 'Aktualneurose' (1898), 'Zwangsneurose' (1894), all were coined by Freud in the 1890s and were clearly in the spirit of *fin-de-siècle* neurology's love affair with neologisms, aimed at capturing supposedly new descriptions of mental states or stable collocations of symptoms.

Words travel, and to very unexpected places: was Gordon R. Gould's paper to the physicists at the 1959 Ann Arbor Conference on Optical Pumping covertly addressed to ten-year-old boys dressing up in intergalactic spacesuits, the double audience determining his choice of the term 'laser' (light amplification by stimulated emission of radiation)? No, of course it wasn't. The audience took hold of the laser (and to a much lesser extent its older brother, the 'maser') and made it travel very far from Bell Labs, just as the general public took hold of 'libido', Freud's borrowing from Krafft-Ebing, and used it as a catch-all term for sex, sometimes knowingly with Freudian connotations, more often unknowingly. Libido did not and has not suffered from its Latinate origins; nor did the travels of the 'Oedipus complex' or 'narcissism' suffer from their recondite Greek origins. Nor has the complete failure of Freud's neologism 'paraphrenia' to make its way, even in the psychoanalytic literature (even Jones failed to include 'paraphrenia' in the 1924 *Glossary*), been a cause for criticism of Freud's hellenizing proclivities. 'Parapraxis' and 'cathexis' have been the cause for much criticism; but the most plausible reason for this is not that

---

[351] Steiner, 'A world wide international trade mark of genuineness?', p. 68n49.
[352] James Strachey, 'Appendix. The emergence of Freud's fundamental hypotheses', *SE* III, 63n2.

the original conception of translation was antipathetic to either the spirit of Freud or the spirit of English.

Two elements in the reception of 'cathexis' need to be distinguished: the reception of the word as an appropriate translation and the reception of the theory of which the term is, according to Strachey, an integral part. Strachey's justification for his choice of the term also had this double character. As Strachey first conceived it in 1921, 'if the "right" translation can be fixed upon as a word with no ostensible meaning at all, people may be induced to try and discover what the meaning really is'[353] (one might add: like Stoney's 'electron', Anderson's 'negatron', Lavoisier's 'caloric', Johannsen's 'gene' or Clausius's hellenism 'Entropie', rendered into English as 'entropy'). This narrow justification, spelt out in a letter to Jones, followed on a detailed exposition of the various contexts in which Freud used the terms 'Besetzung' and 'besetzt' – and Strachey's argument is persuasive in so far as one recognizes how 'technically' complex and protean all these usages were. The need to explain the necessity for Strachey's innovation led to the entry for 'Besetzung' being the longest in the *Glossary* published in 1924:

> *Besetzung*, CATHEXIS (derivative adjective CATHECTIC): If it is desired to discriminate between the process and the state itself the word CATHECT might be used for the latter, but this would have the disadvantage of prematurely forcing a distinction between the two ideas which would be artificial in the present state of our knowledge, and it would seem desirable therefore, at least for the present, to use the term 'cathexis' for both purposes, on the analogy with 'synthesis'. When the term is familiar, the verb cathecticize might perhaps be introduced, though it is not essential.[354] Although the word *Besetzung* can be approximately translated by 'investment' or 'charge' there is a great advantage in reserving a specific and unambiguous term for such a technical concept.[355]

---

[353] Strachey to Jones, 27 November 1921, Vienna, quoted in Ornston, 'The invention of "cathexis" and Strachey's strategy', p. 393.

[354] Strachey's guidelines were subsequently ignored. 'Cathect' never became used for a description of the state itself; instead, a verb form, 'to cathect', was constructed (by back-formation, as the OED puts it) from 'cathexis'; the OED gives Alix Strachey's translation in 1935 of *Inhibition, Symptoms and Anxiety* as the first usage of the past participle form 'cathected', which became the usual verb form. Any usage of the infinitive 'to cathect' is rare until at the earliest the 1950s.

[355] Ernest Jones (ed.), *Glossary for the Use of Translators of Psycho-Analytical Works*, Supplement 1 to the *IJP*, Published for the Institute of Psycho-Analysis, London: Bailliere, Tindall & Cox, 1924, p. 6. This passage from the entry is only partial; a further explanatory part of the entry is completed by a list of twenty-one compound forms and their translations (e.g. 'Besetzungsintensitäten', 'the various intensities (degrees of intensity) of cathexis').

Strachey's argument, endorsed by the Glossary Committee (despite 'some jeers from Rickman'[356]), was that the term is so important for theoretical purposes ('technical') that each of the reasonably plausible translations for a given usage – 'occupation', 'filling', 'investment', 'engagement' – would be unsatisfactory: it would have as an unfortunate but unavoidable consequence the arbitrary accentuation of one particular meaning at the expense of the others. If one were to evade this consequence by varying the translation of the term according to context, a crucial feature of this term would be lost, i.e. the fact that it embodies one of Freud's 'fundamental hypotheses',[357] 'the theory of "cathexis"': 'that in mental functions something is to be distinguished – a quota of affect or sum of excitation – which possesses all the characteristics of a quantity . . . which is capable of increase, diminution, displacement and discharge'.[358]

Despite Strachey's endorsement and well-argued justification, cathexis has not become a treasured part of psychoanalytic vocabulary. While the success or failure of neologisms may not tell us much about their original well-chosenness (so many other factors intervene), Strachey's decades-long attachment and faithfulness to his original choice does tell us something about his reading of Freud. From the very start, his Freud was a medical scientist who, with at least one hand if not the other, was attempting to develop a consistent public terminology for his new science. Bettelheim's anger and sorrow about the misleading terms Strachey chose centres on his conviction that 'the essence of psychoanalysis is to make the unknown known, to make hidden ideas accessible to common understanding'.[359] This characterization of psychoanalysis leaves out the scientific and professional Freud. Was Freud a 'writer' or was he a 'scientist'? He was, quite obviously, both; the perpetually modest Strachey spoke of Joan Riviere's 1922 translation of the *Introductory Lectures* as making it possible 'for the first time for readers of English to realize that Freud was not only a man of science but a master of prose writing'.[360] His publications speak both to a general public and to a specialized audience of professionals, practitioners and followers. The antipathy of Bettelheim to the scientific Freud, the antipathy of other critics to the prosaic business of building up a movement with agreed terminology and rules of practice, conduct and debate is

---

[356] Strachey to Jones, 27 November 1921, Vienna, quoted in Ornston, 'The invention of "cathexis" and Strachey's strategy', p. 393.

[357] Strachey, 'Appendix', *SE* III, 62–8.

[358] *Ibid.*, p. 63 quoting Freud, 'The neuro-psychoses of defence' (1894), *SE* III, 60.

[359] Bruno Bettelheim, *Freud and Man's Soul*, London: Chatto & Windus, 1983, p. 89.

[360] 'Joan Riviere (1883–1962)', pp. 228–9.

expressed in the attack on the Strachey translation, created with the 'serious student' of psychoanalysis in mind, the attack being principally on the translation's purported effect of cutting the general reader off from the realization that psychoanalysis applies first and foremost to that reader.

While it is clear that Strachey's Freud is the 'English man of science' born around 1850, there is another decidedly culturally specific feature of his Freud crafted into English: the literary style which Strachey was quite explicit to uphold as his ideal ('stylistic checker'). This ideal he undoubtedly shared with Alix, Rickman, Sprott and the other Cambridge-educated circles in which he lived his whole life. 'We came out of the same middle-class, professional, cultured, later Victorian, box',[361] as he observed of Joan Riviere at her memorial meeting in 1962. Many criticisms of Strachey are in effect directed at this unmistakeably English voice and literary style. Ornston writes, as if it were self-evidently a criticism, of the *Standard Edition*'s 'urbane style',[362] 'Strachey's self-consciously stiff and masterly voice'[363] and 'Strachey's very British cadence'.[364] Warming to his peroration, he accuses Strachey of traducing Freud by his use of 'formal, Cambridge-Latinate prose';[365] for him, an American, reading the *Standard Edition* is disorienting because 'Cambridge customs can be bewildering.'[366] The light, always amused, touch with which Strachey remembers the early days of deciding on the standard terminology for translating Freud certainly pre-emptively undercuts any attempt to take these terminological issues too seriously. He recalled his first working meetings with Joan Riviere in the summer of 1921, 'at a meeting of a decidedly peculiar institution called the Glossary Committee. It met in Ernest Jones's consulting room in Harley Street and consisted of him, Mrs Riviere, my wife, and me. This quite irresponsible body decided for all time how the technical terms of psycho-analysis were to be translated.'[367] This is Strachey recalling Riviere, but, as he confessed, 'I am afraid there is likely to be almost as much about myself as about her' – and he was, in effect, recalling how this group, who all came from that same middle-class, professional, cultured, later Victorian, box, made Freud English, including the whimsicality with which he could recount his own life's work. Lytton,

---

[361] *Ibid.*, p. 228. [362] Ornston, 'Improving Strachey's Freud', p. 218. [363] *Ibid.*, p. 200.
[364] *Ibid.*, p. 193. [365] Ornston, 'The invention of "cathexis" and Strachey's strategy', pp. 395–6.
[366] *Ibid.*, p. 392.
[367] 'Joan Riviere (1883–1962)', p. 229. 'Scoptophilia' is not only one of the disgraces of the Glossary Committee, but also one of its failures. Google Ngrams show that, after a brief flurry of usage in the 1950s, particularly in the USA, the more sensible, if no less inelegant, 'scopophilia' quickly won out; however, the specifically psychoanalytic literature doggedly continues to use 'scoptophilia' as often as 'scopophilia'.

Virginia, Maynard – all could adopt this self-deprecating, whimsical and charming tone to perfection. All of them acquired their philological skills along with their sophisticated party habits; at country house weekends during the Great War, you could catch Alix 'puzzling over Rabelais with the aid of six dictionaries'.[368] She, James, Riviere and the other members of the Bloomsbury team simply turned these skills to the service of a portrait of Freud, one might say, painted by the 'we' of Jones, the Stracheys, Riviere, Rickman.

Of the four Bloomsbury analysts, the Stephens were to suffer illness and death shockingly early. Adrian's political triumphs in the Society gave him some impetus: to take on the Editorship of the *Journal* in 1945, along with the post of Scientific Secretary; to sit on the Training Committee in 1946–47.[369] But he was unable to fulfil his duties in 1946 because of failing health and he died in 1948. Rickman's (anonymous) obituary spoke of him as a man from another age, that other now-lost England, a lonely man whose life was difficult.[370] Karin suffered much after Adrian's death, though she continued, as she always had done, the struggle to live life to the full. Leonard Woolf's plan to commission a life of Freud from Jones in 1939 had met with no positive response; nor, as we have seen, was his attempt to put together a Freud edition successful. But, clearly giving up on Jones, in 1944 he did successfully persuade Karin to write a book on Freud's contribution to the sum of human knowledge which would contain a comprehensive biography.[371] Karin worked on this manuscript after Adrian's death, the time when her friend and obituarist Marion Milner described her as 'fighting a losing battle'. The book, eventually entitled *Human Misery*,[372] was to take the form of a conversation, with one speaker being a thinly disguised Karin, raising questions and answering them. It remained unfinished at her death, by her own hand, in 1953. And it contained no biography.

Adrian and Karin's elder daughter Ann studied at Cambridge and then in 1943 married Richard Synge, both of them part of the Cambridge circle of

[368] Holroyd, p. 414.
[369] Ernest Jones served as Editor of the *IJP* from 1920 to 1939, to be followed by James Strachey (1940–45), Adrian Stephen (1946), Willi Hoffer, John Rickman and Clifford Scott (1947–48), Willi Hoffer (1949–59), Jock Sutherland (1960–68), Joseph Sandler (1969–78) and Tom Hayley (1978–88).
[370] Anonymous, 'Adrian Leslie Stephen 1883–1948'.
[371] Unpublished letter from Mrs Barbara Halpern to Dr Noel Bradley, *c.* 1980, cited by Gillian Rice, 'The reception of Freud amongst the Bloomsbury Group', MPhil dissertation, University of Cambridge, 1984, p. 57nIII.
[372] Rice, 'The reception of Freud amongst the Bloomsbury Group', p. 58.

Marxist biochemists around Bernal; they had seven children.[373] Late in life, Ann left on record a decidedly jaundiced view of her parents: 'Mother and father were so ill-assorted and so far away from us children that the only things we had in common were the house and the servants . . . There is a lot to be said for the view, put forward by a family friend, that we were orphans from the beginning.'[374] We do not know if Karin took any pleasure in her daughter's growing family; she did live long enough to hear that her son-in-law had been awarded the Nobel Prize in Chemistry in 1952.

James and Alix gave up the lease on 41 Gordon Square in 1956 and went to live at Lord's Wood, the house in the country built by Alix's mother in 1900, where they had spent most of the war years, and bequeathed to Alix on her mother's death in late 1954. James suffered severe problems with his eyes, at one point fearing to lose his sight completely. But he recovered sufficiently to continue his editorial work. His stamina was remarkable: the *Standard Edition*, definitively decided upon in or around 1947, when Strachey turned sixty, was entirely the product of his old age. Strachey, with the help of Jones, Anna Freud, Alan Tyson, Frances Partridge and Alix, brought the twenty-three volumes to completion in 1966. He dedicated the *Standard Edition* 'To the thoughts and words of Sigmund Freud – This their blurred reflection is dedicated by its Contriver.' Even this dedication allows a play between the self-effacing James ('you find the translation contrived?', 'isn't a "contriver" usually an artful plotter?') and the steely-willed and decisive man of letters who allows himself a modest, one-off but unmistakeable identification with the glories of seventeenth- and eighteenth-century English literature where Bloomsbury was most at home. His achievement was recognized with the award of the Schlegel-Tieck prize for translation, which funded a Caribbean escape from the English winter for him and Alix.

Yet in 1962 the old Strachey jumped at the chance of starting a new major project, when the young biographer Michael Holroyd turned up, courtesy of a recommendation from Frances Partridge, at the Stracheys' country house hoping to be able to gain access to Lytton Strachey's correspondence for a major study he had – he so naively thought – already brought close to term. Strachey astutely drew him in to the scheme he promptly hatched, gently but without equivocation, by showing him the

[373] Hugh Gordon, 'Richard Laurence Millington Synge. 28 October 1914–18 August 1994', *BMFRS* 42 (November 1996), 454–79.

[374] Ann Synge, 'Childhood on the edge of Bloomsbury', written for *Charleston Newsletter* 10 (March 1985), 14–15, reprinted in *The Bloomsbury Group: A Collection of Memoirs and Commentary*, ed. Stanford Patrick Rosenbaum, Toronto: University of Toronto Press, 1995, pp. 381–2.

'studio wilderness' at Lord's Wood: 'two great wooden tables piled high with boxes and files, and on the floor were littered innumerable trunks and suitcases – all full of letters, diaries and miscellaneous papers',[375], hoarded for thirty years since Lytton's death and now a challenge like no other for a biographer. And it was James who, having set up Holroyd, did a deal: 'what we agreed to do was to go through the book sentence by sentence trying to hammer out a mutually acceptable text'.[376] Holroyd was even paid a 'bribe' (Strachey's term) to wait till the whole study was complete before publication. Even at the end, James insisted on speaking in his own voice through – where else? – Holroyd's footnotes. Wherever they agreed to disagree, Holroyd would record this, together with James's comments. 'Not the least lively part of this never dull volume consists of pungent footnotes by Strachey's younger brother and literary executor, James, expressing racy disagreement with the biographer',[377] confessed Alan Ryan, reviewing for *The Times* in October 1967.

In this way, Holroyd became the means by which James realized what he himself had contemplated devoting years of the late 1930s to: 'to set about a very long life of Lytton with all the letters',[378] a task he finally shied away from. In this unfulfilled project, James had seen himself as completing what Lytton would, he surmised, have done with the rest of his life if he had not been cut short at the age of fifty-two:

> he would have turned what was implicit in his biographies into an explicit autobiographical campaign to achieve the same treatment under the law for homosexuality as for heterosexuality. I can see now [1994], though I did not see it then, that he would have liked my biography to perform something of the work Lytton might have done between the 1930s and 1950s.[379]

Holroyd was gently but firmly steered towards fulfilling this project. He became a part of Bloomsbury himself, completing what was left unfinished by death. James Strachey died the moment this task (and not the *Standard Edition*) was effectively finished: 'In April 1967, shortly after completing his final notes [on the Lytton manuscript], James died suddenly of a heart attack.'[380] With his shepherding of Holroyd, he had succeeded in completing the Bloomsbury project, as he saw it, of making the life and the work one, making the life into the work, since Bloomsbury had discarded

---

[375] Holroyd, p. xiv.   [376] *Ibid.*, p. xx.

[377] A.P. Ryan, 'Lytton Strachey before he achieved eminence: review of Michael Holroyd, *Lytton Strachey. Vol. I: The Unknown Years (1880–1910)*', *The Times*, 5 October 1967, p. 8.

[378] Virginia Woolf, 5 April 1935, quoted in Holroyd, p. xxiii.   [379] Holroyd, p. xxiii.

[380] *Ibid.*, p. xxi.

all justification for a life through works and made the value of a life in need
of no further justification than itself. What could be more appropriate than
that it was the younger brother, the psychoanalyst Strachey, who brought
this task to a completion? The essence of psychoanalysis is to allow the
interpenetration of a life by a work of analysis, so that, in its purest form,
already achieved in Alix Strachey's modest dream analysis paper of 1922,
the life and the work are united in the unfolding process of analysis. James
had achieved the task of turning his brother's 'life & work' into a form,
Holroyd's biography, that illustrated and thus confirmed the truths sought
by both psychoanalysis and Bloomsbury: the centrality of sex, in all its
forms, imagined, spurned, consummated, bypassed, to a life; the belief in 'a
great deal of a great many kinds of love'.[381] This marked the final victory
over the Victorian ethos – its respectability, earnestness, morality and its
agonized inability to bring into union the private and the public – that
Lytton's *Eminent Victorians* and the exactly contemporaneous 'wild rise of
psychoanalysis' had promised.

Even James Strachey's early and unrequited passion for Noel Olivier had
an unexpected and happy outcome. For Bloomsbury in 1920, James
Strachey still had two 'wives':[382] Noel and Alix. Alix and James married
in June 1920; Noel married a fellow doctor, William Arthur Richards, in
December. Despite Rupert Brooke's reflections on her character in June
1914 after they had become estranged, tinged with his growing compulsive
cattiness – 'Proposed to by me, Adrian Stephen, & James Strachey!
Wouldn't it turn anybody sour. I think she's a miracle to have outlived
that at all'[383] – Virginia Woolf certainly also found her lovable: 'Noel
always enraptures me – she cries over Rupert's letters, she tells me; and
really, I fall in love with her, being so sentimental, for doing it.'[384] But
Noel's practical wisdom shines through the course of her full life. She
established herself as a pioneering female paediatrician at the Westminster
Children's Hospital. She maintained her friendship with James after her
marriage (acting as James's day-to-day doctor for flu injections and other
minor medical needs), and he visited her as she gave birth to five children in
the 1920s and 1930s, reporting to Alix in early 1925 after her first confine-
ment: 'Noel gave me a long account of the business. It sounded decidedly
lustbetont [pleasure-tinged] all through: which is very likely due to her
especially-marked-sense-of-guilt being gratified.'[385] He was godfather to

---

[381] *Ibid.*, p. xxxiii.   [382] *Ibid.*, p. 475.
[383] Harris (ed.), *Song of Love*, p. 273, letter dated 20 June 1914.   [384] VWL III, 93, 8 March 1924.
[385] BF, 16 February 1925, p. 202.

her fourth, Tazza. For the first ten years of her marriage, their correspondence was warm, intimate and friendly; but then in the spring of 1932, she realized that she loved James, was thankful that it wasn't too late, and, very much on her initiative, they began an affair that lasted throughout the 1930s – a love, the 'unattainable romance' as Virginia had called it, rediscovered more than twenty years after they had first become entangled when Noel was close to Rupert Brooke, James's closest friend, and James had fallen in love with her.[386]

Who would have guessed that the directionless thirty-year-old James – 'soft as moss, lethargic as an earthworm' (Virginia Woolf), 'a good fellow but weak and perhaps lacking in tenacity' (Ernest Jones) – would bring two of the great English literary projects of the twentieth century to safe harbour in the final few months before his death at the age of seventy-nine? And the nature of those projects – Michael Holroyd's magnificent biography, the first post-Wolfenden biography in England, of James's brother Lytton; the *Standard Edition* of Freud's works – show how much stamina the great passions of his life elicited from him. Freud, Lytton and Mozart: these were the Masters of his life. Virginia Woolf certainly did not foresee such success; but then, perhaps James Strachey was lucky. His fidelity to all three is only matched by the surprising and admirable success of his marriage – so characteristically inauspicious at first – to Alix.[387] Such an array of achievement, secured by silent and barely recognized hard labour over many decades, is an unusual reminder of the unexpected answers life throws up to that version of Solon's injunction: call no man a failure until he is dead.

What relation did James Strachey's two life projects have to each other? 'My own view is that Lytton's Cambridge years had an important effect on the subsequent mental life in England; especially on the attitude of ordinary people to religion and sex', he told Michael Holroyd in the mid-1960s.

> The young men in my years (though also interested in socialism) were far more open-minded on both those topics than their predecessors – and I believe they handed on what they derived from Lytton, and this (taken in

---

[386] Published correspondences between Brooke and an individual correspondent include those with Strachey and Olivier, though see also the 2015 book *The Second I Saw You: The True Love Story of Rupert Brooke and Phyllis Gardner* by Lorna C. Beckett; these supplement *The Letters of Rupert Brooke, Chosen and Edited by Geoffrey Keynes*, London: Faber & Faber, 1968.

[387] 'Lytton said that James & Alix are to be married in 3 weeks. So after all, she has won. But though satisfactory, I find no excitement in this. They know each other too well to stir one's imagination thinking of their future, as one does with most engagements' (VWD II, 39, 18 May 1920).

conjunction with Freud, who was totally unknown till much later) is, I think, what has resulted in the reform of the general attitude to sex.[388]

So to Lytton and to Freud James attributed the transformation of attitudes to sex and religion – and it was James's life project both to continue their work as a humble amanuensis and to create the monuments recognizing their achievements.[389] But, as Anna Freud said in concluding her obituary of him, he 'was fortunate to find an author and a subject-matter worthy of his efforts'.[390] Anna was, as ever, pressing the case for Freud; but that does not put in doubt her characteristically sharp-eyed observation that, without his encounter with Freud, James Strachey's life would not have been one of such prodigious and unexpected – and of course self-effacing – public achievement.

[388] Holroyd, p. 101.

[389] Annan also attributed much to Lytton and Russell: 'How well known did that pattern of life become! – the English intellectual, in his white-washed country cottage hung with paintings, revolving in a circle of hetero- or homosexual relationships, cultivating the arts and practicing humor and truthfulness in friendship, oblivious to the horrors of politics and divorced from the social structure of country life. And yet who can doubt that Russell and Strachey were the most deadly of all propagandists for the immense revolution in England against the Victorian Ten Commandments?': Noel Annan, 'A very queer gentleman. Review of Holroyd's biography of Lytton Strachey', *New York Review of Books* 10(11), 6 June 1968.

[390] [D.W. Winnicott and Anna Freud], 'James Strachey, 1887–1967', *IJP* 50 (1969), 129–32 at 132.

CHAPTER 10

Freud in Cambridge?

It needs now to be repeated once again: Freud never went to Cambridge. The title of this book is a provocation, a counterfactual.[1] Having read all the chapters, the reader might conclude that, even though the whole weight of evidence we have put forward about the great interest and activity associated with Freud's ideas in Cambridge is substantial, Freud's psychoanalysis never quite arrived in Cambridge. There was never to be a Psychoanalysis Tripos at the University. Promising disciplinary alliances – with psychology, English, anthropology – ultimately failed to materialize. Even the most psychoanalytic of Cambridge groups, the 1925 Group, dispersed with only two members (Strachey and Rickman) remaining analysts, while the others – Penrose, Ramsey and Tansley in particular – stepped off the psychoanalytic path to follow other scientific interests. By the end of the 1920s, the world-changing psychoanalytic enthusiasms of J.D. Bernal had dissipated, to be almost entirely overlaid by another life-consuming commitment, this time to Marxism-Leninism. By 1928, the Malting House School had closed its doors. Even the Congress of the International Psychoanalytical Association planned to take place in Cambridge in 1925 never took place. Psychoanalysis left no institutional trace in Cambridge, no legacy, no research project inspired by psycho-analysis, no clinical grouping of significant analysts, not even a consulting room to be visited by town or gown.[2] The gates of historical forgetting began very quickly to close behind this episode of psychoanalytic enthusiasm.

---

[1] One might even imagine it has a Yiddish question mark omitted – 'Freud in Cambridge?' ('You must be joking!')

[2] The first roughly orthodox analysts to work in Cambridge were Marie Singer and Bernard Zeitlyn in the late 1950s; nor, after the early 1920s, do there seem to have been any 'wilder' therapists based there. In contrast, at Harvard, for example, despite the widespread official and informal hostility to psychoanalysis in the 1930s in the University, by the 1940s those wishing to seek analysis could make use of the analysts associated with the Boston Psychoanalytical Society, who often lived in adjacent Cambridge.

In a wide-ranging and well-informed article, R.D. Hinshelwood distinguished seven points of entry for psychoanalysis into British cultural life in the period 1895 to roughly 1925: the Society for Psychical Research; Havelock Ellis and the development of sexology; psychological developments within psychiatry, associated with W.H.R. Rivers, Bernard Hart, May Sinclair and the Brunswick Square Clinic; the Psycho-Medical Society and the formation of a Medical Section of the British Psychological Society during World War I; the 'literary' strand, dominated by the Bloomsbury Group's intense interest up until the early 1920s; progressive education, including Homer Lane and somewhat later A.S. Neill; discussion in philosophical circles, including the influence of Bergson's and Rivers's work.[3]

The paths of transmission mapped out by *Freud in Cambridge* indicate that this classification scheme, while accurate and pertinent, is insufficient.[4] To the seven paths Hinshelwood lists, we need to add others. There was a wide road to the unconscious which was neither altogether medical nor psychological, neither philosophical nor literary. To map this road, since covered over and silted up, this study examined a number of eminent scientists, nearly all from Cambridge, who, in the early 1920s, were drawn to psychoanalysis. Between them, these individuals explored their dreams, underwent personal analyses, published papers in psychoanalytic journals, wrote popular treatises on psychoanalysis, corresponded with Freud, and were admitted to membership in the BPaS. What they perhaps had most in common to bring them together was 'scientific curiosity'. Amongst them were those scientists of the first or second generation of scientific professionals – that is men (predominantly) who were employed full-time on account of their scientific skills – who regarded the new ideas and practices associated with psychoanalysis as a natural extension of the 'scientific attitude' of careful and empirical inquiry into the nature of human beings. Whether or not they were trained in biology, and regarded psychoanalysis as an extension or supplementary revolution to the Darwinian and other biological advances of the late nineteenth century, is secondary to their self-conception as scientists engaged in a new field of inquiry opened up by Freud. And they were the first to admit that this new

---

[3] R.D. Hinshelwood, 'Psychoanalysis in Britain: points of cultural access, 1893–1918', *IJP* 76 (1995), 135–51.

[4] Tansley himself was certainly aware of this important cultural distinction between the medical and the scientific worlds: in his Memoir concerning his contact with Freud, written in 1953 for Sigmund Freud Archives, he self-consciously described himself as a 'non-medical biologist'.

field of inquiry required 'self-knowledge' – in time-honoured fashion, submitting themselves, first and foremost, to the tools of inquiry.[5]

These representative Cambridge scientists were not the sole conduit for psychoanalytic ideas. From the makers of modern philosophy to the creator of Cambridge English, from the founders of the Malting House School to undergraduate fans, a short-lived culture of enthusiasm existed in Cambridge, both within and outside the University. In 1946, Edward Glover reflected on the dissemination of psychoanalysis in Britain:

> The later course of psycho-analysis in England differed from that followed in other European countries. That it ultimately secured a large measure of acceptance and prestige in lay quarters was due to a considerable extent to the attitudes of certain young intellectual groups. In the early 'twenties, for example, an open-minded attitude to psycho-analysis was an essential part of the equipment of any young Cambridge post-graduate having pretensions to cultural development. Not, by the way, that this was true of his academic mentors, who remained as atrophic in imagination as any other habit-ridden animal. There were, of course, many other groups and many discerning individuals ready to offer intellectual hospitality to the new ideas. And so in course of time Freudian theories percolated in a bowdlerised form, from the gardens of Hampstead and the squares of Bloomsbury, to the drawing-rooms of Kensington. Soon they were to find their way to the maid's pantry. Everywhere and everyday in bus, tube, and the editorial columns of popular daily newspapers a new jargon has come to life – 'wishful thinking,' 'complexes,' 'repressions,' 'inhibitions,' 'sublimations,' 'inferiority feelings,' etc. These terms are lightly and inaccurately bandied about by persons who have no idea to what revolution in thought they owe their origin.[6]

Glover's characterization is amply confirmed by the argument of this book. He had picked out 'any young Cambridge post-graduate having pretensions to cultural development'. He did not mention, for instance, Oxford

---

[5] On the varieties of self-experimentation, see Simon Schaffer, 'Self evidence', *Critical Inquiry* 18 (1992), 327–62.

[6] Edward Glover, 'Eder as psycho-analyst', in J.B. Hobman (ed.), *David Eder: Memoirs of a Modern Pioneer*, London: Gollancz, 1945, pp. 92–3. In an interview for the Columbia Oral History archive in 1965, Glover reiterated this analysis: 'This development [of psychoanalysis] differed in England from that in other countries ... psychoanalysis in this country developed along cultural lines. It began about the middle of the first world war when for the first time Freudian ideas began to percolate here and a number of people having different backgrounds, psychiatric, philosophic, biological or merely cultural, began to take an interest in it ... About 1920 I should have said that in Oxford and Cambridge no young undergraduates who were really interested in cultural matters but would accept or consider Freud; it wasn't a technical psychiatric approach to analysis, it was a cultural approach.' (Edward Glover, Interview by Bluma Swerdloff, 5 August 1965, London, Columbia University Oral History Project, pp. 1–4.)

616 Freud in Cambridge?

or London postgraduates. Just as in Virginia Woolf's famous characterization of the moment in 1910 when human character changed forever, he did not forget to include the maid. Most strikingly, Glover believed, the reception of psychoanalysis differed radically from that in other European countries precisely because of 'the attitudes of certain young intellectual groups'. In England, these groups were above all associated with Cambridge, with established figures like Tansley and Rivers, but also the pre-Great War undergraduate generation of Jeffreys, Rickman, Layard, Farrow, Rees alongside the better known closely knit groups of Apostles-in-Bloomsbury (the Stracheys, the Stephens). To these were added the enthusiasms of the first generation of post-war undergraduates – Ramsey, Penrose, Bernal, Martin, Needham – all nurtured by Ogden's entrepreneurial galaxy of Heretics and book series in psychology, philosophy and science and in time satirized by their Trinity contemporary Nabokov. It is worth highlighting the function of mentor played by key individuals: of Rivers (for Rickman, Layard, Martin and Karin Stephen); of Tansley (for Farrow, Penrose, Needham); of the Pykes for Frank Ramsey; of Hanaghan for Bernal.

Cambridge in the period 1910–30 was to produce many marked for life by the psychoanalytic enthusiasm of the èra. Some of these were to make a mark on the development of psychoanalysis outside Cambridge by themselves becoming analysts in the soon to be orthodox fashion: not only Rickman, Isaacs, the Stracheys, the Stephens, but also Winnicott, Carroll and Bowlby. Of the twenty-eight principal contributors to the Controversial Discussions in the British Society in 1941–44, ten were originally from the Continent (four solitaries – Balint, Foulkes, Heimann, Klein – plus six Viennese émigrés); the Cambridge-linked analysts (Rickman, Isaacs, the Stracheys, the Stephens, Winnicott, Bowlby and we can add Joan Riviere) constituted half of the eighteen 'British analysts' involved.[7] There was no other cohort of comparable coherence within the British Society. They left a distinctive mark of intransigent liberalism on the crisis politics and subsequent governance of British psychoanalysis.

The life and work of A.G. Tansley traces an individual trajectory of the overall reception of psychoanalysis in Cambridge. The ecologist and botanist Tansley had wide scientific interests, including psychology, and pre-war connections, Stopes, Hart and others, through which he could gain a clear sense of the developing field of psychology. Then, during the

[7] FK, pp. xix–xv.

war, his self-help analysis of a critical dream dramatized for him a life-crisis; but he was in the curious position of having analysed his dream without really knowing if he had done so in an orthodox 'Freudian' fashion. So he educated himself more seriously about psychoanalysis and, as a result, wrote a bestseller, *The New Psychology* (1920). His timing was perfect to catch the upsurge of psychoanalytic enthusiasm in Britain, America and the colonies. But the response from readers encountered his doubt about the extent of his own inner understanding of Freudian theory. So he decided to have analysis with Freud. He went to Vienna in the spring of 1922, emerged intent on learning more, resigned his post at Cambridge and completed his analysis in 1924, having changed professional tack. He planned to establish himself as a psychoanalyst in Grantchester, was ready to serve in Ernest Jones's plan that he act as an eminent biological advisor to English psychoanalysis and to establish a psychoanalytic presence in Cambridge. Yet by the beginning of 1927 he had essentially given up his psychoanalytic life in Cambridge and London and rejoined botany as Professor at Oxford. For the rest of his life, his interest in psychoanalysis would remain, as did his immense admiration for Freud, but, from 1927 on, only as a penumbra around his central botanical interests.

Psychoanalysis was a great and surprising adventure in Tansley's life – which is why he is such an exemplary figure. But then psychoanalysis was itself a surprising adventure for European thought and culture in the period 1910–25. Tansley's particular adventure had some distinctive personal characteristics, as well as some shared with others: his personal crisis and dream analysis were the rule rather than the exception, yet they were combined with a serious-minded attempt to render psychoanalysis into his biological language, that of early twentieth-century dynamic botany and evolutionary biology. But the example of Tansley is equally telling because, like many others, there seemed little in his background to prepare for the great interest he took for several years in psychoanalysis.

On the other hand, W.H.R. Rivers was perfectly placed to appreciate Freud's work from within his own scientific areas of expertise. He was committed to working on 'insanity' and 'psychology' in 1892, but his work was rarely clinical from then on. Instead it was perfectly modelled on the research ethos of the University: physiology of vision, comparative psychology of the senses, and classic experiments in neurophysiology and psychopharmacology figured large. At the same time Rivers, the self-made scientist, branched out into ethnography, kinship studies and diffusionist anthropology. There is no reason to doubt Rivers when he later wrote he had 'taken much interest in the general views of Freud before the war'; nor is there reason to doubt that it was the Great War that kindled his entirely different level of interest, both in therapeutics and in dream analysis, leading to his major post-war books (*Instinct and the Unconscious* (1920); *Conflict and Dream* (1923)), both addressing key elements of psychoanalytic theory and practice. As for Tansley, and for many others during the war, Rivers's dream life came to have a significance which became intensely personal and demandingly theoretical. Indeed, this is one of the character-istics of the historical moment: just when the horrors of war for those at the front and those at home were bearing down most crushingly and cruelly, the messages from the individual's dream life became most urgent to under-stand. The influence of the dreams of the shell-shocked surely had some part in this, since the sleeplessness of those poor souls, beset by nightmares from which they could do nothing to escape except refuse sleep, became a distinctive symptom of the condition. What sprang out at these educated scientists was that Freud's psychology was a theory of perpetual mental conflict, most directly accessed through dream analysis: Freud's psychology was a tool-box that fitted perfectly the now revealed stresses and strains of moral, political, personal, not to mention military conflict.

So the dream life that Rivers plunged into was one riven with his personal, professional and patriotic conflicts. He drove himself to pierce the veil of infantile amnesia surrounding those scenes, so crucial in Freud's theory, underpinning his conflicts. He failed – nor do we know how successful his attempts to address any of his own sexual conflicts were. The question whether Rivers was, like Tansley, on the cusp of a pilgrimage to Vienna, as Rickman, whose mentor he was, hoped, or whether his considerable distaste for key features of psychoanalysis – its theory of instincts, its embracing of the transference – was to put him increasingly at that distance where his old colleague and co-founder of Cambridge psychology, C.S. Myers, found himself, is rendered void by his death in June 1922.[8] As close to an English Freud as there could be, his reception of psychoanalysis was more than simply of interest, it was a bellwether. And it was cut short.

A salient characteristic of discipline formation and development in the crucial period of the Great War and its aftermath should not be overlooked here. Raymond Williams made the following dry and astute observation while recalling the Cambridge of his own youth (the 1940s):

---

[8] One can still detect the embers of this hope in Jones's overall sparse account, written in 1957, of the Berlin IPA Congress held in September 1922: 'Rivers of Cambridge had intended to come, but he died suddenly three months before' (J III, 92).

What I have noticed about these memories of groups is that they are at least partly determined by what happens to the people afterwards. The successful, obviously, are more easily remembered; the others, equally important at the time, can be made to fade. There is a close parallel in external images of the intellectual life of the university. People talk of the Cambridge of Moore and Russell, or of Wittgenstein and Richards, and so on. Yet at any time such figures are a tiny minority in the whole intellectual life of the university. It is falsifying, in a particular way, to project the place through these few figures, who are as often as not relatively isolated, or quite uncharacteristic.[9]

Point taken. The Cambridge of disciplinary innovation or success – Rutherford's Cavendish, Hopkins's Dunn Laboratory, Adrian's physiology lab – is in another corner of its activities also the Cambridge of social conservatism, Haldane's defrocking for adultery, intellectual resistance and disdain. Three of our main protagonists, A.G. Tansley, C.S. Myers and Bertrand Russell, abandoned Cambridge in disgust in the early 1920s. For Tansley in 1923, it was the 'conservatives in authority' that propelled him out of botany and towards psychoanalysis; for Myers returning to Cambridge in 1919, it was his increasing revulsion at 'the old academic atmosphere of conservatism and opposition to psychology', best represented by finding that 'the wild rise of psychoanalysis had estranged the Regius Professor of Physic', which led to his resignation in 1922. For Russell, certainly disillusioned by the repudiation he had received from Trinity during the war, it was more his growing lack of confidence in the possibility of his work gaining access to eternal truths of philosophy (a result of reading the manuscript of Wittgenstein's *Tractatus*) and his urgent desire to have children that made him resign his new position at Trinity in the autumn of 1920. (The reason he later gave was what we might call the Haldane reason: that he was 'living in open sin' with Dora.)[10] So the three national leaders in botany, philosophy and psychology, roped together biographically (Tansley and Russell by early friendship; Tansley and Myers as the joint holders of the first Gerstenberg Studentship in 1896), each felt forced to leave Cambridge confronted by the 'conservatives in authority' in the early 1920s. Both Tansley and Myers left Cambridge for the sake of the 'new psychology'; Russell's life tumult was less directly connected to the new psychology, but his principal work at the time, *The Analysis of Mind*, was a major philosophical response to the turbulence in psychology. This tells us a lot both about the weight of conservative authorities in

[9] Ronald Hayman (ed.), *My Cambridge*, London: Robson Books, 1977, p. 57.
[10] MBR I, pp. 594–5.

Cambridge after the war and the revolutionary turmoil in psychology to which they were all responding.

Overall, then, the picture that emerges in Cambridge in the early twentieth century shows both a previously unsuspected covert influence of psychoanalysis and also a failure for it to take root in the rapidly shifting constellation of disciplines and institutional forms. The failure is attributable to many converging causes, but one general one must be emphasized: the fierce hostility to clinical medicine that the University's educational policies embodied in the first half of the twentieth century. Only the scientific foundations of medicine as developed in the laboratories of physiology, biochemistry and pathology were the acceptable disciplines for teaching in Cambridge. Eventually these were deployed as a bulwark against clinical medicine. Long-standing internal political opposition to clinical medicine was bolstered and heavily supported by the inter-war policies of the Medical Research Committee, later MRC, run by Walter Fletcher, Fellow of Trinity, himself a classic product of Cambridge science with long-standing collegial relations with Cambridge science: Fletcher steered the policy of the MRC towards laboratory-based scientific research, systematically avoiding clinical medicine.[11]

Part of the collateral damage of this hostility was the demise in the 1920s of two pre-war initiatives directed at national needs, the Diploma in Public Health and the Diploma in Psychological Medicine.[12] And it was the Diploma in Psychological Medicine that would most obviously have been the platform for the development of the medical aspect of psychoanalysis within the University.[13] Given this deeply ingrained hostility to

---

[11] One substantial indirect consequence of this unfailing indifference of the MRC to clinical medicine was its attitude to the methodology of clinical trials in the inter-war period (see L. Bryder, 'The Medical Research Council and clinical trial methodologies before the 1940s: the failure to develop a "scientific" approach' (2010), JLL Bulletin: Commentaries on the history of treatment evaluation (www .jameslindlibrary.org/articles/the-medical-research-council-and-clinical-trial-methodologies-before-th e-1940s-the-failure-to-develop-a-scientific-approach/)). The principal issue for the MRC in the use of 'controlled trials' – what eventually became the gold standard 'randomised controlled trial' (RCT) – was the perceived need for 'control' of clinical medicine by laboratory medicine and the control of untrustworthy and maverick clinical practitioners, in particular in allocation bias; see M.V. Edwards, 'Control and the therapeutic trial, 1918–1948', MD thesis, University of London, 2004.

[12] The extent of the dogmatic insistence on medical science at the expense of clinical medicine, and the resulting sacrifice on the University's behalf, can be gauged from the fact that the DPH, founded in 1875, had 118 candidates in 1913–14 and the same number in 1920–21 (figures taken from Sir Humphrey Davy Rolleston, The Cambridge Medical School, Cambridge: Cambridge University Press, 1932, p. 43). The London School of Hygiene and Tropical Medicine opened in 1929 and taught for the Cambridge DPH, which provided the University of Cambridge with a convenient excuse for discontinuing the Cambridge degree in 1932.

[13] As the 'Notes' section of the first volume of the IJP recorded, as one more sign of the advance of psychoanalysis, 'At the examination in Psychology for the Cambridge Diploma of Psychological

clinical medicine, the Diploma would have been terminated, even if MacCurdy had fulfilled his contract of employment with the University. As a direct consequence of the particular configuration of disciplines in the first half of the twentieth century, no initiatives within the University for clinical psychoanalysis were feasible.

However, the implanting of psychoanalysis did not necessarily depend upon psychoanalysis being bundled in with clinical medicine. We have singled out three areas of significant interpenetration: the fields of philosophy, literary studies and anthropology – areas where, in Cambridge, new institutions, new specialisms and new examinations were institutionally embedded in the first thirty years of the century. But the influence was singularly indirect and partial – even if critical to the future developments of the disciplines. The dialogue between psychoanalysis and anthropology, through the two leading figures in Britain in the period 1900–40, Rivers and Malinowski, was formative, even if the influence eventually took the form of a reaction – a reaction against any form of psychology, a reaction against any consideration of the power of sexuality. Self-appointedly tasked with the development of an innovative theory and practice of literary criticism for the new initiative of the English Tripos, I.A. Richards drew on his training in the moral sciences, including his time spent in the psychology laboratory, in constructing a pedagogical exercise in the production of error, the exposure of prejudice, 'stock responses' – the psychopathology of poetic reading. This practice, based upon the ideal of scientific self-scrutiny, would later transpose into something rather different, something rather more in keeping with the growing conviction of the moral worth of, the high moral ideals embodied in, the study of literature as compared with all other academic studies. Instead of promoting self-criticism, humility and an almost democratic sensibility in literary reading, as Richards's experiment might have done, Cambridge literary criticism, increasingly at odds with mass culture and with modern technology-infused everyday life, enrolled practical criticism as one method amongst others in the making of guardians of literary values, indeed guardians of civilization itself. The trajectory of philosophy was even more dramatic: from the giddy heights of an ambitious project to provide the foundation of mathematics and the principles of ethics, under the relentlessly critical pressure of Wittgenstein, that other interloper from Vienna, by the 1930s and 1940s philosophy seemed exhausted of anything except a set of tools

Medicine, October 1920, two of the six questions were on Psycho-Analysis' ('Notes', *IJP* 1 (1920), 340–1).

for the dismantling of philosophy itself. This dismantling was conceived of as a therapy. Wittgenstein, vigilantly self-critical at all times, could not escape the analogy between his therapeutic programme for philosophy and the cathartic model provided by psychoanalysis, nor could he, alone of all the major discipline-architects of the era, fail to recognize the dangerous analogy between the formation of his philosophical school and that of the psychoanalytic movement, in which the ascendant leader threatened to play an entirely noxious role in the transmission of knowledge.

But while these psychoanalytic currents were having striking formative effects upon the new disciplines, the core fields of psychology and clinical medicine were undergoing very different changes, only some of which were in any direct response to psychoanalysis. In the development of Cambridge psychology, psychoanalysis did play a striking role, but one which, as in anthropology, was more reactive and negative than positively motivated. Many elements conducive to a relatively rapid incursion of psychoanalysis into an internally reforming psychology disciplinary programme were in place in the early twentieth century. While psychology had been an integral part of the moral sciences since their inception in the 1850s (as logic, 'moral psychology' and 'moral philosophy'), a distinctive semi-autonomous discipline only began to emerge with the introduction of experimental methods and the physiology of the senses – with Rivers's appointment in the 1890s and the development of a research group around him and Haddon in ethnology. This psychology was resolutely experimental at the start, but by the end of the first decade other themes, other methods, observational rather than experimental, other kinds of psychology, were being introduced: animal psychology, child psychology, educational psychology, abnormal psychology. Although Rivers was an unpredictable adventurer, Myers was an eclectic empire-builder increasingly free from any concern about the proper limits of psychology. Give money to build the best psychology lab in Britain; busy himself in the University to introduce a Diploma in Psychological Medicine to make Cambridge the national centre for postgraduate training in a field some way from 'experimental psychology'; encourage innovations in industrial and educational psychology – these he did with energy, generosity and alacrity, both before the war and after. But it was only in the post-war turmoil that one can perceive just how propitious for psychoanalysis was the transformed landscape envisaged by Myers.

Thomas Kuhn's *The Structure of Scientific Revolutions* opened with the suggestion that 'we have been misled by' the histories scientists construct for their fields 'in fundamental ways'. He went on to suggest that the sciences are necessarily founded on a constitutive amnesia about their origins, an amnesia, we can add, whose undoing requires a historian also equipped with the methods of Nietzsche's genealogist, able to see the small, ignoble and implausible beginnings of the glories of the present. The very process by which disciplines crystallize obeys these historical laws, as Simon Schaffer reminds us:

> Yet if, as the philosophers of the *fin-de-siècle* notoriously argued, truths are dead metaphors and scientific instruments are boxed experiments about which one has forgotten that this is what they are, then disciplines are interdisciplines about which the same kind of amnesia has occurred.[14]

Anthropology was long a science before it became a university discipline; and it was a discipline for some decades before, post-World War II, it formed stable professional associations of academic anthropologists. The distinctive genealogy of anthropology, with its roots in the politics of slavery, the administration of the colonies and its late Victorian ethos of salvaging the debris of the primitive before its destruction by the unstoppable forces of modernization, left an ineradicable mark, including the domination of non-professional, non-academic members of the Royal Anthropological Institute well into the middle of the twentieth century. An entirely different yet equally distinctive genealogy for psychology in Britain saw the table-rapping of the spiritualists and the serious-minded toil of the psychical researchers alongside the self-help exhortations and programmes in self-mastery of Arthur Hallam, founder of the London Psycho-Therapeutic Society in 1901, and Emile Coué ('Every day, in every way, I'm getting better and better')[15] and the programmes within universities for detaching from philosophy an experimental psychology, modelled on and authorized by the novel programme of experimental physiology. Nikolas Rose's *The Psychological Complex*, and the social-historical, history-from-below account by Mathew Thomson in *Psychological Subjects*, both paint a picture of inter-war British psychology that is incomparably more extensive than the small pockets of narrowly focused experimental programmes within the universities. Having begun

[14] Simon Schaffer, 'How disciplines look', in Andrew Barry and Georgina Born (eds.), *Interdisciplinarity: Reconfigurations of the Social and Natural Sciences*, London: Routledge, 2013, pp. 57–81 at 58.
[15] Mathew Thomson, *Psychological Subjects: Identity, Culture, and Health in Twentieth-Century Britain*, Oxford: Oxford University Press, 2006; Thomson, 'The popular, the practical and the professional: psychological identities in Britain, 1901–1950', in G.C. Bunn, A.D. Lovie and G.D. Richards (eds.), *Psychology in Britain: Historical Essays and Personal Reflections*, Leicester: BPS Books, 2001, pp. 115–32.

its life in 1901 as a Society dedicated exclusively to experimental psychology and its committed practitioners, Myers's reforms of the British Psychological Society in 1919 were a very unusual reversal of such scientific societies' seemingly inexorable development towards specialization, professionalization and exclusionary rules: the inter-war BPS expressly reached out beyond the laboratories to a far wider membership of doctors, teachers, workers in industry.[16] The inter-war psychoanalysts, led by John Rickman, did play a considerable role in the medical branch of the BPS; Susan Isaacs, elected a member of the BPS in March 1917[17] and active in educational and industrial psychology in the next few years, would be a stalwart of the BPS as her reputation as the country's leading educational psychologist consolidated in the 1930s. It was only after the Second World War, like the social anthropologists with the founding of the professionalizing university-based Association of Social Anthropologists of the UK and Commonwealth (1946), that the academic psychologists once more seized back control of the BPS from the non-academics by further internal reforms slowly initiated and implemented in the 1930s and 40s[18] and the formation of the iconic Experimental Psychology Group (1946), firmly based on the laboratories at the universities of Oxford, Cambridge and London.

In 1919, it was within this ferment of psychology, stirred up by Myers but reflecting new developments within psychological medicine and elsewhere, that Jones chose to found the BPaS with a constitution whose distinctive structure reflected the same forces that had led to the reform of the BPS. While Jones was determined to maintain control and preserve orthodoxy within the Society as dictated by Vienna, he conceded to the general enthusiasm for psychoanalysis of the historical moment and also to the traditional rules of courtesy common to many English scientific societies, whereby interested outsiders were invited to participate and contribute, by creating a special category of associate members.[19]

---

[16] Thanks to Sarah Igo for pointing out how unusual the trajectory of the BPS was in this respect.

[17] BPS Minute Book, p. 96.

[18] Sandy Lovie, 'Three steps to heaven: how the British Psychological Society attained its place in the sun', in Bunn, Lovie and Richards (eds.), *Psychology in Britain*, pp. 95–114.

[19] Freud's conception of the rules governing the early days of the Wednesday Society in Vienna, which became the Viennese Psycho-Analytical Society, were also flexible and open to interested outsiders without relevant education or background, e.g. Max Graf and Otto Rank. He would remain more flexible than Jones well into the 1920s and 1930s, as their interchange over the value of E.P. Farrow's psychoanalytic papers indicates and also over the work of another Tansley acolyte, C.C. Fagg (see Chapter 4, and L. Cameron and J. Forrester, 'Tansley's psychoanalytic network: an episode out of the early history of psychoanalysis in England', *Psychoanalysis and History* 2(2) (2000), 189–256.).

The 1925 Group's interest in psychoanalysis had an entirely different complexion. Each member had had extensive experience of analysis, yet their collective orientation was not 'clinical' – the interest of psychoanalysis was, for them, *scientific* rather than clinical or medical. The 1925 Group was the most distinctive and surprising effect of the swell of Freudianism in Cambridge. A group predominantly composed of first-class scientists, products of the mature scientific culture that developed in Cambridge in the period 1880–1920, exploring an area outside their expertise, but entirely serious in this commitment, as their extended personal analyses and changes of life-course demonstrated, finding a classical Cambridge format for research and inquiry. As an informal group, it was typical of the way in which Cambridge intellectual culture organized itself. But it left no institutional mark. Indeed, the very fact of a mathematician, a brain and logic scientist, a geophysicist, a botanist, a psychiatrist and a translator finding no obstacles to common work and discussion indicates how little disciplinary boundaries mattered at this time – at least to them. However, the culture that gave rise to the very possibility of this group – elite scientific culture, the 'High Science' of Cambridge – had been and indeed was at that time being transformed by the disciplinary innovations of the early twentieth century, within which psychoanalysis would need to find a place, if it were to prolong its life beyond the enthusiasms of hypnotism or of theosophy. Its failure to do so had profound consequences for its subsequent history.

However, the 1925 Group was a striking symptom of the confluence of forces that led to Cambridge being such fertile ground for interest in psychoanalysis: the progressive, liberal, internationalist student culture nurtured by Ogden's Heretics and the *Cambridge Magazine*; the thirst to go out to Vienna after the war seeking analysis, partly in the spirit of a latter-day Grand Tour for modern intellectuals, partly as the committed wing, nurtured by the Apostles (Penrose, Ramsey), of developing Bloomsbury psychoanalytic culture. Young Cambridge was vibrating with post-war revolutionary movements both in science and in politics: not only Bolshevism, the partial success of the suffragists' campaigns and the rapid rise of the Labour Party but also Einstein's revolution in physics, with which Freudian psychoanalysis to many seemed a natural pairing, Wittgenstein's revolution in philosophy (deeply affecting Penrose and Ramsey) and Brouwer's intuitionist attack on formalism in mathematics (affecting Ramsey). To the 1925 Group (at least to Jeffreys, Penrose and Ramsey), these were the revolutions that counted in the early 1920s, as much as if not more so than the modernist achievements of Proust, of

Joyce's *Ulysses*, of Eliot's *The Waste Land* and of Malinowski's *Argonauts of the Western Pacific*. The Group displayed no hint of collective commitment to a psychoanalytic profession: psychoanalysis was simply one of the many intellectual threads that ran through their lives. The invention of the profession of psychoanalyst was simultaneously taking place in Berlin, with the Clinic and its training ethos. Max Eitingon's template, which would have such profound implications for psychoanalysis, was produced for the International Training Committee of the IPA on 3 September 1925; its opening principle was: 'Psycho-analytical training ought no longer to be left to the private initiative of individuals.'[20] A bureaucracy supervising the training of psychoanalysts quickly came into existence, and was made solid in the rapid founding, within the national Psychoanalytical Societies, of Institutes dedicated to training. The category of associate member began to mutate from being a welcoming gesture towards those with a wider interest in psychoanalysis beyond that of professional practice – towards those engaged, like the 1925 Group, in research – to being a 'waiting room' for those wishing to enter, as fully qualified members, into the psychoanalytic profession.[21] Psychoanalysis as a science to be discussed by the 1925 Group and others like it was crowded out by a growing insistence on the profession, on qualifications, on training, and on a politics of exclusion.

A counterfactual may help to clarify these questions. What would have been the chances of establishing a Cambridge Institute of Psycho-Analysis alongside the London Institute established by Ernest Jones in 1926? What would have been necessary for such an initiative? Could one have imagined the University Lecturer in Psychopathology, J.T. MacCurdy, as its

[20] Max Eitingon, 'Preliminary discussion of the question of analytical training. Thursday, September 3, 1925', *BIPA* 7 (1926), 130–43 at 130.

[21] One can catch this movement in the early 1930s in the USA, which had finally followed the IPA (which itself had followed Jones's initiative by introducing the general category of 'associate', without quite accepting its rationale) in eventually introducing associate membership. Thus in 1927, the APA authorized 'the creation of an associate membership to include persons engaged in fields allied to psycho-analysis but who are not engaged in the practice of therapeutic psycho-analysis'. But in 1930, with the issue of lay analysis preoccupying the Americans, a new rule was introduced by the APA that effectively excluded any non-medical 'fields allied to psycho-analysis' from consideration: 'Non-medical applicants for associate membership must have the equivalent of a BA degree. Such applicants under thirty-five years of age shall be required to study medicine. Those over thirty-five, in addition to the educational requirements of a BA degree, shall be required to have had three years of training here and abroad, according to the requirements of the International Training Commission.' The NYPS in 1931 firmed up this revision of the raison d'être of the associate: 'Non-medical applicants shall be considered eligible for associate membership. Such candidates must have a baccalaureate degree from an accredited college or university; they must give adequate reasons for lacking a medical qualification, and they must have received three years' training in psycho-analysis either here or abroad in accordance with the requirements of the International Training Commission.'

Director? Could seminars on 'Psychoanalysis and mind' have been led by Penrose and Ramsey, on 'Philosophy and psychoanalysis' by Ramsey and Jeffreys, on 'Psychoanalysis and biology' by Tansley, on 'Psychoanalysis and education' by Susan Isaacs, on 'The technique of psychoanalysis' by James Strachey? Could the Addenbrooke's Psychological Clinic have been integrated into a clinical teaching programme (an initiative that C.S. Myers, in his expansionist and generous mood of 1919, would obviously have welcomed)?

The idea that a thriving psychoanalytic culture and set of institutions could have developed outside London runs counter to the reality of the history of English psychoanalysis from the 1920s to very late in the twentieth century. The counterfactual question brings into sharp focus the plain fact that the development in Britain of the profession of psychoanalyst was confined to London, and was effectively restricted, in terms of the geography of practitioners, to the consulting rooms of Harley Street, Wimpole Street, Regent's Park and the enclaves of Bloomsbury and Fitzrovia, eventually moving out to Kensington, Hampstead and St John's Wood. This concentration led inevitably to nervous competition for patients, so that, many years too late, Ernest Jones's first thought in 1938 about the influx of immigrants from Hitler's Europe was whether it would be possible for them to find professional lodgings in Britain without disturbing the permanent residents – in Edinburgh, Glasgow, Manchester, Bristol, Oxford and Cambridge, to reproduce his list.[22] The excessive centralization of British psychoanalysis was not repeated in all other countries, and not only in the United States, where large distances encouraged independent psychoanalytic Societies developing – in Buffalo, Baltimore, Cleveland, Washington, Boston, and many other cities, not to mention the short-lived Ward's Island Psychoanalytic Society in the Manhattan borderlands, founded in 1913.[23] The American Psychoanalytic Association was until the 1940s an umbrella organization of wavering and indeterminate powers and policy, one certainly lacking any individual with the iron grip and control fever of Ernest Jones. In Germany, psychoanalytic initiatives sprung up in a number of cities: Groddeck's sanatorium at Baden-Baden; Frieda Reichmann's 'Therapeutikum' in Heidelberg

---

[22] Jones to Anna Freud, 29 April 1938, cited in R. Steiner, "'It is a new kind of diaspora'", *IRP* 16 (1989), 35–72 at 67: 'coming to England is not the same as coming to London ... It will be hard for more than eight analysts at the most to settle in London ... Five provincial towns come into consideration, though we are not yet sure of more than three. They are Edinburgh, Glasgow, Manchester, Bristol, Oxford and Cambridge.' Jones appears to have counted 'Oxford and Cambridge' as one.

[23] JF, 11 November 1913.

(1924); in Leipzig a Society for Psycho-Analytical Research was formed in 1922 at the initiative of Theresa Benedek;[24] a long-standing Stuttgart Study Group of the German Psycho-Analytical Society was still active in 1936.[25]

The most obvious German historical case prompting comparative speculation is the Frankfurt School, based in the Institute for Social Research (founded in 1923) affiliated with the University of Frankfurt, and informally linked to the 'Frankfurt Subsection', founded in 1926,[26] of the German Psycho-Analytical Society. This led to the creation of the South German Institute for Psychoanalysis in Frankfurt-am-Main (1929), which became a guest of the Institute for Social Research. The striking characteristic of the Frankfurt School was the deployment of psychoanalysis within a larger preponderantly academic project, a Marxist-inspired sociology and philosophy, one which was capable of mutating and adapting in another environment, 1940s America, on both coasts, into an influential social psychology and philosophy. What were the preconditions for the success of the Frankfurt School? It could draw on a full generation of foundational sociological theory and empirical studies, of which the work of Max Weber is only the most famous. There was also Karl Mannheim (a life-long admirer of Freud[27]), never a member of the School, who took up a chair in Sociology and Economics at Frankfurt. Most importantly, it was wholly founded by a major philanthropic donor, thus creating long-term resources and intellectual independence, preserved by Max Horkheimer's far-sightedness in moving that financial capital and even official institutional residence out of Germany as early as 1930.

Of course, the constellation of psychoanalytical interests in Cambridge in the 1920s was very different from those in Frankfurt from the early 1920s to 1933, when the Institute was forced to decamp. The most conspicuous difference is the complete absence of a sociological dimension in Cambridge. The only figure in this book with anything close to a modern sociological imagination was J.D. Bernal, with all his passionate

---

[24] 'The psycho-analytical movement', *BIPA* 5 (1924), 409–17.

[25] M. Graber, 'Stuttgart Study Group of the German Psycho-Analytical Society', *BIPA* 17 (1936), 141–2.

[26] S. Radó, 'Frankfurt Subsection', *BIPA* 9 (1928), 390. The members of the Subsection were Frieda Reichmann (Heidelberg), Erich Fromm (Heidelberg), Clara Happel (Frankfurt), Karl Landauer (Chair), Dr Röllenblöck, Heinrich Meng (Stuttgart) and Dr Stein.

[27] Mannheim's wife Julia trained as an analyst, first in the early 1930s in Frankfurt, then in London, qualifying in 1944. Mannheim's Wikipedia entry includes a photograph of the vase containing his ashes on top of a marble monument in Golders Green Columbarium, and notes: 'He was originally placed opposite Sigmund Freud as a planned pairing, but Freud was later relocated' (http://en .wikipedia.org/wiki/Karl_Mannheim, accessed 22 May 2015)

intensity.[28] The contrast with European sociology provided by the example of Rivers is telling: he certainly contributed to sociology (and indeed, much ink has been written on the complex relations between Rivers, his student Radcliffe-Brown and the importing of Durkheim's theories into what became social anthropology), but the one thing that can be said for certain is that Rivers's conception of sociology was entirely continuous with his conception of anthropology. And a 'sociology' that was autonomous, that would have been recognizable as such to Europeans and Americans, would, as we saw, only arrive in Cambridge in the last third of the twentieth century.

Contrasting the aborted psychoanalytic initiatives of 1920s Cambridge with, on the one hand, the institutional and intellectual resilience of the Frankfurt School and, on the other, the sobering failure to establish sociology at Cambridge alongside the relatively effortless creation of the Institute of Criminology courtesy of the consummate outsider-insider Leon Radzinowicz bring to mind Perry Anderson's famous theses, written out of the turbulence of the 1960s. Anderson drew attention to what he judged to be the comparative poverty of English intellectual life after the Great War and its consequent domination by 'White émigrés' – 'White' as in 'White Russians', reflecting the fact that, according to Anderson, it was the 'conservatives' who remained in England, the radicals (amongst whom the Frankfurt School figured prominently) who moved across the Atlantic. Anderson's list of such conservative émigrés, less frayed in the light of subsequent history than one might anticipate, is: Wittgenstein, Malinowski, Namier, Popper, Berlin, Gombrich, Eysenck, Klein, Deutscher.

> The culture of British bourgeois society is organized about an absent centre – a total theory of itself, that should have been either a classical sociology or a national Marxism . . . A White emigration rolled across the flat expanse of English intellectual life, capturing sector after sector, until this traditionally insular culture became dominated by expatriates, of heterogeneous calibre. Simultaneously, there occurred a series of structural distortions in the character and connections of the inherited disciplines. Philosophy was restricted to a technical inventory of language. Political theory was thereby cut off from history. History was divorced from the exploration of political ideas. Psychology was counterposed to them. Economics was dissociated from both political theory and history. Aesthetics was reduced to

[28] Sprott, who later became a founding British sociologist, came late to the field and via social psychology, an unusual route with nothing of the theoretical ambitions associated with Weber, Durkheim and Marx in its tracks.

psychology. The congruence of each sector with its neighbour came to form something like a closed system, in which elements incompatible with the dominant pattern, like psychoanalysis, were quarantined.[29] Suppressed in every obvious sector at home, thought of the totality was painlessly exported abroad, producing the paradox of a major anthropology where there was no such sociology. In the general vacuum thus created, literary criticism usurps ethics and insinuates a philosophy of history.[30]

Two aspects of these famous claims concern us: firstly, the quarantining of psychoanalysis; and secondly, the deepening impoverishment and insularity of English culture.

If there is one conclusion to be drawn from *Freud in Cambridge*, it is that psychoanalysis was very quickly established in certain circles on the 'inside' of respectable English intellectual life, if one takes 'Cambridge' to represent respectability, which is plausible. What did not happen was that psychoanalysis became a University discipline: this needs saying, because Anderson's account is peculiarly oblivious to the university orientation of his account of disciplines (even more important for the 'modern disciplines' like English and anthropology than for his 'inherited disciplines'). So, from one point of view, psychoanalysis definitely was excluded from the ordinary commerce of university interdisciplinary conversation. One of the fruits of this book is the tracking in detail of how wild enthusiasm turned into calculated quarantining, both within the University and without. Anderson concedes that the psychoanalysts (Jones and Glover) played an important part, perhaps even the principal part, in making psychoanalysis into an 'esoteric enclave',[31] in its isolation, exclusion and consequent neutering; but the prehistory of the Institute of Criminology in Cambridge offers a good example, of which there were surely others, of active quarantining from outside by well-entrenched conservative forces, in that instance in the Home Office. These effects became most evident in the period after the Second World War. In the USA, Hollywood, the New York intelligentsia, *Reader's Digest* culture and American psychiatry

---

[29] The original 1968 sentence, rewritten in 1992 in the passage quoted, was: 'The congruence of each sector with its neighbour is circular: together they form something like a closed system. The quarantine of psychoanalysis is an example: it was incompatible with this pattern' (56).

[30] Perry Anderson, 'Components of the national culture', *New Left Review* 50 (1968), 3–57, reprinted (with light revisions) in Anderson, *English Questions*, London: Verso, 1992, pp. 48–104 at 103.

[31] *Ibid.*, p. 42. In the 1968 version: 'It has been sealed off as a technical enclave: an esoteric and specialized pursuit unrelated to any of the central concerns of mainstream "humanistic" culture. There is no Western country where the presence of psychoanalysis in general culture is so vestigial.' In the 1992 version this is recast as: 'Psychoanalysis has been sealed off as an esoteric enclave: a specialized pursuit unrelated to mainstream "humanistic" concerns' (p. 88).

all were permeated with psychoanalytic enthusiasm and orthodoxies. In France, philosophy and the human sciences, Parisian Left Bank intellectuals and French psychiatry all conceded to psychoanalysis an essential role, with Freud firmly established as an 'auteur de gare', a railway station bestseller. In 1968, the contrast with the very limited psychoanalytic successes in Britain, with Bowlbyism and the Tavistock effect making inroads into the new welfare state structures of health and social care, but with no mass psychoanalytic culture – 'the virtually complete absence of any important work by Freud in paperback'[32] – was striking (although Anderson was writing just too early to observe the success in Pelican paperback of John Bowlby's *Child Care and the Growth of Love* (1965) and D.W. Winnicott's *The Child, the Family and the Outside World* (1964)).[33]

But Anderson's key term 'quarantining' points to something very real, with a number of different measures of its validity. The most obvious is that for most of the twentieth century it was virtually impossible to find a practising psychoanalyst outside of the privileged quarters of London. In the most literal sense of the term 'quarantine', the official practice of psychoanalysis was restricted, by seemingly invisible forces, to a few square miles, almost a few streets, in London.[34] If solicitors could thrive in Hastings or Huddersfield, why could not psychoanalysts?

Yet this quarantining, if indeed it was the case for the practice of the orthodox psychoanalysis that was so closely policed by Jones and Glover before the war and increasingly enshrined in post-war Britain in the policies of the BPaS, was famously being undone in post-Second World War Britain by innovations which, for Anderson, did not count. While Klein, the figure that Anderson did pick out, had undoubted great influence within the British Society, the innovations that significantly affected British society at large were associated with the Tavistock Clinic, with Michael Balint's innovations in creating an autonomous and professional

---

[32] *Ibid.*

[33] See Peter Mandler, 'Good reading for the million: the "paperback revolution" and the diffusion of academic knowledges in mid-20th century Britain and America', paper delivered to Cultural History Seminar, Cambridge, 20 May 2015.

[34] The concentration of psychoanalysts in specific exclusive areas was not confined to Britain; the psychoanalysts of Los Angeles 'mainly live and work in Beverly Hills and points west – about 200 psychotherapists operate in "Couch Canyon" located on Bedford Drive' (Douglas Kirsner, *Unfree Associations: Inside Psychoanalytic Institutions*, London: Process, 2000, p. 230); in New York, the main concentration of analysts, from the 1930s on, was to be found on Park Avenue, Central Park West and around Columbia University (N. Hale, 'From Bergasse XIX to Central Park West: the Americanization of psychoanalysis', *Journal of the History of the Behavioral Sciences* 14 (1978), 299–315). The Paris analysts were less geographically concentrated but in the 1930s the majority were to be found in the wealthiest arrondisement (16ème).

practice of group analysis, the Balint Groups, for general practitioners within the newly founded NHS, and then with the innovations associated with the paediatrics of Donald Winnicott and the attachment theory of John Bowlby. These innovations were disseminated not through 'high theory' and the universities, which Anderson assumes is where psycho-analysis might have had an effect, but through the 'caring professions' of clinical psychology, clinical medical practice and social work. Dissemination through widening professional practices, rather than foster-ing of academic reflection and research, was virtually the exclusive path sought and taken.

How are we to weigh Anderson's judgement about the deepening impoverishment and insularity of English culture?[35] After all, the rise of the White émigrés to the respected and recognized heights of disciplinary power and cultural recognition in Britain ('the most conservative major society in Europe [with] a culture in its own image: mediocre and inert') might lead one to the opposite conclusion: the openness and generosity of British culture to immigrants of distinction and flair. More pertinently, the force of Anderson's claim in 1968 is itself most plausibly a symptom of the radical change in British intellectual culture that was taking place in those years, of which Anderson was himself representative. The rise of cultural theory, of social history, of feminism, of radical aesthetics (John Berger's *Ways of Seeing*), of structuralism and post-structuralism – all these would be distinctively British while being constructively open to influences not only from the post-colonial world but also from Europe and especially France; they would also be radically different from the mid-century dis-pensation Anderson deplored. So the judgement concerning the inert mediocrity of British intellectual life should be taken with a pinch of salt, itself expressive of the new forces struggling for expression. Psychoanalysis would take up a new position in the new landscape of the 1970s and after; but that is not part of the history told in this book. The sea-change following the 1960s did, however, make it more difficult to perceive the character of the first British response to psychoanalysis. It was all but impossible in the 1970s to imagine that Cambridge had ever been a hive of psychoanalytic activity.

There is one episode recounted in *Freud in Cambridge* that did, how-ever, engender considerable progeny and have major import in the early

---

[35] Anderson clarified the focus and the limits of his analysis, namely our 'fundamental concepts of man and society', thereby restricting it to 'history, sociology, anthropology, economics, political theory, philosophy, aesthetics, literary criticism, psychology and psychoanalysis' (Anderson, 'Components of the national culture', p. 5).

history of psychoanalysis in Britain – the story of the Malting House School. The experiment of the school also captures an element that has been viewed as distinctively English in the global history of psychoanalysis: the figure of the child. Freud himself had declared in 1925 that:

> None of the applications of psycho-analysis has excited so much interest and aroused so many hopes, and none, consequently, has attracted so many capable workers as its use in the theory and practice of education. It is easy to understand why; for children have become the main subject of psycho-analytic research and have thus replaced in importance the neurotics upon whom its studies began.[36]

Freud was aware, not least because his daughter Anna was a child analyst with no medical qualifications and a teacher, that psychoanalysis was provoking great interest – and opposition – in educational circles both on the continent and in England. Freud, as a keen observer from afar, had no illusions about the independent-mindedness of English analysts in relation to children ('I also knew that the attitude of the English to child analysis was already fixed, even before Anna's appearance'). The first major controversies over child analysis, with the 'Symposium on Child Analysis' held at the BPaS on 4 and 18 May 1927, then published as a series of papers in the *IJP*,[37] opened up a feud between two of the pioneers, Melanie Klein and Anna Freud, which would last their lifetimes. But what 'fixed' the views of the English in relation to the child?

In hindsight the English literature of the early twentieth century for children, imbued with a nostalgic, witty playfulness and openness to high fantasy, is critical – the age after Lewis Carroll, the age of J.M. Barrie and Kenneth Grahame, and in the 1920s A.A. Milne. But this was also the age of the cultural supremacy of the English public school, which had already provoked a reaction in the progressive school movement that, having started in the late nineteenth century with the foundation of Abbotsholme and Bedales, flourished in the 1920s. Progressive education introduced the importance of 'nature', of play, of the imagination, of education arising from the interests of the child. Men such as Pyke and Russell could not countenance having their beloved children educated in the mercilessly bullying atmosphere they themselves had endured. Co-educational Bedales, founded in 1893, was the first such school, and an inordinately large number of protagonists, particularly of the women, in

---

[36] Freud, 'Preface to Aichhorn's *Verwahrloste Jugend*' (1925) *SE* XIX, 273.
[37] The authors of the papers presented were Melanie Klein, Joan Riviere, Nina Searl, Ella Freeman Sharpe, Edward Glover and Ernest Jones.

this book went to that school: Alix Strachey, Lettice Baker, Noel Olivier, Frances Partridge, Margaret Gardiner, Margaret Leathes, as well as John Layard and L.L. Whyte.[38] 'Yes, I went to Cambridge, but I went to Cambridge from Bedales, which gives you a little bit of the flavour',[39] Margaret Gardiner said in an interview in 2000. Whyte recalled: 'At Bedales an advanced rationalism was married to a beautiful theory – implicit in our co-educational schooling – of ideal, whole-natured, relations between man and woman ... This opened grand vistas, wonderful dreams of a radiant humanity of which we Bedalians were the first privileged examples.'[40] Freudianism came naturally to Bedalians in the early 1920s.

Elsewhere, the American Homer Lane founded a community for adults and juvenile delinquents, the Little Commonwealth, during the war (1913–17).[41] Infused with the older American ideas of progressive education together with a radical version of Freud, he established himself as a psychoanalyst in Gordon Square in the early 1920s; it was Lane who transmitted his enthusiasm to A.S. Neill (who, after experiments in Dresden and in Austria, founded Summerhill near Lyme Regis in 1924, moving to its permanent home in Suffolk in 1927[42]), and analysed John Layard in the early 1920s. This was the larger movement to which the Malting House Garden School in Cambridge and the Russells' Beacon Hill School belonged.

Another trail of influence of 1920s progressive education passing via Cambridge leads to an important element in the later psychoanalytic culture of England: the sentimental education of John Bowlby

---

[38] Lancelot Law Whyte (1896–1972) went straight from Bedales into the Army, serving on the front line on the Somme – the core experience of his life. Taking a First in Physics at Cambridge in 1921, he started doctoral work with Rutherford but his internal misery forced him to flee Cambridge and turn to banking and industry, where he found his feet, becoming in 1935 the entrepreneurial energy behind the development of the jet engine by Frank Whittle. Throughout the period from the 1920s to the 1960s, he was also a freelance physicist and philosopher – his theories later influenced Glaister's Order of Woodcraft Chivalry, another 1920s experiment in communal living focused on the child. His book *The Unconscious before Freud* (1960) opens: 'The origins of this work lie back in the years after the first World War, when psychoanalysis was a novelty and I innocently imagined that Freud had just discovered the unconscious mind' (Lancelot Law Whyte, *The Unconscious before Freud*, London: Tavistock, 1962, p. vii). See E. Whyte, 'A biographical note on L.L. Whyte', *Contemporary Psychoanalysis* 10 (1974), 386–8.

[39] Andrew Brown, 'Interview with Margaret Gardiner at her home, 35 Downshire Hill, 24.6.00', transcript courtesy of Andrew Brown.

[40] Lancelot Law Whyte, *Focus and Diversions*, London: The Cresset Press, 1963, pp. 10–11.

[41] W. David Wills, *Homer Lane: A Biography*, London: Allen & Unwin, 1964.

[42] Jonathan Croall, *Neill of Summerhill: The Permanent Rebel*, London: Routledge and Kegan Paul, 1983.

(1907–90).[43] Son of a distinguished surgeon, Bowlby came up to Cambridge in 1925, took a First in Natural Sciences in 1927 and a Second in Moral Sciences – the psychology course – in 1928. Groomed for medicine, Bowlby had acquired an intense interest in psychology – reading Freud's *Introductory Lectures* while taking the Moral Sciences Tripos was a formative experience.[44] A life-changing conversation with John Alford, who 'knew more about what was of interest to me than anyone I had met so far',[45] decided him to try teaching in a progressive school. He spent a few months at Bedales, but then moved to Priory Gate where Alford was working, founded and run by Theodore Faithfull, an unconventional Freudian, a charismatic and unstable stalwart of the progressive moment. Alford had a similar professional background to Bowlby: an undergraduate at King's College, Cambridge, he took a First in Moral Sciences in 1911 then a Second in History in 1912, followed by a year studying psychology at UCL. As an ambulance driver during the war, he lived at Toynbee Hall and worked in the charitable organization the Agenda Club. During this time, like his contemporary from King's John Layard, he had analysis with Homer Lane and, again like Layard, spent time on the continent, in Florence.[46] It was in the late 1920s that he was teaching at Priory Gate School, unpaid, and there influenced Bowlby decisively, not only with his version of psychoanalysis and its relevance to dealing with difficult children, but also in deciding him to study medicine (at UCH) and start a training analysis (he was assigned to Joan Riviere) at the BPaS. The young Bowlby had already developed an interpretation of Freud's trauma theory, viewing separation in childhood as a major trauma. By the end of the 1930s,

---

[43] Important works on Bowlby's early development are: Suzan Van Dijken, *John Bowlby: His Early Life*, London: Free Association Books, 1998; K.S. Van Dijken, R. Van der Veer, M.H. Van IJzendoorn and H.J. Kuipers, 'Bowlby before Bowlby: the sources of an intellectual departure in psychoanalysis and psychology', *Journal of the History of the Behavioral Sciences* 34 (1998), 247–69.

[44] Bowlby Collection, Wellcome Collection, A.1/4, 'An autobiographical note for the Bulletin of the Royal College of Psychiatrists by John Bowlby, October 1980. Eleven books that most influenced my work'. See also 'John Bowlby Interview with Milton Senn, M.D. [1977]', *Beyond the Couch*, the online journal of the American Association for Psychoanalysis in Clinical Social Work 2 (December 2007) at www.beyondthecouch.org/1207/bowlby_int.htm, accessed 22 September 2015.

[45] Alice Smuts, interview with Bowlby, 6 June 1977, p. 2 (Wellcome Collection: CMAC: PP/BOW/ A.5/2), quoted in Van Dijken et al., 'Bowlby before Bowlby', p. 251.

[46] In the early 1930s, Alford was teaching part-time for the Institute of Education and the Courtauld Institute for Art, from where in 1934 he was hired, sight unseen, as the Professor of Fine Arts, at the University of Toronto. He spent the rest of his life in North America, first at Toronto (1934–45), then at the Rhode Island School of Design (1945–53), arriving finally at Tulane College, New Orleans, until his death in 1960. See E. Lisa Panayotidis, 'The Department of Fine Art at the University of Toronto, 1926–1945: institutionalizing the "Culture of the Aesthetic"', *Journal of Canadian Art History / Annales d'histoire de l'art canadien* 25 (2004), 100–22.

he had refined this theory and applied it to criminal behaviour amongst the delinquents with whom he worked for years at the Canonbury Child Guidance Clinic in Islington, London.

Bowlby and 'Bowlbyism' became famous principally after the Second World War, for highlighting the effect of the child's separation from the mother in infancy and early childhood, leading to his theories of maternal deprivation.[47] Combining empirical studies with psychoanalytic theory, his later influential work also involved the joining of ethology with psycho-analysis; this element had an important link with Cambridge, in the person of Robert Hinde, follower of Lorenz and Tinbergen, researching at Cambridge on imprinting in birds, which he extended, with Bowlby's assistance, to rhesus monkeys. Their collaboration, however, did not start with a Cambridge connection: it was Julian Huxley, a friend of Bowlby's wife's family, who in 1951 put Bowlby onto ethology and the work of Lorenz and Tinbergen; it led, somewhat later, to his contact and long-term collaboration with Hinde, Fellow (later Master) of St John's College, Cambridge, and from 1963 Royal Society Professor at the Madingley Sub-Department for the Study of Animal Behaviour in the University. Bowlby's interest in ethology led to the incorporation of imprinting studies into the foundations of his theory of attachment, and Bowlby drew Hinde into participation in the Cases Seminar at the Tavistock. This collabora-tion led eventually to Hinde's own later studies of nursery children.[48]

Bowlby's career is a significant marker of the end-point of this story, precisely because Cambridge in his early days offered him little connection to psychoanalysis. If the enthusiasms of the early 1920s had translated into any kind of institutional or community support for psychoanalysis, one would have expected him to make the most of it. Acquiring an interest in psychology at Cambridge in the late 1920s, he made no connection to the Malting House School, probably because it was foundering while he was intellectually unformed; however, the movement of progressive education of which it was a part would quickly prove decisive, after he had left Cambridge as an undergraduate, in giving him the actual experience that made psychoanalysis his life's vocation. Bowlby was typical in this respect of a steady trickle of other Cambridge undergraduates, who of necessity

[47] The seminal work is of course Denise Riley, *War in the Nursery: Theories of the Child and Mother*, London: Virago, 1983; also Michal Shapira, *The War Inside: Psychoanalysis, Total War, and the Making of the Democratic Self in Postwar Britain*, Cambridge: Cambridge University Press, 2013, esp. chapter 7, pp. 198–237.

[48] 'Robert Hinde: interview with Alan Macfarlane, 20.11.07', at www.alanmacfarlane.com/DO/film show/hinde2_fast.htm, accessed 22 September 2015.

cultivated their embryonic psychoanalytic interests after leaving the University. But these figures do indicate how Cambridge, despite the paucity of its institutional support, still supplied an important flow of students into the world of psychoanalysis, whether in training or in therapy, amongst them Denis Carroll[49] and Henry Dicks.[50]

With the outbreak of the Second World War, Cambridge hosted a sudden efflorescence of psychoanalytic activity, in part because Cambridge was chosen, following government advice, as a site for evacuation of London-based institutions vulnerable to bombing, including Bedford College, Lister Institute, the LSE and the London Child Guidance Clinic. Psychoanalysts with long-standing links to Cambridge did the same: within days of the declaration of war John Rickman was based as a psychiatrist at Haymeads Emergency Hospital, Bishop's Stortford, from where he was coordinating with members of the LSE's Mental Health course now based in Peterhouse, Cambridge[51] to provide emergency services. After consultation with Rickman, Susan Isaacs and Melanie Klein moved to Cambridge, together with Sybil Clement Brown, Susan's close friend and colleague at LSE. The three women lived together for a year at 30 Causewayside, barely a hundred yards from the Malting

---

[49] Carroll (1901–56), a mathematics scholar at Trinity Hall in 1919, took a Double First in the Natural Sciences Tripos in 1922 and 1923, and then was awarded the Michael Foster Studentship in 1923, doing research in physiology under Barcroft. As an undergraduate he was elected a member of the Natural Sciences Club; the talks he gave, one on 'Suggestion in everyday life' (24 February 1923) and another on 'Death' (1 December 1923), gave early indications of his interest in the Freudian field (see *Cambridge University Natural Science Club. Founded March 10th, 1872*, Cambridge: Cambridge University Press, 1982); in the mid-1920s he trained as a doctor at the London Hospital, and held a position at the Maudsley and as a psychoanalyst at the Institute of Psycho-Analysis. He became a close colleague of Edward Glover's and a pioneer in the field of psychoanalytic criminology, through his work at the ISTD, to which he also sponsored Bowlby's affiliation in the mid-1930s.

[50] The best source on Dicks is Daniel Pick, *The Pursuit of the Nazi Mind: Hitler, Hess, and the Analysts*, Oxford: Oxford University Press, 2012, pp. 25–8. Born in Estonia of a timber-trading English father (probably of Jewish extraction) married to a German Protestant, Dicks (1900–77) was educated later in St Petersburg and passed through the fire of the October Revolution. Having volunteered in 1918 for the Artists' Rifles and then served in Russia during the English post-war attempts to combat the stabilization of the Bolshevik Revolution (1918–20), he came up to St John's College, Cambridge as a foundation scholar and took a First in Part I of the Natural Sciences Tripos in 1923. Training in medicine at Bart's and then the Royal Bethlem Hospital, by 1929 he was attached to the Tavistock Clinic, practising in 'psychological medicine' as an eclectic Freudian, with sympathy for Adler and Jung alongside more exotic interests in the occult. To the Tavistock Clinic, Dicks remained faithful throughout his life, writing its authoritative history in 1970. During the Second World War, like Bowlby and several other Tavistock psychiatrists, he served in the Army, assigned (being bilingual in German and English) by his close colleague and friend J.R. Rees, Chief of Army Psychiatry, the task of interviewing Hitler's deputy, Rudolf Hess, after his capture following his enigmatic flight to Scotland in the summer of 1941. In post-war Britain, he created a new Department of Psychiatry at the University of Leeds, before returning to head a marital unit at the Tavistock.

[51] RPBPaS, CRR/F14/04, Margaret Ashdown to John Rickman, 5 September 1939.

House. Klein remained for one year, Isaacs until early 1943.[52] Bowlby's work at the London Child Guidance Clinic meant his clinical work with children shifted to Cambridge, a move he welcomed. The analysts were not alone in evacuating: 3.5 million people evacuated from the large cities in September 1939, most, like Isaacs and Klein, under their own steam; the government provision was principally for schoolchildren and women with young families – over a million were evacuated within days of the declaration of war. Cambridge was a favoured destination for those from Tottenham and Islington, in the north and east of London – 6,700 schoolchildren were evacuated there in September 1939.[53] On 16 October 1939, a meeting was convened at Susan Isaacs's flat in Cambridge: amongst those present were Rickman, Klein, Winnicott, Bowlby, Ruth Darwin (Medical Research Council; creator of the Pinsent-Darwin Fund), Sybil Clement Brown and social workers from Cambridge: 'a group of people representing many aspects of the children-evacuation problem'.[54] Echoing the days of the Malting House School when the Adrians had been so supportive, Hester Adrian, Chairman of the Evacuation Reception Committee, gave full assistance to the Research Committee established by the meeting, led by Isaacs. As Isaacs wrote in the final report, the *Cambridge Evacuation Survey* (1941): 'In the Borough of Cambridge we had at hand not merely a compact field of study, but also many personal contacts which ensured the essential co-operation of officials and voluntary workers.'[55] Thouless, who co-authored the Report with

---

[52] Philip Graham, *Susan Isaacs: A Life Freeing the Minds of Children*, London: Karnac, 2009, p. 299. Isaacs saw several patients in the three and a half years she lived in Cambridge during the war, amongst them William Rushton, physiologist and Fellow of Trinity College, whose research career had stalled and was saved by his five-year psychoanalysis, during which time he shifted from nerve physiology to colour vision, becoming a remarkably sustained and prolific researcher in his new field (see CUEP, W.A.H. Rushton, 'Personal Record of Fellow of the Royal Society' (1969) and Horace Barlow, 'William Rushton. 8 December 1901–21 June 1980', *BMFRS* 32 (1986), 422–59). Another patient was Dr Frances Barnes, daughter of Helen Gardiner (who had been up at Newnham College during the First World War, had an analysis with Ernest Jones and spoke to her daughter of psychoanalysis as 'the coming thing'); Frances was interested in working with children, had contacts with Anna Freud and, while training as a doctor, had analysis with Isaacs at Causewayside in 1942; at the time the issue of marriage was part of her analysis: she married John Barnes, who had studied anthropology at St John's College, Cambridge and in 1969 was appointed first Professor of Sociology at Cambridge. (Laura Cameron interview with Frances Barnes, 24 June 2000.)

[53] Van Dijken, *John Bowlby*, p. 112, citing Angus Calder's *The People's War: Britain 1939–1945*, London: Jonathan Cape, 1969.

[54] RPBPaS, CRR/F14/04, Draft Memorandum, 16 October 1939.

[55] *Cambridge Evacuation Survey: A Wartime Study in Social Welfare and Education*, ed. Susan Isaacs with the co-operation of Sibyl Clement Brown and Robert H. Thouless, London: Methuen, 1941, p. 3.

Isaacs and Clement Brown, was no doubt another of these personal contacts: a student of Rivers with interests in psychology and religion, he had returned to Cambridge in 1938 as an educational psychologist (and therefore excluded from the Department of Experimental Psychology), after having held posts at Manchester and Glasgow.

The *Cambridge Evacuation Survey* was a study of the failure of the evacuation programme: failure because so many of the evacuees straggled back to the danger zones of the cities during the Phoney War. The 6,700 evacuees of September had become 3,650 by November; in July 1940 only 1,624 remained. 'If human nature had been taken into equal account with geography and railway time tables, there would in all likelihood not have been so serious a drift back to the danger areas',[56] Isaacs wrote. In its focus on the importance of addressing failure, the study resonates with that other pioneering research survey conducted in Cambridge a decade earlier, Lella Sargant Florence's account of the introduction of birth control to the working-class districts of Cambridge: its scientific tone, its determination to reveal all the warts, to counteract ignorance, to show how underestimating human needs had led to such mistakes, this time, in wartime, even more on a national scale. But even before the *Survey* was published, a different psychoanalytic note of warning and admonishment had been struck: three of the Cambridge-linked psychoanalysts had very speedily issued a warning about the dangers of the evacuation exercise; Winnicott, Emanuel Miller and Bowlby co-authored a note in the *BMJ* in December 1939 which warned of the dangers of prolonged separation of a small child from its mother, with the consequent juvenile delinquency which would threaten later in life.[57] The *Cambridge Evacuation Report* gave a different, less explicitly psychoanalytic, account of the reasons for the obvious failings in the evacuation exercise, emphasizing 'the crucial importance of family ties and of the feelings of parents and children towards and about each other'.[58] More specifically, the first of three dominant motives for the return home of the evacuated children (the others were complaints about foster homes and the financial burden of the evacuation) was 'family feeling – the anxiety and loneliness of the parents, the homesickness or worry of the children'.[59]

Post-war, the *Survey* did have indirect effects: the Curtis Committee (1946) heard evidence from Bowlby, Isaacs and Winnicott on the effects of

---

[56] *Ibid*, p. 4.
[57] John Bowlby, Emanuel Miller and D.W. Winnicott, 'Evacuation of small children', *BMJ* 2(4119) (1939), 1202–3.
[58] *Cambridge Evacuation Survey*, ed. Isaacs, p. 7.   [59] *Ibid.*, pp. 7–8.

evacuation on the emotional lives of children.[60] With others, their contributions led to the main theme of the ensuing Children Act 1948 being the requirement on Local Authorities 'to exercise their powers with respect to [the child] so as to further his best interests and to afford him opportunity for the proper development of his character and abilities'.[61] The experience of evacuation, together with Bowlby's later studies of deprivation, persuaded the government and Local Authorities of the importance of restricting the amount of time children in need were separated from their families. But this story belongs to the very different post-World War II history of psychoanalysis, with Bowlbyism, the Balint Groups, the consolidation of Kleinianism and the growth of the Middle Group best represented by Winnicott. Those developments are what international psychoanalysis, even fine historians of psychoanalysis such as Shapira and Zaretsky,[62] came to understand as 'British psychoanalysis'. The story told in this book of an earlier moment is forgotten, with whatever historical effects it had slipping into oblivion.

The work devoted to the *Cambridge Evacuation Survey*, indeed its very possibility, was a swan-song of the efflorescence of interest in Cambridge in psychoanalysis during the 1920s, providing some of the key personnel in the study and furnishing the network of contacts upon which the link-up between London and Cambridge was grafted. But long before the accidents of war brought 6,000 children and a core group of analysts to Cambridge, the University- or city-based impetus behind psychoanalytic initiatives no longer existed. The final chapter of this book emphasizes this fact: there is no doubt that Bloomsbury was intimately tied to Cambridge in its inception; the four Bloomsbury analysts prolonged their connection with Cambridge well into the 1920s. But even they let those connections drop in the 1930s, save for Karin Stephen's acceptance of the role of external expert bringing necessary if unconventional views to the medical students. The 'Cambridge' connection went underground, surfacing only indirectly in the crowning achievements of their public careers: instilling a certain political ethos, which was also inherent in their personal conception of psychoanalysis itself, into the constitutional reform of the BPaS and

---

[60] Shapira, *The War Inside*, p. 202; for a full account of the Curtis Committee and the 1948 Children's Act which implemented its recommendations, see Harry Hendrick, *Child Welfare: England 1872–1989*, London: Routledge, 1994, esp. pp. 196–203.

[61] Children Act 1948, Section 12, quoted in Hendrick, *Child Welfare*, p. 199.

[62] Shapira, *The War Inside*; Eli Zaretsky, *Secrets of the Soul: A Social and Cultural History of Psychoanalysis*, New York: Knopf, 2004, chapter 10 'The mother–infant relationship and the postwar Welfare State', pp. 249–75.

impressing upon the English Freud of the *Standard Edition* a quite distinctive literary style and culture.

This study demonstrates not only the range of interest in psychoanalysis in Cambridge in the 1910s and 1920s but also the influx into the early British Society during its formative years of a substantial group of Cambridge-linked analysts. What significance should we attribute to this phenomenon, which speaks both to the character of English culture and to the character of the BPaS? Firstly, it highlights a considerable contrast between the early members of the Viennese and the British Psycho-Analytical Societies. The Viennese Society was nearly completely Jewish, composed of non-elite doctors and a range of adventurous but often marginal cultural entrepreneurs. They were, like Freud himself, part of the first large wave of Jewish immigrants to Vienna, that of the third quarter of the nineteenth century, who successfully entered the professions of law and medicine, the arts and sciences, commerce and banking.[63] Yet the most important vehicle for the wider dissemination of Freud's ideas was not the Viennese Society but the Burghölzli Hospital, directed by Eugen Bleuler, who in the early 1900s directed his assistant, Carl Gustav Jung, to report on and assimilate Freud's ideas. In the period 1906–10, the large numbers of young visiting doctors at the Burghölzli were rapidly informed of the theories and practices of Freud's psychoanalysis. Because the Burghölzli was already one of the principal European hubs, an obligatory passage point, for psychologically oriented psychiatry (the other was Kraepelin's Department in Munich), Freud's ideas were disseminated rapidly throughout some of the best-informed and enterprising of the young psychiatrists and neurologists of the time. The Burghölzli was the launch pad both for a Freudian psychiatry and for the psychoanalytic movement.

If we were to take this twin-centre characteristic for German-speaking psychoanalysis as a model for English psychoanalysis, would Cambridge function as the potential transmitter, the equivalent of the Burghölzli? At first sight, the contrast is greater than the similarity: a cantonal lunatic asylum attached to the University of Zurich that had morphed into a world-renowned centre for treatment and research in psychiatry becoming, for a brief period, the seedbed for a new movement, is very different from an ancient university which, while hostile to clinical medicine, was welcoming of a full range of scientific innovations and whose lack of

[63] Steven Beller, *Vienna and the Jews, 1867–1938: A Cultural History*, Cambridge: Cambridge University Press, 1989; Zaretsky, *Secrets of the Soul*, p. 70.

centralized authority facilitated many diverse intellectual initiatives. But it is precisely what the two institutions had in common as obligatory passage points, one for psychiatry, the other for entry into the academic and social elite of inter-war Britain, that is also striking: the early 1920s at Cambridge accustomed a whole generation of the elite to the reality of Freud's revolution. And, just as the enthusiasm for psychoanalysis cooled quite rapidly at the Burghölzli with Bleuler's withdrawal and Jung's resignation, but left a permanent legacy within German-speaking psychiatry, so the enthusiasm in Cambridge lasted no longer than the post-war period from the Armistice to the General Strike but made 'Freud' a permanent (if often notorious) feature of English intellectual and public life from then on.

Looking at the surge in Cambridge of interest in psychoanalysis from the point of view of English society and culture in general, one difference from Vienna is striking: the relative absence of Jews in the early history of psychoanalysis in Britain, and also locally in Cambridge. None of the 1925 Group was Jewish; none of the undergraduates whose interest pre-Great War and in the early 1920s we have tracked was Jewish. Until the arrival of Melanie Klein in 1926 and Marjorie Franklin's election in 1927, only two members of the early BPaS (David Eder and his sister-in-law Barbara Low) were Jewish.[64] This absence of Jews in early English psychoanalysis, almost unique amongst the different national cultures which took it up, points in two different directions: firstly, towards the excessively closed medical profession, traditionally based in family firms, and pervaded by the burble of routine antisemitism so typical of the English middle and upper classes in the inter-war period; secondly, towards the stable structures of professional power and authority in British life.

Somewhat in contrast to the medics, the legal profession, while never actively welcoming of Jews, had promoted Jews to the highest ranks by the beginning of the twentieth century (the Lord Chief Justice during the First World War, Rufus Isaacs, also a Liberal MP, was Jewish). At this time it was still virtually impossible for a Jew to become a consultant in the prestigious London teaching hospitals until after the Second World War; even then, antisemitism was still active in the higher reaches of the medical profession.[65] In the 1950s, Arthur Rook, dermatologist at Addenbrooke's, advised the highly qualified Maurice Garretts, who had applied for a junior post alongside him, that 'the number of people who are Jewish who live in

[64] Mathew Thomson, '"The solution to his own enigma": connecting the life of Montague David Eder (1865–1936), socialist, psychoanalyst, Zionist and modern saint', *Medical History* 55 (2011), 61–84.
[65] John Cooper, *Pride versus Prejudice: Jewish Doctors and Lawyers in England, 1890–1990*, Portland, OR: The Littman Library of Jewish Civilization, 2003, p. 74.

Cambridge you can count on one hand'; Garretts took the hint and withdrew his application.[66]

The paucity of Jews in British psychoanalysis thus appears to reflect the paucity of Jews in the professions – and that is a striking fact. It tells us that English psychoanalysis was not a profession to which outsiders could gain access precisely because it was new, as it so clearly was in central Europe; it was an insiders' profession, with more than just the Bloomsberries coming out of the 'middle-class, professional, cultured, later Victorian, box'. Even the disproportionate numbers of women who became early members of the BPaS often had distinguished middle-class professional backgrounds.[67] Cambridge was of course simply a concentrated form of English elite culture with respect to its exclusion of Jews. Yet a few Jews did pass through Cambridge and some of them figure significantly in this book: Myers, Pyke, Miller, Bernal (part), Wittgenstein (baptised), Gorer – most of these from monied and privileged backgrounds. And this is the second point: taken together, the relative absence of Jews in English psychoanalysis and the substantial presence of Cambridge (both in general reception and as source of members of the BPaS) tells us much: psychoanalysis was not in any sense subversive or external to the established social and professional order. In the first half of the twentieth century, with its new structures of the professions and new dispensation for the universities becoming well entrenched, English culture had little place for anything disruptive. The cultivation of respectability for psychoanalysis at all costs by Ernest Jones began to irritate younger analysts such as the Stephens in the 1930s and eventually led to the constitutional crisis in the Society of the early 1940s, which, in this light, presages the sea-change that came with the Labour government of 1945 followed by the post-imperial convulsions of the 1960s.

The character of the principal cultural carriers of psychoanalysis in Britain thus looked back to the innovative years of the late nineteenth century, when the formulae for the establishment of the professions were created, modelled principally on law and medicine, and from then on gave

---

[66] *Ibid.*, p. 258.
[67] With her father a vicar, Sylvia Payne trained as a doctor before the Great War and distinguished herself in medical service during it, being awarded a CBE; her brother went to Jesus College, Cambridge and eventually became Governor of Kenya, just when Payne was taking over the Presidency of the BPaS, and then Governor of Ceylon. Ella Freeman Sharpe did attend Nottingham University and would have taken her place at Oxford if she had not been obliged to care for her mother and siblings by the early death of her father. These histories are typical, even of some of the more unconventional members of the Society (as women inevitably were). Both Payne and Sharpe began their psychoanalytic careers training at the Brunswick Square Clinic.

the professions in Britain a unique character: small, compact, imbued with solidarity (self-protection): 'While popularizing the aspiration for autonomy regulated by ethics and self-government, the freedom of the British model [of the professions] from university and bureaucracy proved surprisingly unique.'[68] The component of education which consisted in 'professional training' was placed at arms' length by the universities, in particular by Oxbridge. It is a surprising conclusion, but one historians of education often agree on, that Oxbridge ended up as purer instantiations of the ideals of *Bildung* and *Wissenschaft* than any universities to be found in Germany or the USA – nineteenth-century ideals that survived until the onslaught launched by the Thatcher government in the 1980s. The compact separateness of the professions in Britain mirrored the self-protective autonomy of the universities (led by Oxbridge – though this did not prevent Oxbridge from becoming the breeding ground of the ruling class as represented by the Home and Indian Civil Services). But these two structural features of upper-middle-class institutions form the backdrop to the famous theses of Noel Annan (echoed and revised by Perry Anderson and Stefan Collini) concerning the 'intellectual aristocracy' in England. As Randall Collins glossed the argument:

> [In Britain], leading intellectuals are more elaborately connected by family linkages in the generations from 1840 to 1920 than at virtually any other time in history: there is an intermarrying network that links Russell, Moore, Keynes, Virginia Woolf, and the Bloomsbury circle to the Thackerays, Macaulays, Darwins, Maitlands, Trevelyans, Balfours, and many others. They are genuine cousins, in-laws, and nephews, not merely the metaphorical kinfolk produced by master–pupil lineages.[69]

Annan demonstrated the existence of this network at work within the Colleges of Oxford and Cambridge, then extending out to all the key institutions of English society: parliament, the courts, the Foreign and Colonial Services, the Home Civil Service, the publishing houses, the professions, even the episcopate. Its existence explained, for him, 'the paradox of an intelligentsia which appears to conform rather than rebel against the rest of society.'[70] Whether or not the English were distinctly

[68] Walter Rüegg (ed.), *A History of the Universities in Europe*, Vol III, *Universities in the Nineteenth and Early Twentieth Century*, Cambridge, Cambridge University Press, 2004, p. 380.
[69] Randall Collins, *The Sociology of Philosophies: A Global Theory of Intellectual Change*, Cambridge, MA: Belknap Press, 1998, p. 732.
[70] Noel Annan, 'The intellectual aristocracy', in J. Plumb (ed.), *Studies in Social History: A Tribute to G.M. Trevelyan*, London: Longman, 1955, p. 285, quoted in Stefan Collini, *Absent Minds: Intellectuals in Britain*, Oxford: Oxford University Press, 2006, p. 140.

different from other elites (such as that formed by the Ecole Normale Supérieure in France, also prone to extensive intermarrying and generational consolidation of position),[71] whether or not Annan actually pinpointed an 'intelligentsia' rather than, as Collini argues, simply 'a professional-cum-administrative-cum-academic elite',[72] the extensive influence of a Cambridge route to psychoanalysis in the early twentieth century invites us to view psychoanalysis as being bound into this 'aristocracy'. The two eminent analyst couples, the Stracheys and the Stephens, not to mention the Woolfs who had such a hand in the dissemination of psychoanalysis in Britain before the Second World War, tempt us to this view. Collini rephrases the core of Annan's thesis once he has stripped it of its pompous implausibilities and excesses: 'most of the members of Annan's intellectual aristocracy displayed a considerable social confidence, a familiarity with other members of this elite, and an assumption of access to the wielders of power'.[73] A surprising number of early psychoanalysts, especially those who came via Cambridge, conform to this description.[74] The Apostles, Bedales and Newnham, Bloomsbury: these 'institutions' symbolize the ready overlapping of the progressive wing of the intellectual aristocracy and the advent of a broadly conceived psychoanalytic culture.

The rapid expansionary years of psychoanalysis lasted from the end of the Great War to the founding of the Institute of Psychoanalysis in 1926, when the regulations governing training were introduced. Whether or not

---

[71] Christophe Charle, *Naissance des 'intellectuels' 1880–1900*, Paris: Minuit, 1990; Charle, *Les Intellectuels en Europe au XIXe siècle: essai d'histoire comparé*, Paris: Seuil, 1996; Jean-François Sirinelli, *Génération intellectuelle: Khâgneux et normaliens dans l'entre-deux-guerres*, Paris: Fayard, 1988, discussed in Collini, *Absent Minds*, pp. 267–70.

[72] Collini, *Absent Minds*, p. 142. [73] *Ibid.*, p. 144.

[74] An alternative view of the 'intellectual aristocracy', though one even more committed to looking backward to its Victorian roots, is proposed by anthropologist Adam Kuper in the context of examining 'the end of the great Victorian clans in England' with the First World War against the backdrop of the prevalence of incestuous first-cousin marriage in England during the nineteenth century. The introduction of psychoanalysis into Britain coincided with the moment when these clans lost their power and social significance – Bloomsbury was a last prominent gasp. The steady and momentous decline in the birth rate, which meant there were fewer and fewer cousins or possible suitors in the small number of acceptable or recommended families, together with an ever-reducing need to protect family businesses by 'marrying in', had been overtaken by the rise in limited companies from the 1840s on, by structural changes to capitalism, as well as by the decimating effects of the First World War followed by the early 1920s depression and the Great Depression of the 1930s. These structural changes led to the disappearance of the pattern of intermarrying elite families that was so characteristic of late Victorian and Edwardian England. See Adam Kuper, *Incest and Influence: The Private Life of Bourgeois England*, Cambridge, MA: Harvard University Press, 2009, pp. 251–4. However, Kuper's account relates more to the provincial and urban business classes and is not strictly tailored to the 'intellectual aristocracy', a thesis which is much more closely allied to Francis Galton's *Hereditary Genius* (1869), foundational for eugenics; indeed, some of the same families figure in both works.

the pioneering days of psychoanalysis had really burned themselves out by
1926, as the ebbing of psychoanalytic initiatives in Cambridge and elsewhere indicate, or the new training regulations, preoccupied with professional respectability and intent on exclusion, acted as a sudden inhibition
on the general interest in psychoanalysis amongst enthusiasts, the rate of
expansion of the membership of the British Society slowed dramatically.
At forty-nine in 1924, the British Society was the largest in the world, but
this number had expanded only marginally to sixty by 1934; by 1937 there
were sixty-eight, seven of the additional eight being German-based
refugees.[75] The contrast in the careers of James Strachey and John
Bowlby, Strachey twenty years older than Bowlby but both out of established upper-middle-class professional families, is striking. Strachey mean-
dered for a decade, reviewing for the weeklies, having a brush with medical
school and the Brunswick Square Clinic until, after two years in Vienna,
'without any further ado, I was elected an associate member of the
Society ... So there I was, launched on the treatment of patients, with
no experience, with no supervision, with nothing to help me but some two
years of analysis with Freud.'[76] Bowlby's career started with simultaneous
medical and psychoanalytic training; during the 1930s he was affiliated to
most of the important institutions of the 1930s linked to progressive
psychiatry and psychoanalysis: the Maudsley Hospital, the Department
of Psychology and the Institute of Education at UCL, the London Child
Guidance Clinic, the ISTD. No meander for him: Bowlby had a model
career path and professional training for a psychoanalyst, which means that
the model institutions had already been created ready for him to join.
None of those institutions came out of Cambridge, though many of the
pioneers associated with them did, like Bowlby himself. Tentatively,
Strachey asked in 1963, looking back across the bureaucratic apparatus of
training that had developed since 1920: 'Is it worth-while to leave
a loophole for an occasional maverick? I don't know.'

Meisel and Kendrick made some acutely perceptive remarks about the
early days of psychoanalysis in the 1920s, comparing Alix Strachey's Berlin
with James's London:

---

[75] E. Jones and K. Abraham, 'Report of the Eighth International Psycho-Analytical Congress', *IJP* 5
(1924), 402; 'Reports of Proceedings of Societies', *BIPA* 15 (1934), 525–34 at 526.

[76] Strachey, 'James Strachey's speech at the Jubilee Banquet to celebrate the 50th anniversary of the
British Psycho-Analytic Society on 30 October 1963' (privately printed), quoted in M.R. Khan,
'*The Psychoanalytic Study of the Child*. Vols 34 and 35: Edited by Albert J. Solnit et al. New Haven:
Yale University Press. 1979/1980', *IJP* 63 (1982), 98–100 at 99–100.

in England, psychoanalysis faced greater public resistance, even as it attracted a wider range of intellectual adherents, than it did on the Continent. For all their involvement in the arts and entertainment, the Berlin psychoanalysts enjoyed a greater sense of solidarity and separateness than did their English counterparts. In England psychoanalysis lingered longer in a stage of generous dilettantism; as non-medical practitioners, James and Alix were rather unusual because they made it the principal focus of their lives. Anthropologists, art historians, economists, as well as doctors of various persuasions – even literary types like Lytton – dabbled in psychoanalysis, seeing it as a contribution to their chosen specialities, not a profession in itself. As a result, the British interest in psychoanalysis was strikingly diverse; by contrast, the Berlin Society looks positively parochial.[77]

This 'stage of generous dilettantism' is when activities in Cambridge were at their height; it was also the 'psychological moment in every country when interest in the newness of psycho-analysis became acute', as Ernest Jones, looking back in his autobiography, characterized the coming of psychoanalysis: 'I should date this "moment" in England to be within the first five years after the end of the war, some ten years before it happened in France and Italy.'[78] This 'moment' was when Freud came to Cambridge.

John Bowlby returned to Cambridge in 1977 to receive an honorary DSc, courtesy of his old Director of Studies from the late 1920s, Lord Adrian, retiring as Chancellor of the University, who had nominated him for the degree.[79] Psychoanalysis received its honorary status from the University in this moment, with Bowlby inducted into the Great and the Good alongside the jurist Lord Devlin and Mother Theresa. But Bowlby was also the object of protest and derision. Facing much resistance, the Nursery Action Group (NAG) in Cambridge, inspired by the women's movement, were campaigning for the University to set up child-care facilities for university teachers, workers and students. Bowlbyism, psychoanalysis and Freud represented the conservative ideology that tied women to the family and the home. Consigned to oblivion as the revolutionary representative of sexual immorality and the bearer of a revolution in the values of civilization, Freud returned fifty years on to Cambridge as an unforgiving patriarch guarding the status quo of a conservative University.

Yet while Freud never truly came to Cambridge, his ideas and his movement could never be entirely ignored. Across the late twentieth-century

---

[77] BF, p. 39.
[78] Ernest Jones, *Free Associations: Memories of a Psycho-analyst*, London: Hogarth, 1959, p. 220.
[79] Ursula Bowlby, 'To Cambridge for an Sc.D.', 18 pages, Bowlby Papers, Wellcome Modern Archives, A. 2/12 Cambridge 1977. Laura Cameron interview with Anne Keynes née Adrian.

University, dons found it necessary to teach students the rudiments of those ideas: in the literature of all the major European languages; as scandalous and discarded ideas in psychology; as bearing one of the principal moral theories of the century; as the source for radical interpretations of ancient texts; as laying down a challenge about the existence of 'the unconscious', whether in the paradoxes of self-deception or in the perennial problems of mind–body dualism; as a main vehicle for scientific naturalism and radical atheism for theology to weigh up; as an essential ingredient for theories of child development; even to understand the modernist movement in music associated with the Second Viennese School or to consider the coherence of legal doctrines of guilt and punishment. Psychoanalytic ideas, Freud's ideas, were always on hand to surprise and disconcert. But Cambridge as an institution, as representing the orthodoxies of knowledge, would never accept him. As he himself said in English in the one BBC broadcast he, a refugee seeking to 'die in freedom',[80] gave at the end of his life: 'People did not believe in my facts and thought my theories unsavoury. Resistance was strong and unrelenting. In the end, I succeeded in acquiring pupils and building up an *International Psycho-Analytic Association*. But the struggle is not yet over.'

[80] J III, 240, Freud to Ernst Freud, 12 May 1938.

Bibliography

Archival sources

Birmingham City Library Archives
British Psycho-Analytic Society Archives
British Psychological Society Archives
Cambridge University Archives, University Library
Cambridge University Botanical Garden Archives
Cambridgeshire Collection, Cambridge Central Library
Cambridgeshire Record Office
Croydon Natural History and Scientific Society Archives
Department of Experimental Psychology Archives, University of Cambridge
Frederic E. Clements Papers, American Heritage Center, University of Wyoming
Freud Museum, London
Institute of Education Archives
King's College, Modern Archive, Cambridge
Kingsley Martin Papers, University of Sussex Library
Layard Papers, University of California at San Diego, Mandeville Special Collections
Library of Congress
McMaster University Libraries
Newnham College Archives, Cambridge
Otto Rank Archives, Columbia University, New York
Pyke Papers (held in the private residence of Janet Pyke, London)
St John's College Archives, Cambridge Archives
Sigmund Freud Museum, Vienna
Strachey Papers, British Library
Trinity College Archives, Cambridge
University College London Manuscript and Rare Books Room
Wellcome Library

Theses and other unpublished sources

Buklijas, Tatjana, 'The laboratory and the asylum: the L.C.C. Pathological Laboratory at Claybury, Essex, 1895–1916', MPhil dissertation, University of Cambridge, 1999.

Cameron, Laura, 'Anthropogenic natures: Wicken Fen and histories of disturbance, 1923–1943', PhD dissertation, University of Cambridge, 2001.

Crampton, Colin, 'The Cambridge School. The life, work and influence of James Ward, W.H.R. Rivers, C.S. Myers and Sir Frederic Bartlett', PhD dissertation, University of Edinburgh, 1978.

Edwards, M.V., 'Control and the therapeutic trial, 1918–1948', MD thesis, University of London, 2004.

Howie, D. 'Interpretations of probability 1919–1939: Harold Jeffreys, R.A. Fisher, and the Bayesian controversy', PhD dissertation, University of Pennsylvania, 1999.

Jardine, Boris, 'Scientific moderns', PhD dissertation, University of Cambridge, 2012.

Keeley, James P., '"The coping stone on psychoanalysis": Freud, psychoanalysis and the Society for Psychical Research', PhD dissertation, Columbia University, 2002.

Levy, Marin Katherine, 'Birth control on trial (1930): Lella Secor Florence and the Cambridge Women's Welfare Clinic', unpublished paper.

Mandler, Peter, 'Good reading for the million: the "paperback revolution" and the diffusion of academic knowledges in mid-20th century Britain and America', paper delivered to Cultural History Seminar, Cambridge, 20 May 2015.

Palfrey, D., 'Moral Science at Cambridge University, 1848–1860', PhD dissertation, University of Cambridge, 2001.

Rice, Gillian, 'The reception of Freud amongst the Bloomsbury Group', MPhil dissertation, University of Cambridge, 1984.

Sommer, Andreas, 'Crossing the boundaries of mind and body: psychical research and the origins of modern psychology', PhD dissertation, University of London, 2013.

Sponsel, Alistair, The Cambridge Natural Sciences Tripos, 1915–1949, MSc dissertation, Imperial College, London, 2001.

Stansfield, F.R. 'The growth years of the National Institute of Industrial Psychology, 1921–1930', unpublished paper presented at the British Psychological Society History of Psychology Centre, London, 2 March 2005.

Vincent, J., 'Disestablishing moral science: John Neville Keynes, religion, and the question of cultural authority in Britain 1860–1900', PhD dissertation, University of Cambridge, 2006.

Journals/magazines/newspapers

American Journal of Psychology
Annals of Botany
Biographical Memoirs of Fellows of the Royal Society
Brain
British Journal of Medical Psychology
British Medical Journal
Bulletin of the History of Medicine

Bulletin of the International Psycho-Analytical Association
Cambridge Daily News
Cambridge Magazine
Cambridge University Reporter
Ecology
International Journal of Psycho-Analysis
International Review of Psycho-Analysis
Journal of Ecology
Journal of Mental Science
Journal of the History of the Behavioral Sciences
Mind
New York Review of Books
Philosophical Magazine
Proceedings of the BAAS
Psyche
The Athenaeum
The Cambridge Review
The Lancet
The Nation and Athenæum
The New Phytologist
The Spectator
The Times
University of Cambridge Calendar

Books and articles

Abel, Elizabeth, *Virginia Woolf and the Fictions of Psychoanalysis*, Chicago: University of Chicago Press, 1989.

Abir-Am, P., 'Dorothy Maud Wrinch (1894–1976)', in L.S. Grinstein et al. (eds.), *Women in Chemistry and Physics*, New York: Greenwood, 1993, pp. 605–12.

'Synergy or clash: disciplinary and marital strategies in the career of mathematical biologist Dorothy Wrinch', in Pnina G. Abir-Am and Dorinda Outram (eds.), *Uneasy Careers and Intimate Lives: Women in Science 1789–1979*, with a foreword by Margaret W. Rossiter, New Brunswick: Rutgers University Press, 1987, pp. 239–80, 342–54.

Adrian, E.D., 'Discoveries of the mind', *The Observer*, 28 October 1953.

'The response of human sensory nerves to currents of short duration', *Journal of Physiology, London* 53 (1919), 70–85.

Anderson, Perry, 'Components of the national culture', *New Left Review* 50 (1968), 3–57.

English Questions, London: Verso, 1992.

Anker, Peder, 'The context of ecosystem theory', *Ecosystems* 5 (2002), 611–13.

Imperial Ecology: Environmental Order in the British Empire, 1895–1945, New Haven: Yale University Press, 2001.

Annan, Noel, *The Dons*, London: HarperCollins, 1999.
'The intellectual aristocracy', in J. Plumb (ed.), *Studies in Social History: A Tribute to G.M. Trevelyan*, London: Longmans, Green & Co., 1955.
Anon. 'Obituary [E.P. Farrow]', *Lincolnshire Free Press*, 5 June 1956.
Appignanesi, Lisa and John Forrester, *Freud's Women*, London: Weidenfeld & Nicolson, 1992.
Auden, W.H., 'A consciousness of reality', review of *A writer's diary*, by Virginia Woolf. *The New Yorker*, 6 March 1954.
Forewords and Afterwords, selected by Edward Mendelson, London: Faber & Faber, 1973.
Ayer, A.J. *Bertrand Russell*, Chicago: University of Chicago Press, 1974.
Language, Truth and Logic (1936), London: Penguin, 1971.
Ayres, Peter, *Shaping Ecology: The Life of Arthur Tansley*. Chichester, UK and Hoboken, NJ: Wiley-Blackwell, 2011.
Baldick, Chris, *The Oxford English Literary History*, Vol. X, *1910–1940: The Modern Movement*, Oxford: Clarendon Press, 2004.
The Social Mission of English Criticism, 1848–1932, Oxford: Clarendon Press, 1983.
Balint, M., 'On love and hate' (1951) in Balint, *Primary Love and Psycho-Analytic Technique*, London: Hogarth Press and the Institute of Psycho-Analysis, 1952, pp. 141–56.
Barham, G.F., 'Insanity with myxœdema', *Journal of Mental Science* 58 (1912), 226–35.
Barham, Peter, *Forgotten Lunatics of the Great War*, New Haven: Yale University Press, 2004.
Barkan, E., *The Retreat of Scientific Racism: Changing Concepts of Race in Britain and the United States between the World Wars*, Cambridge: Cambridge University Press, 1992.
Barker, Pat, *The Eye in the Door*. London: Viking, 1993.
The Ghost Road. London: Viking, 1995.
Regeneration. London: Viking, 1991.
Bartlett, F.C., 'Experimental method in psychology', *Nature*, 31 August 1929, pp. 341–5.
'Fifty years of psychology', *Occupational Psychology* 19 (1955), 203–16.
'Remembering Dr. Myers', *Bulletin of the British Psychological Society* 18 (1965), 1–10.
Bartley, William Warren, III, *Wittgenstein*, Philadelphia: Lippincott, 1973.
Bateson, Gregory, 'Experiments in thinking about observed ethnological material', *Philosophy of Science* 8.1 (1941), 53–68.
Naven: A Survey of Problems Suggested by a Composite Picture of the Culture of a New Guinea Tribe Drawn from Three Points of View, Cambridge: Cambridge University Press, 1936.
Bayliss, W.M. (ed.), *Life and its Maintenance: A Symposium on Biological Problems of the Day*, Glasgow: Blackie & Son, 1919.

Beard, Mary, *The Invention of Jane Harrison*, Cambridge, MA: Harvard University Press, 2000.

Beiser, Frederick C., *The German Historicist Tradition*, Oxford: Oxford University Press, 2012.

Bell, Quentin, *Bloomsbury*, London: Weidenfeld & Nicolson, 1968.

Beller, Steven, *Vienna and the Jews, 1867–1938: A Cultural History*, Cambridge: Cambridge University Press, 1989.

Beloff, John, *Parapsychology: A Concise History*. London: The Athlone Press, 1993.

Bennett, Arnold, 'Fiery sign on the horizon', *Evening Standard*, 26 July 1928.

Bernal, J.D., 'Mr. Geoffrey Pyke: an appreciation', *Manchester Guardian*, 25 February 1948.

The World, the Flesh and the Devil, London: Cape, 1968.

Bernal, Martin, *Geography of a Life*, Xlibris, 2012.

Bernfeld, Siegfried, 'On psychoanalytic training', *Psychoanalytic Quarterly* 31 (1962), 453–82.

Berrios, German E., 'British psychopathology since the early 20th century', in German E. Berrios and Hugh Freeman (eds.), *150 Years of British Psychiatry, 1841–1991*, London: Gaskell, 1991, pp. 232–44.

Bettelheim, Bruno, *Freud and Man's Soul*, London: Chatto & Windus, 1983.

Black, Max, 'Relations between logical positivism and the Cambridge school of analysis', *Erkenntnis* 8 (1938/9), 24–35.

Black, Max, John Terence Wisdom and Maurice Cornforth, 'Symposium: Is analysis a useful method in philosophy?', *Proceedings of the Aristotelian Society, Supplementary Volumes*, Vol. XIII, *Modern Tendencies in Philosophy* (1934), pp. 53–118.

Blackett, Patrick, 'Boy Blackett', in Peter Hore (ed.), *Patrick Blackett: Sailor, Scientist and Socialist*, London: Frank Cass, 2003, p. 12.

Blackwell, Stephen H., 'Nabokov's wiener-schnitzel dreams: despair and anti-Freudian poetics', *Nabokov Studies* 7 (2002/3), 129–50.

Blanco, Matte, I, *The Unconscious as Infinite Sets*, London: Duckworth, 1975.

Bland, Lucy, *Banishing the Beast: English Feminism and Sexual Morality 1885–1914* London: Penguin Books, 1995.

Bocking, Stephen, *Ecologists and Environmental Politics: A History of Contemporary Ecology*, New Haven: Yale University Press, 1997.

Boll, Theophilus E. M., 'May Sinclair and the Medico-Psychological Clinic of London', *Proceedings of the American Philosophical Society* 106 (August 1962), 310–26.

Bolt, Bruce A., 'Jeffreys and the Earth', in Keitti Aki and Renata Dmowska (eds.), *Relating Geophysical Structures and Processes: The Jeffreys Volume*, Geophysical Monograph 76, IUGG Volume 16, Washington, DC: International Union of Geodesy and Geophysics and the American Geophysical Union, 1993, pp. 1–10.

Bouveresse, Jacques, *Wittgenstein Reads Freud: The Myth of the Unconscious*, trans. Carol Cosman, foreword Vincent Descombes, Princeton: Princeton University Press, 1995.

Bouwsma, O.K., *Wittgenstein: Conversations, 1949–1951*, ed. J.L. Craft and Ronald E. Hustwit. Indianapolis: Hackett, 1986.

Bowden, P., 'Pioneers in forensic psychiatry. Hamblin Smith: *The Psychoanalytic Panacea*', *Journal of Forensic Psychiatry* 1 (1991), 103–13.

Bowler, Peter J. and John V. Pickstone (eds.), *The Cambridge History of Science*, Vol. VI, *The Modern Biological and Earth Sciences*, Cambridge: Cambridge University Press, 2008.

Boyd, Brian, *Vladimir Nabokov: The Russian Years*, Princeton: Princeton University Press, 1990.

Bradley, Noel, 'Response: Karin Stephen, the superego and internal objects', *Psychoanalysis and History* 4(2) (2002), 225–30.

Brent, Leslie Baruch, 'Susanna Isaacs Elmhirst obituary: child psychiatrist adept at observing disorders from play', *The Guardian*, 29 April 2010.

Broad, C.D., *Scientific Thought*, International Library of Psychology, Philosophy, and Scientific Method, London: Kegan Paul, Trench, Trubner & Co., 1923.

Brooke, Christopher N.L., *A History of the University of Cambridge*, Vol. IV, *1870–1990*, Cambridge: Cambridge University Press, 1993.

Brown, Andrew, *J.D. Bernal: The Sage of Science*, Oxford: Oxford University Press, 2005.

Bulmer, Martin, 'The development of sociology and of empirical social research in Britain', in Bulmer (ed.), *Essays on the History of British Sociological Research*, Cambridge: Cambridge University Press, 1985, pp. 3–36.

'Sociology in Britain in the twentieth century: differentiation and establishment', in A.H. Halsey and W.G. Runciman (eds.), *British Sociology Seen from Within and Without*. Oxford: Oxford University Press for the British Academy, 2005, pp. 36–53.

Bunn, G.C., A.D. Lovie and G.D. Richards (eds.), *Psychology in Britain: Historical Essays and Personal Reflections*, Leicester: BPS Books, 2001.

Burston, Daniel, *The Wing of Madness: The Life and Work of R.D. Laing*, Cambridge, MA: Harvard University Press, 1996.

Buzard, James, *Disorienting Fiction: The Autoethnographic Work of Nineteenth-Century British Novels*, Princeton: Princeton University Press, 2005.

Caine, Barbara, *From Bombay to Bloomsbury: A Biography of the Strachey Family*, Oxford: Oxford University Press, 2005.

'Mothering feminism/mothering feminists: Ray Strachey and *The Cause*', *Women's History Review* 8(2) (1999), 295–310.

'The Stracheys and psychoanalysis', *History Workshop Journal* 45 (1998), 145.

Cambridge University Natural Science Club. Founded March 10th, 1872. Cambridge: Cambridge University Press, 1982.

Cameron, H.C., 'Sir Maurice Craig', *Guy's Hospital Reports* 85 (1935), 251–7.

Cameron, Laura, 'Ecosystems', in Stephan Harrison, Steve Pile and Nigel Thrift, (eds), *Patterned Ground: Entanglements of Nature and Culture*. London: Reaktion Press, 2004, pp. 55–7.

'Histories of disturbance', *Radical History Review* 74 (1999), 2–24.

'Oral history in the Freud Archives: incidents, ethics and relations', *Historical Geography* 29 (2001), 38–44.

'Science, nature and hatred: "finding out" at the Malting House Garden School, 1924–29', *Environment and Planning D: Society and Space* 24 (2006), 851–72.

'Sir Arthur George Tansley', in N. Koertge (ed.), *New Dictionary of Scientific Biography*, Vol. VII, London: Thomson Gale, 2008, pp. 3–10.

Cameron, Laura and Sinead Earley, 'The ecosystem: movements, connections, tensions and translations', *Geoforum* 65 (2015), 473–81.

Cameron, Laura and John Forrester, 'Freud in the field: psychoanalysis, fieldwork and geographical imaginations in interwar Cambridge', in P. Kingsbury and S. Pile (eds.), *Psychoanalytic Geographies*, Farnham: Ashgate, 2014, pp. 41–56.

'"A nice type of the English scientist": Tansley and Freud', *History Workshop Journal* 48 (Autumn 1999), 64–100.

'Tansley's psychoanalytic network: an episode out of the early history of psychoanalysis in England', *Psychoanalysis and History* 2(2) (2000), 189–256.

Cameron, Laura and David Matless, 'Translocal ecologies: the Norfolk Broads, the "natural", and the International Phytogeographical Excursion, 1911', *Journal of the History of Biology* 44(1) (2011), 15–41.

Campbell, Patrick, *Siegfried Sassoon: A Study of the War Poetry*, Jefferson, NC: McFarland, 1999.

Carey, Hugh, *Mansfield Forbes and his Cambridge*, Cambridge: Cambridge University Press, 1984.

Cavell, Stanley, 'The availability of Wittgenstein's later philosophy', *Philosophical Review* 71(1) (1962), 67–93.

Charle, Christophe, *Les Intellectuels en Europe au XIXe siècle: essai d'histoire comparée*, Paris: Seuil, 1996.

Naissance des 'intellectuels', 1880–1900, Paris: Minuit, 1990.

Charlesworth, B., *Evolution in Age-Structured Populations*, Cambridge: Cambridge University Press, 1980, 2nd edition 1994.

Chessick, Richard D., *The Future of Psychoanalysis*, Albany: State University of New York Press, 2007.

Chick, Harriette, Margaret Hume and Marjorie MacFarlane, *War on Disease: A History of the Lister Institute*, London: A. Deutsch, 1971.

Clark, Ronald, *J.B.S.: The Life and Work of J.B.S. Haldane*, London: Bloomsbury, 2011.

Cochrane, Archibald L. with Max Blythe, *One Man's Medicine: An Autobiography of Professor Archie Cochrane*, London: BMJ, The Memoir Club, 1989.

Cohen, Pauline, 'Marie Battle Singer', *Bulletin of the Anna Freud Centre* 8 (1985), 213–15.

Collini, Stefan, *Absent Minds: Intellectuals in Britain*, Oxford: Oxford University Press, 2006.

'The study of English', in Sarah J. Ormrod (ed.), *Cambridge Contributions*, Cambridge: Cambridge University Press, 1998, pp. 42–64.

Collini, Stefan, Donald Wrinch and John Burrow, *That Noble Science of Politics: A Study in Nineteenth-Century Intellectual History*, Cambridge: Cambridge University Press, 1983.

Collins, Randall, *The Sociology of Philosophies: A Global Theory of Intellectual Change*, Cambridge, MA: Belknap Press, 1998.

'Committees at Work', *Bulletin of the American Psychoanalytic Association* 5B (1949), 24–32.

Connell, E.H., 'The use of the method of psychoanalysis in medicine', *Cambridge University Medical Society Magazine* 1 (Easter Term 1923), 97–105.

Constable, J. (ed.), *Selected letters of I.A. Richards*, Oxford: Clarendon, 1990.

Cook, Caldwell, *The Play Way: An Essay in Educational Method*, London: Heineman, 1917.

Cook, Hera, *The Long Sexual Revolution: English Women, Sex, and Contraception, 1800–1975*, Oxford: Oxford University Press, 2004.

Cooper, John, *Pride versus Prejudice: Jewish Doctors and Lawyers in England, 1890–1990*, Oxford and Portland, OR: The Littman Library of Jewish Civilization, 2003.

Copeland, B. Jack, *Colossus: The Secrets of Bletchley Park's Codebreaking Computers*, Oxford: Oxford University Press, 2010.

Costhall, Alan, 'Pear and his peers', in G.C. Bunn, A.D. Lovie and G.D. Richards (eds.), *Psychology in Britain: Historical Essays and Personal Reflections*, Leicester: BPS Books, 2001, pp. 188–204.

Crary, Alice and Rupert Read (eds.), *The New Wittgenstein*, London: Routledge, 2000.

Critchley, Macdonald, 'Remembering Kinnier Wilson', *Movement Disorders* 3(1) (1988), 2–6.

Croall, Jonathan, *Neill of Summerhill: The Permanent Rebel*, London: Routledge and Kegan Paul, 1983.

Crozier, Ivan, '"All the world's a stage": Dora Russell, Norman Haire, and the 1929 London World League for Sexual Reform Congress', *Journal of the History of Sexuality* 12(1) (2003), 16–37.

Crump, Thomas, *The Phenomenon of Money*, London: Routledge and Kegan Paul, 1981.

Danziger, Kurt, *Constructing the Subject: Historical Origins of Psychological Research*, Cambridge: Cambridge University Press, 1990.

'The origins of the psychological experiment as a social Institution', *American Psychologist* 40 (1985), 133–40.

Darling, Elizabeth, *Re-forming Britain: Narratives of Modernity before Reconstruction*, London: Routledge, 2007.

'"The star in the profession she invented for herself": a brief biography of Elizabeth Denby, housing consultant', *Planning Perspectives* 20 (July 2005), 271–300.

Deacon, R., *The Cambridge Apostles: A History of Cambridge University's Elite Intellectual Secret Society*, London: Robert Royce, 1985.

Delany, Paul, *The Neo-Pagans: Friendship and Love in the Rupert Brooke Circle*, London: Macmillan, 1987.

Desrosières, Alain, *The Politics of Large Numbers: A History of Statistical Reasoning*, trans. Camille Naish, Cambridge, MA: Harvard University Press, 1998.

Dicks, H.V., *Fifty Years of the Tavistock Clinic*, London: Routledge and Kegan Paul, 1970.

Dillistone, F.W., *Charles Raven: Naturalist, Historian, Theologian*, London: Hodder and Stoughton, 1975.

Dunbabin, J.P.D., 'Oxford and Cambridge college finances, 1871–1913', *Economic History Review*, New Series 28(4) (1975), 631–47.

Duncan-Jones, E.E., 'The Wizard of Finella. Review, Hugh Carey, *Mansfield Forbes and his Cambridge*', *London Review of Books* 7(1) (24 January 1985), 9.

'Editorial: The nature of desire', *Journal of Neurology, Neurosurgery, and Psychiatry* 3(11) (1922), 274–6.

'Editorial', *Psychic Research Quarterly* 1 (1920–21), 287–8.

Edwards, A.W.F., 'G.H. Hardy (1908) and Hardy-Weinberg equilibrium', *Genetics* 179 (July 2008), 1143–50.

Eissler, K.R., *Medical Orthodoxy and the Future of Psychoanalysis*, New York: International Universities Press, 1965.

Erreygers, G. and G. Di Bartolomeo, 'The debates on Eugenio Rignano's inheritance tax proposals', *History of Political Economy* 39(4) (2007), 605–38.

Etkind, Alexander, *Eros of the Impossible: The History of Psychoanalysis in Russia* (1993), trans. Noah and Maria Rubins, Boulder, CO: Westview Press, 1997.

Fagg, C.C., *Psycho-Analysis: A Select Reading List*, Croydon: Central Library, Town Hall, 1926.

Fairhurst, Susan, A Manchester Girl, 'Cambridge and Manchester', *Manchester University Magazine* 9(8) (1913), 173–5.

Falzeder, E. (ed.), *The Complete Correspondence of Sigmund Freud and Karl Abraham 1907–1925*, London and New York: Karnac, 2002.

Farrow, Ernest Pickworth, 'Experiences with two psycho-analysts', *Psyche* 5 (1925), 234–48.

'Eine Kindheitserinnerung aus dem 6. Lebensmonat', *International Zeitschrift für Psychoanalyse* 12 (1926), 79.

A Practical Method of Self-Analysis, enabling anyone to become deeply psychoanalyzed without a personal analyst, with some results obtained by the Author from early childhood, the earliest memories going back to the age of six months: also recounting the Author's personal experiences with two Psycho-Analysts, with Foreword by the late Professor Sigmund Freud, London: George Allen & Unwin, 1942.

Psychoanalyze Yourself: A Practical Method of Self Treatment, New York: International Universities Press, 1948.

The Study of Vegetation, London: Blackie and Son, 1926.

Federn, Paul, *Zur Psychologie der Revolution: die vaterlose Gesellschaft* (Vienna: Anzengruber, 1919).

Feigl, Herbert, 'Some major issues and developments in the philosophy of science of logical empiricism' (1954), in Herbert Feigl and Michael Scriven (eds.), *The Foundations of Science and the Concepts of Psychology and Psychoanalysis*, Minnesota Studies in the Philosophy of Science 1, Minneapolis: University of Minnesota Press, 1956, pp. 3–38.

Feuer, Lewis S., 'A narrative of personal events and ideas', in S. Hook, W. O'Neill and R. O'Toole (eds.), *Philosophy, History and Social Action: Essays in Honor of Lewis Feuer with an Autobiographic Essay by Lewis Feuer*, Boston Studies in the Philosophy and History of Science, Dordrecht: Springer, 1988, pp. 1–85.

Firkin, Barry G., 'Historical review. Some women pioneers in haematology', *British Journal of Haematology* 108 (2000), 6–12.

Fisher, Kate, *Birth Control, Sex and Marriage in Britain, 1918–1960*, Oxford: Oxford University Press, 2006.

Fisher, Kate and Simon Szreter, *Sex before the Sexual Revolution: Intimate Life in England 1918–1963*, Cambridge: Cambridge University Press, 2010.

'"They prefer withdrawal": the choice of birth control in Britain, 1918–1950', *Journal of Interdisciplinary History* 34(2) (2003), 263–91.

Florence, Barbara Moench (ed.), *Lella Secor: A Diary in Letters 1915–1922*, New York: Burt Franklin, 1978.

Florence, Lella Secor, *Birth Control on Trial*, London: George Allen & Unwin, 1930.

Progress Report on Birth Control, London: William Heinemann, 1956.

Flowers, F.A., III (ed.), *Portraits of Wittgenstein* (4 vols.), Vol. II, Bristol: Thoemmes Press, 1999.

Portraits of Wittgenstein, Vol. III, Bristol: Thoemmes, 1999.

Flügel, J.C., 'A hundred years or so of psychology at University College London', *Bulletin of the British Psychological Society* 23 (1954), 21–30.

Forrester, John, '1919: psychology and psychoanalysis, Cambridge and London: Myers, Jones and MacCurdy', *Psychoanalysis and History* 10(1) (2008), 38–43, 45–7.

'Dream readers', in Forrester (ed.), *Dispatches from the Freud Wars: Psychoanalysis and its Passions*, Cambridge, MA: Harvard University Press, 1997, pp. 138–83.

'The English Freud: W.H.R. Rivers, dreaming and the making of the early twentieth century human sciences', in Sally Alexander and Barbara Taylor (eds.), *History and Psyche: Culture, Psychoanalysis, and the Past*, London: Palgrave, 2012, pp. 72–104.

'Freud in Cambridge', *Critical Quarterly* 46(2) (2004), 1–26.

'Introduction', Sigmund Freud, *Interpreting Dreams*, trans. J.A. Underwood, London: Penguin, 2006, pp. vii–liv.

Language and the Origins of Psychoanalysis, London: Macmillan, 1980.

'The psychoanalytic passion of J.D. Bernal in 1920s Cambridge', *British Journal of Psychotherapy* 26 (2010), 397–404.

'Remembering and forgetting Freud in early twentieth century dreams', *Science in Context* 19(1) (2006), 65–85.

'"A sort of devil" (Keynes on Freud, 1925): reflections on a century of Freud-criticism', *Österreichische Zeitschrift für Geschichtswissenschaften* 14(2) (2003), 70–85.

Fortes, M. 'The first born', *Journal of Child Psychology and Psychiatry* 15 (1974), 81–104.

'Malinowski and Freud', *Psychoanalytic Review* 45A (1958), 127–45.

'The Forty-Second Meeting of The American Psychoanalytic Association, The Netherland Plaza, Cincinnati, Ohio, May 19, 20, 21 and 22, 1940', *Bulletin of the American Psychoanalytic. Association* 3 (1940), 16–65.

Foucault, Michel, *The Order of Things* ([1966] 1970) London: Routledge, 2002.

Franke, Damon, *Modernist Heresies: British Literary History, 1883–1924*, Columbus: Ohio State University Press, 2008.

French, Richard D., *Antivivisection and Medical Science in Victorian Society*, Princeton: Princeton University Press, 1975.

Freud, Anna, *The Ego and the Mechanisms of Defence* (1936), London: Hogarth Press and Institute of Psycho-Analysis, 1968 (revised edition).

Freud, Sigmund and Sándor Ferenczi, *The Correspondence of Sigmund Freud and Sándor Ferenczi*, Vol. III, *1920–1933*, ed. Ernst Falzeder and Eva Brabant, Cambridge, MA: Belknap Press, 2000.

Freud, Sigmund and Oskar Pfister, *Psycho-Analysis and Faith*, London: Hogarth Press, 1963.

G.V.C. (ed.), *Historical Register of the University of Cambridge, Supplement, 1911–1920*, Cambridge: Cambridge University Press, 1922.

Galavotti, Maria Carla (ed.), *Cambridge and Vienna: Frank P. Ramsey and the Vienna Circle*, Vienna Circle Institute Yearbook 12, Dordrecht: Springer, 2004.

Gardiner, Margaret, *A Scatter of Memories*, London: Free Association Books, 1988.

Gardner, D.E.M., *Susan Isaacs*, London: Methuen Educational, 1969.

Garland, David, 'British criminology before 1935', *British Journal of Criminology* 28(2) (1988), 1–17.

Garland, Martha McMackin, *Cambridge before Darwin: The Ideal of Liberal Education, 1800–1860*, Cambridge: Cambridge University Press, 1980.

Gathercole, Peter, Obituary of Reo Fortune, *RAIN* 37 (April 1980), 9.

Gauld, Alan, *The Founders of Psychical Research*, London: Routledge and Kegan Paul, 1968.

Geison, Gerald, *Michael Foster and the Cambridge School of Physiology*, Princeton: Princeton University Press, 1978.

Gere, Cathy and Charlie Gere, 'Introduction: the brain in a vat', *Studies in History and Philosophy of Biological and Biomedical Sciences* 35C(2) (2004), 219–25.

Gilkeson, John S., *Anthropologists and the Rediscovery of America, 1886–1965*, Cambridge: Cambridge University Press, 2014.

Glendinning, Victoria, *Leonard Woolf: A Life*, London: Secker and Warburg, 2006.

Glover, Edward, 'Eder as psycho-analyst', in J.B. Hobman (ed.), *David Eder: Memoirs of a Modern Pioneer*, London: Gollancz, 1945, pp. 92–3.

'In piam memoriam: Emanuel Miller, M.A., F.R.C.P., D.P.M., 1893–1970', *British Journal of Criminology* 11(1) (1971), 8–9.

Godwin, H., *Cambridge and Clare*, Cambridge: Cambridge University Press, 1985.

Goldstein, Jan Ellen, 'The Woolfs' response to Freud – water-spiders, singing canaries, and the second apple', *Psychoanalytic Quarterly* 43 (1974), 438–76.

Goldstein, R.G., 'The higher and lower in mental life: an essay on J. Hughlings Jackson and Freud', *Journal of the American Psychoanalytic Association* 43 (1995), 495–515.

Golley, Frank, *A History of the Ecosystem Concept in Ecology*, New Haven: Yale University Press, 1993.

Gorer, Geoffrey, *The Peoples of Great Russia*, London: The Cresset Press, 1949.

Gorham, Deborah, 'Dora and Bertrand Russell and Beacon Hill School', *Russell: The Journal of Bertrand Russell Studies*, n.s. 25 (summer 2005), 39–76.

Gosselin, R., '*Mental Defect*, by L.S. Penrose', *Psychoanalytic Quarterly* 4 (1935), 529–31.

Graham, Philip, *Susan Isaacs: A Life Freeing the Minds of Children* (London: Karnac, 2009).

Gratton-Guinness, I., *The Search for Mathematical Roots, 1870–1940: Logics, Set Theories and the Foundations of Mathematics from Cantor through Russell to Gödel*, Princeton: Princeton University Press, 2000.

Graves, Richard Perceval, *Robert Graves: The Assault Heroic 1895–1926*. London: Weidenfeld & Nicolson, 1986.

Griffin, Nicolas, *The Selected Letters of Bertrand Russell*, Vol. II, *The Public Years 1914–1970*, London: Routledge, 2002.

Grosskurth, Phyllis, *Melanie Klein: Her World and Her Work*, New York: Knopf, 1986.

The Secret Ring: Freud's Inner Circle and the Politics of Psychoanalysis, London: Cape, 1991.

Grubrick-Simitis, Ilse, *Back to Freud's Texts: Making Silent Documents Speak*, trans. Philip Slotkin, New Haven: Yale University Press, 2006.

Hacking, Ian, *The Taming of Chance*, Cambridge: Cambridge University Press, 1990.

'Telepathy: origins of randomization in experimental design', *Isis* 79(3) (1988), 427–51.

Haffenden, John (ed.), *Selected Letters of William Empson*, Oxford: Oxford University Press, 2006.

Haldane, J.B.S., *Daedelus; or: Science and the Future; a paper read to the Heretics, Cambridge, on February 4th, 1923*, New York: Dutton, 1924.

'Forty years of genetics', in J. Needham and Walter Pagel (eds.), *Background to Modern Science*, Cambridge: Cambridge University Press, 1938.

'A mathematical theory of natural and artificial selection' (Part IV), *Mathematical Proceedings of the Cambridge Philosophical Society* 23 (1927), 607–15.

Hale, Keith (ed.), *Friends and Apostles: The Correspondence of Rupert Brooke and James Strachey, 1905–1914*, New Haven: Yale University Press, 1998.

Hale, N., 'From Bergasse XIX to Central Park West: the Americanization of psychoanalysis', *Journal of the History of the Behavioral Sciences* 14 (1978), 299–315.

Hall, A.R., *The Cambridge Philosophical Society: A History 1819–1969*, Cambridge: Cambridge Philosophical Society, 1969.

Hall, J., 'Psychology and schooling: the impact of Susan Isaacs and Jean Piaget on 1960s science education reform', *History of Education* 29(2) (2000), 514–18.

Hall, Ruth (ed.), *Dear Dr Stopes: Sex in the 1920s*, London: Deutsch, 1978.

Marie Stopes: A Biography, London: Deutsch, 1977.

Halsey, A.H., *A History of Sociology in Britain: Science, Literature and Society*, Oxford: Oxford University Press, 2004.

Harley, David, 'Beacon Hill School', *Russell: The Journal of Bertrand Russell Studies* 35–36 (autumn/winter 1979–80), 5–16.

Harris, Paul L., 'Piaget on causality: the Whig interpretation of cognitive development', *British Journal of Psychology* 100 (2009), 229–32.

Harris, Pippa (ed.), *Song of Love: The Letters of Rupert Brooke and Noel Olivier, 1909–1915*, London: Bloomsbury, 1991.

Harrison, Tom, *Bion, Rickman, Foulkes and the Northfield Experiments: Advancing on a Different Front*, London: Jessica Kingsley, 2000.

Hayman, Ronald (ed.), *My Cambridge*, London: Robson Books, 1977.

Head, Henry, 'Obituary notice of William Halse Rivers Rivers', *Proceedings of the Royal Society* B77(1923), i–iv.

Hearnshaw, L.S., *Cyril Burt, Psychologist*, Ithaca, NY: Cornell University Press, 1979.

A Short History of British Psychology, 1840–1940, London: Methuen, 1964.

Heffernan, Michael and Heike Jöns, 'Research travel and disciplinary identities in the University of Cambridge, 1885–1955', *British Journal for the History of Science* 46(2) (2013), 255–86.

'Heinz Hermann Otto Wolff', *Psychiatric Bulletin* 13 (1989), 584–5.

Hemming, Henry, *Churchill's Iceman*, London: Arrow Books, 2014.

Hendrick, Harry, *Child Welfare: England 1872–1989*, London: Routledge, 1994.

Herle, Anita and Sandra Rouse (eds.), *Cambridge and the Torres Strait: Centenary Essays on the 1898 Anthropological Expedition*, Cambridge: Cambridge University Press, 1998.

Hinshelwood, B., 'Virginia Woolf and psychoanalysis', *International Review of Psycho-Analysis* 17 (1990), 367–71.

Hobsbawm, Eric, 'Preface', in Brenda Swann and Francis Aprahamian (eds.), *J.D. Bernal: A Life in Science and Politics*, London: Verso, 1999.

Hodges, Andrew, *Alan Turing: The Enigma*, London: Burnett Books, 1983.

Holroyd, Michael, *Lytton Strachey: The New Biography*, New York and London: Norton, 1995.

Horn, Lawrence, *A Natural History of Negation*, Chicago: University of Chicago Press, 1989.

Howard, Jane, *Margaret Mead*, New York: Fawcett Columbine, Ballantine, 1984.

Howarth, T.E.B., *Cambridge between Two Wars*, London: Collins, 1978.

Hume, E. Margaret, 'The history of the sieve tubes of *Pteridium aquilinum*, with some notes on *Marsilia quadrifolia* and *Lygodium dichotomum*', *Annals of Botany* 26 (1912), 573–87.

Hume, E. Margaret, 'On the presence of connecting threads in graft hybrids', *New Phytologist* 12(6) (1913), 216–21.

Huxley, A., *Brave New World* (1932), New York: Harper Perennial, 2004.

Isaacs, Susan, *Cambridge Evacuation Survey: A Wartime Study in Social Welfare and Education*, ed. with the co-operation of Sibyl Clement Brown and Robert H. Thouless, London: Methuen, 1941.

  Intellectual Growth in Young Children: With an Appendix on Children's 'Why' Questions by Nathan Isaacs, London: Routledge and Kegan Paul, 1930.

  Social Development in Young Children: A Study of Beginnings (1933), London: George Routledge and Sons, 1946.

James, William, *Essays in Psychical Research* (The Works of William James), Cambridge, MA: Harvard University Press, 1986.

Jebb, Eglantyne, *Cambridge: A Brief Study in Social Questions*, Cambridge: Macmillan & Bowes, 1906.

Jeffreys, H., *Collected Papers of Sir Harold Jeffreys on Geophysics and Other Sciences*, ed. Sir Harold Jeffreys and Bertha Swirles (Lady Jeffreys), London: Gordon & Breach, 1977, Vol. VI.

  'Fisher and inverse probability', *International Statistical Review* 42 (1974), 1–3.

  Scientific Inference, Cambridge: Cambridge University Press, 1931.

Jones, Edgar, Shahina Rahman and Robin Woolven, 'The Maudsley Hospital: design and strategic direction, 1923–1939', *Medical History* 51 (2007), 357–78.

Jones, Ernest, *Free Associations: Memories of a Psycho-analyst*, London: Hogarth, 1959.

  'The nature of desire', *Journal of Neurology and Psychopathology* 3 (1922), 338–41.

  Sigmund Freud: Life and Work, 3 vols., London: Hogarth Press, 1953–57.

  'The theory of symbolism', in *Papers on Psycho-Analysis*, 2nd edition, Published for the Institute of Psycho-Analysis, London: Bailliere, Tindall & Cox, 1918, pp. 129–86.

Jones, Ernest and others (ed.), *Glossary for the Use of Translators of Psycho-analytic Works*, International Journal of Psycho-Analysis Supplement 1, London: Baillière, Tindall & Cox, 1924.

Jöns, Heike, 'Academic travel from Cambridge University and the formation of centres of knowledge, 1885–1954', *Journal of Historical Geography* 34(2) (2008), 338–62.

Jung, C.G. and Sigmund Freud, *The Freud/Jung Letters: The Correspondence between Sigmund Freud and C.G. Jung*, ed. William McGuire, Princeton: Princeton University Press, 1974.

Kamminga, H., 'Hopkins and biochemistry', in Peter Harman and Simon Mitton (eds.), *Cambridge Scientific Minds*, Cambridge: Cambridge University Press, 2002, pp. 172–86.

Kevles, D., *In the Name of Eugenics: Genetics and the Uses of Human Heredity*, Harmondsworth: Penguin, 1985.

Keynes, Geoffrey (ed.), *The Letters of Rupert Brooke*, London: Faber & Faber, 1968.

Keynes, John Maynard, *The Collected Writings of John Maynard Keynes*, Vol. X, *Essays in Biography*, London and Cambridge: Macmillan and Cambridge University Press for the Royal Economic Society, 1972.

The Economic Consequences of the Peace, London: Macmillan, 1919.

Essays in Persuasion, New York: W.W. Norton & Co., 1963.

The General Theory of Employment, Interest and Money, London: Macmillan, 1936.

Treatise on Money, Vol. II, *The Applied Theory of Money* (1930), in *The Collected Writings of John Maynard Keynes*, Vol. VI, London: Macmillan, 1971.

King, Pearl (ed.), *No Ordinary Psychoanalyst: The Exceptional Contributions of John Rickman*, London: Karnac, 2003.

King, Pearl and Alex Holder, 'Great Britain', in Peter Kutter (ed.), *Psychoanalysis International: A Guide to Psychoanalysis throughout the World*, Stuttgart-Bad Cannstatt: Frommann-Holzboorg, 1992, pp. 150–72.

King, Pearl and Riccardo Steiner (eds.), *The Freud–Klein Controversies 1941–45*, London and New York: Tavistock and Routledge, 1991.

Kinkead-Weekes, Mark, *D.H. Lawrence: Triumph to Exile 1912–1922*, Cambridge: Cambridge University Press, 1996.

Kirsner, Douglas, *Unfree Associations: Inside Psychoanalytic Institutions*, London: Process, 2000.

Knights, L.C., *How Many Children Had Lady Macbeth: An Essay on Theory and Practice of Shakespeare Criticism*, Cambridge: The Minority Press, 1933.

Kohler, Robert E., *Lords of the Fly: Drosophila Genetics and the Experimental Life*, Chicago: University of Chicago Press, 1994.

'Walter Fletcher, F.G. Hopkins, and the Dunn Institute of Biochemistry: a case study in the patronage of science', *Isis* 69 (1978), 330–55.

Kolnai, Aurel, *Psychoanalysis and Sociology*, trans. Eden and Cedar Paul, London: George Allen & Unwin, 1921.

Kretschmer, E., *Physique and Character: An Investigation of the Nature of Constitution and of the Theory of Temperament*, London: Kegan Paul, Trench, Trubner & Co., 1925.

Kuhn, Philip, 'Footnotes in the history of psychoanalysis. Observing Ernest Jones discerning the works of Sigmund Freud, 1905–1908', *Psychoanalysis and History* 16(1) (2014), 5–54.

'Subterranean histories. The dissemination of Freud's works into the British discourse on psychological medicine: 1904–1911', *Psychoanalysis and History* 16(2) (2014), 153–214.

Kuhn, Thomas S., *The Structure of Scientific Revolutions*, Chicago: University of Chicago Press, 1962.

Kuklick, Henrika, *The Savage Within: The Social History of British Anthropology, 1885–1945*, Cambridge: Cambridge University Press, 1991.

Kuper, Adam, *Anthropology and Anthropologists: The Modern British School*, 2nd revised edition, London: Routledge and Kegan Paul, 1983.

Incest and Influence: The Private Life of Bourgeois England, Cambridge, MA: Harvard University Press, 2009.

'Psychology and anthropology: the British experience', *History of the Human Sciences* 3 (1990), 397–413.

Kusch, Martin, *Psychologism: A Case Study in the Sociology of Philosophical Knowledge*, London: Routledge, 1995.

Lacan, J., *The Seminar of Jacques Lacan*, Book I, *Freud's Papers on Technique, 1953–1954*, ed. Jacques-Alain Miller, trans. with notes by John Forrester, Cambridge: Cambridge University Press, 1988.

Lampe, David, *Pyke: The Unknown Genius*, London: Evans Brothers, 1959.

Langdon-Brown, Sir Walter, 'Adler's contribution to general medicine', in Philip Mairet and Drs H.C. Squires, Cuthbert Dukes, O.H. Woodcock, Sir Walter Langdon-Brown and others, *The Contribution of Alfred Adler to Psychological Medicine*, London: C.W. Daniel, 1938, pp. 47–60.

'"Just nerves"', in Walter Langdon-Brown, *Thus We Are Men*, London: Kegan Paul, Trench, Trubner & Co., 1938, pp. 77–92.

Langham, Ian, *The Building of British Social Anthropology: W.H.R. Rivers and his Cambridge Disciples in the Development of Kinship Studies, 1898–1931*, Dordrecht: D. Reidel, 1981.

Leavis, F.R., *The Common Pursuit* (1952), Harmondsworth: Penguin, 1962.

Education and the University: A Sketch for an English School, London: Chatto & Windus, 1943.

Lee, Hermione, *Virginia Woolf*, London: Vintage Books, 1997.

Leese, Peter, '"Why are they not cured?" British shellshock treatment during the Great War', in Mark S. Micale and Paul Lerner (eds.), *Traumatic Pasts: History, Psychiatry, and Trauma in the Modern Age, 1870–1930*, Cambridge: Cambridge University Press, 2001, pp. 205–21.

Lemov, Rebecca, *World as Laboratory: Experiments with Mice, Mazes and Men*, New York: Hill and Wang, 2005.

Lévi-Strauss, Claude, "Do dual-organizations exist?" (1956), in *Structural Anthropology* (1958), trans. Claire Jacobson and Brooke Grundfest Schoepf, London: Penguin, 1972, pp. 162–3.

Levy, Paul (ed.), *The Letters of Lytton Strachey*, London: Viking, 2005.

Lewis, Nolan D.C. and Carney Landis, 'Freud's library', *Psychoanalytic Review* 44 (1957), 327–56.

Linehan, Peter (ed.), *St John's College, Cambridge: A History*, Woodbridge: Boydell Press, 2011.

Linstrum, Erik, 'The making of a translator: James Strachey and the origins of British psychoanalysis', *Journal of British Studies* 53(3) (2014), 685–704.

Lipset, D., *Gregory Bateson*, Englewood Cliffs, NJ: Prentice-Hall, 1980.

Lohmann, Roger Ivar, 'Dreams of Fortune: Reo Fortune's psychological theory of cultural ambivalence', *Pacific Studies* 32 (2009), 273–98.

Loughran, Tracey, 'Shell-shock and psychological medicine in First World War Britain', *Social History of Medicine* 22(1) (2009), 88.

Lovie, Sandy, 'Three steps to heaven: how the British Psychological Society attained its place in the sun', in G.C. Bunn, A.D. Lovie and G.D. Richards (eds.), *Psychology in Britain: Historical Essays and Personal Reflections*, Leicester: BPS Books, 2001, pp. 95–114.

Lowe, P. 'Amateurs and professionals: the institutional emergence of British plant ecology', *Journal of the Society for the Bibliograpy of Natural History* 7(4) (1976), 517–35.

Luckhurst, Roger, *The Invention of Telepathy, 1870–1901*, Oxford: Oxford University Press, 2002.

Lyons, Andrew P. and Harriet D. Lyons, *Irregular Connections: A History of Anthropology and Sexuality*, Lincoln: University of Nebraska Press, 2004.

Lythgoe, Katrina, 'Sir Arthur Tansley – the Man', *Bulletin of the British Ecological Society* 36(1) (2005), 26.

MacCurdy, J.T., *War Neuroses, with a Preface by W.H.R. Rivers*, Cambridge: Cambridge University Press, 1918.

Macdonald, Arthur, 'Congress Report', *Science* 20(511) (18 November 1892), 288–90.

Macdonald, Norman, 'Lionel Penrose', *Medical Association for the Prevention of War Proceedings* 2(5) (1972), 117–22.

MacFie Campbell, C., *Problems of Personality: Studies presented to Dr. Morton Prince, Pioneer in American Psychopathology*, London: Kegan Paul, Trench, Trubner & Co., 1925.

MacGibbon, Jean, *There's the Lighthouse*, London: James & James, 1997.

MacLeod, Roy and Russell Moseley, 'The "Naturals" and Victorian Cambridge: reflections on the anatomy of an elite, 1851–1914', *Oxford Review of Education* 6(2) (1980), 177–95.

Maddox, Brenda, *Freud's Wizard*, London: John Murray, 2006.

Malcolm, Norman, *Ludwig Wittgenstein: A Memoir, with a Biographical Sketch by G.H. von Wright*, 2nd edition, Oxford: Clarendon Press, 2001.

Malinowski, Bronislaw, *Sex and Repression in Savage Society*, London: Kegan Paul, Trench, Trubner & Co., 1927.

Mandler, Peter, *Return from the Natives: How Margaret Mead Won the Second World War and Lost the Cold War*, New Haven: Yale University Press, 2013.

Marshall, Alfred, 'The present position of economics' (1885), reprinted in *Memorials of Alfred Marshall*, ed. A.C. Pigou, London: Macmillan, 1925.

Martin, Kingsley, *Father Figures: A First Volume of Autobiography, 1897–1931*, London: Hutchinson, 1966.

'The war generation in England', *Revue des Sciences Politiques* 48 (1925), 211–30.

Martin, Wallace, 'Criticism and the academy', in A. Walton Litz, Louis Menand and Lawrence Rainey (eds.), *The Cambridge History of Literary Criticism*,

Vol. VII, *Modernism and the New Criticism*, Cambridge: Cambridge University Press, 2000, pp. 267–321.

Martindale, Phillippa, '"Against all hushing up and stamping down": the Medico-Psychological Clinic of London and the novelist May Sinclair', *Psychoanalysis and History* 6(2) (2004), 177–200.

Masson, J.M. (ed.), *The Complete Letters of Sigmund Freud to Wilhelm Fliess, 1887–1904*, Cambridge, MA: Harvard University Press, 1984.

McCarthy, Thomas J. and Joseph Needham, 'The excretion of creatinine in nervous diseases', *Quarterly Journal of Experimental Physiology* 13 (1922–23), Supplementary Proceedings of the XIth International Physiological Congress, Edinburgh, July 1923.

McCrea, William, 'Lecture: Sir Ralph Howard Fowler, 1889–1944: a centenary', *Notes and Records, Royal Society of London* 47 (1993), 61–78.

McGuinness, Brian, 'Freud and Wittgenstein', in McGuinness, *Approaches to Wittgenstein: Collected Papers*, London: Routledge, 2002, pp. 224–35.

Wittgenstein: A Life. Young Ludwig (1889–1921), London: Duckworth, 1988.

(ed.), *Wittgenstein in Cambridge: Letters and Documents 1911–1951*, Oxford: Wiley-Blackwell, 2012.

McIntosh, Robert, *The Background of Ecology: Concept and Theory*, Cambridge: Cambridge University Press, 1985.

McWilliams Tullberg, Rita, *Women at Cambridge*, Cambridge: Cambridge University Press, 1998.

Mead, M. 'The swaddling hypothesis: its reception', *American Anthropologist* 56 (1954), 395–409.

Meisel, Perry and Walter Kendrick (eds.), *Bloomsbury/Freud: The Letters of James and Alix Strachey 1924–1925*, London: Chatto & Windus, 1986.

Mellor, D.H., 'Frank Ramsey', *Philosophy* 70 (1995), 243–62.

Meyer-Palmedo, Ingeborg (ed.), *Sigmund Freud, Anna Freud: Correspondence 1904–1938*, Cambridge: Polity, 2013.

Mill, J.S., *System der deduktiven und induktiven Logik: eine Darstellung der Principien wissenschaftlicher Forschung, insbesondere der Naturforschung*, 2 vols., Braunschweig: Vieweg, 1862, 1863.

Miller, E., 'The present discontents in psychopathology', *British Journal of Medical Psychology* 15 (1935), 2–17.

Miller, E. et al., '1950–1970. Retrospects and reflections', *British Journal of Criminology* 10 (1970), 313–23.

Moffat, Wendy, *E.M. Forster: A New Life*, London: Bloomsbury, 2010.

Molnar, Michael (trans., annot. and ed.), *The Diary of Sigmund Freud, 1929–1939: A Record of the Final Decade*, London: Hogarth Press.

Money-Kyrle, R., *The Collected Papers of Roger Money-Kyrle*, ed. Donald Meltzer, Perthshire: Clunie Press, 1978.

Monk, Ray, *Bertrand Russell: The Ghost of Madness, 1921–1970*, London: Jonathan Cape, 2000.

Bertrand Russell: The Spirit of Solitude, London: Jonathan Cape, 1996.

Ludwig Wittgenstein: The Duty of Genius, London: Jonathan Cape, 1990.

Moore, G.E., *Principia Ethica*, Cambridge: Cambridge University Press, 1903.
Selected Writings, ed. T. Baldwin, London: Routledge, 1993.
Mosse, Werner E. et al. (eds.), *Second Chance: Two Centuries of German-speaking Jews in the United Kingdom*. Schriftenreihe wissenschaftlicher Abdhandlungen des Leo Baeck Instituts 48. Tübingen: J.C.B. Mohr (Paul Siebeck), 1991.
Murchison, C. (ed.), *A History of Psychology in Autobiography*, Vol. III, Worcester, MA: Clark University Press, 1936.
Psychological Register, Vol. II, Worcester, MA: Clark University Press, 1929.
Myers, C.S., *Human Personality and its Survival of Bodily Death*, 2 vols., New York: Longmans, Green, and Co., 1903.
'The influence of the late W.H.R. Rivers', in W. H. R. Rivers, *Psychology and Politics and Other Essays*, London: Kegan Paul, Trench, Trubner & Co., 1923, pp. 147–81.
Present-day Applications of Psychology, with Special Reference to Industry, Education and Nervous Breakdown, London: Methuen, 1918.
Shell Shock in France, 1914–1918 (1940), Cambridge: Cambridge University Press, 2012.
Myres, J.L., 'W.H.R. Rivers', *Journal of the Royal Anthropological Institute of Great Britain and Ireland* 53 (January–June 1923), 14–17.
Nabokov, Vladimir, *Lolita* (1955), London: Penguin, 1997.
The Real Life of Sebastian Knight (1941), London: Penguin, 1964.
Speak, Memory: An Autobiography Revisited (1967), New York: Vintage, 1989.
Nagel, Ernest, 'Impressions and appraisals of analytic philosophy in Europe. I', *Journal of Philosophy* 33(1) (1936), 5–24.
Neighbour, Oliver, 'Alan Walker Tyson', *Proceedings of the British Academy* 115 (2002), 367–82.
Neild, Robert, *Riches and Responsibility: The Financial History of Trinity College, Cambridge*, Cambridge: Granta, 2008.
Neill, A.S., *Talking of Summerhill*. London: Victor Gallancz, 1971.
Nepomnyashchy, Catharine Theimer, 'King, queen, sui-mate: Nabokov's defense against Freud's "Uncanny"', *Intertexts (Lubbock)* 12(1/2) (2008), 7–24, 80–1.
Neu, Jerome, *Emotion, Thought and Therapy: A Study of Hume and Spinoza and the Relationship of Philosophical Theories of the Emotions to Psychological Theories of Therapy*, London: Routledge and Kegan Paul, 1977.
New Dictionary of National Biography, ed. Andrew Roberts and Betty Falkenberg, Oxford: Oxford University Press, 2005.
Newnham College Register 1871–1971, Vol. I, *1871–1923*, Cambridge: [The College], 1979.
North, Michael, *Reading 1922: A Return to the Scene of the Modern*, Oxford: Oxford University Press, 1999.
Norton, H.T.J., 'Natural selection and Mendelian variation', *Proceedings of the London Mathematical Society* 28 (1928), 1–45.
Nussbaum, Martha, *The Therapy of Desire: Theory and Practice in Hellenistic Ethics*, Princeton: Princeton University Press, 1994.

Ogden, C.K., I.A. Richards and James Wood, *Foundations of Aesthetics*, London: Allen & Unwin, 1922.

Oliver, F.W. and T.G. Hill, *An Outline of the History of the Botanical Department of University College London*, issued by the Department on the Occasion of the Centenary of the College, 18 June 1927.

Oppenheim, Janet, *The Other World: Spiritualism and Psychical Research in England, 1850–1914*, Cambridge: Cambridge University Press, 1985.

Ornston, D., 'Freud's conception is different from Strachey's', *Journal of the American Psychoanalytic Association* 33 (1985), 379–412.

(ed.) *Translating Freud*, New Haven: Yale University Press, 1992.

Overy, Richard, *The Morbid Age: Britain and the Crisis of Civilization*, London: Allan Lane, 2009.

Oxford Dictionary of National Biography, Oxford: Oxford University Press, 2004.

Panayotidis, E. Lisa, 'The Department of Fine Art at the University of Toronto, 1926–1945: institutionalizing the "culture of the aesthetic"', *Journal of Canadian Art History / Annales d'histoire de l'art canadien* 25 (2004), 100–22.

Parker, Peter, *Isherwood: A Life*, London: Picador, 2004.

Parry, Bronwyn, 'Technologies of immortality: the brain on ice', *Studies in History and Philosophy of Biological and Biomedical Sciences* 35C(2) (2004), 391–413.

Partridge, Frances, *Diaries, 1939–1972*, London: Phoenix, 2001.

Memories (1981), London: Phoenix, 1996.

Paskauskas, R. Andrew (ed.), *The Complete Correspondence of Sigmund Freud and Ernest Jones 1908–1939*, introduction by Riccardo Steiner, Cambridge, MA: The Belknap Press, 1993.

Paul, Margaret, *Frank Ramsey (1903–1930): A Sister's Memoir*, Huntingdon: Smith-Gordon, 2012.

Pear, T.H. 'Some early relations between English ethnologists and psychologists', *Journal of the Royal Anthropological Institute of Great Britain and Ireland* 90 (July–December 1960), 227–37.

Pearson, Jenny, *Analyst of the Imagination: The Life and Work of Charles Rycroft*, London: Karnac, 2004.

Pemberton, J., 'Origins and early history of the Society for Social Medicine in the UK and Ireland', *Journal of Epidemiology and Community Health* 56 (2002), 342–6.

Penrose, L., *Mental Defect*, London: Sidgwick & Jackson, 1933.

On the Objective Study of Crowd Behaviour, London: H.K. Lewis, 1952.

Penrose, Oliver, 'Lionel Penrose, FRS, human geneticist and human being', in *Penrose: Pioneer in Human Genetics. Report on a Symposium held to Celebrate the Centenary of the Birth of Lionel Penrose, 12th & 13th March 1998*, London: Centre for Human Genetics at UCL, 1998.

Lionel S. Penrose FRS: Human Geneticist and Human Being, ed. Peter Cave and Oliver Penrose, Peterborough: Effective Print, 1999.

Penrose, Roger, *Shadows of the Mind*, Oxford: Oxford University Press, 1994.

Pfeiffer, Ernst (ed.), *Sigmund Freud and Lou Andreas-Salomé: Letters*, trans. William and Elaine Robson-Scott, London: Hogarth Press and the Institute of Psycho-Analysis, 1972.

Philips, Ann (ed.), *A Newnham Anthology*, Cambridge: Cambridge University Press, 1979.

Philo, Chris, *A Geographical History of Institutional Provision for the Insane from Medieval Times to the 1860s in England and Wales: The Space Reserved for Insanity*, Lampeter, UK and New York: Edwin Mellen, 2003.

Pick, Daniel, *The Pursuit of the Nazi Mind: Hitler, Hess, and the Analysts*, Oxford: Oxford University Press, 2012.

Pick, D. and L. Roper (eds.), *Dreams and History*, London: Routledge, 2004.

Piketty, Thomas, *Capital in the Twenty-first Century*, Cambridge, MA: The Belknap Press, 2014.

Plowman, D.L. (ed.), *Bridge into the Future: Letters of Max Plowman*, London: A. Dakers, 1944.

Pogson, Beryl Chassereau, *Maurice Nicoll: A Portrait*, London: V. Stuart, 1961.

Preston, Aaron, *Analytic Philosophy: The History of an Illusion*, London: Continuum, 2007.

Pritchard, Jack, *View from a Long Chair: The Memoirs of Jack Pritchard*, London: Routledge and Kegan Paul, 1984.

Prokop, Ursula, *Margaret Stonborough-Wittgenstein: Bauherrin, Intellektuelle, Mäzenin*, 2nd edition, Vienna: Böhlau, 2005.

Provine, W.B., *The Origins of Theoretical Population Genetics*. Chicago: University of Chicago Press, 1971.

Pulman, Bertrand, 'Anthropologie et psychanalyse: "paix et guerre" entre les herméneutiques?', *Connexions* 44 (1984), 81–97.

'Aux origines du débat ethnologie/psychanalyse: W.H.R. Rivers (1864–1922)', *L'Homme* 26(100) (1986), 119–42.

'Le débat anthropologie/psychanalyse et la référence au "terrain"', *Cahiers internationaux de sociologie* 80 (1986), 5–26.

'Malinowski and ignorance of physiological paternity', *Revue française de sociologie* 45, Supplement: An Annual English Selection (2004), 125–46.

Pyke, Richard, *The Lives and Deaths of Roland Greer*, London: R. Cobden-Sanderson, 1928; New York: A. and C. Boni, 1929.

Radcliffe-Brown, A.R., *Structure and Function in Primitive Societies*, Glencoe, IL: The Free Press, 1952.

Radzinowicz, Leon, *Adventures in Criminology*, London: Routledge, 1998.

The Cambridge Institute of Criminology: Its Background and Scope (London: HMSO, 1988).

Raitt, Suzanne, 'Early British psychoanalysis and the Medico-Psychological Clinic', *History Workshop Journal* 58(1)(2004), 63–85.

May Sinclair: A Modern Victorian, Oxford: Oxford University Press, 2000.

Ramsey, F.P., 'Civilisation and happiness' (24 November 1925). In Ramsey, *Notes on Philosophy, Probability and Mathematics*, ed. Maria Carla Galavotti, Naples: Bibliopolis, 1991, pp. 320–4.

Critical Notice. Ludwig Wittgenstein, Tractatus Logico-Philosophicus, with an Introduction by Bertrand Russell. International Library of Psychology, Philosophy and Scientific Method. London: Kegan Paul, Trench, Trubner & Co., 1922.

'The foundations of mathematics' (1926), in Ramsey, *The Foundations of Mathematics and Other Logical Essays*, ed. Richard Braithwaite, London: Routledge, 1931, p. 21.

'An imaginary conversation with John Stuart Mill' (1924), in Ramsey, *Notes on Philosophy, Probability and Mathematics*, ed. Maria Carla Galavotti, Naples: Bibliopolis, 1991, pp. 302–12.

'Mr Keynes on probability', *British Journal for the Philosophy of Science* 40 (1989), 219–22.

On Truth, ed. Nicholas Rescher and Ulrich Majer, Dordrecht: Kluwer, 1991.

Philosophical Papers, ed. D.H. Mellor, Cambridge: Cambridge University Press, 1990.

Rapp, Dean, 'The early discovery of Freud by the British general educated public, 1912–1919', *Social History of Medicine* 3(2) (1990), 217–43.

Richards, Graham, 'C.K. Ogden's basic role in inter-war British psychology', *History and Philosophy of Psychology* 9(1) (2007), 56–65.

Race, Racism and Psychology: Towards a Reflexive History, London: Routledge, 1997.

Richards, I.A., *Principles of Literary Criticism*, London: Kegan Paul, Trench, Trubner & Co., 1924.

Practical Criticism: A Study of Literary Judgment (1929), London: Kegan Paul, Trench, Trubner & Co., 1930.

Rickman, J., *Selected Contributions to Psycho-Analysis*, compiled by W.C.M. Scott, London: Hogarth Press and Institute of Psychoanalysis, 1957.

Riddell, Lord, *Lord Riddell's Intimate Diary of the Peace Conference and After, 1918–1923*, London: Victor Gollancz, 1933.

Riley, Denise, *War in the Nursery: Theories of the Child and Mother*, London: Virago, 1983.

Ritchie, Professor W., *The History of the South African College 1829–1918*, Vol. II, Capetown: T. Maskew Miller, 1918.

Rivers, W.H.R., 'An address on education and mental hygiene' (1922), in Rivers, *Psychology and Politics and Other Essays*, with a Prefatory Note by G. Elliot Smith and an Appreciation by C.S. Myers, London: Kegan Paul, Trench, Trubner & Co., 1923, pp. 95–106.

'Anthropological research outside America', in W.H.R. Rivers, A.E. Jenks and S.G. Morley (eds.), *Reports on the Present Condition and Future Needs of the Science of Anthropology*, Washington, DC: Carnegie Institute, 1913, pp. 5–28.

Conflict and Dream (London: Kegan Paul, Trench, Trubner & Co., 1923).

'Dreams and primitive culture. A lecture delivered in the John Rylands Library on the 10th April 1918', reprinted from *The Bulletin of the John Rylands Library* 4(3–4) (1918), 14.

'A genealogical method of collecting social and vital statistics', *Journal of the Anthropological Institute of Great Britain and Ireland* 30 (1900), 74–82.

The Influence of Alcohol and Other Drugs on Fatigue: The Croonian Lectures Delivered at the Royal College of Physicians in 1906, London: Edward Arnold, 1908.

Instinct and the Unconscious: A Contribution to a Biological Theory of the Psycho-Neuroses, Cambridge: Cambridge University Press, 1920.

'Presidential Address: The symbolism of rebirth', *Folklore* 33(1) (1922), 14–33.

Reports of the Cambridge Anthropological Expedition to Torres Straits, Vol. II, *Physiology and Psychology*, Part I, *Introduction and Vision*, Cambridge: Cambridge University Press, 1901.

'Sociology and psychology', *Sociological Review* 9 (1916), 1–13.

Roach, J.P.C., *A History of the County of Cambridge and the Isle of Ely*, Vol. III, *The City and University of Cambridge*, Oxford: Oxford University Press, 1959.

Robert, Christian P., Nicolas Chopin and Judith Rousseau (2009), 'Harold Jeffreys's theory of probability revisited', *Statistical Science* 24(2) (2009), 141–72.

Robinson, Annabel, *The Life and Works of Jane Ellen Harrison*, Oxford: Oxford University Press, 2002.

Robinson, Ken, 'A portrait of the psychoanalyst as a Bohemian: Ernest Jones and the "lady from Styria"', *Psychoanalysis and History* 15(2) (2013), 165–89.

Rodman, F.R. (ed.), *The Spontaneous Gesture: Selected Letters of D.W. Winnicott*, Cambridge, MA: Harvard University Press, 1987.

Winnicott: His Life and Work, Boston, MA: Da Capo Press, 2004.

Roe, Daphne, 'Lucy Wills (1888–1964): a biographical sketch', *Journal of Nutrition* 108(9) (1978), 1379–83.

Rolleston, Sir Humphrey Davy, *The Cambridge Medical School*, Cambridge: Cambridge University Press, 1932.

Rolph, C.H., *Kingsley: The Life, Letters and Diaries of Kingsley Martin*, London: Victor Gollancz, 1973.

Rook, Arthur, Margaret Carlton and W. Graham Cannon, *The History of Addenbrooke's Hospital*, Cambridge: Cambridge University Press, 2010.

Rose, Nikolas, *Governing the Soul: The Shaping of the Private Self*, London: Routledge, 1990.

The Psychological Complex: Psychology, Politics and Society in England, 1869–1939, London: Routledge and Kegan Paul, 1985.

Rosenbaum, Stanford Patrick (ed.), *The Bloomsbury Group: A Collection of Memoirs and Commentary*, Toronto: University of Toronto Press, 1995.

Rothblatt, Sheldon, 'State and market in British university history', in Stefan Collini, Richard Whatmore and Brian Young (eds.), *Economy, Polity, and Society: British Intellectual History 1750–1950*, Cambridge: Cambridge University Press, 2000, pp. 224–42.

The Revolution of the Dons: Cambridge and Society in Victorian England, London: Faber & Faber, 1968, reprinted Cambridge University Press, 1981.

Rüegg, Walter (ed.), *A History of the University in Europe*, Vol. III, *Universities in the Nineteenth and Early Twentieth Centuries*, Cambridge: Cambridge University Press, 2004.

Russell, Bertrand, *The Analysis of Mind*, London: George Allen & Unwin, 1921.

    The Autobiography of Bertrand Russell, Vol. I, *1872–1914*, London: George Allen & Unwin, 1967.

    The Autobiography of Bertrand Russell, Vol. II, *1914–1944*, London: George Allen & Unwin, 1968.

    The Autobiography of Bertrand Russell, Vol. III, *1944–1967*, London: George Allen & Unwin, 1969.

    The Collected Papers of Bertrand Russell, London: Routledge, 1983.

    'Dr. Schiller's analysis of the *Analysis of Mind*', *Journal of Philosophy* 19(24) (1922), 645–51.

    'Free speech in childhood', *New Statesman and Nation* 1 (30 May 1931), 486.

    'Freudianism', *The Literary Guide and Rationalist Review* 392 (February 1929), 44.

    History of Western Philosophy (1946), London: George Allen & Unwin, new edition, 1961.

    On Education, Especially in Early Childhood (1926), London: George Allen & Unwin, 1966.

    The Scientific Outlook, 2nd edition, London: George Allen & Unwin, 1954.

Russell, Dora, *The Tamarisk Tree*, Vol. I, London: Elek Pemberton, 1975.

Russo, John Paul, *I.A. Richards: His Life and Work*, London: Routledge, 1989.

Rycroft, Charles, 'Where I come from', in Charles Rycroft, *Psychoanalysis and Beyond*, London: Chatto & Windus/The Hogarth Press, 1985, pp. 204–5.

Ryle, Gilbert, *Collected Papers*, Vol. I, London: Hutchinson, 1971.

Sahlins, Marshall, 'The conflicts of the faculty', *Critical Inquiry* 35 (Summer 2009), 997–1017.

Samuels, E., *Bernard Berenson: The Making of a Connoisseur*, Cambridge, MA: The Belknap Press, 1979.

Sargant Florence, Philip and J.R.L. Anderson (eds.), *C.K. Ogden: A Collective Memoir*, London: Elek Pemberton, 1977.

Sassoon, Siegfried, *Sherston's Progress* (1936), London: Folio Society, 1974.

    War Diaries. 1915–1918, ed. and intro. Rupert Hart-Davies, London: Faber & Faber, 1983.

    War Poems, London: Heinemann, 1919.

Saunders, Max, 'Science and futurology in the *To-day and To-Morrow* series: matter, consciousness, time and language', *Interdisciplinary Science Reviews* 34(1) (2009), 68–78.

Saunders, Max and Brian Hurwitz, 'The *To-Day and To-Morrow* series and the popularization of science: an introduction', *Interdisciplinary Science Reviews* 34(1) (2009), 3–8.

Schaffer, Simon, *From Physics to Anthropology and Back Again*, Cambridge: Prickly Pear Pamphlet 3, 1994.

'How disciplines look', in Andrew Barry and Georgina Born (eds.), *Interdisciplinarity: Reconfigurations of the Social and Natural Sciences*, London: Routledge, 2013, pp. 57–81.

'Self evidence', *Critical Inquiry* 18 (1992), 327–62.

Schiller, F.C.S., 'Telepathic communication' (April 1902), *Journal of the Society for Psychical Research* 10 (1901–2), 224.

Searby, Peter, *A History of the University of Cambridge*, Vol. III, *1750–1870*, Cambridge: Cambridge University Press, 1997.

Searle, John, 'Minds, brains and programs', *Behavioral and Brain Sciences* 3 (1980), 417–57.

Shamdasani, Sonu, *Jung and the Making of Modern Psychology: The Dream of a Science*, Cambridge: Cambridge University Press, 2003.

'"Psychotherapy": the invention of a word', *History of the Human Sciences* 18 (2005), 1–22.

Shapira, Michal, *The War Inside: Psychoanalysis, Total War, and the Making of the Democratic Self in Postwar Britain*, Cambridge: Cambridge University Press, 2013.

Shephard, Ben, '"The early treatment of mental disorders": R.G. Rows and Maghull 1914–18', in G. E. Berrios and H. Freeman (eds.), *150 Years of British Psychiatry*, Vol II, *The Aftermath*, London: Athlone, 1996, pp. 434–64.

Head Hunters: The Search for a Science of the Mind, London: Bodley Head, 2014.

A War of Nerves: Soldiers and Psychiatrists 1914–1994, London: Jonathan Cape, 2000.

Sidgwick, A. and E.M. Sidgwick, *Henry Sidgwick: A Memoir*, London: Macmillan, 1906.

Sidgwick, Henry, *The Methods of Ethics* (1874), London: Macmillan, 1907.

Sirinelli, *Génération intellectuelle: Khâgneux et Normaliens dans l'entre-deux-guerres*, Paris: Fayard, 1988.

Skelton, Ross, 'Jonathan Hanaghan. The founder of psychoanalysis in Ireland', *The Crane Bag* 7(2) (1983), 183–90.

Skidelsky, Robert, *John Maynard Keynes*, Vol. I, *Hopes Betrayed 1883–1920*, London: Macmillan, 1983.

John Maynard Keynes, Vol. II, *The Economist as Saviour 1920–1937*, London: Macmillan, 1992.

Slobodin, Richard, *W.H.R. Rivers*, New York: Columbia University Press, 1978, pbk version.

Smith, Lydia, *To Understand and to Help: The Life and Work of Susan Isaacs, 1885–1948*, London: Fairleigh Dickinson University Press, 1985.

Smith, Roger, *The Fontana History of the Human Sciences*, London: Fontana, 1997.

Spitzer, Therese, *Psychobattery: A Chronicle of Psychotherapeutic Abuse*, with medical discussion by Ralph Spitzer and a Foreword by Joseph Needham, Clifton, NJ: Humana Press, 1980.

Sprott, W.J.H., *Social Psychology*, London: Methuen, 1952.

Stansky, Peter, *On or about December 1910: Early Bloomsbury and its Intimate World*, Cambridge, MA: Harvard University Press, 1996.

'Statement of policy', *Analysis* 1(1) (1933), 1.

Steel, D.A., 'Escape and aftermath: Gide in Cambridge 1918', *Yearbook of English Studies* 15, Anglo-French Literary Relations Special Number (1985), 125–59.

'Gide à Cambridge, 1918', *Bulletin des Amis d'André Gide* 125 (janvier 2000), 11–74.

Steel, Ronald, *Walter Lippmann and the American Century* (1980), New York: Little, Brown, 2008.

Steiner, Riccardo, '"Et in arcadia ego . . .?' Some notes on methodological issues in the use of psychoanalytic documents and archives', *IJP* 76 (1995), 739–58.

'A world wide international trade mark of genuineness? – Some observations on the history of the English translation of the work of Sigmund Freud, focusing mainly on his technical terms', *IRP* 14 (1987), 33–102.

Stephen, K., 'Relations between the superego and the ego', *Psychoanalysis and History* 2(1) (2000), 11–28.

Stephenson, William, 'Tribute to Melanie Klein', *Psychoanalysis and History* 12(2) (2010), 245–71.

Stevens, Rosemary, *Medical Practice in Modern England: The Impact of Specialization and State Medicine*, New Haven: Yale University Press, 1966.

Stocking, George W., Jr., *After Tylor: British Social Anthropology, 1888–1951*, Madison: University of Wisconsin Press, 1995.

'Books unwritten, turning points unmarked. Notes for an anti-history of anthropology' (1981), in Stocking, *Delimiting Anthropology: Occasional Essays and Reflections*, Madison: University of Wisconsin Press, 2001, pp. 330–54.

'The ethnographer's magic: fieldwork in British anthropology from Tylor to Malinowski', in Stocking (ed.), *The Ethnographer's Magic and Other Essays in the History of Anthropology*, Madison: University of Wisconsin Press, 1992, pp. 12–59.

(ed.), *Malinowski, Rivers, Benedict and Others: Essays on Culture and Personality*. History of Anthropology 4, Madison: University of Wisconsin Press, 1986.

Stocks, Mary D. (1929) 'The spirit of feminism', *The Woman's Leader*, 13 September, p. 239.

Stolt, Carl-Magnus, 'Why did Freud never receive the Nobel Prize?', in Elisabeth Crawford (ed.), *Historical Studies in the Nobel Archives: The Prizes in Science and Medicine*, Uppsala: Universal Academy Press, 2002, pp. 95–106.

Stone, Martin, 'Shellshock and the psychologists', in W.F. Bynum, Roy Porter and Michael Shepherd (eds.), *The Anatomy of Madness: Essays in the History of Psychiatry*, Vol. II, *Institutions and Society*, London: Tavistock, 1985, pp. 242–71.

Strachey, Barbara, *Remarkable Relations: The Story of the Pearsall Smith Family*, London: Victor Gollancz, 1980.

Strachey, Barbara and Jayne Samuels (eds.), *Mary Berenson: A Self-portrait from her Letters and Diary*, London: Hamish Hamilton, 1983.

Strachey, James (ed.) in collaboration with Anna Freud, assisted by Alix Strachey and Alan Tyson, *The Standard Edition of the Complete Psychological Works of Sigmund Freud* (24 volumes), London: The Hogarth Press and the Institute of Psycho-Analysis, 1953–74.

Strachey, Lytton, 'According to Freud', in Strachey, *The Really Interesting Question, and Other Papers*, ed. with intro. and commentaries by Paul Levy, London: Weidenfeld and Nicolson, 1972.

Queen Victoria (1921), London: Penguin, 1971.

'Supplementary Report of Council', *Supplement to the British Medical Journal*, 29 June 1929, p. 270.

Swann, Brenda and Francis Aprahamian (eds.), *J.D. Bernal: A Life in Science and Politics*, London: Verso, 1999.

Tanner, J.R. (ed.), *Historical Register of the University of Cambridge, being a Supplement to the Calendar with a Record of University Offices, Honours and Distinctions to the Year 1910*, Cambridge: Cambridge University Press, 1917.

Tansley, A.G., *The British Isles and their Vegetation*, Cambridge: Cambridge University Press 1939.

The New Psychology in its Relation to Life, London: George Allen & Unwin, 1920.

Practical Plant Ecology, London: George Allen & Unwin, 1923.

'The psychological connexion of two basic principles of the S.F.S.', *Society for Freedom in Science*, Occasional Pamphlet 12, March 1952.

'Sigmund Freud (1856–1939)', *Obituary Notices of Fellows of The Royal Society* 3 (1939–41), 247–75.

(prepared by Peder Anker), 'The temporal genetic series as a means of approach to philosophy', *Ecosystems* 5 (2002), 614–24.

Types of British Vegetation, Cambridge: Cambridge University Press, 1911.

Taylor, Peter J., Michael Hoyler and David M. Evans, 'A geohistorical study of the rise of modern science: mapping scientific practice through urban networks, 1500–1900', *Minerva* 46 (2008), 391–410.

Thomson, Godfrey, *Education of an Englishman: An Autobiography*, Edinburgh: Moray House College of Education, 1969.

Thomson, Mathew, *Psychological Subjects: Identity, Culture, and Health in Twentieth-Century Britain*, Oxford: Oxford University Press, 2006.

'"The solution to his own enigma": connecting the life of Montague David Eder (1865–1936), socialist, psychoanalyst, Zionist and modern saint', *Medical History* 55 (2011), 61–84.

Tillyard, E.M.W., *The Muse Unchained: An Intimate Account of the Revolution in English Studies at Cambridge*, London: Bowes & Bowes, 1958.

Tomlinson, Margaret, *Three Generations in the Honiton Lace Trade: A Family History*, Exeter: Devon Print Group, 1983.

Toulmin, Stephen, *An Examination of the Place of Reason in Ethics*, Cambridge: Cambridge University Press, 1950.

'The logical status of psycho-analysis', *Analysis* 9(2) (1948), 23–9.

Trist, Eric, 'Guilty of enthusiasm', in Trist, *Management Laureates*, ed. Arthur G. Bedeian, Vol. III, London: JAI, 1993, pp. 191–221.

Tudor-Hart, B.E., 'Are there cases in which lies are necessary?', *Pedagogical Seminary and Journal of Genetic Psychology* 33 (1926), 586–641.

University of Cambridge, *Statutes and Ordinances*, 2011.

University Registry, *Historical Register of the University of Cambridge, Supplement 1921–1930*, Cambridge: Cambridge University Press, 1932.

Urry, James, '"Notes and queries on anthropology" and the development of field methods in British anthropology, 1870–1920', *Proceedings of the Royal Anthropological Institute of Great Britain and Ireland* (1972), 45–57.

Valentine, Elizabeth, 'The other woman', *The Psychologist* 21 (2008), 86–7.

van der Eyken, W. and B. Turner, *Adventures in Education*, London: Allen Lane, The Penguin Press, 1969.

Van Dijken, Suzan, *John Bowlby: His Early Life*, London: Free Association Books, 1998.

Van Dijken, K.S., R. van der Veer, M. H. van IJzendoorn and H. J. Kuipers, 'Bowlby before Bowlby: the sources of an intellectual departure in psycho-analysis and psychology', *Journal of the History of the Behavioral Sciences* 34 (1998), 247–69.

Vickers, Neil, 'Roger Money-Kyrle's *Aspasia: The Future of Amorality* (1932)', *Interdisciplinary Science Reviews* 34(1) (2009), 91–106.

Vincent, 'Les "sciences morales et politiques": de la gloire à l'oubli? Savoirs et politique en Europe au XIXe siècle', *Revue pour l'histoire du CNRS* 18 (automne 2007), 38–43.

Von Wright, G.H. (ed.), *A Portrait of Wittgenstein as a Young Man: From the Diary of David Hume Pinsent 1912–1914*, with an introduction by Anne Keynes, Oxford: Blackwell, 1990.

Vonofakos, Dimitris and Bob Hinshelwood, 'Wilfred Bion's letters to John Rickman (1939–1951)', *Psychoanalysis and History* 14(1) (2012), 53–94.

Wade, N.J., P. Thompson and M. Morgan, 'Guest editorial: the after-effect of Adolf Wohlgemuth's seen motion', *Perception* 43 (2014), 229–34.

Wallace, Stuart, *War and the Image of Germany: British Academics 1914–1918*, London: John Donald, 1988.

Wallerstein, Robert S., *Lay Analysis: Life in the Controversy*, Hillsdale, NJ and London: Analytic, 1998.

Warwick, Andrew, *Masters of Theory: Cambridge and the Rise of Mathematical Physics*, Chicago: University of Chicago Press, 2003.

Waugh, Alexander, *The House of Wittgenstein: A Family at War*, London: Bloomsbury, 2008.

Weatherall, Mark, 'Bread and newspapers: the making of "a revolution in the science of food"', in Harmke Kamminga and Andrew Cunningham (eds.),

The Science and Culture of Nutrition, 1840–1940, Amsterdam and Atlanta, GA: Rodopi, 1995, pp. 179–212.

Gentlemen, Scientists and Doctors: Medicine at Cambridge, 1800–1940, Cambridge: Cambridge University Library, The Boydell Press, 2000.

Wells, H.G. and J. Huxley, *The Science of Life*, London: Hazell, Watson & Viney, 1930.

Werskey, Gary, *The Visible College: A Collective Biography of British Scientists and Socialists of the 1930s* (1979), London: Free Association Books, 1988.

Wharton, C.J. and V. Ovcharenko, 'The history of Russian psychoanalysis and the problem of its periodisation', *Journal of Analytical Psychology* 44 (1999), 341–52.

Whewell, William, *Lectures on the History of Moral Philosophy*, Cambridge: Deighton, Bell, 1862.

Whyte, E., 'A biographical note on L.L. Whyte', *Contemporary Psychoanalysis* 10 (1974), 386–8.

Whyte, Lancelot Law, *Focus and Diversions*, London: The Cresset Press, 1963.

The Unconscious before Freud, London: Tavistock, 1962.

Wilkinson, Greg (ed.), *Talking about Psychiatry*, London: Gaskell, 1993.

Wilkinson, L.P., *A Century of King's, 1873–1972*, Cambridge: King's College, 1980.

Willey, Basil, *Cambridge and Other Memories, 1920–1953*, London: Chatto & Windus, 1968.

Williams, Raymond, *What I Came to Say*, London: Hutchinson Radius, 1989.

Wills, W. David, *Homer Lane: A Biography*, London: Allen & Unwin, 1964.

Wilson, Jean Moorcroft, *Siegfried Sassoon: The Journey from the Trenches: A Biography (1918–1967)*, London: Duckworth, 2003.

Siegfried Sassoon: The Making of a War Poet: A Biography (1886–1918), London: Duckworth, 1998.

Winnicott, D.W., 'A personal view of the Kleinian contribution' (1962) in Winnicott, *The Maturational Processes and the Facilitating Environment: Studies in the Theory of Emotional Development*, London: The Hogarth Press and the Institute of Psycho-Analysis, 1965, pp. 171–8.

Winslow, Edward G., 'Bloomsbury, Freud, and the vulgar passions', *Social Research* 57 (1990), 786–819.

'Keynes and Freud: psychoanalysis and Keynes' account of the "animal spirits of capitalism"', *Social Research* 53 (1986), 549–78.

Winstanley, D.A., *Early Victorian Cambridge*, Cambridge: Cambridge University Press, 1940.

Wisdom, John, *Interpretation and analysis in relation to Bentham's theory of definition*, London: Kegan Paul, Trench, Trubner & Co., 1931.

Philosophy and Psycho-Analysis, Oxford: Blackwell, 1953.

Problems of Mind and Matter, Cambridge: Cambridge University Press, 1934.

Wittgenstein, Ludwig, *Culture and Value*, ed. G.H. von Wright, in collaboration with Heikki Nyman; trans. Peter Winch, Chicago: University of Chicago Press, 1980.

Denkbewegungen: Tagebücher 1930–1932, 1936–1937, hrsg. Ilse Somavilla, Frankfurt: Fischer, 1999.

'A lecture on ethics', *Philosophical Review* 74(1) (1965), 3–12.

Lectures and Conversations on Aesthetics, Psychology and Religious Belief, compiled from notes taken by Yorick Smythies, Rush Rhees and James Taylor; ed. Cyril Barrett, Oxford: Basil Blackwell, 1966.

Letters to C.K. Ogden, with Comments on the English Translation of Tractatus Logico-Philosophicus, ed. with intro. by G.H. von Wright, and an appendix of letters by Frank Plumpton Ramsey, Oxford: Blackwell; Boston: Routledge and Kegan Paul, 1973.

Philosophical Grammar, ed. Rush Rhees, trans. Anthony Kenny, Oxford: Basil Blackwell, 1974.

Preliminary Studies for the 'Philosophical Investigations' Generally Known as The Blue and Brown Books, Oxford: Blackwell, 1958.

Remarks on Colour, ed. G.E.M. Anscombe, Berkeley: University of California Press, 1977, Part 3.

Remarks on the Foundations of Mathematics, ed. G.H. von Wright, R. Rhees and G.E.M. Anscombe, trans. G.E.M. Anscombe, Oxford: Blackwell, 3rd edition, 1978.

Wodehouse, Helen, 'Instincts of death and destruction: a comment on Freud', Presidential Address given at the Froebel Society Annual Meeting, 5 January 1932, published in *Child Life* 35(156) (1932), 5–14.

'Natural selfishness, and its position in the doctrine in Freud', *British Journal of Medical Psychology* 9(1) (1929), 38–59.

Wohlgemuth, A., *A Critical Examination of Psycho-Analysis*, London: George Allen & Unwin; New York: Macmillan, 1923.

'The Freudian psychology', *The Literary Guide and Rationalist Review* 393 (March 1929), 60.

Wooldridge, A., *Measuring the Mind: Education and Psychology in England c. 1860–c.1990*, Cambridge: Cambridge University Press, 1994.

Woolf, Leonard, *Downhill All the Way: An Autobiography of the Years 1919–1939*, London: Hogarth, 1967.

'Everyday life', *New Weekly* 1 (1914), 412.

New Weekly (13 July 1914).

Woolf, Virginia, *The Diary of Virginia Woolf*, Vol. I, *1915–1919*, ed. A.O. Bell, New York: Mariner Press, 1979.

The Diary of Virginia Woolf, Vol. II, *1920–1924*, ed. A.O. Bell, New York: Mariner Press, 1980.

The Diary of Virginia Woolf, Vol. III, *1925–1930*, ed. A.O. Bell, New York: Mariner Press, 1981.

The Diary of Virginia Woolf, Vol. IV, *1931–1935*, ed. A.O. Bell, New York: Mariner Press, 1983.

The Letters of Virginia Woolf, Vol. II, *1912–1922*, ed. N. Nicolson and J. Trautmann, New York: Harcourt Brace Jovanovich, 1976.

The Letters of Virginia Woolf, Vol. III, *1923–1928,* ed. N. Nicolson and J. Trautmann, New York: Harcourt Brace, 1977.

Moments of Being, ed. Jeanne Schulkind, St. Albans: Triad/Panther, 1978.

'Old Bloomsbury', in *Moments of Being,* ed. Jeanne Schulkind, St Albans: Triad/Panther, 1978.

Virginia Woolf: Collected Essays, Vol. III, New York: Harcourt, Brace & World, 1967.

Wrinch, D. and H. Jeffreys, 'On the seismic waves from the Oppau explosion of 1921', *Royal Astronomical Society Monthly Notices,* Geophysics Supplement, January 1923, 15–22.

'The relation of geometry to Einstein's theory of gravitation', *Nature* 106 (1921), 806–9.

Yorke, C., 'Freud's psychology: can it survive?', *Psychoanalytic Study of the Child* 50 (1995), 3–31.

Young, Allan, *The Harmony of Illusion: Inventing Post-Traumatic Stress Disorder,* Princeton: Princeton University Press, 1995.

Young, Michael W., *Malinowski: Odyssey of an Anthropologist, 1884–1920,* New Haven: Yale University Press, 2004.

Zaretsky, Eli, *Secrets of the Soul: A Social and Cultural History of Psychoanalysis,* New York: Knopf, 2004.

'An unpublished letter from Freud on female sexuality', *Gender and Psychoanalysis.* 4 (1999), 99–104.

Zuckerman, Solly, *From Apes to Warlords: the Autobiography (1904–1946) of Solly Zuckerman,* London: Hamilton, 1978.

Websites

Berke, Joseph H., 'Trick or treat: the divided self of R. D. Laing', www.janushead .org/4-1/berke.cfm, accessed 25 November 2005.

Bryder, L. (2010), 'The Medical Research Council and clinical trial methodologies before the 1940s: the failure to develop a "scientific" approach', JLL Bulletin: Commentaries on the history of treatment evaluation, www.jameslindlibrary .org/articles/the-medical-research-council-and-clinical-trial-methodologies-bef ore-the-1940s-the-failure-to-develop-a-scientific-approach/.

Daunton, Martin (2002), 'Equality and incentive: fiscal politics from Gladstone to Brown', www.historyandpolicy.org/archive/pol-paper-print-06.html, accessed 10 September 2006.

http://edge.org/conversation/gregory-bateson-the-centennial, accessed 30 July 2015.

http://en.wikipedia.org/wiki/Karl_Mannheim; accessed 22 May 2015.

http://www.loftyimages.co.uk/gallery_408953.html, accessed 13 April 2015.

http://www.psychoanalysis.org.uk/P17/P17-G-A.htm, accessed 5 December 2014.

'John Bowlby interview with Milton Senn, M.D. [1977]' *Beyond the Couch. The online journal of the American Association for Psychoanalysis in Clinical Social Work* 2 (December 2007), at www.beyondthecouch.org/1207/bowl by_int.htm, accessed 22 September 2015.

Kennedy, S.B., 'The White Stag Group', www.modernart.ie/en/downloads/white
    stagbk.doc, accessed 29 January 2006.
Roach, J.P.C. (ed.), 'The city of Cambridge: economic history', in *A History of
    the County of Cambridge and the Isle of Ely*, Vol. III, *The City and University of
    Cambridge* (Oxford: Oxford University Press, 1959), pp. 86–101,
    www.british-history.ac.uk/vch/cambs/vol3/pp86-101, accessed 6 May 2015.
Robert Hinde: Interview with Alan Macfarlane, 20 November 2007, at www.alan
    macfarlane.com/DO/filmshow/hinde2_fast.htm, accessed 22 September 2015.
Sigel, Lisa Z., 'Censorship and magic tricks in inter-war Britain', *Revue LISA/LISA
    e-journal* [en ligne], 11(1) (2013), mis en ligne le 30 mai 2013, http://lisa.revues
    .org/5211, consulté le 29 mars 2015.
Wilson, Leigh, 'May Sinclair 1863–1946', *The Literary Encyclopedia*, www.litencyc
    .com/index.php.
Young, Robert, 'Melanie Klein 1', lecture in series 'Psychoanalytic Pioneers', given
    at the Tavistock Centre in London, p. 12, http://human-nature.com/rmyou
    ng/papers/pap128.html.

Printed in the United States
by Bookmasters

Printed in the United States
By Bookmasters